Lecture Notes in Computer Science 11760

More information about this series at http://www.springer.com/series/7407

Mário S. Alvim · Kostas Chatzikokolakis ·
Carlos Olarte · Frank Valencia (Eds.)

The Art of Modelling Computational Systems

A Journey from Logic and Concurrency to Security and Privacy

Essays Dedicated to Catuscia Palamidessi on the Occasion of Her 60th Birthday

Springer

Editors
Mário S. Alvim
Universidade Federal de Minas Gerais
Belo Horizonte, Brazil

Kostas Chatzikokolakis
University of Athens
Athens, Greece

Carlos Olarte
Federal University of Rio Grande
do Norte - UFRN
Natal, Brazil

Frank Valencia
CNRS & University Javeriana Cali
Palaiseau, France

ISSN 0302-9743 ISSN 1611-3349 (electronic)
Lecture Notes in Computer Science
ISBN 978-3-030-31174-2 ISBN 978-3-030-31175-9 (eBook)
https://doi.org/10.1007/978-3-030-31175-9

LNCS Sublibrary: SL1 – Theoretical Computer Science and General Issues

Cover illustration: By Nadia Miller

This Springer imprint is published by the registered company Springer Nature Switzerland AG
The registered company address is: Gewerbestrasse 11, 6330 Cham, Switzerland

Preface

This Festschrift has been written in honor of Catuscia Palamidessi on the occasion of her 60th birthday. It features six laudations and 25 papers by close collaborators and friends of hers. These contributions are a tribute to Catuscia's intellectual depth, vision, passion for science, and tenacity in solving technical problems. They also reflect the breadth and impact of her work.

We have known Catuscia for decades. We have had the honor to work closely with her, giving us the opportunity to witness firsthand her amazing intellectual brilliance, dedication, and perseverance, exhibited in both her scientific work and her personal life. Besides her professional success, Catuscia is known for her kindness, for her disposition to improve the communities she is a part of, and for reliably being ready to always support colleagues, students, and friends.

Catuscia's scientific interests include, in chronological order, principles of programming languages, concurrency theory, security, and privacy. She has become one of the most influential scientists in each of these areas and, although they do not seem closely related, Catuscia has moved from one to another by doing seminal work that interconnected them. To trace back her prolific scientific career, we highlight just a few of her most influential papers; it would be impossible to do justice to her entire body of work in a preface.

Catuscia's scientific journey began in the mid-eighties at the University of Pisa in the area of principles of declarative programming languages, in particular logic and functional programming. Some of her most influential papers in this period provided the semantic foundations for some representative declarative languages [17, 18]. She also introduced a notion that became largely studied in the logic programming community: the S-Semantics [16].

Her work on declarative programming led her to the domain of concurrent constraint programming (ccp), a family of declarative languages where concurrency can be naturally expressed. Catuscia published seminal papers on logical and semantic foundations of ccp languages. Her work introduced the first fully abstract model for ccp [9], and provided an elegant proof theory for the correctness of ccp programs [7]. She also worked on a generalization of ccp to account for temporal concurrent behavior [20]. In general, Catuscia's work on ccp helped bring it to the attention of the concurrency theory community.

In the nineties, while working on ccp at CWI in Holland and under the influence of the Dutch process algebra community, Catuscia became interested in process algebra and concurrency theory at large. Her work at the time introduced the notion of embedding as a tool for language comparison [10]. She also helped developed a process algebra theory that can be regarded as a kernel for languages based on asynchronous communication [8]. A few years later at the University of Genoa, she went on to publish one of the most important papers on the expressiveness of process algebra, contradicting conventional wisdom by showing that the π-calculus, a central formalism from concurrency theory, was under certain natural conditions strictly more expressive than its asynchronous version [21]. In the late nineties at Penn State University, she proved that a probabilistic extension of the asynchronous π-calculus was as expressive as the same probabilistic extension of the π-calculus [19]. Catuscia became interested in the interplay between non-deterministic and probabilistic concurrent behavior, which ultimately led her to the next step in her scientific career.

In the early 2000s Catuscia moved to Inria at LIX, École Polytechnique. Her interest in probabilistic concurrency and, more generally, the formalization of probabilistic and non-deterministic concurrent systems, led her to a long and prolific journey on formal approaches to security and privacy. As always, her contributions span a wide variety of topics, to name a few: probabilistic and non-deterministic characterizations of anonymity [6, 22, 23], foundational contributions to information-theoretic approaches to information hiding [11, 13, 14], the influential g-leakage framework [2–4] for quantitative information flow, more recent research on statistical disclosure control, including an insightful metric-based generalization of differential privacy [12] and its very influential application to location privacy, known as geo-indistinguishability [5], applications of game theory [1], and finally applications of machine learning to privacy [15].

Catuscia's scientific journey continues. While developing her ever more sophisticated probabilistic approaches to security and privacy, her interests in probability and statistics themselves deepened, and, as mentioned, she has recently started to focus on the connections between privacy and machine learning. A few months ago, she was awarded an ERC Advanced grant to further explore these connections. We all look forward to the great contributions she will most definitely deliver from this new challenge.

We would like to conclude by expressing our gratitude to all the authors and the reviewers that helped us honor Catuscia with this book. We thank EasyChair for providing excellent tools for paper submission, reviewing, and the production of these proceedings. We thank Springer for their cooperation in publishing this volume. Finally, we are indebted to LIX, École Polytechnique and Inria Saclay for funding the event for the presentation of this book. We thank Catuscia's daughter, Nadia Miller, for drawing a beautiful portrait of her mother for this book.

References

1. Alvim, M.S., Chatzikokolakis, K., Kawamoto, Y., Palamidessi, C.: A game-theoretic approach to information-flow control via protocol composition. Entropy 20(5), 382 (2018)
2. Alvim, M.S., Chatzikokolakis, K., McIver, A., Morgan, C., Palamidessi, C., Smith, G.: Additive and multiplicative notions of leakage, and their capacities. In: CSF. pp. 308–322. IEEE Computer Society (2014)
3. Alvim, M.S., Chatzikokolakis, K., McIver, A., Morgan, C., Palamidessi, C., Smith, G.: Axioms for information leakage. In: CSF. pp. 77–92. IEEE Computer Society (2016)
4. Alvim, M.S., Chatzikokolakis, K., Palamidessi, C., Smith, G.: Measuring information leakage using generalized gain functions. In: CSF. pp. 265–279. IEEE Computer Society (2012)

5. Andrés, M.E., Bordenabe, N.E., Chatzikokolakis, K., Palamidessi, C.: Geo-indistinguishability: differential privacy for location-based systems. In: ACM Conference on Computer and Communications Security. pp. 901–914. ACM (2013)
6. Bhargava, M., Palamidessi, C.: Probabilistic anonymity. In: CONCUR. LNCS, vol. 3653, pp. 171–185. Springer (2005)
7. de Boer, F.S., Gabbrielli, M., Marchiori, E., Palamidessi, C.: Proving concurrent constraint programs correct. ACM Trans. Program. Lang. Syst. 19(5), 685–725 (1997)
8. de Boer, F.S., Klop, J.W., Palamidessi, C.: Asynchronous communication in pro-cess algebra. In: LICS. pp. 137–147. IEEE Computer Society (1992)
9. de Boer, F.S., Palamidessi, C.: A fully abstract model for concurrent constraint programming. In: TAPSOFT, Vol.1. LNCS, vol. 493, pp. 296–319. Springer (1991)
10. de Boer, F.S., Palamidessi, C.: Embedding as a tool for language comparison. Inf. Comput. 108(1), 128–157 (1994)
11. Braun, C., Chatzikokolakis, K., Palamidessi, C.: Quantitative notions of leakage for one-try attacks. Electr. Notes Theor. Comput. Sci. 249, 75–91 (2009)
12. Chatzikokolakis, K., Andrés, M.E., Bordenabe, N.E., Palamidessi, C.: Broadening the scope of differential privacy using metrics. In: Privacy Enhancing Technologies. LNCS, vol. 7981, pp. 82–102. Springer (2013)
13. Chatzikokolakis, K., Palamidessi, C., Panangaden, P.: Anonymity protocols as noisy channels. Inf. Comput. 206(2–4), 378–401 (2008)
14. Chatzikokolakis, K., Palamidessi, C., Panangaden, P.: On the bayes risk in information-hiding protocols. Journal of Computer Security 16(5), 531–571 (2008)
15. Cherubin, G., Chatzikokolakis, K., Palamidessi, C.: F-BLEAU: fast black-box leak-age estimation. In: Symposium on Security and Privacy. pp. 1307–1324. IEEE Computer Society (2019)
16. Falaschi, M., Levi, G., Martelli, M., Palamidessi, C.: A new declarative semantics for logic languages. In: ICLP/SLP. pp. 993–1005. MIT Press (1988)
17. Falaschi, M., Levi, G., Palamidessi, C., Martelli, M.: Declarative modeling of the operational behavior of logic languages. Theor. Comput. Sci. 69(3), 289–318 (1989)
18. Giovannetti, E., Levi, G., Moiso, C., Palamidessi, C.: Kernel-leaf: A logic plus functional language. J. Comput. Syst. Sci. 42(2), 139–185 (1991)
19. Herescu, O.M., Palamidessi, C.: Probabilistic asynchronous pi-calculus. In: FoS-SaCS. LNCS, vol. 1784, pp. 146–160. Springer (2000)
20. Nielsen, M., Palamidessi, C., Valencia, F.D.: Temporal concurrent constraint programming: Denotation, logic and applications. Nord. J. Comput. 9(1), 145–188 (2002)
21. Palamidessi, C.: Comparing the expressive power of the synchronous and asynchronous pi-calculi. Mathematical Structures in Computer Science 13(5), 685–719 (2003)
22. Palamidessi, C.: Anonymity in probabilistic and nondeterministic systems. Electr. Notes Theor. Comput. Sci. 162, 277–279 (2006)
23. Palamidessi, C.: Probabilistic and nondeterministic aspects of anonymity. Electr. Notes Theor. Comput. Sci. 155, 33–42 (2006)

August 2019

Mário S. Alvim
Kostas Chatzikokolakis
Carlos Olarte
Frank Valencia

Organization

Additional Reviewers

Aranda, Jesus
Ayala-Rincón, Maurício
Blanco, Roberto
Bonchi, Filippo
Boreale, Michele
Brodo, Linda
Bruni, Roberto
Cacciagrano, Diletta Romana
Carbone, Marco
Castiglioni, Valentina
Degano, Pierpaolo
Deng, Yuxin
Elsalamouny, Ehab
Falaschi, Moreno
Ferrari, Gianluigi
Gabbrielli, Maurizio
Gadducci, Fabio
Gorla, Daniele
Gorrieri, Roberto
Goubault-Larrecq, Jean
Guzmán, Michell
Kawamoto, Yusuke
Khouzani, Arman Mhr
Knight, Sophia
Laneve, Cosimo
Lanotte, Ruggero
Malacaria, Pasquale
McIver, Annabelle
Morgan, Carroll
Nigam, Vivek
Panangaden, Prakash
Parker, Dave
Perchy, Salim
Pino Duque, Luis Fernando
Rocha, Camilo
Rueda, Camilo
Sangiorgi, Davide
Santini, Francesco
Schwarzentruber, François
Smith, Geoffrey
Tini, Simone
Tiu, Alwen
Toninho, Bernardo
van Glabbeek, Rob
Zanasi, Fabio

Laudations

My Colleague, Wife, and Friend

Dale Miller

Inria Saclay and LIX, École Polytechnique

I remember seeing and speaking with Catuscia for the first time at the Third International Logic Programming Conference in London on June 1986. She was shy and uncomfortable with speaking English, so we had only a brief exchange about the weather. (Many years later I learned that she was one of the reviewers of the paper I presented at that meeting: she suggested accepting it—a good omen.) A couple of years later, about the time she received her Ph.D. from Pisa, Catuscia and I met again briefly at the Advanced School on Foundations of Logic Programming in Alghero, Sardinia in September 1988. This time we spoke even less since she was always disappearing after the lectures with a Dutch boyfriend. During the Logic in Computer Science (LICS) 1991 meeting in Amsterdam, we spoke a bit more, and I even signed the cast that she wore on her leg (injured from playing Frisbee).

It was not until LICS 1993 in Montreal that we had a serious conversation and it was mostly about scientific subjects. We both had an interest in logic and in concurrency theory but from somewhat different directions. Based on a promise that we would explore connections between these two topics, she was able to secure a grant from Italy for us to visit each other in Pisa and in Philadelphia. When she visited me in Philadelphia during the summer of 1994, our relationship moved far beyond the scientific, and our story would no longer be told as a series of meetings during conferences. We married in Pisa in March 1996.

I have shared in a bit of Catuscia's scientific life. During our stay at Penn State University (1997–2002), we managed to find some points of confluence between logic programming and concurrency, and we wrote four papers on how structured operational semantics and labeled transition systems can be encoded into logic. But even from the start of our relationship, Catuscia was following her own internal and demanding research agenda that carried her into and through a series of new research areas. I have vivid memories of us picking a wedding ring for her in Jeweler's Row in Philadelphia: while we had to wait for a jeweler to make some adjustments, she was distracted by showing how to separate the synchronous from the asynchronous π-calculi. Work on that topic required her to move beyond logic and concurrency to distributed computing.

After our move to France and Inria in 2002, we did not work together on research for about 10 years. During that time, her research interests matured and moved on. If a new topic could advance her research interests, she eagerly accepted the challenge to master yet another subject. I remember constantly finding thick textbooks on information theory and statistics under pillows and on top of chairs and sofas. Countless printouts on topics such as the Kantorovich measure, information leakage, differential privacy, geo-indistinguishability, Bayesian inference, and probabilistic model checking were scattered everywhere in the house and the car. I could tell which printouts she had read because those were the ones covered with margin art such as what you see here.

By the next time we collaborated on something scientific—the co-supervising of a Ph.D. student on the topic of differential privacy—I was completely out of my comfort zone. In just a few years, Catuscia moved from being a novice in the area of privacy to being a pioneer in that field. Me, I'm still encoding things in logic.

Although we have not worked on joint scientific problems for years, we regularly discuss a range of professional topics such as advising, teaching, editing, reviewing, hiring, etc. Over dinner, our son—who shows an interest in pursuing academic studies—gets to hear a lot about how academic research is pursued: both the good and bad aspects. Fortunately, all this talk about the not-so-scientific aspects of academics has not dampened his interest in mathematics and science.

Those who know Catuscia well know about her ability to focus and to be absorbed by a research topic. Those strengths that Catuscia brings to research she also brings to

Fig. 1 Photos from our wedding in the Municipality of Pisa in 1996 and of our children with Catuscia in the Roman Colosseum in 2017.

her family. This is particularly true of her role as the mother of our two children, Nadia and Alexis. When our daughter showed an interest in drawing and storytelling, our house gradually filled with novels and comic books that Catuscia and Nadia would read and discuss. A major event in our household was the arrival of a new edition of the French comic book series "Les Nombrils". Our son has a passion for mathematics. When trying to understand some advanced mathematical concepts and problems, he sometimes encounters problems that he cannot solve. But he knows that if he approaches his mother in the evening after dinner with the problem, there is a good chance that she will work throughout the night looking for a solution, which she almost always succeeded in finding by the next morning.

Since she is an avid downhill skier, she realized that to maintain her annual skiing habit, she would need to teach every member of her growing family to ski, starting with me and then with each of the children shortly after they could walk. One of the few times I can remember seeing Catuscia in tears was about 12 years ago in the French Alps. She had been trying to learn to snowboard, but she kept failing. She was so desperate that I found her crying in our apartment one evening. However, she would not give up, and the next day, she bought hip pads that softened the falls while making you look ridiculous on the slopes. After several days with those pads, she mastered the board. Now she can snowboard with the rest of us on skis.

From time-to-time, Catuscia can also get absorbed entirely as a *bricoleuse*—a do-it-yourselfer. For example, a couple of years ago, our dishwasher failed to heat water. After watching a couple shaky videos on YouTube and reading some advice on some web forums, she got the courage to drag me into the maze of electrical wires, plumbing, and motors that is the guts of modern dishwashers. To her credit, we managed to swap out the bad water heater and reassemble the entire washing machine. It is still working today. But fixing a dishwasher is one of her smaller achievements. During the years 2013–2015, she directed the effort to design and build a new home for us. She followed all the construction work and, in the end, she spent months learning how to make some of the finishing touches on the house and built an internal staircase, laid parquet, and cut, painted, and mounted moldings.

Finally, I want to express my delight and appreciation with having Catuscia as a close friend and companion. Her warmth, commitment, humor, and lively energies directed towards her family and me have been an endless source of strength and renewal.

Happy birthday, sweetheart.

The Best Problem-Solver I Know!

Prakash Panangaden

School of Computer Science, McGill University, Montreal, Quebec

Abstract. I briefly describe my long and delightful association with Catuscia.

Keywords: Catuscia Palmidessi is a genius

1 First Section

First, apologies[1] for not producing a technical article. The last few months have been difficult for me and Catuscia will understand. But enough about me. Let us move on the main theme.

1.1. A Subsection Sample

What?! Catuscia is turning 60! What does that make me? Age is in the mind. Catuscia and I are still young and her charm and brilliance are undimmed with the passing of time.

Sample Heading (Third Level) I first met her at an ICALP conference in Warwick in 1990. She was a fresh PhD with a solo paper on something about logic programming. We became friends immediately (after I penetrated the ring of eager males surrounding her) and have been ever since. Though she was shy and not assertive then, her intelligence was obvious[2]. We were both interested in concurrent constraint programming and for the next couple of years were friendly rivals on this topic. Somehow we were rivals without there ever being rancor or hostility.

Over the years we have drifted apart and come back together, research-wise that is. Our friendship has never ebbed. We both (independently!) got interested in probability. She published a sensational paper on the expressive power of asynchronous π-calculus in POPL 1997. Expressiveness was one of my main interests in the late 80s but I definitely missed this result which combined insight and ingenuity in typical Catuscia fashion.

Later she got interested in security and we briefly joined forces to study security from an information theory point of view. Subsequently she build a formidable team of young researchers around herself in Paris who pushed this. She has had–and continues

[1] To Valentina Castiglioni as well as Catuscia.

[2] as was her astonishing capacity for alcohol.

to have–a very successful collaboration with Geoff Smith and others on quantitative information flow which was a boat that I missed.

We have not collaborated on metrics for probability distributions but I like to think that my work on this topic influenced her development of a multiplicative analogue of the Kantorovich metric aimed at understanding differential privacy.

Sample Heading (Fourth Level) But let me not waste my time on stuff one can read in her CV. What I remember is her sensational ability to solve problems. Sure, her papers are filled with ingenious technical tricks but when it came to mathematical puzzles she was amazing. I will recall two that struck me. One was the proof of Sperner's lemma. Take a large triangle and cut it up into smaller triangles in whatever fashion one pleases. The smaller triangles must fit together properly: two adjacent triangles must touch along an edge or only at a vertex. Now label the three vertices of the other triangle 0, 1, 2. The vertices that lie along the edge labelled 01 have to be labelled with 0 or 1 but there is no other constraint. The analogous restriction applies to the other two sides. The interior vertices are labelled arbitrarily. The lemma states that there must be an inner triangle labelled 012. This is a key step in Brouwer's proof of his celebrated fixed point theorem.

Catuscia solved this not with a combinatorial argument but with a clever application of linear algebra which allowed her to extend the result to higher dimensions. I was amazed as I had struggled to produce a hairy proof in the two dimensional case.

Just two years ago there were several of us "old profs" at a summer school. The problem floating around was the following. Consider n marked points on a circle with one of them labelled as the origin. Now start a symmetric random walk at the origin and walk around the circle. Sooner or later every vertex will be hit. What is the probability of the various points, apart from the origin of course, of being the last one visited. In fact the answer is uniform! I totally failed to do this. Catuscia's solution was so ingenious that none of us understood it. Many other luminaries were there as well. Not only did they fail to solve it, we all failed to understand Catuscia's proof and kept trying to poke holes in it. In the end we finally got it and triumphantly "mansplained" her proof to her!

Why am I using these silly heading? Anyway what can I say after all these years? Catuscia is brilliance and grace.

Are we going up? Looking forward to your next tour de force.

1.2 Almost Back to the Top Level

I wish you many more productive and fun years.

Happy Birthday

The Four Senior Academics

Eleanor McIver[1], Mário S. Alvim[2], Annabelle McIver[3], and Carroll Morgan[4,5]

[1] Sydney Girls' High School
[2] Universidade Federal de Minas Gerais
[3] Macquarie University
[4] University of New South Wales
[5] Data 61

Abstract. Catuscia's appreciation of witty humor is well-known among her friends. In this laudation we adapt to an academic scenario one of her favorite sketches, *The Four Yorkshiremen*, which was originally performed on the *At Last the 1948 Show* and, later, by *Monty Python*.

(Four senior academics sitting together at a conference cocktail drinking wine. Symposium on Web-based Artificial Machine Programming (SWAMP '19) poster behind them. Music being played in the background that dies out.)

Mário Alvim: Ahh... very passable, this, very passable.

Annabelle McIver: Nothing like a good glass of Château de Chassilier wine, eh, Carroll?

Carroll Morgan: You're right there, Annabelle.

Eleanor McIver: Who'd a' thought thirty years ago we'd all be sittin' here, invited speakers at the cocktail party of the *Symposium on Web-based Artificial Machine Programming*, drinking Château de Chasillier wine?

MA: Aye. In them days, we'd a' been glad to have been attending the hands-on tutorial.

AM: With a super boring speaker!

EM: Without demo or slides.

CM: *Or* a speaker!

MA: In a filthy, cramped hall.

EM: We never used to have an 'all. We used to have to peek in through the window.

AM: The best *we* could manage was to hand out the name-badges at the welcome desk.

CM: But, you know, we were motivated in those days, though we were only PhD. students.

MA: Aye. *Because* we were PhD students! My supervisor used to say to me: "Publishing a paper should not be your motivation."

EM: She was right. I was more motivated then an' I had *nothin'*. We used to work in this *tiiiny* old lab, with *suuuper* slow Internet.

AM: Lab?! You were lucky to have a *lab*! We used to work at one desk, all hundred and twenty-six of us, one computer. Half the keys were missing from the keyboard, which we were all huddled around, and we could never agree on whose turn it was to type.

CM: You were lucky to have a *desk*! We used to have to work in a corridor!

MA: *Ohhhh,* we used to *dream* of workin' in a corridor! Woulda' been the University of Pisa to us. We used to have to use the photocopier next to the toilets as our working desk. We got disturbed every hour by people wanting to copy hundreds of pages—or worse! Lab?! Hmph.

EM: Well, when I say "lab" it was only our superior's drawer covered by my latest rejected paper. But it was a lab *to us*.

AM: We were evicted from *our* supervisor's drawer; we had to go and work in the lift!

CM: You were lucky to have a *lift*! There were a hundred and sixty of us working in a recycling bin by the coffee cart.

MA: A covered recycling bin?

CM: Aye.

MA: You were lucky! For three months we had to work inside a reusable interoffice envelope. We had to work as the envelope was passed on from office to office, for fourteen hours a day, week-in and week-out. When the envelope got back to our supervisor's desk, she'd tell us our theorems were all trivial!

AM: Luxury. *We* used to have to write our papers on the floor of the lift, prove twenty-seven contradictions every day for a month, and get the attention of our supervisor only when he decided not to use the stairs—if we were lucky!

CM: Well, we had it tough. We had to climb out of the bin at twelve o'clock at night, just before the truck came to haul it away, and all we had for breakfast was the juice that had dribbled out of the plastic bottle, which we had to lick off the roadside. Then we marked assignments for twenty-four hours a day and got one millionth of a bitcoin, but no private key. When we were done, our supervisor would throw all our work away and make us start again.

EM: Right. I had to submit my papers at ten o'clock at night, half an hour before I knew what it was about, eat a handful of moldy pencil shavings, work twenty-nine hours a day in my drawer and pay my supervisor for permission to work at all and, at the end of the day, my supervisor would rip all my work in two and forbid me from ever coming again.

MA: But you try and tell the students of today that… and they won't believe ya'.

ALL: Nope, nope…

To Catuscia: A Beautiful Mind and Spirit

Ehab ElSalamouny[1,2]

[1] Inria and LIX, École Polytechnique, France
[2] Suez Canal University, Egypt

Abstract. It is with great emotion that I write this laudation to Catuscia. Despite the fact I have been, and still, collaborating with her for a decade, it is hard yet to find the suitable words to express the value and beauty of her character. From the perspective of science, Catuscia is a great figure in the areas of quantitative information flow and probabilistic models of computer security and privacy. Also from the social point of view, Catuscia has a wonderful spirit that makes her a really good, if not even the best, friend to people interacting with her. Amazingly, the two perspectives are combined in one person, known as 'Catuscia'.

1 Knowing Catuscia

I had the opportunity to see Catuscia for the first time in 2009, when she invited me to visit her research team 'Cométe' in École Polytechniqe in Paris. At that time I was a PhD student at the university of Southampton, in UK. Here I have to acknowledge the effort of my supervisor, Vladimiro Sassone, to establish this link with Catuscia. Of course I was very happy to collaborate with Catuscia and extend my work on 'computational trust' to the area of privacy and anonymity in which the contributions of Catuscia and her team are significant landmarks. From the first meeting with Catuscia at that time, I loved and enjoyed that collaboration with her, and her wonderful team.

2 Catuscia as a Problem Solver

Catuscia has a sharp and very focused mind. She is amazing at presenting scientific problems and also at giving elegant solutions to them. To Catuscia, as a great professor, research is not only a career, but also a joyful means of entertainment. She loves challenging problems and see them as puzzles to solve, and she does! In many technical problems that we see in our work, Catuscia does not only seek the skeletons of the solutions, but she always has the curiosity to dig into the details of those problems, and exert unlimited energy to chase them to the end.

3 Social Life

I have collaborated with Catuscia as a member of her research group 'Cométe' for many years. Although this group has been always dynamic, with members joining and leaving, Catusica could always preserve an amazing warm familial atmosphere among the members of Cométe. We always socialize, having dinners, watching movies, and in general enjoying life together. One remarkable social event that she is keen to organize every year is the music party, in which friends are invited to show their musical talents and perform pieces of music.

The scientific and social sides of Catuscia's character are always amazingly mixed. It is likely that a conversation about work is intervened with some jokes and puzzles. Also, while socializing, it can very well happen to get into a scientific discussion (sometimes mathematical proofs!). This makes Catuscia very special and wonderful. I really wish her more and more wonderful, productive, and youthful years.

Happy Birthday Catuscia!

Information Leakage in a Music Score

François Fages

Inria Saclay-Île de France, Palaiseau, France

When I was invited to contribute to a Festschrift in honor of Catuscia Palamidessi, I immediately thought I should not miss the opportunity to express my friendly feelings and admiration for the research career of a good friend of 30 years.

Initially, we had some common research axes on Logic Programming and Concurrent Constraint Programming. The GULP conference on these topics in Italy was a high place where Catuscia had no equal to defend, between Pisa and Florence, what should be the true Italian pronunciation, but with too subtle arguments for a foreigner. In 1990, the ICLP conference in Jerusalem and Eilat was a memorable mark of Catuscia's passion for Concurrent Logic Programming. In 2000, our small CSE Penn State-Inria workshop on "Concurrency and Logic" was premonitory of a new life for Catuscia at Inria. In the last decade however, our research interests evolved in different directions, respectively on quantitative information flows and quantitative biology.

When reading the recent papers of Catuscia on information leakage in information flows, in the quest of finding something to write related to those topics, a music unexpectedly came to my mind, with two concurrent voices playing obsessively on some jazzy rhythm, responding each other, and producing some bits of information about a secret message subject to interpretation, as always in music.

It is my pleasure to dedicate that little piece of music, called *"Concur in C"*, to Catuscia for her 60th birthday. The music score is given below[1].

But could we try to apply Catuscia's theory on information leakage in information flows to the notes of each voice, in the hope of revealing something more on the meaning of this mysterious musical message remained enigmatic to the composer?

In information flow theory, a *channel*, or a *voice* in our musical interpretation, is a triple (X, Y, C), where X and Y are finite sets (of secret input values and produced output music notes) and the channel matrix C is an $|X| \times |Y|$ matrix giving the probability of getting output y when the input is x. Given a prior distribution π on X, the joint distribution $p(x,y) = \pi[x]C[x,y]$ on random variables $X \times Y$ is the (unique) joint distribution that gives marginal probabilities $p(x) = \sum_y p(x,y) = \pi[x]$, conditional probabilities $p(y|x) = p(x,y)/p(x) = C[x,y]$ (if $p(x)$ is nonzero), and similarly $p(y)$ and $p(x|y)$.

The marginal distribution of the music notes in a voice are given by the frequencies of the notes in that voice. In *Concur in C*, the upper and lower voices happen to play the same number of notes, 186, with slightly different distributions and slightly higher

[1] *"Concur in C"* can be heard with several orchestrations (piano, flute-saxophone, flute-organ-bass, big band) on http://lifeware.inria.fr/fages/music/.

Shannon's entropy $(-\sum_{i=1}^{186} p_i(y).\log_2(p_i(y)))$, i.e. more information, in the upper voice:

Number of occurrences	C	C# Db	D	D# Eb	E	F	F# Gb	G	G# Ab	A	A# Bb	B	Total
Upper voice $p^u(y)$	32 0.17	0 0	0 0	54 0.29	2 0.01	14 0.07	20 0.11	40 0.22	0 0	0 0	22 0.12	2 0.01	186 entropy 2.56
Lower voice $p^l(y)$	41 0.22	0 0	0 0	36 0.19	0 0	22 0.12	27 0.15	39 0.21	0 0	0 0	21 0.11	0 0	186 entropy 2.53

In music, it is expected that the upper voice contains more information than the lower voice usually dedicated to accompaniment. But here, the two voices play similar melodies in counterpoint which results in similar entropy values for both voices. It would be interesting to test this approach on more elaborate scores.

Thank you Catuscia for having once again inspired me, in an unexpected fashion, this self-surprising gift with curious questions about the quantification of information in music.

There has always been something magic in you.

Happy 60th Birthday Catuscia!

Acknowledgments. Special thanks go to Hubert Garavel for having improved the layout of my music score, and for his tolerance regarding my parallel fifths.

Concur in C

To Catuscia Palamidessi for her 60th birthday

F. Fages

13
15
18
20
23

An Italian Renaissance

Elena Marchiori and Jan Rutten

Dear Catuscia, Your stay in Amsterdam signalled the beginning of an *Italian Renaissance* at CWI, during which many Italian guests came to work there. It all had begun with the Esprit 415 project, in which both Pisa and CWI and many other institutes were involved. At CWI, you were a co-founder of the *Gaudi* group, consisting of yourself, Frank, Joost and Jan. The members of this group shared a passion not only for science but also for travelling, going out, dining and much more. As a consequence, the decision of the group where to submit its papers was not only based on the quality of the conference but, in equal measure, on the attractiveness of its location.

Often enough, this worked extremely well: it has generated scientific output and, at the same time, it has brought the group to lovely places all around the world. Images come to mind of its members floating in the Dead Sea and climbing Masada, of champagne and caviar in the Moscovite Hotel Ukraina, of nightclubs in London and Paris, and many more such.

Such are the wonderful fringe benefits of our profession.

Another important aspect of your presence in Amsterdam was your introduction of the three Dutch boys mentioned above to the finer aspects of Italian cuisine. Various of the recipes that you taught them have become permanent classics on their repertoire. On one occasion, you asked your mother to come over and cook for them, which turned into an unforgettable feast. Another highlight was the visit to your parents house, and the joint consumption of one of their chickens, which had been baptised *Giovanni* for the occasion. Best chicken ever.

Your amazing combination of scientific and social skills has helped new Italians at CWI to become fully integrated in a short time and enjoy work and life in Amsterdam. You have many and diverse qualities which provided examples and inspiration for trying new experiences. For instance, you could play squash and at the same time have an interesting conversation. Also, after a long social event until late evening you could go back to your place, sit at the table with a 'quartino di vino' and enjoy doing research. Magic Catuscia, Happy 60th Birthday!

Acknowledgement. The authors want to thank Catuscia also for having introduced them to each other, thereby making a substantial contribution to both their happiness.

Contents

Concurrency

Logic and Constraint Programming

Security and Privacy

Models and Puzzles

Concurrency

Variations of the Itai-Rodeh Algorithm for Computing Anonymous Ring Size

Wan Fokkink(✉) and Guus Samsom

Vrije Universiteit Amsterdam, Amsterdam, The Netherlands
w.j.fokkink@vu.nl, guus_samsom@hotmail.com

Abstract. We propose two adaptations of the probabilistic Itai-Rodeh algorithm for computing the size of an anonymous asynchronous ring. This Monte Carlo algorithm (inevitably) allows for wrong outcomes. Our adaptations reduce the chance that this happens. Furthermore, we propose a new algorithm that has a better message complexity.

1 Introduction

In an anonymous network, the nodes do not carry a unique ID. Typically, this is the case if there are no unique hardware IDs (for example, LEGO Mindstorms). Also, when each node has a unique ID but cannot reveal it to the other nodes, this is similar to having no unique IDs at all. For instance, nodes may want to hide such information because of security concerns (e.g. [1]), or because transmitting and storing IDs may be deemed too expensive, as is the case in the IEEE 1394 serial bus [10].

Itai and Rodeh [11] proposed probabilistic distributed algorithms for anonymous rings with asynchronous message-passing communication. One algorithm elects a leader in such networks, another computes its number of nodes. The leader election algorithm is inevitably Las Vegas, meaning that it contains infinite executions in which no leader is ever elected. Moreover, it must require that the nodes know the ring size. The ring size algorithm is inevitably Monte Carlo, meaning that it contains finite executions in which a wrong ring size is computed. The general idea of both algorithms is that nodes repeatedly choose a random ID and then perform an election or ring size algorithm for non-anonymous rings. If a conflict due to identical IDs by different nodes is detected, a new election or size estimation round may be started.

Here we propose some adaptations to the Itai-Rodeh ring size algorithm. First, we let nodes stick to the first random ID they choose, which reduces the chance of a wrong outcome, because each new choice of random IDs may contain an undesirable symmetry that leads to a premature termination of the execution. Second, we prioritize the order in which a node treats incoming messages, to push estimates of the ring size at the nodes upward as quickly as possible. Finally, we propose an alternative ring size algorithm that has a worst-case message complexity of $O(N^2)$, as opposed to the message complexity $O(N^3)$ of the Itai-Rodeh ring size algorithm.

M. S. Alvim et al. (Eds.): Palamidessi Festschrift, LNCS 11760, pp. 3–13, 2019.
https://doi.org/10.1007/978-3-030-31175-9_1

It is with great pleasure that we make this contribution to the Festschrift on the occasion of the 60th birthday of Catuscia Palamidessi. She has made important research contributions on probabilistic systems and anonymity, especially with regard to security and in the context of the π-calculus.

2 Election in Anonymous Rings

We say that a distributed computer network is *anonymous* if its nodes do not carry a unique ID. Some problems that are easily solved in networks with unique node IDs turn out to be insurmountable in anonymous networks. Boldi and Vigna [4] provided effective characterizations of the relations that can be computed on anonymous networks, if a bound on the network size is known.

We first consider election algorithms, which let the nodes in a network choose one leader among them. This first part is included here because it emphasizes the importance of the second part: We will see that precomputing the size of an anonymous ring network plays an important role in electing a leader in such a network. Moreover, several features of the Itai-Rodeh election algorithm, which will be explained below, are carried over to the Itai-Rodeh ring size algorithm, which is the main focus of this paper.

Angluin [2] showed that no election algorithm for anonymous asynchronous networks always terminates. The idea is that if the initial network configuration is symmetric (typically, a ring network in which all nodes are in the same state and all channels are empty), then there is always an infinite execution that cannot escape this symmetry, meaning that no leader is ever elected. Vice versa, if one leader has been elected, all nodes can be given a unique ID using a traversal algorithm (e.g. a depth-first search) initiated by the leader.

In a *probabilistic* algorithm, a node may flip a coin and perform an event based on the outcome of this coin flip. For probabilistic algorithms, one can calculate the probability that an execution from some given set of possible executions will occur. Probabilistic algorithms for which all executions terminate in a correct configuration are in general not so interesting, because any deterministic version of such an algorithm (for example, let the coin flip always yield heads) produces a correct nonprobabilistic algorithm. Therefore generally two classes of probabilistic algorithms are considered. A probabilistic algorithm is *Las Vegas* if the probability that it terminates is greater than 0 and all terminal configurations are correct. It is *Monte Carlo* if it always terminates and the probability that a terminal configuration is correct is greater than 0.

From now on we will consider only anonymous directed ring networks, with asynchronous communication. That is, nodes do not carry a unique ID and the network topology is a ring structure in which messages can only travel in a clockwise direction. Rings have the symmetric topology required for the impossibility results regarding election (explained above) and computing network size (which will be discussed in due course). It is worth noting that for acyclic anonymous networks, an algorithm exists for computing the network size, whereby the execution is started at the leaves of the network and works its way toward the center of the network.

Itai and Rodeh [11] proposed a Las Vegas election algorithm for anonymous directed ring networks in which the probability mass of the infinite executions is 0. In other words, the algorithm terminates with probability 1. (Note that this does allow the presence of infinite executions.) Their algorithm is based on the Chang-Roberts algorithm [5] for non-anonymous rings. Bakhshi, Fokkink, Pang and van de Pol [3] proposed a Las Vegas election algorithm for anonymous rings based on Franklin's algorithm [9] for non-anonymous rings.

In the Itai-Rodeh election algorithm, nodes choose a random ID and then select the node with the largest ID as the leader. Since multiple nodes may choose this largest ID, multiple election rounds may be necessary. In more detail, the algorithm works as follows. Initially, all nodes are active. At the start of each round, the active nodes randomly select an ID and send this ID to their clockwise next neighbor, with a hop counter set to 1. Since nodes are supposed to know the ring size N, a node can recognize from the hop counter when its own message returns after completing a round trip. Each message carries an additional bit that is dirtied if the message visits a node with the same ID that is not its originator.

Passive nodes simply pass on incoming messages, with the hop counter increased by 1. An active node that receives a message (for its current round) compares the ID of the message with its own randomly chosen ID for this round. There are three cases:

- If the message ID is smaller than the node ID, the message is purged.
- If the message ID is larger than the node ID, the node becomes passive and the message is passed on, with the hop counter increased by 1.
- If the message ID is equal to the node ID but its hop counter is smaller than N, then the message is passed on, again with the hop counter increased by 1, but also with a dirtied bit.

If a message returns to its originator with the hop counter N, the recipient checks whether the bit still clean. If so, the node becomes the leader (and it is certain that all other nodes are by now passive). If on the other hand the bit has been dirtied, then the node proceeds to a next election round (because another node chose the same ID in this round) and chooses a new random ID.

Messages carry the round number of the sender to avoid confusion, in case a message of an earlier round reaches its destination after some delay. (A variant of the algorithm without round numbers, in the case of FIFO channels, was proposed in [7,8].)

3 Computing the Size of an Anonymous Ring

As mentioned above, in the Itai-Rodeh election algorithm it is required that nodes know the ring size. This requirement is crucial, because there is no Las Vegas algorithm to compute the size of anonymous rings; every probabilistic algorithm for computing the size of an anonymous ring must allow for incorrect outcomes. This implies that there is no Las Vegas algorithm for election in anonymous rings if nodes do not know the ring size, because when there is one leader, network size can be computed using a traversal algorithm initiated by the leader.

The proof that there is no Las Vegas algorithm to compute the size of an anonymous ring goes roughly as follows. Suppose that such an algorithm does exist. We apply it on an anonymous ring of size $N > 2$. Consider an execution E that terminates with the correct outcome N. We cut the ring open between two of its nodes, make a copy of the resulting line of nodes, and glue the two parts together, yielding an anonymous ring of size $2N$. Now we can perform the execution E twice, on the two halves of this ring. Hereby it is crucial that at the two places where the two halves were glued together, the recipient of a message cannot recognize that the message now originates from a node in the other half, due to anonymity. This execution on the ring of size $2N$ terminates with the incorrect outcome N.

The Itai-Rodeh ring size algorithm targets anonymous directed rings. We have seen that it must be a Monte Carlo algorithm, meaning that it must allow for incorrect outcomes. However, in the Itai-Rodeh ring size algorithm, the probability of an erroneous outcome can be arbitrarily close to 0 by letting the nodes randomly select IDs from a sufficiently large domain.

Each node p maintains an estimate est_p of the ring size; initially $est_p = 2$. During any execution of the algorithm, est_p will never exceed the correct estimate N. The algorithm proceeds in estimate rounds. Every time a node finds that its estimate is too conservative, it moves to another round. That is, each node p initiates an estimate round at the start of the algorithm as well as at every update of est_p. The following detailed description of the algorithm is based on the ones in [6,12].

In each round, p randomly selects an ID id_p from $\{1, \ldots, R\}$ for some positive number R and sends the message $(est_p, id_p, 1)$ to its next neighbor. The third value is a hop count, which is increased by 1 every time the message is forwarded.

Now p waits for a message (est, id, h) to arrive. An invariant of such messages is that always $h \leq est$. When a message arrives, p acts as follows, depending on the parameter values in this message:

- $est < est_p$:
  The estimate of the message is more conservative than p's estimate, so p dismisses the message.
- $est > est_p$:
  The estimate of the message improves on p's estimate, so p increases its estimate. We distinguish between two cases:
  - $h < est$:
    The estimate est may be correct. So p sends $(est, id, h + 1)$ to give the message the chance to complete its round trip. Moreover, p performs $est_p \leftarrow est$.
  - $h = est$:
    The estimate est is too conservative because the message traveled est hops but did not complete its round trip. Therefore p performs $est_p \leftarrow est + 1$.

- $est = est_p$:
  The estimate of the message and that of p agree. We distinguish between two cases:
  - $h < est$:
    p sends $(est, id, h + 1)$ to give the message the chance to complete its round trip.
  - $h = est$:
    We again distinguish between two cases:
    * $id \neq id_p$:
      The estimate *est* is too conservative because the message traveled *est* hops but did not complete its round trip. Therefore p performs $est_p \leftarrow est + 1$.
    * $id = id_p$:
      Possibly p's own message returned (or a message originating from another node *est* hops before p that unfortunately happened to select the same ID as p in this estimate round). In this case, p dismisses the message.

When the algorithm terminates, $est_p \leq N$ for all nodes p, because a node increases its estimate only when it is certain that its current estimate is too conservative. Furthermore, est_p converges to the same value at all nodes p. If this were not the case, clearly there would be nodes p and q where p is q's predecessor in the ring and p's final estimate is larger than that of q. But then p's message in its final estimate round would have increased q's estimate to p's estimate.

The Itai-Rodeh ring size algorithm is a Monte Carlo algorithm: it may terminate with an estimate smaller than N. This can happen if in a round with an estimate $est < N$ all nodes at distance *est* from each other happen to select the same ID. The probability that the algorithm terminates with an incorrect outcome clearly becomes smaller when the domain $\{1, \ldots, R\}$ from which random IDs are drawn is made larger. This probability tends to 0 when R tends to infinity, for a fixed N.

In particular, if N is a prime number, it is not hard to compute the probability of a correct outcome, under the simplifying assumption that nodes never skip an estimate round. Since N is prime, the nodes only terminate with an estimate $1 < est < N$ if they have all chosen the same identity. Hence, the correct ring size N is computed if in each election round $e = 2, \ldots, N - 1$, the nodes do not all select the same identity. In other words, the probability that ring size N is computed is

$$\left(1 - \frac{1}{R^{N-1}}\right)^{N-2}$$

The side condition that we only consider executions in which no node skips a round is essential here (and means the probability above is in fact a bit too conservative). Namely, a node p skips an estimate round if it receives a message (est, id, h) with either $h = est = est_p + 1$ or $est > est_p + 1$. In those cases p skips (at least) the estimate round $est_p + 1$.

The worst-case message complexity of the Itai-Rodeh ring size algorithm is $O(N^3)$: each node starts at most $N - 1$ estimate rounds, and during each round it sends out one message, which takes at most N steps. The worst-case time complexity is $O(N^2)$, under the assumption that messages take at most one time unit to reach their destination, since then the ith estimate round completes after at most $i{\cdot}N$ time units, for $i = 1, \ldots, N-1$.

4 Adaptations of the Itai-Rodeh Ring Size Algorithm

We now propose some adaptations of the Itai-Rodeh ring size algorithm, with the aims to increase the chance of a correct outcome and to decrease its message complexity. We have also performed some simulations of implementations of the original Itai-Rodeh ring size algorithm and our adaptations, for small values of N, to get an impression of the impact of the adaptations.

4.1 Choose a Random ID Only Once

A first, simple adaptation of the Itai-Rodeh ring size algorithm is to let the nodes stick to the first random ID they choose, instead of selecting a new random ID at each increased estimate. That this is beneficial for the chance of a correct outcome is clear in the case that N is prime. In that case the only possibility for a wrong outcome is if all nodes choose the same ID at the start. In other words, the probability that ring size N is computed is

$$1 - \frac{1}{R^{N-1}}$$

Basically, the principle that is essential for the Itai-Rodeh election algorithm, letting a node choose a new random ID at each round, is harmful for the Itai-Rodeh ring size algorithm. The latter algorithm may terminate with a wrong outcome if the chosen IDs in an estimate round display a certain symmetry. Letting the nodes choose a new ID at each round increases the chance that in some round the chosen IDs contain a detrimental symmetry.

4.2 Prioritization of Received Messages

For simplicity, the analysis of the probability that the correct ring size is computed, at the end of Sect. 3, assumed that nodes never skip an estimate round. Since estimations of the ring size are guaranteed to be conservative, the possibility that a node skips an estimate round increases the chance that an execution of the Itai-Rodeh ring size algorithm terminates with the correct outcome. To increase the chance that estimate rounds are skipped, it is beneficial to define a prioritization on the order in which a node p treats concurrently received messages in its buffer.

- First of all, the priority is based on the value of *est*.
- Second, the priority is based on the value of h.
- Third (so if two received messages carry the same estimate and hop counter), a message with an identity different from id_p has a higher priority than a message with the identity id_p.

In all three cases the priority aims to increase the chance that est_p skips a value.

4.3 Each Node Sends Out Only One Message

To decrease the worst-case message complexity from $O(N^3)$ to $O(N^2)$, we let each node send out only one message, which in the worst case (from the point of message complexity) completes the entire round trip. We exclude the value est_p of the sender p from the message, so that it only contains the ID of p and a hop count. The intention is that each of these messages completes the round trip and helps to increase estimates of visited nodes on the way.

Again, initially $est_p = 2$ at each node p, and est_p will never exceed N. Each node p randomly selects an ID id_p from $\{1, \ldots, R\}$ for some positive number R and sends the message $(id_p, 1)$ to its next neighbor. Now p waits for a message (id, h) to arrive.

When p receives a message $m = (id_p, h)$ with $h \geq est_p$, it assumes that this message originates from p itself, so (only) in that case the received message is not forwarded, while est_p is updated to h. If p later receives another message showing that the value $est_p = h$ is too conservative, then p must forward the message m after all. Therefore a Boolean variable $passive_p$ is set when p receives m, to recall that p must send the message $(id_p, est_p + 1)$ if the value of est_p is increased at some later moment in time. Initially the value of $passive_p$ is $false$.

When node p receives a message (id, h), it acts as follows, depending on the parameter values in this message:

- $h < est_p$:
  p forwards the message with the hop count increased by 1, i.e., $(id, h + 1)$.
- $h \geq est_p$:
  If $passive_p = true$, then p forwards $(id_p, est_p{+}1)$, because p earlier stopped the incoming message (id_p, est_p) by mistake. (We note that in this case $h > est_p$, because a node never receives two different messages with the same hop count.) Moreover, p then performs $passive_p \leftarrow false$. We distinguish between two cases:
  * $id \neq id_p$:
    Since the message does not originate from p, the value of est_p is too conservative. Therefore p performs $est_p \leftarrow h + 1$. Furthermore, it forwards $(id, h + 1)$.
  * $id = id_p$:
    Possibly p's own message returned (or a message originating from another node h hops before p that unfortunately happened to select the same ID as p). Then p performs $est_p \leftarrow h$. Moreover, p performs $passive_p \leftarrow true$, to recall that it did not forward the message (id_p, est_p).

Since no message can be forwarded beyond the node it originated from, it is not hard to see that the algorithm terminates, and that then $est_p \leq N$ at all nodes p. Moreover, the $passive_p$ flags ensure that each node at any moment has stopped at most one message. Since each node sends out one message, this implies that each node stops exactly one message permanently. Clearly, the IDs of a node p and of the message m that is stopped permanently at p coincide, and the final value of est_p coincides with the value of the hop counter of m. The fact that the predecessor q of p in the ring forwarded m implies that upon termination, $est_q \geq est_p$. Since this inequality holds for each pair of neighbors in the ring, it follows that upon termination, all nodes carry the same estimate.

This algorithm has worst-case message complexity $O(N^2)$ (compared to $O(N^3)$ for the Itai-Rodeh ring size algorithm), because each node sends out only one message, which takes at most N steps.

4.4 Implementation and Simulation Results

We implemented in Java the original Itai-Rodeh ring size algorithm, as well as the two adaptations proposed in Sects. 4.1 and 4.2, and the following pseudocode description of our algorithm proposed in Sect. 4.3 which describes how a node p acts when it receives a message (id, h). (Initially est_p has the value 2, $passive_p$ has the value *false*, and p sends the message $(id_p, 1)$ to its successor $next_p$ in the ring.)

```
if h < est_p then
|  send (id, h + 1) to next_p;
else
|  if passive_p = true then
|  |  send (id_p, est_p + 1) to next_p;
|  |  passive_p ← false;
|  end
|  if id ≠ id_p then
|  |  est_p ← h + 1;
|  |  send (id, h + 1) to next_p;
|  else
|  |  est_p ← h;
|  |  passive_p ← true;
|  end
```

We ran a million simulations for both $N = 12$ and $N = 13$, to get an impression of the impact of the adaptations on the performance. These numbers were chosen because 12 has relatively many divisors while 13 is prime. The outcomes of these experiments with regard to the Itai-Rodeh ring size algorithm and the adaptations proposed in Sects. 4.1 and 4.2 are plotted in Fig. 1. The horizontal axis charts the possible ring sizes, ranging from 2 up to and including $N-1$, while the vertical axis expresses how many of the experiments produced a certain ring size; in the plot for $N = 12$, these numbers must be multiplied with 10^4. On the horizontal axis, the correct ring size N has been excluded because this column would dwarf the other ones. Already with these small values for N, the simulations took a significant amount of time, due to the message complexity of $O(N^3)$.

For $N = 12$ the impact of sticking to the first chosen ID is already significant, and for $N = 13$ it makes the probability of computing a size between 2 and

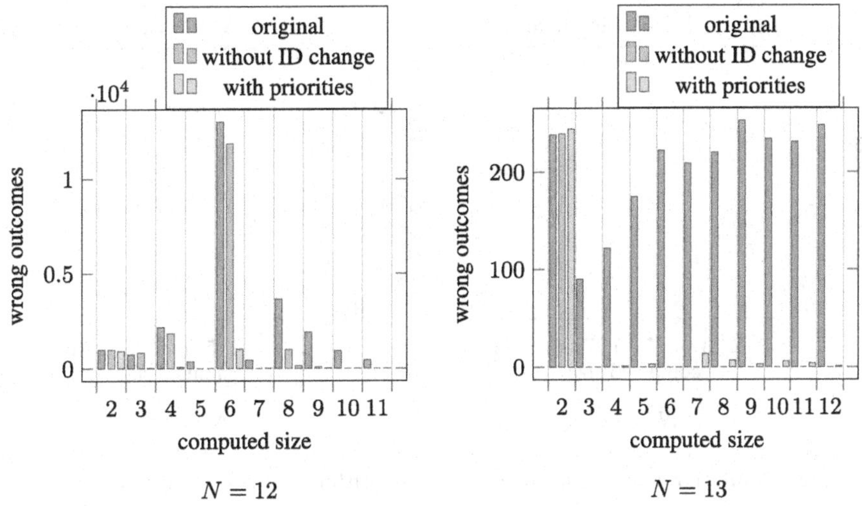

Fig. 1. Simulation results for three variants of the Itai-Rodeh ring size algorithm, with $N = 12$ and $N = 13$.

13 drop to 0, because 13 is prime. The impact of imposing priorities on the order in which concurrently received messages are treated is rather dramatic in our experiments. (In this implementation we included ID changes, to clearly distinguish the effects the two optimizations have on the performance of the Itai-Rodeh algorithm.) This dramatic effect is due to the fact that in our simulations, channel delays were chosen to be negligible, so that in-buffers at nodes will often contain multiple messages. In case of a larger channel delay, in-buffers will mostly contain no more than one message, in which case clearly prioritization has no effect at all. Finally, we note that for $N = 13$, the dip in wrong outcomes for the values 3, 4, and 5 for the original Itai-Rodeh algorithm is caused by the fact that nodes may skip estimate rounds, as explained at the end of Sect. 3. The chance that this happens is larger for small values of est_p, because then it is more likely that a message (est, id, h) arrives at p with $h = est = est_p + 1$ or $est > est_p + 1$.

In Fig. 2 the simulation results of our ring size algorithm from Sect. 4.3 are plotted, again for $N = 12$ and $N = 13$. For the sake of a clear comparison with the original Itai-Rodeh algorithm, we refrained from using prioritization of received messages. It can be observed that the probability of computing a wrong outcome is comparable to the original Itai-Rodeh algorithm without ID changes. However, these simulations took much less time, owing to the message complexity of $O(N^2)$.

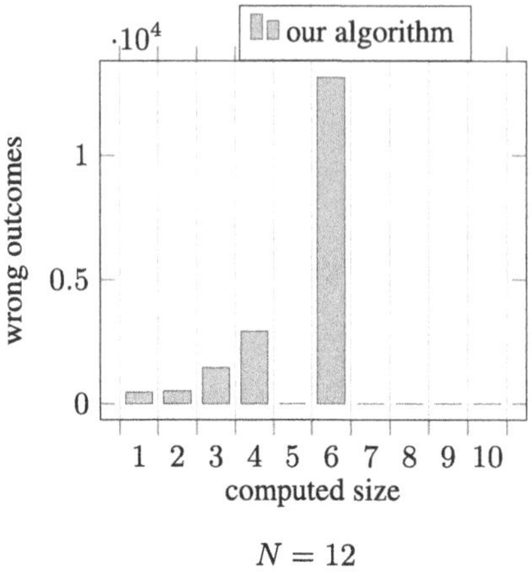

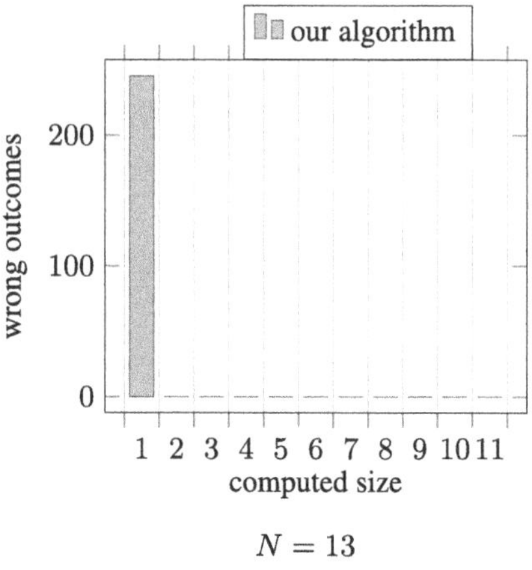

Fig. 2. Simulation results for our ring size algorithm, with $N = 12$ and $N = 13$.

5 Conclusion

We proposed two optimizations of the Itai-Rodeh algorithm for computing the size of an anonymous ring: nodes stick to the first random ID they select, and the treatment of received messages is prioritized to stimulate the fast propagation of larger estimates through the ring. Furthermore, we proposed a new algorithm for computing the size of an anonymous ring, in which each node sends out only one message, so that the worst-case message complexity is $O(N^2)$ (compared to $O(N^3)$ for the Itai-Rodeh algorithm).

The implementations and the simulation experiments are available at https://github.com/gsamsom/Itai-Rodeh-simulatie.

In [8], the probabilistic model checker PRISM was used to analyze a finite-state version of the Itai-Rodeh leader election algorithm. Moreover, this algorithm served as a benchmark for different optimization techniques for probabilistic model checking, e.g. in [13]. Likewise, to complement our simulation results, it would be interesting to apply probabilistic model checking to analyze the probability that the different variations of the Itai-Rodeh ring size algorithm terminate correctly, on anonymous rings of varying sizes. Moreover, these variations could be employed for a benchmark in probabilistic model checking.

References

1. Andrés, M., Palamidessi, C., van Rossum, P., Sokolova, A.: Information hiding in probabilistic concurrent systems. Theoret. Comput. Sci. **412**(28), 3072–3089 (2011)
2. Angluin, D.: Local and global properties in networks of processors. In: Proceedings of the STOC 1980, pp. 82–93. ACM (1980)

3. Bakhshi, R., Fokkink, W., Pang, J., van de Pol, J.: Leader election in anonymous rings: Franklin goes probabilistic. In: Proceedings of the IFIP TCS 2008, pp. 57–72. IFIP (2008)
4. Boldi, P., Vigna, S.: An effective characterization of computability in anonymous networks. In: Welch, J. (ed.) DISC 2001. LNCS, vol. 2180, pp. 33–47. Springer, Heidelberg (2001). https://doi.org/10.1007/3-540-45414-4_3
5. Chang, E., Roberts, R.: An improved algorithm for decentralized extrema-finding in circular configurations of processes. Commun. ACM **22**, 281–283 (1979)
6. Fokkink, W.: Distributed Algorithms: An Intuitive Approach. MIT Press, Cambridge (2013)
7. Fokkink, W., Pang, J.: Simplifying Itai-Rodeh leader election for anonymous rings. In: Proceedings of the AVOCS 2004. Electronic Notes in Theoretical Computer Science, vol. 128, no. 6, pp. 53–68. Elsevier (2005)
8. Fokkink, W., Pang, J.: Variations on Itai-Rodeh leader election for anonymous rings and their analysis in PRISM. J. Univ. Comput. Sci. **12**, 981–1006 (2006)
9. Franklin, W.R.: On an improved algorithm for decentralized extrema-finding in circular configurations of processes. Commun. ACM **25**(5), 336–337 (1982)
10. IEEE Computer Society: IEEE standard for a high performance serial bus. Technical report, IEEE (1994)
11. Itai, A., Rodeh, M.: Symmetry breaking in distributed networks. Inf. Comput. **88**(1), 60–87 (1990)
12. Tel, G.: Introduction to Distributed Algorithms, 2nd edn. Cambridge University Press, Cambridge (2000)
13. Timmer, M., Katoen, J.P., van de Pol, J., Stoelinga, M.: Confluence reduction for Markov automata. Theoret. Comput. Sci. **655**, 193–219 (2016)

Axiomatizing Team Equivalence for Finite-State Machines

Roberto Gorrieri(✉)

Dipartimento di Informatica — Scienza e Ingegneria,
Università di Bologna, Mura A. Zamboni, 7, 40127 Bologna, Italy
roberto.gorrieri@unibo.it

Abstract. Finite-state machines, a simple class of finite Petri nets, were equipped in [10] with a truly concurrent, bisimulation-based, behavioral equivalence, called *team equivalence*, which conservatively extends bisimulation equivalence on labeled transition systems and which is checked in a distributed manner, without necessarily building a global model of the overall behavior. The process algebra CFM [9] is expressive enough to represent all and only the finite-state machines, up to net isomorphism. Here we first prove that this equivalence is a congruence for the operators of CFM, then we show some algebraic properties of team equivalence and, finally, we provide a finite, sound and complete, axiomatization for team equivalence over CFM.

1 Introduction

By finite-state machine (FSM, for short) we mean a simple type of finite Petri net [9,16,19] whose transitions have singleton pre-set and singleton, or empty, post-set. The name originates from the fact that an unmarked net of this kind is essentially isomorphic to a nondeterministic finite automaton (NFA), usually called a finite-state machine as well. However, semantically, our FSMs are richer than NFAs because, as their initial marking may be not a singleton, these nets can also exhibit concurrent behavior, while NFAs are strictly sequential. FSMs are also similar to finite-state, labeled transition systems (LTSs, for short) [11], a class of models that are suitable for describing sequential, nondeterministic systems, and are also widely used as a semantic model for process algebras (see, e.g., [8]). On this class of models, there is widespread agreement that a natural and convenient equivalence relation is bisimilarity [13,15].

In [10] we defined a new truly-concurrent equivalence relation over FSMs, called *team equivalence*, that can be computed in a distributed manner, without resorting to a global model of the overall behavior of the analyzed marked net. Since an FSM is so similar to an LTS, the basic idea we started with was to define bisimulation equivalence directly over the set of places of the *unmarked* net. The advantage is that bisimulation equivalence is a relation on places, rather than on markings (as it is customary for Petri nets), and so much more easily computable; more precisely, if m is the number of net transitions and n is the number of places,

M. S. Alvim et al. (Eds.): Palamidessi Festschrift, LNCS 11760, pp. 14–32, 2019.
https://doi.org/10.1007/978-3-030-31175-9_2

checking whether two places are bisimilar can be done in $O(m \log (n+1))$ time, by adapting the optimal algorithm in [17] for standard bisimulation on LTSs. After the bisimulation equivalence over the set of places has been computed, we can define, in a purely structural way, that two markings m_1 and m_2 are *team equivalent* if they have the same cardinality, say $|m_1| = k = |m_2|$, and there is a bisimulation-preserving, bijective mapping between the two markings, so that each of the k pairs of places (s_1, s_2), with $s_1 \in m_1$ and $s_2 \in m_2$, is such that s_1 and s_2 are bisimilar. Once bisimilarity on places has been computed, checking whether two markings of size k are team equivalent can be computed in $O(k^2)$ time.

Note that to check whether two markings are team equivalent we need not to construct an LTS describing the global behavior of the whole system, but only to find a suitable, bisimulation-preserving match among the local, sequential states (i.e., the elements of the markings). Nonetheless, we proved that team equivalence is coherent with the global behavior of the net. More precisely, we showed in [10] that team equivalence is finer than interleaving bisimilarity (so it respects the token game), actually it coincides with *strong place bisimilarity* [1] (and so it respects the causal semantics of nets, more precisely, it is finer than fully-concurrent bisimilarity [3]).

In [9] we proved that the class of FSMs can be "alphabetized" by means of the process algebra CFM: not only the net semantics of each CFM process term is an FSM, but also, given an FSM N, we can single out a CFM process term p_N such that its net semantics is an FSM isomorphic to N. This means that we can define team equivalence also over CFM process terms. CFM is a simple process algebra: it is actually a slight extension to finite-state CCS [13] and a subcalculus of both regular CCS and BPP [8].

The goals of this paper are three: we want (i) to prove that team equivalence is a congruence for the CFM operators, (ii) to study its algebraic properties and, finally, (iii) to provide a sound and complete, finite axiomatization of team equivalence for CFM. This axiomatization is not very surprising: it is enough to add to (a slighty revised version of) the finite axiomatization of interleaving bisimulation for finite-state CCS [12], three axioms for parallel composition stating that it is associative, commutative and with $\mathbf{0}$ as neutral element. However, the technical treatment is simpler than [12], as we base our axiomatization on guarded process constants (e.g., $C \doteq a.C$) rather than on the recursive constructs (with possible unguarded variables; e.g., $\mu X a.X + X$). Moreover, this is the first finite, sound and complete, axiomatization of a truly-concurrent behavioral semantics for a process algebra admitting recursive behavior.

The paper is organized as follows. Section 2 introduces the basic definitions about FSMs, bisimulation on places and team equivalence. Section 3 introduces the process algebra CFM, its syntax, its net semantics and recalls the so-called *representability theorem* from [9]. Section 4 shows that team equivalence is a congruence for the CFM operators and studies the algebraic properties of this equivalence. Section 5 presents the finite axiomatization of team equivalence for CFM, proving that it is sound and complete. Finally, Sect. 6 discusses some related literature and future research.

2 Finite-State Machines and Team Equivalence

Definition 1 (Multiset). *Let* $\mathbb{N}$ *be the set of natural numbers. Given a finite set* S*, a* multiset *over* S *is a function* $m : S \to \mathbb{N}$*. The* support *set* $dom(m)$ *of* m *is* $\{s \in S \mid m(s) \neq 0\}$*. The set of all multisets over* S*, denoted by* $\mathscr{M}(S)$*, is ranged over by* m*. We write* $s \in m$ *if* $m(s) > 0$*. The* multiplicity *of* s *in* m *is given by the number* $m(s)$*. The* size *of* m*, denoted by* $|m|$*, is the number* $\sum_{s \in S} m(s)$*, i.e., the total number of its elements. A multiset* m *such that* $dom(m) = \emptyset$ *is called* empty *and is denoted by* θ*. We write* $m \subseteq m'$ *if* $m(s) \leq m'(s)$ *for all* $s \in S$*.*

Multiset union $\_ \oplus \_$ *is defined as follows:* $(m \oplus m')(s) = m(s) + m'(s)$*; it is commutative, associative and has* θ *as neutral element.* Multiset difference $\_ \ominus \_$ *is defined as follows:* $(m_1 \ominus m_2)(s) = max\{m_1(s) - m_2(s), 0\}$*. The* scalar product *of a number* j *with* m *is the multiset* $j \cdot m$ *defined as* $(j \cdot m)(s) = j \cdot (m(s))$*. By* s_i *we also denote the multiset with as only element* s_i*. Hence, a multiset* m *over* $S = \{s_1, \ldots, s_n\}$ *can be represented as* $k_1 \cdot s_1 \oplus k_2 \cdot s_2 \oplus \ldots \oplus k_n \cdot s_n$*, where* $k_j = m(s_j) \geq 0$ *for* $j = 1, \ldots, n$*.* □

Definition 2 (Finite-state machine). *A labeled* finite-state machine *(FSM, for short) is a tuple* $N = (S, A, T)$*, where*

- S *is the finite set of* places*, ranged over by* s *(possibly indexed),*
- A *is the finite set of* labels*, ranged over by* ℓ *(possibly indexed), and*
- $T \subseteq S \times A \times (S \cup \{\theta\})$ *is the finite set of* transitions*, ranged over by* t*.*

Given a transition $t = (s, \ell, m)$*, we use the notation* $^\bullet t$ *to denote its* pre-set s *(which is a single place) of tokens to be consumed;* $l(t)$ *for its* label ℓ*, and* $t^\bullet$ *to denote its* post-set m *(which is a place or the empty multiset* θ*) of tokens to be produced. Hence, transition* t *can be also represented as* $^\bullet t \xrightarrow{l(t)} t^\bullet$*.* □

Graphically, a place is represented by a circle, a transition by a box, connected by a directed arc from the place in its pre-set and to the place in its post-set, if any.

Definition 3 (Marking, FSM net system). *A multiset over* S *is called a* marking*. Given a marking* m *and a place* s*, we say that the place* s *contains* $m(s)$ tokens*, graphically represented by* $m(s)$ *bullets inside place* s*. An* FSM net system $N(m_0)$ *is a tuple* (S, A, T, m_0)*, where* (S, A, T) *is an FSM and* m_0 *is a marking over* S*, called the* initial marking*. We also say that* $N(m_0)$ *is a* marked *net. An FSM net system* $N(m_0) = (S, A, T, m_0)$ *is* sequential *if* m_0 *is a singleton, i.e.,* $|m_0| = 1$*; while it is* concurrent *if* m_0 *is arbitrary.* □

Definition 4 (Token game, firing sequence, transition sequence, reachable markings). *Given an FSM* $N = (S, A, T)$*, a transition* t *is* enabled *at marking* m*, denoted by* $m[t\rangle$*, if* $^\bullet t \subseteq m$*. The execution (or* firing*) of* t *enabled at* m *produces the marking* $m' = (m \ominus {^\bullet t}) \oplus t^\bullet$*. This is written as* $m[t\rangle m'$*. This procedure is called the* token game*. A* firing sequence *starting at* m *is defined inductively as follows:*

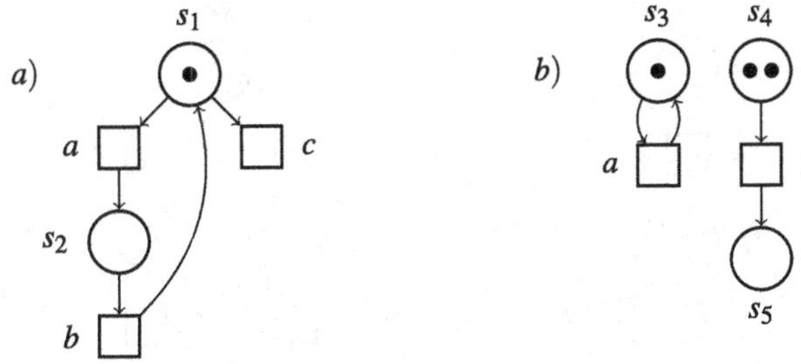

Fig. 1. A sequential finite-state machine in (a), and a concurrent finite-state machine in (b)

- $m[\epsilon\rangle m$ *is a firing sequence (where* ϵ *denotes an empty sequence of transitions) and*
- *if* $m[\sigma\rangle m'$ *is a firing sequence and* $m'[t\rangle m''$, *then* $m[\sigma t\rangle m''$ *is a firing sequence.*

If $\sigma = t_1 \dots t_n$ *(for* $n \geq 0$*) and* $m[\sigma\rangle m'$ *is a firing sequence, then there exist* $m_1, \dots, m_{n+1}$ *such that* $m = m_1[t_1\rangle m_2[t_2\rangle \dots m_n[t_n\rangle m_{n+1} = m'$, *and* $\sigma = t_1 \dots t_n$ *is called a* transition sequence *starting at* m *and ending at* m'. *The set of* reachable markings *from* m *is* $reach(m) = \{m' \mid \exists \sigma. m[\sigma\rangle m'\}$. □

Definition 5 (Dynamically reduced). *An FSM net system* $N(m_0) = (S, A, T, m_0)$ *is* dynamically reduced *if* $\forall s \in S\, \exists m \in reach(m_0).m(s) \geq 1$ *and* $\forall t \in T\, \exists m, m' \in reach(m_0)$ *such that* $m[t\rangle m'$. □

Example 1. By using the usual drawing convention for Petri nets, Fig. 1 shows in (a) a sequential FSM, which performs a, possibly empty, sequence of a's and b's, until it performs one c and then stops *successfully* (the token disappears in the end). Note that a sequential FSM is such that any reachable marking is a singleton or empty. Hence, a sequential FSM is a *safe* (or 1-bounded) net: each place in any reachable marking can hold one token at most. In (b), a concurrent FSM is depicted: it can perform a forever, interleaved with two occurrences of b, only: the two tokens in s_4 will eventually reach s_5, which is a place representing unsuccessful termination (deadlock). Note that a concurrent FSM is a *k-bounded* net, where k is the size of the initial marking: each place in any reachable marking can hold k tokens at most. Hence, the set $reach(m)$ is finite for any m. As a final comment, we mention that for each FSM $N = (S, A, T)$ and each place $s \in S$, the set $reach(s)$ is a subset of $S \cup \{\theta\}$. □

We recall the definition of (strong) bisimulation on places for unmarked FSMs, originally introduced in [10]. In this definition (and in the following ones), the markings m_1 and m_2 can only be either the empty marking θ or a single place, because of the shape of FSM transitions.

Definition 6 (Bisimulation on places). *Let* $N = (S, A, T)$ *be an FSM. A* bisimulation on places *is a relation* $R \subseteq S \times S$ *such that if* $(s_1, s_2) \in R$ *then for all* $\ell \in A$

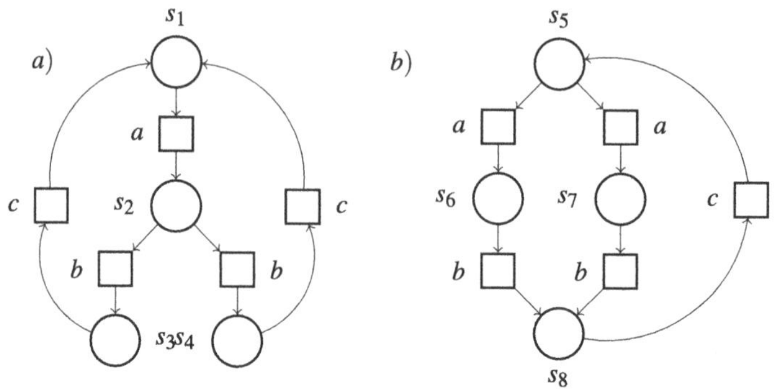

Fig. 2. Two bisimilar FSMs

- *$\forall m_1$ such that $s_1 \xrightarrow{\ell} m_1$, $\exists m_2$ such that $s_2 \xrightarrow{\ell} m_2$ and either $m_1 = \theta = m_2$ or $(m_1, m_2) \in R$,*
- *$\forall m_2$ such that $s_2 \xrightarrow{\ell} m_2$, $\exists m_1$ such that $s_1 \xrightarrow{\ell} m_1$ and either $m_1 = \theta = m_2$ or $(m_1, m_2) \in R$.*

Two places s and s' are bisimilar *(or* bisimulation equivalent*), denoted $s \sim s'$, if there exists a bisimulation R such that $(s, s') \in R$.* □

Example 2. Consider the nets in Fig. 2. It is not difficult to realize that relation $R = \{(s_1, s_5), (s_2, s_6), (s_2, s_7), (s_3, s_8), (s_4, s_8)\}$ is a bisimulation. □

We argued in [10] that, if m is the number of net transitions and n of places, checking whether two places of an FSM are bisimilar can be done in $O(m \log (n+1))$ time, by adapting the algorithm in [17] for ordinary bisimulation on LTSs. Moreover, bisimulation on places enjoys the same properties of bisimulation on LTSs, i.e., it is coinductive and equipped with a fixed-point characterization. Bisimulation equivalence is an equivalence relation, indeed, and can be used in order to minimize the net [10].

2.1 Team Equivalence

Definition 7 (Additive closure). *Given an FSM net $N = (S, A, T)$ and a* place relation *$R \subseteq S \times S$, we define a* marking relation *$R^\oplus \subseteq \mathscr{M}(S) \times \mathscr{M}(S)$, called the* additive closure *of R, as the least relation induced by the following axiom and rule.*

$$\frac{}{(\theta, \theta) \in R^\oplus} \qquad \frac{(s_1, s_2) \in R \quad (m_1, m_2) \in R^\oplus}{(s_1 \oplus m_1, s_2 \oplus m_2) \in R^\oplus}$$

□

Note that, by definition, two markings are related by $R^\oplus$ only if they have the same size; in fact, the axiom states that the empty marking is related to

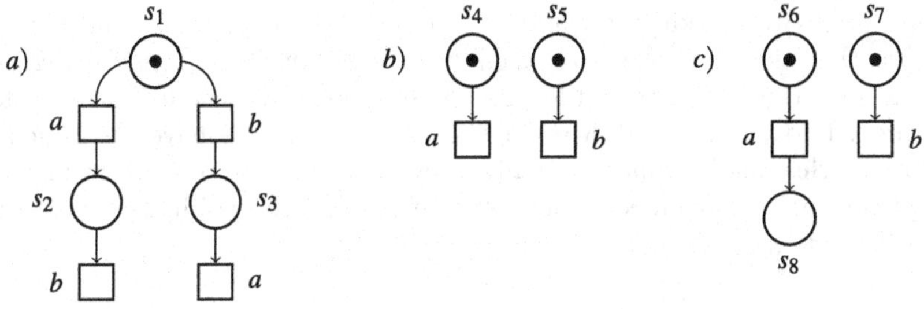

Fig. 3. Three non-team equivalent net systems: $a.b.\mathbf{0} + b.a.\mathbf{0}$, $a.\mathbf{0} \,|\, b.\mathbf{0}$ and $a.C \,|\, b.\mathbf{0}$ (with $C \doteq \mathbf{0}$)

itself, while the rule, assuming by induction that m_1 and m_2 have the same size, ensures that $s_1 \oplus m_1$ and $s_2 \oplus m_2$ have the same size. An alternative way to define that two markings m_1 and m_2 are related by $R^{\oplus}$ is to state that m_1 can be represented as $s_1 \oplus s_2 \oplus \ldots \oplus s_k$, m_2 can be represented as $s'_1 \oplus s'_2 \oplus \ldots \oplus s'_k$ and $(s_i, s'_i) \in R$ for $i = 1, \ldots, k$. Note that if R is an equivalence relation, then $R^{\oplus}$ is also an equivalence relation. The complexity of checking if two markings m_1 and m_2 of equal size are related by $R^{\oplus}$ is very low. In fact, if R is implemented as an adjacency matrix, then the complexity of checking if two markings m_1 and m_2 (represented as a list of places with multiplicities) are related by $R^{\oplus}$ is $O(k^2)$, where k is the size of the markings, since the problem is essentially that of finding for each element s_1 of m_1 a matching, R-related element s_2 of m_2.

Definition 8 (Team equivalence). *Given an FSM net $N = (S, A, T)$ and the bisimulation equivalence $\sim \subseteq S \times S$, we define* team equivalence $\sim^{\oplus} \subseteq \mathscr{M}(S) \times \mathscr{M}(S)$, *as the additive closure of $\sim$.* □

Proposition 1. *For any FSM $N = (S, A, T)$, if $m_1 \sim^{\oplus} m_2$, then $|m_1| = |m_2|$.* □

Example 3. Continuing Example 2 about Fig. 2, we have, for instance, that the marking $s_1 \oplus 2 \cdot s_2 \oplus s_3 \oplus s_4$ is team equivalent to any marking with one token in s_5, two tokens distributed over s_6 and s_7, and two tokens in s_8; e.g., $s_5 \oplus s_6 \oplus s_7 \oplus 2 \cdot s_8$ or $s_5 \oplus 2 \cdot s_7 \oplus 2 \cdot s_8$. Note that $m_1 = s_1 \oplus 2 \cdot s_2$ has the same size of $m_2 = 2 \cdot s_5 \oplus s_6$, but the two are not team equivalent; in fact, we can first match s_1 with one instance of s_5; then, one instance of s_2 with s_6; but, now, we are unable to find a match for the second instance of s_2, because the only element left in m_2 is s_5, and s_2 is not bisimilar to s_5. □

Example 4. Team equivalence is a truly concurrent equivalence. According to [9], the sequential FSM in Fig. 3(a) denotes (actually, it is isomorphic to) the net for the sequential CFM (see next Section) process term $a.b.\mathbf{0} + b.a.\mathbf{0}$, which can perform the two actions a and b in either order. On the contrary, the concurrent FSM in (b) denotes the net for the parallel CFM process term $a.\mathbf{0} | b.\mathbf{0}$. Note that s_1 is not team equivalent to $s_4 \oplus s_5$, because the two markings have different size. Nonetheless, s_1 and $s_4 \oplus s_5$ are interleaving bisimilar [9,10]. □

Example 5. If two markings m_1 and m_2 are interleaving bisimilar and have the same size, then they may be not team equivalent. For instance, consider Fig. 3(c), which denotes the net for the CFM process term $a.C \mid b.\mathbf{0}$, where C is a constant with empty body, i.e., $C \doteq \mathbf{0}$. Markings $s_4 \oplus s_5$ and $s_6 \oplus s_7$ have the same size, they are interleaving bisimilar (actually, they are even fully concurrent bisimilar [3]), but they are not team equivalent: even if $s_5 \sim s_7$, the residuals are not team bisimilar: $s_4 \nsim s_6$. □

3 CFM: Syntax and Net Semantics

Let *Act* be a finite set of actions, ranged over by μ, and let $\mathscr{C}$ be a finite set of constants, disjoint from *Act*, ranged over by $A, B, C, \ldots$. The size of the sets *Act* and $\mathscr{C}$ is not important: we assume that they can be chosen as large as needed. The CFM *terms* (where CFM is the acronym of *Concurrent Finite-state Machines*) are generated from actions and constants by the following abstract syntax:

$$
\begin{array}{lll}
s ::= \mathbf{0} \mid \mu.q \mid s+s & & \textit{guarded processes} \\
q ::= s \mid C & & \textit{sequential processes} \\
p ::= q \mid p \mid p & & \textit{parallel processes}
\end{array}
$$

where $\mathbf{0}$ is the empty process, $\mu.q$ is a process where action μ prefixes the residual q ($\mu.-$ is the *action prefixing* operator), $s_1 + s_2$ denotes the alternative composition of s_1 and s_2 ($-+-$ is the *choice* operator), $p_1 \mid p_2$ denotes the asynchronous parallel composition of p_1 and p_2 and C is a constant. A constant C may be equipped with a definition, but this must be a guarded process, i.e., $C \doteq s$. A term p is a CFM *process* if each constant in $Const(p)$ (the set of constants used by p; see [9] for details) is equipped with a defining equation (in category s). The set of CFM processes is denoted by $\mathscr{P}_{CFM}$, the set of its sequential processes, i.e., those in syntactic category q, by $\mathscr{P}_{CFM}^{seq}$ and the set of its guarded processes, i.e., those in syntactic category s, by $\mathscr{P}_{CFM}^{grd}$.

The net for CFM is such that the set of places S_{CFM} is the set of the sequential CFM processes, without $\mathbf{0}$, i.e., $S_{CFM} = \mathscr{P}_{CFM}^{seq} \setminus \{\mathbf{0}\}$. The decomposition function $dec : \mathscr{P}_{CFM} \to \mathscr{M}(S_{CFM})$, mapping process terms to markings, is defined in Table 1. An easy induction proves that for any $p \in \mathscr{P}_{CFM}$, $dec(p)$ is a finite multiset of sequential processes. Note that, if $C \doteq \mathbf{0}$, then $\theta = dec(\mathbf{0}) \neq dec(C) = \{C\}$.

Table 1. Decomposition function

$dec(\mathbf{0}) = \theta$	$dec(\mu.p) = \{\mu.p\}$
$dec(p+p') = \{p+p'\}$	$dec(C) = \{C\}$
$dec(p \mid p') = dec(p) \oplus dec(p')$	

Table 2. Denotational net semantics

$$
\begin{array}{rl}
[\![\mathbf{0}]\!]_I = (\emptyset,\emptyset,\emptyset,\emptyset) & \\
[\![\mu.p]\!]_I = (S,A,T,\{\mu.p\}) & \text{given } [\![p]\!]_I = (S',A',T',dec(p)) \text{ and where} \\
 & S=\{\mu.p\}\cup S',\ A=\{\mu\}\cup A',\ T=\{(\{\mu.p\},\mu,dec(p))\}\cup T' \\
[\![p_1+p_2]\!]_I = (S,A,T,\{p_1+p_2\}) & \text{given } [\![p_i]\!]_I = (S_i,A_i,T_i,dec(p_i)) \text{ for } i=1,2, \text{ and where} \\
 & S=\{p_1+p_2\}\cup S'_1\cup S'_2, \text{ with, for } i=1,2, \\
 & S'_i = \begin{cases} S_i & \exists t\in T_i \text{ such that } t^\bullet(p_i)>0 \vee p_i=\mathbf{0}, \\ S_i\setminus\{p_i\} & \text{otherwise} \end{cases} \\
 & A=A_1\cup A_2,\ T=T'\cup T'_1\cup T'_2, \text{ with, for } i=1,2, \\
 & T'_i = \begin{cases} T_i & \exists t\in T_i.t^\bullet(p_i)>0 \vee p_i=\mathbf{0}, \\ T_i\setminus\{t\in T_i \mid {}^\bullet t(p_i)>0\} & \text{otherwise} \end{cases} \\
 & T'=\{(\{p_1+p_2\},\mu,m) \mid (\{p_i\},\mu,m)\in T_i, i=1,2\} \\
[\![C]\!]_I = (\{C\},\emptyset,\emptyset,\{C\}) & \text{if } C\in I \\
[\![C]\!]_I = (S,A,T,\{C\}) & \text{if } C\notin I, \text{ given } C\doteq p \text{ and } [\![p]\!]_{I\cup\{C\}} = (S',A',T',dec(p)) \\
 & A=A', S=\{C\}\cup S'', \text{ where} \\
 & S'' = \begin{cases} S' & \exists t\in T'.t^\bullet(p)>0 \vee p=\mathbf{0}, \\ S'\setminus\{p\} & \text{otherwise} \end{cases} \\
 & T=\{(\{C\},\mu,m) \mid (\{p\},\mu,m)\in T'\}\cup T'' \text{ where} \\
 & T'' = \begin{cases} T' & \exists t\in T'.t^\bullet(p)>0 \vee p=\mathbf{0}, \\ T'\setminus\{t\in T' \mid {}^\bullet t(p)>0\} & \text{otherwise} \end{cases} \\
[\![p_1 \mid p_2]\!]_I = (S,A,T,m_0) & \text{given } [\![p_i]\!]_I = (S_i,A_i,T_i,m_i) \text{ for } i=1,2, \text{ and where} \\
 & S=S_1\cup S_2,\ A=A_1\cup A_2,\ T=T_1\cup T_2,\ m_0=m_1\oplus m_2
\end{array}
$$

Now we provide a construction of the net system $[\![p]\!]_\emptyset$ associated with process p, which is compositional and denotational in style. The details of the construction are outlined in Table 2. The mapping is parametrized by a set of constants that has already been found while scanning p; such a set is initially empty and it is used to avoid looping on recursive constants. The definition is syntax driven and also the places of the constructed net are syntactic objects, i.e., CFM sequential process terms. For instance, the net system $[\![a.\mathbf{0}]\!]_\emptyset$ is a net composed of one single marked place, namely process $a.\mathbf{0}$, and one single transition $(\{a.\mathbf{0}\}, a, \emptyset)$. A bit of care is needed in the rule for choice: in order to include only strictly necessary places and transitions, the initial place p_1 (or p_2) of the subnet $[\![p_1]\!]_I$ (or $[\![p_2]\!]_I$) is to be kept in the net for $p_1 + p_2$ only if there exists a transition reaching place p_1 (or p_2) in $[\![p_1]\!]_I$ (or $[\![p_2]\!]_I$), otherwise p_1 (or p_2) can be safely removed in the new net. Similarly, for the rule for constants.

Example 6. Consider constant $B \doteq b.A$, where $A \doteq a.b.A$. By using the definitions in Table 2, $[\![A]\!]_{\{A,B\}} = (\{A\}, \emptyset, \emptyset, \{A\})$. Then, by action prefixing,

$[\![b.A]\!]_{\{A,B\}} = (\{b.A, A\}, \{b\}, \{(\{b.A\}, b, \{A\})\}, \{b.A\})$. Again, by action prefixing,

$[\![a.b.A]\!]_{\{A,B\}} = (\{a.b.A, b.A, A\}, \{a, b\}, \{(\{a.b.A\}, a, \{b.A\}), (\{b.A\}, b, \{A\})\}, \{a.b.A\})$.

Now, the rule for constants ensures that

$[\![A]\!]_{\{B\}} = (\{b.A, A\}, \{a, b\}, \{(\{A\}, a, \{b.A\}), (\{b.A\}, b, \{A\})\}, \{A\})$.

Note that place $a.b.A$ has been removed, as no transition in $[\![a.b.A]\!]_{\{A,B\}}$ reaches that place. By action prefixing,

$[\![b.A]\!]_{\{B\}} = (\{b.A, A\}, \{a, b\}, \{(\{A\}, a, \{b.A\}), (\{b.A\}, b, \{A\})\}, \{b.A\})$,

i.e., this operation changes only the initial marking, but does not affect the underlying net! Finally,

$[\![B]\!]_\emptyset = (\{B, b.A, A\}, \{a, b\}, \{(\{B\}, b, \{A\}), (\{A\}, a, \{b.A\}), (\{b.A\}, b, \{A\})\}, \{B\})$.

Note that place $b.A$ has been kept, because there is a transition in the net $[\![b.A]\!]_{\{B\}}$ that reaches that place. □

Example 7. Consider the CFM process $B \mid b.A \mid C \mid C$, where $B \doteq b.A$, $A \doteq a.b.A$ and $C \doteq a.C$. The nets for the processes $b.A$ and B are described in Example 6. The concurrent FSM associated with $B \mid b.A$ is

$[\![B \mid b.A]\!]_\emptyset =$

$(\{B, b.A, A\}, \{a, b\}, \{(\{B\}, b, \{A\}), (\{A\}, a, \{b.A\}), (\{b.A\}, b, \{A\})\}, \{B, b.A\})$,

where the only addition to the net for B is one token in place $b.A$. The sequential FSM for C is $[\![C]\!]_\emptyset = (\{C\}, \{a\}, \{(\{C\}, a, \{C\})\}, \{C\})$.

The concurrent FSM for $C \mid C$ is

$[\![C \mid C]\!]_\emptyset = (\{C\}, \{a\}, \{(\{C\}, a, \{C\})\}, \{C, C\})$.

And the whole net for $B \mid b.A \mid C \mid C$ is $[\![B \mid b.A \mid C \mid C]\!]_\emptyset = (S, A, T, m_0)$, where

$S = \{B, b.A, A, C\}$,

$A = \{a, b\}$,

$T = \{(\{B\}, b, \{A\}), (\{A\}, a, \{b.A\}), (\{b.A\}, b, \{A\}), (\{C\}, a, \{C\})\}$,

$m_0 = \{B, b.A, C, C\}$.

The resulting net is depicted in Fig. 4. □

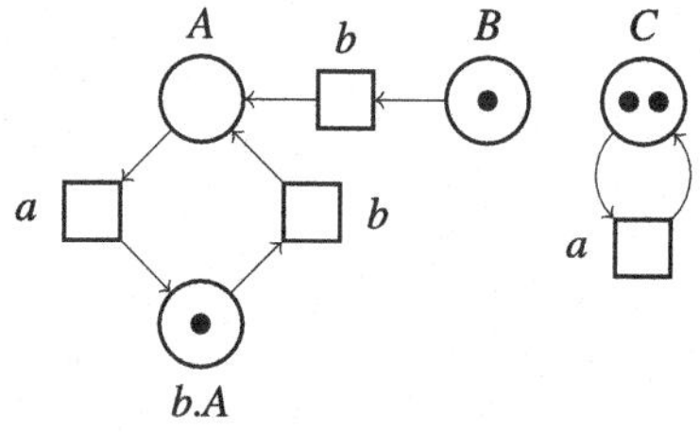

Fig. 4. The concurrent finite-state machine for $B \mid b.A \mid C \mid C$ of Example 7

We now list some properties of the semantics, whose proofs are in [9], which state that CFM really represents the class of FSMs.

Theorem 1 (Only Concurrent FSMs). *For each CFM process p, $[\![p]\!]_\emptyset$ is a concurrent finite-state machine.* □

Definition 9 (Translating Concurrent FSMs into CFM Process Terms). *Let $N(m_0) = (S, A, T, m_0)$—with $S = \{s_1, \ldots, s_n\}$, $A \subseteq Act$, $T = \{t_1, \ldots, t_k\}$, and $l(t_j) = \mu_j$—be a concurrent finite-state machine. Function $\mathscr{T}_{CFM}(-)$, from concurrent finite-state machines to CFM processes, is defined as*

$$\mathscr{T}_{CFM}(N(m_0)) = \underbrace{C_1|\cdots|C_1}_{m_0(s_1)}|\cdots|\underbrace{C_n|\cdots|C_n}_{m_0(s_n)}$$

where each C_i is equipped with a defining equation $C_i \doteq c_i^1 + \cdots + c_i^k$ (with $C_i \doteq \mathbf{0}$ if $k = 0$), and each summand c_i^j, for $j = 1, \ldots, k$, is equal to

- $\mathbf{0}$, *if $s_i \notin {}^\bullet t_j$;*
- $\mu_j.\mathbf{0}$, *if ${}^\bullet t_j = \{s_i\}$ and $t_j^\bullet = \emptyset$;*
- $\mu_j.C_h$, *if ${}^\bullet t_j = \{s_i\}$ and $t_j^\bullet = \{s_h\}$.* □

Theorem 2 (All Concurrent FSMs). *Let $N(m_0) = (S, A, T, m_0)$ be a dynamically reduced, concurrent finite-state machine such that $A \subseteq Act$, and let $p = \mathscr{T}_{CFM}(N(m_0))$. Then, $[\![p]\!]_\emptyset$ is isomorphic to $N(m_0)$.* □

4 Congruence and Algebraic Properties

Definition 10. *Two CFM processes p and q are team bisimilar, denoted $p \sim^\oplus q$, if, by considering the (union of the) nets $[\![p]\!]_\emptyset$ and $[\![q]\!]_\emptyset$, we have that $dec(p) \sim^\oplus dec(q)$.* □

For sequential CFM processes, team equivalence $\sim^\oplus$ coincides with bisimilarity $\sim$. Now we show that team equivalence is a congruence for all the CFM operators.

Proposition 2. *(1) For each $p, q \in \mathscr{P}_{CFM}^{seq}$, if $p \sim q$ (or $p = q = \mathbf{0}$), then $\mu.p \sim \mu.q$ for all $\mu \in Act$.*

(2) For each $p, q \in \mathscr{P}_{CFM}^{grd}$, if $p \sim q$ (or $p = q = \mathbf{0}$), then $p + r \sim q + r$ for all $r \in \mathscr{P}_{CFM}^{grd}$.

Proof. Assume R is a bisimulation such that $(p, q) \in R$ (or $R = \emptyset$ in case $p = q = \mathbf{0}$). For case 1, consider, for each $\mu \in Act$, relation $R_\mu = \{(\mu.p, \mu.q)\} \cup R$. It is very easy to check that R_μ is a bisimulation on places. For case 2, it is very easy to check that, for each $r \in \mathscr{P}_{CFM}^{grd}$, the relation $R_r = \{(p{+}r, q{+}r)\} \cup R \cup \mathscr{I}_r$ is a bisimulation, where $\mathscr{I}_r = \{(r', r') \mid r' \in reach(r), r' \neq \theta\}$ if $r \neq \mathbf{0}$, otherwise $\mathscr{I}_r = \emptyset$. □

Proposition 3. *For every $p, q, r \in \mathscr{P}_{CFM}$, if $p \sim^\oplus q$, then $p\,|\,r \sim^\oplus q\,|\,r$.*

Proof. By induction on the size of $dec(p)$. If $|dec(p)| = 0 = |dec(q)|$, then $p = \mathbf{0} = q$. Hence, $dec(p\,|\,r) = dec(r) = dec(q\,|\,r)$ and the thesis follows trivially, because $\sim^\oplus$ is reflexive. Since $dec(p) \sim^\oplus dec(q)$, if $|dec(p)| = k + 1$ for some

$k \geq 0$, then by Definition 7, there exist p_1, p_2, q_1, q_2 such that $p_1 \sim q_1$, $dec(p_2) \sim^{\oplus} dec(q_2)$, $dec(p) = p_1 \oplus dec(p_2)$ and $dec(q) = q_1 \oplus dec(q_2)$. Since $|dec(p_2)| = k = |dec(q_2)|$ and $p_2 \sim^{\oplus} q_2$, by induction, we have that $p_2 \,|\, r \sim^{\oplus} q_2 \,|\, r$. Since $p_1 \sim q_1$, by Definition 7, we have that $dec(p \,|\, r) = p_1 \oplus dec(p_2 \,|\, r) \sim^{\oplus} q_1 \oplus dec(q_2 \,|\, r) = dec(q \,|\, r)$. Hence, $p \,|\, r \sim^{\oplus} q \,|\, r$. □

Note that the symmetric cases $r + p \sim r + q$ and $r \,|\, p \sim^{\oplus} r \,|\, q$ are implied by the fact that the operators of choice and parallelism are commutative w.r.t. $\sim$ and $\sim^{\oplus}$, respectively (see Propositions 4 and 6).

Still there is one construct missing: recursion, defined over guarded terms only. Here we simply sketch the issue. Consider an extension of CFM where terms can be constructed using variables, such as $x, y, \ldots$ (which are in syntactic category q): this defines an "open" CFM, where terms may be not given a complete semantics. For instance, $p_1(x) = a.(b.\mathbf{0} + c.x)$ and $p_2(x) = a.(c.x + b.\mathbf{0})$ are open guarded CFM terms. An open term $p(x_1, \ldots, x_n)$ can be *closed* by means of a substitution as follows: $p(x_1, \ldots, x_n)\{r_1/x_1, \ldots, r_n/x_n\}$, with the effect that each occurrence of the free variable x_i is replaced by the *closed* CFM sequential process r_i, for $i = 1, \ldots, n$. For instance, $p_1(x)\{d.\mathbf{0}/x\} = a.(b.\mathbf{0} + c.d.\mathbf{0})$. A natural extension of bisimulation equivalence $\sim$ over open *guarded* terms is as follows: $p(x_1, \ldots, x_n) \sim q(x_1, \ldots, x_n)$ if for all tuples of (closed) CFM sequential terms $(r_1, \ldots, r_n)$,

$$p(x_1, \ldots, x_n)\{r_1/x_1, \ldots, r_n/x_n\} \sim q(x_1, \ldots, x_n)\{r_1/x_1, \ldots, r_n/x_n\}.$$

E.g., it is easy to see that $p_1(x) \sim p_2(x)$. As a matter of fact, for all r, $p_1(x)\{r/x\} = a.(b.\mathbf{0} + c.r) \sim a.(c.r + b.\mathbf{0}) = p_2(x)\{r/x\}$, which can be easily proved by means of the algebraic properties (to be proved below) and the congruence ones of $\sim$. For simplicity's sake, let us now restrict our attention to open guarded terms using a single undefined variable. We can *recursively close* an open term $p(x)$ by means of a recursively defined constant. For instance, $A \doteq p(x)\{A/x\}$. The resulting process constant A is a closed CFM sequential process. By saying that strong bisimilarity is a congruence for recursion we mean the following: If $p(x) \sim q(x)$ and $A \doteq p(x)\{A/x\}$ and $B \doteq q(x)\{B/x\}$, then $A \sim B$. The following theorem states this fact.

Theorem 3. *Let p and q be two open guarded CFM terms, with one variable x at most. Let $A \doteq p\{A/x\}$, $B \doteq q\{B/x\}$ and $p \sim q$. Then $A \sim B$.*

Proof. *Consider $R = \{(r\{A/x\}, r\{B/x\}) \mid r \in reach(p) \cup reach(q), r \neq \theta\}$. Note that when r is x,*[1] *we get $(A, B) \in R$. The proof that R is a strong bisimulation up to $\sim$ [8, 13] is not difficult. By symmetry, it is enough to prove that if $r\{A/x\} \xrightarrow{\mu} p'$, then $r\{B/x\} \xrightarrow{\mu} q'$ with $p' \sim R \sim q'$ (or $p' = \theta = q'$). The proof proceeds by induction on the definition of the net for $r\{A/x\}$.*

[1] The *open* net semantics for open CFM extends the net semantics in Table 2 with $[\![x]\!]_I = (\{x\}, \emptyset, \emptyset, \{x\})$, so that, e.g., the semantics of $a.x$ is $(\{a.x\}, \{a\}, \{(a.x, a, x)\}, a.x)$.

- $r = \mu.r'$. *In this case,* $r\{A/x\} = \mu.r'\{A/x\} \xrightarrow{\mu} r'\{A/x\}$ *(or* $dec(r') = \theta$*). Similarly, transition* $r\{B/x\} = \mu.r'\{B/x\} \xrightarrow{\mu} r'\{B/x\}$ *(or* $dec(r') = \theta$*) is derivable, and* $(r'\{A/x\}, r'\{B/x\}) \in R$ *(or they both reach* θ*).*
- $r = r_1 + r_2$. *In this case,* $r\{A/x\} = r_1\{A/x\} + r_2\{A/x\}$. *A transition from* $r\{A/x\}$*, e.g.,* $r_1\{A/x\} + r_2\{A/x\} \xrightarrow{\mu} p'$*, is derivable only if* $r_i\{A/x\} \xrightarrow{\mu} p'$ *for some* $i = 1, 2$. *Without loss of generality, assume the transition is due to* $r_1\{A/x\} \xrightarrow{\mu} p'$. *Since* r_1 *is guarded,* $r_1\{A/x\} \xrightarrow{\mu} p'$ *is derivable only if* $r_1 \xrightarrow{\mu} \overline{r}$*, with* $p' = \overline{r}\{A/x\}$. *Therefore, we can derive* $r_1\{B/x\} \xrightarrow{\mu} \overline{r}\{B/x\}$ *and so also* $r\{B/x\} = r_1\{B/x\} + r_2\{B/x\} \xrightarrow{\mu} \overline{r}\{B/x\}$*, with* $(\overline{r}\{A/x\}, \overline{r}\{B/x\}) \in R$ *(or* $\overline{r} = \theta$*).*
- $r = D$*, with* $D \doteq s$. *So,* $r\{A/x\} \doteq s\{A/x\}$ *and* $r\{B/x\} \doteq s\{B/x\}$. *If* $r\{A/x\} \xrightarrow{\mu} p'$*, then this is possible only if* $s\{A/x\} \xrightarrow{\mu} p'$. *Since* s *is guarded,* $s\{A/x\} \xrightarrow{\mu} p'$ *is possible only if* $s \xrightarrow{\mu} \overline{s}$ *with* $p' = \overline{s}\{A/x\}$. *Therefore,* $s\{B/x\} \xrightarrow{\mu} \overline{s}\{B/x\}$ *is derivable, too, as well as* $r\{B/x\} \xrightarrow{\mu} \overline{s}\{B/x\}$*, with* $(\overline{s}\{A/x\}, \overline{s}\{B/x\}) \in R$.
- $r = x$. *Then, we have* $r\{A/x\} = A$ *and* $r\{B/x\} = B$. *We want to prove that for each* $A \xrightarrow{\mu} p'$*, there exists a process* q' *such that* $B \xrightarrow{\mu} q'$ *with* $p' \sim R \sim q'$ *(or* $p' = \theta = q'$*). By hypothesis,* $A \doteq p\{A/x\}$*, hence also* $p\{A/x\} \xrightarrow{\mu} p'$ *is a transition in the net for* $p\{A/x\}$. *Since* p *is guarded,* $p\{A/x\} \xrightarrow{\mu} p'$ *is derivable only if* $p \xrightarrow{\mu} \overline{p}$ *with* $p' = \overline{p}\{A/x\}$. *Hence,* $p\{B/x\} \xrightarrow{\mu} \overline{p}\{B/x\}$ *is derivable, too. Since* $p \sim q$*, we have that* $q \xrightarrow{\mu} \overline{q}$ *with* $\overline{p} \sim \overline{q}$ *(or* $\overline{p} = \theta = \overline{q}$*). Hence,* $q\{B/x\} \xrightarrow{\mu} \overline{q}\{B/x\}$ *is derivable, too, with* $\overline{p}\{B/x\} \sim \overline{q}\{B/x\}$ *(or* $\overline{p} = \theta = \overline{q}$*). Since* $B \doteq q\{B/x\}$*,* $B \xrightarrow{\mu} \overline{q}\{B/x\}$ *is derivable, too, with* $\overline{p}\{A/x\} \sim \overline{p}\{A/x\} R \overline{p}\{B/x\} \sim \overline{q}\{B/x\}$ *(or* $\overline{p} = \theta = \overline{q}$*), as required. This concludes the proof that* R *is a strong bisimulation up to* $\sim$. □

The extension to the case of open terms with multiple undefined variables, e.g., $p(x_1, \ldots, x_n)$ can be obtained in a standard way [8,13].

Now we list the algebraic properties of team equivalence. On sequential processes we have the following algebraic laws.

Proposition 4 (Laws of the choice op.). *For each* $p, q, r \in \mathscr{P}_{CFM}^{grd}$*, the following hold:*

$p + (q + r) \sim (p + q) + r$		*(associativity)*
$p + q \sim q + p$		*(commutativity)*
$p + \mathbf{0} \sim p$	*if* $p \neq \mathbf{0}$	*(identity)*
$p + p \sim p$	*if* $p \neq \mathbf{0}$	*(idempotency)*

Proof. For each law, it is enough to exhibit a suitable bisimulation relation. For instance, for idempotency, for each p *in syntactic category* s *(*$p \neq \mathbf{0}$*), take* $R_p = \{(p + p, p)\} \cup \mathscr{I}_p$ *where* $\mathscr{I}_p = \{(q, q) \mid q \in reach(p), q \neq \theta\}$ *is the identity relation. It is an easy exercise to check that* R_p *is a bisimulation on the places of* $[\![p + p]\!]_\emptyset$ *and* $[\![p]\!]_\emptyset$. □

Proposition 5 (Laws of the constant). *For each $p \in \mathscr{P}^{grd}_{CFM}$, and each $C \in \mathscr{C}$, the following hold:*

if $C \doteq \mathbf{0}$, *then*	$C \sim \mathbf{0} + \mathbf{0}$	*(stuck)*
if $C \doteq p$ *and* $p \neq \mathbf{0}$, *then*	$C \sim p$	*(unfolding)*
if $C \doteq p\{C/x\}$ *and* $q \sim p\{q/x\}$ *then*	$C \sim q$	*(folding)*

where in the last law p is actually also open on x (while q is closed).

Proof. *The stuck property is trivial: since the decomposition of a constant is a place, if the body is stuck, it corresponds to a stuck place, such as $\mathbf{0} + \mathbf{0}$. The required bisimulation on places proving the unfolding property is $R_{C,p} = \{(C,p)\} \cup \mathscr{I}_C$, where $\mathscr{I}_C = \{(q,q) \mid q \in reach(C), q \neq \theta\}$ is the identity relation. For the folding property, note that this is implied by the following: if $q_1 \sim p\{q_1/x\}$ and $q_2 \sim p\{q_2/x\}$ then $q_1 \sim q_2$. In fact, if we choose $q_1 = C$, then $C = q_1 \sim p\{q_1/x\} = p\{C/x\}$ (which holds by hypothesis, due to the unfolding property) and $C = q_1 \sim q_2$, which is the thesis. This statement can be easily proven by showing that the relation $R = \{(r\{q_1/x\}, r\{q_2/x\}) \mid r \in reach(p) \cup \{x\}, r \neq \theta\}$ is a bisimulation up to $\sim$ [8,13]. Clearly, when $r = x$, we have that $(q_1, q_2) \in R$. So, it remains to prove the bisimulation (up to) conditions. If $r\{q_1/x\} \xrightarrow{\mu} t$, this can be due to one of the following:*

- *$r \xrightarrow{\mu} r'$ and so $t = r'\{q_1/x\}$. Also $r\{q_2/x\} \xrightarrow{\mu} r'\{q_2/x\}$ is derivable and the pair $(r'\{q_1/x\}, r'\{q_2/x\})$ is in R (if $r' \neq \theta$; otherwise, the condition is trivially satisfied).*
- *$r = x$ and $q_1 \xrightarrow{\mu} q_1'$, and so $t = q_1'$. Assuming $q_1' \neq \theta$, since $q_1 \sim p\{q_1/x\}$ and p is guarded, we have that there exists p' such that $p \xrightarrow{\mu} p'$ and $p\{q_1/x\} \xrightarrow{\mu} p'\{q_1/x\}$ with $q_1' \sim p'\{q_1/x\}$. Therefore, $p\{q_2/x\} \xrightarrow{\mu} p'\{q_2/x\}$ is derivable, too. Since $q_2 \sim p\{q_2/x\}$, it follows that there exists a place q_2' such that $q_2 \xrightarrow{\mu} q_2'$ with $q_2' \sim p'\{q_2/x\}$. Summing up, if $x\{q_1/x\} = q_1 \xrightarrow{\mu} q_1'$, then $x\{q_2/x\} = q_2 \xrightarrow{\mu} q_2'$ such that $q_1' \sim p'\{q_1/x\}$, $(p'\{q_1/x\}, p'\{q_2/x\}) \in R$ and $p'\{q_2/x\} \sim q_2'$, as required by the bisimulation up to condition. Finally, if $q_1' = \theta$, then also $p'\{q_1/x\} = \theta$, as well as $p'\{q_2/x\}$ and q_2', so that the thesis follows trivially.*

Simmetrically, if $r\{q_2/x\}$ moves first. Hence, R is a bisimulation up to $\sim$. □

Proposition 6 (Laws of the parallel operator). *For each $p, q, r \in \mathscr{P}_{CFM}$, the following hold:*

$$
\begin{array}{rcll}
p \mid (q \mid r) & \sim^{\oplus} & (p \mid q) \mid r & \textit{(associativity)} \\
p \mid q & \sim^{\oplus} & q \mid p & \textit{(commutativity)} \\
p \mid \mathbf{0} & \sim^{\oplus} & p & \textit{(identity)}
\end{array}
$$

Proof. *To prove that each law is sound, it is enough to observe that the net for the process in the left-hand-side is exactly the same as the net for the process in the right-hand-side. For instance, $[\![p \mid q]\!]_\emptyset = [\![q \mid p]\!]_\emptyset$. In fact, $dec(p \mid q) = dec(p) \oplus dec(q) = dec(q) \oplus dec(p) = dec(q \mid p)$ and the resulting net is obtained by simply joining the net for p with the net for q. Therefore, the identity relation on places, which is a bisimulation, is enough to prove that $dec(p \mid q) \sim^{\oplus} dec(q \mid p)$.* □

5 Axiomatization

In this section we provide a sound and complete, finite axiomatization of team equivalence over CFM. For simplicity's sake, the syntactic definition of open CFM is given with only one syntactic category, but each ground instantiation of an axiom must respect the syntactic definition of CFM given in Sect. 3; this means that we can write the axiom $x + (y + z) = (x + y) + z$, but it is invalid to instantiate it to $C + (a.\mathbf{0} + b.\mathbf{0}) = (C + a.\mathbf{0}) + b.\mathbf{0}$ because these are not legal CFM processes (the constant C cannot be used as a summand).

The set of axioms are outlined in Table 3. We call E the set of axioms {**A1, A2, A3, A4, R1, R2, R3, P1, P2, P3**}. By the notation $E \vdash p = q$ we mean that there exists an equational deduction proof of the equality $p = q$, by using the axioms in E. Besides the usual equational deduction rules of reflexivity, symmetry, transitivity, substitutivity and instantiation (see, e.g., [8]), in order to deal with constants we need also the following recursion congruence rule:

$$\frac{p = q \ \wedge \ A \doteq p\{A/x\} \ \wedge \ B \doteq q\{B/x\}}{A = B}$$

The axioms **A1**–**A4** are the usual axioms for choice where, however, **A3**–**A4** have the side condition $x \neq \mathbf{0}$; hence, it is not possible to prove $E \vdash \mathbf{0} + \mathbf{0} = \mathbf{0}$, as expected, because these two terms have a completely different semantics. The conditional axioms **R1**–**R3** are about process constants. Note that axiom **R2** requires that p is not (equal to) $\mathbf{0}$ (condition $p \neq \mathbf{0}$). Note also that these conditional axioms are actually a finite collection of axioms, one for each constant definition: since the set $\mathscr{C}$ of process constants is finite, the instances of **R1**–**R3** are finitely many. Finally, we have axioms **P1**–**P3** for parallel composition.

Table 3. Axioms for team equivalence

A1	Associativity	$x+(y+z) = (x+y)+z$	
A2	Commutativity	$x+y = y+x$	
A3	Identity	$x+\mathbf{0} = x$	if $x \neq \mathbf{0}$
A4	Idempotence	$x+x = x$	if $x \neq \mathbf{0}$
R1	Stuck	if $C \doteq \mathbf{0}$, then $C = \mathbf{0}+\mathbf{0}$	
R2	Unfolding	if $C \doteq p \wedge p \neq \mathbf{0}$, then $C = p$	
R3	Folding	if $C \doteq p\{C/x\} \wedge q = p\{q/x\}$, then $C = q$	
P1	Associativity	$x \mid (y \mid z) = (x \mid y) \mid z$	
P2	Commutativity	$x \mid y = y \mid x$	
P3	Identity	$x \mid \mathbf{0} = x$	

Theorem 4 (Soundness). *For every $p, q \in \mathscr{P}_{CFM}$, if $E \vdash p = q$, then $p \sim^{\oplus} q$.*

Proof. *The proof is by induction on the proof of $E \vdash p = q$. The thesis follows by observing that all the axioms in E are sound by Propositions 4, 5 and 6 and that $\sim^{\oplus}$ is a congruence.* □

Proposition 7 (Unique solution). *Let $\widetilde{X} = (x_1, x_2, \ldots, x_n)$ be a tuple of variables and let $\widetilde{p} = (p_1, p_2, \ldots, p_n)$ be a tuple of open guarded CFM terms (in syntactic category s), using the variables in $\widetilde{X}$. Then, there exists a tuple $\widetilde{q} = (q_1, q_2, \ldots, q_n)$ of closed sequential CFM terms such that*

$E \vdash q_i = p_i\{\widetilde{q}/\widetilde{X}\}$ *for* $i = 1, \ldots, n$.

Moreover, if the same property holds for $\widetilde{q'} = (q'_1, q'_2, \ldots, q'_n)$, then

$E \vdash q'_i = q_i$ *for* $i = 1, \ldots, n$.

Proof. *By induction on n. For $n = 1$, if $p_1 = \mathbf{0}$, then $p_1\{q_1/x_1\} = \mathbf{0}$ and so the unique value for q_1 is $q_1 = \mathbf{0}$. Otherwise, we choose $q_1 = C_1$, and we close this constant with the definition $C_1 \doteq p_1\{C_1/x_1\}$, and so the result follows immediately using axiom* **R2**. *This solution is unique: if $E \vdash r_1 = p_1\{r_1/x_1\}$, since $C_1 \doteq p_1\{C_1/x_1\}$, by axiom* **R3** *we get $E \vdash C_1 = r_1$.*

Now assume a tuple $\widetilde{p} = (p_1, p_2, \ldots, p_n)$ and the term p_{n+1}, so that they are all open on $\widetilde{X} = (x_1, x_2, \ldots, x_n)$ and the additional x_{n+1}. Assume, w.l.o.g., that x_{n+1} occurs in p_{n+1}. First, define $C_{n+1} \doteq p_{n+1}\{C_{n+1}/x_{n+1}\}$, so that C_{n+1} is now open on $\widetilde{X}$. Therefore, also for $i = 1, \ldots, n$, each $p_i\{C_{n+1}/x_{n+1}\}$ is now open on $\widetilde{X}$. Thus, we are now able to use induction on $\widetilde{X}$ and $(p_1\{C_{n+1}/x_{n+1}\}, \ldots, p_n\{C_{n+1}/x_{n+1}\})$, to conclude that there exists a tuple $\widetilde{q} = (q_1, q_2, \ldots, q_n)$ of closed sequential CFM terms such that

$E \vdash q_i = (p_i\{C_{n+1}/x_{n+1}\})\{\widetilde{q}/\widetilde{X}\} = p_i\{\widetilde{q}/\widetilde{X}, C_{n+1}\{\widetilde{q}/\widetilde{X}\}/x_{n+1}\}$ *for* $i = 1, \ldots, n$.

Note that above by $C_{n+1}\{\widetilde{q}/\widetilde{X}\}$ we have implicitly closed the definition of C_{n+1} as

$C_{n+1} \doteq p_{n+1}\{C_{n+1}/x_{n+1}\}\{\widetilde{q}/\widetilde{X}\} = p_{n+1}\{\widetilde{q}/\widetilde{X}\}\{C_{n+1}/x_{n+1}\}$,

so that C_{n+1} can be chosen as q_{n+1}. By axiom **R2**, *$E \vdash C_{n+1} = p_{n+1}\{\widetilde{q}/\widetilde{X}\}\{C_{n+1}/x_{n+1}\}$.*

Unicity of the tuple $(\widetilde{q}, q_{n+1})$ can be proved by using axiom **R3**. *Assume to have another solution tuple $(\widetilde{q'}, q'_{n+1})$. This means that*

$E \vdash q'_i = p_i\{\widetilde{q'}/\widetilde{X}, q'_{n+1}/x_{n+1}\}$ *for* $i = 1, \ldots, n+1$.

By induction, we can assume that $E \vdash q_i = q'_i$, for $i = 1, \ldots, n$.

Since $E \vdash C_{n+1} = p_{n+1}\{\widetilde{q}/\widetilde{X}\}\{C_{n+1}/x_{n+1}\}$ by axiom **R2**, *by substitutivity we get $E \vdash C_{n+1} = p_{n+1}\{\widetilde{q'}/\widetilde{X}\}\{C_{n+1}/x_{n+1}\}$. Let F be a constant defined as follows: $F \doteq p_{n+1}\{\widetilde{q'}/\widetilde{X}\}\{F/x_{n+1}\}$. Then, by axiom* **R3**, *$E \vdash C_{n+1} = F$. Hence, since*

$E \vdash q'_{n+1} = p_{n+1}\{\widetilde{q'}/\widetilde{X}\}\{q'_{n+1}/x_{n+1}\}$

by axiom **R3**, *we get $E \vdash F = q'_{n+1}$; so the thesis $E \vdash C_{n+1} = q'_{n+1}$ by transitivity.* □

Proposition 8 (Equational characterization). *For every $p \in \mathscr{P}_{CFM}^{seq}$, there exist $k \geq 1$ and $p_1, p_2, \ldots, p_k \in \mathscr{P}_{CFM}^{seq}$ such that, for $i = 1, \ldots, k$, $E \vdash p_i = p'_i$ where p'_i can be either $\mathbf{0}$ or $\mathbf{0}+\mathbf{0}$ or $\sum_{j=1}^{n(i)} a_{ij}.p_{f(i,j)}$ (with $n(i) \geq 1$ and $f(i,j) \in \{1, 2, \ldots, k\}$), and $E \vdash p = p_1$.*

Proof. By induction on the syntactic definition of p. If $p = \mathbf{0}$, then $k = 1$ and $p_1 = \mathbf{0} = p'_1$. If $p = \mu.q$, then by induction there exist $k \geq 1$ and $q_1, q_2, \ldots, q_k$ such that for $i = 1, \ldots, k$, $E \vdash q_i = q'_i$ where q'_i can be either $\mathbf{0}$ or $\mathbf{0} + \mathbf{0}$ or a sumform $\sum_{j=1}^{n(i)} a_{ij}.q_{f(i,j)}$, and $E \vdash q = q_1$. Then, $p_1 = \mu.q_1$, $p_2 = q_1, \ldots, p_{k+1} = q_k$. And so, $p'_1 = \mu.p_2$ (i.e., a singleton sumform), $p'_2 = q'_1, \ldots, p'_{k+1} = q'_k$.

If $p = r_1 + r_2$, then, by induction there exist $r_1^1, \ldots r_{k_1}^1$ and $r_1^2, \ldots r_{k_2}^2$, such that $E \vdash r_i^j = s_i^j$, where s_i^j can be either $\mathbf{0}$ or $\mathbf{0}+\mathbf{0}$ or a sumform, and, moreover, $E \vdash r_1 = r_1^1 = s_1^1$ and $E \vdash r_2 = r_1^2 = s_1^2$. We can take $p_1 = r_1^1 + r_1^2$ and the other $k_1 + k_2$ terms p_i are $r_1^1, \ldots r_{k_1}^1$ and $r_1^2, \ldots r_{k_2}^2$. Since for each r_i^j there is already a suitable s_i^j, it remains to define p'_1. If s_1^1 is $\mathbf{0}$ or $\mathbf{0} + \mathbf{0}$ and s_1^2 is $\mathbf{0}$ or $\mathbf{0} + \mathbf{0}$, then $k = 1$ and $p'_1 = \mathbf{0} + \mathbf{0}$ (by axiom **A3**). If s_1^1 is $\mathbf{0}$ or $\mathbf{0} + \mathbf{0}$ and $s_1^2 = \sum_{j=1}^{n_2(1)} a_{1j}.r^2_{f_2(1,j)}$, then $p'_1 = s_1^2$ (by axiom **A3**). Symmetrically, if s_1^2 is $\mathbf{0}$ or $\mathbf{0} + \mathbf{0}$ and $s_1^1 = \sum_{j=1}^{n_1(1)} a_{1j}.r^1_{f_1(1,j)}$. If $s_1^1 = \sum_{j=1}^{n_1(1)} a_{1j}.r^1_{f_1(1,j)}$ and $s_1^2 = \sum_{j=1}^{n_2(1)} a'_{1j}.r^2_{f_2(1,j)}$, then $p'_1 = \sum_{j=1}^{n_1(1)} a_{1j}.r^1_{f_1(1,j)} + \sum_{j=1}^{n_2(1)} a'_{1j}.r^2_{f_2(1,j)}$.

If $p = C$ and $C \doteq r$, then[2] by induction, for r there exist $k \geq 1$ and $r_1, \ldots, r_k$ such that for $i = 1, \ldots, k$, $E \vdash r_i = r'_i$ where r'_i can be either $\mathbf{0}$ or $\mathbf{0} + \mathbf{0}$ or a sumform $\sum_{j=1}^{n(i)} a_{ij}.r_{f(i,j)}$, and $E \vdash r = r_1$. If r'_1 is $\mathbf{0}$, then $k = 1$, $r = r_1 = \mathbf{0}$ and $p_1 = p'_1 = \mathbf{0}+\mathbf{0}$ by axiom **R1**. If r'_1 is $\mathbf{0}+\mathbf{0}$, then $k = 1$ and $p_1 = p'_1 = \mathbf{0}+\mathbf{0}$. If r'_1 is $\sum_{j=1}^{n(1)} a_{1j}.r_{f(1,j)}$, then we get $E \vdash C = r'_1$, because $E \vdash r = r_1$, $E \vdash r_1 = r'_1$ and by axiom **R2** $E \vdash C = r$; hence, $p_i = r_i$ and $p'_i = r'_i$ for $i = 1, \ldots, k$. □

Lemma 1 (Completeness for sequential terms). *For every $p, p' \in \mathscr{P}_{CFM}^{seq}$, if $p \sim p'$ (or $p = \mathbf{0} = p'$), then $E \vdash p = p'$.*

Proof. If $p = \mathbf{0} = p'$, then $E \vdash p = p'$ by reflexivity. Otherwise, by Proposition 8, we have that there exist $p_1, p_2, \ldots, p_k$ and $r_1, r_2, \ldots, r_k$ such that $E \vdash p = p_1$, for $i = 1, \ldots, k$, $E \vdash p_i = r_i$ and r_i is either $\mathbf{0}$ or $\mathbf{0} + \mathbf{0}$ or a sumform $\sum_{j=1}^{n(i)} a_{ij}.p_{f(i,j)}$. Similarly, there exist $p'_1, p'_2, \ldots, p'_h$ and $r'_1, r'_2, \ldots, r'_h$ such that $E \vdash p' = p'_1$, for $i = 1, \ldots, h$, $E \vdash p'_i = r'_i$ and r'_i is either $\mathbf{0}$ or $\mathbf{0} + \mathbf{0}$ or a sumform $\sum_{j=1}^{n'(i)} a'_{ij}.p'_{f'(i,j)}$.

By Theorem 4, we have that $p \sim p_1 \sim r_1$ and $p' \sim p'_1 \sim r'_1$; by transitivity, we have that $p_1 \sim p'_1$ and $r_1 \sim r'_1$. If $r_1 = \mathbf{0} + \mathbf{0}$, then also $r'_1 = \mathbf{0} + \mathbf{0}$, and so $E \vdash r_1 = r'_1$ by reflexivity and $E \vdash p = p'$ by transitivity. Otherwise, let $r_1 = \sum_{j=1}^{n(1)} a_{1j}.p_{f(1,j)}$ and $r'_1 = \sum_{j=1}^{n'(1)} a'_{1j}.p'_{f'(1,j)}$. Now, let $I = \{(i, i') \mid p_i \sim$

[2] Note that if C has been already encountered while scanning the original term p, then the resulting term is C itself, and so the proof does not loop; indeed, to be precise, induction is on the syntactic definition parametrized w.r.t. a set of already encountered constants (initially empty); however, this parametrization is not explicit in the proof.

$p'_{i'}\}$. *Clearly,* $(1,1) \in I$. *Moreover, since* p_i *and* $p'_{i'}$ *are bisimilar when* $(i,i') \in I$, *the following hold: for each* $(i,i') \in I$, *there exists a total surjective relation* $J_{ii'}$ *between* $\{1,2,\ldots n(i)\}$ *and* $\{1,2,\ldots n'(i')\}$ *given by* $J_{ii'} = \{(j,j') \mid a_{ij} = a'_{i'j'} \wedge (f(i,j), f'(i',j')) \in I\}$. *Now, for each* $(i,i') \in I$, *let us consider the open terms*

$$t_{ii'} = \sum_{(j,j') \in J_{ii'}} a_{ij}.x_{f(i,j),f'(i',j')}$$

By Proposition 7, for each $(i,i') \in I$, *there exists* $s_{ii'}$ *such that* $E \vdash s_{ii'} = t_{ii'}\{\widetilde{s}/\widetilde{X}\}$, *where* $\widetilde{s}$ *denotes the tuple of terms* $s_{ii'}$ *for each* $(i,i') \in I$, *and* $\widetilde{X}$ *denotes the tuple of variables* $x_{ii'}$ *for each* $(i,i') \in I$. *In particular, we have* $E \vdash s_{ii'} = \sum_{(j,j') \in J_{ii'}} a_{ij}.s_{f(i,j),f'(i',j')}$. *If we close each* $t_{ii'}$ *by replacing* $x_{f(i,j),f'(i',j')}$ *with* $p_{f(i,j)}$, *we get*

$\sum_{(j,j') \in J_{ii'}} a_{ij}.p_{f(i,j)}$

which is equal, via axioms **A1-A4**, *to* r_i: *in fact,* $J_{ii'}$ *is surjective so that the two summations differ only for possible repeated summands. Since* $E \vdash p_i = r_i$ *for* $i = 1, \ldots, k$, *we get that* $E \vdash r_i = \sum_{(j,j') \in J_{ii'}} a_{ij}.r_{f(i,j)}$. *Therefore, we note that* r_i *is such that* $E \vdash r_i = t_{ii'}\{\widetilde{r}/\widetilde{X}\}$ *and so, by Proposition 7, we have that* $E \vdash s_{ii'} = r_i$. *Since* $(1,1) \in I$, *we have that* $E \vdash s_{11} = r_1$. *Similarly, if we close each* $t_{ii'}$ *by replacing* $x_{f(i,j),f'(i',j')}$ *with* $p'_{f'(i',j')}$, *we get*

$\sum_{(j,j') \in J_{ii'}} a_{ij}.p'_{f'(i',j')}$

which is equal, via axioms **A1-A4**, *to* $r'_{i'}$. *Since* $E \vdash p'_i = r'_i$ *for* $i = 1, \ldots, h$, *we get that* $E \vdash r'_{i'} = \sum_{(j,j') \in J_{ii'}} a_{ij}.r'_{f'(i',j')}$. *Therefore, we note that* $r'_{i'}$ *is such that* $E \vdash r'_{i'} = t_{ii'}\{\widetilde{r'}/\widetilde{X}\}$ *and so, by Proposition 7, we have that* $E \vdash s_{ii'} = r'_{i'}$. *Since* $(1,1) \in I$, *we have that* $E \vdash s_{11} = r'_1$; *by transitivity, it follows that* $E \vdash r_1 = r'_1$, *and so that* $E \vdash p = p'$. □

Theorem 5 (Completeness). *For every* $p, q \in \mathscr{P}_{CFM}$, *if* $p \sim^{\oplus} q$, *then* $E \vdash p = q$.

Proof. The proof is by induction on the size of dec(p). If $|dec(p)| = 0$, *then* $p = \mathbf{0} = q$, *and* $E \vdash p = q$ *by reflexivity. If* $|dec(p)| = k+1$, *then there exist* p_1, p_2, q_1, q_2 *such that* $dec(p) = p_1 \oplus dec(p_2)$, $dec(q) = q_1 \oplus dec(q_2)$, $p_1 \sim q_1$ *and* $dec(p_2) \sim^{\oplus} dec(q_2)$. *By the definition of the decomposition function and by axioms* **P1-P3**, *this means that* $E \vdash p = p_1 \,|\, p_2$ *and* $E \vdash q = q_1 \,|\, q_2$. *By Lemma 1 we have that* $E \vdash p_1 = q_1$. *By induction, we have that* $E \vdash p_2 = q_2$. *By substitutivity we get* $E \vdash p_1 \,|\, p_2 = q_1 \,|\, q_2$ *and so the thesis follows by transitivity.* □

6 Conclusion and Future Research

Team equivalence is a truly concurrent equivalence which seems the most natural, intuitive and simple extension of LTS bisimulation equivalence to FSMs. It also has a very low complexity: checking whether two places are bisimilar can be done in $O(m \log (n+1))$ time, where m is the number of transitions and n

the number of places, and then, once bisimilarity on places has been computed, checking whether two markings of size k are team equivalent can be computed in $O(k^2)$ time. As, in order to perform team equivalence checking, there is no need to compute the LTSs of the global behavior of the systems under scrutiny, our proposal seems a natural solution to solving the state-space explosion problem for FSMs.

Moreover, team equivalence is intuitively appealing as it coincides with *strong place bisimilarity* [1] and is coarser than the branching-time semantics of *isomorphism of (nondeterministic) occurrence nets* (or unfoldings) [6] and finer than the linear-time semantics of *isomorphism of causal (or deterministic occurrence) nets* [2,14]; moreover, it is finer than *history-preserving bisimilarity* (hpb for short) [5,7,18], which on nets takes the form of so-called *fully concurrent bisimilarity* [3]. Hence, strong team equivalence does respect the causal behavior of FSMs.

The axiomatization we have provided for team equivalence is the first *finite*, sound and complete, axiomatization of a truly-concurrent equivalence for a process algebra admitting recursive behavior. It is based on Milner's finite axiomatization of interleaving bisimilarity for finite-state CCS [12], even if our technical treatment, based on guarded constants, is simpler than that. An example of a non-finite axiomatization of a non-interleaving equivalence for a generous process algebra is [4].

Our axiomatization can be adapted to characterize *finitely* also hpb over FSMs: we claim that it is enough to remove the side-condition $x \neq \mathbf{0}$ in axioms **A3-A4**, to drop axiom **R1** and to remove the premise $p \neq \mathbf{0}$ in the conditional axiom **R2**. This explains the essence of the difference between these two equivalences: hpb is slightly coarser than team equivalence because only the former may relate markings with different size where, however, the possible additional components are all stuck. Moreover, to get the axiomatization for the linear-time versions of team equivalence and history-preserving bisimilarity for FSMs, it is enough to add the distributivity axiom: $\mu.(x + y) = \mu.x + \mu.y$.

We plan to prove that the proposed axiomatization of team equivalence is sound and complete also for the more generous class of nets whose only constraint is that transitions have singleton pre-set, namely the class of BPP nets [9], whose language is the BPP process algebra.

Acknowledgments. The anonymous referees are thanked for their useful comments and suggestions. Catuscia Palamidessi is thanked for her illuminating scientific work.

References

1. Autant, C., Belmesk, Z., Schnoebelen, P.: Strong bisimilarity on nets revisited. In: Aarts, E.H.L., van Leeuwen, J., Rem, M. (eds.) PARLE 1991. LNCS, vol. 506, pp. 295–312. Springer, Heidelberg (1991). https://doi.org/10.1007/3-540-54152-7_71
2. Best, E., Devillers, R.: Sequential and concurrent behavior in petri net theory. Theoret. Comput. Sci. **55**(1), 87–136 (1987)
3. Best, E., Devillers, R., Kiehn, A., Pomello, L.: Concurrent bisimulations in petri nets. Acta Inf. **28**(3), 231–264 (1991)

4. Bravetti, M., Gorrieri, R.: Deciding and axiomatizing weak ST bisimulation for a process algebra with recursion and action refinement. ACM Trans. Comput. Log. **3**(4), 465–520 (2002)
5. Degano, P., De Nicola, R., Montanari, U.: Partial orderings descriptions and observations of nondeterministic concurrent processes. In: de Bakker, J.W., de Roever, W.-P., Rozenberg, G. (eds.) REX 1988. LNCS, vol. 354, pp. 438–466. Springer, Heidelberg (1989). https://doi.org/10.1007/BFb0013030
6. van Glabbeek, R.J., Vaandrager, F.W.: Petri net models for algebraic theories of concurrency. In: de Bakker, J.W., Nijman, A.J., Treleaven, P.C. (eds.) PARLE 1987. LNCS, vol. 259, pp. 224–242. Springer, Heidelberg (1987). https://doi.org/10.1007/3-540-17945-3_13
7. van Glabbeek, R.J., Goltz, U.: Equivalence notions for concurrent systems and refinement of actions. In: Kreczmar, A., Mirkowska, G. (eds.) MFCS 1989. LNCS, vol. 379, pp. 237–248. Springer, Heidelberg (1989). https://doi.org/10.1007/3-540-51486-4_71
8. Gorrieri, R., Versari, C.: Introduction to Concurrency Theory: Transition Systems and CCS. TTCSAES. Springer, Cham (2015). https://doi.org/10.1007/978-3-319-21491-7
9. Gorrieri, R.: Process Algebras for Petri Nets: The Alphabetization of Distributed Systems. EATCS Monographs in Theoretical Computer Science. Springer, Cham (2017). https://doi.org/10.1007/978-3-319-55559-1
10. Gorrieri, R.: Verification of finite-state machines: a distributed approach. J. Logic Algebraic Methods Program. **96**, 65–80 (2018)
11. Keller, R.: Formal verification of parallel programs. Commun. ACM **19**(7), 561–572 (1976)
12. Milner, R.: A complete inference systems for a class of regular behaviors. J. Comput. Syst. Sci. **28**, 439–466 (1984)
13. Milner, R.: Communication and Concurrency. Prentice-Hall, Upper Saddle River (1989)
14. Olderog, E.R.: Nets, Terms and Formulas. Cambridge Tracts in Theoretical Computer Science, vol. 23. Cambridge University Press, Cambridge (1991)
15. Park, D.M.R.: Concurrency and automata on infinite sequences. In: Deussen, P. (ed.) GI-TCS 1981. LNCS, vol. 104, pp. 167–183. Springer, Heidelberg (1981). https://doi.org/10.1007/BFb0017309
16. Peterson, J.L.: Petri Net Theory and the Modeling of Systems. Prentice-Hall, Upper Saddle River (1981)
17. Paige, R., Tarjan, R.E.: Three partition refinement algorithms. SIAM J. Comput. **16**(6), 973–989 (1987)
18. Rabinovich, A., Trakhtenbrot, B.A.: Behavior structures and nets. Fundam. Inf. **11**(4), 357–404 (1988)
19. Reisig, W.: Understanding Petri Nets: Modeling Techniques, Analysis Methods, Case Studies. Springer, Heidelberg (2013)

Asynchronous π-calculus at Work: The Call-by-Need Strategy

Davide Sangiorgi(✉)

Focus Team, University of Bologna and Inria, Bologna, Italy
davide.sangiorgi@gmail.com

Abstract. In a well-known and influential paper [17] Palamidessi has shown that the expressive power of the Asynchronous π-calculus is strictly less than that of the full (synchronous) π-calculus. This gap in expressiveness has a correspondence, however, in sharper semantic properties for the former calculus, notably concerning algebraic laws. This paper substantiates this, taking, as a case study, the encoding of call-by-need λ-calculus into the π-calculus. We actually adopt the *Local* Asynchronous π-calculus, that has even sharper semantic properties. We exploit such properties to prove some instances of validity of β-reduction (meaning that the source and target terms of a β-reduction are mapped onto behaviourally equivalent processes). Nearly all results would fail in the ordinary synchronous π-calculus. We show that however the full β-reduction is not valid. We also consider a refined encoding in which some further instances of β-validity hold. We conclude with a few questions for future work.

1 Introduction

Since the introduction of the π-calculus, a lot of effort has been devoted to its comparison with the λ-calculus, beginning with Milner's seminar work on functions as processes [14]. The attention has gone mostly to *call-by-name* and *call-by-value* λ-calculi [19], and the main results concern operational correspondence, validity of β-reduction, characterisation of the equivalence induced on λ-terms by the π-calculus encoding [6,14,21,22,27]. In particular, the call-by-name encoding, for its simplicity, is often presented as *the* π-calculus representation of functions.

In a call-by-name reduction, the redex contracted is the leftmost one; the reduction occurs regardless of whether the argument of the function is a value (as in call-by-value). As a consequence, if the argument is not a value and will be used several times, its evaluation will be repeated the same number of times. In implementation of programming languages following call-by-name, this repetition of evaluation is avoided: evaluation occurs only once, the first time the term is used, and the value so obtained is recorded for future uses. This implementation technique is referred to as *call-by-need* evaluation (or strategy) [28]. Thus call-by-need uses explicit environments and β-reduction does not require

M. S. Alvim et al. (Eds.): Palamidessi Festschrift, LNCS 11760, pp. 33–49, 2019.
https://doi.org/10.1007/978-3-030-31175-9_3

substituting a term for a variable, as in call-by-name (or call-by-value)—just substituting a reference to a term for a variable. In this sense call-by-need is closer to the π-calculus than call-by-name, as substitutions in the π-calculus only involve names. Again, the modifications that take us from call-by-name to call-by-need can be easily represented in a π-calculus encoding [24].

The π-calculus, having a rich and well-developed theory, as well as a remarkable expressiveness, has been advocated as a foundational model for reasoning about higher-order languages, including equivalence between programs and correctness of compilers and compiler optimisations [25,26]. Indeed, the π-calculus and related languages have been used, via appropriate encodings, as a target language of compilers, for a number of experimental programming languages, beginning with Pict [18] and Join [7].

The above raises the question about how, and at which extent, the π-calculus and its current theory can be used to prove the correctness of call-by-need as an optimised implementation strategy for call-by-name. The only work on the correctness of the π-calculus representation of call-by-need is by Brock and Ostheimer [5]. The paper considers operational correspondence, between reduction in a call-by-need system and in the encoding π-calculus terms. However there are foundametal semantic issues that remain unexplored. A major one is the validity of β-reduction, namely the property that the processes encoding β-convertible λ-terms are behaviourally iundistinguishable. The property holds in call-by-name (and it is at the heart of its theory), as well as in the π-calculus encoding of call-by-name. One would therefore hope to find analoguos results for call-by-need. The correctness of the process representation of call-by-need is the topic of the present paper, focusing on the validity of β-reduction.

In a well-known and influential paper [17] Palamidessi has shown that the expressive power of the asynchronous π-calculus is strictly less than that of the full (synchronous) π-calculus. This gap in expressiveness has a correspondence, however, in sharper semantic properties for the former calculus, notably concerning algebraic laws. This paper may be seen as a demonstration of this, since most the proofs are carried out using algebraic laws that are only valid in the asynchronous π-calculus—precisely in the Asynchronous Local π-calculus, ALπ, [12], where only the output capability of names may be exported.

In Sect. 2 we present ALπ and some of its laws. In Sect. 3 we briefly recall the call-by-name and call-by-need λ-calculus. In Sect. 4 we consider two encodings of call-by-need. We show that limited forms of validity β-reduction hold, and that the general property fails. The questions that follow from this, discussed in Sect. 5, may contribute to open some interesting directions for future work, which may also shed further light on the theory of the π-calculus and similar name-passing calculi.

2 The Asynchronous Local π-calculus

2.1 Syntax

Small letters $a, b, \ldots, x, y, \ldots$ range over the infinite set of names, and $P, Q, R, \ldots$ over the set of all processes. A tilde represents a tuple. The i-th elements of a tuple $\widetilde{E}$ is referred to as $\widetilde{E}_i$. Our notations are extended to tuples componentwise.

The Asynchronous Local π-calculus (ALπ) [12] is built from the operators of inaction, input prefix, output, parallel composition, restriction, and replication:

$$P := \mathbf{0} \;\Big|\; a(\widetilde{b}).P \;\Big|\; \overline{a}\langle\widetilde{b}\rangle \;\Big|\; P_1 \mid P_2 \;\Big|\; \boldsymbol{\nu} a\, P \;\Big|\; !a(\widetilde{b}).P\,.$$

with the *syntactic constraint* that in processes $a(\widetilde{b}).P$ and $!a(\widetilde{b}).P$ names $\widetilde{b}$ may not occur free in P in input position.

When the tilde is empty, the surrounding brackets () and ⟨⟩ will be omitted. $\mathbf{0}$ is the inactive process. An input-prefixed process $a(\widetilde{b}).P$, where $\widetilde{b}$ has pairwise distinct components, waits for a tuple of names $\widetilde{c}$ to be sent along a and then behaves like $P\{\widetilde{c}/\widetilde{b}\}$, where $\{\widetilde{c}/\widetilde{b}\}$ is the simultaneous substitution of names $\widetilde{b}$ with names $\widetilde{c}$. An output particle $\overline{a}\langle\widetilde{b}\rangle$ emits names $\widetilde{b}$ at a. Parallel composition is to run two processes in parallel. The restriction $\boldsymbol{\nu} a\, P$ makes name a local, or private, to P. A replication $!P$ stands for a countable infinite number of copies of P in parallel. We assign parallel composition the lowest precedence among the operators.

2.2 Terminologies and Notations

We write $\overline{a}(b).b(\widetilde{c}).Q$ as an abbreviation for $\boldsymbol{\nu} b\,(\overline{a}b \mid b(\widetilde{c}).Q)$, and similarly for $\overline{a}(b).!b(\widetilde{c}).Q$. The prefix '$\overline{a}(b)$' is called a *bound output.* In prefixes $a(\widetilde{b})$ and $\overline{a}\langle\widetilde{b}\rangle$, we call a the *subject* and $\widetilde{b}$ the *object.* We use α to range over prefixes. We often abbreviate $\alpha.\mathbf{0}$ as α, and $\boldsymbol{\nu} a\, \boldsymbol{\nu} b\, P$ as $(\boldsymbol{\nu} a, b)\, P$. An input prefix $a(\widetilde{b}).P$ and a restriction $\boldsymbol{\nu} b\, P$ are binders for names $\widetilde{b}$ and b, respectively, and give rise in the expected way to the definition of *free names* (fn), *bound names* (bn) and *names* (n) of a term or a prefix, and *alpha conversion.* We identify processes or actions that only differ on the choice of the bound names. The symbol $=$ will mean "syntactic identity modulo alpha conversion". Sometimes, we use $\stackrel{\text{def}}{=}$ as abbreviation mechanism, to assign a name to an expression to which we want to refer later. In a statement, a name declared *fresh* is supposed to be different from any other name appearing in the objects of the statement, like processes or substitutions. Substitutions are of the form $\{\widetilde{b}/\widetilde{c}\}$, and are finite assignments of names to names. A context is a process expression with a hole $[\cdot]$ in it. We use C to range over contexts; then $C[P]$ is the process obtained from C by filling its hole with P.

2.3 Sorting

Following Milner [13], we only admit *well-sorted agents*, that is agents obeying a predefined *sorting* discipline in their manipulation of names. The sorting prevents arity mismatching in communications, like in $\overline{a}\langle b, c\rangle \mid a(x).Q$. A sorting is an assignment of *sorts* to names, which specifies the arity of each name and, recursively, of the names carried by that name. We do not present the formal system of sorting because it is not essential in the exposition of the topics in the present paper.

We will however allow sorting to identify *linear* names, that is, names that are supposed to be used only once. Linearity will be used a few times to replace input-replicated prefixes with ordinary input prefixes. Again, we omit the details of linearity in type systems (sorting is a form of type system), as they are by now standard [9,24].

$$\text{INP: } a(\widetilde{b}).\,P \xrightarrow{a(\widetilde{b})} P \qquad\qquad \text{OUT: } \overline{a}\langle\widetilde{b}\rangle \xrightarrow{\overline{a}\langle\widetilde{b}\rangle} \mathbf{0}$$

$$\text{REP: } \frac{P \mid !P \xrightarrow{\mu} P'}{!P \xrightarrow{\mu} P'} \qquad\qquad \text{PAR: } \frac{P \xrightarrow{\mu} P'}{P \mid Q \xrightarrow{\mu} P' \mid Q} \text{ if } \mathsf{bn}(\mu) \cap \mathsf{fn}(Q) = \emptyset$$

$$\text{COM: } \frac{P \xrightarrow{a(\widetilde{c})} P' \qquad Q \xrightarrow{(\boldsymbol{\nu}\,\widetilde{d})\,\overline{a}\langle\widetilde{b}\rangle} Q'}{P \mid Q \xrightarrow{\tau} \boldsymbol{\nu}\widetilde{d}\,(P'\{\widetilde{b}/\widetilde{c}\} \mid Q')} \text{ if } \widetilde{d} \cap \mathsf{fn}(P) = \emptyset$$

$$\text{RES: } \frac{P \xrightarrow{\mu} P'}{\boldsymbol{\nu}a\,P \xrightarrow{\mu} \boldsymbol{\nu}a\,P'} \; a \notin \mathsf{n}(\mu) \qquad \text{OPEN: } \frac{P \xrightarrow{(\boldsymbol{\nu}\,\widetilde{d})\,\overline{a}\langle\widetilde{b}\rangle} P'}{\boldsymbol{\nu}c\,P \xrightarrow{(\boldsymbol{\nu}\,c,\widetilde{d})\,\overline{a}\langle\widetilde{b}\rangle} P'} \; c \in \widetilde{b} - \widetilde{d},\; a \neq c.$$

Fig. 1. The transition system for ALπ

2.4 Relations

A process has three possible forms of action. A *silent action* $P \xrightarrow{\tau} P'$ represents an interaction, i.e. an internal activity in P. *Input* and *output actions* are, respectively, of the form $P \xrightarrow{a(\widetilde{d})} P'$ and $P \xrightarrow{(\boldsymbol{\nu}\,\widetilde{d})\,\overline{a}\langle\widetilde{b}\rangle} P'$. In both cases, the action occurs at a—the *subject* of the action. In the output action, $\widetilde{b}$ is the tuple of names which are emitted, and $\widetilde{d} \subseteq \widetilde{b}$ are private names which are carried out from their current scope. We use μ to represent the label of a generic action (not to be confused with α, which represents prefixes). In an input action $a(\widetilde{d})$ and in an output action $(\boldsymbol{\nu}\,\widetilde{d})\,\overline{a}\langle\widetilde{b}\rangle$, names $\widetilde{d}$ are *bound*, the remaining ones free. Bound and free names of an action μ, respectively written $\mathsf{bn}(\mu)$ and $\mathsf{fn}(\mu)$, are defined accordingly. The *names* of μ, briefly $\mathsf{n}(\mu)$, are $\mathsf{bn}(\mu) \cup \mathsf{fn}(\mu)$. The transition system of the calculus is presented in Fig. 1. We have omitted the symmetric versions of rules PAR and COM. Alpha convertible processes have deemed to have the same transitions. We often abbreviate $P \xrightarrow{\tau} Q$ with $P \longrightarrow Q$. The 'weak' arrow $\Longrightarrow$ is the reflexive and transitive closure of $\longrightarrow$.

We use the symbol $\equiv$ to denote *structural congruence*, a relation used to rearrange the structure of processes [13]. We shall also use it to represent garbage-collection of restrictions and of inert terms.

Definition 1 (Structural congruence). Structural congruence, $\equiv$, *is the smallest congruence relation satisfying the axioms below:*

- $P \mid \mathbf{0} \equiv P,\ \ P \mid Q \equiv Q \mid P,\ \ P \mid (Q \mid R) \equiv (P \mid Q) \mid R$
- $!a(x).P \equiv a(x).P \mid !a(x).P;$
- $\boldsymbol{\nu} a\, \mathbf{0} \equiv \mathbf{0},\ \ \boldsymbol{\nu} a\, \boldsymbol{\nu} b\, P \equiv \boldsymbol{\nu} b\, \boldsymbol{\nu} a\, P,\ \ \boldsymbol{\nu} a\, (P \mid Q) \equiv P \mid \boldsymbol{\nu} a\, Q$ *if* $a \notin \mathsf{fn}(P)$;
- $\boldsymbol{\nu} a\, (P \mid !a(\tilde{b}).Q) \equiv P$ *and* $\boldsymbol{\nu} a\, (P \mid a(\tilde{b}).Q) \equiv P$, *if* $a \notin \mathsf{fn}(P)$.

(A derivable law is $\boldsymbol{\nu} a\, P \equiv P$, for a not free in P.) A standard behavioural equivalence for the π-calculus is *barbed congruence*. Barbed congruence can be defined in any calculus possessing: (i) an *interaction relation* (the τ-steps in the π-calculus), modelling the evolution of the system; and (ii) an *observability predicate* $\downarrow_a$ for each name a, which detects the possibility of a process of accepting a communication with the environment at a. More precisely, we write $P \downarrow_a$ if P can make an output action whose subject is a, that is, if there are P' and an output action μ with subject a such that $P \xrightarrow{\mu} P'$. We write $P \Downarrow_a$ if $P \Longrightarrow P'$ and $P' \downarrow_a$. Unlike synchronous π-calculus, in asynchronous calculi it is natural to restrict the observation to output actions [1]. The reason is that in asynchronous calculi the observer has no direct way of knowing when a message emitted is received.

Definition 2 (Barbed congruence). *A symmetric relation $\mathcal{S}$ on π-calculus processes is a* barbed bisimulation *if $P\ \mathcal{S}\ Q$ implies:*

1. *If $P \xrightarrow{\tau} P'$ then there exists Q' such that $Q \Longrightarrow Q'$ and $P'\ \mathcal{S}\ Q'$.*
2. *If $P \downarrow_a$ then $Q \Downarrow_a$.*

Two π-calculus processes P and Q are barbed bisimilar *if $P\ \mathcal{S}\ Q$ for some barbed bisimulation $\mathcal{S}$. We say that P and Q are* barbed congruent, *written $P \approx Q$, if for each π-calculus context C, processes $C[P]$ and $C[Q]$ are barbed bisimilar.*

Strong barbed congruence, written $\sim$, is defined analogously but replacing the weak arrows $\Longrightarrow$ and $\Downarrow_a$ with the strong arrows $\longrightarrow$ and $\downarrow_a$. As expected, we have $\equiv\ \subseteq\ \sim\ \subseteq\ \approx$; each containment is strict.

2.5 Further Algebraic Laws

Most of the proofs in the paper are carried out using algebraic reasoning. We report here some important laws. First some simple laws that are valid in the full (synchronous) π-calculus (Lemma 1). Then laws that are specific to ALπ.

Lemma 1. *1.* $\boldsymbol{\nu} a\ (\overline{a}(b).P \mid a(y).Q) \approx (\boldsymbol{\nu} a, b)\ (P \mid Q\{^b\!/\!_y\})$ *and* $\boldsymbol{\nu} a\ (\overline{a}(b).P \mid !a(y).Q) \approx (\boldsymbol{\nu} a, b)\ (P \mid Q\{^b\!/\!_y\} \mid !a(y).Q)$;

2. $\boldsymbol{\nu}a\,(\alpha.Q \mid !a(\widetilde{x}).P) \sim \alpha.\boldsymbol{\nu}a\,(Q \mid !a(\widetilde{x}).P)$, *if* $\mathsf{bn}(\alpha) \cap \mathsf{fn}(a(\widetilde{x}).P) = \emptyset$ *and* $a \notin \mathsf{n}(\alpha)$ *(a similar law holds without the replication);*
3. $a(\widetilde{x}).(P \mid !a(\widetilde{x}).P) \sim !a(\widetilde{x}).P$.

Important laws of ALπ are the following ones. Their validity hinges on the asynchronous and output-capability properties of ALπ. For simplicity we present them on monadic prefixes.

Lemma 2. *We have* $\overline{a}b \approx \boldsymbol{\nu}c\,(\overline{a}c \mid !c(x).\overline{b}x)$. *Moreover, if b is linear, then the replication can be removed thus:* $\overline{a}b \approx \boldsymbol{\nu}c\,(\overline{a}c \mid c(x).\overline{b}x)$.

Next, we report some distributivity laws for private replications, i.e., for systems of the form

$$\boldsymbol{\nu}y\,(P \mid !y(\widetilde{q}).Q)$$

in which y may occur free in P and Q only in output position. One should think of Q as a private resource of P, for P is the only process who can access Q; indeed P can activate as many copies of Q as needed. One such law has already been given as Lemma 1(2). (The laws can be generalised to the full π-calculus, but need stronger assumptions.)

Lemma 3. *Suppose a occurs free in P, R, Q only in output position. Then:*

1. $\boldsymbol{\nu}a\,(P \mid R \mid !a(\widetilde{b}).Q) \sim \boldsymbol{\nu}a\,(P \mid !a(\widetilde{b}).Q) \mid \boldsymbol{\nu}a\,(R \mid !a(\widetilde{b}).Q)$;
2. $\boldsymbol{\nu}a\,((!P) \mid !a(\widetilde{b}).Q) \sim !\boldsymbol{\nu}a\,(P \mid !a(\widetilde{b}).Q)$;
3. $\boldsymbol{\nu}a\,(\alpha.P \mid !a(\widetilde{b}).Q) \sim \alpha.\boldsymbol{\nu}a\,(P \mid !a(\widetilde{b}).Q)$, *if* $\mathsf{bn}(\alpha) \cap \mathsf{fn}(a(\widetilde{b}).Q) = \emptyset$ *and* $a \notin \mathsf{n}(\alpha)$;
4. $\boldsymbol{\nu}a\,((\boldsymbol{\nu}c\,P) \mid !a(\widetilde{b}).Q) \sim \boldsymbol{\nu}c\,\boldsymbol{\nu}a\,(P \mid !a(\widetilde{b}).Q)$ *if* $c \notin \mathsf{fn}(a(\widetilde{b}).Q)$.

ALπ has also sharper properties concerning labelled characterisation of bisimilarity and associated congruence properties [3,12].

3 The λ-calculus

We use M, N to range over the set Λ of λ-terms, and x, y, z to range over variables. The set Λ of λ-terms is given by the grammar:

$$M ::= x \;\Big|\; \lambda x.M \;\Big|\; MN$$

A *redex* is a term of the form $(\lambda x.M)N$, and then its *contractum* is $M\{N/x\}$. In *call-by-name evaluation* [19], the redex is always at the extreme left of a term. We omit the standard evaluation rules.

Call-by-need [2,28] optimises call-by-name as follows, so to guarantee that in the contractum $M\{N/x\}$ the evaluation of N is not performed more than once. Roughly, N is placed in an environment, and the evaluation continues on M. When x is needed (i.e., x reaches the leftmost position), then N is evaluated and, if a value (i.e., an abstraction) is obtained, say V, then V replaces x (value V can replace all occurrences of x or, more commonly, only the leftmost occurrence, and then other occurrences of x when they reach the outermost position). Call-by-need is best presented in a graph; or in a system with a `let` construct to represent sharing. We refer to Ariola et al. [2] for details, as they are not essential for understanding the remainder of the paper; see also the references in Sect. 5.

We sometimes omit λ in nested abstractions, thus for example, $\lambda x_1 x_2.M$ stands for $\lambda x_1.\lambda x_2.M$. We assume the standard concepts of free and bound variables and substitutions, and identify α-convertible terms. Thus, throughout the paper '=' is syntactic equality modulo α-conversion.

Following the call-by-value terminology, the set of abstractions and variables are the *values.* (Indeed, call-by-need may also be thought of as a modified form of call-by-value, in which the evaluation of the argument of a function $\lambda x.M$ is made only when x is used for the first time, rather than before performing the reduction.)

4 The Encoding and Its Properties

4.1 Background Material

Figure 2 presents the call-by-name and call-by-need encodings [16,24]. The call-by-name one is a variant of the original encoding by Milner [14], with the advantage that it can be written in ALπ and can be easily modified to follow call-by-need.

We explain the encodings. The important part is the treatment of application. Both in call-by-name and in call-by-need, a function located at q (its 'location') is a process that signals to be a function on q, and then receives a pointer x to the argument N together with the location p for the next interaction. Now the evaluation of M continues. The difference between call-by-name and call-by-need arises when the argument N is needed. This is signaled by an output at x that also provides the location for the evaluation of a copy of N. In call-by-name, every output at x triggers the evaluation of a new copy of N. In call-by-need, in contrast, the evaluation is made only the first time. Precisely, in call-by-need N is evaluated at the first request and, when it becomes a value, a pointer to this value is returned (instantiating w, in the table). This pointer is returned to the process that requested N. When further requests for N are made, the pointer is returned immediately. Thus, for instance, in the call-by-name encoding of $(\lambda.xx)(II)$ term II is evaluated twice, whereas in the call-by-need encoding only once. In all encodings, the location names (in the table, the names ranged over by p, q, r) are linear.

Correcteness of call-by-name has been studied in depth. In particular, it has been shown that β-reduction is validated by the encoding, that the encoding gives rise to a term model for the λ-calculus, and that the equivalence on λ-terms induced by the encoding corresponds to the best tree-structures of the λ-calculus—which are also at the heart of its denotational semantics—namely Böhm Trees and Lévy-Longo Trees [14,23] Correctness of the call-by-need encoding has been studied only by Brock and Ostheimer [5], and only for operational correspondence with respect to Ariola et al.'s system [2]. (The encoding in Fig. 2

call-by-name encoding

$$\mathcal{M}[\![\lambda x.\, M]\!]p \stackrel{\mathrm{def}}{=} \overline{p}(v).\, !v(x,q).\, \mathcal{M}[\![M]\!]q$$

$$\mathcal{M}[\![x]\!]p \stackrel{\mathrm{def}}{=} \overline{x}p$$

$$\mathcal{M}[\![MN]\!]p \stackrel{\mathrm{def}}{=} (\boldsymbol{\nu} q\,)\Big(\mathcal{M}[\![M]\!]q \mid q(v).\, \boldsymbol{\nu} x\, \overline{v}\langle x,p\rangle.\, !x(r).\, \mathcal{M}[\![N]\!]r\Big)$$

call-by-need encoding

$$\mathcal{N}[\![\lambda x.\, M]\!]p \stackrel{\mathrm{def}}{=} \overline{p}(v).\, !v(x,q).\, \mathcal{N}[\![M]\!]q$$

$$\mathcal{N}[\![x]\!]p \stackrel{\mathrm{def}}{=} \overline{x}p$$

$$\mathcal{N}[\![MN]\!]p \stackrel{\mathrm{def}}{=}$$

$$(\boldsymbol{\nu} q\,)\Big(\mathcal{N}[\![M]\!]q \mid q(v).\, \boldsymbol{\nu} x\, \overline{v}\langle x,p\rangle.\, x(r).\, \boldsymbol{\nu} q'\, (\mathcal{N}[\![N]\!]q' \mid q'(w).\, (\overline{r}w \mid !x(r').\, \overline{r'}w))\Big)$$

Fig. 2. The encoding of call-by-name and call-by-need

is actually a minor improvement over that in [5]—avoiding one reduction step during a β-reducition—and maintains the results of operational correspondence in [5] recalled below.) Following Ariola et al.'s system [2] we write $M \Downarrow$ if the call-by-need computation of M terminates, and $M \Uparrow$ it the computation does not terminate.

Theorem 1 (Brock and Ostheimer [5]**).** *We have, for M closed:*

1. *$M \Downarrow$ iff $\mathcal{N}[\![M]\!]p \Downarrow_p$;*
2. *$M \Uparrow$ iff $\mathcal{N}[\![M]\!]p \Uparrow$.*

The proof in [5] considers an extended version of the call-by-need system in [2], one that yields a closer (nearly one-to-one) correspondence between reductions in the call-by-need system and reductions on the encoding π-calculus processes.

Note that, since M is closed, the only free name of $\mathcal{N}[\![M]\!]p$ is p; and since p is used in $\mathcal{N}[\![M]\!]p$ in output, the first visible action of $\mathcal{N}[\![M]\!]p$ (if there is one) is an output at p.

However, operational correspondence alone is not fully satisfactory as a criterium for correctness. It does not ensure foundamental semantic properties of the source language terms. In the following sections we focus on the validity of β-reduction.

4.2 β-validity

We consider in this section a few cases of validity of β-reduction; that is, the property that a β-redex $(\lambda x.M)N$ and its contractum $M\{^N\!/x\}$ are barbed congruent when represented as ALπ processes.

A form of β-reduction that is straightforward to handle is one in which the argument is never used.

Theorem 2. *If* $x \notin \mathsf{fv}(M)$ *then* $\mathcal{N}[\![(\lambda x.M)N]\!]p \approx \mathcal{N}[\![M\{^N\!/\!_x\}]\!]p$.

A more interesting form deals with β-reduction between closed values.

Theorem 3. $\mathcal{N}[\![(\lambda x.M)(\lambda y.N)]\!]p \approx \mathcal{N}[\![M\{^{\lambda y.N}\!/\!_x\}]\!]p$.

Proof. Using algebraic reasoning, we first derive:

$$
\begin{aligned}
&\mathcal{N}[\![(\lambda x.M)(\lambda y.N)]\!]p \\
&= (\boldsymbol{\nu} q\,)\Big(\mathcal{N}[\![\lambda x.M]\!]q \mid \\
&\qquad q(v).\boldsymbol{\nu} x\,\overline{v}\langle x,p\rangle.x(r).\boldsymbol{\nu} q'\,(\mathcal{N}[\![\lambda y.N]\!]q' \mid \\
&\qquad\qquad q'(w).(\overline{r}w \mid !x(r').\overline{r'}w))\Big) \\
&= (\boldsymbol{\nu} q\,)\Big(\overline{q}(v).!v(x,q').\mathcal{N}[\![M]\!]q' \mid \\
&\qquad q(v).\boldsymbol{\nu} x\,\overline{v}\langle x,p\rangle.x(r).\boldsymbol{\nu} q'\,(\mathcal{N}[\![\lambda y.N]\!]q' \mid \\
&\qquad\qquad q'(w).(\overline{r}w \mid !x(r').\overline{r'}w))\Big) \\
&\approx (\boldsymbol{\nu} x\,)\Big(\mathcal{N}[\![M]\!]p \mid \\
&\qquad x(r).\boldsymbol{\nu} q'\,(\mathcal{N}[\![\lambda y.N]\!]q' \mid \\
&\qquad\qquad q'(w).(\overline{r}w \mid !x(r').\overline{r'}w))\Big) \\
&= (\boldsymbol{\nu} x\,)\Big(\mathcal{N}[\![M]\!]p \mid \\
&\qquad x(r).\boldsymbol{\nu} q'\,(\overline{q'}(v).!v(y,q'').\mathcal{N}[\![N]\!]q'' \mid \\
&\qquad\qquad q'(w).(\overline{r}w \mid !x(r').\overline{r'}w))\Big) \\
&\approx (\boldsymbol{\nu} x\,)\Big(\mathcal{N}[\![M]\!]p \mid \\
&\qquad x(r).\boldsymbol{\nu} v\,(!v(y,q'').\mathcal{N}[\![N]\!]q'' \mid \\
&\qquad\qquad (\overline{r}v \mid !x(r').\overline{r'}v))\Big) \\
&\sim (\boldsymbol{\nu} x,v\,)\Big(\mathcal{N}[\![M]\!]p \mid \\
&\qquad !v(y,q'').\mathcal{N}[\![N]\!]q'' \mid \\
&\qquad x(r).(\overline{r}v \mid !x(r').\overline{r'}v)\Big) \\
&\sim (\boldsymbol{\nu} x,v\,)\Big(\mathcal{N}[\![M]\!]p \mid \\
&\qquad !v(y,q'').\mathcal{N}[\![N]\!]q'' \mid \\
&\qquad !x(r').\overline{r'}v\Big)
\end{aligned}
$$

where the two occurrences of $\approx$ represent applications of law (1) of Lemma 1, and the two occurrences of $\sim$ are due to laws (2) and (3) of the same lemma, respectively.

Now we proceed by induction on the structure of M. If M is variable different from x then the two replications at v and x can be garbage-collected and we are done. If $M = x$, then

$$
\begin{aligned}
(\boldsymbol{\nu} x, v)(\mathcal{N}[\![M]\!]p|!v(y,q'').\mathcal{N}[\![N]\!]q''|!x(r').\overline{r'}v) &= \\
(\boldsymbol{\nu} x, v)(\overline{x}p|!v(y,q'').\mathcal{N}[\![N]\!]q''|!x(r').\overline{r'}v) &\approx \\
(\boldsymbol{\nu} x, v)(\overline{p}v|!v(y,q'').\mathcal{N}[\![N]\!]q''|!x(r').\overline{r'}v) &\equiv \\
(\boldsymbol{\nu} v)(\overline{p}v|!v(y,q'').\mathcal{N}[\![N]\!]q'') &= \\
\mathcal{N}[\![\lambda y.N]\!]p &
\end{aligned}
$$

where $\approx$ is obtained from law (1) of Lemma 1, and $\equiv$ from the garbage-collection laws of Definition 1.

When M is an abstraction or an application, we proceed by induction and exploit the distributivity properties of private replications in Lemma 3.

Finally we consider the case when the argument of the function is divergent—a form of β-reduction that is not valid in call-by-value.

Theorem 4. *Suppose $(\lambda x.M)N$ is closed. If $N \Uparrow$ then we have $\mathcal{N}[\![(\lambda x.M)N]\!]p \approx \mathcal{N}[\![M\{^N\!/\!_x\}]\!]p$.*

Proof. Using Theorem 1 we have $\mathcal{N}[\![N]\!]q \Uparrow$, for any q, hence $\mathcal{N}[\![N]\!]q \approx \mathbf{0}$. As a consequence, using algebraic reasoning similar to that in the proof of Theorem 3, we obtain

$$\mathcal{N}[\![(\lambda x.M)N]\!]p \approx \boldsymbol{\nu} x\, (\mathcal{N}[\![M]\!]p \mid x(r).\mathbf{0})$$

Now, since x occurs in $\mathcal{N}[\![M]\!]p$ only in output subject position, each output at x, say $\overline{x}r$, can be removed, or replaced by $\mathcal{N}[\![N]\!]r$ (because in the relation $\approx$ with $\mathbf{0}$), up-to $\approx$. This yields $\mathcal{N}[\![M\{^N\!/\!_x\}]\!]p$.

4.3 Failure of General β-validity

However, in call-by-name and call-by-need β-reduction is not confined to values. We show that, in the call-by-need encoding, the general β-reduction fails. Notably β-reduction fails when the argument of the function is a variable. For this we show that

$$\mathcal{N}[\![yy]\!]p \not\approx \mathcal{N}[\![(\lambda z.zz)y]\!]p \tag{1}$$

While for simplicity this counterexample is shown for open terms, a similar one can be given for closed terms, by closing the two terms with an abstraction, i.e.,

$$\mathcal{N}[\![\lambda y.(yy)]\!]p \not\approx \mathcal{N}[\![\lambda y.((\lambda z.zz)y)]\!]p$$

The remainder of the section is devoted to the proof of (1). We first unroll the initial traces of the two processes (the only traces that they can perform) We have:

$$
\begin{aligned}
&\mathcal{N}[\![yy]\!]p \\
&= (\boldsymbol{\nu} q')\Big(\overline{y}q' \mid q'(w).\boldsymbol{\nu} x\, \overline{w}\langle x', p\rangle.x'(r).\boldsymbol{\nu} q''\, (\overline{y}q'' \mid \\
&\qquad\qquad q''(w').(\overline{r}w' \mid !x(r').\overline{r'}w'))\Big) \\
&\xrightarrow{\overline{y}(q')}\xrightarrow{q'(w)}\xrightarrow{\boldsymbol{\nu} x'\, \overline{w}\langle x',p\rangle}\xrightarrow{x'(r)} \boldsymbol{\nu} q''\, (\overline{y}q'' \mid \\
&\qquad\qquad q''(w').(\overline{r}w' \mid !x'(r').\overline{r'}w')) \\
&\qquad \xrightarrow{\overline{y}(q'')}\xrightarrow{q''(w')} \quad \overline{r}w' \mid !x'(r').\overline{r'}w'
\end{aligned}
$$

Since the above is the only possible trace of the term, we have

$$\mathcal{N}[\![yy]\!]p \sim \overline{y}(q).q(v).\boldsymbol{\nu}x\,\overline{v}\langle x,p\rangle.x(r).\overline{y}(q').q'(w).(\overline{r}w \mid !x(r').\overline{r'}w) \qquad (2)$$

We now consider the analogous trace for $\mathcal{N}[\![(\lambda z.zz)y]\!]p$. Below, the uses of $\equiv$ are due to some garbage-collection of restrictions and private inputs (possibly replicated), and some rearrangements of the scopes of some restrictions; the use of $\sim$ is due to (2).

$$\begin{array}{ll}
& \mathcal{N}[\![(\lambda z.zz)y]\!]p \\
= & (\boldsymbol{\nu}q\,)\Big(\mathcal{N}[\![\lambda z.zz]\!]q \mid \\
& \quad q(v).\boldsymbol{\nu}x\,\overline{v}\langle x,p\rangle.x(r).\boldsymbol{\nu}q'\,(\mathcal{N}[\![y]\!]q' \mid \\
& \qquad\qquad q'(w).(\overline{r}w \mid !x(r').\overline{r'}w))\Big) \\
= & (\boldsymbol{\nu}q\,)\Big(\overline{q}(v).!v(z,q').\mathcal{N}[\![zz]\!]q' \mid \\
& \quad q(v).\boldsymbol{\nu}x\,\overline{v}\langle x,p\rangle.x(r).\boldsymbol{\nu}q'\,(\overline{y}q' \mid \\
& \qquad\qquad q'(w).(\overline{r}w \mid !x(r').\overline{r'}w))\Big) \\
\xrightarrow{\tau}\xrightarrow{\tau}\equiv & \boldsymbol{\nu}x\,\Big(\mathcal{N}[\![xx]\!]p \mid \\
& \quad x(r).\boldsymbol{\nu}q'\,(\overline{y}q' \mid \\
& \qquad q'(w).(\overline{r}w \mid !x(r').\overline{r'}w))\Big) \\
\sim & \boldsymbol{\nu}x\,\Big(\\
& \overline{x}(q).q(v).\boldsymbol{\nu}x'\,\overline{v}\langle x',p\rangle.x'(r).\overline{x}(q').q'(w).(\overline{r}w \mid !x'(r').\overline{r'}w) \mid \\
& x(r).\boldsymbol{\nu}q'\,(\overline{y}q' \mid \\
& \qquad q'(w).(\overline{r}w \mid !x(r').\overline{r'}w))\Big) \\
\xrightarrow{\tau}\equiv & (\boldsymbol{\nu}x,q,q') \\
& \Big(q(v).\boldsymbol{\nu}x'\,\overline{v}\langle x',p\rangle.x'(r).\overline{x}(q').q'(w).(\overline{r}w \mid !x'(r').\overline{r'}w) \mid \\
& \overline{y}q' \mid \\
& q'(w).(\overline{q}w \mid !x(r').\overline{r'}w)\Big) \\
\xrightarrow{\overline{y}(q')}\xrightarrow{q'(w)} & (\boldsymbol{\nu}x,q) \\
& \Big(q(v).\boldsymbol{\nu}x'\,\overline{v}\langle x',p\rangle.x'(r).\overline{x}(q').q'(w).(\overline{r}w \mid !x'(r').\overline{r'}w) \mid \\
& (\overline{q}w \mid !x(r').\overline{r'}w)\Big) \\
\xrightarrow{\tau} & (\boldsymbol{\nu}x)\,\Big(\boldsymbol{\nu}x'\,\overline{w}\langle x',p\rangle.x'(r).\overline{x}(q').q'(w).(\overline{r}w \mid !x'(r').\overline{r'}w) \mid \\
& \quad !x(r').\overline{r'}w\Big) \\
\xrightarrow{\boldsymbol{\nu}x'\,\overline{w}\langle x',p\rangle}\xrightarrow{x'(r)} & (\boldsymbol{\nu}x)\,\Big(\overline{x}(q').q'(w).(\overline{r}w \mid !x'(r').\overline{r'}w) \mid \\
& \quad !x(r').\overline{r'}w\Big) \\
\xrightarrow{\tau}\xrightarrow{\tau}\equiv & \overline{r}w \mid !x'(r').\overline{r'}w
\end{array}$$

Again, the above is the only possible trace for the term. Up-to some renaming, the final derivative is the same as the final derivative of the trace emanating from $\mathcal{N}[\![yy]\!]p$ examined earlier. However, the two traces are different—the first is longer. As a consequence, the two terms can be distinguished, for instance in the context

$$C \stackrel{\text{def}}{=} \boldsymbol{\nu} y\,([\cdot] \mid y(q).\overline{q}(v).v(x,p).\boldsymbol{\nu} r\,(\overline{x}r \mid y(q').\overline{h}))$$

The observable output at h becomes visible only when the context is filled with the first term, $\mathcal{N}[\![yy]\!]p$ (the one that produces the longer trace).

In contrast, in the call-by-name encoding the validity of the full β-reduction holds [23]. Therefore the above counterexample not only shows that the call-by-need encoding is observably different from the call-by-name one; it also tells us that the properties that the two encoding satisfy are quite different.

4.4 A Refined Encoding

In this section we experiment with a refinement $\mathcal{R}$ of the encoding so to improve on the problems described in Sect. 4.3.

Such encoding is shown in Fig. 3. In the definition of application for call-by-need, the argument of a function is interrogated only once, the first time the argument is used. Future uses of the argument will directly use the answer so received, without repeating the interrogation—this is indeed the essence of the call-by-need optimisation over call-by-name. To mirror this policy we modify the encoding of an abstraction $\lambda x.M$, so that the body M will interrogate the parameter x only once. As a consequence, in this refined encoding when the head 'λx' of the function is consumed a local entry is left that takes care of the dialogue with x; in particular the local entry makes sure that x is consulted only once. The refined encoding, while it exhibits more interactions than the original one, in a distributed setting may be thought of as an optimisation of the latter, as the interactions with the new local entry replace interactions with the possibly remote actual parameter for x. We write $\mathtt{LE}(x,y)$ for a local entry in which the internal resource is x and the external one is y; it will be convenient to break its definition into two parts, using the auxiliary local entry $\mathtt{LE}'(s,r,x)$; see Fig. 3.

The local entry is unnecessary if the internal resource is used only once.

Lemma 4. *Suppose z' appears only once in M. Then*

$$\boldsymbol{\nu} z'\,(\mathcal{R}[\![M]\!]p \mid \mathtt{LE}(z',z)) \approx \mathcal{R}[\![M\{z/z'\}]\!]p$$

Proof. By induction on the structure of M. The most interesting case is when $M = z'$, in which case we exploit the (linear) law of ALπ in Lemma 2. When M is an application we exploit the hypothesis (z' occurring only once), and simple algebraic manipulation, so to be able to carry out the induction.

The next lemma shows that local entries compose.

Lemma 5. $\boldsymbol{\nu} x\,(\mathtt{LE}(z,x) \mid \mathtt{LE}(x,y)) \approx \mathtt{LE}(z,y)$.

Proof. We use laws (2) and (1) of Lemma 1, and the garbage-collection laws of structural congruence.

$$\mathcal{R}[\![\lambda x.\, M]\!]p \stackrel{\text{def}}{=} \overline{p}(v).\, !v(x', q).\, \boldsymbol{\nu} x\, (\; \mathcal{R}[\![M]\!]q \mid \mathtt{LE}(x, x'))$$

$$\mathcal{R}[\![x]\!]p \stackrel{\text{def}}{=} \overline{x}p$$

$$\mathcal{R}[\![MN]\!]p \stackrel{\text{def}}{=} (\boldsymbol{\nu} q)\Big(\mathcal{R}[\![M]\!]q \mid q(v).\, \boldsymbol{\nu} x\, \overline{v}\langle x, p\rangle.\, x(r).\, \boldsymbol{\nu} q'\, (\mathcal{R}[\![N]\!]q' \mid \mathtt{LE}'(q', r, x))\Big)$$

where

$$\mathtt{LE}(x, y) \stackrel{\text{def}}{=} x(r).\, \overline{y}(s).\, \mathtt{LE}'(s, r, x)$$

$$\mathtt{LE}'(s, r, x) \stackrel{\text{def}}{=} s(v).\, (\overline{r}v \mid !x(r').\, \overline{r'}v)$$

Fig. 3. The refined call-by-need encoding

We revisite the counterexample of Sect. 4.3, that involves terms yy and $(\lambda z.zz)y$, under the refined encoding $\mathcal{R}$. All free variables should be protected under a local entry, except for the variables that occur only once (by Lemma 4). We begin by examining $\boldsymbol{\nu} y'\, (\mathcal{R}[\![y'y']\!]q \mid \mathtt{LE}(y', y))$. We have:

$$\begin{array}{ll}
& \boldsymbol{\nu} y'\, (\mathcal{R}[\![y'y']\!]q \mid \mathtt{LE}(y', y)) \\
\sim & \boldsymbol{\nu} y'\, (\overline{y'}(q).q(v).\boldsymbol{\nu} x\, \overline{v}\langle x, p\rangle.x(r).\overline{y'}(q').q'(w).(\overline{r}w \mid !x(r').\overline{r'}w) \\
& \quad \mid y'(r).\overline{y}(r').r'(v).(\overline{r}v \mid !y'(r).\overline{r}v)) \\
\approx & (\boldsymbol{\nu} y', q)\, (q(v).\boldsymbol{\nu} x\, \overline{v}\langle x, p\rangle.x(r).\overline{y'}(q').q'(w).(\overline{r}w \mid !x(r').\overline{r'}w) \\
& \quad \mid \overline{y}(r').r'(v).(\overline{q}v \mid !y'(q).\overline{q}v)) \\
\xrightarrow{\overline{y}(r')}\xrightarrow{r'(v)} & (\boldsymbol{\nu} y', q)\, (q(v).\boldsymbol{\nu} x\, \overline{v}\langle x, p\rangle.x(r).\overline{y'}(q').q'(w).(\overline{r}w \mid !x(r').\overline{r'}w) \\
& \quad \mid \overline{q}v \mid !y'(q).\overline{q}v) \\
\xrightarrow{\tau} & (\boldsymbol{\nu} y')\, (\boldsymbol{\nu} x\, \overline{v}\langle x, p\rangle.x(r).\overline{y'}(q').q'(w).(\overline{r}w \mid !x(r').\overline{r'}w) \\
& \quad \mid !y'(q).\overline{q}v) \\
\xrightarrow{\boldsymbol{\nu} x\, \overline{v}\langle x, p\rangle}\xrightarrow{x(r)} & (\boldsymbol{\nu} y')\, (\overline{y'}(q').q'(w).(\overline{r}w \mid !x(r').\overline{r'}w) \\
& \quad \mid !y'(q).\overline{q}v) \\
\xrightarrow{\tau}\xrightarrow{\tau} & \overline{r}v \mid !x(r').\overline{r'}v
\end{array}$$

where the occurrence of $\sim$ is justified by (2) (the two encodings coincide on terms that do not contain abstractions) and definition of $\mathtt{LE}(y', y)$, and the occurrence of $\approx$ comes from law (1) of Lemma 1.

We now consider the second term, $(\lambda z.zz)y$, under the refined encoding. First we note that, if the argument of a function L is a variable, then the refined encoding can be simplified thus:

$$\begin{aligned}\mathcal{R}[\![Ly]\!]p &= (\boldsymbol{\nu} q\,)\Big(\mathcal{R}[\![L]\!]q \mid \\ &\qquad q(v).\boldsymbol{\nu} x\,\overline{v}\langle x,p\rangle.x(r).\boldsymbol{\nu} q'\,(\mathcal{N}[\![y]\!]q' \mid \\ &\qquad\qquad \mathtt{LE}'(q',r,x))\Big) \\ &= (\boldsymbol{\nu} q\,)\Big(\mathcal{R}[\![L]\!]q \mid \\ &\qquad q(v).\boldsymbol{\nu} x\,\overline{v}\langle x,p\rangle.\mathtt{LE}(x,y)\Big)\end{aligned}$$

Using this property, we have:

$$\begin{aligned}\mathcal{R}[\![(\lambda z.zz)y]\!]p &= (\boldsymbol{\nu} q\,)\Big(\mathcal{R}[\![\lambda z.zz]\!]q \mid \\ &\qquad q(v).\boldsymbol{\nu} x\,\overline{v}\langle x,p\rangle.\mathtt{LE}(x,y)\Big) \\ &= (\boldsymbol{\nu} q\,)(\overline{q}(v).!v(z',q).\boldsymbol{\nu} z\,(\mathcal{R}[\![zz]\!]q \mid \mathtt{LE}(z,z')) \\ &\qquad \mid q(v).\boldsymbol{\nu} x\,\overline{v}\langle x,p\rangle.\mathtt{LE}(x,y)) \\ &\xrightarrow{\tau}\xrightarrow{\tau} \boldsymbol{\nu} z\,\boldsymbol{\nu} x\,(\mathcal{R}[\![zz]\!]p \mid \mathtt{LE}(z,x) \mid \mathtt{LE}(x,y)) \\ &\approx \boldsymbol{\nu} z\,(\mathcal{R}[\![zz]\!]p \mid \mathtt{LE}(z,y))\end{aligned}$$

where $\approx$ is obtained by composition of local entries (Lemma 5).

The above reasoning shows that the behaviour of the initial and refined encodings are the same on the term $(\lambda z.zz)y$:

$$\mathcal{R}[\![(\lambda z.zz)y]\!]p \approx \mathcal{N}[\![(\lambda z.zz)y]\!]p$$

And it is also the same as that of the refined encoding on the β-contractum yy, when the free variables y are protected under the appropriate local entry, i.e.,

$$\mathcal{R}[\![(\lambda zz)y]\!]p \approx \boldsymbol{\nu} y'\,(\mathcal{R}[\![y'y']\!]p \mid \mathtt{LE}(y',y))$$

The result also holds by closing up the terms:

$$\mathcal{R}[\![\lambda y.(yy)]\!]p \approx \mathcal{R}[\![\lambda y.((\lambda z.zz)y)]\!]p$$

In the refined encoding $\mathcal{R}$, the local entry (necessary for confining the behaviour of yy) is produced by the encoding of the abstraction.

More generally, proceeding as above one can show that, on the encoding $\mathcal{R}$, β-reduction is valid when the argument of a function is a variable. Moreover, β-reduction is also valid when the argument is itself a function, reasoning as in Theorem 3. We may therefore conclude that β-reduction is valid when the argument is a value (as it is the case for the encoding of the call-by-value strategy [24]).

We write $\mathtt{LE}(\widetilde{x},\widetilde{y})$ for $\mathtt{LE}(x_1,y_1) \mid \ldots \mid \mathtt{LE}(x_n,y_n)$ where n is the length of the tuples $\widetilde{x}$ and $\widetilde{y}$. We use V to range over values.

Theorem 5. *Suppose* $\mathsf{fv}((\lambda z.MV)) \subseteq \widetilde{x}$. *Then, for any* y *and fresh* $\widetilde{y}$, *we have*

$$\boldsymbol{\nu}\widetilde{x}\,(\mathcal{R}[\![(\lambda z.M)V]\!]p \mid \mathtt{LE}(\widetilde{x},\widetilde{y})) \approx \boldsymbol{\nu}\widetilde{x}\,(\mathcal{R}[\![M\{{}^{V}\!/\!_{z}\}]\!]p \mid \mathtt{LE}(\widetilde{x},\widetilde{y}))$$

Similarly to what done earlier, a local entry $\mathtt{LE}(x_i, y_i)$ may be removed when the variable x_i appears at most once.

However, even in the encoding $\mathcal{R}$, the full β-reduction is unvalid. As a counterexample we can use the terms $(xz)(xz)$ and $(\lambda y.yy)(xz)$. Indeed we have:

$$\boldsymbol{\nu} xz\ (\mathcal{R}[\![(xz)(xz)]\!]p \mid \mathtt{LE}(x, x') \mid \mathtt{LE}(z, z') \not\approx \mathcal{R}[\![(\lambda y.yy)(xz)]\!]p$$

We omit the calculations, which are rather involved. Intuitively, the difference appears because the full trace of the process computing the application xz is visible twice in the first process, whereas, in the second process, the second time, part of the trace is concealed.

5 Conclusions

Call-by-need was proposed by Wadsworth [28] as an implementation technique. Formalizations of call-by-need on a λ-calculus with a `let` construct or with environments include Ariola et al. [2], Launchbury [11], Purushothaman and Seaman [20], and Yoshida [29]. The uniform encoding in Sect. 4 is from [16]. A study of the correctness of the call-by-need encoding in Fig. 2 is in [5]. Encodings of graph reductions, related to call-by-need, into π-calculus were given in [4,8] but their correctness was not studied. Niehren [15] used encodings of call-by-name, call-by-value and call-by-need λ-calculi into π-calculus to compare the time complexity of the strategies.

In the paper we have used the theory of the Asynchronous Local π-calculus (ALπ) [12] to reason about the encoding of the call-by-need λ-calculus strategy as processes. We have mainly focused on the validity of β-reduction. We have showed that various instances of the property on closed terms hold, though the general property fails. We have also considered a refined encoding in which β-reduction on arbitrary values (though not on arbitrary terms) holds. All this leaves us with some challenging questions, that we leave for future work:

1. In the refined encoding, we use special processes called *local entries* to protect the formal parameter of the function, thus improving the results about β-validity. Is it possible to further protect variables (or terms) so to recover the full β-validity?
2. Is there a different form of behavioural equivalence under which the full β-validity holds, in the initial or in the refined encoding?
3. What is an appropriate process preorder under which call-by-need can indeed be proved to be an optimisation of call-by-name?
4. What is the equivalence on λ-terms induced by the call-by-need encoding? Following the results for call-by-name and call-by-value, one expects to recover some kind of tree structure (Böhm Trees and Lévy-Longo Tree for call-by-name, Lassen's trees [10] for call-by-value). We are not aware of similar tree structures for call-by-need. Hence investigating this question may also shed light on what should the appropriate forms of trees for call-by-need be.

In questions (2) and (3), 'easy' answers may be obtained by confining the testing contexts to be encodings of λ-calculus contexts. The challenge is to find more general and useful answers, with applications outside the realm of pure λ-calculi. One may consider forms of behavioural types.

In questions (1) and (2), perhaps requiring the validity of the full β-reduction, in the same way as for call-by-name, is too demanding. Indeed in this way probably the tree structure referred to in question (4) is likely to be the same as that for call-by-name. One may find it acceptable to limit β-validity to reductions between closed terms.

Acknowledgments. Thanks to the reviewers for their careful reading of the paper and their suggestions. Research partly supported by the H2020-MSCA-RISE project ID 778233 "Behavioural Application Program Interfaces (BEHAPI)".

References

1. Amadio, R., Castellani, I., Sangiorgi, D.: On bisimulations for the asynchronous π-calculus. Theoret. Comput. Sci. **195**, 291–324 (1998)
2. Ariola, Z., Felleisen, M., Maraist, J., Odersky, M., Wadler, P.: A call-by-need λ-calculus. In: Proceedings of the 22th POPL. ACM Press (1995)
3. Boreale, M., Sangiorgi, D.: Some congruence properties for π-calculus bisimilarities. Theoret. Comput. Sci. **198**, 159–176 (1998)
4. Boudol, G.: Some chemical abstract machines. In: de Bakker, J.W., de Roever, W.-P., Rozenberg, G. (eds.) REX 1993. LNCS, vol. 803, pp. 92–123. Springer, Heidelberg (1994). https://doi.org/10.1007/3-540-58043-3_18
5. Brock, S., Ostheimer, G.: Process semantics of graph reduction. In: Lee, I., Smolka, S.A. (eds.) CONCUR 1995. LNCS, vol. 962, pp. 471–485. Springer, Heidelberg (1995). https://doi.org/10.1007/3-540-60218-6_36
6. Durier, A., Hirschkoff, D., Sangiorgi, D.: Eager functions as processes. In: 33nd Annual ACM/IEEE Symposium on Logic in Computer Science, LICS 2018. IEEE Computer Society (2018)
7. Fournet, C., Gonthier, G.: The join calculus: a language for distributed mobile programming. In: Barthe, G., Dybjer, P., Pinto, L., Saraiva, J. (eds.) APPSEM 2000. LNCS, vol. 2395, pp. 268–332. Springer, Heidelberg (2002). https://doi.org/10.1007/3-540-45699-6_6
8. Jeffrey, A.: A chemical abstract machine for graph reduction extended abstract. In: Brookes, S., Main, M., Melton, A., Mislove, M., Schmidt, D. (eds.) MFPS 1993. LNCS, vol. 802, pp. 293–303. Springer, Heidelberg (1994). https://doi.org/10.1007/3-540-58027-1_14
9. Kobayashi, N., Pierce, B., Turner, D.: Linearity and the pi-calculus. TOPLAS **21**(5), 914–947 (1999). Preliminary summary appeared in Proceedings of POPL'96
10. Lassen, S.B.: Eager normal form bisimulation. In: Proceedings of the 20th IEEE Symposium on Logic in Computer Science (LICS 2005), Chicago, IL, USA, 26–29 June 2005, pp. 345–354 (2005)
11. Launchbury, J.: A natural semantics for lazy evaluation. In: Proceedings of the 20th POPL. ACM Press (1993)
12. Merro, M., Sangiorgi, D.: On asynchrony in name-passing calculi. J. Math. Struct. Comput. Sci. **14**(5), 715–767 (2004)

13. Milner, R.: The polyadic π-calculus: a tutorial. Technical report, ECS-LFCS-91-180, LFCS, Department of Computer Science, Edinburgh University, October 1991. Also in Bauer, F.L., Brauer, W., Schwichtenberg, H. (eds.) Logic and Algebra of Specification. Springer, Heidelberg (1991)
14. Milner, R.: Functions as processes. J. Math. Struct. Comput. Sci. **2**(2), 119–141 (1992)
15. Niehren, J.: Functional computation as concurrent computation. In: Proceedings of the 23th POPL. ACM Press (1996)
16. Ostheimer, G., Davie, A.: π-calculus characterisations of some practical λ-calculus reductions strategies. Technical report, CS/93/14, St. Andrews (1993)
17. Palamidessi, C.: Comparing the expressive power of the synchronous and asynchronous pi-calculi. J. Math. Struct. Comput. Sci. **13**(5), 685–719 (2003)
18. Pierce, B.C., Turner, D.N.: Pict: a programming language based on the pi-calculus. In: Plotkin, G., Stirling, C., Tofte, M. (eds.) Proof, Language and Interaction: Essays in Honour of Robin Milner. MIT Press, Cambridge (2000)
19. Plotkin, G.: Call by name, call by value and the λ-calculus. Theoret. Comput. Sci. **1**, 125–159 (1975)
20. Purushothaman, S., Seaman, J.: An adequate operational semantics of sharing in lazy evaluation. In: Krieg-Brückner, B. (ed.) ESOP 1992. LNCS, vol. 582, pp. 435–450. Springer, Heidelberg (1992). https://doi.org/10.1007/3-540-55253-7_26
21. Sangiorgi, D.: An investigation into functions as processes. In: Brookes, S., Main, M., Melton, A., Mislove, M., Schmidt, D. (eds.) MFPS 1993. LNCS, vol. 802, pp. 143–159. Springer, Heidelberg (1994). https://doi.org/10.1007/3-540-58027-1_7
22. Sangiorgi, D.: From λ to π, or: Rediscovering continuations. J. Math. Struct. Comput. Sci. **9**(4), 367–401 (1999). Special Issue on "Lambda-Calculus and Logic" in Honour of Roger Hindley
23. Sangiorgi, D.: Lazy functions and mobile processes. In: Plotkin, G., Stirling, C., Tofte, M. (eds.) Proof, Language and Interaction: Essays in Honour of Robin Milner. MIT Press, Cambridge (2000)
24. Sangiorgi, D., Walker, D.: The π-calculus: A Theory of Mobile Processes. Cambridge University Press, Cambridge (2001)
25. Sangiorgi, D.: Typed pi-calculus at work: a correctness proof of Jones's parallelisation transformation on concurrent objects. TAPOS **5**(1), 25–33 (1999)
26. Turner, N.: The polymorphic pi-calculus: theory and implementation. Ph.D. thesis, Department of Computer Science, University of Edinburgh (1996)
27. Vasconcelos, V.T.: Lambda and pi calculi, CAM and SECD machines. J. Funct. Program. **15**(1), 101–127 (2005)
28. Wadsworth, C.P.: Semantics and pragmatics of the lambda calculus. Ph.D. thesis, University of Oxford (1971)
29. Yoshida, N.: Optimal reduction in weak lambda-calculus with shared environments. In: Proceedings of the FPCA 1993, Functional Programming and Computer Architecture, pp. 243–252 (1993)

Deadlock Analysis of Wait-Notify Coordination

Cosimo Laneve[1] and Luca Padovani[2](✉)

[1] Department of Computer Science and Engineering, University of Bologna – Inria Focus, Bologna, Italy
[2] Dipartimento di Informatica, Università di Torino, Turin, Italy
luca.padovani@unito.it

Abstract. Deadlock analysis of concurrent programs that contain coordination primitives (`wait`, `notify` and `notifyAll`) is notoriously challenging. Not only these primitives affect the scheduling of processes, but also notifications unmatched by a corresponding `wait` are silently lost. We design a behavioral type system for a core calculus featuring shared objects and `Java`-like coordination primitives. The type system is based on a simple language of object protocols – called *usages* – to determine whether objects are used *reliably*, so as to guarantee deadlock freedom.

1 Introduction

Locks and condition variables [10] are established mechanisms for process coordination that are found, in one form or another, in most programming languages. `Java` provides `synchronized` blocks for enforcing exclusive access to shared objects and `wait`, `notify` and `notifyAll` primitives to coordinate the threads using them: a thread performing a `wait` operation on an object releases the lock on that object and is suspended; a `notify` operation performed on an object awakens a thread suspended on it, if there is one; `notifyAll` is similar to `notify`, except that it awakens all suspended threads. Writing correct concurrent programs using these primitives is notoriously difficult. For example, a thread may block indefinitely if it attempts to lock an object that is permanently owned by a different thread or if it suspends waiting for a notification that is never sent.

We see an instance of non-trivial concurrent `Java` program in Fig. 1, which models a coordination problem whereby a single *consumer* retrieves items from two *producers*. Each producer repeatedly generates and stores a new item into a buffer (line 5), notifies the consumer that the item is available (line 6) and waits until the consumer has received the item (line 7). At each iteration, the consumer waits for an item from each buffer (lines 12 and 15) and notifies the corresponding producer that the item has been processed (lines 14 and 17). The main thread of the program forks the producers (lines 25–26) and then runs as

C. Laneve—Research partly supported by the H2020-MSCA-RISE project ID 778233 (BEHAPI).

M. S. Alvim et al. (Eds.): Palamidessi Festschrift, LNCS 11760, pp. 50–67, 2019.
https://doi.org/10.1007/978-3-030-31175-9_4

```
1  public static void producer(Buffer x) {
2      int item = 0;
3      while (true)
4          synchronized (x) {
5              x.Put(item++);
6              x.notify();
7              x.wait();
8          }
9  }
10 public static void consumer(Buffer x, Buffer y) {
11     while (true) {
12         x.wait();
13         System.out.println(x.Get());
14         x.notify();
15         y.wait();
16         System.out.println(y.Get());
17         y.notify();
18     }
19 }
20 public static void main(String[] args) {
21     Buffer x = new Buffer();
22     Buffer y = new Buffer();
23     synchronized (x) {
24         synchronized (y) {
25             new Thread(() -> producer(x)).start();
26             new Thread(() -> producer(y)).start();
27             consumer(x, y);
28         }
29     }
30 }
```

Fig. 1. Multiple-producers/single-consumer coordination in Java.

the consumer (line 27). In this example, the fact that producers and consumer use the two buffers in mutual exclusion is guaranteed by the syntactic structure of the code, since all accesses to x and y occur within synchronized blocks. However, understanding whether the three threads coordinate correctly so as to realize the desired continuous flow of information is not as obvious. This difficulty is largely due to the *ephemeral* nature of notifications: a notification sent to a shared object has an effect only if there is another thread waiting to be notified on that object. Otherwise, the notification is lost, with likely undesired implications. For example, suppose to change the program in Fig. 1 so that the synchronized blocks now in the main method (lines 23–24) are moved into the consumer method, each protecting accesses to the corresponding buffer. This change would give producer and consumer a visually appealing symmetric structure, but the correctness of the program would be fatally compromised. Now a producer could lock x before the consumer and notify x at a time when the consumer is not yet waiting for a notification. Eventually, the consumer would block waiting for a notification on `x` that will never arrive, leading to a deadlock.

The contribution of this paper is a behavioral type system ensuring that well-typed programs using shared objects and coordination primitives are deadlock free. The type system combines two features inspired by previous works on the static analysis of concurrent programs. First, we use a formulation of the typing rule for parallel compositions akin to that found in linear logic interpretations of session types [3,18]. Unlike these works, where session endpoints are linear resources, here we apply the typing rule in a setting with shared (hence, non-linear) objects. Second, we rely on behavioral types – called *usages* – to make sure that objects are used *reliably*, ruling out deadlocks due to missing notifications. Kobayashi [12] has already studied a type system based on usages for the deadlock analysis of pi-calculus processes. He shows how to reduce usage reliability to a reachability problem in a Petri net. As is, this reduction does not apply to our setting because the encoding of usages with coordination primitives requires the use of Petri nets with *inhibitor arcs* [2].

The rest of the paper is structured as follows. Section 2 presents a core calculus of concurrent programs featuring threads, shared objects and coordination primitives. Section 3 defines types, usages and the key notion of *reliability*. Typing rules and properties of well-typed programs are given in Sect. 4. Section 5 discusses related work in more detail and Sect. 6 concludes. Proofs and the handling of the `notifyAll` primitive can be found in the full paper [15].

2 Language Syntax and Semantics

We define a core language of concurrent programs featuring threads, shared objects and a minimal set of coordination primitives inspired to those of `Java`. We formalize our language as a process calculus comprising standard constructs (termination, conditional behavior, object creation, parallel composition, recursion) in which actions represent coordination primitives on objects. Instead of providing a `synchronized` construct to enforce mutually exclusive access to a shared object, we use explicit acquire and release operations on the object.

Formally, our calculus makes use of a countable set of *variables and object names*, ranged over by x, y, z, and a set of *procedure names*, ranged over by A. A *program* is a pair $(\mathcal{D}, P)$, where $\mathcal{D}$ is a *finite set* of *procedure definitions* of the form $A(\overline{x}) = P_A$, with $\overline{x}$ and P_A respectively being the *formal parameters* and the *body* of A. In a program $(\mathcal{D}, P)$ we say that P is the *main process*. Hereafter we write $\overline{\alpha}$ for possibly empty, finite sequences $\alpha_1, \dots, \alpha_n$ of various entities. The syntax of processes, expressions and actions is given in Table 1.

Expressions comprise integer constants, variables and object names, and an unspecified set of operators op such as $+$, $\leq$, and so forth. Expressions are evaluated by means of a total function $[\![\cdot]\!]$ such that $[\![x]\!] = x$.

The process done performs no action. The process $\pi.P$ performs the action π and continues as P. The conditional process $\mathsf{if}\ e\ \mathsf{then}\ P\ \mathsf{else}\ Q$ behaves as P if $[\![e]\!] \neq 0$ or else as Q. The process $\mathsf{new}\ x\ \mathsf{in}\ P$ creates a new object x with scope

Table 1. Syntax of the language with runtime syntax marked by †.

Expression	e ::=	n	(constant)
	|	x	(variable)
	|	e op e	(operator)
Process	P, Q ::=	done	(termination)
	|	$\pi.P$	(action prefix)
	|	if e then P else Q	(conditional)
	|	new x in P	(new object)
	|	fork$\{P\}Q$	(new process)
	|	$A(\overline{e})$	(invocation)
Action	π ::=	acq(x)	(acquire)
	|	rel(x)	(release)
	|	wait(x)	(wait)
	|	wait(x, n)	(waiting)†
	|	notify(x)	(notify)

P. The process fork$\{P\}Q$ forks P and continues as Q. Finally, $A(\overline{e})$ denotes the invocation of the process corresponding to A with actual parameters $\overline{e}$.

An action is either acq(x), which acquires the lock on x, or rel(x), which releases the lock on x, or wait(x), meaning that the process unlocks x and suspends waiting for a notification, or notify(x) that notifies a process waiting on x, if there is any. Objects are reentrant and can be locked multiple times by the same process. The prefix wait(x, n) is a runtime version of wait(x) that keeps track of the number of times (n) x has been locked before wait(x) was performed. User programs are not supposed to contain wait(x, n) prefixes. We write acq$(x)^n.P$ in place of acq$(x)\cdots$acq$(x).P$ where there are n subsequent acquisitions of x.

To define the operational semantics of a program we make use of an infinite set of *process identifiers*, ranged over by t and s, and *states*, which are pairs of the form $\mathcal{H} \Vdash \mathcal{P}$ made of a *heap* $\mathcal{H}$ and a *process pool* $\mathcal{P}$. Heaps are finite maps from object names to pairs of the form t, n where t identifies the process that has locked an object x and n is the number of times x has been acquired. We allow t to be the distinguished name $\bullet$ and n to be 0 when x is unlocked. Process pools are finite maps from process identifiers to processes and represent the set of processes running at a given time. In the following we occasionally write $\mathcal{H}$ as $\{x_i : t_i, n_i\}_{i \in I}$ and $\mathcal{P}$ as $\{t_i : P_i\}_{i \in I}$. We also write $\mathcal{H}, \mathcal{H}'$ for the union of $\mathcal{H}$ and $\mathcal{H}'$ when $dom(\mathcal{H}) \cap dom(\mathcal{H}') = \emptyset$. Similarly for $\mathcal{P}, \mathcal{P}'$.

The operational semantics of a program $(\mathcal{D}, P)$ is determined by the transition relation $\longrightarrow_{\mathcal{D}}$ defined in Table 2 applied to the initial state $\emptyset \Vdash main : P$. To reduce clutter, for each rule we only show those parts of the heap and of the process pool that are affected by the rule. For example, the verbose version of [R-DONE] is $\mathcal{H} \Vdash t :$ done$, \mathcal{P} \longrightarrow_{\mathcal{D}} \mathcal{H} \Vdash \mathcal{P}$. Rules [R-ACQ-*] and [R-REL-*] model the acquisition and release of a lock. There are two versions of each rule to account

Table 2. Reduction rules.

[R-ACQ-1]	$x : \bullet, 0 \Vdash t : \mathsf{acq}(x).P \longrightarrow_{\mathcal{D}} x : t, 1 \Vdash t : P$	
[R-ACQ-2]	$x : t, n \Vdash t : \mathsf{acq}(x).P \longrightarrow_{\mathcal{D}} x : t, n+1 \Vdash t : P$	if $n > 0$
[R-REL-1]	$x : t, 1 \Vdash t : \mathsf{rel}(x).P \longrightarrow_{\mathcal{D}} x : \bullet, 0 \Vdash t : P$	
[R-REL-2]	$x : t, n+1 \Vdash t : \mathsf{rel}(x).P \longrightarrow_{\mathcal{D}} x : t, n \Vdash t : P$	if $n > 0$
[R-WAIT]	$x : t, n \Vdash t : \mathsf{wait}(x).P \longrightarrow_{\mathcal{D}} x : \bullet, 0 \Vdash t : \mathsf{wait}(x, n).P$	if $n > 0$
[R-NFY-1]	$\Vdash t : \mathsf{wait}(x, n).P, s : \mathsf{notify}(x).Q \longrightarrow_{\mathcal{D}} \Vdash t : \mathsf{acq}(x)^n.P, s : Q$	
[R-NFY-2]	$\Vdash t : \mathsf{notify}(x).P, \mathcal{P} \longrightarrow_{\mathcal{D}} \Vdash t : P, \mathcal{P}$	if $\mathsf{wait}(x) \notin \mathcal{P}$
[R-DONE]	$\Vdash t : \mathsf{done} \longrightarrow_{\mathcal{D}} \Vdash$	
[R-IF-1]	$\Vdash t : \mathsf{if}\ e\ \mathsf{then}\ P\ \mathsf{else}\ Q \longrightarrow_{\mathcal{D}} \Vdash t : P$	if $[\![e]\!] \neq 0$
[R-IF-2]	$\Vdash t : \mathsf{if}\ e\ \mathsf{then}\ P\ \mathsf{else}\ Q \longrightarrow_{\mathcal{D}} \Vdash t : Q$	if $[\![e]\!] = 0$
[R-FORK]	$\Vdash t : \mathsf{fork}\{Q\}P, \mathcal{P} \longrightarrow_{\mathcal{D}} \Vdash t : P, s : Q, \mathcal{P}$	
[R-NEW]	$\mathcal{H} \Vdash t : \mathsf{new}\ x\ \mathsf{in}\ P \longrightarrow_{\mathcal{D}} \mathcal{H}, x : \bullet, 0 \Vdash t : P$	
[R-CALL]	$\Vdash t : A(\overline{e}) \longrightarrow_{\mathcal{D}} \Vdash t : P\{[\![\overline{e}]\!]/\overline{x}\}$	if $A(\overline{x}) = P \in \mathcal{D}$

for the fact that locks are reentrant and can be acquired multiple times. Rule [R-WAIT] models a process that unlocks an object x and suspends waiting for a notification on it. The number n of acquisitions is stored in the runtime prefix $\mathsf{wait}(x, n)$ so that, by the time the process is awoken by a notification, it re-acquires x the appropriate number of times. Rules [R-NFY-*] model notifications. The two rules differ depending on whether or not there exists a process that is waiting for such notification. In [R-NFY-1], one waiting process is awoken and guarded by the appropriate number of acquisitions. In [R-NFY-2], the side condition $\mathsf{wait}(x) \notin \mathcal{P}$ means that $\mathcal{P}$ does not contain a process of the form $t : \mathsf{wait}(x, n).P$, implying that no process is currently suspended waiting for a notification on x. In this case, the notification is simply lost. Rules [R-DONE], [R-IF-*], [R-FORK], [R-NEW] and [R-CALL] model terminated processes, conditional processes, forks, object creation and procedure calls as expected. Notice that [R-FORK] has an implicit assumption stating that the name s of the process being created is fresh. This is because the composition $\mathcal{P}, \mathcal{P}'$ is well defined only provided that $dom(\mathcal{P}) \cap dom(\mathcal{P}') = \emptyset$. Similarly, rule [R-NEW] implicitly assumes that the object x being created is fresh.

We write $\Longrightarrow_{\mathcal{D}}$ for the reflexive, transitive closure of $\longrightarrow_{\mathcal{D}}$ and $\mathcal{H} \Vdash \mathcal{P} \not\longrightarrow_{\mathcal{D}}$ if there exist no $\mathcal{H}'$ and $\mathcal{P}'$ such that $\mathcal{H} \Vdash \mathcal{P} \longrightarrow_{\mathcal{D}} \mathcal{H}' \Vdash \mathcal{P}'$. With this notation in place, we formalize the property that we aim to ensure with the type system:

Definition 1 (deadlock-free program). *We say that* $(\mathcal{D}, P)$ *is* deadlock free *if* $\emptyset \Vdash main : P \Longrightarrow_{\mathcal{D}} \mathcal{H} \Vdash \mathcal{P} \not\longrightarrow_{\mathcal{D}}$ *implies* $\mathcal{H} = \{x_i : \bullet, 0\}_{i \in I}$ *and* $\mathcal{P} = \emptyset$.

In words, a program is deadlock free if every maximal, finite computation starting from the initial state – in which the heap is empty and there is only one running process *main* : P – ends in a state in which all the objects in the heap are unlocked and all processes have terminated.

We conclude the section showing how to model the producer-consumer coordination program of Fig. 1 in our calculus.

Table 3. Syntax of types and usages.

Type	$\mathsf{T} ::= \mathsf{int} \mid n \cdot U$
Usage	$U ::= \mathbf{0} \mid \kappa.U \mid U \mid V \mid U + V \mid \alpha \mid \mu\alpha.U$
Usage prefix	$\kappa ::= \mathit{acq} \mid \mathit{rel} \mid \mathit{wait} \mid \mathit{waiting} \mid \mathit{notify}$

Example 1 (multiple-producers/single-consumer coordination). We model the program in Fig. 1 by means of the following procedure definitions, which make use of two objects x and y to coordinate producer and consumer:

$$\begin{aligned}
\mathsf{Main} &= \mathsf{new}\ x\ \mathsf{in}\ \mathsf{new}\ y\ \mathsf{in}\ \mathsf{acq}(x).\mathsf{acq}(y).\mathsf{fork}\{\mathsf{P}(x)\}\ \mathsf{fork}\{\mathsf{P}(y)\}\ \mathsf{C}(x,y)\\
\mathsf{P}(x) &= \mathsf{acq}(x).\mathsf{notify}(x).\mathsf{wait}(x).\mathsf{rel}(x).\mathsf{P}(x)\\
\mathsf{C}(x,y) &= \mathsf{wait}(x).\mathsf{notify}(x).\mathsf{wait}(y).\mathsf{notify}(y).\mathsf{C}(x,y)
\end{aligned}$$

The overall structure of these procedures matches quite closely that of the correspondings methods in Fig. 1. Let us discuss the differences. First of all, in the calculus we focus on coordination primitives. All other operations performed on objects – notably, `Put` and `Get` in Fig. 1 – that do not affect coordination are not modeled explicitly. Second, we model `while` loops using recursion. Third, we model `synchronized` blocks in Fig. 1 as (matching) pairs of acquire-release operations on objects. Notice that the consumer never performs explicit releases, for x and y are always released by $\mathsf{wait}(x)$ and $\mathsf{wait}(y)$. This corresponds to the fact that, in Fig. 1, the whole code of the consumer runs within a block that synchronizes on x and y.

It is worth noting that, even after all the background noise that is present in Fig. 1 has been removed, understanding whether the program deadlocks is not trivial. As anticipated in the introduction, the most critical aspect is determining whether x and y are always notified at a time when there is a process waiting to be awoken. ■

3 Types and Usages

Our type system rules out deadlocks using two orthogonal mechanisms. The first one makes sure that well-typed programs do not contain cycles of parallel processes linked by shared objects. This mechanism suffices to avoid circular waits, but does not guarantee that each process suspended on a wait operation is awoken by a notification. To rule out these situations, we also associate each shared object with a simple protocol description, called *usage*, that specifies the operations performed by processes on it. Then, we make sure that usages are *reliable*, namely that each wait operation is matched by (at least) one notification. The rest of this section is devoted to the formal definition of types, usages and related notions, leading to the formalization of reliability. The description of the actual typing rules is deferred to Sect. 4.

The syntax of types and usages is shown in Table 3. A type is either int, denoting an integer value, or an object type $n \cdot U$ where n is a natural number called *counter* and U is a *usage*. The counter indicates the number of times the object has been acquired. The usage describes the combined operations performed by processes on the object. The usage $\mathbf{0}$ describes an object on which no operations are performed. The usage $\kappa.U$ describes an object that is used to perform the operation κ and then according to U. Usage prefixes *acq*, *rel*, *wait*, *waiting* and *notify* are in direct correspondence with the actions in Table 1. In particular, *waiting* is a "runtime version" of *wait* and describes an object on which the wait operation has been performed and is waiting to be notified. Usages of user programs are not supposed to contain the *waiting* prefix. The usage $U \mid V$ describes an object that is used by concurrent processes according to U and V, whereas the usage $U + V$ describes an object that is used either according to U or to V. Terms of the form $\mu\alpha.U$ and α are used to describe recursive usages.

We adopt standard conventions concerning (recursive) usages: we assume contractiveness, namely that, in a usage $\mu\alpha.U$, the variable α occurring in U is always guarded by a usage prefix; we identify usages modulo (un)folding of recursions; we omit trailing $\mathbf{0}$'s.

To illustrate usages before describing the typing rules, consider the process

$$\mathsf{fork}\{\mathsf{acq}(x).\mathsf{rel}(x)\}\ \mathsf{acq}(y).\mathsf{acq}(x).\mathsf{rel}(x).\mathsf{rel}(y)$$

which uses two objects x and y. Notice that x is acquired by two concurrent subprocesses and that, in one case, this acquisition is nested within the acquisition and release of y. The operations performed on x are described by the usage $acq.rel \mid acq.rel$ whereas those performed on y are described by the usage $acq.rel$.

We now proceed to define a series of auxiliary notions related to types and usages and that play key roles in the type system and the proofs of its soundness. To begin with, we formalize a predicate on usages that allows us to identify those objects on which there is no process waiting for a notification:

Definition 2 (wait-free usage). *Let* $\mathsf{wf}(\cdot)$ *be the least predicate on usages defined by the axioms and rules below:*

$$\mathsf{wf}(\mathbf{0}) \qquad \mathsf{wf}(acq.U) \qquad \frac{\mathsf{wf}(U) \qquad \mathsf{wf}(V)}{\mathsf{wf}(U \mid V)} \qquad \mathsf{wf}(U+V)$$

Note that $\mathsf{wf}(U + V)$ holds regardless of U and V because the usage $U + V$ describes an object on which a process behaves according to either U or V, but the process has not committed to any such behavior yet. Hence, the process cannot be waiting for a notification on the object.

Next, we define a reduction relation $\rightsquigarrow$ describing the evolution of the type of an object as the object is used by processes. As we have anticipated earlier, the effect of operations depends on whether and how many times the object has been locked. For this reason, the reduction relation we are about to define concerns types and not just usages.

Table 4. Reduction of types.

[U-ACQ-1]
$$0 \cdot acq.U \mid V \rightsquigarrow 1 \cdot U \mid V$$

[U-ACQ-2]
$$n+1 \cdot U \rightsquigarrow n+2 \cdot U$$

[U-REL-1]
$$1 \cdot rel.U \mid V \rightsquigarrow 0 \cdot U \mid V$$

[U-REL-2]
$$n+2 \cdot U \rightsquigarrow n+1 \cdot U$$

[U-WAIT]
$$n+1 \cdot wait.U \mid V \rightsquigarrow 0 \cdot waiting.U \mid V$$

[U-CHOICE]
$$n \cdot (U + U') \mid V \rightsquigarrow n \cdot U \mid V$$

[U-NFY-1]
$$n+1 \cdot waiting.U \mid notify.U' \mid V \rightsquigarrow n+1 \cdot acq.U \mid U' \mid V$$

[U-NFY-2]
$$\frac{\mathsf{wf}(V)}{n+1 \cdot notify.U \mid V \rightsquigarrow n+1 \cdot U \mid V}$$

[U-CONG]
$$\frac{U \equiv U' \qquad n \cdot U' \rightsquigarrow n' \cdot V' \qquad V' \equiv V}{n \cdot U \rightsquigarrow n' \cdot V}$$

Definition 3 (type reduction). *Let $\equiv$ be the least congruence on usages containing commutativity and associativity of $\mid$ with identity* $\mathbf{0}$, *commutativity, associativity, and idempotency of* $+$. *The reduction relation* $\mathtt{T} \rightsquigarrow \mathtt{T}'$ *is the least relation defined by the axioms and rules in Table 4. As usual, we write* $\rightsquigarrow^*$ *for the reflexive and transitive closure of* $\rightsquigarrow$.

Rules [U-ACQ-1] and [U-REL-1] model the acquisition and the release of an object; in the first case, the object must be unlocked (the counter is 0) whereas, in the second case, the object must have been locked exactly once. Rules [U-ACQ-2] and [U-REL-2] model nested acquisitions and releases on an object. These rules simply change the counter to reflect the actual number of (nested) acquisitions and do not correspond to an actual prefix in the usage. Rule [U-WAIT] models a wait operation performed on an object: the counter is set to 0 and the *wait* prefix is replaced by *waiting*, indicating that the object has been unlocked waiting for a notification. This makes it possible for another process to acquire the object and eventually notify the one who waits. Rules [U-NFY-1] and [U-NFY-2] model notifications. In [U-NFY-1], the notification occurs at a time when there is indeed another process that is waiting to be notified. In this case, the waiting process attempts to acquire the object. In [U-NFY-2], the notification occurs at a time when no other process is waiting to be notified. In this case, the notification has no effect whatsoever and is lost. Finally, rule [U-CHOICE] (in conjunction with commutativity of $+$) models the nondeterministic choice between two possible usages of an object and [U-CONG] closes reductions under usage congruence.

Not all types are acceptable for our type system. More specifically, there are two properties that we wish to be guaranteed:

- whenever there is a pending acquisition operation on a locked object, the object is eventually unlocked;

– whenever there is a pending wait operation on an object, the object is eventually notified.

A type that satisfies these two properties is said to be *reliable*:

Definition 4 (type reliability). *We say that* T *is* reliable, *written* $rel(\mathtt{T})$, *if the following conditions hold for all* n, U *and* V:

1. $\mathtt{T} \rightsquigarrow^* n \cdot acq.U \mid V$ *implies* $n \cdot V \rightsquigarrow^* 0 \cdot V'$ *for some* V', *and*
2. $\mathtt{T} \rightsquigarrow^* 0 \cdot waiting.U \mid V$ *implies* $0 \cdot V \rightsquigarrow^* n \cdot notify.U' \mid V'$ *for some* U', V'.

For example, it is easy to verify that $0 \cdot acq.(wait.rel \mid acq.notify.rel)$ is reliable whereas $0 \cdot acq.wait.rel \mid acq.notify.rel$ is not. In the latter usage, the object may be acquired by a process that notifies the object at a time when there is no other process waiting to be notified. Eventually, the object is acquired again but the awaited notification is lost.

Example 2. Consider the type $0 \cdot acq.(U \mid V)$ where $U \stackrel{\text{def}}{=} \mu\alpha.acq.notify.wait.rel.\alpha$ and $V \stackrel{\text{def}}{=} \mu\alpha.wait.notify.\alpha$. To prove that $0 \cdot acq.(U \mid V)$ is reliable, we derive

$$
\begin{array}{rll}
0 \cdot acq.(U \mid V) & \rightsquigarrow 1 \cdot U \mid V & \text{[U-ACQ-1]} \\
& \rightsquigarrow 0 \cdot U \mid waiting.notify.V & \text{[U-WAIT]}\ (\star) \\
& \rightsquigarrow 1 \cdot notify.wait.rel.U \mid waiting.notify.V & \text{[U-ACQ-1]} \\
& \rightsquigarrow 1 \cdot wait.rel.U \mid acq.notify.V & \text{[U-NFY-1]} \\
& \rightsquigarrow 0 \cdot waiting.rel.U \mid acq.notify.V & \text{[U-WAIT]} \\
& \rightsquigarrow 1 \cdot waiting.rel.U \mid notify.V & \text{[U-ACQ-1]} \\
& \rightsquigarrow 1 \cdot acq.rel.U \mid V & \text{[U-NFY-1]} \\
& \rightsquigarrow 0 \cdot acq.rel.U \mid waiting.notify.V & \text{[U-WAIT]} \\
& \rightsquigarrow 1 \cdot rel.U \mid waiting.notify.V & \text{[U-ACQ-1]} \\
& \rightsquigarrow 0 \cdot U \mid waiting.notify.V & \text{[U-REL-1]}\ (\star)
\end{array}
$$

with the obvious applications of [U-CONG] not reported. Observe that no other reductions are possible apart from those shown above, that the two types labelled $(\star)$ are equal, and that both conditions of Definition 4 are satisfied for each reachable state. As we shall see in Example 3, the type $0 \cdot acq.(U \mid V)$ describes the behavior of the main thread of Example 1 with respect to each buffer. ■

4 Static Semantics

4.1 Type Environments

The type system uses *type environments*, ranged over by Γ, which are finite sets of associations on variables and procedure names defined by the grammar below:

$$\textbf{Type environment} \quad \Gamma \quad ::= \quad \emptyset \mid x : \mathtt{T}, \Gamma \mid A : [\overline{\mathtt{T}}], \Gamma$$

An association $x : \mathtt{T}$ indicates that x has type T, whereas an association $A : [\overline{\mathtt{T}}]$ indicates that A is a procedure accepting parameters of type $\overline{\mathtt{T}}$.

We write $dom(\Gamma)$ for the set of variable/procedure names for which there is an association in Γ and $\Gamma(x)$ for the type associated with $x \in dom(\Gamma)$. With an abuse of notation, we write Γ, Γ' for the union of Γ and Γ' when $dom(\Gamma) \cap dom(\Gamma') = \emptyset$. In addition: we write $live(\Gamma)$ for the subset of $dom(\Gamma)$ of *live object references* on which there are pending operations, that is $live(\Gamma) \stackrel{\text{def}}{=} \{x \in dom(\Gamma) \mid \Gamma(x) = n \cdot U \wedge (n > 0 \vee U \not\equiv \mathbf{0})\}$; we write $iszero(\Gamma)$ if $\Gamma(x) = 0 \cdot U$ for every $x \in dom(\Gamma)$; we write $noAct(\Gamma)$ if $live(\Gamma) = \emptyset$.

The same object may be used in different ways in different parts of a program. In order to track the combined usage of the object we inductively define two operators $\mid$ and $+$ on type environments with the same domain. Intuitively, $(\Gamma \mid \Gamma')(x)$ is the type of an object that is used *both* as specified in Γ *and also* as specified in Γ' whereas $(\Gamma + \Gamma')(x)$ is the type of an object that is used *either* as specified in Γ *or* as specified in Γ'. The former case happens if x is shared by two concurrent processes respectively typed by Γ and Γ'. The latter case happens if x is used in different branches of a conditional process. Formally:

$$
\begin{aligned}
\emptyset \mid \emptyset &= \emptyset \\
x : \mathsf{int}, \Gamma \mid x : \mathsf{int}, \Gamma' &= x : \mathsf{int}, (\Gamma \mid \Gamma') \\
A : [\overline{\mathsf{T}}], \Gamma \mid A : [\overline{\mathsf{T}}], \Gamma' &= A : [\overline{\mathsf{T}}], (\Gamma \mid \Gamma') \\
n \cdot U, \Gamma \mid m \cdot V, \Gamma' &= n + m \cdot U \mid V, (\Gamma \mid \Gamma') \qquad n = 0 \vee m = 0
\end{aligned}
$$

$$
\begin{aligned}
\emptyset + \emptyset &= \emptyset \\
x : \mathsf{int}, \Gamma + x : \mathsf{int}, \Gamma' &= x : \mathsf{int}, (\Gamma + \Gamma') \\
A : [\overline{\mathsf{T}}], \Gamma + A : [\overline{\mathsf{T}}], \Gamma' &= A : [\overline{\mathsf{T}}], (\Gamma + \Gamma') \\
n \cdot U, \Gamma + n \cdot V, \Gamma' &= n \cdot U + V, (\Gamma + \Gamma')
\end{aligned}
$$

Note that both $\mid$ and $+$ for environments are partial operators and that the former enforces the property that the same object cannot be owned by more than one process at any given time (at least one of the counters must be 0). It is easy to see that $\mid$ on environments is commutative and associative (modulo $\equiv$ on usages). In the following we write $\prod_{i=1..n} \Gamma_i$ in place of $\Gamma_1 \mid \cdots \mid \Gamma_n$.

As usual for behavioral type systems, the type environment used for typing a process is an abstraction of the behavior of the process projected on the objects it uses. In particular, a *live* object association $x : n \cdot U$ in the type environment of a process means that the process uses x as specified by U, whereas an association such as $x : 0 \cdot \mathbf{0}$ means that the process does not use x at all. To prevent circular waits between parallel processes, we forbid the existence of cycles in the corresponding type environments:

Definition 5 (acyclic type environments). *We say that a family* $\{\overline{\Gamma}\}$ *of type environments has a* cycle $x_1, \dots, x_n$ *of* $n \geq 2$ *pairwise distinct names if there exist* $\Gamma_1, \dots, \Gamma_n \in \{\overline{\Gamma}\}$ *such that* $x_i \in live(\Gamma_i) \cap live(\Gamma_{(i \bmod n)+1})$ *for all* $1 \leq i \leq n$. *We say that* $\{\overline{\Gamma}\}$ *is* acyclic *if it has no cycle.*

Observe that the acyclicity of a family of type environments can be established efficiently, for example by means of a standard union-find algorithm that stores in the same partition two environments if they share a live object.

Table 5. Typing rules for user syntax.

Typing rules for (sequences of) expressions $\boxed{\Gamma \vdash \overline{e} : \overline{\mathtt{T}}}$

$$\text{[T-CONST]}\ \frac{noAct(\Gamma)}{\Gamma \vdash n : \mathsf{int}} \qquad \text{[T-VAR]}\ \frac{noAct(\Gamma)}{\Gamma, x : \mathtt{T} \vdash x : \mathtt{T}} \qquad \text{[T-OP]}\ \frac{\Gamma \vdash e : \mathsf{int} \qquad \Gamma \vdash e' : \mathsf{int}}{\Gamma \vdash e \;\mathsf{op}\; e' : \mathsf{int}}$$

$$\text{[T-SEQ]}\ \frac{\Gamma_i \vdash e_i : \mathtt{T}_i \ {}^{(i=1..n)}}{\Gamma_1 \mid \cdots \mid \Gamma_n \vdash e_1, \ldots, e_n : \mathtt{T}_1, \ldots, \mathtt{T}_n}$$

Typing rules for processes $\boxed{\Gamma \vdash P}$

$$\text{[T-DONE]}\ \frac{noAct(\Gamma)}{\Gamma \vdash \mathsf{done}} \qquad \text{[T-ACQ-1]}\ \frac{\Gamma, x : 1 \cdot U \vdash P}{\Gamma, x : 0 \cdot acq.U \vdash \mathsf{acq}(x).P} \qquad \text{[T-ACQ-2]}\ \frac{\Gamma, x : n+2 \cdot U \vdash P}{\Gamma, x : n+1 \cdot U \vdash \mathsf{acq}(x).P}$$

$$\text{[T-REL-1]}\ \frac{\Gamma, x : 0 \cdot U \vdash P}{\Gamma, x : 1 \cdot rel.U \vdash \mathsf{rel}(x).P} \qquad \text{[T-REL-2]}\ \frac{\Gamma, x : n+1 \cdot U \vdash P}{\Gamma, x : n+2 \cdot U \vdash \mathsf{rel}(x).P}$$

$$\text{[T-WAIT]}\ \frac{\Gamma, x : n+1 \cdot U \vdash P}{\Gamma, x : n+1 \cdot wait.U \vdash \mathsf{wait}(x).P} \qquad \text{[T-NOTIFY]}\ \frac{\Gamma, x : n+1 \cdot U \vdash P}{\Gamma, x : n+1 \cdot notify.U \vdash \mathsf{notify}(x).P}$$

$$\text{[T-IF]}\ \frac{\Gamma \vdash e : \mathsf{int} \qquad \Gamma_i \vdash P_i \ {}^{(i=1,2)}}{\Gamma \mid (\Gamma_1 + \Gamma_2) \vdash \mathsf{if}\ e\ \mathsf{then}\ P_1\ \mathsf{else}\ P_2} \qquad \text{[T-CALL]}\ \frac{\Gamma \vdash \overline{e} : \overline{\mathtt{T}}}{\Gamma, A : [\overline{\mathtt{T}}] \vdash A(\overline{e})}$$

$$\text{[T-FORK]}\ \frac{\Gamma_i \vdash P_i \ {}^{(i=1,2)} \qquad iszero(\Gamma_1) \qquad \{\Gamma_1, \Gamma_2\}\ \text{acyclic}}{\Gamma_1 \mid \Gamma_2 \vdash \mathsf{fork}\{P_1\}P_2} \qquad \text{[T-NEW]}\ \frac{\Gamma, x : 0 \cdot U \vdash P \qquad rel(0, U)}{\Gamma \vdash \mathsf{new}\ x\ \mathsf{in}\ P}$$

Typing rule for programs $\boxed{\Gamma \vdash (\mathcal{D}, P)}$

$$\text{[T-PROGRAM]}\ \frac{\Gamma = A_i : [\overline{\mathtt{T}}_i] \ {}^{(i=1..n)} \qquad \Gamma, \overline{x}_i : \overline{\mathtt{T}}_i \vdash P_i \ {}^{(i=1..n)} \qquad \Gamma \vdash P}{\Gamma \vdash (\{A_i(\overline{x}_i) = P_i\}_{i=1..n}, P)}$$

4.2 Typing Rules for User Syntax

The typing rules for the language in Sect. 2 are defined in Table 5 and derive three kinds of judgments. A judgment $\Gamma \vdash e : \mathtt{T}$ means that the expression e is well typed in Γ and has type $\mathtt{T}$. A judgment $\Gamma \vdash P$ means that the process P is well typed in Γ. In particular, P uses each object $x \in dom(\Gamma)$ according to the usage in $\Gamma(x)$. Finally, a judgment $\Gamma \vdash (\mathcal{D}, P)$ means that the program $(\mathcal{D}, P)$ is well typed in Γ. We now describe the typing rules.

The typing rules for (sequences of) expressions are unremarkable except for the fact that the unused part of the type environment cannot contain live associations, hence the premise $noAct(\Gamma)$ in [T-CONST] and [T-VAR].

Rule [T-DONE] states that the terminated process is well typed in an environment without live associations, because done does not perform any operation.

Rules [T-ACQ-1] and [T-ACQ-2] concern a process $\mathsf{acq}(x).P$ that acquires the lock on x and then continues as P. The difference between the two rules is that in [T-ACQ-1] the process is attempting to acquire the lock for the first time (the counter of the object is 0), whereas in [T-ACQ-2] the process is performing a reentrant acquisition, having already acquired the lock $n + 1$ times. The continuation P is typed in an environment that reflects the (possibly reentrant) acquision of x. Note that the *acq* action occurs in the usage of x only in the case of [T-ACQ-1]. As we have anticipated in Sect. 3, in usages we only keep track of non-reentrant acquisitions and releases.

Rules [T-REL-1] and [T-REL-2] concern a process $\mathsf{rel}(x).P$. As in the case of [T-ACQ-*] rules, they differ depending on whether the lock is actually released ([T-REL-1]) or not ([T-REL-2]). Only in the first case the release action is noted in the usage of x. Besides that, the rules update the environment for typing the continuation P.

Rules [T-WAIT] and [T-NOTIFY] concern the coordination primitives. In both cases, the object must have been previously acquired (the counter is strictly positive) and the number of acquisitions does not change.

Rule [T-IF] is essentially standard. Since only one of the two continuations P_1 and P_2 executes, the respective type environments Γ_1 and Γ_2 are composed using the appropriate disjunctive operator.

Rule [T-FORK] types a parallel composition of two processes P_1 and P_2. The objects used by the parallel composition are used *both* by P_1 and also by P_2. For this reason, the respective type environments are combined using $\mid$. The $iszero(\Gamma_1)$ premise enforces the property that the process P_1 being forked off does not own any lock. The last premise requires that the family $\{\Gamma_1, \Gamma_2\}$ be acyclic, which is equivalent to checking that $live(\Gamma_1) \cap live(\Gamma_2)$ contains at most one element. This prevents circular waits between P_1 and P_2 as discussed earlier.

Rule [T-NEW] concerns the creation of a new object. The object is initially unlocked (its counter is 0) and its type must be reliable (Definition 4).

Rule [T-CALL] is unremarkable and types a process invocation. The only standard requirement is for the types of the arguments to match those expected in the corresponding process declaration.

The typing rule [T-PROGRAM] ensures that all process names have a corresponding definition and verifies that the main process is itself well typed.

Example 3. Let us show that the `Main` process in Example 1 is well typed. To do that, consider the type environment $\Gamma = \mathtt{P} : [0 \cdot U], \mathtt{C} : [1 \cdot V, 1 \cdot V]$ where U and V are the usages defined in Example 2 and observe that $noAct(\Gamma)$ holds.

For the two invocations $\mathtt{P}(x)$ and $\mathtt{P}(y)$ we easily derive

$$(1)\quad \dfrac{\overline{\Gamma, x : 0 \cdot U, y : 0 \cdot \mathbf{0} \vdash x : 0 \cdot U}}{\Gamma, x : 0 \cdot U, y : 0 \cdot \mathbf{0} \vdash \mathtt{P}(x)} \qquad (2)\quad \dfrac{\overline{\Gamma, x : 0 \cdot \mathbf{0}, y : 0 \cdot U \vdash y : 0 \cdot U}}{\Gamma, x : 0 \cdot \mathbf{0}, y : 0 \cdot U \vdash \mathtt{P}(y)}$$

using [T-VAR] and [T-CALL]. Then we have

$$\dfrac{\dfrac{\dfrac{\dfrac{(1) \quad \dfrac{(2) \quad \dfrac{\dfrac{}{\Gamma, x : 1 \cdot V, y : 1 \cdot V \vdash x, y : 1 \cdot V, 1 \cdot V}\,\text{[T-SEQ]}}{\Gamma, x : 1 \cdot V, y : 1 \cdot V \vdash \mathsf{C}(x, y)}\,\text{[T-CALL]}}{\Gamma, x : 1 \cdot V, y : 1 \cdot U \mid V \vdash \mathsf{fork}\{\mathtt{P}(y)\}\ \mathsf{C}(x, y)}\,\text{[T-FORK]}}{\Gamma, x : 1 \cdot U \mid V, y : 1 \cdot U \mid V \vdash \mathsf{fork}\{\mathtt{P}(x)\}\ \mathsf{fork}\{\mathtt{P}(y)\}\ \mathsf{C}(x, y)}\,\text{[T-FORK]}}{\Gamma, x : 1 \cdot U \mid V, y : 0 \cdot acq.(U \mid V) \vdash \mathsf{acq}(y).\mathsf{fork}\{\mathtt{P}(x)\}\ \mathsf{fork}\{\mathtt{P}(y)\}\ \mathsf{C}(x, y)}\,\text{[T-ACQ-1]}}{\dfrac{\Gamma, x : 0 \cdot acq.(U \mid V), y : 0 \cdot acq.(U \mid V) \vdash \mathsf{acq}(x) \cdots}{\dfrac{\Gamma, x : 0 \cdot acq.(U \mid V) \vdash \mathsf{new}\ y\ \mathsf{in}\ \cdots}{\Gamma \vdash \mathsf{new}\ x\ \mathsf{in}\ \mathsf{new}\ y\ \mathsf{in}\ \mathsf{acq}(x).\mathsf{acq}(y).\mathsf{fork}\{\mathtt{P}(x)\}\ \mathsf{fork}\{\mathtt{P}(y)\}\ \mathsf{C}(x, y)}\,\text{[T-NEW]}}\,\text{[T-NEW]}}\,\text{[T-ACQ-1]}$$

where the reliability of $0 \cdot acq.(U \mid V)$ needed in the applications of [T-NEW] has already been proved in Example 2. In the applications of [T-FORK], the acyclicity of the involved environments is easily established since in each conclusion of (1) and (2) there is only one live association, for x and y respectively. ∎

4.3 Typing Rules for Runtime Syntax and States

The soundness proof of our type system follows a standard structure and includes a subject reduction result stating that typing (but not necessarily types) are preserved by reductions. Since the operational semantics of a program makes use of constructs that occur at runtime only (notably, waiting processes and states) we must extend the typing rules to these constructs before we can formulate the properties of the type system. The additional typing rules are given in Table 6.

Rule [T-WAITING] accounts for a process waiting for a notification, after which it will attempt to acquire the lock on x. Once x is notified and the process awakened, the process will acquire the lock n times, reflecting the state of acquisitions at the time the process performed the $\mathsf{wait}(x)$ operation (see [R-WAIT]).

The rule [T-STATE] for states looks more complex than it actually is. For the most part, this rule is a generalization of [T-FORK] and [T-NEW] to an arbitrary number of concurrent processes (indexed by $i \in I$) and of objects (indexed by $j \in J$). From left to right, the premises of the rule ensure that:

- each process P_i is well typed in its corresponding environment Γ_i;
- the type of each object x_j, which describes the overall usage of x_j by all the processes, is reliable;

Table 6. Typing rules for runtime syntax.

[T-WAITING]

$$\frac{\Gamma, x : n \cdot U \vdash P}{\Gamma, x : 0 \cdot waiting.U \vdash \mathsf{wait}(x, n).P}$$

[T-STATE]

$$\frac{\begin{array}{cccc}\Gamma = \prod_{i \in I} \Gamma_i & \Gamma_i \vdash P_i^{\ (i \in I)} & rel(\Gamma(x_j))^{\ (j \in J)} & \{\Gamma_i\}_{i \in I} \text{ acyclic} \\ \Gamma_i(x_j) = n+1, U \iff t_i = s_j^{\ (i \in I, j \in J)} & & \Gamma(x_j) = 0 \cdot U \iff s_j = \bullet^{\ (j \in J)} & \end{array}}{\Gamma \vdash \{x_j : s_j, n_j\}_{j \in J} \Vdash \prod_{i \in I} t_i : P_i}$$

- the family $\{\Gamma_i\}_{i \in I}$ of type environments is acyclic;
- the process t_i owns the object x_j if and only if the counter for x_j in the type environment Γ_i of P_i is strictly positive. Because of the definition of $\mid$ for type environments, this implies that no other process owns x_j;
- no process owns the object x_j if and only if the counter for x_j in the overall environment is zero.

4.4 Properties of Well-Typed Programs

As usual, the key lemma for proving soundness of the type system is subject reduction, stating that a well-typed state reduces to well-typed state. In our case, this result guarantees the preservation of typing, but not necessarily the preservation of types. Indeed, as a program reduces and operations are performed on objects, the type of such objects changes consequently. To account for these changes, we lift reduction of types to type environments, thus:

Definition 6 (environment reduction). *The reduction relation for environments, noted $\rightsquigarrow$, is the least relation such that:*

$$\Gamma \rightsquigarrow \Gamma, x : \mathrm{T} \qquad \frac{\mathrm{T} \rightsquigarrow \mathrm{T}'}{\Gamma, x : \mathrm{T} \rightsquigarrow \Gamma, x : \mathrm{T}'}$$

As usual, $\rightsquigarrow^*$ denotes the reflexive, transitive closure of $\rightsquigarrow$.

The first rule accounts for the possibility that a new object x is created. We can now formally state subject reduction, which shows that typing is preserved for any reduction of a well-typed program:

Lemma 1 (subject reduction). *Let $\Gamma \vdash (\mathcal{D}, P)$ and $\emptyset \Vdash main : P \longrightarrow_{\mathcal{D}}^{*} \mathcal{H}' \Vdash \mathcal{P}'$. Then $\Gamma' \vdash \mathcal{H}' \Vdash \mathcal{P}'$ for some Γ' such that $\Gamma \rightsquigarrow^* \Gamma'$.*

The soundness theorem states that well-typed programs are deadlock free:

Theorem 1 (soundness). *If $\Gamma \vdash (\mathcal{D}, P)$, then $(\mathcal{D}, P)$ is deadlock free.*

5 Related Work

Despite the number of works on deadlock analysis of concurrent programs, only a few address coordination primitives. Below we discuss the most relevant ones.

Static techniques typically employ control-flow analysis to build a dependency graph between objects and enforce its acyclicity. These techniques may adopt some heuristics to remove likely false positives, but they are necessarily conservatives. For instance, Deshmukh *et al.* [5] analyze libraries of concurrent objects looking for deadlocks that may manifest for *some* clients of such objects, by considering all possible aliasing between the locks involved in the objects. The technique of von Praun [17] is based on the detection of particular patterns in the code, such as two threads that perform $\mathsf{wait}(x)$ and $\mathsf{wait}(y)$ in different orders, which does not necessarily lead to a deadlock. Naik *et al.* [16] combine different static analyses that correspond to different conditions that are necessary to yield a deadlock. Their technique concerns lock acquisition and release, but not coordination primitives. Williams *et al.* [19] build a *lock-order graph* that describes the sequences of lock acquisitions in a library of Java classes. In particular, they consider implicit acquisitions due to wait operations, but not deadlocks caused by missed notifications. Hamin and Jacobs [8] present a refinement of separation logic to reason on locks and wait/notify coordination primitives. Their technique ensures deadlock freedom by imposing an ordering on the use of locks and by checking that each wait is matched by at least one notification. The logic allows them to address single wait operations within loops, which is something our type system is unable to handle. On the other hand, the use of a lock ordering limits the technique in presence of loops and recursion whereby blocking operations on several locks are interleaved, as in our running example (Fig. 1).

Dynamic techniques perform deadlock detection by analyzing the log or scheduling of a program execution [1,6,11]. By considering actual program runs, these techniques potentially offer better precision, at the cost of delayed deadlock detection. Agarwal and Stoller [1] define feasible sequences, called *traces*, that are consistent with the original order of events from each thread and with constraints imposed by synchronization events. By analyzing all the possible traces, they verify that a wait operation always *happens before* a notify operation. Joshi *et al.* [11] extract a simple multi-threaded program from the source code that records relevant operations for finding deadlocks. Then, they consider any possible interleavings of the simple program by means of a model checker in search of deadlocks. The technique returns both false positives (the simple program manifests a deadlock that never occurs in the source program) and false negatives (the simple program is defined by observing a single execution and the deadlock may occur in another execution). Deadlocks due to coordination primitives are not covered by the technique of Eslamimehr and Palsberg [6].

Demartini *et al.* [4] translate `Java` into the Promela language, for which the SPIN model checker verifies deadlock freedom. Their analysis reports all deadlock possibilities so long as the program does not exceed the maximum number of modeled objects or threads. `Java` Pathfinder, a well-known tool that is used to analyze execution traces, also performs model checking by translating

Java to Promela [9]. When checking matches between wait and notify operations, this technique may require the analysis of a large number of traces.

Kobayashi and Laneve [13] present a deadlock analysis for the pi-calculus which provided the initial inspiration for this work. In fact, attempts were made to encode the language of Sect. 2 in the pi-calculus so as to exploit the technique of Kobayashi and Laneve. The encoding approach proved to be unsatisfactory because of the many false positives it triggered. Notably, Kobayashi and Laneve [13], following [12], define a notion of usage reliability which can be reduced to a reachability problem in Petri nets. However, the encoding of usages with `wait-notify` primitives requires the use of Petri nets with *inhibitor arcs* [2], which are more expressive than standard Petri nets.

Our type system does not impose an order on the usage of locks. Rather, it adopts a typing rule for parallel threads ([T-FORK]) inspired by session type systems based on linear logic [3,18]. The key idea is to require that, whenever two threads are combined together in a parallel composition, they can only interact through at most one object, as suggested by the structure of the cut and tensor rules in (classical) linear logic [7]. This approach results in a simple and expressive type system which can deal with recursive processes interleaving blocking actions on different objects (Fig. 1). The downside is that [T-FORK] imposes well-typed programs to exhibit a forest-like topology, ruling out some interesting programs which are in the scope of other techniques, such as those based on lams [13].

6 Concluding Remarks

We have described a deadlock analysis technique for concurrent programs using Java-like coordination primitives `wait` and `notify`. Our technique is based on behavioral types, called usages, that may be encoded as Petri nets with inhibitor arcs [2]. Thereby, we reduce deadlock freedom to the reachability problem in this class of Petri nets that is *partially decidable*. The details, as well as the extension of our technique to the `notifyAll` primitive can be found in the full paper [15].

Our technique is unable to address programs where two threads share more than one object because of the acyclicity constraint in rule [T-FORK]. A more fine-grained approach for tracking object dependencies has been developed by Laneve [14] and is based on lams [13]. However, this approach does not consider coordination primitives. We initially tried to combine lams and usages with coordination primitives, but the resulting type system proved to be overly restrictive with respect to recursive processes: the amount of dependencies prevented the typing of any recursive process interleaving blocking operations on two or more objects (such as the consumer in Fig. 1). Whether lams and usages with coordination primitives can be reconciled is still to be determined.

Another limitation of the type system is that it assumes precise knowledge of the number of acquisitions for each shared object, to the point that types contain a *counter* for this purpose. However, this information is not always statically available. It may be interesting to investigate whether this limitation can be lifted by allowing a form of *counter polymorphism*.

Finally, our model ignores object fields and methods for the sake of simplicity. Coping with these features might require some non-trivial extensions of the type system, such as effect annotations on method types, in order to keep track of the type of fields as they are updated.

References

1. Agarwal, R., Stoller, S.D.: Run-time detection of potential deadlocks for programs with locks, semaphores, and condition variables. In: Proceedings of PADTAD 2006, pp. 51–60. ACM (2006). https://doi.org/10.1145/1147403.1147413
2. Busi, N.: Analysis issues in petri nets with inhibitor arcs. Theor. Comput. Sci. **275**(1–2), 127–177 (2002). https://doi.org/10.1016/S0304-3975(01)00127-X
3. Caires, L., Pfenning, F., Toninho, B.: Linear logic propositions as session types. Math. Struct. Comput. Sci. **26**(3), 367–423 (2016). https://doi.org/10.1017/S0960129514000218
4. Demartini, C., Iosif, R., Sisto, R.: A deadlock detection tool for concurrent java programs. Softw. Pract. Exper. **29**(7), 577–603 (1999)
5. Deshmukh, J.V., Emerson, E.A., Sankaranarayanan, S.: Symbolic modular deadlock analysis. Autom. Softw. Eng. **18**(3–4), 325–362 (2011)
6. Eslamimehr, M., Palsberg, J.: Sherlock: scalable deadlock detection for concurrent programs. In: Proceedings of FSE 2014, pp. 353–365. ACM (2014)
7. Girard, J.: Linear logic. Theor. Comput. Sci. **50**, 1–102 (1987). https://doi.org/10.1016/0304-3975(87)90045-4
8. Hamin, J., Jacobs, B.: Deadlock-free monitors. In: Ahmed, A. (ed.) ESOP 2018. LNCS, vol. 10801, pp. 415–441. Springer, Cham (2018). https://doi.org/10.1007/978-3-319-89884-1_15
9. Havelund, K.: Using runtime analysis to guide model checking of Java programs. In: Havelund, K., Penix, J., Visser, W. (eds.) SPIN 2000. LNCS, vol. 1885, pp. 245–264. Springer, Heidelberg (2000). https://doi.org/10.1007/10722468_15
10. Hoare, C.A.R.: Monitors: an operating system structuring concept. Commun. ACM **17**(10), 549–557 (1974). https://doi.org/10.1145/355620.361161
11. Joshi, P., Naik, M., Sen, K., Gay, D.: An effective dynamic analysis for detecting generalized deadlocks. In: Proceedings of FSE 2010, pp. 327–336. ACM (2010). https://doi.org/10.1145/1882291.1882339
12. Kobayashi, N.: Type-based information flow analysis for the pi-calculus. Acta Informatica **42**(4–5), 291–347 (2005). https://doi.org/10.1007/s00236-005-0179-x
13. Kobayashi, N., Laneve, C.: Deadlock analysis of unbounded process networks. Inf. Comput. **252**, 48–70 (2017). https://doi.org/10.1016/j.ic.2016.03.004
14. Laneve, C.: A lightweight deadlock analysis for programs with threads and reentrant locks. In: Havelund, K., Peleska, J., Roscoe, B., de Vink, E. (eds.) FM 2018. LNCS, vol. 10951, pp. 608–624. Springer, Cham (2018). https://doi.org/10.1007/978-3-319-95582-7_36
15. Laneve, C., Padovani, L.: Deadlock analysis of wait-notify coordination. Technical report, Computer Science Department, University of Bologna (2019). https://hal.inria.fr/hal-02166082
16. Naik, M., Park, C., Sen, K., Gay, D.: Effective static deadlock detection. In: Proceedings of ICSE 2009, pp. 386–396. IEEE (2009). https://doi.org/10.1109/ICSE.2009.5070538

17. von Praun, C.: Detecting synchronization defects in multi-threaded object-oriented programs. Ph.D. thesis, Swiss Federal Institute of Technology, Zurich (2004)
18. Wadler, P.: Propositions as sessions. J. Funct. Program. **24**(2–3), 384–418 (2014). https://doi.org/10.1017/S095679681400001X
19. Williams, A., Thies, W., Ernst, M.D.: Static deadlock detection for Java libraries. In: Black, A.P. (ed.) ECOOP 2005. LNCS, vol. 3586, pp. 602–629. Springer, Heidelberg (2005). https://doi.org/10.1007/11531142_26

Enhancing Reaction Systems: A Process Algebraic Approach

Linda Brodo[1(✉)], Roberto Bruni[2], and Moreno Falaschi[3]

[1] Dipartimento di Scienze economiche e aziendali, Università di Sassari, Sassari, Italy
brodo@uniss.it
[2] Dipartimento di Informatica, Università di Pisa, Pisa, Italy
bruni@di.unipi.it
[3] Dipartimento di Ingegneria dell'Informazione e Scienze Matematiche, Università di Siena, Siena, Italy
moreno.falaschi@unisi.it

Abstract. In the area of Natural Computing, reaction systems are a qualitative abstraction inspired by the functioning of living cells, suitable to model the main mechanisms of biochemical reactions. This model has already been applied and extended successfully to various areas of research. Reaction systems interact with the environment represented by the context, and pose problems of implementation, as it is a new computation model. In this paper we consider the `link`-calculus, which allows to model multiparty interaction in concurrent systems, and show that it allows to embed reaction systems, by representing the behaviour of each entity and preserving faithfully their features. We show the correctness and completeness of our embedding. We illustrate our framework by showing how to embed a *lac* operon regulatory network. Finally, our framework can contribute to increase the expressiveness of reaction systems, by exploiting the interaction among different reaction systems.

Keywords: Process algebras · Reaction systems · Multi-party interaction

1 Introduction

Natural Computing is an emerging area of research which has two main aspects: human designed computing inspired by nature, and computation performed in nature. Reaction Systems (RSs) [9] are a rewriting formalism inspired by the way biochemical reactions take place in living cells. This theory has already shown to be relevant in several different fields, such as computer science [17], biology [1–3,15], and molecular chemistry [18]. Reaction Systems formalise the mechanisms of biochemical systems, such as *facilitation* and *inhibition*. As a qualitative approximation of the real biochemical reactions, they consider if a necessary reagent is or not present, and likewise they consider if an inhibiting molecule is or not present. The possible reactants and inhibitors are called

M. S. Alvim et al. (Eds.): Palamidessi Festschrift, LNCS 11760, pp. 68–85, 2019.
https://doi.org/10.1007/978-3-030-31175-9_5

'entities'. RSs model in a direct way the interaction of a living cell with the environment (called 'context'). However, two RSs are seen as independent models and do not interact.

In this paper, we present an encoding from RSs, to the open multiparty process algebra cCNA[1], a variant of the `link`-calculus [5,8] without name mobility. This formalism allows several processes to synchronise and communicate altogether, at the same time, with a new communicating mechanism based on links and link chains. Our initial motivation for introducing this mechanism was to encode Mobile Ambients [12], obtaining a much stronger operational correspondence than any one available in the literature, such as the one in [10]. This allowed us to easily encode calculi for biology equipped with membranes, as in [6]. Process calculi have been used successfully to model biological processes, see [4] for a recent survey. We illustrate our embedding by means of some simple basic examples, and then we consider a more complex example, by modeling a RS representing a regulatory network for *lac* operon, presented in [15]. We also show that our embedding preserves the main features of RSs, and prove its correctness and completeness. Our main contributions are as follows:

- the behaviour of the context, for each single entity, can be specified in a recursive way as an ordinary process;
- our translation results in a cCNA system where from each state only one transition can be generated, thus the cCNA computation is fully deterministic as that of the encoded RS;
- we can express the behaviour of entity mutation, in such a way that the mutated entity s' can take part to only a subset of rules requiring entity s;
- with a little coding effort, two RSs can communicate; i.e. a subset of those entities that the context can provide, are then provided by a second RS.

The main drawback of our proposal, is that the cCNA translation is verbose. Nevertheless it is clear that our translation can be automatised by means of a proper front-end in an implementation of the `link`-calculus.

As we have remarked, in our translation, Reaction Systems get the ability to interact between them in a synchronized manner. This interaction is not foreseen in the basic RS framework, as it can only happen with the context. By exploiting recursion, the kind of interactions which can be defined can be complex and expressive. Example 3 and more in general the discussion in Sect. 6 show that the interaction between RSs can help to model new scenarios.

Structure of the Paper. Section 2 describes RSs and their semantics (interactive processes). Section 3 describes briefly the cCNA process algebra and its operational semantics. Section 4 defines the embedding of RSs in cCNA processes and shows some simple examples to illustrate it. Section 5 shows a more complex example taken from the literature on RSs and illustrates a *lac* operon. Section 6 presents some features and advantages of our embedding for the compositionality of RSs. Finally, Sect. 7 discusses future work and concludes.

[1] After 'chained Core Network Algebra'.

2 Reaction Systems

Natural Computing is concerned with human-designed computing inspired by nature as well as with computation taking place in nature. The theory of Reaction Systems [9] was born in the field of Natural Computing to model the behaviour of biochemical reactions taking place in living cells. Despite its initial aim, this formalism has shown to be quite useful not only for modeling biological phenomena, but also for the contributions given to computer science [17], theory of computing, mathematics, biology [1–3,15], and molecular chemistry [18]. Here we briefly review the basic notions of RSs, see [9] for more details.

The mechanisms that are at the basis of biochemical reactions and thus regulate the functioning of a living cell, are *facilitation* and *inhibition*. These mechanisms are reflected in the basic definitions of Reaction Systems.

Definition 1 (Reaction). *A reaction is a triplet* $a = (R, I, P)$*, where* R, I, P *are finite, non empty sets and* $R \cap I = \emptyset$*. If* S *is a set such that* $R, I, P \subseteq S$*, then* a *is a reaction in* S*.*

The sets R, I, P are also written R_a, I_a, P_a and called the *reactant set* of a, the *inhibitor set* of a, and the *product set* of a, respectively. All reactants are needed for the reaction to take place. Any inhibitor blocks the reaction if it is present. Products are the outcome of the reaction. Also, $R_a \cup I_a$ is the set of the resources of a and $rac(S)$ denotes the set of all reactions in S. Because R and I are non empty, all products are produced from at least one reactant and every reaction can be inhibited in some way. Sometimes artificial inhibitors are used that are never produced by any reaction. For the sake of simplicity, and without loss of generality, in some examples, we will allow I to be empty.

Definition 2 (Reaction System). *A Reaction System (RS) is an ordered pair* $\mathcal{A} = (S, A)$ *such that* S *is a finite set, and* $A \subseteq rac(S)$*.*

The set S is called the *background set* of $\mathcal{A}$, its elements are called *entities*, and they represent molecular substances (e.g., atoms, ions, molecules) that may be present in the states of a biochemical system. The set A is the set of *reactions* of $\mathcal{A}$. Since S is finite, so is A: we denote by $|A|$ the number of reactions in A.

Definition 3 (Reaction Result). *Given a finite set of entities* S*, let* $T \subseteq S$*.*

1. *Let* $a \in rac(S)$ *be a reaction. Then* a *is enabled by* T*, denoted by* $en_a(T)$*, if* $R_a \subseteq T$ *and* $I_a \cap T = \emptyset$*.*
2. *Let* $a \in rac(S)$ *be a reaction. The result of* a *on* T*, denoted by* $res_a(T)$*, is defined by:* $res_a(T) = P_a$ *if* $en_a(T)$*, and* $res_a(T) = \emptyset$ *otherwise.*
3. *Let* $A \subseteq rac(S)$ *be a finite set of reactions. The result of* A *on* T*, denoted by* $res_A(T)$*, is defined by:* $res_A(T) = \bigcup_{a \in A} res_a(T)$*.*

The theory of Reaction Systems is based on the following assumptions.

- **No permanency.** An entity of a set T vanishes unless it is sustained by a reaction. This reflects the fact that a living cell would die for lack of energy, without chemical reactions.

- **No counting.** The basic model of RSs is very abstract and qualitative, i.e. the quantity of entities that are present in a cell is not taken into account.
- **Threshold nature of resources.** From the previous item, we assume that either an entity is available and there is enough of it (i.e. there are no conflicts), or it is not available at all.

The dynamic behaviour of a RS is formalized in terms of *interactive processes.*

Definition 4 (Interactive Process). *Let $\mathcal{A} = (S, A)$ be a RS and let $n \geq 0$. An n-step* interactive process *in $\mathcal{A}$ is a pair $\pi = (\gamma, \delta)$ of finite sequences s.t. $\gamma = \{C_i\}_{i\in[0,n]}$ and $\delta = \{D_i\}_{i\in[0,n]}$ where $C_i, D_i \subseteq S$ for any $i \in [0, n]$, $D_0 = \emptyset$, and $D_i = res_{\mathcal{A}}(D_{i-1} \cup C_{i-1})$ for any $i \in [1, n]$.*

Living cells are seen as open systems that continuously react with the external environment, in discrete steps. The sequence γ is the *context sequence* of π and represents the influence of the environment on the Reaction System. The sequence δ is the *result sequence* of π and it is entirely determined by γ and A. The sequence $\tau = W_0, \ldots, W_n$ with $W_i = C_i \cup D_i$, for any $i \in [0, n]$ is called a *state sequence*. Each state W_i in a state sequence is the union of two sets: the context C_i at step i and the result of the previous step.

For technical reasons, we extend in a straightforward way the notion of an interactive process to deal with infinite sequences.

Definition 5 (extended interactive process). *Let $\mathcal{A} = (S, A)$ be a RS, and let $\pi = (\gamma, \delta)$ be an n-step interactive process, with $\gamma = \{C_i\}_{i\in[0,n]}$ and $\delta = \{D_i\}_{i\in[0,n]}$. Then, let $\pi' = (\gamma', \delta')$ be the extended interactive process of $\pi = (\gamma, \delta)$, defined as $\gamma' = \{C'_i\}_{i\in\mathbb{N}}$, $\delta' = \{D'_i\}_{i\in\mathbb{N}}$, where $C'_j = C_j$ for $j \in [0, n]$ and $C'_j = \emptyset$ for $j > n$, $D'_0 = D_0$ and $D'_j = res_A(D'_{j-1} \cup C'_{j-1})$ for $j \geq 1$.*

The next example shows that the functioning of the interacting processes is deterministic, once the context is fixed.

Example 1. Let $\mathcal{A} = (S, A)$ be a RS with $S = \{s_1, s_2, s_3, s_4\}$, and $A = \{a_1, a_2\}$, where $a_1 = (\{s_1\}, \{s_3\}, \{s_1\})$, $a_2 = (\{s_2\}, \{s_4\}, \{s_2\})$. Now, we show the first three steps of an interacting process $\pi = (\gamma, \delta)$ where the context provides entities as follows: $\gamma = \{C_0 = \{s_1, s_2\}, C_1 = \{s_3\}, C_2 = \{s_1, s_4\}, C_3 = \emptyset, C_4, \ldots, C_n\}$. Initally $D_0 = \emptyset$ and from $C_0 \cup D_0 = \{s_1, s_2\}$ we get to $C_1 \cup D_1 = \{s_1, s_2, s_3\}$ by applying a_1, a_2 (they both are enabled); after that we get to $C_2 \cup D_2 = \{s_1, s_2, s_4\}$, by applying a_2 only (a_1 is not enabled); finally we get to $C_3 \cup D_3 = \{s_1\}$ by applying a_1 (a_2 is not enabled). Thus $\delta = \{D_0 = \emptyset, D_1 = \{s_1, s_2\}, D_2 = \{s_2\}, D_3 = \{s_1\}, D_4, \ldots, D_n\}$. We remark that at every state $C_i \cup D_i$ all the reactions which are enabled are applied, so the computation is deterministic.

3 Chained CNA (cCNA)

In this section we introduce the syntax and operational semantics of a variant of the `link`-calculus [5], the cCNA (chained CNA), where the prefixes are link chains.

Link Chains. Let $\mathcal{C}$ be the set of channels, ranged over by $a, b, ...$, and let $\mathcal{N} = \mathcal{C} \cup \{\tau\} \cup \{\square\}$ be the set of actions, ranged over by $\alpha, \beta, ...$, where the symbol τ denotes a *silent* action, while the symbol $\square$ denotes a *virtual* (non-specified) action. A *link* is a pair $\ell = {}^{\alpha}\backslash_{\beta}$; it is *solid* if $\alpha, \beta \neq \square$; the link ${}^{\square}\backslash_{\square}$ is called *virtual*; the link ${}^{\tau}\backslash_{\tau}$ is called *silent*. A link is *valid* if it is solid or virtual. A *link chain* is a finite sequence $v = \ell_1 ... \ell_n$ of valid links $\ell_i = {}^{\alpha_i}\backslash_{\beta_i}$ such that:

1. for any $i \in [1, n-1]$, $\begin{cases} \beta_i, \alpha_{i+1} \in \mathcal{C} & \text{implies } \beta_i = \alpha_{i+1} \\ \beta_i = \tau & \text{iff } \alpha_{i+1} = \tau \end{cases}$
2. $\exists i \in [1, n].\ \ell_i \neq {}^{\square}\backslash_{\square}$.

A link chain whose links are silent is also called silent. Virtual links ${}^{\square}\backslash_{\square}$ represent missing elements of a chain. The equivalence $\bowtie$ models expansion and contraction of virtual links to adjust the length of a link chain.

Definition 6 (Equivalence $\bowtie$). *We let $\bowtie$ be the least equivalence relation over link chains closed under the axioms (whenever both sides are well defined):*

$$\begin{array}{llll} v\,{}^{\square}\backslash_{\square} \bowtie v & v_1\,{}^{\square}\backslash^{\square}_{\square}\backslash_{\square}\,v_2 \bowtie v_1\,{}^{\square}\backslash_{\square}\,v_2 \\ {}^{\square}\backslash_{\square}\,v \bowtie v & v_1\,{}^{\alpha}\backslash^{a}_{a}\backslash_{\beta}\,v_2 \bowtie v_1\,{}^{\alpha}\backslash^{\square}_{a}\backslash^{a}_{\square}\backslash_{\beta}\,v_2 \end{array}$$

Two link chains of equal length can be merged whenever each position occupied by a solid link in one chain is occupied by a virtual link in the other chain and solid links in adjacent positions match. Positions occupied by virtual links in both chains remain virtual. Merging is denoted by $v_1 \bullet v_2$. For example, given $v_1 = {}^{a}\backslash^{\square}_{b}\backslash^{\square}_{\square}\backslash_{\square}$ and $v_2 = {}^{\square}\backslash^{b}_{\square}\backslash^{\square}_{c}\backslash_{\square}$ we have $v_1 \bullet v_2 = {}^{a}\backslash^{b}_{b}\backslash^{\square}_{c}\backslash_{\square}$.

Some names in a link chain can be restricted as non observable and transformed into silent actions τ. This is possible only if they are matched by some adjacent link. Restriction is denoted by $(\nu\, a)v$. For example, given $v = {}^{a}\backslash^{b}_{b}\backslash^{\square}_{c}\backslash_{\square}$ we have $(\nu\, b)v = {}^{a}\backslash^{\tau}_{\tau}\backslash^{\square}_{c}\backslash_{\square}$.

Syntax. The cCNA processes are generated by the following grammar:

$$P, Q ::= \textstyle\sum_{i \in I} v_i.P_i \ \mid\ P|Q \ \mid\ (\nu\, a)P \ \mid\ P[\phi] \ \mid\ A$$

where v_i is a link chain, ϕ is a channel renaming function, and A is a process identifier for which we assume a definition $A \triangleq P$ is available in a given set Δ of (possibly recursive) process definitions. We let $\mathbf{0}$, the inactive process, denote the empty summation.

The syntax of cCNA extends that of CNA [8] by allowing to use link chains as prefixes instead of links. This extension was already discussed in [8] and it preserves all the main formal properties of CNA. For the rest it features non-deterministic choice, parallel composition, restriction, relabelling and possibly recursive definitions of the form $A \triangleq P$ for some constant A. Here we do not consider name mobility, which is present instead in the `link`-calculus.

$$\frac{v \bowtie v_j \quad j \in I}{\sum_{i\in I} v_i.P_i \xrightarrow{v} P_j}\ (Sum) \qquad \frac{P \xrightarrow{v} P' \qquad (A \triangleq P) \in \Delta}{A \xrightarrow{v} P'}\ (Ide)$$

$$\frac{P \xrightarrow{v} P'}{P[\phi] \xrightarrow{\phi(v)} P'[\phi]}\ (Rel) \qquad \frac{P \xrightarrow{v} P'}{(\nu\, a)P \xrightarrow{(\nu\, a)v} (\nu\, a)P'}\ (Res)$$

$$\frac{P \xrightarrow{v} P'}{P|Q \xrightarrow{v} P'|Q}\ (Lpar) \qquad \frac{P \xrightarrow{v'} P' \qquad Q \xrightarrow{v} Q'}{P|Q \xrightarrow{v \bullet v'} P'|Q'}\ (Com)$$

Fig. 1. SOS semantics of the cCNA (rule (*Rpar*) omitted).

Semantics. The operational semantics of cCNA is defined in the SOS style by the inference rules in Fig. 1. The rules are reminiscent of those for Milner's CCS and they essentially coincide with those of CNA in [8], except for the presence of prefixes that are link chains instead of single links. Briefly: rule (*Sum*) selects one alternative and uses, as a label, a possible contraction/expansion v of the link chain v_j in the selected prefix; rule (*Ide*) selects one transition of the process defined by a constant; rule (*Rel*) renames the channels in the label as indicated by ϕ; rule (*Res*) restricts some names in the label (it cannot be applied when $(\nu\, a)v$ is not defined); rules (*Lpar*) and (*Rpar*) account for interleaving in parallel composition; rule (*Com*) synchronises interactions (it cannot be applied when $v \bullet v'$ is not defined). Analogously to CNA, the operational semantics of cCNA satisfies the so called Accordion Lemma: whenever $P \xrightarrow{v} Q$ and $v' \bowtie v$ then $P \xrightarrow{v'} Q$. As a matter of notation, we write $P \to Q$ when $P \xrightarrow{v} Q$ for some silent link chain v and call it a *silent transition*. Similarly, a sequence of j silent transitions is denoted $P \to^j Q$.

3.1 Notation for Link Chains

Hereafter we make use of some new notations for link chains that will facilitate the presentation of our translation.

Definition 7 (Replication). *Let v be a link chain. Its n times replication v^n is defined recursively by letting $v^0 = \epsilon$ (i.e. the empty chain) and $v^n = v^{n-1}v$, with the hypothesis that all the links in the resulting link chains match.*

For example, the expression $({}^{a}\backslash^{\square}_{b}\backslash_{\square})^3$ denotes the chain ${}^{a}\backslash^{\square}_{b}\backslash^{a}_{\square}\backslash^{\square}_{b}\backslash^{a}_{\square}\backslash^{\square}_{b}\backslash_{\square}$. We introduce the *half link* that will be used in conjunction with the *open block of chain* to form regular link chains. Let ${}^{a}\backslash$ denote the *half left link* of a link ${}^{a}\backslash_{x}$, and conversely let $\backslash_{a}$ denote the *half right link* of ${}^{x}\backslash_{a}$.

Definition 8 (Open block). *Let R be a totally ordered, finite set of names. We define an* open block *as* $(\mathop{\backslash\!\!\backslash}_{c \in R} {}^{\square}_{c_i}\backslash^{c_o}_{\square})$, *where c_i and c_o are annotated version of the name c, as follows*

set	*open block expression*	*result*
$R = \emptyset$	$\left(\backslash\!\!\backslash_{c \in R}\, {}^{\square}_{c_i}\backslash^{c_o}_{\square}\right)$	ϵ
$R = \{a\}$	$\left(\backslash\!\!\backslash_{c \in R}\, {}^{\square}_{c_i}\backslash^{c_o}_{\square}\right)$	${}^{\square}_{a_i}\backslash^{a_o}_{\square}$
$R = \{a\} \uplus R'$ *with* $a = \min R$	$\left(\backslash\!\!\backslash_{c \in R}\, {}^{\square}_{c_i}\backslash^{c_o}_{\square}\right)$	${}^{\square}_{a_i}\backslash^{a_o}_{\square}\backslash\left(\backslash\!\!\backslash_{c \in R'}\, {}^{\square}_{c_i}\backslash^{c_o}_{\square}\right)$

We then combine half links and open blocks to form regular link chains. For example, for $R = \{a, b\}$ the expression $\left(\backslash\!\!\backslash_{c \in R}\, {}^{\square}_{c_i}\backslash^{c_o}_{\square}\right)$ denotes the block of chains ${}^{\square}_{a_i}\backslash^{a_o}_{\square}\backslash^{\square}_{b_i}\backslash^{b_o}_{\square}$; and the expression ${}^{r_1}\backslash\left(\backslash\!\!\backslash_{c \in R}\, {}^{\square}_{c_i}\backslash^{c_o}_{\square}\right)\backslash_{r_2}$ denotes the chain ${}^{r_1}\backslash^{\square}_{a_i}\backslash^{a_o}_{\square}\backslash^{\square}_{b_i}\backslash^{b_o}_{\square}\backslash_{r_2}$.

4 From Reaction Systems to cCNA

Here we present a translation from Reaction Systems to cCNA. The idea is to define separated processes for representing the behaviour of each entity, each reaction, and for the provisioning of each entity by the context.

Processes for Entities. Given an entity $s \in S$, we exploit four different names for the interactions over s: names s_i, s_o are used to test the presence of s in the system; names $\hat{s}_i$, $\hat{s}_o$ are used to test the provisioning of s from the context; names $\tilde{s}_i$, $\tilde{s}_o$ are used to test the production of s by some reaction; names $\overline{s}_i$, $\overline{s}_o$ are used to test the absence of s in the system; and names $\underline{s}_i$, $\underline{s}_o$ are used to test the absence of s from the context. We let P_s be the process implementing the presence of s in the system, and $\overline{P_s}$ its absence. They are defined below:

$$
\begin{aligned}
P_s \triangleq\; & \textstyle\sum_{h\geq 0,k\geq 0}({}^{s_i}\backslash^{\square}_{s_o}\backslash_{\square})^h\; {}^{\hat{s}_i}\backslash^{\square}_{\hat{s}_o}\backslash_{\square}({}^{\tilde{s}_i}\backslash^{\square}_{\tilde{s}_o}\backslash_{\square})^k.P_s\\
& +\\
& \textstyle\sum_{h\geq 0,k\geq 1}({}^{s_i}\backslash^{\square}_{s_o}\backslash_{\square})^h\; {}^{\underline{s}_i}\backslash^{\square}_{\underline{s}_o}\backslash_{\square}({}^{\tilde{s}_i}\backslash^{\square}_{\tilde{s}_o}\backslash_{\square})^k.P_s\\
& +\\
& \textstyle\sum_{h\geq 0}({}^{s_i}\backslash^{\square}_{s_o}\backslash_{\square})^h\; {}^{\underline{s}_i}\backslash_{\underline{s}_o}.\overline{P_s}
\end{aligned}
$$

$$
\begin{aligned}
\overline{P_s} \triangleq\; & \textstyle\sum_{h\geq 0,k\geq 0}({}^{\overline{s}_i}\backslash^{\square}_{\overline{s}_o}\backslash_{\square})^h\; {}^{\hat{s}_i}\backslash^{\square}_{\hat{s}_o}\backslash_{\square}({}^{\tilde{s}_i}\backslash^{\square}_{\tilde{s}_o}\backslash_{\square})^k.P_s\\
& +\\
& \textstyle\sum_{h\geq 0,k\geq 1}({}^{\overline{s}_i}\backslash^{\square}_{\overline{s}_o}\backslash_{\square})^h\; {}^{\underline{s}_i}\backslash^{\square}_{\underline{s}_o}\backslash_{\square}({}^{\tilde{s}_i}\backslash^{\square}_{\tilde{s}_o}\backslash_{\square})^k.P_s\\
& +\\
& \textstyle\sum_{h\geq 0}({}^{\overline{s}_i}\backslash^{\square}_{\overline{s}_o}\backslash_{\square})^h\; {}^{\underline{s}_i}\backslash_{\underline{s}_o}.\overline{P_s}
\end{aligned}
$$

The first line of P_s accounts for the case where s is tested for presence by h reactions and produced by k reactions, while being provided by the context (${}^{\hat{s}_i}\backslash_{\hat{s}_o}$). Thus, s will be present at the next step (the continuation is P_s). Here h and k are not known a priori and therefore any combination is possible. By knowing the number of reactions that test s, we can bound the maximum values of h and k. The second line accounts for the analogous case where s is not provided by the context (${}^{\underline{s}_i}\backslash_{\underline{s}_o}$). The condition $k \geq 1$ guarantees that s will remain present (the continuation is P_s). The third line accounts for the case where s is tested for presence by h reactions, but it is neither produced nor provided by the context. Therefore, in the next step s will be absent in the system (the continuation is $\overline{P_s}$). Note that in the case of $\overline{P_s}$ the test for presence of s in the system is just replaced by the test for its absence. As before, s will be present again in the system if s will be produced (rows 1 and 2 in $\overline{P_s}$ code), or if the context will provide it (row 1 in $\overline{P_s}$ code).

Processes for Reactions. We assume that each reaction a is assigned a progressive number j. The process for reaction $a_j = (R_j, I_j, P_j)$ must assert either the possibility to apply the reaction or its impossibility. The first case happens when all its reactants are present (the link ${}^{s_i}\backslash_{s_o}$ is requested for any $s \in R_j$) and all its inhibitors are absent (the link ${}^{\overline{e}_i}\backslash_{\overline{e}_o}$ is requested for any $e \in I_j$), then the product set is released (the link ${}^{\tilde{c}_i}\backslash_{\tilde{c}_o}$ is requested for any $c \in P_j$). The next case can happen for two reasons: one of the reactants is absent (the link ${}^{\overline{s}_i}\backslash_{\overline{s}_o}$ is requested for some $s \in R_j$) or one of the inhibitors is present (the link ${}^{e_i}\backslash_{e_o}$ is requested for some $e \in I_j$). The process is recursive so that reactions can be applied at any step.

$$
\begin{aligned}
P_{a_j} \triangleq\ & {}^{r_j}\backslash \Big(\bigparallel_{s \in R_j} {}^{\Box}_{s_i}\backslash^{s_o}_{\Box}\Big)\backslash \Big(\bigparallel_{e \in I_j} {}^{\Box}_{\overline{e}_i}\backslash^{\overline{e}_o}_{\Box}\Big)\backslash^{\Box}_{r_{j+1}}\backslash^{p_j}_{\Box}\backslash \Big(\bigparallel_{c \in P_j} {}^{\Box}_{\tilde{c}_i}\backslash^{\tilde{c}_o}_{\Box}\Big)\backslash_{p_{j+1}}.P_{a_j} && \{a_j \text{ is applicable}\} \\
& + \\
& \textstyle\sum_{s \in R_j} {}^{r_j}\backslash^{\Box}_{\overline{s}_i}\backslash^{\overline{s}_o}_{\Box}\backslash^{\Box}_{r_{j+1}}\backslash^{p_j}_{\Box}\backslash_{p_{j+1}}.P_{a_j} && \{a_j \text{ is not applicable}\} \\
& + \\
& \textstyle\sum_{e \in I_j} {}^{r_j}\backslash^{\Box}_{e_i}\backslash^{e_o}_{\Box}\backslash^{\Box}_{r_{i+1}}\backslash^{p_j}_{\Box}\backslash_{p_{j+1}}.P_{a_j} && \{a_j\text{is not applicable}\}
\end{aligned}
$$

Formally speaking, the definition of the case when a_j is applicable (row 1) requires R_j, I_j and P_j to be totally ordered, because open block expressions are used. It is worth noting that the chosen orders are inessential for exploiting P_{a_j} in the encoding of a RS, but they are needed to disambiguate its definition. Without loss of generality, we assume that all the entities are enumerated, likewise reactions, so that R_j, I_j and P_j inherit the same order. We exploit names r_j, p_j to join the chains provided by the application of all the reactions. Channels r_j and r_{j+1} enclose the enabling/disabling condition of reaction a_j. Channels p_i and p_{j+1} enclose the links related to the entities produced by a_j. We will see that all the link chain labels of transitions follow the same schema: first we find all the reactions limited to the reactants and inhibitors (chained using r_j channels), then all the supplies by the contexts (chained using cxt_j channels, to be introduced next), and finally the products for all the reactions (chained using p_j channels). In the following there is an example explaining this schema.

Processes for Contexts. For each entity $s \in S$, we introduce another process Cxt_s, participating in each transition and determining whether the entity s is provided by the context or not. As already said, we assume that entities are enumerated and use the names cxt_j to concatenate the chains formed by the application of all the contexts. For each entity s with number j, at step $n > 0$ there are two possible behaviours:

$$Cxt_s^n \triangleq \begin{cases} {}^{cxt_j}\backslash^{\square}_{\hat{s}_i}\backslash^{\hat{s}_o}_{\square}\backslash_{cxt_{j+1}}.Cxt_s^{n+1} & \text{if the context provides } s \text{ at the n-th step} \\ {}^{cxt_j}\backslash^{\square}_{\underline{s}_i}\backslash^{\underline{s}_o}_{\square}\backslash_{cxt_{j+1}}.Cxt_s^{n+1} & \text{otherwise} \end{cases}$$
$$Cxt_s \triangleq Cxt_s^1$$

We only consider Cxt_s^n with $n > 0$, as the entities such that $s \in C_0$ (resp. $s \notin C_0$) are modeled by the P_s processes (resp. $\overline{P_s}$) in the initial cCNA process. The intrinsic modularity of cCNA allows us to run different context behaviours for the same system by changing the definitions of Cxt_s^n.

Definition 9 (Translation). *Let $\mathcal{A} = (S, A)$ be a RS, and let $\pi = (\gamma, \delta)$ be an extended interactive process in $\mathcal{A}$, with $\gamma = \{C_i\}_{i \in \mathbb{N}}$. We define its cCNA translation $[\![\mathcal{A}, \gamma]\!]$ as follows:*

$$[\![\mathcal{A}, \gamma]\!] = (\nu\, reacts, cxts, ents, prods)(\Pi_{s \in C_0} P_s | \Pi_{s \notin C_0} \overline{P_s} | \Pi_{a \in A} P_a | \Pi_{s \in S} Cxt_s),$$

with reacts be the set of reaction names r_j, cxts the set of context names cxt_j, ents the set of decorated entity names $\{s_i, s_o, \hat{s}_i, \hat{s}_o, \tilde{s}_i, \tilde{s}_o, \overline{s}_i, \overline{s}_o, \underline{s}_i, \underline{s}_o | s \in S\}$, and prods be the set of names p_j associated to each reaction. In the following, we set names = reacts ∪ cxts ∪ ents ∪ prods. For notational convenience, we fix that $r_1 = \tau$, $r_{u+1} = cxt_1$ for u the number of reacts, and $cxt_{w+1} = p_1$ $p_{u+1} = \tau$ for w the number of entities.

It is important to observe that, for each transition, our cCNA encoding requires all the processes P_a, with $a \in A$, and Cxt_s and P_s, with $s \in S$, be interacting in that transition. This is due to the fact that all the channels r_j, p_j, cxt_j, and s_i, and s_o are restricted. Each reaction defines a pattern to be satisfied, i.e. each reaction inserts as many virtual links as the number of reactants, inhibitors, and products, as required by the corresponding reaction.

Lemma 1. *Let $\mathcal{A} = (S, A)$ be a RS and let $\pi = (\gamma, \delta)$ be an extended interactive process in $\mathcal{A}$. Let $P = [\![\mathcal{A}, \gamma]\!]$ its cCNA translation. If there exists P' such that $t = (P \xrightarrow{(\nu\, names)v} P')$ is a transition of P, then*

1. *for each reaction $a_j \in A$, the corresponding channels r_j and p_j appear in v; for each entity $s_h \in S$ (where h is the identifying number of s), the corresponding channel s_h (suitably decorated), and the corresponding channel cxt_h appear in v;*
2. *for each reaction $a \in A$ and each entity $s \in S$, each virtual link offered by processes P_a and Cxt_s is overlapped by exactly one solid link offered by processes representing entities.*

Example 2. Let $\mathcal{A}$ be a RS whose specification contains two entities, $s1$ and $s2$, and, among the others, the reaction $a = (\{s1\}, \{\dots\}, \{s1\})$ that guarantees the persistence of the entity $s1$ once it is present in the system. Note that we use here a dummy inhibitor $(\dots)$ which will never be present. Then, we assume an extended interactive process $\pi = (\gamma, \delta)$ where the context γ provides $s1$ and $s2$.

Our translation includes the processes:

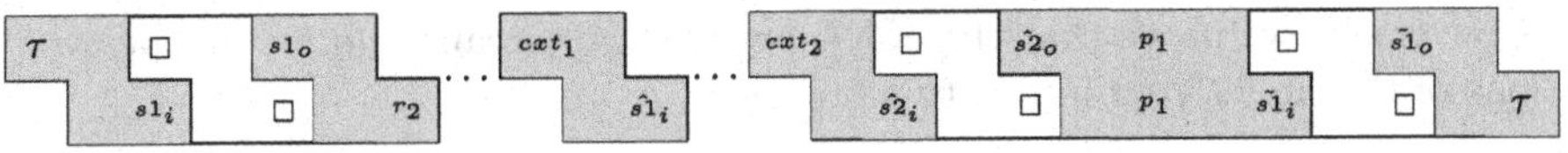

Fig. 2. The link chain structure arising from reactions and context processes. (Color figure online)

$$P_a \triangleq \tau\backslash^{\square}_{s1_i}\backslash^{s1_o}_{\square}\backslash^{\square}_{r_2}\backslash^{p_1}_{\square}\backslash^{\square}_{\tilde{s1}_i}\backslash^{\tilde{s1}_o}_{\square}\backslash_{p_2}.P_a + \dots;$$

$$P_{s1} \triangleq s1_i\backslash^{\square}_{s1_o}\backslash^{\hat{s1}_i}_{\square}\backslash^{\square}_{\hat{s1}_o}\backslash^{\tilde{s1}_i}_{\square}\backslash_{\tilde{s1}_o}.P_{s1} + \dots; \qquad P_{s2} \triangleq \hat{s2}_i\backslash_{\hat{s2}_o}.P_{s2} + \dots;$$

$$Cxt_{s1} \triangleq cxt_1\backslash^{\square}_{\hat{s1}_i}\backslash^{\hat{s1}_o}_{\square}\backslash_{cxt_2}.Cxt_{s1} \qquad Cxt_{s2} \triangleq cxt_2\backslash^{\square}_{\hat{s2}_i}\backslash^{\hat{s2}_o}_{\square}\backslash_{p_1}.Cxt_{s2}$$

Now, we assume that $s1$ is in the initial state of $\mathcal{A}$, and in Fig. 2 we show the structure of a link chain label related to the execution of a transition of the cCNA system: $(\nu\, names)(P_{s1}|P_{s2}|P_a|\dots|Cxt_{s1}|Cxt_{s2})$. The yellow blocks are referred to the processes encoding the reactions (P_a, in our case) and the contexts (Cxt_{s1} and Cxt_{s2}). As the figure puts in evidence, these two kinds of processes determine the structure of the link chain, from end to end, i.e. from the left τ to the right one. We could say that these processes form the *backbone* of the interaction. In contrast, the processes encoding the entities (P_{s1}, and P_{s2}, in our case) provide the solid links to overlap the virtual links of the backbone.

Example 2 outlines two different roles of the processes defining the translation of an interactive process: those processes encoding the reactions and the context provide the backbone of each transition, whereas the processes encoding the entities provide the resources needed for the communication to take place.

With the next proposition, we analyse the structure of a cCNA process encoding of a reactive process after one step transition. In the following four statements, for brevity, we let $\mathcal{A} = (S, A)$ be a RS, and let $\pi = (\gamma, \delta)$ be an extended interactive process in A, with $\gamma = \{C_i\}_{i\in\mathbb{N}}$ and $\delta = \{D_i\}_{i\in\mathbb{N}}$. Moreover, we denote by π^j the shift of π starting at the j-th state sequence; formally we let $\pi^j = (\gamma^j, \delta^j)$ with $\gamma^j = \{C_i'\}_{i\in\mathbb{N}}$, $\delta^j = \{D_i'\}_{i\in\mathbb{N}}$ such that $C_0' = C_j \cup D_j$, $D_0' = \emptyset$, and $C_i' = C_{i+j}$, $D_i' = D_{i+j}$ for any $i \geq 1$.

Proposition 1 (Correctness 1). *Let* $P = [\![\mathcal{A}, \gamma]\!]$ *with*

$$P = (\nu\, names)(\Pi_{a\in A} P_a | \Pi_{s\in S} Cxt_s^1 | \Pi_{s\in C_0} P_s | \Pi_{s\notin C_0} \overline{P}_s).$$

If $P \xrightarrow{v} Q$, *then* v *is a silent action and* $Q = [\![\mathcal{A}, \gamma^1]\!]$, *namely*

$$Q = (\nu\, names)(\Pi_{a\in A} P_a | \Pi_{s\in S} Cxt_s^2 | \Pi_{s\in C_1\cup D_1} P_s | \Pi_{s\notin C_1\cup D_1} \overline{P}_s).$$

Now, we extend the previous result to a series of transitions.

Corollary 1 (Correctness 2). *Let* $P = [\![\mathcal{A}, \gamma]\!]$ *and* $j \geq 1$. *If there exists* Q *such that* $P \rightarrow^j Q$, *then* $Q = [\![\mathcal{A}, \gamma^j]\!]$.

Conversely, we prove that the *c*CNA process $[\![\mathcal{A},\gamma]\!]$ can simulate all the evolutions of the underlying interactive process.

Proposition 2 (Completeness 1). $[\![\mathcal{A},\gamma]\!] \to [\![\mathcal{A},\gamma^1]\!]$.

Now, we extend the previous result to a series of transitions.

Corollary 2 (Completeness 2). $[\![\mathcal{A},\gamma]\!] \to^j [\![\mathcal{A},\gamma^j]\!]$.

5 Example: *lac* operon

In this section we present the encoding of a RS example taken from [15].

5.1 The *lac* operon

An operon is a cluster of genes under the control of a single promoter. The *lac* operon is involved in the metabolism of lactose in *Escherichia coli* cells; it is composed by three adjacent structural genes (plus some regulatory components): *lacZ*, *lacY* and *lacA* encoding for two enzymes Z and A, and a transporter Y, involved in the digestion of the *lactose*. The main regulations are:

- the gene *lacI* encodes a repressor protein I;
- the DNA sequence, called *promoter*, is recognised by a RNA polymerase to initiate the transcription of the genes *lacZ*, *lacY* and *lacA*;
- a DNA segment, called the *operator* (OP), obstructs the RNA polymerase functionality when the repressor protein I is bound to it forming I-OP;
- a short DNA sequence, called the *CAP-binding site*, when it is bound to the complex composed by the protein CAP and the signal molecule *cAMP*, acts as a promoter for the interaction between the RNA polymerase and the promoter.

The functionality of the *lac* operon depends on the integration of two control mechanisms, one mediated by *lactose*, and the other one mediated by *glucose*.

In the first control mechanism, an effect of the absence of the *lactose* is that I is able to bind the operator sequence preventing the *lac* operon expression. If lactose is available, I is unable to bind the operator sequence, and the *lac* operon can be potentially expressed.

In the second control mechanism, when glucose is absent, the molecule *cAMP* and the protein CAP increase the *lac* operon expression, thanks to the fact that the binding between the molecular complex *cAMP*-CAP and the *CAP-binding site* increases. In summary, the condition promoting the operon gene expression is when the *lactose* is present and the *glucose* is absent.

In the following we report the description of the *lac* operon mechanism in the reaction system formalism and then show its encoding in cCNA.

5.2 The RS Formalization

The reaction system for the *lac* operon is defined as $\mathcal{A}_{lac} = (S, A)$, where the set S represents the main biochemical components involved in this genetic system, while the reaction set A contains the biochemical reactions involved in the regulation of the *lac* operon expression. Formally, the *lac* operon reaction system is defined as follows: S is the set

$$\{lac, Z, Y, A, lacI, I, I\text{-}OP, cya, cAMP, crp, CAP, cAMP\text{-}CAP, lactose, glucose\},$$

and A consists of the following 10 reactions:

$$\begin{array}{ll}
a_1 = (\{lac\}, \{...\}, \{lac\}), & a_6 \ = (\{cya\}, \{...\}, \{cAMP\}), \\
a_2 = (\{lacI\}, \{...\}, \{lacI\}), & a_7 \ = (\{crp\}, \{...\}, \{crp\}), \\
a_3 = (\{lacI\}, \{...\}, \{I\}), & a_8 \ = (\{crp\}, \{...\}, \{CAP\}), \\
a_4 = (\{I\}, \{lactose\}, \{I\text{-}OP\}), & a_9 \ = (\{cAMP, CAP\}, \{glucose\}, \{cAMP\text{-}CAP\}), \\
a_5 = (\{cya\}, \{...\}, \{cya\}), & a_{10} = (\{lac, cAMP\text{-}CAP\}, \{I\text{-}OP\}, \{Z, Y, A\}).
\end{array}$$

The default context (DC) is composed by those entities that are always present in the system $DC = \{lac, lacI, I, cya, cAMP, crp, CAP\}$, whereas the *lactose* and the *glucose* are given non-deterministically by the context.

5.3 The RS Encoding

For the sake of readability, the encoding we propose exploits the specific features of the example in hand to perform some simplifications:

- for the entities in the default context, $s \in DC$, as they are persistent, we do not provide the $\overline{P_s}$ processes and the Cxt_s processes;
- for the reactions requiring the presence of entities $s \in DC$, we do not provide the reaction alternative behaviour when s is absent;
- the Cxt_s processes are specified only for those entitiess that are really provided by the context.

Moreover, we do not model the *dummy* entity that is specified by dots (...) by the RS reactions in Sect. 5.2. Finally, we exclude the *duplication reactions* (a_1, a_2, a_5, a_7), and renumber the remaining reactions :

old	new	reactions
a_3	a_1	$= (\{lacI\}, \{...\}, \{I\})$,
a_4	a_2	$= (\{I\}, \{lactose\}, \{I\text{-}OP\})$,
a_6	a_3	$= (\{cya\}, \{...\}, \{cAMP\})$,
a_8	a_4	$= (\{crp\}, \{...\}, \{CAP\})$,
a_9	a_5	$= (\{cAMP, CAP\}, \{glucose\}, \{cAMP\text{-}CAP\})$,
a_{10}	a_6	$= (\{lac, cAMP\text{-}CAP\}, \{I\text{-}OP\}, \{Z, Y, A\})$.

Expression Reactions. First we define the parametric process

$$P_i(s1, s2) \triangleq {}^{r_i}\backslash^{\square}_{s1_i}\backslash^{s1_o}_{\square}\backslash^{\square}_{r_{i+1}}\backslash^{p_i}_{\square}\backslash^{\square}_{\tilde{s2}_i}\backslash^{\tilde{s2}_o}_{\square}\backslash_{p_{i+1}}.P_i(s1, s2)$$

Then, we let $P_{a1} \triangleq P_1(lacI, I)$, $P_{a3} \triangleq P_3(cya, cAMP)$, and $P_{a4} \triangleq P_4(crp, CAP)$.

Regulation Reactions.

$$
\begin{aligned}
P_{a2} \triangleq\ & {}^{r_2}\backslash^{\square}_{I_i}\backslash^{I_o}_{\square}\backslash^{\square}_{\overline{lactose}_i}\backslash^{\overline{lactose}_o}_{\square}\backslash^{\square}_{r_3}\backslash^{p_2}_{\square}\backslash^{\square}_{\widetilde{I\text{-}OP}_i}\backslash^{\widetilde{I\text{-}OP}_o}_{\square}\backslash_{p_3}.P_{a2}\\
&+\\
& {}^{r_2}\backslash^{\square}_{lactose_i}\backslash^{lactose_o}_{\square}\backslash^{\square}_{r_3}\backslash^{p_2}_{\square}\backslash_{p_3}.P_{a2} \ +\ {}^{r_2}\backslash^{\square}_{\overline{I}_i}\backslash^{\overline{I}_o}_{\square}\backslash^{\square}_{r_3}\backslash^{p_2}_{\square}\backslash_{p_3}.P_{a2}
\end{aligned}
$$

$$
\begin{aligned}
P_{a5} \triangleq\ & {}^{r_5}\backslash^{\square}_{cAMP_i}\backslash^{cAMP_o}_{\square}\backslash^{\square}_{CAP_i}\backslash^{CAP_o}_{\square}\backslash^{\square}_{\overline{glucose}_i}\backslash^{\overline{glucose}_o}_{\square}\backslash^{\square}_{r_6}\backslash^{p_5}_{\square}\backslash^{\square}_{\widetilde{cAMP\text{-}CAP}_i}\backslash^{\widetilde{cAMP\text{-}CAP}_o}_{\square}\backslash_{p_6}.P_{a5}\\
&+\\
& {}^{r_5}\backslash^{\square}_{glucose_i}\backslash^{glucose_o}_{\square}\backslash^{\square}_{r_6}\backslash^{p_5}_{\square}\backslash_{p_6}.P_{a5}\\
&+\\
& {}^{r_5}\backslash^{\square}_{\overline{cAMP}_i}\backslash^{\overline{cAMP}_o}_{\square}\backslash^{\square}_{r_6}\backslash^{p_5}_{\square}\backslash_{p_6}.P_{a5} \ +\ {}^{r_5}\backslash^{\square}_{\overline{CAP}_i}\backslash^{\overline{CAP}_o}_{\square}\backslash^{\square}_{r_6}\backslash^{p_5}_{\square}\backslash_{p_6}.P_{a5}
\end{aligned}
$$

$$
\begin{aligned}
P_{a6} \triangleq\ & {}^{r_6}\backslash^{\square}_{lac_i}\backslash^{lac_o}_{\square}\backslash^{\square}_{cAMP\text{-}CAP_i}\backslash^{cAMP\text{-}CAP_o}_{\square}\backslash^{\square}_{\overline{I\text{-}OP}_i}\backslash^{\overline{I\text{-}OP}_o}_{\square}\backslash^{\square}_{cxt_1}\backslash^{p_6}_{\square}\backslash^{\square}_{\tilde{z}_i}\backslash^{\tilde{z}_o}_{\square}\backslash^{\square}_{\tilde{y}_i}\backslash^{\tilde{y}_o}_{\square}\backslash^{\square}_{\tilde{A}_i}\backslash^{\tilde{A}_o}_{\square}\backslash_{\tau}.P_{a6}\\
&+\\
& {}^{r_6}\backslash^{\square}_{I\text{-}OP_i}\backslash^{I\text{-}OP_o}_{\square}\backslash^{\square}_{cxt_1}\backslash^{p_6}_{\square}\backslash_{\tau}.P_{a6}\\
&+\\
& {}^{r_6}\backslash^{\square}_{\overline{lac}_i}\backslash^{\overline{lac}_o}_{\square}\backslash^{\square}_{cxt_1}\backslash^{p_6}_{\square}\backslash_{\tau}.P_{a6} \ +\ \backslash^{\square}_{\overline{cAMP\text{-}CAP}_i}\backslash^{\overline{cAMP\text{-}CAP}_o}_{\square}\backslash^{\square}_{cxt_1}\backslash^{p_6}_{\square}\backslash_{\tau}.P_{a6}
\end{aligned}
$$

Processes for the Entities. We exploit the specificity of the example in hand to optimise the code, and we specify exactly the number of solid links that each process encoding an entity must offer. For the always present entities we let:

$$
\begin{array}{ll}
P_{cya} \triangleq {}^{cya_i}\backslash_{cya_o}.P_{cya} & \quad P_{crp} \triangleq {}^{crp_i}\backslash_{crp_o}.P_{crp}\\
P_{lacI} \triangleq {}^{lacI_i}\backslash_{lacI_o}.P_{lacI} & \quad P_{lac} \triangleq {}^{lac_i}\backslash_{lac_o}.P_{lac}
\end{array}
$$

For the entities always produced (i.e. not present only at the first step), we provide a parametric definition $P_e(s) \triangleq {}^{s_i}\backslash^{\square}_{s_o}\backslash^{\tilde{s}_i}_{\square}\backslash_{\tilde{s}_o}.P_e(s) + {}^{\tilde{s}_i}\backslash_{\tilde{s}_o}.P_e(s)$. There are three entities of the second type:

$$
P_{cAMP} \triangleq P_e(cAMP) \qquad P_{CAP} \triangleq P_e(CAP) \qquad P_I \triangleq P_e(I).
$$

The entity *I-OP* can be either produced (by a_2) or tested for absence (by a_6). Correspondingly, the process $P_{I\text{-}OP}$ is defined as follows:

$$
\begin{aligned}
P_{I\text{-}OP} &\triangleq \textstyle\sum_{h=0}^{1}({}^{I\text{-}OP_i}\backslash^{\square}_{I\text{-}OP_o}\backslash_{\square})^h\ {}^{\widetilde{I\text{-}OP}_i}\backslash_{\widetilde{I\text{-}OP}_o}.P_{I\text{-}OP} \ +\ {}^{I\text{-}OP_i}\backslash_{I\text{-}OP_o}.\overline{P_{I\text{-}OP}}\\
\overline{P_{I\text{-}OP}} &\triangleq \textstyle\sum_{h=0}^{1}({}^{\overline{I\text{-}OP}_i}\backslash^{\square}_{\overline{I\text{-}OP}_o}\backslash_{\square})^h\ {}^{\widetilde{I\text{-}OP}_i}\backslash_{\widetilde{I\text{-}OP}_o}.P_{I\text{-}OP} \ +\ {}^{\overline{I\text{-}OP}_i}\backslash_{\overline{I\text{-}OP}_o}.\overline{P_{I\text{-}OP}}
\end{aligned}
$$

The process $P_{cAMP\text{-}CAP}$ is similar to $P_{I\text{-}OP}$, as it is produced by a_5 and tested for presence by a_6. Its code is omitted.

The *lactose* is provided by the context and tested for absence by a_2.

$$
\begin{aligned}
P_{lactose} \triangleq\ & \textstyle\sum_{h=0}^{1}({}^{lactose_i}\backslash^{\square}_{lactose_o}\backslash_{\square})^h\ {}^{\widehat{lactose}_i}\backslash_{\widehat{lactose}_o}.P_{lactose}\\
&+\\
& \textstyle\sum_{h=0}^{1}({}^{lactose_i}\backslash^{\square}_{lactose_o}\backslash_{\square})^h\ {}^{\underline{lactose}_i}\backslash_{\underline{lactose}_o}.\overline{P_{lactose}}
\end{aligned}
$$

$$
\begin{aligned}
\overline{P_{lactose}} \triangleq\ & \textstyle\sum_{h=0}^{1}({}^{\overline{lactose}_i}\backslash^{\square}_{\overline{lactose}_o}\backslash_{\square})^h\ {}^{\widehat{lactose}_i}\backslash_{\widehat{lactose}_o}.P_{lactose}\\
&+\\
& \textstyle\sum_{h=0}^{1}({}^{\overline{lactose}_i}\backslash^{\square}_{\overline{lactose}_o}\backslash_{\square})^h\ {}^{\underline{lactose}_i}\backslash_{\underline{lactose}_o}.\overline{P_{lactose}}
\end{aligned}
$$

The process $P_{glucose}$ is similar to $P_{lactose}$ and tested for absence by a_5. Its code is omitted.

The entity z can only be produced by rule a_6, while it is never provided by the context. Moreover, there is no rule for testing its presence or absence.

$$P_z \triangleq {}^{\tilde{z}_i}\backslash^{\square}_{\tilde{z}_o}\backslash^{z_i}_{\square}\backslash_{\underline{z_o}}.P_z \;+\; {}^{\underline{z_i}}\backslash_{\underline{z_o}}.\overline{P_z} \qquad\qquad \overline{P_z} \triangleq {}^{\tilde{z}_i}\backslash^{\square}_{\tilde{z}_o}\backslash^{z_i}_{\square}\backslash_{\underline{z_o}}.\overline{P_z} \;+\; {}^{\underline{z_i}}\backslash_{\underline{z_o}}.\overline{P_z}$$

The entities y and A are treated likewise z. Their processes are omitted.

Context. The entities in DC are assumed always present by default, so no context process is needed for them. The entities z, y, and A are assumed never provided by the context:

$$Cxt_z \triangleq {}^{cxt_1}\backslash^{\square}_{\underline{z}_i}\backslash^{\underline{z}_o}_{\square}\backslash_{cxt_2}.Cxt_z \qquad\qquad Cxt_y \triangleq {}^{cxt_2}\backslash^{\square}_{\underline{y}_i}\backslash^{\underline{y}_o}_{\square}\backslash_{cxt_3}.Cxt_y$$
$$Cxt_A \triangleq {}^{cxt_3}\backslash^{\square}_{\underline{A}_i}\backslash^{\underline{A}_o}_{\square}\backslash_{cxt_4}.Cxt_A$$

Also, for the sake of presentation, we assume that the *glucose* is never provided and *lactose* is always provided by the context:

$$Cxt_{lactose} \triangleq {}^{cxt_4}\backslash^{\square}_{\widehat{lactose}_i}\backslash^{\widehat{lactose}_o}_{\square}\backslash_{cxt_5}.Cxt_{lactose}$$
$$Cxt_{glucose} \triangleq {}^{cxt_5}\backslash^{\square}_{\underline{glucose}_i}\backslash^{\underline{glucose}_o}_{\square}\backslash_{p_1}.Cxt_{glucose}$$

In the following we let $CXT \triangleq Cxt_z|Cxt_y|Cxt_A|Cxt_{lactose}|Cxt_{glucose}$ be the processes for context. The whole system is as follows:

$$lacOp \triangleq (\nu\, names)(\Pi_{i=1}^{6} P_{ai}|\Pi_{s\in DC} P_s|\Pi_{s\in S\setminus DC}\overline{P}_s|CXT)$$

Execution. Now, we show the execution of two transitions. After the first transition, the entity *cAMP-CAP* is produced due to the absence of *glucose*(see P_{a5}), while the presence of *lactose* inhibits the production of *I-OP* (see P_{a2}): $lacOp \xrightarrow{(\nu\, names)v} lacOp'$, where

$$v = {}^{\tau}\backslash_{lac_i} \cdots {}^{lactose_o}\backslash_{r3} \cdots {}^{r_5}\backslash_{cAMP_i} \cdots {}^{\overline{glucose}_o}\backslash_{r_6} \cdots {}^{p_5}\backslash_{c\widetilde{AMP\text{-}C}AP_i} \cdots {}^{p_6}\backslash_{\tau}$$
$$lacOp' \triangleq (\nu\, names)(\Pi_{i=1}^{6} P_{ai}|\Pi_{s\in AP} P_s|\Pi_{s\in S\setminus AP}\overline{P}_s|CXT)$$

with $AP = DC \cup \{cAMP\text{-}CAP\}$ the actual context.
After the second step the entities z, y and A are produced, due to the presence of *cAMP-CAP* and the absence of *I-OP* (see P_{a6}), thus $lacOp' \xrightarrow{(\nu\, names)v'} lacOp''$ where:

$$v' = {}^{\tau}\backslash_{lac_i} \cdots {}^{lac_o}\backslash_{cAMP-CAP_i} \cdots {}^{\overline{I-OP}_o}\backslash^{\dots}_{cxt_1}\, \underline{glucose}_o\backslash^{p_1}_{p_1}\backslash^{p_6}_{\dots}\backslash^{\tilde{z}_i}_{\tilde{z}_i}\backslash^{\tilde{z}_o}_{\tilde{z}_o}\backslash^{\tilde{y}_i}_{\tilde{y}_i}\backslash^{\tilde{y}_o}_{\tilde{y}_o}\backslash^{\tilde{A}_i}_{\tilde{A}_i}\backslash^{\tilde{A}_o}_{\tilde{A}_o}\backslash_{\tau}$$
$$lacOp'' \triangleq (\nu\, names)(\Pi_{i=1}^{6} P_{ai}|\Pi_{s\in AP'} P_s|\Pi_{s\in S\setminus AP'}\overline{P}_s|CXT)$$

with $AP' = DC \cup \{z, y, A\}$.

6 Enhanced Reaction Systems

Our encoding increases the expressivity of RS concerning: the behaviour of the context, the possibility of alternative behaviour of mutated entities and the communication between two different reaction systems. It is important to note that our encoding guarantees that from each state, in the cCNA transition system, only one transition comes out, as the dynamics is totally deterministic.

6.1 Recursive Contexts

In RS, the behaviour of the context is finite. For the first n steps, it is specified which are the entities that are provided from the context. Using cCNA we can describe in a natural way the behaviour of the context in a recursive way. Then, the context behaviour would not necessarily end after n steps, and could be infinite. For example, in an extended interactive process, we may want that the entity s is intermittently provided by the context every two steps:

$$\begin{array}{lll} Cxt_s & \triangleq {}^{cxt_j}\backslash^{\square}_{\hat{s}_i}\backslash^{\hat{s}_o}_{\square}\backslash_{cxt_{j+1}}.Cxt_{s'}, & \text{context provides } s; \\ Cxt_{s'} & \triangleq {}^{cxt_j}\backslash^{\square}_{\underline{s}_i}\backslash^{\underline{s}_o}_{\square}\backslash_{cxt_{j+1}}.Cxt_{s''}, & \text{context doesn't provide } s; \\ Cxt_{s''} & \triangleq {}^{cxt_j}\backslash^{\square}_{\underline{s}_i}\backslash^{\underline{s}_o}_{\square}\backslash_{cxt_{j+1}}.Cxt_{s}, & \text{context doesn't provide } s. \end{array}$$

6.2 Mutating Entities

In RS, when an entity is present, it can potentially be involved in each reaction where it is required. With a few more lines of code, in cCNA it is possible to describe the behaviour of a mutation of an entity, in a way that the mutated version of the entity can take part to only a subset of the rules requiring the *normal version* of the entity. For example, let us assume that entity $s1$ is consumed by reactions $a1$ and $a2$. Reaction $a1$ produces $s1$ if $s2$ is present, otherwise $a1$ produces a mutated version of $s1$, say $s1'$. When $s1'$ is produced, reaction $a1$ behaves in the same way as if $s1$ would be absent, whereas $a2$ recognises the presence of $s1'$ and behaves in the same way as if $s1$ would be present. Technically, in both cases it is enough to add one more non deterministic choice in the code of P_{a1} and P_{a2}.

Table 1. The two reaction systems $rs1$ and $rs2$.

$rs1$	$rs2$
$a_1 = (s, \ , x)$	$a_2 = (y, \ , s)$

6.3 Communicating Reaction Systems

We sketch how it is possible to program two RS encodings, in a way that the entities that usually come from context of one RS will be provided instead from the other RS.

Example 3. Let $rs1$ and $rs2$ be two RSs, defined by the rules in Table 1. Now, we set our example such that the two contexts, for $rs1$ and $rs2$, do not provide any entities. We also assume that entity s in $rs1$ is provided by $rs2$, as $rs2$ produces a quantity of s that is enough for $rs1$ and $rs2$. For technical reasons, we can not use the same name for s in both the two RSs, then we use the name ss in $rs2$. We need to modify our translation technique to suite this new setting. As we do not model contexts, we introduce *dummy* channel names dx and dss to model the case x and ss are not produced. Also, thanks to the simplicity of the example, we can leave out the use of the p_i channels. This streamlining does not affect the programming technique we propose to make two RSs communicate. First, we translate the reaction in $rs1$:

$$[\![a_1]\!] \triangleq P_{a_1} \triangleq {}^{\tau}\backslash^{\square}_{s_i}\backslash^{s_o}_{\square}\backslash^{\square}_{\tilde{x}_i}\backslash^{\tilde{x}_o}_{\square}\backslash_{a_2}.P_{a_1} + {}^{\tau}\backslash^{\square}_{\overline{s}_i}\backslash^{\overline{s}_o}_{\square}\backslash^{\square}_{dx_i}\backslash^{dx_o}_{\square}\backslash_{a_2}.P_{a_1}$$

Please note, that prefixes of process P_{a_1} end with the channel name a_2, as the link chain is now connected with the reaction of $rs2$. The translation for the entities follows.

$$[\![s]\!] \triangleq P_s \triangleq {}^{s_i}\backslash^{\square}_{s_o}\backslash^{\hat{s}_i}_{\square}\backslash_{\hat{s}_o}.P_s + {}^{s_i}\backslash_{s_o}.\overline{P_s} \qquad [\![x]\!] \triangleq P_x \triangleq {}^{\tilde{x}_i}\backslash_{\tilde{x}_o}.P_x + {}^{dx_i}\backslash_{dx_o}.\overline{P_x}$$
$$\overline{P_s} \triangleq {}^{\overline{s}_i}\backslash^{\square}_{\overline{s}_o}\backslash^{\hat{s}_i}_{\square}\backslash_{\hat{s}_o}.P_s + {}^{\overline{s}_i}\backslash_{\overline{s}_i}.\overline{P_s} \qquad \overline{P_x} \triangleq {}^{\tilde{x}_i}\backslash_{\tilde{x}_o}.P_x + {}^{dx_i}\backslash_{dx_o}.\overline{P_x}$$

The translation for the $rs2$ follows.

$$[\![a_2]\!] \triangleq P_{a_2} \triangleq {}^{a_2}\backslash^{\square}_{y_i}\backslash^{y_o}_{\square}\backslash^{\square}_{\tilde{s}s_i}\backslash^{\tilde{s}s_o}_{\square}\backslash_{\tau}.P_{a_2} + {}^{a_2}\backslash^{\square}_{\overline{y}_i}\backslash^{\overline{y}_o}_{\square}\backslash^{\square}_{dss_i}\backslash^{dss_o}_{\square}\backslash_{\tau}.P_{a_2}$$

In the translation of the entities in $rs2$, we introduce the mechanism that allows the entity s (ss in $rs2$) to be provided in $rs1$. Every time ss is produced in $rs2$, a virtual link is created to synchronise with $rs1$ on link ${}^{\hat{s}_i}\backslash_{\hat{s}_o}$:

$$[\![ss]\!] \triangleq P_{ss} \triangleq {}^{\tilde{s}s_i}\backslash^{\square}_{\hat{s}_i}\backslash^{\hat{s}_o}_{\square}\backslash_{\tilde{s}s_o}.P_{ss} + {}^{dss_i}\backslash_{dss_o}.\overline{P_{ss}} \qquad [\![y]\!] \triangleq P_y \triangleq {}^{y_i}\backslash_{y_o}.\overline{P_y}$$
$$\overline{P_{ss}} \triangleq {}^{\tilde{s}s_i}\backslash^{\square}_{\hat{s}_i}\backslash^{\hat{s}_o}_{\square}\backslash_{\tilde{s}s_o}.P_{ss} + {}^{dss_i}\backslash_{dss_o}.\overline{P_{ss}} \qquad \overline{P_y} \triangleq {}^{\overline{y}_i}\backslash_{\overline{y}_o}.\overline{P_y}$$

We now assume that the initial system is $S \triangleq (\nu\, names)(P_{a_1}|P_{a_2}|P_s|P_y|\overline{P_x}|\overline{P_{ss}})$, i.e. only entities s and y are present. Now, the only possible transition has the following label (that we report without restriction):

$${}^{\tau}\backslash^{s_i}_{s_i}\backslash^{s_o}_{s_o}\backslash^{\tilde{x}_i}_{\tilde{x}_i}\backslash^{\tilde{x}_o}_{\tilde{x}_o}\backslash^{a_2}_{a_2}\backslash^{y_i}_{y_i}\backslash^{y_o}_{y_o}\backslash^{\tilde{s}s_i}_{\tilde{s}s_i}\backslash^{\hat{s}_i}_{\hat{s}_i}\backslash^{\hat{s}_o}_{\hat{s}_o}\backslash^{\tilde{s}s_o}_{\tilde{s}s_o}\backslash_{\tau},$$

where the black links belong to the prefixes of P_{a_1}, and P_{a_2}, the blue links belong to P_s, the gray links belong to P_y, and $\overline{P_x}$ and the red links belong to $\overline{P_{ss}}$. After the execution, the entity s is still present in $rs1$ as it has been provided by $rs2$.

As we have briefly sketched, our model of two *communicating reaction systems* can enable the study of the behaviour of one RS in relation to another one. Thus, the products of the reactions of one RS can become the input for another one. This could allow for a modular approach to modeling complex systems, by composing different Reaction Systems.

7 Conclusion

In this paper we have exploited a variant of the `link`-calculus where prefixes are link chains and no more single links. This variant was already briefly discussed by the end of [8]. This variant allowed us to define an elegant embedding of reaction systems, an emerging formalism to model computationally biochemical systems. This translation shows several benefits. For instance, the context behaviour can also be expressed recursively; entity mutations can be expressed easily; reaction systems can communicate between them.

We believe that our embedding can contribute to extend the applications of reaction systems to diverse fields of computer science, and life sciences. As we have already mentioned, the evolution of each process resulting from our embedding is deterministic, thus we do not have the problem of having infinitely many transitions in the produced labelled transition system. In any case, we can exploit the implementation of the symbolic semantics of the `link`-calculus [11] that is available at [19].

As future work, we plan to implement a prototype of our framework, with an automatic translation from RSs to the `link`-calculus. We believe that our work can also help to extend the framework of RSs towards a model which can improve the communication between different RSs. We also believe that our work can make possible to investigate how to apply formal techniques to prove properties of the modeled systems [7,13,14,16,20].

Acknowledgments. We thank the anonymous reviewers for their detailed and very useful criticisms and recommendations that helped us to improve our paper.

References

1. Azimi, S.: Steady states of constrained reaction systems. Theor. Comput. Sci. **701**(C), 20–26 (2017). https://doi.org/10.1016/j.tcs.2017.03.047
2. Azimi, S., Iancu, B., Petre, I.: Reaction system models for the heat shock response. Fundam. Inf. **131**(3–4), 299–312 (2014). https://doi.org/10.3233/FI-2014-1016
3. Barbuti, R., Gori, R., Levi, F., Milazzo, P.: Investigating dynamic causalities in reaction systems. Theor. Comput. Sci. **623**, 114–145 (2016)
4. Bernini, A., Brodo, L., Degano, P., Falaschi, M., Hermith, D.: Process calculi for biological processes. Nat. Comput. **17**(2), 345–373 (2018)
5. Bodei, C., Brodo, L., Bruni, R.: Open multiparty interaction. In: Martí-Oliet, N., Palomino, M. (eds.) WADT 2012. LNCS, vol. 7841, pp. 1–23. Springer, Heidelberg (2013). https://doi.org/10.1007/978-3-642-37635-1_1

6. Bodei, C., Brodo, L., Bruni, R., Chiarugi, D.: A flat process calculus for nested membrane interactions. Sci. Ann. Comp. Sci. **24**(1), 91–136 (2014)
7. Bodei, C., Brodo, L., Gori, R., Levi, F., Bernini, A., Hermith, D.: A static analysis for Brane Calculi providing global occurrence counting information. Theoret. Comput. Sci. **696**, 11–51 (2017)
8. Bodei, C., Brodo, L., Bruni, R.: A formal approach to open multiparty interactions. Theoret. Comput. Sci. **763**, 38–65 (2019)
9. Brijder, R., Ehrenfeucht, A., Main, M., Rozenberg, G.: A tour of reaction systems. Int. J. Found. Comput. Sci. **22**(07), 1499–1517 (2011)
10. Brodo, L.: On the expressiveness of pi-calculus for encoding mobile ambients. Math. Struct. Comput. Sci. **28**(2), 202–240 (2018)
11. Brodo, L., Olarte, C.: Symbolic semantics for multiparty interactions in the link-calculus. In: Steffen, B., Baier, C., van den Brand, M., Eder, J., Hinchey, M., Margaria, T. (eds.) SOFSEM 2017. LNCS, vol. 10139, pp. 62–75. Springer, Cham (2017). https://doi.org/10.1007/978-3-319-51963-0_6
12. Cardelli, L., Gordon, A.D.: Mobile ambients. Theoret. Comput. Sci. **240**(1), 177–213 (2000)
13. Chiarugi, D., Falaschi, M., Hermith, D., Olarte, C., Torella, L.: Modelling non-markovian dynamics in biochemical reactions. BMC Syst. Biol. **9**(S–3), S8 (2015)
14. Chiarugi, D., Falaschi, M., Olarte, C., Palamidessi, C.: Compositional modelling of signalling pathways in timed concurrent constraint programming. In: Proceedings of ACM BCB 2010, pp. 414–417. ACM, New York (2010)
15. Corolli, L., Maj, C., Marinia, F., Besozzi, D., Mauri, G.: An excursion in reaction systems: from computer science to biology. Theoret. Comput. Sci. **454**, 95–108 (2012)
16. Falaschi, M., Olarte, C., Palamidessi, C.: Abstract interpretation of temporal concurrent constraint programs. Theory Pract. Logic Program. **15**(3), 312–357 (2015)
17. Męski, A., Penczek, W., Rozenberg, G.: Model checking temporal properties of reaction systems. Inf. Sci. **313**, 22–42 (2015). https://doi.org/10.1016/j.ins.2015.03.048
18. Okubo, F., Yokomori, T.: The computational capability of chemical reaction automata. Natural Comput. **15**(2), 215–224 (2016). https://doi.org/10.1007/s11047-015-9504-7
19. Olarte, C.: SiLVer: Symbolic links verifier, December 2018. http://subsell.logic.at/links/links-web/index.html
20. Olarte, C., Chiarugi, D., Falaschi, M., Hermith, D.: A proof theoretic view of spatial and temporal dependencies in biochemical systems. Theor. Comput. Sci. **641**, 25–42 (2016)

Checking the Expressivity of Firewall Languages

Lorenzo Ceragioli[1](✉), Pierpaolo Degano[1], and Letterio Galletta[2]

[1] Dipartimento di Informatica, Università di Pisa, Pisa, Italy
lorenzo.ceragioli@phd.unipi.it, degano@di.unipi.it
[2] IMT School for Advanced Studies, Lucca, Italy
letterio.galletta@imtlucca.it

Abstract. Designing and maintaining firewall configurations is hard, also for expert system administrators. Indeed, policies are made of a large number of rules and are written in low-level configuration languages that are specific to the firewall system in use. As part of a larger group, we have addressed these issues and have proposed a semantic-based transcompilation pipeline. It is supported by FWS, a tool that analyses a real configuration and ports it from a firewall system to another. To our surprise, we discovered that some configurations expressed in a real firewall system cannot be ported to another system, preserving the semantics. Here we outline the main reasons for the detected differences between the firewall languages, and describe F2F, a tool that checks if a given configuration in a system can be ported to another system, and reports its user on which parts cause problems and why.

1 Introduction

Firewalls are the basic mechanisms for the protection of computer networks. Their effectiveness heavily depends on the correctness of configurations, since even a small flaw may break up completely the security or the functionality of an entire network.

Policies are typically written in low-level configuration languages that are specific to the firewall system in use and support non-trivial control flow constructs, such as *call* and *goto*. A configuration usually consists of a large number of rules interacting with each other, often in some obscure manner. Indeed, the way rules are used depends on the firewall system in hand, and also on contextual information, e.g. the order in which rules appear in the configuration. Also, the way in which these systems work has to be inferred from manuals and by experiencing with them, since almost always they have no formal semantics. It is therefore hard to understand the firewall behaviour, also because of the different ways in which packets are processed by the network stack of the operating system

The first two authors have been partially supported by project PRA_2018_66 *DECLware: Declarative methodologies for designing and deploying applications* of the Università di Pisa; the third author by IMT project *PAI VeriOSS*.

M. S. Alvim et al. (Eds.): Palamidessi Festschrift, LNCS 11760, pp. 86–100, 2019.
https://doi.org/10.1007/978-3-030-31175-9_6

running on the firewall. An additional source of mess is Network Address Translation (NAT) that translates IP addresses and performs port redirection while packets traverse the firewall. In cooperation with a larger group of researchers, we have proposed a semantic-based transcompilation pipeline [3,4] to support system administrators. The pipeline is implemented by the tool FWS, described in [4]. It is available online[1] and it provides several utilities. Its first stages support a system administrator in analysing the current firewall configuration, by first extracting a formal description of the firewall behaviour. Then, the user can port a configuration from a system to another, preserving its meaning. Crucial to all the pipeline is the use of the Intermediate Firewall Configuration Language (IFCL) that has a formal semantics. This intermediate language has been used in [3,4] as a common framework to encode firewalls written in `iptables` [12], `ipfw` [13] and `pf` [11], which are the most used languages in Unix and Linux.

To our surprise, while implementing the pipeline, we discovered that some configurations could not be compiled to a language different from the source one. To investigate on these corner cases, we used the new denotational semantics of IFCL and the new algorithm for porting configurations given in [6]. The expressivity of real firewall systems is then formally compared, and we proved that their languages originate a partial order [5]. In particular there exists a firewall expressible in `iptables`, but neither in `ipfw` nor in `pf`; also, `iptables` is incomparable and `ipfw` dominates `pf`.

Here, we describe F2F, a tool that given a configuration c for a source system $\mathcal{L}$ and a target system $\mathcal{L}'$, checks if there exists a configuration c' in $\mathcal{L}'$ such that the behaviour is the same. If instead there is none, it reports to the user on which packets cause problems and why. The information provided by F2F can guide the system administrator in the choice of another target system, but could also be used to make an informed decision on how to change the configuration in order to make it expressible by the target system. We also present a simple yet realistic use case for F2F, that is available online[2] and takes a few seconds to perform such a check on medium size configurations, although it is in a preliminary version.

Our purpose here is to complement the papers [3–6] that have the full details about the theoretical basis of the tool; in particular, we refer the interested reader to [5,6] for the new denotational semantics of IFCL and for an analysis on the expressive power of various firewall languages.

Related Work. To the best of our knowledge the expressive power of firewall systems is formally investigate only in [5], which provides us with firm guidelines for the development of F2F.

In the literature there are many proposals for modeling and analyzing firewall configurations. Some of them, e.g. see [7,9,10], do not rely on a formal semantics and typically either compile from a high level to a low level language or check if configurations comply with given specifications.

1 https://github.com/secgroup/fws.
2 https://github.com/lceragioli/F2F.

The following instead provide firewall systems with a formal semantics. Diekmann et al. [8] consider a subset of `iptables` without packet transformations, and mechanize it using Isabelle/HOL. Furthermore, they define and prove correct a simplification procedure that aims to make configurations easier to be analyzed by automatic tools. Adão et al. [1] introduce Mignis, a high level firewall language for specifying abstract filtering policies, and a compiler from it into `iptables`, which is formally proved correct. The paper formalizes the semantics of Mignis and gives an operational semantics of `iptables`, including packet filtering and translations. Successively, Adão et al. [2] propose a denotational semantics for Mignis that maps a configuration into a packet transformer, representing all the accepted packets with the corresponding translations, similarly to the one in [5], which is the basis of our tool.

Plan of the Paper. Section 2 surveys the firewall system we consider in this paper, the IFCL semantics and the transcompilation pipeline of [4].

Section 3 presents an `iptables` configuration, implementing a policy to be applied in a simple yet realistic scenario, and shows how our tool detects that neither `ipfw` nor `pf` can express this configuration. The internals of the tool F2F are described in Sect. 4 with the help of some examples that illustrate the main reasons for the differences between the expressive power of the considered firewall languages. In Sect. 5 we conclude and discuss some future work.

2 Background

In the following we survey the most used firewall languages, the intermediate language IFCL and the way it is used to encode them. Finally, we briefly present the stages of the transcompilation pipeline.

2.1 `iptables`

It is the default tool for packet filtering in Linux and it operates on top of Netfilter, the framework for packets processing of the Linux kernel [12].

An `iptables` configuration is built on *tables* and *chains*. Intuitively, each table has a specific purpose and is made of a collection of chains. The most commonly used tables are: `Filter` for packet filtering; `Nat` for network address translation; `Mangle` for packet alteration. Chains are actually rulesets, and are ordered lists of policy rules that are inspected to find the first one matching the packet under evaluation. There are five predefined chains, and the user can extend them defining its own. Chains are inspected by the Linux kernel at specific moments of the packet life cycle [14]: `PreRouting`, when a packet reaches the host; `Forward`, when a packet is routed through the host; `PostRouting`, when a packet is about to leave the host; `Input`, when a packet is routed to the host; `Output`, when a packet is generated by the host.

Each rule specifies a *condition* and an action to apply to the matching packet, called *target*. The most commonly used targets are: ACCEPT, to accept the packet;

DROP, to discard the packet; DNAT, to perform destination NAT, i.e., a translation of the destination address; SNAT, to perform source NAT, i.e., a translation of the source address. There are also targets that allow implementing mechanisms similar to jumps and procedure calls, but since they can be macro-expanded (see below), we ignore them and refer the reader to the technical documentation [12]. Built-in chains have a user-configurable default policy (ACCEPT or DROP) to be applied when the evaluation reaches the end of a built-in chain.

2.2 ipfw

It is the standard firewall for FreeBSD [13]. A configuration consists of a single list of rules (called ruleset) that is inspected twice, when the packet enters the firewall and when it exits. It is possible to specify when a certain rule has only to be applied in either case by using the keywords `in` and `out`. Similarly to `iptables`, rules are inspected sequentially until the first match occurs and the corresponding action is taken. The packet is dropped if there is no matching rule. The sequential order of inspection is altered by special targets that jump to a rule that follows the current one.

2.3 pf

This is the standard firewall of OpenBSD [11] and MacOS since version 10.7. A `pf` configuration consists of a single ruleset, inspected when the packet enters and exits the host. Similarly to the other systems, each rule consists of a condition and a target that specifies how to process the packets matching the condition. The most common targets are: `pass` and `block` to accept and reject packets, respectively; `rdr` and `nat` to perform destination and source NAT, respectively. The rule to apply is the *last matched rule* (unless otherwise specified). Moreover, when a packet enters the host, DNAT rules are examined first and filtering is performed after the address translation. Similarly, when a packet leaves the host, its source address is translated by the relevant SNAT rules, and then the resulting packet is filtered.

2.4 The Intermediate Language IFCL

We use the Intermediate Firewall Configuration Language (IFCL) [3,4] as a common setting in which we encode the firewall systems above. It has been originally equipped with an operational semantics [3].

A firewall configuration in IFCL consists of a set of rulesets and a control diagram. As usual, a ruleset is a list of rules that are inspected sequentially to find the first rule matching a packet. A *control diagram* is a graph C that deterministically describes the order in which rulesets are applied by the network stack of the operating system. Every node q represents thus a processing phase and it is associated with the ruleset to apply when the control reaches q. Arcs are labeled with predicates that encode routing decisions taken by the firewall, e.g., depending on whether a packet comes from the external world. Intuitively, a packet p is accepted if there exists a path π from the initial to the final node of C such that p passes the checks of (and is possibly transformed by) the rulesets associated with the nodes of π.

A firewall rule consists of a predicate ϕ over packets and an action t, called *target*, defining how packets matching ϕ are processed. For our purposes, it suffices to consider the following subset of targets considered in [4] (note that we can safely neglect targets altering the control, like *call*, *goto* and *return*, because they can suitably be macro-expanded)

`ACCEPT`	the packet is accepted
`DROP`	the packet is discarded
`NAT`(n_d, n_s)	apply NAT

In the `NAT` action, n_d and n_s specify how to translate the destination and source addresses/ports of a packet and, as done before, we use $-$ to denote the identity translation. For instance, $n_d = n : -$ means that the destination address of a packet is translated according to n, while the port is kept unchanged.

Formally, an IFCL firewall is defined as follows:

Definition 1 (Firewall). *An IFCL* firewall *is a pair* (C, Σ) *where* C *is a control diagram and* Σ *is a configuration, defined below.*

Let $\mathcal{P} \subseteq \mathbb{P}$ *be a set of predicates over packets* p. *A* deterministic control diagram C *is a tuple* (Q, A, q_i, q_f), *where*

- Q *is the set of nodes;*
- $A \subseteq Q \times \mathcal{P} \times Q$ *is the set of arcs, such that whenever* $(q, \psi, q'), (q, \psi', q'') \in A$ *and* $q' \neq q''$ *then* $\neg(\psi \wedge \psi')$*;*
- $q_i, q_f \in Q$ *are special nodes denoting the start of elaboration of an incoming packet* p *and the end if* p *is accepted.*

A configuration *is a pair* $\Sigma = (\rho, c)$, *where*

- ρ *a set of rulesets;*
- $c \colon Q \to \rho$ *is a function assigning a ruleset to each node* $q \in Q$.

2.5 Modeling `iptables`, `ipfw` and `pf` in IFCL

We now survey the encoding of the firewall systems mentioned above into IFCL. The full definitions are in [3,4], and we only report here the relevant information for our treatment. We remark that the behaviour of those firewall systems is only informally defined by manuals, except for `iptables` for which [1] introduced a formal semantics. However they all inherit the formal semantics of IFCL, via the encodings below (for `iptables` the inherited semantics and that of [1] coincide).

Hereafter let $\mathcal{S}$ be the set of the addresses of the firewall interfaces; let $p \in \mathbb{P}$ be a packet; let $d(p)$ and $s(p)$ denote the destination address and source address

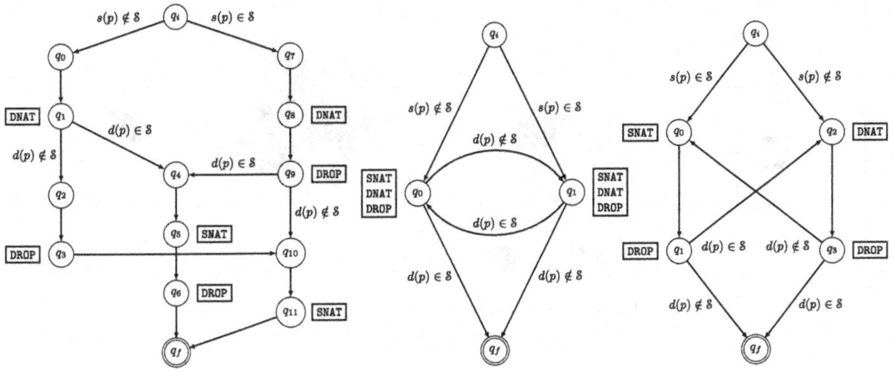

Fig. 1. Control diagrams of `iptables`, `ipfw` and `pf`.

of p, respectively; and let $d(p) \in \mathbb{S}$ ($s(p) \in \mathbb{S}$, respectively) specify that p is for (comes from, respectively) the firewall. In the control diagrams of Fig. 1 we label arcs with predicates expressing constraints on the header of packets according to $\mathbb{S}$; arcs with no label implicitly carry "*true*".

***iptables*.** Figure 1a shows the control diagram $C_{\texttt{iptables}}$ of `iptables` (recall that its tables contain predefined chains). The encoding into IFCL associates these predefined chains with the nodes of the control diagram. For example, the table `Nat` contains the `PostRouting` chain that is associated with q_{11}. It is important to note that in `iptables` a `DNAT` is only performed in nodes q_1, q_8, whereas `SNAT` only in nodes q_5, q_{11}. Similarly, DROP can only be applied when the control is in either nodes q_3, q_6, q_9. As we will see later on, these capabilities are represented by the labels in the boxes.

***ipfw*.** The control diagram of `ipfw` is in Fig. 1b. The encoding of [4] splits the single `ipfw` ruleset in two rulesets containing each the rules annotated with the keyword `in` and `out`, respectively (if not annotated, the rule goes in both). Both rulesets can filter packets and transform them. The node q_0 is associated with the ruleset applied when an IP packet reaches the host from the net, whereas, q_1 is for when the packet leaves the host. Note that the loop between the nodes q_0 and q_1 causes no problem, because firewalls detects cycles and drop the packet causing it.

***pf*.** Figure 1c displays the control diagram of `pf`, where the nodes q_2 and q_3 are associated with the rulesets applied to packets that reach the firewall, while q_0 and q_1 are for when they leave the firewall. Also in this case, the single ruleset of `pf` is split in different rulesets. A source NAT can only be applied to packets leaving the firewall, and destination NAT only those reaching it. Importantly NAT rules (in nodes q_0 and q_2) are evaluated *before* filtering (in nodes q_1 and q_3). To represent the *last matching-rule* policy of `pf`, it suffices to reverse the order of rules inside rulesets.

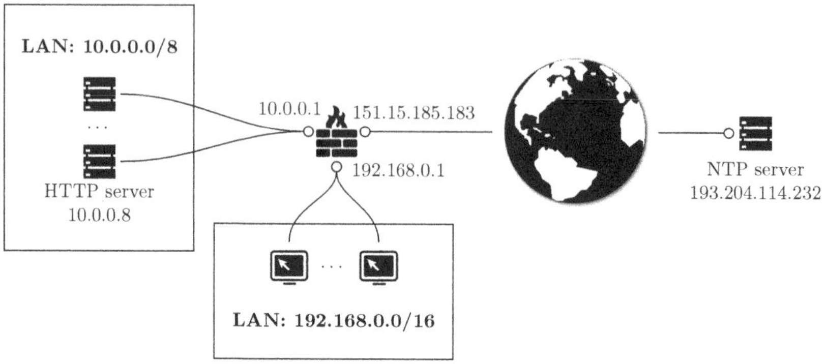

Fig. 2. A simple scenario.

2.6 Transcompilation Pipeline

The transcompilation pipeline of [3,4] supports system administrators in (i) *decompiling* real configurations into abstract specifications representing the set of the permitted connections; (ii) performing *policy refactoring* by supporting configuration updates and elimination of redundant rules, thus obtaining minimal and clean configurations; (iii) automatically *porting* a configuration written for a system into the language of another system.

The proposed transcompiling pipeline is made of the following stages:

1. decompile a policy from the source language to IFCL;
2. extract the meaning of the policy as a table describing how the accepted packets are translated;
3. compile the semantic table into the target language.

All the steps above are supported by the tool FWS, described in [4] and available online.[3] Experiments have been made on te languages mentioned above.

3 An Example Illustrating the Expressivity Problem

We present below a specific network with a firewall, and a specific configuration, expressed in `iptables`. Then, using F2F we illustrate the reasons why this configuration can neither be ported in `ipfw` nor in `pf`, without resorting to ad hoc extensions of these languages.

Consider the firewall connected to the Internet using the IP 151.15.185.183, in the network in Fig. 2. It protects two LANs with addresses ranging in 10.0.0.0/8 and 192.168.0.0/16,[4] respectively, because all the connections from and to the Internet pass through the firewall. The first LAN hosts various servers, and the

[3] https://github.com/secgroup/fws.

[4] We use the standard CIDR notation to denote the range of IP addresses.

```
*nat
:PREROUTING ACCEPT [0:0]
:INPUT ACCEPT [0:0]
:OUTPUT ACCEPT [0:0]
:POSTROUTING ACCEPT [0:0]

-A PREROUTING -p udp --dport 123 -j DNAT --to 193.204.114.232
-A OUTPUT -p udp --dport 123 -j DNAT --to 193.204.114.232
-A PREROUTING -p tcp -d 151.15.185.183 --dport 80 -j DNAT --to 10.0.0.8
-A OUTPUT -p tcp -d 151.15.185.183 --dport 80 -j DNAT --to 10.0.0.8

-A POSTROUTING -d 192.168.0.0/16 -j ACCEPT
-A INPUT -d 192.168.0.0/16 -j ACCEPT
-A POSTROUTING -d 10.0.0.0/8 -j ACCEPT
-A INPUT -d 10.0.0.0/8 -j ACCEPT
-A POSTROUTING -j SNAT --to 151.15.185.183
-A INPUT -j SNAT --to 151.15.185.183

COMMIT

*filter
:INPUT DROP [0:0]
:FORWARD DROP [0:0]
:OUTPUT DROP [0:0]

-A INPUT -m state --state ESTABLISHED -j ACCEPT
-A INPUT -p tcp -d 10.0.0.8 --dport 80 -j ACCEPT
-A INPUT -s 10.0.0.0/8 -d 10.0.0.0/8 -j ACCEPT
-A INPUT -s 192.168.0.0/16 ! -d 10.0.0.0/8 -j ACCEPT
-A INPUT -p udp -d 193.204.114.232 --dport 123 -j ACCEPT

-A FORWARD -m state --state ESTABLISHED -j ACCEPT
-A FORWARD -p tcp -d 10.0.0.8 --dport 80 -j ACCEPT
-A FORWARD -s 10.0.0.0/8 -d 10.0.0.0/8 -j ACCEPT
-A FORWARD -s 192.168.0.0/16 ! -d 10.0.0.0/8 -j ACCEPT
-A FORWARD -p udp -d 193.204.114.232 --dport 123 -j ACCEPT

-A OUTPUT -m state --state ESTABLISHED -j ACCEPT
-A OUTPUT -p tcp -d 10.0.0.8 --dport 80 -j ACCEPT
-A OUTPUT -s 10.0.0.0/8 -d 10.0.0.0/8 -j ACCEPT
-A OUTPUT -s 192.168.0.0/16 ! -d 10.0.0.0/8 -j ACCEPT
-A OUTPUT -p udp -d 193.204.114.232 --dport 123 -j ACCEPT

COMMIT
```

Fig. 3. A simple configuration in `iptables`.

second one common users. The firewall interacts with the LANs through the interfaces with the addresses 10.0.0.1 and 192.168.0.1, respectively.

We assume that hosts in the LANs have private addresses that cannot be routed on the Internet, hence the source address of outgoing packets and the destination address of incoming packets are translated using NAT. Also these hosts are not allowed to directly communicate with each other, and thus their messages always pass through the firewall.

The configuration in Fig. 3 enforces the following behaviour. The hosts in the LANs can freely communicate each other, but servers cannot start a connection towards a common user. When the protocol is NTP (Network Time Protocol) on port 123, the connection is redirected to the remote host 193.204.114.232, via a destination NAT (`DNAT`). The connections from the Internet towards the hosts in

```
(venv) user@here:~/$ fwp iptables ~/interfaces ~/iptables.conf ipfw
Solving: [##################################################] (  36/  36) 100.00%

PROBLEM FOUND!
In ipfw the following rule schema is not expressible!
=============================================================================
| sIp   | dIp   ||  tr_sIp      : tr_sPort |  tr_dIp      : tr_dPort |
=============================================================================
| Self  | Self  || SNAT ( Self)  : id      | DNAT (~Self)  : id       |
=============================================================================
Hence the following is impossible to achieve:
=====================================================================================================
||    sIp      | sPort |      dIp        | dPort | prot ||         tr_src          |          tr_dst          ||
=====================================================================================================
|| 192.168.0.1 |   *   |    127.0.0.1    |  123  | udp  || 151.15.185.183 : -  | 193.204.114.232 : -  ||
||             |       | 151.15.185.183  |       |      ||                     |                      ||
||             |       |    10.0.0.1     |       |      ||                     |                      ||
||             |       |   192.168.0.1   |       |      ||                     |                      ||
=====================================================================================================

PROBLEM FOUND!
In ipfw the following rule schema is not expressible!
=============================================================================
| sIp   | dIp   ||  tr_sIp      : tr_sPort |  tr_dIp      : tr_dPort |
=============================================================================
| Self  | ~Self || SNAT ( Self) : id       | DNAT (~Self) : id        |
=============================================================================
Hence the following is impossible to achieve:
=====================================================================================================================
||    sIp      | sPort |              dIp                | dPort | prot ||        tr_src        |         tr_dst         ||
=====================================================================================================================
|| 192.168.0.1 |   *   |      0.0.0.0 - 10.0.0.0         |  123  | udp  || 151.15.185.183 : -  | 193.204.114.232 : -  ||
||             |       |     10.0.0.2 - 127.0.0.0        |       |      ||                     |                      ||
||             |       |   127.0.0.2 - 151.15.185.182    |       |      ||                     |                      ||
||             |       | 151.15.185.184 - 192.168.0.0    |       |      ||                     |                      ||
||             |       | 192.168.0.2 - 255.255.255.255   |       |      ||                     |                      ||
=====================================================================================================================
```

Fig. 4. Checking the portability of the configuration in `ipfw`.

the LANs are blocked, except than those to the external address 151.15.185.183 on port 80 (written 151.15.185.183 : 80), that are redirected to the HTTP server at 10.0.0.8, also via `DNAT`. Every host can connect to 193.204.114.232, while only common users can connect to other Internet addresses. The source address of these outgoing packets get their source address translated to the external address of the firewall, via a `SNAT`.

Assume we want to port this configuration into `ipfw`, and for that we invoke F2F to check if this is possible. Intuitively, the configuration is internally represented as a table whose rows represent the accepted packets and how they are translated. For example, consider a UDP packet with destination 151.15.185.184 : 123 and source 192.168.0.1 : $*$ ($*$ stands for any port). It is accepted when its destination is translated to 193.204.114.232 : 123 and its source to 151.15.185.183 : $-$ ($-$ stands for "the port remains the same"), because its destination is first translated by rule 10, then it passes through the `OUTPUT` chain in the *filter* table and is forwarded by rule 42, finally it is subject to `SNAT` in rule 16 that is the first matching rule of the `POSTROUTING` chain in *nat* table. Roughly, the relevant portion of the corresponding row will look as follows:

sIp	sPort	dIp	dPort	prot	+tr_scr+	+tr_dst+
192.168.0.1	$*$	151.15.185.184	123	udp	151.15.185.183 : $-$	193.204.114.232:123

```
(venv) user@here:~/$ fwp iptables ~/interfaces ~/iptables.conf pf

PROBLEM FOUND!
In pf the following rule schema is not expressible!
==============================================================================
| sIp   | dIp   ||   tr_sIp     : tr_sPort  |    tr_dIp      : tr_dPort  |
==============================================================================
| Self  | Self  || id             : id        | DNAT (~Self) : id          |
==============================================================================
Hence the following is impossible to achieve:
=====================================================================================
||       sIp       | sPort |       dIp       | dPort | prot || tr_src |     tr_dst     ||
=====================================================================================
||    127.0.0.1    |   *   | 151.15.185.183  |  80   | tcp  || - : -  | 10.0.0.8 : -   ||
|| 151.15.185.183  |       |                 |       |      ||        |                ||
||     10.0.0.1    |       |                 |       |      ||        |                ||
||   192.168.0.1   |       |                 |       |      ||        |                ||
=====================================================================================

PROBLEM FOUND!
In pf the following rule schema is not expressible!
==============================================================================
| sIp   | dIp   ||   tr_sIp     : tr_sPort  |    tr_dIp      : tr_dPort  |
==============================================================================
| Self  | Self  || SNAT ( Self) : id        | DNAT (~Self) : id          |
==============================================================================
Hence the following is impossible to achieve:
=====================================================================================================
||     sIp     | sPort |       dIp       | dPort | prot ||        tr_src        |         tr_dst         ||
=====================================================================================================
|| 192.168.0.1 |   *   |    127.0.0.1    |  123  | udp  || 151.15.185.183 : -   | 193.204.114.232 : -    ||
||             |       | 151.15.185.183  |       |      ||                      |                        ||
||             |       |     10.0.0.1    |       |      ||                      |                        ||
||             |       |   192.168.0.1   |       |      ||                      |                        ||
=====================================================================================================

PROBLEM FOUND!
In pf the following rule schema is not expressible!
==============================================================================
| sIp   | dIp   ||   tr_sIp     : tr_sPort  |    tr_dIp      : tr_dPort  |
==============================================================================
| Self  | ~Self || SNAT ( Self) : id        | DNAT (~Self) : id          |
==============================================================================
Hence the following is impossible to achieve:
=====================================================================================================================
||     sIp     | sPort |                dIp                | dPort | prot ||        tr_src        |         tr_dst         ||
=====================================================================================================================
|| 192.168.0.1 |   *   |       0.0.0.0 - 10.0.0.0          |  123  | udp  || 151.15.185.183 : -   | 193.204.114.232 : -    ||
||             |       |       10.0.0.2 - 127.0.0.0        |       |      ||                      |                        ||
||             |       |    127.0.0.2 - 151.15.185.182     |       |      ||                      |                        ||
||             |       |   151.15.185.184 - 192.168.0.0    |       |      ||                      |                        ||
||             |       |  192.168.0.2 - 255.255.255.255    |       |      ||                      |                        ||
=====================================================================================================================
```

Fig. 5. Checking the portability of the configuration in `pf`.

where `sIp` and `dIp` stand for source and destination IP address, `sPort` and `dPort` for their ports, `prot` for the protocol, `tr_scr`, `tr_dst` for the transformations applied (for brevity the IP address and the port are separated by a ":").

Starting from this tabular representation, F2F checks if the same transformations can be obtained by a configuration of the target system. Roughly, to do that, it visits the target control diagram and verifies if it is possible to apply the relevant transformations in the nodes of the paths followed by a packet. If this is not possible, then the counterexamples are reported in tabular form.

Back to our example, Figs. 4 and 5 display the output of F2F on the configuration in Fig. 3 when the targets are `ipfw` and `pf`, respectively. The parameters of the tool are the source firewall system (here `iptables`); a file with the firewall interfaces and their addresses (`interfaces`); the input configuration (`iptables.conf`); and the target system. Some problems are detected because of a different expressivity power of `iptables` and `ipfw`, and thus porting this configuration cannot preserve the semantics. For each of them the tool displays the reason why, and the rules of the source configuration that cannot be ported, in tabular form.

The first problem with `ipfw` arises when (i) a packet has source and destination addresses of the firewall; (ii) the source is transformed by `SNAT` to an address of the firewall itself (`Self`); (iii) the destination is transformed `DNAT` to an address not of the firewall (`~Self`).The reason is that `ipfw` cannot apply both transformations in this case.The second table lists those packets that cannot be correctly handled. For example, the packet with source IP 192.168.0.1 and any source port, with destination IP 10.0.0.1 and port 123, and protocol UDP cannot be translated to a packet with source 151.15.185.183 and destination 193.204.114.232, with no changes to the ports.

Also the second problem with `ipfw` arises because of a pair `SNAT` and `DNAT` on a packet generated by the firewall but this time with another host as destination. E.g., the UDP packet with source 192.168.0.1 : $*$ and destination 151.15.185.184 : 123 shown in Fig. 3 cannot be transformed in the packet with source 151.15.185.183 : $*$ and destination 193.204.114.232 : 123.

Figure 5 shows that the configuration in Fig. 3 cannot be ported to `pf`, as well. Besides the same problems already discussed for `ipfw`, the tool reports in the first subtable that a packet generated by the firewall and directed to one of its interfaces cannot be redirected to another external host, e.g., the remote NTP, by applying a `DNAT`. Note that this case implicitly shows that there is a packet expressible in `ipfw`, but not in `pf`.

4 Why and How It Works

This section describes the internals of our tool that works in two steps. The first extracts the meaning of a configuration in a given language $\mathcal{L}$ as a function from packets to transformations, which has been intuitively represented as a table in Sect. 3. The second step uses the semantics just obtained and checks if the configuration can be expressed in a target language $\mathcal{L}'$.

4.1 The Denotational Semantics of Configurations in a Nutshell

Given a configuration in $\mathcal{L}$, FWS decompiles it into a set of rules in IFCL, associated with the relevant control diagram of $\mathcal{L}$ [4].

This intermediate representation of the configuration is then associated with a function $\mathcal{F}$ that maps packets into (the sets of) transformations they are subject to while the kernel of the operating system processes them. (Recall that packets belonging to established connections are not treated here, since they are accepted by default and usually never translated[5]). A *transformation* can be the discard

[5] Actually, some translations may occur, typically `SNAT`, but these are performed by other components of the operating system at run-time.

of a packet (represented by $\lambda_\perp$) or a specification of what happens to each of its fields: either it is left unchanged (id) or it is transformed by NAT to a specific value a (λ_a). The semantics of a single ruleset and then of an entire IFCL firewall are defined in [5] as the composition of those functions, collected while packets traverse the paths of the control diagram. In Sect. 2 the meaning of a firewall configuration is expressed as a table, which is a succinct representation of its denotational semantics, where each row corresponds to an equivalence class of packets associated to the same transformation.

For simplifying the treatment of the expressivity of firewall languages it is convenient to annotate each node in a control diagram with the kind of transformations that are allowed in that node. This is because in the translation to IFCL some types of rules cannot be associated with all the nodes. For example, in `pf` rules containing the DROP target can only be associated with nodes q_1 and q_3. Graphically, in Fig. 1, we include these annotations, or *capability labels*, in boxes, and when the only allowed transformation is the identity `ID`, we omit the box. We dispense the actual reasons of the exact association of nodes with labels in the three control diagrams,[6] because not necessary for the present treatment; the interest reader can find the complete presentation of them in [5].

4.2 Checking Portability of Configurations

By exploiting the semantic function $\mathcal{F}$ and the control diagram of the target language $\mathcal{L}'$, we check if a configuration having the same semantics is expressible in $\mathcal{L}'$. The underlying idea of the checking algorithm follows. Given a packet p, we follow the path in the control diagram of $\mathcal{L}'$ from the initial node to the final one, if p is accepted, or to the node that drops p. Along the path, we collect the list of the capabilities T_j associated with each node q_j, i.e. the transformations that can be applied when the control reaches q and graphically included in boxes. Now, let $\mathcal{F}(p)$ be the sequence of transformations $t_1, \cdots, t_n$ that transform p to $\tilde{p}$. If for all the packets and for all j it is $t_j \in T_j$, then the configuration can be ported from $\mathcal{L}$ to $\mathcal{L}'$ (recall that t_j can be the identity).

Consider again the UDP packet p with source IP 192.168.0.1 : $*$ and destination IP 10.0.0.1 : 123. As discussed in Sect. 3, it is accepted by the configuration in Fig. 3, traversing the following path in the control diagram of `iptables`:

$$q_i \rightarrow q_7 \rightarrow q_8 \rightarrow q_9 \rightarrow q_{10} \rightarrow q_{11} \rightarrow q_f$$

where node q_8 transforms 10.0.0.1:123 to 193.204.114.232: 123 by a `DNAT`, and the node q_{11} transforms 192.168.0.1 : $*$ to 151.15.185.183 : $*$, the interface towards the Internet. Indeed, `DNAT` is a capability of q_8 and `SNAT` of q_{11}.

Instead, p can only follow the two following paths in the control diagram of `ipfw`, attempting to obtain the same behavior of `iptables`:

[6] As a matter of fact, it is not `ipfw` that actually translates address, but it demands this task to other lower level components, possibly to the operating system kernel itself. For the sake of generality, we have only modelled such calls, because the actual translations heavily depend on the specific setting of the system hosting the firewall.

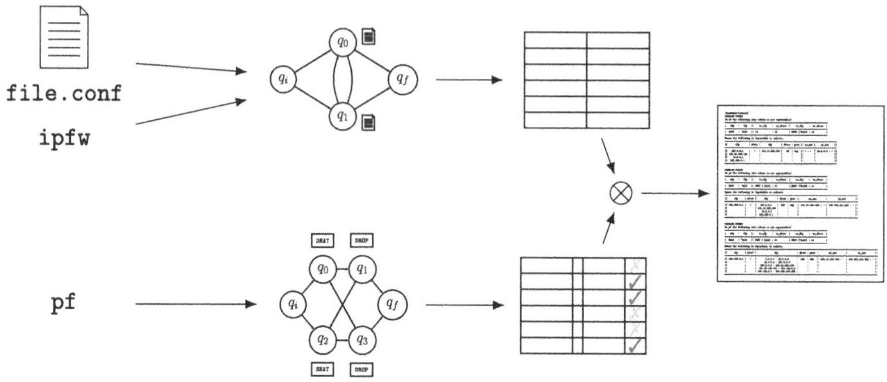

Fig. 6. How F2F checks the portability of an `ipfw` configuration (`file.conf` in the figure) into `pf`.

$$q_i \rightarrow q_1 \rightarrow q_f \qquad\qquad q_i \rightarrow q_1 \rightarrow q_0 \rightarrow q_f$$

We have the following two cases

1. in the leftmost path a `DNAT` occurs in q_1 and p is accepted with no source address transformation applied, violating the intended semantics;
2. in the rightmost path an `SNAT` applies, then in q_0 we do not apply the `DNAT` because otherwise the predicate of the arc leading to q_f would be falsified. If instead we apply `DNAT` in q_0, the control moves back to q_1 and a loop will be detected and the packet will be dropped.

Consider now p' with with source IP 192.168.0.1 : $*$ and destination IP 151.15.185.182 : 123 that is transformed to one with source 151.15.185.183 : $*$ and destination 193.204.114.232 : 123 (of the NTP server) along the following accepting path in `iptables` (the same just seen above):

$$q_i \rightarrow q_7 \rightarrow q_8 \rightarrow q_9 \rightarrow q_{10} \rightarrow q_{11} \rightarrow q_f$$

In the control diagram of `pf` there only is one path that p' can follow:

$$q_i \rightarrow q_0 \rightarrow q_1 \rightarrow q_f$$

Note that while the source address can be transformed by node q_0, it is not possible to perform `DNAT` because the only node with capability `DNAT` is q_2, which is not reachable since the predicate of the arc requires the packet destination to be an interface of the firewall. Hence when the packet is accepted by the final node its fields do not contain the expected values.

Given a control diagram with its capability labels, the algorithm of [5] returns a table in which each row represents both the expressible and the inexpressible transformations for a given class of packets. The procedure for building these table is similar to the one sketched above for checking a single packet p. Rather than on single packets, our algorithm actually works on (a representative of the) equivalence classes of packets that cover every possible behaviour of the firewall in hand. The tool F2F matches this expressivity table with the one representing $\mathfrak{F}$ and computed by FWS, looking for clashes. In Fig. 6 we recapitulate the steps needed for evaluating the portability of a configuration: (i) first translate the source policy into IFCL; use the control diagram of the source language (with its capability labels) and perform a syntactic translation; and then compute the semantic function $\mathfrak{F}$ in a tabular form (top of the figure); (ii) similarly, take the control diagram of the target language with its capability labels, and derive the table representing its expressive power (bottom of the figure); (iii) finally compare the two tables to produce the final result (right part of the figure).

5 Conclusions

We have briefly introduced the pipeline of [3,4] that supports system administrators in better understanding, in analysing and maintaining a firewall configuration, and in porting it from one firewall configuration language to another. It is folklore that firewall configuration languages only differ in pragmatic aspects and in syntactic sugar. To our surprise, we discovered that this is not the case, and we proved in [5] that the most used ones, `iptables`, `ipfw` and `pf`, form a hierarchy of expressivity.

We have described F2F, a tool available online[7] that checks if a configuration is expressible in a given language. We have also discussed some revealing cases in which this is not possible, by inspecting the outputs of F2F. This results have the form of a table showing the shape of the packets that can be filtered and redirected in one language, but not in another. Here, we have mainly concentrated on the form of the source and destination addresses before and after possible translations. Although in a preliminary version, our tool can be used efficiently, because it is a matter of seconds checking the expressivity of a medium size configuration.

Network administrators can use F2F to choose the firewall system more appropriate to the current situation, or at least to evaluate if their preferred one, as well as which of the available tools are the more suited for implementing the needed configuration. Also, a system administrator can overcome the detected limitation of the firewall systems in use and patch the configuration in hand in the very troubling points, by resorting to calls to procedures in some programming language.

Future work will consider different firewall systems, like Cisco-IOS, which is particularly challenging because the control diagram is also affected by routing choices. We plan to involve systems administrators in experimenting F2F on real configurations, and get feedback from them. This of course requires to improve the interface of F2F.

[7] https://github.com/lceragioli/F2F.

References

1. Adão, P., Bozzato, C., Rossi, G.D., Focardi, R., Luccio, F.L.: Mignis: a semantic based tool for firewall configuration. In: IEEE 27th Computer Security Foundations Symposium, CSF 2014, pp. 351–365 (2014)
2. Adão, P., Focardi, R., Guttman, J.D., Luccio, F.L.: Localizing firewall security policies. In: Proceedings of the 29th IEEE CSF, Lisbon, Portugal, 27 June–1 July 2016, pp. 194–209 (2016)
3. Bodei, C., Degano, P., Focardi, R., Galletta, L., Tempesta, M.: Transcompiling firewalls. In: Bauer, L., Küsters, R. (eds.) POST 2018. LNCS, vol. 10804, pp. 303–324. Springer, Cham (2018). https://doi.org/10.1007/978-3-319-89722-6_13
4. Bodei, C., Degano, P., Focardi, R., Galletta, L., Tempesta, M., Veronese, L.: Language-independent synthesis of firewall policies. In: Proceedings of the 3rd IEEE European Symposium on Security and Privacy (2018)
5. Ceragioli, L., Degano, P., Galletta, L.: Are All Firewall Systems Equally Powerful? Submitted for publication. https://sysma.imtlucca.it/wp-content/uploads/2019/03/firewall-expressivity.pdf
6. Ceragioli, L., Galletta, L., Tempesta, M.: From firewalls to functions and back. In: Italian Conference on Cybersecurity ITASEC 2019. CEUR Proceedings, vol. 2315 (2019). http://ceur-ws.org/Vol-2315/paper04.pdf
7. Cuppens, F., Cuppens-Boulahia, N., Sans, T., Miège, A.: A formal approach to specify and deploy a network security policy. In: Dimitrakos, T., Martinelli, F. (eds.) Formal Aspects in Security and Trust. IIFIP, vol. 173, pp. 203–218. Springer, Boston, MA (2005). https://doi.org/10.1007/0-387-24098-5_15
8. Diekmann, C., Hupel, L., Michaelis, J., Haslbeck, M.P.L., Carle, G.: Verified iptables firewall analysis and verification. J. Autom. Reason. **61**(1–4), 191–242 (2018)
9. Foley, S.N., Neville, U.: A firewall algebra for openstack. In: 2015 IEEE Conference on Communications and Network Security, CNS 2015, pp. 541–549 (2015)
10. Martínez, S., Cabot, J., Garcia-Alfaro, J., Cuppens, F., Cuppens-Boulahia, N.: A model-driven approach for the extraction of network access-control policies. In: Proceedings of the MDSec 2012, pp. 5:1–5:6. ACM (2012)
11. Packet Filter (PF). https://www.openbsd.org/faq/pf/
12. Russell, R.: Linux 2.4 packet filtering HOWTO (2002). http://www.netfilter.org/documentation/HOWTO/packet-filtering-HOWTO.html
13. The IPFW Firewall. https://www.freebsd.org/doc/handbook/firewalls-ipfw.html
14. The Netfilter Project. https://www.netfilter.org/

Polymorphic Session Processes as Morphisms

Bernardo Toninho[1(✉)] and Nobuko Yoshida[2]

[1] Universidade Nova de Lisboa and NOVA-LINCS, Lisbon, Portugal
btoninho@fct.unl.pt
[2] Imperial College London, London, UK

Abstract. The study of expressiveness of concurrent processes via session types opens a connection between linear logic and mobile processes, grounded in the rigorous logical background of propositions-as-types. One such study includes a notion of parametric session polymorphism, which connects session typed processes with rich higher-order functional computations. This work proposes a novel and non-trivial application of session parametricity – an encoding of inductive and coinductive session types, justified via the theory of initial algebras and final co-algebras using a processes-as-morphisms viewpoint. The correctness of the encoding (i.e. universality) relies crucially on parametricity and the associated relational lifting of sessions.

Keywords: Expressiveness · Session types · π-calculus · F-Algebra · Linear logic

1 Introduction

The study of expressiveness of the π-calculus by Palamidessi [19,20] opened a new field of process calculi, linking the areas of distributed computation and distributed algorithms with the π-calculus. Around the same period, *session types* [13] were introduced as a typing discipline that is able to ensure communication safety and deadlock-freedom of communication protocols among two or more parties [14]. In particular, a tight connection between session types and linear logic (a *propositons-as-session types* correspondence) [6,7] has produced several new developments and logically-motivated techniques [15,27,32] to augment both the theory and practice of session-based message-passing concurrency. Notably, parametric session polymorphism [5] (in the sense of Reynolds [24]) has been proposed and a corresponding Abstraction theorem has been established. Despite its practical significance [22], the expressiveness of parametric session polymorphism has been relatively unexplored.

In this paper, we study the expressiveness of parametric session polymorphism in a logical setting and the induced parametric equivalence on session-typed processes. More precisely, we study a notion of *processes-as-morphisms* and *session type operators-as-functors* (in the categorical sense) to develop an

M. S. Alvim et al. (Eds.): Palamidessi Festschrift, LNCS 11760, pp. 101–117, 2019.
https://doi.org/10.1007/978-3-030-31175-9_7

encoding of inductive and coinductive session types in the style of initial algebras and final coalgebras in System F [1,11,23,31] but in a process setting, introducing new interesting reasoning techniques such as a relational lifting of sessions, crucial for the correctness of the encoding of coinductive session types.

To develop the encoding of inductive and coinductive types (Sects. 3.3 and 3.4) we introduce the notion of a functorial action on typed processes (Sect. 3.2), showing it satisfies functoriality up-to parametric equivalence (Theorem 3.3), and enabling us to represent the necessary morphisms as session-typed processes (Theorems 3.7 and 3.9). Crucially, to show the universality of our encoding, i.e. the uniqueness clauses of initiality (Theorem 3.8) and finality (Theorem 3.12), we rely on fundamental properties of parametricity applied to a process setting, such as relational liftings of sessions.

Our results are yet another testament to the strength of propositions-as-sessions: by relying on a rigorous logical background we are able to produce rigorous results such as universality in a session-typed process setting using elegant logical techniques, which give rise to novel reasoning techniques in a process setting such as relational liftings (used in the arguably non-trivial proof of Lemma 3.10). Just as the categorical view point reveals new approaches to functional programming, this work aims to be a stepping stone towards a general notion of algebraic programming [2,3] in a concurrent setting, exploiting the algebraic structures that naturally arise via propositions-as-sessions in order to provide new abstraction mechanisms for concurrent programming. By barely scratching the surface, we can already study a concurrent notion of functor, initial algebra and final co-algebra. The study of further abstraction techniques such as natural transformations is the obvious continuation of this work.

Contributions. We summarise the key contributions of this work:

- We develop a notion of functorial action of a type operator on session-typed processes, satisfying the necessary functor laws;
- We study encodings of inductive session types as initial algebras (Sect. 3.3) of a given type operator and, dually, coinductive session types as final coalgebras (Sect. 3.4).
- Using parametricity, we show our encodings satisfy the necessary initiality (Theorems 3.7 and 3.8) and finality (Theorems 3.9 and 3.12) properties, and are thus correct wrt the semantics of inductive and coinductive types.

We conclude with discussion of related work (Sect. 4). The technical report [30] includes the omitted definitions and detailed proofs.

2 Polymorphic Session π-Calculus

This section summarises the polymorphic session π-calculus dubbed Polyπ [5], arising as a process assignment to second-order linear logic [9], its typing system and behavioural equivalences.

The calculus is a synchronous π-calculus with binary guarded choice, input-guarded replication, channel links and prefixes for type input and output.

2.1 Processes and Typing

Definition 2.1 (Process and Type Syntax). Syntax. *Given an infinite set Λ of names x, y, z, u, v, the grammar of processes P, Q, R and session types A, B, C is defined by:*

$$\begin{array}{rl} P, Q, R ::= & x\langle y\rangle.P \mid x(y).P \mid P \mid Q \mid (\nu y)P \mid [x \leftrightarrow y] \mid \mathbf{0} \\ \mid & x\langle A\rangle.P \mid x(Y).P \mid x.\mathsf{inl}; P \mid x.\mathsf{inr}; P \mid x.\mathsf{case}(P, Q) \mid !x(y).P \\ A, B ::= & \mathbf{1} \mid A \multimap B \mid A \otimes B \mid A \,\&\, B \mid A \oplus B \mid !A \mid \forall X.A \mid \exists X.A \mid X \end{array}$$

$x\langle y\rangle.P$ denotes the output of channel y on x with continuation process P; $x(y).P$ denotes an input along x, bound to y in P; $P \mid Q$ denotes parallel composition; $(\nu y)P$ denotes the restriction of name y to the scope of P; $\mathbf{0}$ denotes the inactive process; $[x \leftrightarrow y]$ denotes the linking of the two channels x and y (implemented as renaming); $x\langle A\rangle.P$ and $x(Y).P$ denote the sending and receiving of a *type* A along x bound to Y in P of the receiver process; $x.\mathsf{inl}; P$ and $x.\mathsf{inr}; P$ denote the emission of a selection between the left or right branch of a receiver $x.\mathsf{case}(P, Q)$ process; $!x(y).P$ denotes an input-guarded replication, that spawns replicas upon receiving an input along x. We often abbreviate $(\nu y)x\langle y\rangle.P$ to $\overline{x}\langle y\rangle.P$ and omit trailing $\mathbf{0}$ processes. By convention, we range over linear channels with x, y, z and shared channels with u, v, w. We write $P_{x,y}$ to single out that x and y are free in P.

Session types denote the communication behaviour that takes place along a given channel between communicating processes. Our syntax of session types is that of (intuitionistic) linear logic propositions which are assigned to channels according to their usages in processes: $\mathbf{1}$ denotes the type of a channel along which no further behaviour occurs; $A \multimap B$ denotes a session that waits to receive a channel of type A and will then proceed as a session of type B; dually, $A \otimes B$ denotes a session that sends a channel of type A and continues as B; $A \,\&\, B$ denotes a session that offers a choice between proceeding as behaviours A or B; $A \oplus B$ denotes a session that internally chooses to continue as either A or B, signaling appropriately to the communicating partner; $!A$ denotes a session offering an unbounded (but finite) number of behaviours of type A; $\forall X.A$ denotes a polymorphic session that receives a type B and behaves uniformly as $A\{B/X\}$; dually, $\exists X.A$ denotes an existentially typed session, which emits a type B and behaves as $A\{B/X\}$.

Operational Semantics. The operational semantics of our calculus are presented as a standard labelled transition system (Fig. 1) modulo $\equiv_\alpha$ congruence, in the style of the *early* system for the π-calculus [26]. In the remainder of this work we write $\equiv$ for the standard π-calculus structural congruence, extended with the clause $[x \leftrightarrow y] \equiv [y \leftrightarrow x]$. We write $\equiv_!$ for structural congruence extended with the so-called sharpened replication axioms [26].

A transition $P \xrightarrow{\alpha} Q$ denotes that P may evolve to Q by performing the action represented by label α. An action α ($\overline{\alpha}$) requires a matching $\overline{\alpha}$ (α) in the environment to enable progress. Labels of our transition semantics include: the silent internal action τ, output and bound output actions ($\overline{x\langle y\rangle}$ and $\overline{(\nu z)x\langle z\rangle}$);

input action $x(y)$; labels for pertaining to the binary choice actions ($x.\mathsf{inl}$, $\overline{x.\mathsf{inl}}$, $x.\mathsf{inr}$, and $\overline{x.\mathsf{inr}}$); and output and input actions of types ($\overline{x\langle A\rangle}$ and $x(A)$).

Definition 2.2 (Labeled Transition System). The relation *labeled transition* ($P \xrightarrow{\alpha} Q$) is defined by the rules in Fig. 1, subject to the side conditions: in rule (res), we require $y \notin fn(\alpha)$; in rule (par), we require $bn(\alpha) \cap fn(R) = \emptyset$; in rule (close), we require $y \notin fn(Q)$. We omit the symmetric versions of rules (par), (com), (lout), (lin), (close) and closure under α-conversion.

We write $\rho_1\rho_2$ for the composition of relations ρ_1, ρ_2. We write $\rightarrow$ to stand for $\xrightarrow{\tau}\equiv$. Weak transitions are defined as usual: we write $\Longrightarrow$ for the reflexive, transitive closure of $\xrightarrow{\tau}$. Given $\alpha \neq \tau$, notation $\stackrel{\alpha}{\Longrightarrow}$ stands for $\Longrightarrow\xrightarrow{\alpha}\Longrightarrow$ and $\stackrel{\tau}{\Longrightarrow}$ stands for $\Longrightarrow$.

(out) $\quad x\langle y\rangle.P \xrightarrow{\overline{x\langle y\rangle}} P$

(in) $\quad x(y).P \xrightarrow{x(z)} P\{z/y\}$

(outT) $\quad x\langle A\rangle.P \xrightarrow{\overline{x\langle A\rangle}} P$

(inT) $\quad x(Y).P \xrightarrow{x(B)} P\{B/Y\}$

(lout) $\quad x.\mathtt{inl}; P \xrightarrow{\overline{x.\mathsf{inl}}} P$

(id) $\quad (\nu x)([x \leftrightarrow y] \mid P) \xrightarrow{\tau} P\{y/x\}$

(lin) $\quad x.\mathtt{case}(P, Q) \xrightarrow{x.\mathsf{inl}} P$

(rep) $\quad !x(y).P \xrightarrow{x(z)} P\{z/y\} \mid !x(y).P$

(open) $$\frac{P \xrightarrow{\overline{x\langle y\rangle}} Q}{(\nu y)P \xrightarrow{\overline{(\nu y)x\langle y\rangle}} Q}$$

(close) $$\frac{P \xrightarrow{\overline{(\nu y)x\langle y\rangle}} P' \quad Q \xrightarrow{x(y)} Q'}{P \mid Q \xrightarrow{\tau} (\nu y)(P' \mid Q')}$$

(par) $$\frac{P \xrightarrow{\alpha} Q}{P \mid R \xrightarrow{\alpha} Q \mid R}$$

(com) $$\frac{P \xrightarrow{\overline{\alpha}} P' \quad Q \xrightarrow{\alpha} Q'}{P \mid Q \xrightarrow{\tau} P' \mid Q'}$$

(res) $$\frac{P \xrightarrow{\alpha} Q}{(\nu y)P \xrightarrow{\alpha} (\nu y)Q}$$

Fig. 1. π-calculus labeled transition system.

Typing System. The typing system $\Omega; \Gamma; \Delta \vdash P :: z{:}A$ is given in Fig. 2. The judgement means that process P offers a session of type A along channel z, using the *linear* sessions in Δ, (potentially) using the unrestricted or *shared* sessions in Γ, with polymorphic type variables maintained in Ω. We use a well-formedness judgment $\Omega \vdash A\,\mathsf{type}$ which states that A is well-formed wrt the type variable environment Ω (i.e. $fv(A) \subseteq \Omega$). The rules are defined up-to structural congruence $\equiv$. We often write T for the right-hand side typing $z{:}A$, $\cdot$ for the empty context and Δ, Δ' for the union of contexts Δ and Δ', only defined when Δ and Δ' are disjoint. We write $\cdot \vdash P :: T$ to denote that P is *closed*.

As in [6,7,21,32], the typing discipline enforces that outputs always have as object a *fresh* name, in the style of the internal mobility π-calculus [25]. The typing rules of Fig. 2 can be divided into three classes: *right rules* (marked with R), which explicate how a process can *offer* a session of a given type; *left rules* (marked with L), which define how a process can *use* or interact with a session of a given type; and *judgmental rules* which define basic logical principles such as composition of proofs (rule cut) and identity (rule id). Rule $\forall$R defines the

(id)
$$\frac{}{\Omega;\Gamma;x{:}A \vdash [x \leftrightarrow z] :: z{:}A}$$

(cut)
$$\frac{\Omega;\Gamma;\Delta_1 \vdash P :: x{:}A \quad \Omega;\Gamma;\Delta_2, x{:}A \vdash Q :: z{:}C}{\Omega;\Gamma;\Delta_1,\Delta_2 \vdash (\nu x)(P \mid Q) :: z{:}C}$$

($\multimap$L)
$$\frac{\Omega;\Gamma;\Delta_1 \vdash P :: y{:}A \quad \Omega;\Gamma;\Delta_2, x{:}B \vdash Q :: z{:}C}{\Omega;\Gamma;\Delta_1,\Delta_2, x{:}A \multimap B \vdash (\nu y)x\langle y\rangle.(P \mid Q) :: z{:}C}$$

($\forall$R)
$$\frac{\Omega, X;\Gamma;\Delta \vdash P :: z{:}A}{\Omega;\Gamma;\Delta \vdash z(X).P :: z{:}\forall X.A}$$

($\forall$L)
$$\frac{\Omega \vdash B \quad \Omega;\Gamma;\Delta, x{:}A\{B/X\} \vdash P :: z{:}C}{\Omega;\Gamma;\Delta, x{:}\forall X.A \vdash x\langle B\rangle.P :: z{:}C}$$

($\exists$R)
$$\frac{\Omega \vdash B \quad \Omega;\Gamma;\Delta \vdash P :: z{:}A\{B/X\}}{\Omega;\Gamma;\Delta \vdash z\langle B\rangle.P :: z{:}\exists X.A}$$

($\exists$L)
$$\frac{\Omega, X;\Gamma;\Delta, x{:}A \vdash P :: z{:}C}{\Omega;\Gamma;\Delta, x{:}\exists X.A \vdash x(X).P :: z{:}C}$$

($\multimap$R)
$$\frac{\Omega;\Gamma;\Delta, x{:}A \vdash P :: z{:}B}{\Omega;\Gamma;\Delta \vdash z(x).P :: z{:}A \multimap B}$$

Fig. 2. Type system (selected rules)

meaning of (impredicative) universal quantification over session types, stating that a session of type $\forall X.A$ inputs a type and then behaves uniformly as A; dually, to use such a session (rule $\forall$L), a process must output a type B which then warrants the use of the session as type $A\{B/X\}$. Rule $\multimap$R types session input, where a session of type $A \multimap B$ inputs a session of type A which is used to produce a session of type B; dually, one uses such a session (rule $\multimap$L) by producing a fresh session of type A (that uses a disjoint set of sessions to those of the continuation) and outputting the fresh session along x, which is then used as a type B.

The typing system ensures strong correctness properties on typed processes. As shown in [5], typing entails Subject Reduction, Global Progress, and Termination.

2.2 Observational Equivalences

We briefly summarise the typed congruence and logical equivalence with polymorphism, giving rise to a suitable notion of relational parametricity in the sense of Reynolds [24], defined as a contextual logical relation on typed processes [5]. The logical relation is reminiscent of a typed bisimulation. However, extra care is needed to ensure well-foundedness due to impredicative type instantiation. As a consequence, the logical relation allows us to reason about process equivalences where type variables are not necessarily instantiated with *the same*, but rather *related* types.

Typed Barbed Congruence ($\cong$). We use the typed contextual congruence from [5], which preserves *observable* actions, called barbs. Formally, *barbed congruence*, noted $\cong$, is the largest equivalence on well-typed processes that is τ-closed, barb preserving, and contextually closed under typed contexts; see [5] for the full definition.

Logical Equivalence. The definition of logical equivalence is no more than a typed contextual bisimulation with the following reading: given two open processes P and Q (i.e. processes with non-empty left-hand side typings), we define their equivalence by inductively closing out the context, composing with equivalent processes offering appropriately typed sessions; when processes are closed, we have a single distinguished session channel along which we can perform observations, and proceed inductively on the structure of the specified session type. We can then show that such an equivalence satisfies the necessary fundamental properties (Theorem 2.4).

Formally, the logical relation is defined using the candidates technique of Girard [8,10]. In this setting, an *equivalence candidate* is an equivalence relation on well-typed processes satisfying basic closure conditions: an equivalence candidate must be compatible with barbed congruence and closed under forward and backward reduction. We write $\mathcal{R} :: z{:}A{\Leftrightarrow}B$ for such a candidate relation, such that $(P, Q) \in \mathcal{R} :: z{:}A{\Leftrightarrow}B$ further requires that $\cdot \vdash P :: z{:}A$ and $\cdot \vdash Q :: z{:}B$.

To define the logical relation we rely on some auxiliary notation. We write $\omega : \Omega$ to denote that a type substitution ω assigns a closed type to the type variables in Ω. Given two substitutions $\omega : \Omega$ and $\omega' : \Omega$, we define a candidate assignment η between ω and ω' as a mapping of candidate $\eta(X) :: -{:}\omega(X){\Leftrightarrow}\omega'(X)$ to the type variables in Ω, where the particular choice of a distinguished right-hand side channel is *delayed* (i.e. instantiated later on). We write $\eta(X)(z)$ for the instantiation of the (delayed) candidate with the name z. We write $\eta : \omega{\Leftrightarrow}\omega'$ to denote that η is a candidate assignment between ω and ω'; and $\hat{\omega}(P)$ to denote the application of substitution ω to P.

We define a sequent-indexed family of process relations, that is, a set of pairs of processes (P, Q), written $\Gamma; \Delta \vdash P \approx_{\mathrm{L}} Q :: T[\eta : \omega{\Leftrightarrow}\omega']$, satisfying some conditions, is assigned to sequents of the form $\Omega; \Gamma; \Delta \vdash T$, with $\omega : \Omega$, $\omega' : \Omega$ and $\eta : \omega{\Leftrightarrow}\omega'$. Logical equivalence is defined inductively on the size of the typing contexts and then on the structure of the right-hand side type.

Definition 2.3 (Logical Equivalence). ***(Base Case)*** *Given a type A and mappings ω, ω', η, we define* logical equivalence, *noted $P \approx_L Q :: z{:}A[\eta : \omega{\Leftrightarrow}\omega']$, as the smallest binary relation containing all pairs of processes (P, Q) such that (i) $\cdot \vdash \hat{\omega}(P) :: z{:}\hat{\omega}(A)$; (ii) $\cdot \vdash \hat{\omega}'(Q) :: z{:}\hat{\omega}'(A)$; and (iii) satisfies the conditions given below:*

$$P \approx_L Q :: z{:}X[\eta : \omega{\Leftrightarrow}\omega'] \ \textit{iff}\ (P, Q) \in \eta(X)(z)$$

$$P \approx_L Q :: z{:}\forall X.A[\eta : \omega{\Leftrightarrow}\omega'] \ \textit{iff}\ \forall B_1, B_2, P', \mathcal{R} :: -{:}B_1{\Leftrightarrow}B_2.\ (P \xrightarrow{z(B_1)} P')\ \textit{implies}$$
$$\exists Q'.Q \stackrel{z(B_2)}{\Longrightarrow} Q',\ P' \approx_L Q' :: z{:}A[\eta[X \mapsto \mathcal{R}] : \omega[X \mapsto B_1]{\Leftrightarrow}\omega'[X \mapsto B_2]]$$

$$P \approx_L Q :: z{:}\exists X.A[\eta : \omega{\Leftrightarrow}\omega'] \ \textit{iff}\ \exists B_1, B_2, \mathcal{R} :: -{:}B_1{\Leftrightarrow}B_2.\ (P \xrightarrow{\overline{z\langle B_1\rangle}} P')\ \textit{implies}$$
$$\exists Q'.Q \stackrel{\overline{z\langle B_2\rangle}}{\Longrightarrow} Q',\ P' \approx_L Q' :: z{:}A[\eta[X \mapsto \mathcal{R}] : \omega[X \mapsto B_1]{\Leftrightarrow}\omega'[X \mapsto B_2]]$$

(Inductive Case) *Let Γ, Δ be non empty. Given $\Omega; \Gamma; \Delta \vdash P :: T$ and $\Omega; \Gamma; \Delta \vdash Q :: T$, the binary relation on processes $\Gamma; \Delta \vdash P \approx_L Q :: T[\eta : \omega{\Leftrightarrow}\omega']$ (with $\omega, \omega' : \Omega$ and $\eta : \omega{\Leftrightarrow}\omega'$) is inductively defined as:*

$$\Gamma; \Delta, y : A \vdash P \approx_L Q :: T[\eta : \omega \Leftrightarrow \omega'] \text{ iff } \forall R_1, R_2.\ s.t.\ R_1 \approx_L R_2 :: y{:}A[\eta : \omega \Leftrightarrow \omega'],$$
$$\Gamma; \Delta \vdash (\nu y)(\hat{\omega}(P) \mid \hat{\omega}(R_1)) \approx_L (\nu y)(\hat{\omega}'(Q) \mid \hat{\omega}'(R_2)) :: T[\eta : \omega \Leftrightarrow \omega']$$
$$\Gamma, u : A; \Delta \vdash P \approx_L Q :: T[\eta : \omega \Leftrightarrow \omega'] \text{ iff } \forall R_1, R_2.\ s.t.\ R_1 \approx_L R_2 :: y{:}A[\eta : \omega \Leftrightarrow \omega'],$$
$$\Gamma; \Delta \vdash (\nu u)(\hat{\omega}(P) \mid !u(y).\hat{\omega}(R_1)) \approx_L (\nu u)(\hat{\omega}'(Q) \mid !u(y).\hat{\omega}'(R_2)) :: T[\eta : \omega \Leftrightarrow \omega']$$

For the sake of readability we often omit the $\eta : \omega \Leftrightarrow \omega'$ portion of $\approx_{\mathrm{L}}$, which is henceforth implicitly universally quantified. Thus, we write $\Omega; \Gamma; \Delta \vdash P \approx_{\mathrm{L}} Q :: z{:}A$ (or $P \approx_{\mathrm{L}} Q$) iff the two given processes are logically equivalent for all consistent instantiations of its type variables.

It is instructive to inspect the clauses for type input and output ($\forall X.A$ and $\exists X.A$, respectively): in the former, the two processes must be able to match inputs of any pair of *related* types (i.e. types related by a candidate), such that the continuations are related at the open type A with the appropriate type variable instantiations, following Girard [10]. Dually, for type output we require the existence of a pair of output types and candidate relation such that the continuation processes are related. The power of this style of logical relation comes from the fact that polymorphic equivalences do not require the same type to be instantiated in both processes, but rather that the types are *related* (via a suitable equivalence candidate relation).

Theorem 2.4 (Properties of Logical Equivalence [5]).
Parametricity: *If* $\Omega; \Gamma; \Delta \vdash P :: z{:}A$ *then, for all* $\omega, \omega' : \Omega$ *and* $\eta : \omega \Leftrightarrow \omega'$, *we have* $\Gamma; \Delta \vdash \hat{\omega}(P) \approx_L \hat{\omega'}(P) :: z{:}A[\eta : \omega \Leftrightarrow \omega']$; **Soundness:** *If* $\Omega; \Gamma; \Delta \vdash P \approx_L Q :: z{:}A$ *then* $\mathcal{C}[P] \cong \mathcal{C}[Q] :: z{:}A$, *for any closing* $\mathcal{C}[-]$; **Completeness:** *If* $\Omega; \Gamma; \Delta \vdash P \cong Q :: z{:}A$ *then* $\Gamma; \Delta \vdash P \approx_L Q :: z{:}A$.

3 F-Algebras in Polymorphic Sessions

This section shows how to interpret diagrams and type operators in Polyπ (Sect. 3.2), which are then used to represent inductive (Sect. 3.3) and coinductive (Sect. 3.4) session types.

In order to present our encoding of inductive and coinductive sessions through session polymorphism (Sects. 3.3 and 3.4), we first summarise the standard interpretation of inductive and coinductive types as algebras of a type operator (Sect. 3.1). We then show how to interpret diagrams, type operators (and their associated functorial action) at the level of the session calculus (Sect. 3.2), which enables us to carry out our main development.

3.1 Inductive and Coinductive Session Types

The study of polymorphism in the λ-calculus [1,4,11,23] has shown that parametric polymorphism is expressive enough to encode both inductive and coinductive types in a very precise sense, via a faithful representation of initial and final (co)algebras [16] which does not require extending the syntax nor the semantics of the calculus.

The polymorphic session calculus can express fairly intricate communication behaviours, including generic protocols through both existential and universal polymorphism (i.e. protocols that are parametric in their sub-protocols). The introduction of recursive behaviours in the logical-based session typing framework has been addressed through the introduction of explicit inductive and coinductive session types [15,28] and the corresponding process constructs, preserving the good properties of the framework such as strong normalisation and absence of deadlocks.

Given the logical foundation of the polymorphic session calculus it is natural to wonder if polymorphic sessions are powerful enough to represent inductive and coinductive behaviours in a systematic way.

Inductive and Coinductive Types in System F. Exploring an algebraic interpretation of polymorphism where types are interpreted as functors, it can be shown that given a type F with a free variable X that occurs only positively (i.e. occurrences of X are on the left-hand side of an even number of function arrows), the polymorphic type $\forall X.((F(X) \to X) \to X)$ forms an initial F-algebra [1] (we write $F(X)$ to denote that X occurs in F).

An algebra of a type operator F is a pairing of a type X and a function $f : F(X) \to X$. For an algebra (X, f) to be initial, it must be the case that for any other algebra (X', f') there must exist a *unique* function $h : X \to X'$ such that $h \circ f = f' \circ F(h)$, where $F(h)$ stands for the functorial action of F applied to h. This enables the representation of *inductively* defined structures using an algebraic or categorical justification. For instance, the natural numbers can be seen as the initial F-algebra of $F(X) = \mathbf{1} + X$ (where $\mathbf{1}$ is the unit type and $+$ is the coproduct type), and are thus *already present* in System F, in a precise sense, as the type $\forall X.((\mathbf{1} + X) \to X) \to X$ (noting that both $\mathbf{1}$ and $+$ can also be faithfully encoded in System F). A similar story can be told for *coinductively* defined structures, which correspond to final F-coalgebras and are representable with the polymorphic type $\exists X.(X \to F(X)) \times X$, where $\times$ is the product type. Final coalgebras consist of a pair (X, f) where X is a type and $f : X \to F(X)$ such that for any other coalgebra (X', f') there exists a *unique* $h : X' \to X$ satisfying $f \circ h = F(h) \circ f'$. In the remainder of this section we assume the positivity requirement on F mentioned above. While the complete formal development of the representation of inductive and coinductive types in System F would lead us to far astray, we summarise here the key concepts as they apply to the λ-calculus (the interested reader can refer to [11] for the full categorical details).

To show that the polymorphic type $T_i \triangleq \forall X.((F(X) \to X) \to X)$ is an initial F-algebra, one exhibits a pair of λ-terms, often dubbed fold and in, such that the diagram in Fig. 3(a) commutes (for any A, where $F(f)$, where f is a λ-term, denotes the functorial action of F applied to f), and, crucially, that fold is *unique*. When these conditions hold, we are justified in saying that T_i is a least fixed point of F. Through a fairly simple calculation, it is easy to see that:

$$\begin{aligned}\mathsf{fold} &\triangleq \Lambda X.\lambda x{:}F(X) \to X.\lambda t{:}T_i.t[X](x)\\ \mathsf{in} &\triangleq \lambda x{:}F(T_i).\Lambda X.\lambda y{:}F(X) \to X.y\,(F(\mathsf{fold}[X](x))(x))\end{aligned}$$

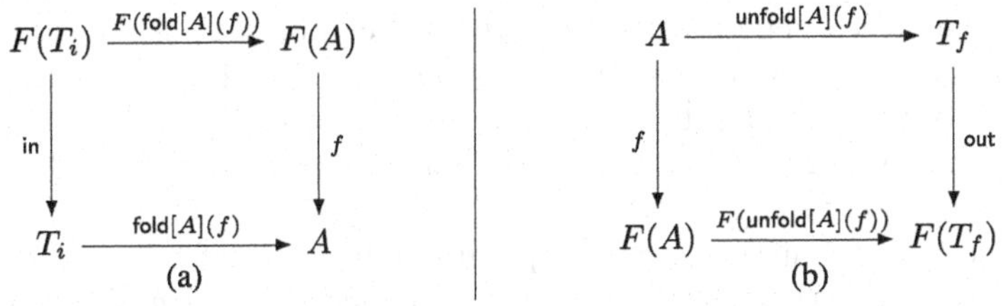

Fig. 3. Diagrams for initial F-algebras and final F-coalgebras

satisfy the necessary equalities. To show uniqueness one appeals to *parametricity*, which allows us to prove that any function of the appropriate type is equivalent to fold. This uniqueness property is often dubbed *initiality* or *universality*.

The construction of final F-coalgebras and their justification as *greatest* fixed points is dual. Assuming the existence of products in the calculus and taking $T_f \triangleq \exists X.(X \to F(X)) \times X$, we produce the λ-terms

$$\begin{aligned} \mathsf{unfold} &\triangleq \Lambda X.\lambda f{:}X \to F(X).\lambda x{:}T_f.\mathsf{pack}\, X\, \mathsf{with}\, (f, x) \\ \mathsf{out} &\triangleq \lambda t : T_f.\mathsf{let}\, (X, (f, x)) = t\, \mathsf{in}\, F(\mathsf{unfold}[X](f))\, (f(x)) \end{aligned}$$

such that the diagram in Fig. 3(b) commutes and unfold is unique up to parametricity.

3.2 Diagrams, Functors and Functorial Action

In order to represent diagrams in our session framework, we have that the nodes are *session types* and the arrows are session-typed *processes* that are able to map a session of a given type to one of another. However, given that typing assigns session types to *channels* and not processes, a session transformer from A to B is parameterised by the channel that offers the A behaviour and the channel along which B is to be offered.

Thus, with a pair of types $x{:}A$ and $z{:}B$ and a process P of type $x{:}A \vdash P :: z{:}B$, we can schematically represent the arrow from $x{:}A$ to $z{:}B$ as $x{:}A \xrightarrow{P} z{:}B$. Given another process Q such that $z{:}B \vdash Q :: y{:}A$, we can represent arrow composition as the diagram of Fig. 4, denoting that we can transform a session $x{:}A$ into $y{:}A$ by either "following the diagram" along the P and Q arrows or we may compose (i.e. cut) the two arrows outright to form $(\nu z)(P \mid Q)$.

We note that the diagram in Fig. 4 commutes up-to structural congruence, since composition of P and Q is structurally equivalent to $(\nu x)(P \mid Q)$. This is in contrast with the diagram in Fig. 5, where we must show that $(\nu z)(P \mid Q) = [x \leftrightarrow y]$, for which $\equiv$ is insufficient, but we can use $\approx_{\mathtt{L}}$ of Sect. 2.2 as $(\nu z)(P \mid Q) \approx_{\mathtt{L}} [x \leftrightarrow y]$. Hence, unless otherwise stated, we assume that equality in diagrams is $\approx_{\mathtt{L}}$.

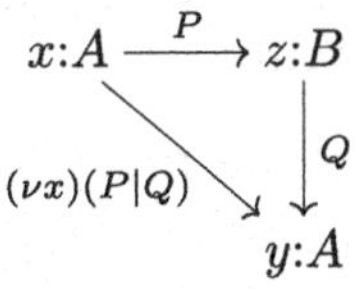

Fig. 4. Composition (up-to $\equiv$)

Type Operators. Type operators, or *functors*, can be seen as transformations on (session) types, in the sense that given some type A and a type operator $X.B$ you can apply the operator to A to produce the type $B\{A/X\}$. For instance, the type operator $X.\mathbf{1} \multimap (X \otimes \mathbf{1})$ can be applied to some well-formed type A to produce the session type $\mathbf{1} \multimap (A \otimes \mathbf{1})$.

As is the case in functional generic programming, a key feature of type operators as functors is that they act as type-directed transformations in a very precise sense: given a type that has the same "shape" of the type operator, with some instantiation of the type variable (e.g. for the operator $X.\mathbf{1} \multimap (X \otimes \mathbf{1})$, the type $\mathbf{1} \multimap (A \otimes \mathbf{1})$, for some A), the functorial action of the type operator allows us to *automatically* transform a program (in our case, a session) of type $\mathbf{1} \multimap (A \otimes \mathbf{1})$ into one of, say, type $\mathbf{1} \multimap (B \otimes \mathbf{1})$ by simply providing a process that coerces a session of type A into one of type B – the functorial action takes care of all the additional boilerplate.

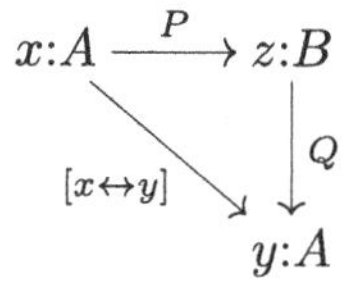

Fig. 5. Composition (up-to $\approx_{\mathsf{L}}$)

In a functional setting, the functorial action of a type operator is formalised as a structure preserving map, defined inductively on the structure of the type operator, such that $\mathsf{map}_{X.F(X)}(\lambda x.M)(N)$ takes a function $\lambda x.M$ of type $A \to B$ and a (to be transformed) term N of type $F(A)$, producing a term of type $F(B)$. In our session-typed setting, the natural representation $\mathsf{map}_{X.F(X)}$ must consist of a type directed *process transformation* that given a session $F(A)$ and a way to transform sessions of type A into sessions of type B produces a session $F(B)$. That is, the map construction acts as a *modular* adapter that given a process that transforms protocols of type A to protocols of type B, can perform the adaptation from A to B compositionally in any protocol where A occurs as a *sub-protocol.*

Given that our typing judgment is of the form $\Omega; \Gamma; \Delta \vdash P :: z{:}A$, where $\{z\} \cap dom(\Delta) \cap dom(\Gamma) = \emptyset$, denoting that P *offers* a session of type A along channel z when composed with processes that offer the (linear) session channels specified in Δ and the (shared) session channels specified in Γ, our session transformers must be *parameterized* with the appropriate channel names. To account for the full generality of positive type operators and their associated *covariant* and *contravariant* functorial actions we define the two map processes, $\mathsf{map}^{+}_{z:X.A}(Q_{x,y})(z')$ and $\mathsf{map}^{-}_{z:X.A}(Q_{x,y})(z')$. The former applies to covariant functors, taking a session z' of type $A\{B'/X\}$ into $A\{B/X\}$ (offered along z) via the transformation process $Q_{x,y}$ (typed as $x{:}B' \vdash Q :: y{:}B$). The latter applies to contravariant functors, taking a session z' of type $A\{B/X\}$ into $A\{B'/X\}$.

Definition 3.1 (Functorial Action of a Type Operator). *Given a type operator $X.A$ and process $x{:}B' \vdash Q_{x,y} :: y{:}B$, we define its covariant and contravariant functorial action $\mathsf{map}^{+}_{z:X.A}(Q_{x,y})(z')$ and $\mathsf{map}^{-}_{z:X.A}(Q_{x,y})(z')$ inductively on the structure of $X.A$, where map^{+} applies to positive operators and map^{-} to negative ones, as follows (we write $\mathsf{map}^{\dagger}$ for either map^{+} or map^{-} and $\bar{\dagger}$ to stand for the reverse of $\dagger$):*

$$\begin{aligned}
\mathsf{map}^{+}_{z:X.Y}(Q_{x,y})(z') &\triangleq [z' \leftrightarrow z] \quad (Y \neq X)\\
\mathsf{map}^{\dagger}_{z:X.\mathbf{1}}(Q_{x,y})(z') &\triangleq [z' \leftrightarrow z]\\
\mathsf{map}^{+}_{z:X.X}(Q_{x,y})(z') &\triangleq (\nu y)(Q\{z'/x\} \mid [y \leftrightarrow z])\\
\mathsf{map}^{\dagger}_{z:X.A\otimes B}(Q_{x,y})(z') &\triangleq z'(a').\overline{z}\langle a\rangle.(\mathsf{map}^{\dagger}_{a:X.A}(Q_{x,y})(a') \mid \mathsf{map}^{\dagger}_{z:X.B}(Q_{x,y})(z'))\\
\mathsf{map}^{\dagger}_{z:X.A\multimap B}(Q_{x,y})(z') &\triangleq z(a).\overline{z}'\langle a'\rangle.(\mathsf{map}^{\overline{\dagger}}_{a:X.A}(Q_{x,y})(a') \mid \mathsf{map}^{\dagger}_{z:X.B}(Q_{x,y})(z'))\\
\mathsf{map}^{\dagger}_{z:X.\forall Y.A}(Q_{x,y})(z') &\triangleq z(Y).z'\langle Y\rangle.\mathsf{map}^{\dagger}_{z:X.A}(Q_{x,y})(z')\\
\mathsf{map}^{\dagger}_{z:X.\exists Y.A}(Q_{x,y})(z') &\triangleq z'(Y).z'\langle Y\rangle.\mathsf{map}^{\dagger}_{z:X.A}(Q_{x,y})(z')
\end{aligned}$$

Lemma 3.2. *For all type operators $X.F(X)$, well-formed types B, B' and $x{:}B' \vdash Q :: y{:}B$:*

- *If $X.F(X)$ is positive and $\Omega; \Gamma; \Delta \vdash P :: w{:}F(B')$ then $\Omega; \Gamma; \Delta \vdash (\nu w)(P \mid \mathsf{map}^{+}_{z:X.F(X)}(Q_{x,y})(w)) :: z{:}F(B)$.*
- *If $X.F(X)$ is negative and $\Omega; \Gamma; \Delta \vdash P :: w{:}F(B)$, then $\Omega; \Gamma; \Delta \vdash (\nu w)(P \mid \mathsf{map}^{-}_{z:X.F(X)}(Q_{x,y})(w)) :: z{:}F(B')$.*

Functor Laws. We prove that our development of functors is *canonical* by showing that the *functor laws* are preserved by map. Specifically, map preserves identity and composition up to $\approx_{\mathtt{L}}$.

Theorem 3.3 (Functor Laws). *For any type operator $X.A$:*

Identity: *If $\Omega; \Gamma; \Delta \vdash P :: z' : A\{B/X\}$ then*

$$\Omega; \Gamma; \Delta \vdash (\nu z')(P \mid \mathsf{map}^{\dagger}_{z:X.A}([x \leftrightarrow y])(z')) \approx_{L} (\nu z')(P \mid [z' \leftrightarrow z]) :: z{:}A\{B/X\}$$

Associativity: *Let $\Omega; \Gamma; x{:}B_1 \vdash P :: a{:}B_2$, $\Omega; \Gamma; a{:}B_2 \vdash Q :: y{:}B_3$ and $\Omega; \Gamma; \Delta \vdash R :: z'{:}A_0$. If $X.A$ is positive then:*

$$\begin{aligned}
\Omega; \Gamma; z'{:}A\{B_1/X\} \vdash \mathsf{map}^{+}_{z:X.A}((\nu a)(P \mid Q)_{x,y})(z') \approx\\
(\nu z'')(\mathsf{map}^{+}_{z'':X.A}(P_{x,a})(z') \mid \mathsf{map}^{+}_{z:X.A}(Q_{a,y})(z'')) :: z{:}A\{B_3/X\}
\end{aligned}$$

If $X.A$ is negative then

$$\begin{aligned}
\Omega; \Gamma; z'{:}A\{B_3/X\} \vdash \mathsf{map}^{-}_{z:X.A}((\nu a)(P \mid Q)_{x,y})(z') \approx\\
(\nu z'')(\mathsf{map}^{-}_{z'':X.A}(Q_{a,y})(z') \mid \mathsf{map}^{-}_{z:X.A}(P_{x,y})(z'')) :: z{:}A\{B_1/X\}
\end{aligned}$$

Henceforth we write map to stand for map^{+}.

3.3 Encoding Inductive Types

An inductive type $\mu_i X.F(X)$, where $X.F(X)$ is a positive type operator is encoded as $[\![\mu_i X.F(X)]\!] \triangleq \forall X.!(F(X) \multimap X) \multimap X$. As discussed in Sect. 3.1, inductive types can be understood as initial F-algebras. Let $T_i = [\![\mu_i X.F(X)]\!]$. We must show that the F-algebra $(T_i, \alpha_{x,y})$, with $x{:}F(T_i) \vdash \alpha_{x,y} :: y{:}T_i$, is initial. Thus, we must show that the diagram of Fig. 6 commutes and that fold is unique (we write β_{x_2,y_2} for a process $x_2{:}F(X) \vdash \beta_{x_2,y_2} :: y_2{:}X$ and $!\beta$ for $!u(y_2).y_2(x_2).\beta_{x_2,y_2}$). We define fold_{y_1,y_2} and $\alpha_{x,y}$ as follows:

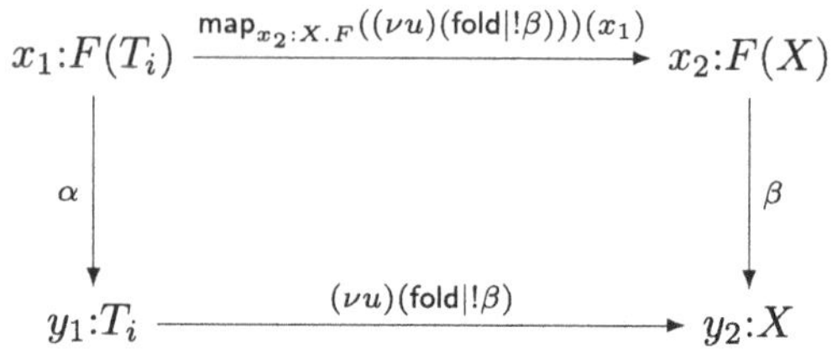

Fig. 6. Weak Initiality

$$\begin{array}{ll} \mathsf{fold}_{y_1,y_2} & \triangleq y_1\langle X\rangle.\overline{y_1}\langle b\rangle.(!b(w).\overline{u}\langle c\rangle.[c \leftrightarrow w] \mid [y_1 \leftrightarrow y_2]) \\ \alpha_{x_1,y_1} & \triangleq y_1(X).y_1(u).\overline{u}\langle f\rangle.\overline{f}\langle b\rangle(\mathsf{map}_{b:X.F}(\mathsf{fold}_{y_1,y_2})(x_1) \mid [f \leftrightarrow y_1]) \end{array}$$

where $X; u{:}F(X) \multimap X; y_1{:}T_i \vdash \mathsf{fold}_{y_1,y_2} :: y_2{:}X$ and $\cdot;\cdot; x_1{:}F(T_i) \vdash \alpha_{x_1,y_1} :: y_1{:}\forall X.!(F(X) \multimap X) \multimap X$.

To show that Fig. 6 commutes, we prove the observational equivalence stated in Theorem 3.7, using the three lemmas below.

Lemma 3.4. *Let $X.A$ be a positive type operator. If $\Omega; \Gamma, u{:}B; x{:}B_1 \vdash P :: y{:}B_2$ and $\Omega; \Gamma; \cdot \vdash Q :: w{:}B$ then:*

$$\Omega; \Gamma; z'{:}A\{B_1/X\} \vdash \mathsf{map}_{z:X.A}((\nu u)(!u(w).Q \mid P))(z') \approx_L (\nu u)(!u(w).Q \mid \mathsf{map}_{z:X.A}(P)(z')) :: z{:}A\{B_2/X\}$$

Lemma 3.5. *For any $\Omega; \Gamma; \cdot \vdash P :: x{:}A$ we have that:*

$$\Omega; \Gamma; \cdot \vdash (\nu u)(!b(w).\overline{u}\langle c\rangle.[c \leftrightarrow w] \mid !u(y_2).P) \approx_L !b(y_2).P :: b{:}!A$$

Lemma 3.6. *Let $\Omega; \Gamma; \cdot \vdash U_1 \approx_L U_2 :: x{:}A$ and $\Omega; \Gamma; \cdot \vdash R_1 \approx_L R_2 :: u{:}!A$ then: (1) $\Omega; \Gamma; \cdot \vdash !u(x).U_1 \approx_L R_1 :: u{:}!A$; (2) $\Omega; \Gamma; \cdot \vdash !u(x).U_2 \approx_L R_2 :: u{:}!A$*

Theorem 3.7 (Weak Initiality). *Let $X.F(X)$ be a positive type operator, then for any type X and morphism β such that $x_2{:}F(X) \vdash \beta_{x_2,y_2} :: y_2{:}X$ we have that:*

$$X; \cdot; x_1{:}F(T_i) \vdash (\nu y_1)(\alpha_{x_1,y_1} \mid (\nu u)(\mathsf{fold}_{y_1,y_2} \mid !\beta)) \approx_L (\nu x_2)(\mathsf{map}_{x_2:X.F}((\nu u)(\mathsf{fold}_{y_1,y_2} \mid !\beta))(x_1) \mid \beta_{x_2,y_2}) :: y_2{:}X$$

Initiality. Having shown that the diagram of Fig. 6 commutes, we have that the F-algebra (T_i, α) is weakly initial in the sense that there exists a morphism (constructable with fold) from it to any other such F-algebra. We now show that the algebra is indeed initial. In other words, we show that fold is the *only* morphism (up-to $\approx_\mathrm{L}$) that makes the diagram of Fig. 6 commute.

Theorem 3.8 (Initiality of T_i). *Let $X.F$ be a positive type operator. For any F-algebra (X, β), we have that for all H such that $X; u{:}F(X) \multimap X; y_1{:}T_i \vdash H :: y_2{:}X$ the following holds: $X; u{:}F(X) \multimap X; y_1{:}T_i \vdash H \approx_L \mathsf{fold}_{y_1,y_2} :: y_2{:}X$.*

Proof. We only sketch the key elements of the proof. The proof of initiality requires showing an equivalence at the (open) type variable X. In particular, we must show that for all pairs of closed types (and respective admissible relations) that may instantiate X, fold and H are equivalent. By making crucial use of parametricity (i.e. that fold and H are equivalent to themselves for any such admissible relations) we can construct an admissible relation that allows us to discharge the main proof obligation.

3.4 Encoding Coinductive Types

We perform a similar development to that of Sect. 3.3 but for coinductive types. A coinductive type $\mu_f X.F(X)$, where $X.F(X)$ is a positive type operator is encoded as: $[\![\mu_f X.F(X)]\!] \triangleq \exists X.!(X \multimap F(X)) \otimes X$. Coinductive types are interpreted as final F-coalgebras. Let $T_f = [\![\mu_f X.F(X)]\!]$, we must show that the F-coalgebra $(T_f, \alpha_{x,y})$, with $x{:}T_i \vdash \alpha_{x,y} :: y{:}F(T_i)$ is final. As before, we show that the diagram of Fig. 7 commutes and that unfold is unique (we write β for a process $x_1{:}X \vdash \beta_{x_1,y_1} :: y_1{:}F(X)$ and $!\beta$ as before). We define $\mathsf{unfold}_{x_1,x_2}$ and $\alpha_{x,y}$ as:

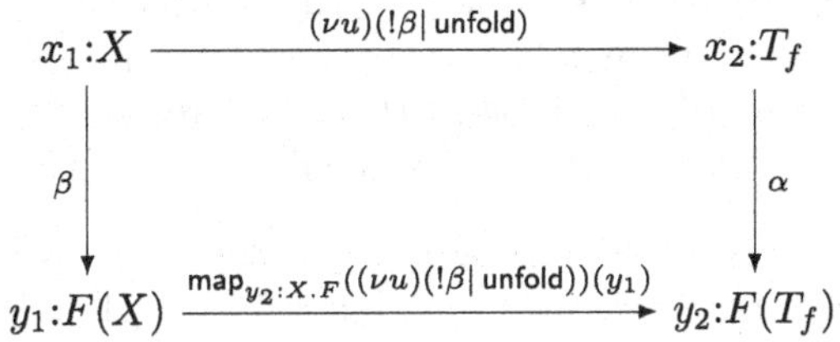

Fig. 7. Weak Finality

$$\begin{aligned}
\mathsf{unfold}_{x_1,x_2} &\triangleq x_2\langle X\rangle.\overline{x_2}\langle c\rangle.(!c(v).\overline{u}\langle c'\rangle.[c' \leftrightarrow v] \mid [x_1 \leftrightarrow x_2]) \\
\alpha_{x_2,y_2} &\triangleq x_2(X).x_2(u).\overline{u}\langle c\rangle.\overline{c}\langle c'\rangle.([x_2 \leftrightarrow c'] \mid \mathsf{map}_{y_2:X.F}(\mathsf{unfold}_{x_1,x_2})(c))
\end{aligned}$$

where $X; u{:}X \multimap F(X); x_1{:}X \vdash \mathsf{unfold}_{x_1,x_2} :: x_2{:}T_f$ and $\cdot;\cdot; x_2{:}T_f \vdash \alpha_{x_2,y_2} :: y_2{:} F(T_f)$. We show that diagram above commutes by proving Lemma 3.9.

Theorem 3.9 (Weak Finality). *Let $X.F(X)$ be a positive type operator. For any type X and morphism β such that $x_1{:}X \vdash \beta_{x_1,y_1} :: y_1{:}F(X)$ we have that: $X; \cdot; x_1{:}X \vdash$*

$$\begin{aligned}
&(\nu x_2)((\nu u)(!\beta \mid \mathsf{unfold}_{x_1,x_2}) \mid \alpha_{x_2,y_2}) \\
&\quad \approx_L (\nu y_1)(\beta_{x_1,y_1} \mid \mathsf{map}_{y_2:X.F}((\nu u)(!\beta \mid \mathsf{unfold}))(y_1)) :: y_2{:}F(T_f)
\end{aligned}$$

Finality. To show the universality of $(T_f, \alpha_{x,y})$ we establish a dual property to Theorem 3.8, proving that unfold is the *unique* morphism for which the diagram of Fig. 7 commutes. However, unlike in the development of Theorem 3.8, the existential flavour of the encoding requires a slightly more intricate formal argument.

We first establish the following two equivalences: Lemma 3.10 states that using unfold to produce a session morphism from A to T_f is equivalent to first using a coalgebra morphism H from A to A' and then using unfold at type A' to ultimately produce a session of type T_f. The proof of this lemma is more challenging than the development of Sect. 3.3 due to the encoding using an existentially quantified type (i.e., parametricity provides a less general property). The proof showcases an instance of a particularly powerful style of reasoning that is not usually available in a process setting. We establish the result by performing the *relational lifting* of *process* H, viewed as a relation between A and A', which is the *graph relation* of H (i.e. $(d, d') \in \mathcal{R}$ iff $d_y \cong (\nu x)(d'_x \mid H_{x,y})$); Lemma 3.11 shows that unfolding the coalgebra morphism α is an identity at type T_f. The proof of Lemma 3.10 is the most intricate constructing a graph relation between A and A' with process H. The corresponding diagram is given in Fig. 8. These two lemmas conclude the finality result (Theorem 3.12).

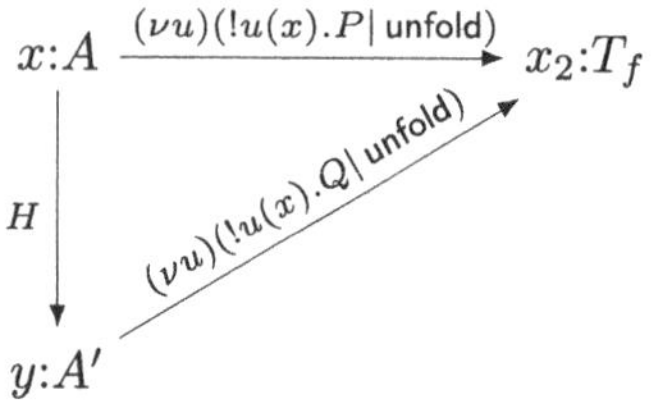

Fig. 8. Unfolding

Lemma 3.10. *Let $X.F$ be a positive type operator. Let $x{:}A \vdash H :: y{:}A'$ be a morphism of coalgebras from $P :: x{:}A \multimap F(A)$ to $Q :: x{:}A' \multimap F(A')$. The equivalence holds:*

$$x{:}A \vdash (\nu y)(H \mid (\nu u)(!u(x).Q \mid \mathsf{unfold}_{y,x_2}\{A'/X\})) \\ \approx_L (\nu u)(!u(x).P \mid \mathsf{unfold}_{x,x_2}\{A/X\}) :: x_2{:}T_f$$

Proof. Since we must establish an equivalence at an existential type, we need to exhibit one particular type instantiation of the existential variable (and respective admissible relation). Noting that reasoning directly from the parametricity of H and unfold does not enable us to fulfill this proof obligation, we proceed by lifting the morphism H to the relational level, showing that the two processes are equivalent when we consider the *graph* relation of the coalgebra morphism explained above (relating processes that are observationally equivalent when one of them is composed with H). By appealing to this lifting via parametricity we conclude using the coalgebraic properties of H.

Lemma 3.11. $x{:}T_f \vdash (\nu u)(!u(x).x(y).\alpha_{y,x} \mid \mathsf{unfold}_{x,z}\{T_f/X\}) \approx_L [x \leftrightarrow z] :: z{:}T_f$

Theorem 3.12 (Finality of T_f). *Let $X.F$ be a positive type operator, $x{:}A \vdash P_{x,y} :: y{:}F(A)$ an F-coalgebra and $x{:}A \vdash H_{x,y} :: y{:}T_f$ be a coalgebra map from P to α:*

$$\cdot; \cdot; x{:}A \vdash H_{x,x_2} \approx_L (\nu u)(!u(y).y(x).P_{x,y} \mid \mathsf{unfold}\{A/X\}) :: x_2{:}T_f$$

Proof. The proof combines the relational lifting of the coalgebra morphism H (Lemma 3.10) and Lemma 3.11 to discharge the main proof obligation.

Remark 1 (Relational Lifting). The development of the encoding of coinductive types enables us to exhibit the full expressive power of parametric equivalence for typed processes, allowing us to reason about process equivalences at *different* types. This is made explicit in the proof of Lemma 3.10, where we use parametric equivalence to perform a relational lifting of a process (i.e. a coalgebra process morphism between A and A'): the two given processes are equivalent when the one on the left-hand side instantiates the existential type with A and the one on the right with A', given that the relation between the types is precisely the composition with the coalgebra process morphism on the right-hand side of the equivalence (i.e. the relational lifting of the morphism).

4 Conclusion

This work has explored a *processes-as-morphisms* approach to demonstrate the expressiveness of polymorphic mobile processes, showing an encoding of inductive and coinductive session types, justified via the theory of initial algebras and final co-algebras. Our work gives a direct account of encodability comparing to the method in [29] which *indirectly* proves an identical result using fully abstract encodings from/to linear System F [33]. Our work crucially uses a rigorous proof-theoretic correspondence between (intuitionistic) linear-logic and session types. The notion of "expressiveness" in our paper differs from the standard (positive and negative) encodability criteria defined by e.g. Palamidessi [19,20] for the π-calculus. Our hope is that this new algebraic encodability will potentially enable an exploration of algebraic constructs beyond initial and final co-algebras in a session programming setting. We wish to further study the meaning of functors and natural transformations in a logic-based session-typed setting, both from a more fundamental viewpoint but also in terms of practical programming patterns. This also relates to Miller's work on logic programming languages [12,17,18] where a proof-theoretic foundation based on (intuitionistic) logic gives a uniform, clear basis for (logic) programming languages.

Acknowledgements. The authors would like to thank Dominic Orchard and the anonymous reviewers for their comments and suggestions. This work is partially supported by EPSRC EP/K034413/1, EP/K011715/1, EP/L00058X/1, EP/N027833/1, EP/N028201/1 and NOVA LINCS (UID/CEC/04516/2019).

References

1. Bainbridge, E.S., Freyd, P.J., Scedrov, A., Scott, P.J.: Functorial polymorphism. Theor. Comput. Sci. **70**(1), 35–64 (1990)
2. Bird, R., de Moor, O.: Algebra of Programming. Prentice-Hall Inc., Upper Saddle River (1997)
3. Bird, R.S., de Moor, O., Hoogendijk, P.F.: Generic functional programming with types and relations. J. Funct. Program. **6**(1), 1–28 (1996)
4. Birkedal, L., Møgelberg, R.E., Petersen, R.L.: Linear Abadi and Plotkin Logic. Logical Methods Comput. Sci. **2**(5) (2006). https://lmcs.episciences.org/2233

5. Caires, L., Pérez, J.A., Pfenning, F., Toninho, B.: Behavioral polymorphism and parametricity in session-based communication. In: Felleisen, M., Gardner, P. (eds.) ESOP 2013. LNCS, vol. 7792, pp. 330–349. Springer, Heidelberg (2013). https://doi.org/10.1007/978-3-642-37036-6_19
6. Caires, L., Pfenning, F.: Session types as intuitionistic linear propositions. In: Gastin, P., Laroussinie, F. (eds.) CONCUR 2010. LNCS, vol. 6269, pp. 222–236. Springer, Heidelberg (2010). https://doi.org/10.1007/978-3-642-15375-4_16
7. Caires, L., Pfenning, F., Toninho, B.: Linear logic propositions as session types. Math. Struct. Comput. Sci. **26**(3), 367–423 (2016)
8. Girard, J.: Une extension de l'interprétation de Gödel à l'analyse, et son application à l'élimination de coupures dans l'analyse et la théorie des types. In: Proceedings of the 2nd Scandinavian Logic Symposium, pp. 63–92 (1971)
9. Girard, J.: Linear logic. Theor. Comput. Sci. **50**, 1–102 (1987)
10. Girard, J., Lafont, Y., Taylor, P.: Proofs and Types. Cambridge University Press, Cambridge (1989)
11. Hasegawa, R.: Categorical data types in parametric polymorphism. Math. Struct. Comput. Sci. **4**(1), 71–109 (1994)
12. Hodas, J., Miller, D.: Logic programming in a fragment of intuitionistic linear logic. Inf. Comput. **110**, 327–365 (1994)
13. Honda, K., Vasconcelos, V.T., Kubo, M.: Language primitives and type discipline for structured communication-based programming. In: Hankin, C. (ed.) ESOP 1998. LNCS, vol. 1381, pp. 122–138. Springer, Heidelberg (1998). https://doi.org/10.1007/BFb0053567
14. Honda, K., Yoshida, N., Carbone, M.: Multiparty asynchronous session types. In: POPL 2008, pp. 273–284 (2008)
15. Lindley, S., Morris, J.G.: Talking bananas: structural recursion for session types. In: ICFP 2016, pp. 434–447 (2016)
16. Mendler, N.P.: Recursive types and type constraints in second-order lambda calculus. In: LICS, pp. 30–36 (1987)
17. Miller, D.: A logic programming language with lambda-abstraction, function variables, and simple unification. J. Logic Comput. **1**(4), 497–536 (1991)
18. Miller, D., Nadathur, G., Pfenning, F., Scedrov, A.: Uniform proofs as a foundation for logic programming. Ann. Pure Appl. Logic **51**, 125–157 (1991)
19. Palamidessi, C.: Comparing the expressive power of the synchronous and the asynchronous π-calculus. In: Proceedings of the 24th ACM SIGPLAN-SIGACT Symposium on Principles of Programming Languages, POPL 1997, pp. 256–265. ACM, New York (1997)
20. Palamidessi, C.: Comparing the expressive power of the synchronous and asynchronous π-calculi. Math. Struct. Comput. Sci. **13**(5), 685–719 (2003)
21. Pérez, J.A., Caires, L., Pfenning, F., Toninho, B.: Linear logical relations for session-based concurrency. In: Seidl, H. (ed.) ESOP 2012. LNCS, vol. 7211, pp. 539–558. Springer, Heidelberg (2012). https://doi.org/10.1007/978-3-642-28869-2_27
22. Pierce, B.C., Sangiorgi, D.: Behavioral equivalence in the polymorphic pi-calculus. J. ACM **47**(3), 531–584 (2000)
23. Plotkin, G., Abadi, M.: A logic for parametric polymorphism. In: Bezem, M., Groote, J.F. (eds.) TLCA 1993. LNCS, vol. 664, pp. 361–375. Springer, Heidelberg (1993). https://doi.org/10.1007/BFb0037118
24. Reynolds, J.C.: Types, abstraction and parametric polymorphism. In: IFIP Congress, pp. 513–523 (1983)

25. Sangiorgi, D.: Pi-calculus, internal mobility, and agent-passing calculi. Theor. Comput. Sci. **167**(1&2), 235–274 (1996)
26. Sangiorgi, D., Walker, D.: The Pi-Calculus: A Theory of Mobile Processes. Cambridge University Press, Cambridge (2001)
27. Toninho, B., Caires, L., Pfenning, F.: Functions as session-typed processes. In: Birkedal, L. (ed.) FoSSaCS 2012. LNCS, vol. 7213, pp. 346–360. Springer, Heidelberg (2012). https://doi.org/10.1007/978-3-642-28729-9_23
28. Toninho, B., Caires, L., Pfenning, F.: Corecursion and non-divergence in session-typed processes. In: Maffei, M., Tuosto, E. (eds.) TGC 2014. LNCS, vol. 8902, pp. 159–175. Springer, Heidelberg (2014). https://doi.org/10.1007/978-3-662-45917-1_11
29. Toninho, B., Yoshida, N.: On polymorphic sessions and functions. In: Ahmed, A. (ed.) ESOP 2018. LNCS, vol. 10801, pp. 827–855. Springer, Cham (2018). https://doi.org/10.1007/978-3-319-89884-1_29
30. Toninho, B., Yoshida, N.: Polymorphic session processes as morphisms. Technical report 02/2019, Department of Computing, Imperial College London, July 2019. https://www.doc.ic.ac.uk/research/technicalreports/2019/DTRS19-2.pdf
31. Wadler, P.: Recursive types for free! http://homepages.inf.ed.ac.uk/wadler/papers/free-rectypes/free-rectypes.txt
32. Wadler, P.: Propositions as sessions. J. Funct. Program. **24**(2–3), 384–418 (2014)
33. Zhao, J., Zhang, Q., Zdancewic, S.: Relational parametricity for a polymorphic linear lambda calculus. In: Ueda, K. (ed.) APLAS 2010. LNCS, vol. 6461, pp. 344–359. Springer, Heidelberg (2010). https://doi.org/10.1007/978-3-642-17164-2_24

Guess Who's Coming: Runtime Inclusion of Participants in Choreographies

Maurizio Gabbrielli[1,2(✉)], Saverio Giallorenzo[3], Ivan Lanese[1,2], and Jacopo Mauro[3]

[1] University of Bologna, Bologna, Italy
maurizio.gabbrielli@unibo.it
[2] INRIA, Sophia Antipolis, France
[3] University of Southern Denmark, Odense, Denmark

Abstract. In Choreographic Programming, a choreography specifies in a single artefact the expected behaviour of all the participants in a distributed system. The choreography is used to synthesise correct-by-construction programs for each participant.

In previous work, we defined Dynamic Choreographies to support the update of distributed systems at runtime.

In this work, we extend Dynamic Choreographies to include new participants at runtime, capturing those use cases where the system might be updated to interact with new, unforeseen stakeholders. We formalise our extension, prove its correctness, and present an implementation in the AIOCJ choreographic framework.

Keywords: Choreographic programming · Adaptation of distributed systems · Dynamic inclusion of new software components

1 To Go or Not to Go ... Almost an Introduction

"Would you go to Padova?" Bob asked while typing on his old computer.

"Well, it depends" answered Alice, without stopping to stare at the paper on her desk. She was puzzled by the proof she was reading. The more she was going deep into it, the less she was convinced about its correctness.

"Depends on what?"

"Oh, on many factors: time and effort required, money, you know, the usual things ... I don't understand how it's been possible to publish this proof. It's completely hand-waved! And you know what the author told me when I met him at a conference in Jerusalem? He said that the proof was not accurate because ... we're not mathematicians but computer scientists! Can you believe that? Computer scientists ... then, my dear hand-waver, what is computer science? A

Research partially supported by the EU H2020 RISE programme under the Marie Skłodowska-Curie grant agreement No 778233.

M. S. Alvim et al. (Eds.): Palamidessi Festschrift, LNCS 11760, pp. 118–138, 2019.
https://doi.org/10.1007/978-3-030-31175-9_8

branch of astrology or, perhaps, by any chance, something that happens to use also math?" Alice was getting visibly nervous about that paper and its author. The questions from Bob did not help in relaxing her.

"Hum, let's try this package" mumbled Bob after a short silence "this should fix these nasty Latex problems ... I believe everything in Padova is pretty standard. Nothing sensationally different from other places. So all in all those factors you mentioned do not help in the decision."

"Then, perhaps, it could be worth considering if you really want to go there. I mean, of course it could be nice, but is it really necessary?" commented Alice.

"Indeed, that is the right question. Is it necessary, or perhaps could one avoid it? Perhaps one could do something else or go to some other place. All in all, there are many places in Italy ... it's a really difficult choice."

The conversation continued for a while, with Alice and Bob working at their own desks. Then, the door of the office slammed open. "Guess who's coming!" shouted Charlie entering the room "How are things going? How many papers have you written in the last week?"

Charlie was the kind of office-mate who continuously asks questions. Always anxious about his performance, compared to the ones of his colleagues. He was a nice guy, and often his questions raised interesting problems. However, when he was at the office, it was simply impossible to do serious technical work.

"What are you talking about? Some new full abstraction proof?" pressed Charlie.

Alice raised her head, stretched her legs, looked for a while out of the window, and then answered "Oh, nothing serious, we are talking about going to Padova." Bob was baffled: "Well, if you allow me, my dear, it's a serious matter. It's actually quite important for my life!"

"Come on Bob, it's just a trip. Moreover, I was not invited to Padova. So, all in all, it's not nice of you to ask me all these questions!"

Bob stopped typing on his computer. He sat still for a few seconds. Then, he slowly turned his head towards Alice. He was genuinely surprised. He was used to some strange remarks from Alice but this time she was beyond her standards.

"Why in the world should they invite you to Padova, Alice!? You have already accepted a position at the University of Genova. Of course the University of Padova did not consider you."

"Position? University? Wait a minute! What are you talking about? Work has nothing to do with this trip to Padova. It's just a trip to go and visit that exhibition George has been talking about for months. He could have invited me too ... but that's OK, after all I've a lot of work to do in these days, so, no problem, really, no problem" replied Alice.

Bob's head was spinning "Exhibition? Why in the universe should I go and visit an exhibition in Padova? And why should I ask your opinion about going to an exhibition? I was talking about accepting or not a position at the University of Padova, that's what I was talking about!" Silence fell like a hammer.

"Well colleagues," broke Charlie "see how my simple question has solved a problem that otherwise could have lasted, unresolved, for hours? It was a kind of deadlock! So both of you owe me a beer. See you!" And with the usual slam of the door Charlie exited the room leaving Alice and Bob speechless.

In the story above, the reader might have recognised Catuscia, who played the role of Alice. Indeed this is a true anecdote, dating back to the Pisa times "some" years ago. Also the story about the wrong proof is true, but we leave to the reader the task of finding out who the author of the proof was (difficult) or guess the identity of Bob (easy). Charlie also corresponds to a real person, even though his character was changed a bit for narrative reasons. And, yes, George is ... George!

At the time of this story choreographic languages had not been invented yet, however, in our opinion, the story has four morals: (i) formalising conversations among computer scientists, as if they were distributed applications, can sometimes be useful; (ii) allowing the inclusion of new participants at runtime, interacting in unforeseen ways, can also be useful; (iii) as a consequence of (i) and (ii) dynamic choreographic languages with runtime inclusion of new participants—as presented in this paper—can be useful not only for computer applications, but also for computer scientists; (iv) most importantly, never, ever, ask a question to Catuscia if she is working on a proof!

2 Introduction

Today, applications are often distributed and involve multiple participants which interact by exchanging messages. Programming the intended behaviour of such applications requires understanding how the behaviour of a participant program combines with that of others, to produce the global behaviour of the application. There is a tension between the *global* desired behaviour of a distributed application and the fact that it is programmed by developing *local* programs. Choreographic Programming [1,2] provides a good trade-off by allowing developers on the one hand to program the global behaviour directly and, on the other hand, to automatically generate the local programs that implement the global behaviour. As an example, consider the following choreography that describes the behaviour of an application composed of one `client` and one `seller`:

```
1  product@client = getInput( "Insert product name" );
2  quote: client( product ) -> seller( order )
```

The execution starts with an action performed by the `client`: an input request to the local user (line 1). The semicolon at the end of the line is a sequential composition operator, hence the user input should complete before execution proceeds to line 2. Then, a communication between the `client` and the `seller` takes place: the `client` sends a message and the `seller` receives it.

This code specifies the global behaviour and can be compiled automatically into two different local programs, one for the client and one for the seller.

Choreographies avoid by construction the presence of common errors like communication deadlocks and message races and are therefore useful to implement correct distributed systems [3]. Moreover, as testified by the dynamic choreographic language AIOCJ [2], choreographies can be extended to support the adaptation of running distributed applications. This is indeed an important task for the development of modern cloud native applications since, due to the widely adoption of DevOps techniques for Continuous Integration and Continuous Deployment [4–6], the entire application needs to be updated, upgraded with new features, or patched guaranteeing no downtimes.

AIOCJ proposes a general mechanism to structure application updates. Inside applications, blocks of code delimited by *scopes* may be dynamically replaced by new blocks of code, called *updates*.

For instance, consider the previous example and assume that the interaction between the `client` and `seller` may change. This is enabled by replacing the second line with the following code.

```
scope @client{
  quote: client( product ) -> seller( order ) }
```

In essence, a scope is a delimiter that defines which part of the application can be updated. Each scope identifies a *coordinator* of the update (`client` in this case). The coordinator is the participant responsible for ensuring that the distributed adaptation of the code within the scope, if and when needed, is performed successfully. The code change is specified by program rules that target the desired scope. For instance, assuming that the seller requires not only the name of the product but also the customer location to make a targeted offer, the previous code can be changed by applying the following rule.

```
rule { do {
  loc@client = getInput("Insert your location" );
  quote: client( product + loc ) -> seller( order ) }
}
```

Notably, rules can be defined and inserted in the system while it is running, without downtimes. At a choreographic level, updates are applied atomically to all the involved participants. The AIOCJ framework guarantees that the compilation and the application of the adaptation rules generate correct behaviours, avoiding the inconsistencies typical of distributed code updates.

While AIOCJ [2] provides a safe and reliable mechanism to adapt the code of running applications, it has one major weakness: it does not support the introduction of new participants at run time. Unfortunately, due to the change of business or legal needs, the need to introduce another entity (e.g., auditing program, logging system) that interacts with an existing and running system may arise.

In this work, we address this issue by proposing an extension of the Dynamic Choreographic AIOCJ framework which allows AIOCJ rules to add new participants to the choreography. In particular:

- we formalise the extension of AIOCJ rules to add new participants;
- we prove that the extension with new participants satisfies the properties of deadlock freedom and (for finite traces) of correctness of compilation and adaptation;

- we extend the AIOCJ development and deployment framework to support the addition of new participants.

Structure of the Paper. To exemplify the approach we start by presenting in Sect. 3 a simple use case. In Sect. 4 we formalise the extension proving that it preserves the correctness properties. In Sect. 5 we present the implementation strategy. Section 6 discusses related work and Sect. 7 draws some concluding remarks.

3 A Client-Seller Use Case

In this section we present a simple use case exemplifying how new participants can be added at run time.

For simplicity, we consider a client-seller system for online shopping where the client sends messages to the seller that processes them and returns an answer to the client. Originally the program performs just this activity. Unfortunately, due to legislation changes, the requests by the client could go towards an auditing process and therefore they must be logged by an independent authority. Hence, logging can not be performed by the seller but requires a new and dedicated service, physically deployed in a location different from the seller one.

Adaptation is performed in two stages:

1. when writing the original AIOCJ program, one should foresee which parts of the code could be adapted in the future (but not which new behaviour will be required by the adaptation), and enclose them into `scopes`;
2. while the AIOCJ program is running, one should write adaptation rules to introduce the desired new behaviour.

```
1 product@client = getInput( "Insert product name" );
2 request: client( product ) -> seller ( order );
3 scope @seller{
4   // process the order and compute result, here XXX
5   result@seller = "XXX";
6   response: seller( result ) -> client ( result )
7 }
```

Let us suppose that the main deployed and running application was created by using the choreographic program above. The original programmer foresaw the possibility that the code run by the seller to compute the answer could be subject to changes. For this reason, the code at lines 4–6, where the answer is computed and sent back to the client, is enclosed in a scope.

To enforce the new legislation, the seller now needs to log all the incoming requests from the client. The logging entity is a new participant in the choreography. Using the extension for AIOCJ that we propose, the rule triggering the adaptation can be written as shown in Listing 1.1.

The rule introduces a new participant (also called a role) `logger` by using the keywords `newRoles` (line 4). New roles in AIOCJ rules are different from roles

```
1  rule {
2   include log from "socket://localhost:8002"
3   include getTime from "socket://localhost:8003"
4   newRoles: logger
5   location@logger: "socket://independent_autority.com:8080"
6   on {} // applicability conditions, irrelevant in this example
7   do {
8    log: seller( order ) -> logger( entry );
9    {
10     time@logger = getTime();
11     _@logger = log( time + ": " + entry )
12   } | {
13     // process the order and compute result, here XXX
14     result@seller = "XXX";
15     response: seller( result ) -> client ( result )
16   }}
17 }
```

Listing 1.1. Adaptation rule adding new role.

previously involved in the target AIOCJ program (if needed they are renamed to avoid clashes of names) and take part to the choreography only while the body of the rule executes. As for normal roles, the URI of new roles is declared using the keyword `location` (line 5). In this particular case we suppose that the code will be deployed at the location reachable at URL `independent_autority.com` on port 8080.

Lines 2–3 define two external services `log` and `getTime` that are used, respectively, to store a message in the database and to get the current time. These two services are supposed to be available on the facility where the `logger` code will run. The new code, as specified in lines 8–15, requires the server to send relevant information on each transaction to the logger (in parallel to the answer to the client). In particular, the server computes a timestamp for each request (line 10) and logs the transaction information together with the timestamp (line 11).

To conclude this section, we remark that for presentation purposes the running example uses only a limited subset of the AIOCJ functionalities not including, e.g., the mechanism for fine-grained rule applicability (provided by the `on` block) which can be used to specify when and whether a rule applies to a given scope. We refer the reader to [2,7] for further details and to the AIOCJ website[1] for the full code of the example (choreography and rule) and the executable programs generated by their compilation.

[1] http://www.cs.unibo.it/projects/jolie/aiocj_examples/external_roles.html.

4 Theoretical Model

In this section, we give a brief overview of the theory of Dynamic Choreographies [2] and we present the changes needed to support the inclusion of new roles in runtime updates. For the sake of presentation we do not report all the formal definitions detailed in [2], but we just give a general intuition of the theory presented in [2] focusing only on the formalisation of our extension. We also prove that some correctness results from [2] are preserved by the extension.

A graphical representation explaining the key elements of Dynamic Choreographies is depicted below.

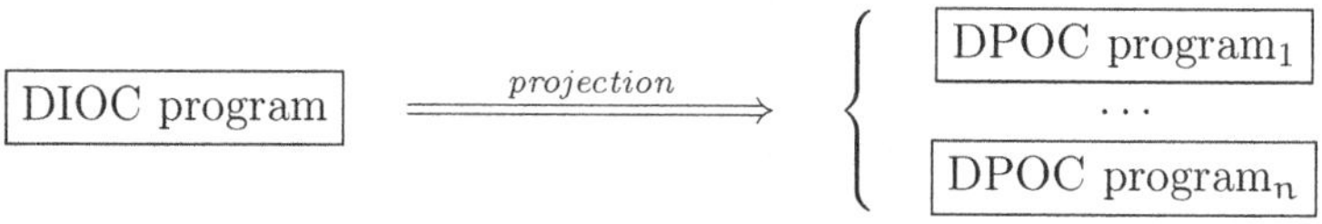

From the left, we have DIOC, which stands for Dynamic Interaction-Oriented Choreographies. This is the high-level language that programmers use to specify the behaviour of the whole distributed system and models the one we have used in the code snippets in the previous sections. It will be presented in Sect. 4.1. The second element is the *projection* procedure which transforms a DIOC into a set of executable programs, one for every participant in the DIOC code. Since the extension in this paper does not affect the projection from [2], we just illustrate the intuition behind it in Sect. 4.3. Finally, we have the target language of the projection, DPOC, standing for Dynamic Process-Oriented Choreographies. DPOC is a calculus inspired by process calculi like the π-calculus and CCS, and it is equipped with primitives for runtime code update. It is designed as a common abstraction for real languages (C, Java, Python) to make the theory more general avoiding to target a specific language. DPOC is presented in Sect. 4.2.

4.1 Dynamic Interaction-Oriented Choreographies

We start by presenting the language to program applications, called DIOC.

DIOCs rely on a set of *Roles*, ranged over by $\mathbf{R}, \mathbf{S}, \ldots$, to identify the participants in the choreography. Each role has its own local state. Roles exchange messages over labels, called *operations* and ranged over by o. We require expressions, ranged over by e, to include at least values, belonging to a set *Val* ranged over by v, and variables, belonging to a set *Var* ranged over by $\mathsf{x}, \mathsf{y}, \ldots$. *DIOC processes* are ranged over by $\mathcal{I}, \mathcal{I}', \ldots$. We present below the DIOC production rules.

$$
\begin{array}{llllllll}
\mathcal{I} ::= & \mid \mathsf{o} : \mathbf{R}(\mathsf{e}) \rightarrow \mathbf{S}(\mathsf{x}) & \textit{(interact)} & \mid & \mathsf{x}@\mathbf{R} = \mathsf{e} & \textit{(assign)} & \mid & \mathcal{I}; \mathcal{I}' \ \textit{(seq)} \\
& \mid \mathbf{if}\ \mathsf{b}@\mathbf{R}\ \{\mathcal{I}\}\ \mathbf{else}\ \{\mathcal{I}'\} & \textit{(cond)} & \mid & \mathbf{scope}\ @\mathbf{R}\ \{\mathcal{I}\} & \textit{(scope)} & \mid & \mathcal{I} | \mathcal{I}' \ \textit{(par)} \\
& \mid \mathbf{while}\ \mathsf{b}@\mathbf{R}\ \{\mathcal{I}\} & \textit{(while)} & \mid & \mathbf{0} & \textit{(end)} & \mid & \mathbf{1} \ \ \textit{(skip)}
\end{array}
$$

Interaction $\mathsf{o} : \mathbf{R}(\mathsf{e}) \rightarrow \mathbf{S}(\mathsf{x})$ means that role $\mathbf{R}$ sends a message on operation o to role $\mathbf{S}$. The sent value is obtained by evaluating expression e in the local state of **R** (evaluation of expressions is always atomic) and it is then stored in the local variable $\mathtt{x}$ of **S**. Assignment $\mathtt{x}@\mathbf{R} = \mathtt{e}$ assigns the evaluation of expression $\mathtt{e}$ in the local state of **R** to its local variable $\mathtt{x}$. Processes $\mathcal{I};\mathcal{I}'$ and $\mathcal{I}|\mathcal{I}'$ denote the standard sequential and parallel composition of processes. The conditional **if** $\mathtt{b}@\mathbf{R}\ \{\mathcal{I}\}$ **else** $\{\mathcal{I}'\}$ and the iteration **while** $\mathtt{b}@\mathbf{R}\ \{\mathcal{I}\}$ are guarded by the evaluation of the boolean expression $\mathtt{b}$ in the local state of **R**. The construct **scope** $@\mathbf{R}\ \{\mathcal{I}\}$ delimits a subterm $\mathcal{I}$ of the DIOC process that may be updated in the future. In **scope** $@\mathbf{R}\ \{\mathcal{I}\}$, role **R** is the coordinator of the update, which ensures that either none of the participants update, or they all apply the same update. Finally, **1** defines a DIOC process that can only terminate while **0** represents a terminated DIOC. The latter is needed for the definition of the operational semantics and not intended to be used by the programmer. We call initial a DIOC where **0** never occurs.

DIOC processes $\mathcal{I}$ execute within a DIOC system $\langle \Sigma, \mathbf{I}, \mathcal{I}\rangle$, which pairs them with a *global state* Σ (disjoint union of the local states) and a set of available updates **I**, i.e., a set of DIOCs that may replace scopes. Set **I** may change at runtime. The semantics of DIOC systems is defined in terms of a labelled transition system (LTS) of the shape $\langle \Sigma, \mathbf{I}, \mathcal{I}\rangle \xrightarrow{\mu} \langle \Sigma', \mathbf{I}', \mathcal{I}'\rangle$ where we use μ to range over the labels. Notably, changes of set **I** are visible in the labels, thus allowing to track or restrict the set of available updates if needed.

To support inclusion of new roles in DIOCs we need to change one rule in their semantics, namely rule $\lfloor^{\mathrm{DIOC}}|_{\mathrm{UP}}\rceil$ (reported below). As done in [2], we annotate the **scope** (as well as other constructs) with an index $\mathtt{i} \in \mathbb{N}$.[2] The only requirement over annotated DIOCs is that indexes within the same program must to be distinct.

To apply an update, we need to make sure that the roles marked as new in the update are not present in the running DIOC. While in principle one could consider as new all roles not present in the target scope, in practice one needs to declare the location only for new roles, hence they need to be distinguished. For this reason, from now on updates include a set of roles expected to be new. Thus, we introduce function $\mathsf{newRoles}$ that, given an update, returns the set of new roles. Function $\mathsf{newRoles}$ is similar to function roles (cf. [2]), which instead extracts the set of roles present in a given DIOC $\mathcal{I}$.

We report below the form of rule $\lfloor^{\mathrm{DIOC}}|_{\mathrm{UP}}\rceil$ used by our extension.

$$\frac{\mathcal{I}' \in \mathbf{I} \quad \mathsf{roles}(\mathcal{I}') \setminus \mathsf{newRoles}(\mathcal{I}') \subseteq \mathsf{roles}(\mathcal{I}) \quad \mathcal{I}'' = \mathsf{fresh}(\mathcal{I}') \quad \mathsf{connected}(\mathcal{I}') \quad \mathsf{freshIndexes}(\mathcal{I}')}{\langle \Sigma, \mathbf{I}, \mathtt{i}\text{: } \mathbf{scope}\ @\mathbf{R}\ \{\mathcal{I}\}\rangle \xrightarrow{\mathcal{I}''} \left\langle \Sigma, \mathbf{I}, \mathcal{I}''\right\rangle}\ \lfloor^{\mathrm{DIOC}}|_{\mathrm{UP}}\rceil$$

First, condition $\mathcal{I}' \in \mathbf{I}$ selects an update in the set of updates. Then, condition $\mathsf{roles}(\mathcal{I}') \setminus \mathsf{newRoles}(\mathcal{I}') \subseteq \mathsf{roles}(\mathcal{I})$ ensures that roles which are not declared as new were already present in the scope. This weakens the condition $\mathsf{roles}(\mathcal{I}') \subseteq \mathsf{roles}(\mathcal{I})$

[2] Annotated DIOC constructs are useful in the definition of the projection and the related proofs to avoid interference between different constructs.

in [2], which required *all* roles to be already present in the scope. The update which is actually applied (and advertised in the label of the transition), $\mathcal{I}''$, is obtained by renaming new roles so that they are fresh for the whole DIOC. Renaming is performed by function fresh that generates new names for roles w.r.t. the whole DIOC.[3] Finally, the predicate $\mathsf{connected}(\mathcal{I}')$ checks that the DIOC is well formed while the predicate $\mathsf{freshIndex}(\mathcal{I}')$ checks that there are no interaction interferences between the rule and the original DIOC. We refer to [2] for the detailed description of both the predicates, since their behaviour is orthogonal to the addition of new roles.

4.2 Dynamic Process-Oriented Choreographies

DPOC is the abstract and formally defined language used as target by the projection function. Hence, this is the language of the programs that implement the DIOC specification.

DPOCs include *processes*, ranged over by $\mathsf{P}, \mathsf{P}', \ldots$, describing the behaviour of participants. A process P for *DPOC role* $\mathbf{R}$ executing in a local state Γ is denoted as $(\mathsf{P}, \Gamma)_{\mathbf{R}}$. A collection of executing processes for different roles is a *Network*, ranged over by $\mathcal{N}$, $\mathcal{N}'$. Finally, a DPOC system is a DPOC network equipped with a set of updates $\mathbf{I}$, namely a pair $\langle \mathbf{I}, \mathcal{N} \rangle$.

Like in DIOCs, DPOC processes communicate over operations o. Among all the communications, there are some auxiliary ones that the projection uses to implement the synchronisation mechanisms needed to realise the global choreography. We use o^* to range over auxiliary operations and $\mathsf{o}^?$ to range over both normal and auxiliary operations.

Following [2], for technical reasons, DPOC constructs are annotated using indexes $\mathsf{i}, \iota \in \mathbb{N}$. Indexes are also used to disambiguate operation names. The syntax of DPOCs is the following.

$$
\begin{array}{llllllll}
\mathsf{P} ::= & \iota\colon \mathsf{i}.\mathsf{o}^? : \mathsf{x}\ \mathbf{from}\ \mathbf{R} & \textit{(receive)} & \mid \mathsf{i}\colon \mathbf{if}\ \mathsf{b}\ \{\mathsf{P}\}\ \mathbf{else}\ \{\mathsf{P}'\} & \textit{(cond)} & \mid \mathsf{P}; \mathsf{P}' & \textit{(seq)} \\
& \mid \iota\colon \mathsf{i}.\mathsf{o}^? : \mathsf{e}\ \mathbf{to}\ \mathbf{R} & \textit{(send)} & \mid \iota\colon \mathbf{scope}\ @\mathbf{R}\ \{\mathsf{P}\}\ \mathbf{roles}\ \{\mathsf{S}\} & \textit{(coord)} & \mid \mathsf{P} \mid \mathsf{P}' & \textit{(par)} \\
& \mid \iota\colon \mathsf{i}.\mathsf{o}^* : \mathsf{X}\ \mathbf{to}\ \mathbf{R} & \textit{(send-up)} & \mid \iota\colon \mathbf{scope}\ @\mathbf{R}\ \{\mathsf{P}\} & \textit{(scope)} & \mid \mathbf{1} & \textit{(skip)} \\
& \mid \iota\colon \mathsf{x} = \mathsf{e} & \textit{(assign)} & \mid \iota\colon \mathbf{spawn}\ @\mathbf{R}\ \{\mathsf{P}\} & \textit{(spawn)} & \mid \mathbf{0} & \textit{(end)} \\
& \mid \mathsf{i}\colon \mathbf{while}\ \mathsf{b}\ \{\mathsf{P}\} & \textit{(while)} & & & &
\end{array}
$$

$$
\mathsf{X} ::= \quad \mathbf{no} \quad \mid \quad \mathsf{P} \qquad\qquad \mathcal{N} ::= \quad (\mathsf{P}, \Gamma)_{\mathbf{R}} \quad \mid \quad \mathcal{N} \parallel \mathcal{N}'
$$

[3] A compositional formalisation of function fresh would require to check freshness at all the steps of the derivation of the transition, to avoid name clashes with roles which are not in the scope but only in the context. To simplify the presentation, here we assume that function fresh has access to the set of roles of the whole DIOC.

DPOC processes include the action of receiving a message written as ι: $\mathtt{i.o}^?$: $\mathtt{x}$ **from R** and meaning that a message from role **R** is received on a specific operation $\mathtt{i.o}^?$ (either normal or auxiliary) and the value stored in variable $\mathtt{x}$. Similarly, the send action ι: $\mathtt{i.o}^?$: $\mathtt{e}$ **to R** sends the value of an expression $\mathtt{e}$ to operation $\mathtt{i.o}^?$ of role **R**. DPOC offers also a higher-order send action that instead of sending a value sends a process. This operation, written as ι: $\mathtt{i.o}^*$: X **to R**, means that the higher-order argument X is sent to role **R** and is used to distribute the updated code. In particular, X may be either the new DPOC process P that **R** has to execute or a token no notifying that no update is needed and the execution can continue with the pre-existing code. Processes also feature assignment ι: $\mathtt{x} = \mathtt{e}$ of the value of expression $\mathtt{e}$ to the variable $\mathtt{x}$.

As standard for process calculi, $\mathsf{P};\mathsf{P}'$ and $\mathsf{P}|\mathsf{P}'$ denote the sequential and parallel composition of P and P'. DPOC processes also include conditionals $\mathtt{i}$: **if** $\mathtt{b}$ $\{\mathsf{P}\}$ **else** $\{\mathsf{P}'\}$ and loops $\mathtt{i}$: **while** $\mathtt{b}$ $\{\mathsf{P}\}$. We also have the process **1** that can only successfully terminate, and the terminated process **0**. Peculiar for DPOC is instead the scope constructor for delimiting a block of code. There are two versions of it: one for the process leading the possible adaptation and one for a process involved in the adaptation but not leading it. Construct ι: **scope @R** $\{\mathsf{P}\}$ **roles** $\{\mathsf{S}\}$ defines a scope with body P and set of participants S, and may occur only inside role **R**, which acts as coordinator of the update. The shorter version ι: **scope @R** $\{\mathsf{P}\}$ is used instead inside the code of some role $\mathbf{R}_1$, which is not the coordinator **R** of the update. The difference is due to the fact that the coordinator **R** needs to know the set S of involved roles to be able to send to them their updates.

All these constructs were already present in the original DPOC described in [2]. For our extension, we introduce only the new construct ι: **spawn @R** $\{\mathsf{P}\}$, which indicates the runtime creation of a new role **R** running behaviour P.

The semantics of DPOCs is defined in terms of a labelled transition system composed of two layers. One is the semantics of DPOC roles, which specifies the local actions of each process and has the shape $(\mathsf{P},\Gamma)_{\mathbf{R}} \xrightarrow{\delta} (\mathsf{P}',\Gamma')_{\mathbf{R}}$. The second is the semantics of DPOC systems, which defines how roles interact with each other and has the shape $\langle \mathbf{I}, \mathcal{N}\rangle \xrightarrow{\eta} \langle \mathbf{I}', \mathcal{N}'\rangle$.

We now present the changes introduced in the semantics of DPOCs to support runtime role inclusion. We updated three rules and introduced two new ones. We start describing the revised rule $\lfloor^{\text{DPOC}}|_{\text{LEAD-UP}}\rceil$, which defines the semantics of the process coordinating the update in the case where an update takes place. Main novelties are: (*i*) it supports the application of updates with roles not present in the body of the **scope** and (*ii*) it includes in the behaviour of the coordinator the **spawn** instructions needed to create the new participants (if any) present in the rule. Below, we use greyed-out circled number to ease the description of the reductum.

$$\frac{\mathsf{Q} = \mathsf{newRoles}(\mathcal{I}) \quad \mathsf{roles}(\mathcal{I}) \setminus \mathsf{Q} \subseteq \mathsf{S} \quad \mathsf{connected}(\mathcal{I}) \quad \mathsf{freshIndexes}(\mathcal{I})}{(\iota\text{: } \mathbf{scope}\ @\mathbf{R}\ \{\mathsf{P}\}\ \mathbf{roles}\ \{\mathsf{S}\}, \Gamma)_{\mathbf{R}} \xrightarrow{\mathcal{I}}} \lfloor^{\text{DPOC}}|_{\text{LEAD-UP}}\rceil$$

$$\begin{pmatrix} \textcircled{1} \prod_{\mathbf{R}_j \in \mathsf{Q}} \iota\text{: } \mathbf{spawn}\ @\mathsf{R}_j\ \{\iota\text{: } \mathbf{scope}\ @\mathsf{R}\ \{\mathbf{1}\}\}; \\ \textcircled{2} \prod_{\mathsf{R}_j \in (\mathsf{S} \cup \mathsf{Q}) \setminus \{\mathbf{R}\}} \iota\text{: } \iota.\mathsf{sb}^*_\iota : \pi(\mathcal{I}, \mathbf{R}_j)\ \mathbf{to}\ \mathbf{R}_j; \\ \textcircled{3}\ \pi(\mathcal{I}, \mathbf{R}); \\ \textcircled{4} \prod_{\mathsf{R}_j \in (\mathsf{S} \cup \mathsf{Q}) \setminus \{\mathbf{R}\}} (\iota\text{: } \iota.\mathsf{se}^*_\iota : \_\ \mathbf{from}\ \mathbf{R}_j), \Gamma \end{pmatrix}_{\mathbf{R}}$$

As done in $\lfloor^{\text{DIOC}}|_{\text{UP}}\rceil$ for DIOCs, in the updated version of $\lfloor^{\text{DPOC}}|_{\text{LEAD-UP}}\rceil$ we make sure that the roles included in an update which are not new (i.e., not contained in Q) are already present in the scope (i.e., contained in S). We also check that the update satisfies the predicates connected and freshIndexes. In DIOCs, rule $\lfloor^{\text{DIOC}}|_{\text{UP}}\rceil$ also α-converts new roles to ensure their freshness. In DPOCs the same conversion is performed at the level of DPOC networks, in rule $\lfloor^{\text{DPOC}}|_{\text{LIFT-UP}}\rceil$, described at the end of this section.

In the reductum of rule $\lfloor^{\text{DPOC}}|_{\text{LEAD-UP}}\rceil$, described below, π is the process-projection function (see Sect. 4.3, cf. [2]) that generates the code from choreography $\mathcal{I}$ for role **R**. In the reductum, first ① the coordinator requires the creation of all the new roles. All new spawned roles have the same behaviour ι: **scope** @R {**1**}, which means that, when started, the new roles wait to receive from the coordinator of the update the code they need to execute (exactly as the old roles do when reaching a scope). Then, ② the coordinator sends, using a high-order communication on auxiliary operation sb^*_ι, the new code that the other roles need to execute. After having updated all the coordinated roles, ③ also the coordinator executes its own part of the update $\pi(\mathcal{I}, \mathbf{R})$. Finally ④ the coordinator waits for a message from each role involved in the update to make sure that the updated code has been completed.

To support the **spawn** construct, we define its semantics at the level of DPOC roles with the two rules denoted $\lfloor^{\text{DPOC}}|_{\text{SPAWN}}\rceil$ and $\lfloor^{\text{DPOC}}|_{\text{NEWROLE}}\rceil$. The former triggers the **spawn** at the level of DPOC roles, while the latter creates the new role at the level of DPOC systems.

$$\frac{}{(\iota\text{: } \mathbf{spawn}\ @\mathbf{R}'\ \{\mathsf{P}\}, \Gamma)_{\mathbf{R}} \xrightarrow{\mathbf{R}'\{\mathsf{P}\}} (\mathbf{1}, \Gamma)_{\mathbf{R}}} \lfloor^{\text{DPOC}}|_{\text{SPAWN}}\rceil$$

$$\frac{\mathcal{N} \xrightarrow{\mathbf{R}'\{\mathsf{P}\}} \mathcal{N}'}{\langle \mathbf{I}, \mathcal{N} \rangle \xrightarrow{\tau} \langle \mathbf{I}, \mathcal{N} \parallel (\mathsf{P}, \Gamma_0)_{\mathbf{R}'} \rangle} \lfloor^{\text{DPOC}}|_{\text{NEWROLE}}\rceil$$

Note that new roles are paired with an empty state Γ_0.

To complete the update on the semantics of DPOCs, we need to revise the original lifting rule to avoid capturing the new label $\mathbf{R}'\{\mathrm{P}\}$, which is instead handled by rule $\lfloor^{\mathrm{DPOC}}|_{\mathrm{NewRole}}\rceil$.

$$\frac{\mathcal{N} \xrightarrow{\delta} \mathcal{N}' \quad \delta \notin \{\mathcal{I},\ \mathbf{R}'\{\mathrm{P}\}\}}{\langle \mathbf{I}, \mathcal{N}\rangle \xrightarrow{\delta} \langle \mathbf{I}, \mathcal{N}'\rangle} \lfloor^{\mathrm{DPOC}}|_{\mathrm{Lift}}\rceil$$

Finally, we need to change rule $\lfloor^{\mathrm{DPOC}}|_{\mathrm{Lift\text{-}Up}}\rceil$ to perform the α-conversion of role names. Note that, while the rule selects $\mathcal{I}' \in \mathbf{I}$ as the DIOC update, reductions happen on its α-converted version $\mathcal{I}$.

$$\frac{\mathcal{I}' \in \mathbf{I} \quad \mathcal{N} \xrightarrow{\mathcal{I}} \mathcal{N}' \quad \mathcal{I} = \mathsf{fresh}(\mathcal{I}')}{\langle \mathbf{I}, \mathcal{N}\rangle \xrightarrow{\mathcal{I}} \langle \mathbf{I}, \mathcal{N}'\rangle} \lfloor^{\mathrm{DPOC}}|_{\mathrm{Lift\text{-}Up}}\rceil$$

4.3 Projection

Given a DIOC program, the projection function proj returns a DPOC network (i.e., a combination of interacting DPOC processes) that implements the semantics of the originating DIOC program. Each process is obtained by projecting the DIOC behaviour on a specific role using the process projection function π. Since **spawn** constructs are introduced during the execution of scopes, there was no need to extend the original definition of the projection function.

For the sake of presentation here we provide just an example of application of the projection, referring the interested reader to [2] for the full definition. In particular, in the following we show the application of π to DIOC interactions. Function π, given an annotated DIOC process and a role $\mathbf{R}$, returns a DPOC process for role $\mathbf{R}$. Below, in case ① the input role is the sender $\mathbf{R}_1$ and π returns the corresponding send action (indexed by i and on operation i.o) towards the receiver $\mathbf{R}_2$. Case ② is complementary to ① for the reception while case ③ produces $\mathbf{1}$, i.e., the role has no part in the interaction and skips that action.

$$
\begin{array}{ll}
① \ \pi(\ \mathrm{i}\colon \mathrm{o} : \mathbf{R}_1(\mathrm{e}) \rightarrow \mathbf{R}_2(\mathrm{x}), \mathbf{R}_1\ ) = \mathrm{i}\colon \mathrm{i.o} : \mathrm{e}\ \textbf{to}\ \mathbf{R}_2 & \\
② \ \pi(\ \mathrm{i}\colon \mathrm{o} : \mathbf{R}_1(\mathrm{e}) \rightarrow \mathbf{R}_2(\mathrm{x}), \mathbf{R}_2\ ) = \mathrm{i}\colon \mathrm{i.o} : \mathrm{x}\ \textbf{from}\ \mathbf{R}_1 & \\
③ \ \pi(\ \mathrm{i}\colon \mathrm{o} : \mathbf{R}_1(\mathrm{e}) \rightarrow \mathbf{R}_2(\mathrm{x}), \mathbf{R}'\ ) = \mathbf{1} & \text{if } \mathbf{R}' \notin \{\mathbf{R}_1, \mathbf{R}_2\}
\end{array}
$$

4.4 Example of Projection and Adaptation

To better clarify how our framework works, we provide here a minimal example of the projection and the adaptation step for an excerpt of the use case in Sect. 3, where the new role **logger** enters the choreography upon adaptation. We start by annotating the code from Sect. 3, to be able to project it since the projection function requires well-annotated DIOCs. For brevity, we consider the excerpt from lines 3–7 where the **seller** role coordinates the adaptation of the scope. We monotonically annotate every instruction starting from 1. This results in the well-annotated DIOC below.

$$\begin{array}{l} 1\text{: } \textbf{scope } @\textbf{seller}\{ \\ \quad 2\text{: } \mathtt{result}@\textbf{seller} = \texttt{"XXX"}; \\ \quad 3\text{: } \mathtt{response} : \textbf{seller}(\mathtt{result}) \rightarrow \textbf{client}(\mathtt{result})\} \end{array}$$

From the annotated DIOC, the projection generates two DPOC processes, one for the **seller** and one for the **client**. We report below the projection on the **seller**, together with a step derived using rule $\lfloor^{\text{DPOC}}|_{\text{LEAD-UP}}\rceil$, where new role **logger** enters the system. In the projection of the program, for simplicity, we omit the indexes that prefix the operations. On the left, we show the DPOC code that is obtained by projecting the previous choreography on the **seller** role. On the right, we present the DPOC process of the **seller** after the adaption step which applies the rule in Listing 1.1, whose body is denoted in the following by $\mathcal{I}$.

$$\begin{array}{l} 1\text{: } \textbf{scope } @\textbf{seller}\{ \\ \quad \mathtt{result} = \texttt{"XXX"}; \\ \quad \mathtt{response} : \mathtt{result} \textbf{ to client} \\ \} \textbf{ roles } \{ \textbf{ client } \} \end{array} \xrightarrow{\mathcal{I}} \begin{array}{l} 1\text{: } \textbf{spawn } @\textbf{logger}\{1 : \textbf{scope}@\textbf{seller}\{1\}\}; \\ \{1\text{: } \mathtt{sb}_1^* : \pi(\mathcal{I}, \textbf{client}) \textbf{ to client} \mid \\ \ 1\text{: } \mathtt{sb}_1^* : \pi(\mathcal{I}, \textbf{logger}) \textbf{ to logger}\}; \\ \pi(\mathcal{I}, \textbf{seller}); \\ \{1\text{: } \mathtt{se}_1^* : \_ \textbf{ from client} \mid \\ \ 1\text{: } \mathtt{se}_1^* : \_ \textbf{ from logger}\} \end{array}$$

As can be seen, the projection of a DIOC scope is a DPOC scope. If no adaptation rule is applied, the body of the scope is executed, hence the **seller** simply computes the answer storing it in its variable $\mathtt{result}$ and then sends it to the **client**. If, instead, an adaptation rule is applied, the **seller** first spawns new roles (if any) found in the adaptation rule. In the case of the adaptation rule in Listing 1.1, the only new role is **logger**. Once spawned, the **logger** executes the code **scope@seller**{1}. Hence, the **logger** waits to get from the **seller** the code to be executed. It is up to the seller, as leader of the adaptation, to send the new code, obtained from the body of the adaptation rule using the π operator, to both the **client** and the **logger**. This is done by performing the communications on operations 1: $\mathtt{sb}_1^*$. Then, the **seller** executes its own adapted code $\pi(\mathcal{I}, \textbf{seller})$. When the **seller** terminates the execution of its adapted code, it waits for all the led roles to notify that they also terminated the execution of their adapted code (this is done by communications on operations 1: $\mathtt{se}_1^*$) before starting to execute the rest of the choreography.

4.5 Properties

Our model, extended to support the inclusion of new roles in runtime updates, preserves the correctness properties of [2]. In particular, we will show in Theorem 1 that a DIOC system and its projection are weak trace equivalent (for finite traces), that is they have the same behaviour up to internal τ actions and communications on auxiliary operations. As shown in [2], this property implies deadlock freedom, which is indeed formalised by requiring finite internal traces (an internal trace is obtained by removing all transitions with label $\mathcal{I}$ from a trace) to end with $\surd$.

Definition 1 (DIOC traces). *A* (strong) trace *of a DIOC system* $\langle \Sigma_1, \mathbf{I}_1, \mathcal{I}_1 \rangle$ *is a sequence (finite or infinite) of labels* $\mu_1, \mu_2, \ldots$ *such that there is a sequence of DIOC system transitions* $\langle \Sigma_1, \mathbf{I}_1, \mathcal{I}_1 \rangle \xrightarrow{\mu_1} \langle \Sigma_2, \mathbf{I}_2, \mathcal{I}_2 \rangle \xrightarrow{\mu_2} \ldots$.

A weak trace *of a DIOC system* $\langle \Sigma_1, \mathbf{I}_1, \mathcal{I}_1 \rangle$ *is a sequence of labels* $\mu_1, \mu_2, \ldots$ *obtained by removing all silent labels* τ *from a trace of* $\langle \Sigma_1, \mathbf{I}_1, \mathcal{I}_1 \rangle$.

Definition 2 (DPOC traces). *A* (strong) trace *of a DPOC system* $\langle \mathbf{I}_1, \mathcal{N}_1 \rangle$ *is a sequence (finite or infinite) of labels* $\eta_1, \eta_2, \ldots$ *with*

$$\eta_i \in \{\tau, o^? : \mathbf{R}_1(v) \to \mathbf{R}_2(x), o^* : \mathbf{R}_1(X) \to \mathbf{R}_2(), \surd, \mathcal{I}, \textbf{no-up}, \mathbf{I}\}$$

such that there is a sequence of transitions $\langle \mathbf{I}_1, \mathcal{N}_1 \rangle \xrightarrow{\eta_1} \langle \mathbf{I}_2, \mathcal{N}_2 \rangle \xrightarrow{\eta_2} \ldots$.
A weak trace *of a DPOC system* $\langle \mathbf{I}_1, \mathcal{N}_1 \rangle$ *is a sequence of labels* $\eta_1, \eta_2, \ldots$ *obtained by removing all the labels corresponding to auxiliary communications, i.e., of the form* $o^* : \mathbf{R}_1(v) \to \mathbf{R}_2(x)$ *or* $o^* : \mathbf{R}_1(X) \to \mathbf{R}_2()$*, and the silent labels* τ*, from a trace of* $\langle \mathbf{I}_1, \mathcal{N}_1 \rangle$.

DPOC traces do not allow send and receive actions. Indeed these actions represent incomplete interactions, thus they are needed for compositionality reasons, but they do not represent relevant behaviours of complete systems. Note also that these actions have no correspondence at the DIOC level, where only whole interactions are allowed.

Definition 3 (Finite trace equivalence). *Two DIOC systems, or two DPOC systems, or a DIOC and a DPOC system are* finite (weak) trace equivalent *iff their sets of finite (weak) traces do coincide.*

Lemma 1 (Projection preserves weak traces [2, Corollary 7.5]). *Consider the semantics without new roles. For each initial, connected DIOC process* $\mathcal{I}$*, each state* Σ*, each set of updates* $\mathbf{I}$*, the DIOC system* $\langle \Sigma, \mathbf{I}, \mathcal{I} \rangle$ *and the DPOC system* $\langle \mathbf{I}, \mathsf{proj}(\mathcal{I}, \Sigma) \rangle$ *are weak trace equivalent.*

We now show that a similar result holds in the system with new roles. We will show this only for finite traces. We conjecture that this holds also for infinite traces, however, while the proof for finite traces can be done using the result in [2, Corollary 7.5] as a black box, we have not been able to do the same for infinite traces. The alternative of redoing all the proofs in [2, Corollary 7.5] (including the preliminary results therein) is beyond the scope of this paper. This last option would also allow us to prove race freedom and orphan message freedom.

Theorem 1 (Projection preserves finite weak traces, with new roles). *For each initial, connected DIOC process* $\mathcal{I}$*, each state* Σ*, each set of updates* $\mathbf{I}$*, the DIOC system* $\langle \Sigma, \mathbf{I}, \mathcal{I} \rangle$ *and the DPOC system* $\langle \mathbf{I}, \mathsf{proj}(\mathcal{I}, \Sigma) \rangle$ *are finite weak trace equivalent.*

Proof (sketch). For each finite weak trace we provide a translation from the DIOC $\mathcal{I}$ generating it to a DIOC $\mathcal{I}'$ where the same finite weak trace does not require new roles. We show that $\mathcal{I}$ and $\mathcal{I}'$ are finite weak trace equivalent, and the same holds for their projections. Hence, given a finite weak trace of $\mathcal{I}$, $\mathcal{I}'$ also has it. Thanks to Lemma 1 this means that the projection of $\mathcal{I}'$ has the trace thus implying that the trace also belongs to the traces of the projection of the original system $\mathcal{I}$. Dually, given a finite weak trace of the projection of the original system $\mathcal{I}$, then the projection of $\mathcal{I}'$ has it thus implying that $\mathcal{I}'$ and $\mathcal{I}$ have the same trace.

Let us consider a finite fixed weak trace. It is clear that the trace uses only a finite amount of new roles. The translation takes a DIOC $\mathcal{I}$ with these new roles and replaces each scope **scope** $@\mathbf{R}$ $\{\mathcal{I}\}$ with

$$\mathbf{scope}\ @\mathbf{R}\ \{\mathbf{if}\ \mathtt{false}@\mathbf{R}\ \{\prod_{\mathbf{R}'} \mathtt{unused}@\mathbf{R}' = 0\}\ \mathbf{else}\ \{\mathbf{1}\}; \mathcal{I}\}$$

where the product stands for n-ary parallel composition, and ranges over all new roles that are created by instantiating the considered scope in the trace under analysis. Variable $\mathtt{unused}$ is an unused variable.

The evaluation of the conditional always selects the else branch, hence it only causes one additional τ step. Apart from this, DIOC $\mathcal{I}$ and its translation generate the same traces. Also, the trace under analysis can be generated by the new DIOC without the need for dynamically generating new roles.

Let us now consider the projections. The projection of $\mathcal{I}$ does not have the auxiliary steps needed for the conditional (a τ to evaluate the guard and some auxiliary communications and τ steps to notify the result of the evaluation to all the involved roles), and has some additional τ steps to spawn the new roles. However, these are all auxiliary steps, hence the weak traces do coincide. □

We remark that even if the proof above relies on the system without new roles, this does not mean that the two systems are equivalent. Indeed, the proof requires one system without new roles for each possible trace, hence infinitely many of them. One single system with new roles captures all these behaviours.

Finally, we note that the DIOC and DPOC are not image-finite [8]. Indeed, both of them allow for changing in an arbitrary way the set of available updates in one step. Hence, each system can have an infinite amount of successors. If we allow only for a finite number of rules at a time, given that the evaluation of expressions is deterministic, the system becomes image-finite and therefore the (infinite) weak trace equivalence can be derived directly from the finite weak trace equivalence [8].

5 Implementation

In this section we overview the components and functionalities of the AIOCJ runtime support and present the main design and implementation choices made to support the inclusion of new roles in choreographies.

AIOCJ is a framework that allows the development of choreographies and update rules that can be projected to runnable and deployable distributed programs. To improve usability, as exemplified in Sect. 3, the AIOCJ syntax used to define the choreographies is not the formal DIOC one, but an embellished version. The target language for the projection function is Jolie [9]. This language, often used to develop microservices [10], was adopted because it offers programming primitives very close to the DPOC ones, thus making the definition of the projection function easier.

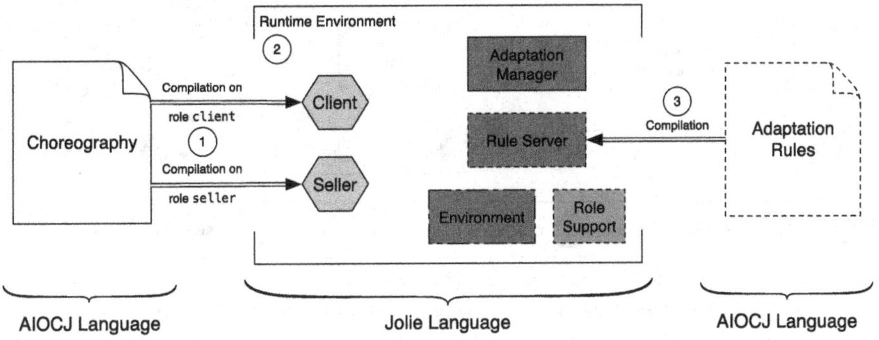

Fig. 1. AIOCJ components. Left and right, choreographic artefacts. Centre, executable components: projected roles (hexagons) and runtime support programs (rectangles).

The basic AIOCJ runtime support includes three kinds of components: the Adaptation Manager, the Rule Server, and the Environment. These components are depicted as rectangles in Fig. 1. Additionally, to support inclusion of new roles at runtime, we introduced a new component, called the Role Supporter; depicted in Fig. 1 as a rectangle with a darker colour.

All runtime support components are optional, meaning that an AIOCJ choreography without adaptation scopes does not need the presence of any of the runtime support components to execute. When scopes are present, the only mandatory component is the Adaptation Manager that interacts with the projected code to find possible applicable rules and retrieve their code. In turn, the Adaptation Manager works as registry for Rule Servers, which store the adaptation rules. Every time a new set of adaptation rules is projected with AIOCJ, the AIOCJ compiler synthesises a new Rule Server that contains the executable code corresponding to the projection of the body of the rule on each participant occurring there. A Rule Server registers to the Adaptation Manager, making its rules available.

When managing an adaptation step, the Adaptation Manager invokes the registered Rule Servers to check which rules are applicable. If at least one rule is applicable, the Adaptation Manager selects one and obtains the code update for the selected rule. Since in AIOCJ applicability conditions of rules may refer to properties of the execution environment (e.g., time, temperature), the AIOCJ runtime offers an Environment service that stores and publishes data on the status of the execution environment.

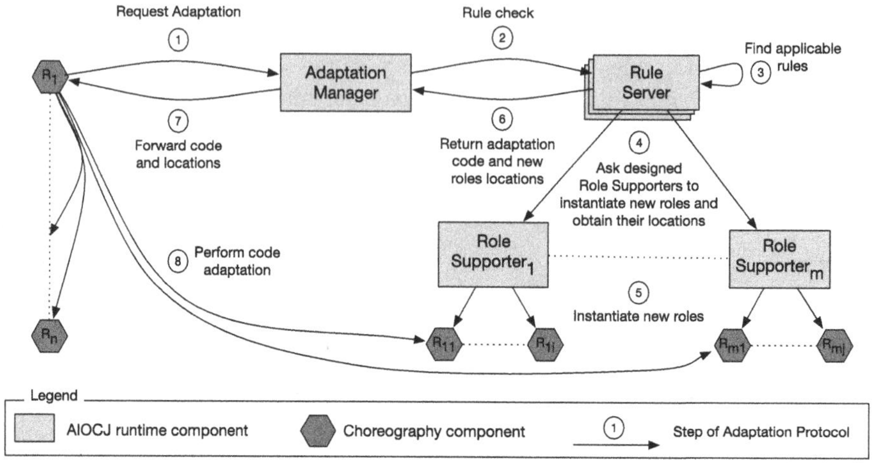

Fig. 2. Representation of the steps of adaptation with new roles.

The Role Supporter component, as the name entails, supports the deployment of new roles. New roles are meant to add a new participant into a pre-existing choreography accessing new functionalities and useful, e.g., for system integration and evolution [11]. A Role Supporter component has to be deployed in the premises of the location of each new participant. The deployment information that is abstracted away in DIOCs, is instead explicitly defined in AIOCJ rules. New roles are marked with the keyword `newRoles`. Their location is stated using the syntax shown at line 5 of Listing 1.1 (`location@logger:"..."`).

The protocol to coordinate the instantiation of new roles is depicted in Fig. 2. The protocol starts with the request made by an adaptation leader (R_1) to the Adaptation Manager ①. After receiving the request, the Adaptation Manager contacts its registered Rule Servers to look for applicable rules ②. If there is an adaptation rule that is applicable, the Rule Servers find it ③. Now, if the selected rule contains new roles, the Rule Server checks if the new roles can be deployed (i.e., there are Role Supporters available at the locations of the new roles).[4] Assuming new roles are needed, at step ④ the Rule Server contacts the interested Role Supporters and invokes the instantiation of dedicated new roles (e.g., $R_{11} \dots R_{mj}$). The new roles are instantiated at step ⑤. Each new role is located at a unique, fresh address (prefixed by the location defined in the rule). In this way, parallel executions of the same rule can run without the need for any coordination among parallel rule applications. After each new role

[4] Note that in case the new roles cannot be instantiated, the adaptation rule is considered not applicable and the process of rule selection proceeds discarding this rule.

has started, its Role Supporter responds to the invoking Rule Server with the address of the instantiated role. Hence, the Rule Server collects the locations of all new roles, which the adaptation leader (R_1) will use to contact them to finalise the adaptation process. After this, or if no new roles are needed, the protocol continues with steps ⑥ and ⑦ that forward the adaptation code of each participant back to the Adaptation Manager and, immediately after, to the adaptation leader. Finally, at step ⑧, the adaptation leader (R_1) distributes the adaptation code to the participants (the previously present $\mathsf{R}_1, \cdots, \mathsf{R}_n$ and the newly instantiated $\mathsf{R}_{11} \cdots \mathsf{R}_{mj}$), to proceed executing the updated behaviour.

All the components for managing adaptation described above are written in Jolie [9] and can be easily deployed using some scripts we provide. For the extension, the code of the Rule Server has been updated to take into account the possibility to use Role Supporters. This last component has been instead created from scratch. The code of the Adaptation Manger and Environment did not require any changes. For more technical details on AIOCJ, an explanation on how to deploy and use AIOCJ with the new extension and its actual implementation we refer the interested reader to [7,12].

6 Related Work

This paper extends dynamic choreographic programming [2] to support the introduction at runtime of new participants. While referring to the related work in [2] for further details, in this section we describe the main distinctive features of our approach and the work closest to it. As far as we know, the approach in [2] is the only one encompassing (i) adaptation for distributed systems, (ii) guarantees of relevant correctness properties by construction, and (iii) a working implementation. To the best of our knowledge, existing proposals share only two of those qualities.

In the literature we can find several middlewares and architectures enabling run-time adaptation [13–17] (see also the related survey [18]). These proposals provide tools for programming adaptive systems, but they do not offer by construction correctness guarantees on the behaviour of the system during and after adaptation. Some of them, however, such as [17], allow one to check correctness properties using techniques such as model checking. In order to do this, they assume knowledge of all the possible available adaptations at the moment of writing the adaptable application.

Other approaches are based on session types [19–22], choreography languages [23,24], behavioural contracts [25], and ad-hoc scripting languages [26]. Those works provide high-level specifications to describe the expected behaviour of a distributed system, ensuring relevant correctness properties, however they assume a static system and are not suitable for runtime adaptation.

There are also approaches based on adaptive choreographies [27–29], however they are not implemented and they concentrate on correctness checking more than on code generation. In particular, [27] concentrates on systems that autonomously switch among a set of pre-defined behaviours, [28] supports system update when no protocol is ongoing, and [29] requires to check global conditions on the system to ensure correctness of adaptation and therefore it is not suitable for large and complex distributed systems.

Finally, among the existing proposals based on choreographic programming, we note [30], where the authors define a compositionality mechanism for choreographies that, although not specifically targeted for adaptation, constitutes a first technical step to support it.

7 Conclusion and Future Work

We presented an extension of dynamic choreographic programming [2] to support the runtime introduction of new roles, extending both the related theory and the AIOCJ programming language [7].

Directions for future research include optimizing code generation to reduce the number of auxiliary communications, introducing in choreographic programs more structured forms of adaptation such as aspects, and developing or exploiting state-of-the-art DevOps tools to automatise the deployment of the different services generated by the AIOCJ framework.

References

1. Montesi, F.: Kickstarting choreographic programming. In: Hildebrandt, T., Ravara, A., van der Werf, J.M., Weidlich, M. (eds.) WS-FM 2014-2015. LNCS, vol. 9421, pp. 3–10. Springer, Cham (2016). https://doi.org/10.1007/978-3-319-33612-1_1
2. Dalla Preda, M., Gabbrielli, M., Giallorenzo, S., Lanese, I., Mauro, J.: Dynamic choreographies: theory and implementation. Logical Methods Comput. Sci. **13**(2) (2017)
3. Lu, S., Park, S., Seo, E., Zhou, Y.: Learning from mistakes: a comprehensive study on real world concurrency bug characteristics. In: ASPLOS, pp. 329–339. ACM (2008)
4. Humble, J., Farley, D.: Continuous Delivery: Reliable Software Releases through Build, Test, and Deployment Automation. Pearson Education, London (2010)
5. Gabbrielli, M., Giallorenzo, S., Guidi, C., Mauro, J., Montesi, F.: Self-reconfiguring microservices. In: Ábrahám, E., Bonsangue, M., Johnsen, E.B. (eds.) Theory and Practice of Formal Methods. LNCS, vol. 9660, pp. 194–210. Springer, Cham (2016). https://doi.org/10.1007/978-3-319-30734-3_14
6. Bravetti, M., Giallorenzo, S., Mauro, J., Talevi, I., Zavattaro, G.: Optimal and automated deployment for microservices. In: Hähnle, R., van der Aalst, W. (eds.) FASE 2019. LNCS, vol. 11424, pp. 351–368. Springer, Cham (2019). https://doi.org/10.1007/978-3-030-16722-6_21
7. AIOCJ website. http://www.cs.unibo.it/projects/jolie/aiocj.html
8. Bergstra, J.A., Ponse, A., Smolka, S.A. (eds.): Handbook of Process Algebra. Elsevier Science Inc., New York (2001)
9. Montesi, F., Guidi, C., Zavattaro, G.: Service-oriented programming with Jolie. In: Bouguettaya, A., Sheng, Q., Daniel, F. (eds.) Web Services Foundations, pp. 81–107. Springer, New York (2014). https://doi.org/10.1007/978-1-4614-7518-7_4

10. Dragoni, N., et al.: Microservices: yesterday, today, and tomorrow. Present and Ulterior Software Engineering, pp. 195–216. Springer, Cham (2017). https://doi.org/10.1007/978-3-319-67425-4_12
11. Giallorenzo, S., Lanese, I., Russo, D.: ChIP: a choreographic integration process. In: Panetto, H., Debruyne, C., Proper, H., Ardagna, C., Roman, D., Meersman, R. (eds.) OTM 2018. Lecture Notes in Computer Science, vol. 11230, pp. 22–40. Springer, Cham (2018). https://doi.org/10.1007/978-3-030-02671-4_2
12. Giallorenzo, S., Lanese, I., Mauro, J., Gabbrielli, M.: Programming adaptive microservice applications: an AIOCJ tutorial. In: Behavioural Types: from Theory to Tools, pp. 147–167. River Publishers (2017)
13. Bucchiarone, A., Marconi, A., Pistore, M., Raik, H.: Dynamic adaptation of fragment-based and context-aware business processes. In: ICWS, pp. 33–41. IEEE (2012)
14. Chen, W.-K., Hiltunen, M.A., Schlichting, R.D.: Constructing adaptive software in distributed systems. ICDCS. LNCS, vol. 6084, pp. 635–643. Springer, Heidelberg (2001). https://doi.org/10.1109/ICDSC.2001.918994
15. Ghezzi, C., Pradella, M., Salvaneschi, G.: An evaluation of the adaptation capabilities in programming languages. In: SEAMS, pp. 50–59. ACM (2011)
16. Lanese, I., Bucchiarone, A., Montesi, F.: A framework for rule-based dynamic adaptation. In: Wirsing, M., Hofmann, M., Rauschmayer, A. (eds.) TGC 2010. LNCS, vol. 6084, pp. 284–300. Springer, Heidelberg (2010). https://doi.org/10.1007/978-3-642-15640-3_19
17. Zhang, J., Goldsby, H., Cheng, B.H.C.: Modular verification of dynamically adaptive systems. In: AOSD, pp. 161–172. ACM (2009)
18. Leite, L.A.F., et al.: A systematic literature review of service choreography adaptation. SOCA **7**(3), 199–216 (2013)
19. Carbone, M., Honda, K., Yoshida, N.: Structured communication-centered programming for web services. ACM Trans. Program. Lang. Syst. **34**(2), 8 (2012)
20. Carbone, M., Montesi, F.: Deadlock-freedom-by-design: multiparty asynchronous global programming. In: POPL, pp. 263–274. ACM (2013)
21. Castagna, G., Dezani-Ciancaglini, M., Padovani, L.: On global types and multiparty session. Logical Methods Comput. Sci. **8**(1) (2012)
22. Honda, K., Yoshida, N., Carbone, M.: Multiparty asynchronous session types. In: POPL, pp. 273–284. ACM (2008)
23. Basu, S., Bultan, T., Ouederni, M.: Deciding choreography realizability. In: POPL, pp. 191–202. ACM (2012)
24. Lanese, I., Guidi, C., Montesi, F., Zavattaro, G.: Bridging the gap between interaction- and process-oriented choreographies. In: SEFM, pp. 323–332. IEEE (2008)
25. Bravetti, M., Zavattaro, G.: Towards a unifying theory for choreography conformance and contract compliance. In: Lumpe, M., Vanderperren, W. (eds.) SC 2007. LNCS, vol. 4829, pp. 34–50. Springer, Heidelberg (2007). https://doi.org/10.1007/978-3-540-77351-1_4
26. Bergstra, J.A., Klint, P.: The discrete time TOOLBUS - a software coordination architecture. Sci. Comput. Program. **31**(2–3), 205–229 (1998)
27. Coppo, M., Dezani-Ciancaglini, M., Venneri, B.: Self-adaptive multiparty sessions. SOCA **9**(3–4), 249–268 (2015)
28. Di Giusto, C., Pérez, J.A.: Event-based run-time adaptation in communication-centric systems. Formal Asp. Comput. **28**(4), 531–566 (2016)

29. Anderson, G., Rathke, J.: Dynamic software update for message passing programs. In: Jhala, R., Igarashi, A. (eds.) APLAS 2012. LNCS, vol. 7705, pp. 207–222. Springer, Heidelberg (2012). https://doi.org/10.1007/978-3-642-35182-2_15
30. Montesi, F., Yoshida, N.: Compositional choreographies. In: D'Argenio, P.R., Melgratti, H. (eds.) CONCUR 2013. LNCS, vol. 8052, pp. 425–439. Springer, Heidelberg (2013). https://doi.org/10.1007/978-3-642-40184-8_30

A Complete Axiomatization of Branching Bisimilarity for a Simple Process Language with Probabilistic Choice (Extended Abstract)

Rob J. van Glabbeek[1,2], Jan Friso Groote[3], and Erik P. de Vink[3](✉)

[1] Data61, CSIRO, Sydney, Australia
rvg@cs.stanford.edu
[2] Computer Science and Engineering, University of New South Wales, Sydney, Australia
[3] Department of Mathematics and Computer Science, Eindhoven University of Technology, Eindhoven, The Netherlands
J.F.Groote@tue.nl, evink@win.tue.nl

Abstract. This paper proposes a notion of branching bisimilarity for non-deterministic probabilistic processes. In order to characterize the corresponding notion of rooted branching probabilistic bisimilarity, an equational theory is proposed for a basic, recursion-free process language with non-deterministic as well as probabilistic choice. The proof of completeness of the axiomatization builds on the completeness of strong probabilistic bisimilarity on the one hand and on the notion of a concrete process, i.e. a process that does not display (partially) inert τ-moves, on the other hand. The approach is first presented for the non-deterministic fragment of the calculus and next generalized to incorporate probabilistic choice, too.

1 Introduction

In [11], in a setting of a process language featuring both non-deterministic and probabilistic choice, Yuxin Deng and Catuscia Palamidessi propose an equational theory for a notion of weak bisimilarity and prove its soundness and completeness. Not surprisingly, the axioms dealing with a silent step are reminiscent to the well-known τ-laws of Milner [26,27]. The process language treated in [11] includes recursion, thereby extending the calculus and axiomatization of [6]. While the weak transitions of [11] can be characterized as finitary, infinitary semantics is treated in [15], providing a sound and complete axiomatization also building on the seminal work of Milner [27].

In this paper we focus on branching bisimilarity in the sense of [17], rather than on weak bisimilarity as in [6,11,15]. In the non-probabilistic setting branching bisimilarity has the advantage over weak bisimilarity that it has far more efficient algorithms [18,19]. Furthermore, it has a strong logical underpinning [9].

M. S. Alvim et al. (Eds.): Palamidessi Festschrift, LNCS 11760, pp. 139–162, 2019.
https://doi.org/10.1007/978-3-030-31175-9_9

It would be very attractive to have these advantages available also in the probabilistic case, where model checking is more demanding. See also the initial work reported in [21].

For a similarly basic process language as in [11], without recursion though, we propose a notion of branching probabilistic bisimilarity as well as a sound and complete equational axiomatization. Hence, instead of lifting all τ-laws to the probabilistic setting, we only need to do this for the B-axiom of [17], the single axiom capturing inert silent steps. For what is referred to as the alternating model [22], branching probabilistic bisimilarity has been studied in [2,3]. Also [28] discusses branching probabilistic bisimilarity. However, the proposed notions of branching bisimilarity are either no congruence for the parallel operator, or they invalidate the identities below which we desire. The paper [1] proposes a complete theory for a variant of branching bisimilarity that is not consistent with the first τ-law unfortunately.

Our investigation is led by the wish to identify the three processes below, involving as a subprocess a probabilistic choice between P and Q. Essentially, ignoring the occurrence of the action a involved, the three processes represent (i) a probabilistic choice of weight $\frac{3}{4}$ between to instances of the subprocess mentioned, (ii) the subprocess on its own, and (iii) a probabilistic choice of weight $\frac{1}{3}$ for the subprocess and a rescaling of the subprocess, in part overlapping.

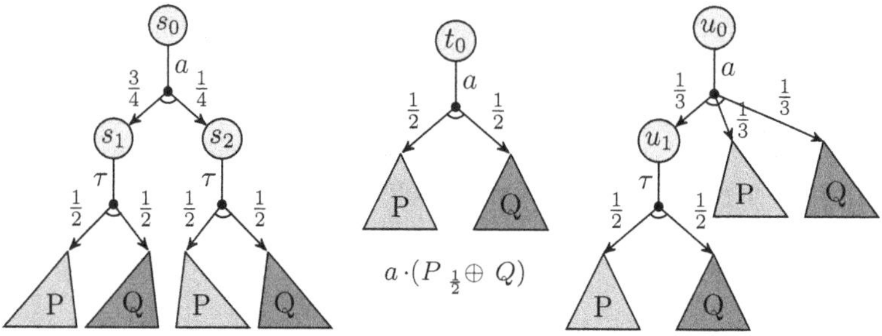

$a\cdot\big(\partial(\tau\cdot(P\ {}_{\frac{1}{2}}\!\oplus\ Q))\ {}_{\frac{3}{4}}\!\oplus\ \partial(\tau\cdot(P\ {}_{\frac{1}{2}}\!\oplus\ Q))\big)$

$a\cdot(P\ {}_{\frac{1}{2}}\!\oplus\ Q)$

$a\cdot\big(\partial(\tau\cdot(P\ {}_{\frac{1}{2}}\!\oplus\ Q))\ {}_{\frac{1}{3}}\!\oplus\ (P\ {}_{\frac{1}{2}}\!\oplus\ Q)\big)$

In our view, all three processes starting from s_0, t_0, and u_0 are equivalent. The behavior that can be observed from them when ignoring τ-steps and coin tosses to resolve probabilistic choices is the same. This leads to a definition of probabilistic branching bisimilarity that hitherto was not proposed in the literature and appears to be the pendant of weak distribution bisimilarity defined by [13].

As for [11] we seek to stay close to the treatment of the non-deterministic fragment of the process calculus at hand. However, as an alternate route in proving completeness, we rely on the definition of a concrete process. We first apply the approach for strictly non-deterministic processes and *mutatis mutandis* for the whole language allowing processes that involve both non-deterministic and probabilistic choice. For now, let's call a process concrete if it doesn't exhibit inert transitions, i.e. τ-transitions that don't change the potential behavior of the process essentially. The approach we follow first establishes soundness for branching (probabilistic) bisimilarity and soundness and completeness for strong (probabilistic) bisimilarity. Because of the non-inertness of the silent steps involved, strong and branching bisimilarity coincide for concrete processes. The trick then is to relate a pair of branching (probabilistically) bisimilar processes to a corresponding pair of concrete processes. Since these are also branching (probabilistically) bisimilar as argued, they are consequently strongly (probabilistically) bisimilar, and, voilà, provably equal by the completeness result for strong (probabilistic) bisimilarity.

The remainder of the paper is organized as follows. In Sect. 2 we gather some notation regarding probability distributions. For illustration purposes Sect. 3 treats the simpler setting of non-deterministic processes reiterating the completeness proof for the equational theory of [17] for rooted branching bisimilarity. Next, after introducing branching probabilistic bisimilarity and some of its fundamental properties in Sects. 4 and 5, respectively, in Sect. 6 we prove the main result, viz. the completeness of an equational theory for rooted branching probabilistic bisimilarity, following the same lines set out in Sect. 3. In Sect. 7 we wrap up and make concluding remarks.

2 Preliminaries

Let $\mathit{Distr}(X)$ be the set of distributions over the set X of finite support. The support of a distribution μ is denoted as $\mathit{spt}(\mu)$. Each distribution $\mu \in \mathit{Distr}(X)$ can be represented as $\mu = \bigoplus_{i \in I} p_i * x_i$ when $\mu(x_i) = p_i$ for $i \in I$ and $\sum_{i \in I} p_i = 1$. We assume I to be a finite index set. In concrete cases, when no confusion arises, the separator $*$ is omitted from the notation. For convenience later, we do not require $x_i \neq x_{i'}$ for $i \neq i'$ nor $p_i > 0$ for $i, i' \in I$.

We use $\delta(x)$ to denote the Dirac distribution for $x \in X$. For $\mu, \nu \in \mathit{Distr}(X)$ and $r \in [0,1]$ we define $\mu \,{}_r\!\oplus\, \nu \in \mathit{Distr}(X)$ by $(\mu \,{}_r\!\oplus\, \nu)(x) = r \cdot \mu(x) + (1-r) \cdot \nu(x)$. By definition $\mu \,{}_0\!\oplus\, \nu = \nu$ and $\mu \,{}_1\!\oplus\, \nu = \mu$. For an index set I, $p_i \in [0,1]$ and $\mu_i \in \mathit{Distr}(X)$, we define $\bigoplus_{i \in I} p_i * \mu_i \in \mathit{Distr}(X)$ by $(\bigoplus_{i \in I} p_i * \mu_i)(x) = \sum_{i \in I} p_i \cdot \mu_i(x)$ for $x \in X$. For $\mu = \bigoplus_{i \in I} p_i * \mu_i$, $\nu = \bigoplus_{i \in I} p_i * \nu_i$, and $r \in [0,1]$ it holds that $\mu \,{}_r\!\oplus\, \nu = \bigoplus_{i \in I} (\mu_i \,{}_r\!\oplus\, \nu_i)$.

For a binary relation $\mathcal{R} \subseteq \mathit{Distr}(X) \times \mathit{Distr}(X)$ we use $\mathcal{R}^\dagger$ to denote its symmetric closure.

3 Completeness: The Non-deterministic Case

In this section we present an approach to prove completeness of an axiomatic theory for branching bisimilarity exploiting the notion of a concrete process in the setting of a basic process language. In the remainder of the paper we extend the approach to a process language involving probabilistic choice.

We assume to be given a set of actions $\mathcal{A}$ including the so-called silent action τ. The process language we consider is called a Minimal Process Language in [4]. It provides inaction $\mathbf{0}$, a prefix construct for each action $a \in \mathcal{A}$, and non-deterministic choice.

Definition 3.1 (Syntax). The class $\mathcal{E}$ of non-deterministic processes over $\mathcal{A}$, with typical element E, is given by

$$E ::= \mathbf{0} \mid \alpha \cdot E \mid E + E$$

with actions α from $\mathcal{A}$.

The process $\mathbf{0}$ cannot perform any action, $\alpha \cdot E$ can perform action α and subsequently behave as E, and $E_1 + E_2$ represents the choice in behavior between E_1 and E_2.

For $E \in \mathcal{E}$ we define its complexity $c(E)$ by $c(\mathbf{0}) = 0$, $c(\alpha \cdot E) = c(E) + 1$, and $c(E + F) = c(E) + c(F)$.

The behavior of processes in $\mathcal{E}$ is given by a structured operational semantics going back to [24].

Definition 3.2 (Operational semantics). The transition relation $\rightarrow \subseteq \mathcal{E} \times \mathcal{A} \times \mathcal{E}$ is given by

$$\frac{}{\alpha \cdot E \xrightarrow{\alpha} E}\ (\text{PREF})$$

$$\frac{E_1 \xrightarrow{\alpha} E_1}{E_1 + E_2 \xrightarrow{\alpha} E_1}\ (\text{ND-CHOICE 1}) \qquad \frac{E_2 \xrightarrow{\alpha} E_2}{E_1 + E_2 \xrightarrow{\alpha} E_2}\ (\text{ND-CHOICE 2})$$

We have auxiliary definitions and relations derived from the transition relation of Definition 3.2. A process $E' \in \mathcal{E}$ is called a derivative of a process $E \in \mathcal{E}$ iff $E_0, \dots, E_n \in \mathcal{E}$ and $\alpha_1, \dots, \alpha_n$ exist such that $E \equiv E_0, E_{i-1} \xrightarrow{\alpha_i} E_i$, and $E_n \equiv E'$. We define $der(E) = \{ E' \in \mathcal{E} \mid E' \text{ derivative of } E \}$. Furthermore, for $E, E' \in \mathcal{E}$ and $\alpha \in \mathcal{A}$ we write $E \xrightarrow{(\alpha)} E'$ iff $E \xrightarrow{\alpha} E'$, or $\alpha = \tau$ and $E = E'$. We use $\Rightarrow$ to denote the reflexive transitive closure of $\xrightarrow{(\tau)}$.

The definitions of strong and branching bisimilarity for $\mathcal{E}$ are standard and adapted from [17,26].

Definition 3.3 (Strong and branching bisimilarity).

(a) A symmetric relation $\mathcal{R} \subseteq \mathcal{E} \times \mathcal{E}$ is called a *strong bisimulation relation* iff for all $E, E', F \in \mathcal{E}$ if $E \,\mathcal{R}\, F$ and $E \xrightarrow{\alpha} E'$ then there is an $F' \in \mathcal{E}$ such that

$$F \xrightarrow{\alpha} F' \text{ and } E' \,\mathcal{R}\, F'.$$

(b) A symmetric relation $\mathcal{R} \subseteq \mathcal{E} \times \mathcal{E}$ is called a *branching bisimulation relation* iff for all $E, E', F \in \mathcal{E}$ if $E \,\mathcal{R}\, F$ and $E \xrightarrow{\alpha} E'$, then there are $\bar{F}, F' \in \mathcal{E}$ such that

$$F \Rightarrow \bar{F},\ \bar{F} \xrightarrow{(\alpha)} F',\ E \,\mathcal{R}\, \bar{F}, \text{ and } E' \,\mathcal{R}\, F'.$$

(c) Strong bisimilarity, denoted by $\leftrightarroweq \subseteq \mathcal{E} \times \mathcal{E}$, and branching bisimilarity, written as $\leftrightarroweq_b \subseteq \mathcal{E} \times \mathcal{E}$, are defined as the largest strong bisimulation relation on $\mathcal{E}$ and the largest branching bisimulation relation on $\mathcal{E}$, respectively.

Clearly, in view of the definitions, strong bisimilarity between two processes implies branching bisimilarity between the two processes.

If for a transition $E \xrightarrow{\tau} E'$ we have that $E \leftrightarroweq_b E'$, the transition is called *inert*. A process $\bar{E}$ is called *concrete* iff it has no inert transitions, i.e., if $E' \in der(\bar{E})$ and $E' \xrightarrow{\tau} E''$, then $E' \not\leftrightarroweq_b E''$. We write $\mathcal{E}_{cc} = \{ \bar{E} \in \mathcal{E} \mid \bar{E} \text{ concrete} \}$.

Next we introduce a restricted form of branching bisimilarity, called rooted branching bisimilarity, instigated by the fact that branching bisimilarity itself is not a congruence for the choice operator. This makes branching bisimilarity unsuitable for equational reasoning where it is natural to replace subterms by equivalent terms. Note that weak bisimilarity has the same problem [26].

For example, we have for any process E that E and $\tau \cdot E$ are branching bisimilar, but in the context of a non-deterministic alternative they may not, i.e., it is not necessarily the case $E + F \leftrightarroweq_b \tau \cdot E + F$. More concretely, although $\mathbf{0} \leftrightarroweq_b \tau \cdot \mathbf{0}$, it does not hold that $\mathbf{0} + b \cdot \mathbf{0} \leftrightarroweq_b \tau \cdot \mathbf{0} + b \cdot \mathbf{0}$. The τ-move of $\tau \cdot \mathbf{0} + b \cdot \mathbf{0}$ to $\mathbf{0}$ has no counterpart in $\mathbf{0} + b \cdot \mathbf{0}$ because $\mathbf{0} + b \cdot \mathbf{0} \not\leftrightarroweq_b \mathbf{0}$.

Definition 3.4. A symmetric $\mathcal{R} \subseteq \mathcal{E} \times \mathcal{E}$ is called a rooted branching bisimulation relation iff for all $E, F \in \mathcal{E}$ such that $E \,\mathcal{R}\, F$ it holds that if $E \xrightarrow{\alpha} E'$ for $\alpha \in \mathcal{A}$, $E' \in \mathcal{E}$ then $F \xrightarrow{\alpha} F'$ and $E' \leftrightarroweq_b F'$ for some $F' \in \mathcal{E}$. Rooted branching bisimilarity, denoted by $\leftrightarroweq_{rb} \subseteq \mathit{Distr}(\mathcal{E}) \times \mathit{Distr}(\mathcal{E})$, is defined as the largest rooted branching bisimulation relation.

The definition of rooted branching bisimilarity boils down to calling processes $E, F \in \mathcal{E}$ rooted branching bisimilar, notation $E \leftrightarroweq_{rb} F$, iff (i) $E \xrightarrow{\alpha} E'$ implies $F \xrightarrow{\alpha} F'$ and $E' \leftrightarroweq_b F'$ for some $F' \in \mathcal{E}$ and, vice versa, (ii) $F \xrightarrow{\alpha} F'$ implies $E \xrightarrow{\alpha} E'$ and $E' \leftrightarroweq_b F'$ for some $E' \in \mathcal{E}$. The formulation of Definition 3.4 for the nondeterministic processes of this section corresponds directly to the definition of rooted branching *probabilistic* bisimulation that we will introduce in Sect. 4, see Definition 4.5.

Direct from the definitions we see $\leftrightarroweq \subseteq \leftrightarroweq_{rb} \subseteq \leftrightarroweq_b$. As implicitly announced we have a congruence result for rooted branching bisimilarity.

Lemma 3.5 ([17]). $\leftrightarroweq_{rb}$ is a congruence on $\mathcal{E}$ for the operators $\cdot$ and $+$.

It is well-known that strong and branching bisimilarity for $\mathcal{E}$ can be equationally characterized [4,17,26].

Definition 3.6 (Axiomatization of $\leftrightarroweq$ and $\leftrightarroweq_{rb}$). The theory AX is given by the axioms A1 to A4 listed in Table 1. The theory AX^b contains in addition the axiom B.

If two processes are provably equal, they are rooted branching bisimilar.

Lemma 3.7 (Soundness). For all $E, F \in \mathcal{E}$, if $AX^b \vdash E = F$ then $E \leftrightarroweq_{rb} F$.

Proof (Sketch). First one shows that the left-hand side and the right-hand side of the axioms of AX^b are rooted branching bisimilar. Next, one observes that rooted branching bisimilarity is a congruence. □

Table 1. Axioms for strong and branching bisimilarity

A1	$E + F = F + E$
A2	$(E + F) + G = E + (F + G)$
A3	$E + E = E$
A4	$E + \mathbf{0} = E$
B	$\alpha \cdot (F + \tau \cdot (E + F)) = \alpha \cdot (E + F)$

Strong bisimilarity is equationally characterized by the axioms A1 to A4 of Table 1. See for example [26, Sect. 7.4] for a proof.

Theorem 3.8 (AX sound and complete for $\leftrightarrows$). For all processes $E, F \in \mathcal{E}$ it holds that $AX \vdash E = F$ iff $E \leftrightarrows F$.

For concrete processes that have no inert transitions, branching bisimilarity and strong bisimilarity coincide. Hence, in view of Theorem 3.8, branching bisimilarity implies equality for AX.

Lemma 3.9. For all concrete $\bar{E}, \bar{F} \in \mathcal{E}_{cc}$, if $\bar{E} \leftrightarrows_b \bar{F}$ then both $\bar{E} \leftrightarrows \bar{F}$ and $AX \vdash \bar{E} = \bar{F}$.

Proof (Sketch). Consider $\bar{E}, \bar{F} \in \mathcal{E}_{cc}$ such that $\bar{E} \leftrightarrows_b \bar{F}$. Let $\mathcal{R}$ be a branching bisimulation relation relating $\bar{E}$ and $\bar{F}$. Define $\mathcal{R}'$ as the restriction of $\mathcal{R}$ to the derivatives of $\bar{E}$ and $\bar{F}$, i.e., $\mathcal{R}' = \mathcal{R} \cap ((der(\bar{E}) \times der(\bar{F})) \cup (der(\bar{F}) \times der(\bar{E})))$. Then $\mathcal{R}'$ is a strong bisimulation relation, since none of the processes involved admits an inert τ-transition. By the completeness of AX, see Theorem 3.8, it follows that $AX \vdash \bar{E} = \bar{F}$. □

We are now in a position to prove the main technical result of this section, viz. that branching bisimilarity implies equality under a prefix. In the proof the notion of a concrete process plays a central role.

Lemma 3.10.

(a) For all processes $E \in \mathcal{E}$, a concrete process $\bar{E} \in \mathcal{E}_{cc}$ exists such that $E \leftrightarrows_b \bar{E}$ and $AX^b \vdash \alpha \cdot E = \alpha \cdot \bar{E}$ for all $\alpha \in \mathcal{A}$.

(b) For all processes $F, G \in \mathcal{E}$, if $F \leftrightarrows_b G$ then $AX^b \vdash \alpha \cdot F = \alpha \cdot G$ for all $\alpha \in \mathcal{A}$.

Proof. We prove statements (a) and (b) by simultaneously induction on $c(E)$ and $\max\{c(F), c(G)\}$, respectively.

Basis, $c(E) = 0$. We have that $E = \mathbf{0} + \cdots + \mathbf{0}$. Hence, take $\bar{E} = \mathbf{0}$. Clearly, part (a) of the lemma holds as $\mathbf{0}$ is concrete, $E \leftrightarrows_b \mathbf{0}$ and $AX^b \vdash \alpha \cdot E = \alpha \cdot \mathbf{0}$ for all $\alpha \in \mathcal{A}$.

Induction step for (a): $c(E) > 0$. The process E can be written as $\sum_{i \in I} \alpha_i \cdot E_i$ for some finite I and suitable $\alpha_i \in \mathcal{A}$ and $E_i \in \mathcal{E}$.

First suppose that for some $i_0 \in I$ we have $\alpha_{i_0} = \tau$ and $E_{i_0} \leftrightarrows_b E$. Then $AX \vdash E = H + \tau.E_{i_0}$, where $H := \sum_{i \in I \setminus \{i_o\}} \alpha_i \cdot E_i$. By the induction hypothesis (a), there is a term $\bar{E}_{i_0} \in \mathcal{E}_{cc}$ such that $E_{i_0} \leftrightarrows_b \bar{E}_{i_0}$. We claim that $\bar{E}_{i_0} \leftrightarrows_b E_{i_0} + H$.

For suppose $\bar{E}_{i_0} \xrightarrow{\alpha} F$. Then $E_{i_0} \Rightarrow E'_{i_0} \xrightarrow{(\alpha)} G$ where $\bar{E}_{i_0} \leftrightarrows_b E'_{i_0}$ and $F \leftrightarrows_b G$. In case $E_{i_0} = E'_{i_0}$ it follows that $E_{i_0} \xrightarrow{(\alpha)} G$. Since $\bar{E}_{i_o}$ is concrete, either $\alpha \neq \tau$ or $F \not\leftrightarrows_b \bar{E}_{i_0}$. Hence, $\alpha \neq \tau$ or $G \not\leftrightarrows_b E_{i_0}$. So $E_{i_0} \xrightarrow{\alpha} G$. Consequently, $E_{i_0} + H \xrightarrow{\alpha} G$. In case $E_{i_0} \neq E'_{i_0}$ we have $E_{i_0} + H \Rightarrow E'_{i_0} \xrightarrow{(\alpha)} G$.

Now suppose $E_{i_0} + H \xrightarrow{\alpha} F$. Then either $E_{i_0} \xrightarrow{\alpha} F$ or $H \xrightarrow{\alpha} F$. In the first case we have $\bar{E}_{i_0} \Rightarrow \bar{E}'_{i_0} \xrightarrow{(\alpha)} G$ where $E_{i_0} \leftrightarrows_b \bar{E}'_{i_0}$ and $F \leftrightarrows_b G$, while in the latter case $E \xrightarrow{\alpha} F$, and since $E \leftrightarrows_b E_{i_0} \leftrightarrows_b \bar{E}_{i_0}$ we have $\bar{E}_{i_0} \Rightarrow \bar{E}'_{i_0} \xrightarrow{(\alpha)} G$ where $E \leftrightarrows_b \bar{E}'_{i_0}$ and $F \leftrightarrows_b G$. Because $\bar{E}_{i_0}$ is concrete, $\bar{E}'_{i_o} = \bar{E}_{i_0}$. Thus $\bar{E}_{i_0} \xrightarrow{(\alpha)} G$ with $F \leftrightarrows_b G$, which was to be shown.

Hence $E_{i_0} \leftrightarrows_b \bar{E}_{i_0} \leftrightarrows_b E_{i_0} + H$. Clearly $c(E_{i_0}), c(E_{i_0} + H) < c(E)$. Therefore, by the induction hypothesis (b), $AX^b \vdash \tau \cdot E_{i_0} = \tau \cdot (E_{i_0} + H)$. By the induction hypothesis (a), there is a term $\bar{E} \in \mathcal{E}_{cc}$ such that $\bar{E} \leftrightarrows_b E_{i_0} + H$ and $AX^b \vdash \alpha \cdot \bar{E} = \alpha \cdot (E_{i_0} + H)$. Now we have $E \leftrightarrows_b E_{i_0} \leftrightarrows_b E_{i_0} + H \leftrightarrows_b \bar{E}$. Therefore,

$$\begin{aligned} AX^b \vdash \alpha \cdot E &= \alpha \cdot (H + \tau \cdot E_{i_0}) \\ &= \alpha \cdot (H + \tau \cdot (E_{i_0} + H)) && (\text{since } AX^b \vdash \tau \cdot E_{i_0} = \tau \cdot (E_{i_0} + H)) \\ &= \alpha \cdot (E_{i_0} + H) && (\text{use axiom B}) \\ &= \alpha \cdot \bar{E} && (\text{by the choice of } \bar{E}). \end{aligned}$$

Hence, we have shown the existence of a desired process $\bar{E}$ with the required properties.

Now suppose, for all $i \in I$ we have $\alpha_i \neq \tau$ or $E_i \not\leftrightarrows_b E$. Clearly $c(E_i) < c(E)$ for all $i \in I$. By the induction hypothesis we can find, for all $i \in I$, concrete $\bar{E}_i$ such that $\bar{E}_i \leftrightarrows_b E_i$ and $AX^b \vdash \alpha \cdot \bar{E}_i = \alpha \cdot E_i$ for all $\alpha \in \mathcal{A}$. Define $\bar{E} = \sum_{i \in I} \alpha_i \cdot \bar{E}_i$. Then $\bar{E} \leftrightarrows_b E$ and $\bar{E}$ is concrete too, since $\bar{E}_i \leftrightarrows_b E_i \not\leftrightarrows_b E \leftrightarrows_b \bar{E}$ for $i \in I$ in case $\alpha_i = \tau$. Moreover, $AX^b \vdash E = \bar{E}$, since $E = \sum_{i \in I} \alpha_i \cdot E_i = \sum_{i \in I} \alpha_i \cdot \bar{E}_i = \bar{E}$. Hence, for $\alpha \in \mathcal{A}$, $AX^b \vdash \alpha \cdot E = \alpha \cdot \bar{E}$.

Both the base and the induction step for (b): $\max\{c(F), c(G)\} \geqslant 0$. Suppose $F \leftrightarrows_b G$. Pick $\bar{F}, \bar{G} \in \mathcal{E}_{cc}$ such that $F \leftrightarrows_b \bar{F}$ and $AX^b \vdash \alpha \cdot F = \alpha \cdot \bar{F}$ for all $\alpha \in \mathcal{A}$, and similarly for G and $\bar{G}$. Then we have $\bar{F} \leftrightarrows_b \bar{G}$. Since $\bar{F}$ and $\bar{G}$ are concrete it follows that $AX \vdash \bar{F} = \bar{G}$, see Lemma 3.9. Now pick any $\alpha \in \mathcal{A}$. Then we have $AX^b \vdash \alpha \cdot F = \alpha \cdot \bar{F} = \alpha \cdot \bar{G} = \alpha \cdot G$. □

By now we have gathered sufficient building blocks to prove the main result of this section.

Theorem 3.11 (*AX^b* **sound and complete for** $\leftrightarrows_{rb}$**).** For all processes $E, F \in \mathcal{E}$ it holds that $E \leftrightarrows_{rb} F$ iff $AX^b \vdash E = F$.

Proof. In view of Lemma 3.7 we only need to prove completeness of AX^b for rooted branching bisimilarity. Suppose $E, F \in \mathcal{E}$ and $E \leftrightarrows_{rb} F$. Let $E = \sum_{i \in I} \alpha_i \cdot E_i$ and $F = \sum_{j \in J} \beta_j \cdot F_j$ for suitable index sets I and J, $\alpha_i, \beta_j \in \mathcal{A}$, $E_i, F_j \in \mathcal{E}$. Since $E \leftrightarrows_{rb} F$ we have (i) for all $i \in I$ there is a $j \in J$ such that

$\alpha_i = \beta_j$ and $E_i \leftrightarroweq_b F_j$, and, symmetrically, (ii) for all $j \in J$ there is an $i \in I$ such that $\alpha_i = \beta_j$ and $E_i \leftrightarroweq_b F_j$. Put $K = \{ (i,j) \in I \times J \mid (\alpha_i = \beta_j) \wedge (E_i \leftrightarroweq_b F_j) \}$. Define the processes $G, H \in \mathcal{E}$ by

$$G = \textstyle\sum_{k \in K} \gamma_k \cdot G_k \quad \text{and} \quad H = \textstyle\sum_{k \in K} \zeta_k \cdot H_k$$

where, for $i \in I$, $\gamma_k = \alpha_i$ and $G_k \equiv E_i$ if $k = (i,j)$ for some $j \in J$, and, similarly for $j \in J$, $\zeta_k = \beta_j$ and $H_k \equiv F_j$ if $k = (i,j)$ for some $i \in I$. Then G and H are well-defined. Moreover, $AX \vdash E = G$ and $AX \vdash F = H$.

For $k \in K$, say $k = (i,j)$, it holds that $\gamma_k = \alpha_i = \beta_j = \zeta_k$ and $G_k \equiv E_i \leftrightarroweq_b F_j \equiv H_k$, by definition of K. By Lemma 3.10b we obtain, for all $k \in K$, $AX^b \vdash \gamma_k \cdot G_k = \zeta_k \cdot H_k$. From this we get $AX^b \vdash E = \sum_{i \in I} \alpha_i \cdot E_i = \sum_{k \in K} \gamma_k \cdot G_k = \sum_{k \in K} \zeta_k \cdot H_k = \sum_{j \in J} \beta_j \cdot F_j = F$ which concludes the proof of the theorem. □

4 Branching Bisimilarity for Probabilistic Processes

In this section we define branching bisimilarity for probabilistic processes. Following [6], we start with adapting the syntax of processes, now distinguishing non-deterministic processes $E \in \mathcal{E}$ and probabilistic processes $P \in \mathcal{P}$.

Definition 4.1 (Syntax). The classes $\mathcal{E}$ and $\mathcal{P}$ of non-deterministic and probabilistic processes over $\mathcal{A}$, respectively, ranged over by E and P, are given by

$$E ::= \mathbf{0} \mid \alpha \cdot P \mid E + E$$
$$P ::= \partial(E) \mid P \;{}_r\!\oplus P$$

with actions α from $\mathcal{A}$ where $r \in (0,1)$.

The probabilistic process $P_1 \;{}_r\!\oplus P_2$ executes the behavior of P_1 with probability r and the behavior P_2 with probability $1 - r$. By convention, $P \;{}_1\!\oplus Q$ denotes P and $P \;{}_0\!\oplus Q$ denotes Q.

We again introduce the complexity measure c, now for non-deterministic and probabilistic processes, based on the depth of a process. The complexity measure $c : \mathcal{E} \cup \mathcal{P} \to \mathbb{N}$ is given by $c(\mathbf{0}) = 0$, $c(\alpha \cdot P) = c(P) + 1$, $c(E + F) = c(E) + c(F)$, and $c(\partial(E)) = c(E) + 1$, $c(P \;{}_r\!\oplus Q) = c(P) + c(Q)$.

As usual SOS semantics for $\mathcal{E}$ and $\mathcal{P}$ makes use of two types of transition relations [6,22].

Definition 4.2 (Operational semantics).

(a) The transition relations $\rightarrow \subseteq \mathcal{E} \times \mathcal{A} \times \mathit{Distr}(\mathcal{E})$ and $\mapsto \subseteq \mathcal{P} \times \mathit{Distr}(\mathcal{E})$ are given by

$$\frac{P \mapsto \mu}{\alpha \cdot P \xrightarrow{\alpha} \mu}\ (\text{PREF})$$

$$\frac{E_1 \xrightarrow{\alpha} \mu_1}{E_1 + E_2 \xrightarrow{\alpha} \mu_1}\ (\text{ND-CHOICE 1}) \qquad \frac{E_2 \xrightarrow{\alpha} \mu_2}{E_1 + E_2 \xrightarrow{\alpha} \mu_2}\ (\text{ND-CHOICE 2})$$

$$\frac{}{\partial(E) \mapsto \delta(E)}\ (\text{DIRAC}) \qquad \frac{P_1 \mapsto \mu_1 \quad P_2 \mapsto \mu_2}{P_1 \;{}_r\!\oplus P_2 \mapsto \mu_1 \;{}_r\!\oplus \mu_2}\ (\text{P-CHOICE})$$

(b) The transition relation $\rightarrow \subseteq Distr(\mathcal{E}) \times \mathcal{A} \times Distr(\mathcal{E})$ is such that $\mu \xrightarrow{\alpha} \mu'$ whenever $\mu = \bigoplus_{i \in I} p_i * E_i$, $\mu' = \bigoplus_{i \in I} p_i * \mu'_i$, and $E_i \xrightarrow{\alpha} \mu'_i$ for all $i \in I$.

With $[\![P]\!]$, for $P \in \mathcal{P}$, we denote the unique distribution μ such that $P \mapsto \mu$.

The transition relation $\rightarrow$ on distributions allows for a probabilistic combination of non-deterministic alternatives resulting in a so-called combined transition, cf. [28,29]. For example, for $E \equiv a \cdot (P \,{}_{1/2}\!\oplus Q) + a \cdot (P \,{}_{1/3}\!\oplus Q)$, the Dirac process $\delta(E) \equiv \delta(a \cdot (P \,{}_{1/2}\!\oplus Q) + a \cdot (P \,{}_{1/3}\!\oplus Q))$ provides an a-transition to $[\![P \,{}_{1/2}\!\oplus Q]\!]$ as well as an a-transition to $[\![P \,{}_{1/3}\!\oplus Q]\!]$. However, since for distribution $\delta(E)$ it holds that $\delta(E) = \frac{1}{2}\delta(E) \oplus \frac{1}{2}\delta(E)$ there is also a transition

$$\delta(E) = \tfrac{1}{2}\delta(E) \oplus \tfrac{1}{2}\delta(E) \xrightarrow{a} \tfrac{1}{2}[\![P \,{}_{1/2}\!\oplus Q]\!] \oplus \tfrac{1}{2}[\![P \,{}_{1/3}\!\oplus Q]\!] = [\![P \,{}_{5/12}\!\oplus Q]\!].$$

As noted in [30], the ability to combine transitions is crucial for obtaining transitivity of probabilistic process equivalences that take internal actions into account.

Referring to the example in the introduction, the processes of t_0 and u_0 will be identified. However, without the splitting of the source distribution μ as provided by Definition 4.2, we are not able to relate t_0 and u_0 directly, or rather their direct derivatives, while meeting the natural transfer conditions (see Definition 4.4). The difficulty arises when both P and Q can do a τ-transition to non-bisimilar processes.

In preparation to the definition of the notion of branching probabilistic bisimilarity below we introduce some notation.

Definition 4.3. For $\mu, \mu' \in Distr(\mathcal{E})$ and $\alpha \in \mathcal{A}$ we write $\mu \xrightarrow{(\alpha)} \mu'$ iff (i) $\mu \xrightarrow{\alpha} \mu'$, or (ii) $\alpha = \tau$ and $\mu = \mu_1 \,{}_r\!\oplus \mu_2$, $\mu' = \mu'_1 \,{}_r\!\oplus \mu'_2$ such that $\mu_1 \xrightarrow{\tau} \mu'_1$ and $\mu_2 \underset{(\tau)}{=} \mu'_2$ for some $r \in [0,1]$. We use $\Rightarrow$ to denote the reflexive transitive closure of $\xrightarrow{(\tau)}$.

Thus, for example,

$$\tfrac{1}{3}\delta(\tau \cdot (P \,{}_{1/2}\!\oplus Q)) \oplus \tfrac{2}{3}[\![P \,{}_{1/2}\!\oplus Q]\!] \xrightarrow{(\tau)} [\![P \,{}_{1/2}\!\oplus Q]\!] \quad \text{and}$$
$$\tfrac{1}{2}\delta(\tau \cdot \partial(\tau \cdot P)) \oplus \tfrac{1}{3}\delta(\tau \cdot P) \oplus \tfrac{1}{6}[\![P]\!] \Rightarrow [\![P]\!].$$

We are now in a position to define strong probabilistic bisimilarity and branching probabilistic bisimilarity. Note that the notion of strong probabilistic bisimilarity is the variant with combined transitions as defined in [6,29].

Definition 4.4 (Strong and branching probabilistic bisimilarity).

(a) A symmetric relation $\mathcal{R} \subseteq Distr(\mathcal{E}) \times Distr(\mathcal{E})$ is called *decomposable* iff for all $\mu, \nu \in Distr(\mathcal{E})$ such that $\mu \,\mathcal{R}\, \nu$ and $\mu = \bigoplus_{i \in I} p_i * \mu_i$ there are $\nu_i \in Distr(\mathcal{E})$, for $i \in I$, such that

$$\nu = \textstyle\bigoplus_{i \in I} p_i * \nu_i \text{ and } \mu_i \,\mathcal{R}\, \nu_i \text{ for all } i \in I.$$

(b) A decomposable relation $\mathcal{R} \subseteq Distr(\mathcal{E}) \times Distr(\mathcal{E})$ is called a *strong probabilistic bisimulation relation* iff for all $\mu, \nu \in Distr(\mathcal{E})$ such that $\mu \mathcal{R} \nu$ and $\mu \xrightarrow{\alpha} \mu'$ there is a $\nu' \in Distr(\mathcal{E})$ such that

$$\nu \xrightarrow{\alpha} \nu' \text{ and } \mu' \mathcal{R} \nu'.$$

(c) A symmetric relation $\mathcal{R} \subseteq Distr(\mathcal{E}) \times Distr(\mathcal{E})$ is called *weakly decomposable* iff for all $\mu, \nu \in Distr(\mathcal{E})$ such that $\mu \mathcal{R} \nu$ and $\mu = \bigoplus_{i \in I} p_i * \mu_i$ there are $\bar{\nu}, \nu_i \in Distr(\mathcal{E})$, for $i \in I$, such that

$$\nu \Rightarrow \bar{\nu},\ \mu \mathcal{R} \bar{\nu},\ \bar{\nu} = \textstyle\bigoplus_{i \in I} p_i * \nu_i, \text{ and } \mu_i \mathcal{R} \nu_i \text{ for all } i \in I.$$

(d) A weakly decomposable relation $\mathcal{R} \subseteq Distr(\mathcal{E}) \times Distr(\mathcal{E})$ is called a *branching* probabilistic bisimulation relation iff for all $\mu, \nu \in Distr(\mathcal{E})$ such that $\mu \mathcal{R} \nu$ and $\mu \xrightarrow{\alpha} \mu'$, there are $\bar{\nu}, \nu' \in Distr(\mathcal{E})$ such that

$$\nu \Rightarrow \bar{\nu},\ \bar{\nu} \xrightarrow{(\alpha)} \nu',\ \mu \mathcal{R} \bar{\nu}, \text{ and } \mu' \mathcal{R} \nu'.$$

(e) Strong probabilistic bisimilarity, denoted by $\leftrightarroweq \subseteq Distr(\mathcal{E}) \times Distr(\mathcal{E})$, and branching probabilistic bisimilarity, written as $\leftrightarroweq_b \subseteq Distr(\mathcal{E}) \times Distr(\mathcal{E})$, are respectively defined as the largest strong probabilistic bisimulation relation on $Distr(\mathcal{E})$ and as the largest branching probabilistic bisimulation relation on $Distr(\mathcal{E})$.

By comparison, on finite processes, as used in this paper, the branching probabilistic bisimilarity of Segala and Lynch [29] can be defined in our framework exactly as in (d) and (e) above, but taking a decomposable instead of a weakly decomposable relation. This yields a strictly finer equivalence, distinguishing the processes s_0, t_0 and u_0 from the introduction.

The notion of decomposability has been adopted from [23] and weak decomposability from [25]. The underlying idea stems from [10]. These notions provide a convenient dexterity to deal with behavior of sub-distributions, e.g., to distinguish $\frac{1}{2}\partial(a \cdot \partial(\mathbf{0})) \oplus \frac{1}{2}\partial(b \cdot \partial(\mathbf{0}))$ from $\partial(\mathbf{0})$, as well as combined behavior.

Our definition of branching probabilistic bisimilarity is based on distributions rather than on states and has similarity with the notion of weak distribution bisimilarity proposed by Eisentraut et al. in [13]. Consider the running example of [13], reproduced in Fig. 1 and reformulated in terms of the process language at hand. The states ① and ⑥ are identified with respect to weak distribution bisimilarity as detailed in [13]. Correspondingly, putting

$$\begin{aligned} E_1 &= \tau \cdot \big(\partial(\tau \cdot \partial(\tau \cdot P + c \cdot Q + \tau \cdot R) + c \cdot Q + \tau \cdot R)\ {}_{1/2}{\oplus} \\ &\qquad \partial(\tau \cdot (\partial(\tau \cdot P + c \cdot Q + \tau \cdot R)\ {}_{1/2}{\oplus}\ \partial(\mathbf{0})))\big) \\ E_6 &= \tau \cdot (\partial(\tau \cdot P + c \cdot Q + \tau \cdot R)\ {}_{3/4}{\oplus}\ \partial(\mathbf{0})) \end{aligned}$$

the non-deterministic processes E_1 and E_6 are identified with respect to branching probabilistic bisimilarity.

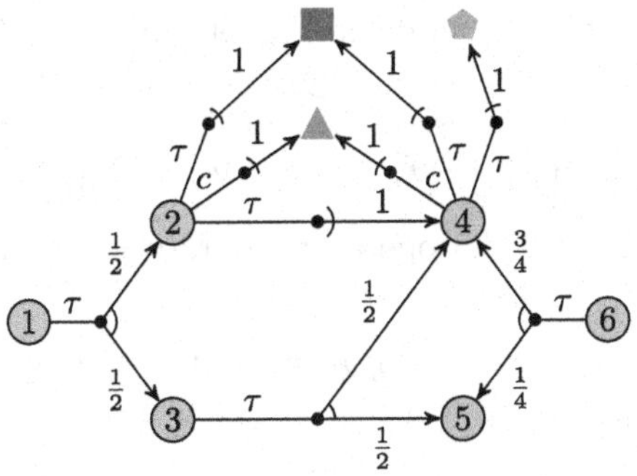

Fig. 1. Probabilistic automaton of [13]

Note that strong and branching probabilistic bisimilarity are well-defined since any union of strong or branching probabilistic bisimulation relations is again a strong or branching probabilistic bisimulation relation. In particular, (weak) decomposability is preserved under arbitrary unions.

As we did for the non-deterministic setting, we introduce a notion of rooted branching probabilistic bisimilarity for distributions over processes.

Definition 4.5. A symmetric and decomposable relation $\mathcal{R} \subseteq \mathit{Distr}(\mathcal{E}) \times \mathit{Distr}(\mathcal{E})$ is called a rooted branching probabilistic bisimulation relation iff for all $\mu, \nu \in \mathit{Distr}(\mathcal{E})$ such that $\mu \, \mathcal{R} \, \nu$ it holds that if $\mu \xrightarrow{\alpha} \mu'$ for $\alpha \in \mathcal{A}$, $\mu' \in \mathit{Distr}(\mathcal{E})$ then $\nu \xrightarrow{\alpha} \nu'$ and $\mu' \leftrightarroweq_b \nu'$ for some $\nu' \in \mathit{Distr}(\mathcal{E})$. Rooted branching probabilistic bisimilarity, denoted by $\leftrightarroweq_{rb} \subseteq \mathit{Distr}(\mathcal{E}) \times \mathit{Distr}(\mathcal{E})$, is defined as the largest rooted branching probabilistic bisimulation relation.

Since any union of rooted branching probabilistic bisimulation relations is again a rooted branching probabilistic bisimulation relation, rooted branching probabilistic bisimilarity $\leftrightarroweq_{rb}$ is well-defined.

Note, the two probabilistic processes $P = \partial(\tau \cdot \partial(a \cdot \partial(\mathbf{0})))\,{}_{1/2}{\oplus}\, \partial(b \cdot \partial(\mathbf{0}))$ and $Q = \partial(a \cdot \partial(\mathbf{0}))\,{}_{1/2}{\oplus}\, \partial(b \cdot \partial(\mathbf{0}))$ are *not* rooted branching probabilistically bisimilar. Any rooted branching probabilistic bisimulation relation is by decomposability required to relate the respective probabilistic components $\partial(\tau \cdot \partial(a \cdot \partial(\mathbf{0})))$ and $\partial(a \cdot \partial(\mathbf{0}))$, which clearly do not meet the transfer condition. Thus, since $\partial(\tau \cdot \partial(a \cdot \partial(\mathbf{0}))) \not\leftrightarroweq_{rb} \partial(a \cdot \partial(\mathbf{0}))$ also $P \not\leftrightarroweq_{rb} Q$.

Two non-deterministic processes are considered to be strongly, rooted branching, or branching probabilistically bisimilar iff their Dirac distributions are, i.e., $E \leftrightarroweq F$ iff $\delta(E) \leftrightarroweq \delta(F)$, $E \leftrightarroweq_{rb} F$ iff $\delta(E) \leftrightarroweq_{rb} \delta(F)$, and $E \leftrightarroweq_b F$ iff $\delta(E) \leftrightarroweq_b \delta(F)$. Two probabilistic processes are considered to be strongly, rooted branching, or branching probabilistically bisimilar iff their associated distributions over $\mathcal{E}$ are.

We show that branching probabilistic bisimilarity, although not a congruence for non-deterministic choice, is a congruence for probabilistic choice. We first need a technical result, the proof of which is omitted here, see [16].

Lemma 4.6. Let I and J be finite index sets, $p_i, q_j \in [0,1]$ and $\xi, \mu_i, \nu_j \in \mathit{Distr}(\mathcal{E})$, for $i \in I$ and $j \in J$, with $\xi = \bigoplus_{i\in I} p_i * \mu_i$ and $\xi = \bigoplus_{j\in J} q_j * \nu_j$. Then there are $r_{ij} \in [0,1]$ and $\varrho_{ij} \in \mathit{Distr}(\mathcal{E})$ such that $\sum_{i\in I} r_{ij} = q_j$, $\sum_{j\in J} r_{ij} = p_i$, $p_i * \mu_i = \bigoplus_{j\in J} r_{ij} * \varrho_{ij}$ for all $i \in I$, and $q_j * \nu_j = \bigoplus_{i\in I} r_{ij} * \varrho_{ij}$ for all $j \in J$.

The next result states that the operator ${}_r\!\oplus$ respects branching probabilistic bisimilarity.

Lemma 4.7. Let $\mu_1, \mu_2, \nu_1, \nu_2 \in \mathit{Distr}(\mathcal{E})$ and $r \in (0,1)$. If $\mu_1 \leftrightarrows_b \nu_1$ and $\mu_2 \leftrightarrows_b \nu_2$ then $\mu_1 \,{}_r\!\oplus\, \mu_2 \leftrightarrows_b \nu_1 \,{}_r\!\oplus\, \nu_2$.

Also for a proof of Lemma 4.7 we refer to [16]. A direct consequence of the lemma is that if $P_1 \leftrightarrows_b Q_1$ and $P_2 \leftrightarrows_b Q_2$ then $P_1 \,{}_r\!\oplus\, P_2 \leftrightarrows_b Q_1 \,{}_r\!\oplus\, Q_2$.

Lemma 4.8 (Congruence). The relations $\leftrightarrows$, $\leftrightarrows_{rb}$, and $\leftrightarrows_b$ on $\mathcal{E}$ and $\mathcal{P}$ are equivalence relations, and the relations $\leftrightarrows$ and $\leftrightarrows_{rb}$ are congruences on $\mathcal{E}$ and $\mathcal{P}$.

Proof. The proof of $\leftrightarrows$, $\leftrightarrows_{rb}$, and $\leftrightarrows_b$ being equivalence relations involves a number of straightforward auxiliary results, in particular for the case of transitivity, and are omitted here.

Regarding congruence the interesting cases are for non-deterministic and probabilistic choice with respect to rooted branching probabilistic bisimilarity. Suppose $E_1 \leftrightarrows_{rb} F_1$ and $E_2 \leftrightarrows_{rb} F_2$. Then $\mathcal{R} = \{\langle \delta(E_1 + E_2), \delta(F_1 + F_2)\rangle\}^\dagger$ is a rooted branching probabilistic bisimulation relation. Clearly, $\mathcal{R}$ is symmetric and decomposable. Moreover, if $\delta(E_1 + E_2) \xrightarrow{\alpha} \mu'$, then either $\delta(E_1) \xrightarrow{\alpha} \mu'$, $\delta(E_2) \xrightarrow{\alpha} \mu'$, or $\delta(E_1) \xrightarrow{\alpha} \mu'_1$, $\delta(E_2) \xrightarrow{\alpha} \mu'_2$ and $\mu' = \mu'_1 \,{}_r\!\oplus\, \mu'_2$ for suitable $\mu'_1, \mu'_2 \in \mathit{Distr}(\mathcal{E})$ and $r \in (0,1)$. We only consider the last case, as the first two are simpler. Hence, we can find $\nu'_1, \nu'_2 \in \mathit{Distr}(\mathcal{E})$ such that $\delta(F_1) \xrightarrow{\alpha} \nu'_1$, $\delta(F_2) \xrightarrow{\alpha} \nu'_2$, $\mu'_1 \leftrightarrows_b \nu'_1$, and $\mu'_2 \leftrightarrows_b \nu'_2$. From this it follows that $\delta(F_1 + F_2) \xrightarrow{\alpha} \nu'$ and $\mu' \leftrightarrows_b \nu'$ for $\nu' = \nu'_1 \,{}_r\!\oplus\, \nu'_2$ using Lemma 4.7.

Suppose $P_1 \leftrightarrows_{rb} Q_1$ and $P_2 \leftrightarrows_{rb} Q_2$ with $\mathcal{R}_1$ and $\mathcal{R}_2$ rooted branching probabilistic bisimulation relations relating $[\![P_1]\!]$ with $[\![Q_1]\!]$, and $[\![P_2]\!]$ with $[\![Q_2]\!]$, respectively, and fix some $r \in (0,1)$. Then $\mathcal{R} = \{\langle \mu_1 \,{}_r\!\oplus\, \mu_2, \nu_1 \,{}_r\!\oplus\, \nu_2\rangle \mid \mu_1 \mathcal{R}_1 \nu_1,\ \mu_2 \mathcal{R}_2 \nu_2\}$ is a rooted branching probabilistic bisimulation relation relating $[\![P_1 \,{}_r\!\oplus\, P_2]\!]$ with $[\![Q_1 \,{}_r\!\oplus\, Q_2]\!]$. Symmetry is straightforward and decomposability can be shown by application of Lemma 4.6.

So we are left to prove the transfer property. Suppose $(\mu_1 \,{}_r\!\oplus\, \mu_2)\mathcal{R}(\nu_1 \,{}_r\!\oplus\, \nu_2)$, thus $\mu_1 \mathcal{R}_1 \mu_2$ and $\nu_1 \mathcal{R}_2 \nu_2$. If $\mu_1 \,{}_r\!\oplus\, \mu_2 \xrightarrow{\alpha} \mu'$ then $\mu_1 \xrightarrow{\alpha} \mu'_1, \mu_2 \xrightarrow{\alpha} \mu'_2$ and $\mu' = \mu'_1 \,{}_r\!\oplus\, \mu'_2$ for suitable $\mu'_1, \mu'_2 \in \mathit{Distr}(\mathcal{E})$. By assumption, ν'_1 and ν'_2 exist such that $\nu_1 \xrightarrow{\alpha} \nu'_1, \nu_2 \xrightarrow{\alpha} \nu'_2$, $\mu'_1 \leftrightarrows_b \nu'_1$, and $\mu'_2 \leftrightarrows_b \nu'_2$. From this we obtain $\nu_1 \,{}_r\!\oplus\, \nu_2 \xrightarrow{\alpha} \nu'$ and by Lemma 4.7 $\mu' \leftrightarrows_b \nu'$ for $\nu' = \nu'_1 \,{}_r\!\oplus\, \nu'_2$. □

5 A Few Fundamental Properties of Branching Bisimilarity

In this section we show two fundamental properties of branching probabilistic bisimilarity that we need further on: the stuttering property, known from [17] for non-deterministic processes, and cancellativity of probabilistic choice with respect to $\leftrightarrow_b$.

Lemma 5.1 (Stuttering Property). If $\mu \Rightarrow \bar{\mu} \Rightarrow \nu$ and $\mu \leftrightarrow_b \nu$ then $\mu \leftrightarrow_b \bar{\mu}$.

Proof. We show that the relation $\leftrightarrow_b \cup \{(\mu, \bar{\mu}), (\bar{\mu}, \mu)\}$ is a branching probabilistic bisimulation.

First suppose $\mu \xrightarrow{\alpha} \mu'$. Then there are $\bar{\nu}, \nu' \in \mathit{Distr}(\mathcal{E})$ such that

$$\nu \Rightarrow \bar{\nu},\ \bar{\nu} \xrightarrow{(\alpha)} \nu',\ \mu \leftrightarrow_b \bar{\nu},\ \text{and}\ \mu' \leftrightarrow_b \nu'.$$

Since $\bar{\mu} \Rightarrow \nu$, we have $\bar{\mu} \Rightarrow \bar{\nu}$, which had to be shown. Now suppose $\bar{\mu} \xrightarrow{\alpha} \mu'$. Then certainly $\mu \Rightarrow \bar{\mu} \xrightarrow{\alpha} \mu'$.

To show weak decomposability, suppose $\mu = \bigoplus_{i \in I} p_i * \mu_i$. Then there are $\bar{\nu}, \nu_i \in \mathit{Distr}(\mathcal{E})$, for $i \in I$, such that

$$\nu \Rightarrow \bar{\nu},\ \mu \mathcal{R} \bar{\nu},\ \bar{\nu} = \textstyle\bigoplus_{i \in I} p_i * \nu_i,\ \text{and}\ \mu_i \mathcal{R} \nu_i\ \text{for all}\, i \in I.$$

Again it suffices to point out that $\bar{\mu} \Rightarrow \bar{\nu}$. Conversely, suppose $\bar{\mu} = \bigoplus_{i \in I} p_i * \bar{\mu}_i$. Then $\mu \Rightarrow \bar{\mu} = \bigoplus_{i \in I} p_i * \bar{\mu}_i$. □

Lemma 5.2 (Cancellativity). Let $\mu, \mu', \nu, \nu' \in \mathit{Distr}(\mathcal{E})$. If $\mu\,{}_r{\oplus}\,\nu \leftrightarrow_b \mu'\,{}_r{\oplus}\,\nu'$ with $r \in (0, 1]$ and $\nu \leftrightarrow_b \nu'$, then $\mu \leftrightarrow_b \mu'$.

A proof can be found in [16].

6 Completeness: The Probabilistic Case

In this section we provide a sound and complete equational characterization of rooted branching probabilistic bisimilarity. The completeness result is obtained along the same lines as the corresponding result for branching bisimilarity for the non-deterministic processes in Sect. 3. We extend and adapt the non-deterministic theories *AX* and AX^b of Sect. 3.

Definition 6.1 (Axiomatization of $\leftrightarrow$ and $\leftrightarrow_{rb}$). The theory AX_p is given by the axioms A1 to A4, the axioms P1 to P3 and C listed in Table 2. The theory AX_p^b contains in addition the axioms BP and G.

The axioms A1–A4 for non-deterministic processes are as before. Regarding probabilistic processes, for the axioms P1 and P2 dealing with commutativity and associativity, we need to take care of the probabilities involved. For P2, it follows from the given restrictions that also $(1{-}r)s = (1{-}r')s'$, i.e., the probability for Q to execute is equal for the left-hand and right-hand side of the equation. Axiom P3 expresses that a probabilistic choice between equal processes can be eliminated. Axiom C expresses that any two nondeterministic transitions can be executed in a combined fashion: one with probability r and one with the complementary probability $1{-}r$.

The axioms P1 and P2 allow us to write each probabilistic process P as

$$\partial(E_1)\ {}_{r_1}\!\oplus\ (\partial(E_2)\ {}_{r_2}\!\oplus\ (\partial(E_3)\ {}_{r_3}\!\oplus \cdots))$$

for non-deterministic processes E_i. In the sequel we denote such a process by $\bigoplus_{i\in I} p_i * E_i$ with $p_i = r_i \prod_{j=1}^{i-1}(1-r_j)$. More specifically, if a probabilistic process P corresponds to a distribution $\bigoplus_{i\in I} p_i * E_i$, then we have $AX_p \vdash P = \bigoplus_{i\in I} p_i * E_i$, as can be shown by induction on the structure of P.

For axioms BP and G of Table 2 we introduce the notation $E \sqsubseteq P$ for $E \in \mathcal{E}$, $P \in \mathcal{P}$. We define

$$\begin{aligned} E \sqsubseteq P \quad \text{iff} \quad \forall \alpha \in \mathcal{A}, \mu \in \mathit{Distr}(\mathcal{E})\colon E \xrightarrow{\alpha} \mu \quad &\Longrightarrow \\ \exists \nu \in \mathit{Distr}(\mathcal{E})\colon [\![P]\!] &\xrightarrow{(\alpha)} \nu \wedge \mu \leftrightarroweq_b \nu\,. \end{aligned}$$

Thus, we require that every transition of the non-deterministic process E can be directly matched by the probabilistic process P. Note, if $E \sqsubseteq P$ and $\delta(E) \xrightarrow{\alpha} \mu$, then $[\![P]\!] \xrightarrow{(\alpha)} \nu$ for some $\nu \in \mathit{Distr}(\mathcal{E})$ such that $\mu \leftrightarroweq_b \nu$: If $\delta(E) \xrightarrow{\alpha} \mu$, then $\mu = \bigoplus_{i\in I} p_i * \mu_i$ and $E \xrightarrow{\alpha} \mu_i$ for suitable $p_i \geqslant 0$, $\mu_i \in \mathit{Distr}(\mathcal{E})$. Since $E \sqsubseteq P$, we have for each $i \in I$ that $[\![P]\!] \xrightarrow{(\alpha)} \nu_i$ for some $\nu_i \in \mathit{Distr}(\mathcal{E})$ satisfying $\mu_i \leftrightarroweq_b \nu_i$. Hence $[\![P]\!] \xrightarrow{(\alpha)} \nu := \bigoplus_{i\in I} p_i * \nu_i$ and $\mu \leftrightarroweq_b \nu$ by Lemma 4.7.

Axiom BP is an adaptation of axiom B of the theory AX^b to the probabilistic setting of AX^b_p. In the setting of non-deterministic processes the implication $F \xrightarrow{\alpha} F' \Longrightarrow E \xrightarrow{\alpha} E' \wedge F' \leftrightarroweq_b E'$ for some E' is captured by $E + F \leftrightarroweq_{rb} E$. If we reformulate axiom B as $E + F = E \Longrightarrow \alpha\cdot(F + \tau\cdot E) = \alpha\cdot E$, then it becomes more similar to axiom BP in Table 2.

As to BP, in the context of a preceding action α and a probabilistic process Q, a non-deterministic alternative E that is also offered by a probabilistic process after a τ-prefix can be dispensed with, together with the prefix τ. In a formulation without the prefix α and the probabilistic alternative Q, but with the specific condition $E \sqsubseteq P$, and retaining the τ-prefix on the right-hand side, the axiom BP shows similarity with axioms T2 and T3 in [15] which, in turn, are reminiscent of axioms T1 and T2 of [11]; these axioms stem from Milner's second τ-law [26].

Let us illustrate the working of axiom BP. Consider the non-deterministic process $E = b\cdot\partial(\mathbf{0})$ and the probabilistic process $P = \partial(a\cdot\partial(\mathbf{0}) + b\cdot\partial(\mathbf{0}))\ {}_{1/2}\!\oplus\ \partial(b\cdot\partial(\mathbf{0}))$. Then we have $E \sqsubseteq P$, i.e.

$$b\cdot\partial(\mathbf{0}) \sqsubseteq \partial(a\cdot\partial(\mathbf{0}) + b\cdot\partial(\mathbf{0}))\ {}_{1/2}\!\oplus\ \partial(b\cdot\partial(\mathbf{0}))\,.$$

Therefore, we have by application of axiom BP the provable equality

$$\begin{aligned} AX^b_p \vdash \alpha\cdot\big(\partial(b\cdot\partial(\mathbf{0}) + \tau\cdot(\partial(a\cdot\partial(\mathbf{0}) + b\cdot\partial(\mathbf{0}))\ {}_{1/2}\!\oplus\ \partial(b\cdot\partial(\mathbf{0}))))\ {}_{r}\!\oplus\ Q\big) &= \\ \alpha\cdot\big((\partial(a\cdot\partial(\mathbf{0}) + b\cdot\partial(\mathbf{0}))\ {}_{1/2}\!\oplus\ \partial(b\cdot\partial(\mathbf{0})))\ {}_{r}\!\oplus\ Q\big)&\,. \end{aligned}$$

Another example is $a\cdot(P_1\ {}_r\!\oplus\ P_2) \sqsubseteq \partial(b\cdot R + a\cdot P_1)\ {}_r\!\oplus\ \partial(c\cdot S + a\cdot P_2)$, so

$$\begin{aligned} AX^b_p \vdash \alpha\cdot\big(\partial(a\cdot(P_1\ {}_r\!\oplus\ P_2) + \tau\cdot(\partial(b\cdot R + a\cdot P_1)\ {}_r\!\oplus\ \partial(c\cdot S + a\cdot P_2)))\ {}_r\!\oplus\ Q\big) &= \\ \alpha\cdot\big((\partial(b\cdot R + a\cdot P_1)\ {}_r\!\oplus\ \partial(c\cdot S + a\cdot P_2))\ {}_r\!\oplus\ Q\big)&. \end{aligned}$$

Table 2. Axioms for strong and rooted branching probabilistic bisimilarity

A1	$E + F = F + E$
A2	$(E + F) + G = E + (F + G)$
A3	$E + E = E$
A4	$E + \mathbf{0} = E$
P1	$P \,{}_r\!\oplus\, Q = Q \,{}_{1-r}\!\oplus\, P$
P2	$P \,{}_r\!\oplus\, (Q \,{}_s\!\oplus\, R) = (P \,{}_{\bar{r}}\!\oplus\, Q) \,{}_{\bar{s}}\!\oplus\, R$ where $r = \bar{r}\bar{s}$ and $(1-r)(1-s) = 1-\bar{s}$
P3	$P \,{}_r\!\oplus\, P = P$
C	$\alpha \cdot P + \alpha \cdot Q = \alpha \cdot P + \alpha \cdot (P \,{}_r\!\oplus\, Q) + \alpha \cdot Q$
BP	if $E \sqsubseteq P$ then $\alpha \cdot (\partial(E + \tau \cdot P) \,{}_r\!\oplus\, Q) = \alpha \cdot (P \,{}_r\!\oplus\, Q)$
G	if $E \sqsubseteq \partial(F)$ then $\alpha \cdot (\partial(E + F) \,{}_r\!\oplus\, Q) = \alpha \cdot (\partial(F) \,{}_r\!\oplus\, Q)$

An example illustrating the use of (α), rather than α, as label of the matching transition of $[\![P]\!]$ in the definition of $\sqsubseteq$ is

$$\tau \cdot (\partial(b \cdot P + \tau \cdot Q) \,{}_r\!\oplus\, Q) \sqsubseteq \partial(b \cdot P + \tau \cdot Q)$$

from which we obtain

$$AX_p^b \vdash \alpha \cdot \big(\partial(\tau \cdot (\partial(b \cdot P + \tau \cdot Q) \,{}_r\!\oplus\, Q) + \tau \cdot (\partial(b \cdot P + \tau \cdot Q))) \,{}_r\!\oplus\, R\big) = \alpha \cdot \big((\partial(b \cdot P + \tau \cdot Q)) \,{}_r\!\oplus\, R\big).$$

The axiom G roughly is a variant of BP without the τ prefixing the process P. A typical example, matching the one above, is

$$AX_p^b \vdash \alpha \cdot \big(\partial(\tau \cdot (\partial(b \cdot P + \tau \cdot Q) \,{}_r\!\oplus\, Q) + (b \cdot P + \tau \cdot Q)) \,{}_r\!\oplus\, R\big) = \alpha \cdot \big((\partial(b \cdot P + \tau \cdot Q)) \,{}_r\!\oplus\, R\big)\,.$$

The occurrences of the prefix $\alpha \cdot \_$ in BP and G are related to the root condition for non-deterministic processes, cf. axiom B in Sect. 3.

Lemma 6.2. The following simplifications of the axiom BP are derivable:

$$(i)\ AX_p^b \vdash \alpha \cdot \partial(E + \tau \cdot P) = \alpha \cdot P \quad \text{if } E \sqsubseteq P,$$
$$(ii)\ AX_p^b \vdash \alpha \cdot (\partial(\tau \cdot P) \,{}_r\!\oplus\, R) = \alpha \cdot (P \,{}_r\!\oplus\, R) \text{ and}$$
$$(iii)\ AX_p^b \vdash \alpha \cdot \partial(\tau \cdot P) = \alpha \cdot P\,.$$

See [16] for a proof of the above lemma. Similar simplifications of axiom G can be found. Using the properties given by Lemma 6.2 the process identities mentioned in the introduction can easily be proven. Returning to the processes E_1 and E_6 related to Fig. 1, we have

$$\begin{aligned}
AX_p^b \vdash E_1 \quad &= \quad \tau\cdot\big(\partial(\tau\cdot\partial(\tau\cdot P + c\cdot Q + \tau\cdot R) + c\cdot Q + \tau\cdot R)\\
&\qquad {}_{1/2}\oplus\ \partial(\tau\cdot(\partial(\tau\cdot P + c\cdot Q + \tau\cdot R)\ {}_{1/2}\oplus\ \partial(\mathbf{0})))\big)\\
&\overset{\text{BP}}{=} \quad \tau\cdot\big(\partial(\tau\cdot P + c\cdot Q + \tau\cdot R)\\
&\qquad {}_{1/2}\oplus\ \partial(\tau\cdot(\partial(\tau\cdot P + c\cdot Q + \tau\cdot R)\ {}_{1/2}\oplus\ \partial(\mathbf{0})))\big)\\
&\overset{6.2\,(ii),\,\text{P1}}{=} \quad \tau\cdot\big(\partial(\tau\cdot P + c\cdot Q + \tau\cdot R)\\
&\qquad {}_{1/2}\oplus\ (\partial(\tau\cdot P + c\cdot Q + \tau\cdot R)\ {}_{1/2}\oplus\ \partial(\mathbf{0}))\big)\\
&\overset{\text{P2}}{=} \quad \tau\cdot\big((\partial(\tau\cdot P + c\cdot Q + \tau\cdot R)_{2/3}\oplus\\
&\qquad \partial(\tau\cdot P + c\cdot Q + \tau\cdot R))\ {}_{3/4}\oplus\ \partial(\mathbf{0})\big)\\
&\overset{\text{P3}}{=} \quad \tau\cdot(\partial(\tau\cdot P + c\cdot Q + \tau\cdot R)\ {}_{3/4}\oplus\ \partial(\mathbf{0})) = E_6\,.
\end{aligned}$$

Soundness of the theory AX_p for strong probabilistic bisimilarity and of the theory AX_p^b for rooted branching probabilistic bisimilarity is straightforward.

Lemma 6.3 (Soundness). For all $P, Q \in \mathcal{P}$, if $AX_p \vdash P = Q$ then $P \leftrightarroweq Q$, and if $AX_p^b \vdash P = Q$ then $P \leftrightarroweq_{rb} Q$.

Proof. As usual, in view of $\leftrightarroweq$ and $\leftrightarroweq_{rb}$ being congruences, one only needs to prove the left-hand and right-hand sides of the axioms to be strongly or rooted branching probabilistically bisimilar. We only treat the cases of the axioms BP and G with respect to rooted branching probabilistic bisimilarity.

For BP, by Definition 4.5 and Lemma 4.7, it suffices to show that $P \leftrightarroweq_b \partial(E + \tau\cdot P)$ if $E \sqsubseteq P$. Suppose $\delta(E + \tau\cdot P) \xrightarrow{\alpha} \mu$. We distinguish two cases: (i) $\delta(E) \xrightarrow{\alpha} \mu$; (ii) $\alpha = \tau$, $\delta(E) \xrightarrow{\tau} \mu'$ and $\mu = [\![P]\!]\ {}_r\oplus\ \mu'$ for some $r \in (0,1]$. For (i), by definition of $E \sqsubseteq P$, we have $[\![P]\!] \xrightarrow{(\alpha)} \nu$ and $\mu \leftrightarroweq_b \nu$ for suitable $\nu \in Distr(\mathcal{E})$. For (ii), again by $E \sqsubseteq P$, we have $[\![P]\!] \xrightarrow{(\tau)} \nu'$ and $\mu' \leftrightarroweq_b \nu'$. Thus $[\![P]\!] = [\![P]\!]\,{}_r\oplus\,[\![P]\!] \xrightarrow{(\tau)} \nu$ and $\mu \leftrightarroweq_b \nu$ for $\nu = [\![P]\!]\ {}_r\oplus\ \nu'$, as was to be shown. Conversely, $[\![P]\!] \xrightarrow{\alpha} \mu$ trivially implies that $\delta(E + \tau\cdot P) \Rightarrow [\![P]\!] \xrightarrow{(\alpha)} \mu$. The requirement on weak decomposability also holds trivially.

For G, by Definition 4.5 and Lemma 4.7, it suffices to show that $E + F \leftrightarroweq_b F$ if $E \sqsubseteq \partial(F)$. Put $\mathcal{R} = \{\langle E + F, F\rangle\}^\dagger \cup \leftrightarroweq_b$. We verify that $\mathcal{R}$ is a branching probabilistic bisimulation. Naturally, $\delta(E) \xrightarrow{\alpha} \mu$ implies $\delta(E{+}F) \xrightarrow{\alpha} \mu$, and also weak decomposability is easy. Finally, suppose $\delta(E + F) \xrightarrow{\alpha} \mu$. Since $E \sqsubseteq \partial(F)$ now we have $\delta(E) \xrightarrow{(\alpha)} \nu$ for some ν with $\mu \leftrightarroweq_b \nu$. □

As for the process language with non-deterministic processes only, we aim at a completeness proof that is built on completeness of strong bisimilarity and the notion of a concrete process. Equational characterization of strong probabilistic bisimilarity has been addressed by various authors. The theory AX_p provides a sound and complete theory. For a proof, see e.g. [23].

Lemma 6.4. The theory AX_p is sound and complete for strong bisimilarity.

The next lemma provides a more state-based characterization of strong probabilistic bisimilarity.

Lemma 6.5. Let $\mathcal{R} \subseteq \mathit{Distr}(\mathcal{E}) \times \mathit{Distr}(\mathcal{E})$ be a decomposable relation such that

$$\mu_1 \,\mathcal{R}\, \nu_1 \text{ and } \mu_2 \,\mathcal{R}\, \nu_2 \quad \text{implies} \quad (\mu_1 \,{}_r{\oplus}\, \mu_2) \,\mathcal{R}\, (\nu_1 \,{}_r{\oplus}\, \nu_2) \tag{1}$$

and for each pair $E, F \in \mathcal{E}$

$$\delta(E) \,\mathcal{R}\, \delta(F) \text{ and } E \xrightarrow{\alpha} \mu' \quad \text{implies} \quad \delta(F) \xrightarrow{\alpha} \nu' \text{ and } \mu' \,\mathcal{R}\, \nu' \tag{2}$$

for a suitable $\nu' \in \mathit{Distr}(\mathcal{E})$. Then $\mu \,\mathcal{R}\, \nu$ implies $\mu \leftrightarroweq \nu$.

Proof. We show that $\mathcal{R}$ is a strong probabilistic bisimulation relation. So, let $\mu, \nu \in \mathit{Distr}(\mathcal{E})$ such that $\mu \,\mathcal{R}\, \nu$ and $\mu \xrightarrow{\alpha} \mu'$. By Definition 4.2(b) we have $\mu = \bigoplus_{i \in I} p_i * E_i$, $\mu' = \bigoplus_{i \in I} p_i * \mu'_i$, and $E_i \xrightarrow{\alpha} \mu'_i$ for all $i \in I$. Since $\mathcal{R}$ is decomposable, there are $\nu_i \in \mathit{Distr}(\mathcal{E})$, for $i \in I$, such that

$$\nu = \textstyle\bigoplus_{i \in I} p_i * \nu_i \quad \text{and} \quad \delta(E_i) \,\mathcal{R}\, \nu_i \text{ for all } i \in I.$$

Let, for each $i \in I$, $\nu_i = \bigoplus_{j \in J_i} p_{ij} * F_{ij}$. Since $\mathcal{R}$ is decomposable, there are $\mu_{ij} \in \mathit{Distr}(\mathcal{E})$, for $j \in J_i$, such that

$$\delta(E_i) = \textstyle\bigoplus_{i \in J_i} p_{ij} * \mu_{ij} \quad \text{and} \quad \mu_{ij} \,\mathcal{R}\, \delta(F_{ij}) \text{ for all } j \in J_i.$$

Here $\mu_{ij} = \delta(E_i)$. Writing $E_{ij} := E_i$, $q_{ij} := p_i \cdot p_{ij}$ and $K = \{(i,j) \mid i \in I \wedge j \in J_i\}$ we obtain

$$\mu = \textstyle\bigoplus_{k \in K} q_k * E_k \;,\; \nu = \bigoplus_{k \in K} q_k * F_k \quad \text{and} \quad \delta(E_k) \,\mathcal{R}\, \delta(F_k) \text{ for all } k \in K.$$

Let $\mu'_{ij} := \mu'_i$ for all $i \in I$ and $j \in J_i$. Then $\mu' = \bigoplus_{k \in K} q_k * \mu'_k$. Using that $E_k \xrightarrow{\alpha} \mu'_k$ for all $k \in K$, there must be distributions ν'_k for $k \in K$ such that

$$\delta(F_k) \xrightarrow{\alpha} \nu'_k \quad \text{and} \quad \mu'_k \,\mathcal{R}\, \nu'_k.$$

By Definition 4.2(b) this implies $\nu \xrightarrow{\alpha} \nu'$, for $\nu' := \bigoplus_{k \in K} q_k * \nu'_k$. Moreover, (1) yields $\mu' \,\mathcal{R}\, \nu'$. □

The following technical lemma expresses that two rooted branching probabilistically bisimilar processes can be represented in a similar way.

Lemma 6.6. For all $Q, R \in \mathcal{P}$, if $Q \leftrightarroweq_{rb} R$ then there are an index set I as well as for all $i \in I$ suitable $p_i > 0$ and $F_i, G_i \in \mathcal{E}$ such that $F_i \leftrightarroweq_{rb} G_i$, $AX_p \vdash Q = \bigoplus_{i \in I} p_i * F_i$ and $AX_p \vdash R = \bigoplus_{i \in I} p_i * G_i$.

Proof. Suppose $AX_p \vdash Q = \bigoplus_{j \in J} q_j * F'_j$. Since $Q \leftrightarroweq_{rb} R$ and $\leftrightarroweq_{rb}$ is decomposable, the process R can be written as $\bigoplus_{j \in J} q_j * R_j$ with $R_j \in \mathcal{P}$ for $j \in J$, such that $\partial(F_j) \leftrightarroweq_{rb} R_j$. Therefore, each distribution R_j can be written as $\bigoplus_{k \in K_j} r_{jk} * G_{jk}$ where $\partial(F'_j) \leftrightarroweq_{rb} \partial(G_{jk})$ for all $j \in J$ and $k \in K_j$. We now define $F_{jk} = F'_j$ for $j \in J$ and $k \in K_j$. Then, using the axioms P1, P2 and P3 we can derive

$$AX_p \vdash Q = \bigoplus_{j \in J} \bigoplus_{k \in K_j} q_j r_{jk} * F_{jk}$$
$$AX_p \vdash R = \bigoplus_{j \in J} \bigoplus_{k \in K_j} q_j r_{jk} * G_{jk}$$

with $F_{jk} \leftrightarroweq_b G_{jk}$ for $j \in J$ and $k \in K_j$. This proves the lemma. □

Similar to the non-deterministic case, a transition $E \xrightarrow{\tau} \mu$ is called *inert* iff $\delta(E) \leftrightarroweq_b \mu$. Typical cases of inert transitions include

$$\tau \cdot P \xrightarrow{\tau} [\![P]\!] \quad \text{and} \quad E + \tau \cdot \partial(E) \xrightarrow{\tau} \delta(E) .$$

Furthermore, a transition $E \xrightarrow{\tau} \mu_1 \,{}_r\!\oplus\, \mu_2$ with $r \in (0,1]$ and $\delta(E) \leftrightarroweq_b \mu_1$ is called *partially inert*. A typical case is

$$\tau \cdot (\partial(b \cdot P + \tau \cdot Q) \,{}_r\!\oplus\, Q) + b \cdot P + \tau \cdot Q \xrightarrow{\tau} \delta(b \cdot P + \tau \cdot Q) \,{}_r\!\oplus\, [\![Q]\!] .$$

Here $\delta\big(\tau \cdot (\partial(b \cdot P + \tau \cdot Q) \,{}_r\!\oplus\, Q) + b \cdot P + \tau \cdot Q\big) \leftrightarroweq_b \delta(b \cdot P + \tau \cdot Q)$ because $\delta(b \cdot P + \tau \cdot Q) \xrightarrow{(\tau)} \delta(b \cdot P + \tau \cdot Q) \,{}_r\!\oplus\, [\![Q]\!]$.

In Sect. 3 a process is called concrete if it does not exhibit an inert transition. In the setting with probabilistic choice we need to be more careful. For example, we also want to exclude processes of the form

$$\partial(\tau \cdot P) \,{}_{1/2}\!\oplus\, \partial(a \cdot Q)) \quad \text{and} \quad \partial(a \cdot P) \,{}_{1/2}\!\oplus\, \partial(b \cdot (\partial(\tau \cdot Q) \,{}_{1/3}\!\oplus\, Q))$$

from being concrete, although they cannot perform a transition by themselves at all. Therefore, we define the *derivatives* $der(P) \subseteq \mathcal{E}$ of a probabilistic process $P \in \mathcal{P}$ by

$$der(P \,{}_r\!\oplus\, Q) := der(P) \cup der(Q)$$
$$der(\partial(\textstyle\sum_{i \in I} \alpha_i \cdot P_i)) := \{ \textstyle\sum_{i \in I} \alpha_i \cdot P_i \} \cup \textstyle\bigcup_{i \in I} der(P_i)$$

and define a process $\bar{P} \in \mathcal{P}$ to be *concrete* iff none of its derivatives can perform a partially inert transition, i.e., if there is no transition $E \xrightarrow{\tau} \mu_1 \,{}_r\!\oplus\, \mu_2$ with $E \in der(\bar{P})$, $r \in (0,1]$ and $\delta(E) \leftrightarroweq_b \mu_1$. A non-deterministic process $\bar{E}$ is called concrete if the probabilistic process $\partial(\bar{E})$ is. Moreover, we define two sets of concrete processes:

$$\mathcal{E}_{cc} = \{ \bar{E} \in \mathcal{E} \mid \bar{E} \text{ is concrete} \} \quad \text{and} \quad \mathcal{P}_{cc} = \{ \bar{P} \in \mathcal{P} \mid \bar{P} \text{is concrete} \}.$$

Furthermore, we call a process $E \in Distr(\mathcal{E})$ *rigid* iff there is no inert transition $E \xrightarrow{\tau} \mu$, and write $\mathcal{E}_r = \{ \bar{E} \in \mathcal{E} \mid \bar{E} \text{ is rigid} \}$. Naturally, $\mathcal{E}_{cc} \subseteq \mathcal{E}_r$.

We use concrete and rigid processes to build the proof of the completeness result for rooted branching probabilistic bisimilarity on top of the completeness proof of strong probabilistic bisimilarity. The following lemma lists all properties of concrete and rigid processes we need in our completeness proof.

Lemma 6.7.

(a) If $E = \sum_{i\in I} \alpha_i \cdot P_i$ with $P_i \in \mathcal{P}_{cc}$ and, for all $i \in I$, $\alpha_i \neq \tau$ or $[\![P_i]\!]$ cannot be written as $\mu_1 \,{}_r\!\oplus\, \mu_2$ with $r \in (0,1]$ and $\delta(E) \leftrightarroweq_b \mu_1$, then $E \in \mathcal{E}_{cc}$.
(b) If $P_1, P_2 \in \mathcal{P}_{cc}$ then $P_1 \,{}_r\!\oplus\, P_2 \in \mathcal{P}_{cc}$.
(c) If $\mu = \bigoplus_{i\in I} p_i * \mu_i \in \mathit{Distr}(\mathcal{E}_{cc})$ with each $p_i > 0$, then each $\mu_i \in \mathit{Distr}(\mathcal{E}_{cc})$.
(d) If $\mu \in \mathit{Distr}(\mathcal{E}_{cc})$ and $\mu \xrightarrow{(\alpha)} \mu'$ then $\mu' \in \mathit{Distr}(\mathcal{E}_{cc})$.
(e) If $\mu \in \mathit{Distr}(\mathcal{E}_r)$ and $\mu \Rightarrow \mu'$ with $\mu \leftrightarroweq_b \mu'$ then $\mu = \mu'$.
(f) If $E \in \mathcal{E}_{cc}$, $F \in \mathcal{E}$, and $\mu, \nu \in \mathit{Distr}(\mathcal{E})$ are such that $E \leftrightarroweq_b F$, $E \xrightarrow{\alpha} \mu$, $\delta(F) \xrightarrow{(\alpha)} \nu$ and $\mu \leftrightarroweq_b \nu$, then $\delta(F) \xrightarrow{\alpha} \nu$.

Proof. Properties (a), (b), (c) and (d) follow immediately from the definitions, in the case of (d) also using Definition 4.2(b).

For (e), let $\mu \in \mathit{Distr}(\mathcal{E}_r)$ and $\mu \Rightarrow \mu'$ with $\mu \leftrightarroweq_b \mu'$. Towards a contradiction, suppose $\mu \neq \mu'$. Then there must be a distribution $\bar{\mu} \neq \mu$ such that $\mu \xrightarrow{(\tau)} \bar{\mu}$ and $\bar{\mu} \Rightarrow \mu'$. We may even choose $\bar{\mu}$ such that the transition $\mu \xrightarrow{(\tau)} \bar{\mu}$ acts on only one (rigid) state in the support of μ, i.e. there are $E \in \mathcal{E}$, $r \in (0,1]$ and $\rho, \nu \in \mathit{Distr}(\mathcal{E})$ such that $\mu = \delta(E) \,{}_r\!\oplus\, \rho$, $E \xrightarrow{\tau} \nu$ and $\bar{\mu} = \nu \,{}_r\!\oplus\, \rho$. By Lemma 5.1 $\delta(E) \,{}_r\!\oplus\, \rho = \mu \leftrightarroweq_b \bar{\mu} = \nu \,{}_r\!\oplus\, \rho$. Hence by Lemma 5.2 $\delta(E) \leftrightarroweq_b \nu$. So the transition $E \xrightarrow{\tau} \nu$ is inert, contradicting $E \in \mathcal{E}_r$.

To establish (f), suppose $E \in \mathcal{E}_{cc}$, $F \in \mathcal{E}$, and $\mu, \nu \in \mathit{Distr}(\mathcal{E})$ are such that $E \leftrightarroweq_b F$, $E \xrightarrow{\alpha} \mu$, $\delta(F) \xrightarrow{(\alpha)} \nu$ and $\mu \leftrightarroweq_b \nu$. Assume $\alpha = \tau$, for otherwise the statement is trivial. Then $\delta(F) \xrightarrow{\tau} \nu_1$ and $\nu = \nu_1 \,{}_r\!\oplus\, \delta(F)$ for some $\nu_1 \in \mathit{Distr}(\mathcal{E})$ and $r \in [0,1]$. Since $\leftrightarroweq_b$ is weakly decomposable, there are $\bar{\mu}, \mu_1, \mu_2 \in \mathit{Distr}(\mathcal{E})$ such that $\mu \Rightarrow \bar{\mu}$, $\mu \leftrightarroweq_b \bar{\mu}$, $\bar{\mu} = \mu_1 \,{}_r\!\oplus\, \mu_2$, $\mu_1 \leftrightarroweq_b \nu_1$ and $\mu_2 \leftrightarroweq_b \delta(F)$. Since μ is concrete, using case (d) of the lemma, $\bar{\mu} = \mu$ by case (e). Thus $E \xrightarrow{\tau} \mu_1 \,{}_r\!\oplus\, \mu_2$ with $\delta(E) \leftrightarroweq_b \delta(F) \leftrightarroweq_b \mu_2$. Since E is concrete, this transition cannot be partially inert. Thus, we must have $r = 1$. It follows that $\delta(F) \xrightarrow{\alpha} \nu$. □

Lemma 6.8. For all $\bar{P}, \bar{Q} \in \mathcal{P}_{cc}$, if $\bar{P} \leftrightarroweq_b \bar{Q}$ then $\bar{P} \leftrightarroweq \bar{Q}$ and $AX_p \vdash \bar{P} = \bar{Q}$.

Proof. Let $\mathcal{R} := \leftrightarroweq_b \cap\, (\mathit{Distr}(\mathcal{E}_{cc}) \times \mathit{Distr}(\mathcal{E}_{cc}))$. Then, by Lemma 6.7(c)–(d), $\mathcal{R}$ is a branching probabilistic bisimulation relation relating $\bar{P}$ and $\bar{Q}$. We show that $\mathcal{R}$ moreover satisfies the conditions of Lemma 6.5. Condition (1) is a direct consequence of Lemmas 4.7 and 6.7(b). That $\mathcal{R}$ is decomposable follows since it is weakly decomposable, in combination with Lemma 6.7(e). Now, in order to verify condition (2), suppose $\delta(E) \,\mathcal{R}\, \delta(F)$ and $E \xrightarrow{\alpha} \mu$. Then $\delta(F) \Rightarrow \bar{\nu} \xrightarrow{(\alpha)} \nu$ for some $\bar{\nu}, \nu \in \mathcal{P}$ with $\delta(E) \,\mathcal{R}\, \bar{\nu}$ and $\mu \,\mathcal{R}\, \nu$. By Lemma 6.7(e) we have $\bar{\nu} = \delta(F)$. Thus $\delta(F) \xrightarrow{(\alpha)} \nu$. Hence, using Lemma 6.7(f) it follows that $\delta(F) \xrightarrow{\alpha} \nu$. With $\mathcal{R}$ satisfying conditions (1) and (2), Lemma 6.5 yields $\bar{P} \leftrightarroweq \bar{Q}$. By Lemma 6.4 we obtain $AX_p \vdash \bar{P} = \bar{Q}$. □

Before we are in a position to prove our main result we need one more technical lemma. Here we write $AX_p^b \vdash P_1 \approx P_2$ as a shorthand for

$$\forall \alpha \in \mathcal{A}\, \forall Q \in \mathcal{P}\, \forall r \in (0,1)\colon\; AX_p^b \vdash \alpha \cdot (P_1 \,{}_r\!\oplus\, Q) = \alpha \cdot (P_2 \,{}_r\!\oplus\, Q)\,.$$

For example, using axiom BP, if $E \sqsubseteq P$ then $AX_p^b \vdash \partial(E + \tau \cdot P) \approx P$. Likewise, using G, if $E \sqsubseteq \partial(F)$ then $\partial(E + F) \approx \partial(F)$. As in the proof of Lemma 6.2(i), from $AX_p^b \vdash P_1 \approx P_2$ it also follows that $AX_p^b \vdash \alpha \cdot P_1 \approx \alpha \cdot P_2$ for all $\alpha \in \mathcal{A}$.

For a complete proof of the lemma we refer the reader to [16]. Note, in the proof we rely on the axioms BP and G.

Lemma 6.9.

(a) For each non-deterministic process $E \in \mathcal{E}$ there is a concrete probabilistic process $\bar{P} \in \mathcal{P}_{cc}$ such that $AX_p^b \vdash \partial(E) \approx \bar{P}$.
(b) For each probabilistic process $P \in \mathcal{P}$ there is a concrete probabilistic process $\bar{P} \in \mathcal{P}_{cc}$ such that $AX_p^b \vdash \alpha \cdot P = \alpha \cdot \bar{P}$ for all $\alpha \in \mathcal{A}$.
(c) For all probabilistic processes $Q, R \in \mathcal{P}$, if $Q \leftrightarrow_b R$ then $AX_p^b \vdash \alpha \cdot Q = \alpha \cdot R$ for all $\alpha \in \mathcal{A}$.

Proof (Sketch). We proceed by simultaneous induction on $c(E)$, $c(P)$, and $\max\{c(Q), c(R)\}$. For part (a) we write a process E with $c(E) > 0$ as $\sum_{i \in I} \alpha_i \cdot P_i$ for some index set I and suitable $\alpha_i \in \mathcal{A}$ and $P_i \in \mathcal{P}$. By the induction hypothesis (b), we can pick for each $i \in I$ a concrete probabilistic process $\bar{P}_i \in \mathcal{P}_{cc}$ with $AX_p^b \vdash \alpha_i \cdot \bar{P}_i = \alpha_i \cdot P_i$. Now $AX_p \vdash E = \bar{E}$ for $\bar{E} := \sum_{i \in I} \alpha_i \cdot \bar{P}_i$. We then distinguish two subcases: (i) for some $i_0 \in I$, $\alpha_{i_0} = \tau$ and $\bar{P}_{i_0} \leftrightarrow_b \partial(\bar{E})$; (ii) for all $i \in I$, $\alpha_i \neq \tau$ or $P_i \not\leftrightarrow_b \partial(\bar{E})$, i.e., $\bar{E}$ is rigid.

Subcase (i) is relatively easy. We make use of the axiom BP. For the axiom to apply we need to verify that $H \sqsubseteq \bar{P}_{i_0}$ where $H := \sum_{i \in I \setminus \{i_0\}} \alpha_i \cdot \bar{P}_i$.

For subcase (ii) we show that a concrete process $\bar{C} \in \mathcal{E}_{cc}$ exists such that $AX_p^b \vdash \partial(\bar{E}) \approx \partial(\bar{C})$ for which we proceed by induction on the number of indices $k \in I$ such that $\alpha_k = \tau$ and $[\![P_k]\!]$ can be written as $\mu_1 \,_r\!\oplus\, \mu_2$ with $r \in (0,1)$ and $\delta(\bar{E}) \leftrightarrow_b \mu_1$. For the induction step we prove that $\tau \cdot P_{i_0} \sqsubseteq H$ and we make use of axiom G. As an intermediate result we show $\bar{E} \leftrightarrow_b H$.

Parts (b) and (c) follow directly from parts (a) and (b), respectively, with the help of Lemmas 6.7 and 6.8. □

By now we have gathered all ingredients for showing that the theory AX_p^b is an equational characterization of rooted branching probabilistic bisimilarity. In the proof of the theorem we make use of axiom C.

Theorem 6.10 (AX_p^b sound and complete for $\leftrightarrow_{rb}$). For all non-deterministic processes $E, F \in \mathcal{E}$ and all probabilistic processes $P, Q \in \mathcal{P}$ it holds that $E \leftrightarrow_{rb} F$ iff $AX_p^b \vdash E = F$ and $P \leftrightarrow_{rb} Q$ iff $AX_p^b \vdash P = Q$.

Proof. As we have settled the soundness of AX_p^b in Lemma 6.3, it remains to show that AX_p^b is complete. So, let $E, F \in \mathcal{E}$ such that $E \leftrightarrow_{rb} F$. Suppose $E = \sum_{i \in I} \alpha_i \cdot P_i$ and $F = \sum_{j \in J} \beta_j \cdot Q_j$ for suitable index sets I, J, actions α_i, β_j, and probabilistic processes P_i, Q_j.

Since, for each $i \in I$, $E \xrightarrow{\alpha_i} [\![P_i]\!]$ we have $\delta(F) \xrightarrow{\alpha_i} \bigoplus_{j \in J_i} q_{ij} * Q_j$ and $P_i \leftrightarrow_b \bigoplus_{j \in J_i} q_{ij} * Q_j$ for some subset $J_i \subseteq J$ and suitable $q_{ij} \geqslant 0$. Similarly, there exist for $j \in J$ a subset $I_j \subseteq I$ and $p_{ij} \geqslant 0$ such that $\delta(E) \xrightarrow{\beta_j} \bigoplus_{i \in I_j} p_{ij} * P_i$ and $Q_j \leftrightarroweq_b \bigoplus_{i \in I_j} p_{ij} * P_i$. By $|J|+|I|$ series of applications of axiom C we obtain

$$AX_p \vdash E = \textstyle\sum_{i \in I} \alpha_i \cdot P_i + \sum_{j \in J} \beta_j \cdot (\bigoplus_{i \in I_j} p_{ij} * P_i), \text{ and} \tag{3}$$
$$AX_p \vdash F = \textstyle\sum_{j \in J} \beta_j \cdot Q_j + \sum_{i \in I} \alpha_i \cdot (\bigoplus_{j \in J_i} q_{ij} * Q_j) . \tag{4}$$

Since $P_i \leftrightarroweq_b \bigoplus_{j \in J_i} q_{ij} * Q_j$ and $Q_j \leftrightarroweq_b \bigoplus_{i \in I_j} p_{ij} * P_i$ we obtain by Lemma 6.9

$$AX_p^b \vdash \alpha_i \cdot P_i = \alpha_i \cdot \textstyle\bigoplus_{j \in J_i} q_{ij} * Q_j \text{ and } AX_p^b \vdash \beta_j \cdot Q_j = \beta_j \cdot \bigoplus_{i \in I_j} p_{ij} * P_i$$

for $i \in I$, $j \in J$. Combining this with Eqs. (3) and (4) yields $AX_p^b \vdash E = F$.

Now, let $P, Q \in \mathcal{P}$ such that $P \leftrightarroweq_{rb} Q$. By Lemma 6.6 we have

$$AX_p \vdash P = \textstyle\bigoplus_{i \in I} p_i * E_i \qquad AX_p \vdash Q = \bigoplus_{i \in I} p_i * F_i \qquad \forall i \in I \colon E_i \leftrightarroweq_{rb} F_i$$

for a suitable index set I, $p_i > 0$, $E_i, F_i \in \mathcal{E}$, for $i \in I$. By the conclusion of the first paragraph of this proof we have $AX_p^b \vdash E_i = F_i$ for $i \in I$. Hence $AX_p^b \vdash P = Q$. □

7 Concluding Remarks

We presented an axiomatization of rooted branching probabilistic bisimilarity and proved its soundness and completeness. In doing so, we aimed to stay close to a straightforward completeness proof for the axiomatization of rooted branching bisimilarity for non-deterministic processes that employed concrete processes, which is also presented in this paper. In particular, the route via concrete processes guided us to find the right formulation of the axioms BP and G for branching bisimilarity in the probabilistic case.

Future work will include the study of the extension of the setting of the present paper with a parallel operator [12]. In particular a congruence result for the parallel operator should be obtained, which for the mixed non-deterministic and probabilistic setting can be challenging. Also the inclusion of recursion [11, 15] is a clear direction for further research.

The present conditional form of axioms BP and G is only semantically motivated. However, the axiom G has a purely syntactic counterpart of the form

$$\begin{aligned} \alpha \cdot \big(\partial(\textstyle\sum_{i \in I} \tau \cdot (P_i \,{}_{r_i}\!\oplus\, \partial(E + \sum_{i \in I} \tau \cdot P_i)) + E + \sum_{i \in I} \tau \cdot P_i) \,{}_r\!\oplus\, Q \big) \\ = \alpha \cdot (\partial(E + \textstyle\sum_{i \in I} \tau \cdot P_i) \,{}_r\!\oplus\, Q) \end{aligned}$$

whereas BP can be written as

$$\alpha \cdot \big(\partial(\textstyle\sum_{i \in I} \alpha_i \cdot (P_i \,{}_{r_i}\!\oplus\, P) + \sum_{i \in J} \alpha_i \cdot P_i + \tau \cdot P) \,{}_r\!\oplus\, Q \big) \;=\; \alpha \cdot (P \,{}_r\!\oplus\, Q)$$

with $I \cap J = \emptyset$, where $\alpha_i = \tau$ for all $i \in I$, $P_i = \bigoplus_{k \in K} r_k * P_{ik}$ for all $i \in I \cup J$, and

$$P = \textstyle\bigoplus_{k \in K} r_k * (F_k + \sum_{i \in I \cup J} \alpha_i \cdot \bigoplus_{k \in K} P_{ik}) .$$

Admittedly, this form is a bit complicated to work with. An alternative approach could be to axiomatize the relation $\sqsubseteq$, or perhaps to introduce and axiomatize an auxiliary process operator $+'$ such that $E \sqsubseteq P$ can be translated into the condition $E +' P = P$ or similar.

Also, we want to develop a minimization algorithm for probabilistic processes modulo branching probabilistic bisimilarity. Eisentraut et al. propose in [13] an algorithm for deciding equivalence with respect to weak distribution bisimilarity relying on a state-based characterization, a result presently not available in our setting. Other work and proposals for weak bisimilarity include [8,14,31], but these do not fit well with the installed base of our toolset [7]. For the case of strong probabilistic bisimilarity without combined transitions we recently developed in [20] an algorithm improving upon the early results of [5]. In [31] a polynomial algorithm for Segala's probabilistic branching bisimilarity, which differs from our notion of probabilistic branching bisimilarity, is defined. We hope to arrive at an efficient algorithm by combining ideas from [31–33] and of [18,19].

References

1. Andova, S., Georgievska, S.: On compositionality, efficiency, and applicability of abstraction in probabilistic systems. In: Nielsen, M., Kučera, A., Miltersen, P.B., Palamidessi, C., Tůma, P., Valencia, F. (eds.) SOFSEM 2009. LNCS, vol. 5404, pp. 67–78. Springer, Heidelberg (2009). https://doi.org/10.1007/978-3-540-95891-8_10
2. Andova, S., Georgievska, S., Trcka, N.: Branching bisimulation congruence for probabilistic systems. Theoret. Comput. Sci. **413**, 58–72 (2012)
3. Andova, S., Willemse, T.A.C.: Branching bisimulation for probabilistic systems: characteristics and decidability. Theoret. Comput. Sci. **356**, 325–355 (2006)
4. Baeten, J.C.M., Basten, T., Reniers, M.A.: Process Algebra: Equational Theories of Communicating Processes. Cambridge Tracts in Theoretical Computer Science, vol. 50. CUP, Cambridge (2010)
5. Baier, C., Engelen, B., Majster-Cederbaum, M.E.: Deciding bisimilarity and similarity for probabilistic processes. J. Comput. Syst. Sci. **60**, 187–231 (2000)
6. Bandini, E., Segala, R.: Axiomatizations for probabilistic bisimulation. In: Orejas, F., Spirakis, P.G., van Leeuwen, J. (eds.) ICALP 2001. LNCS, vol. 2076, pp. 370–381. Springer, Heidelberg (2001). https://doi.org/10.1007/3-540-48224-5_31
7. Bunte, O., et al.: The mCRL2 toolset for analysing concurrent systems. In: Vojnar, T., Zhang, L. (eds.) TACAS 2019. LNCS, vol. 11428, pp. 21–39. Springer, Cham (2019). https://doi.org/10.1007/978-3-030-17465-1_2
8. Cattani, S., Segala, R.: Decision algorithms for probabilistic bisimulation*. In: Brim, L., Křetínský, M., Kučera, A., Jančar, P. (eds.) CONCUR 2002. LNCS, vol. 2421, pp. 371–386. Springer, Heidelberg (2002). https://doi.org/10.1007/3-540-45694-5_25
9. De Nicola, R., Vaandrager, F.W.: Three logics for branching bisimulation. J. ACM **42**(2), 458–487 (1995)
10. Deng, Y., van Glabbeek, R.J., Hennessy, M., Morgan, C.C.: Testing finitary probabilistic processes. In: Bravetti, M., Zavattaro, G. (eds.) CONCUR 2009. LNCS, vol. 5710, pp. 274–288. Springer, Heidelberg (2009). https://doi.org/10.1007/978-3-642-04081-8_19

11. Deng, Y., Palamidessi, C.: Axiomatizations for probabilistic finite-state behaviors. Theoret. Comput. Sci. **373**, 92–114 (2007)
12. Deng, Y., Palamidessi, C., Pang, J.: Compositional reasoning for probabilistic finite-state behaviors. In: Middeldorp, A., van Oostrom, V., van Raamsdonk, F., de Vrijer, R. (eds.) Processes, Terms and Cycles: Steps on the Road to Infinity. LNCS, vol. 3838, pp. 309–337. Springer, Heidelberg (2005). https://doi.org/10.1007/11601548_17
13. Eisentraut, C., Hermanns, H., Krämer, J., Turrini, A., Zhang, L.: Deciding bisimilarities on distributions. In: Joshi, K., Siegle, M., Stoelinga, M., D'Argenio, P.R. (eds.) QEST 2013. LNCS, vol. 8054, pp. 72–88. Springer, Heidelberg (2013). https://doi.org/10.1007/978-3-642-40196-1_6
14. Ferrer Fioriti, L.M., Hashemi, V., Hermanns, H., Turrini, A.: Deciding probabilistic automata weak bisimulation: theory and practice. Formal Aspects Comput. **28**, 09–143 (2016)
15. Fisher, N., van Glabbeek, R.J.: Axiomatising infinitary probabilistic weak bisimilarity of finite-state behaviours. J. Logical Algebraic Methods Program. **102**, 64–102 (2019)
16. van Glabbeek, R.J., Groote, J.F., de Vink, E.P.: A complete axiomatization of branching bisimilarity for a simple process language with probabilistic choice. Technical report, Eindhoven University of Technology (2019). http://rvg.web.cse.unsw.edu.au/pub/AxiomProbBranchingBis.pdf
17. van Glabbeek, R.J., Weijland, W.P.: Branching time and abstraction in bisimulation semantics. J. ACM **43**, 555–600 (1996)
18. Groote, J.F., Jansen, D.N., Keiren, J.J.A., Wijs, A.: An $O(m \log n)$ algorithm for computing stuttering equivalence and branching bisimulation. ACM Trans. Comput. Logic **18**(2), 13:1–13:34 (2017)
19. Groote, J.F., Vaandrager, F.: An efficient algorithm for branching bisimulation and stuttering equivalence. In: Paterson, M.S. (ed.) ICALP 1990. LNCS, vol. 443, pp. 626–638. Springer, Heidelberg (1990). https://doi.org/10.1007/BFb0032063
20. Groote, J.F., Rivera Verduzco, H.J., de Vink, E.P.: An efficient algorithm to determine probabilistic bisimulation. Algorithms **11**(9), 131, 1–22 (2018)
21. Groote, J.F., de Vink, E.P.: Problem solving using process algebra considered insightful. In: Katoen, J.-P., Langerak, R., Rensink, A. (eds.) ModelEd, TestEd, TrustEd. LNCS, vol. 10500, pp. 48–63. Springer, Cham (2017). https://doi.org/10.1007/978-3-319-68270-9_3
22. Hansson, H., Jonsson, B.: A calculus for communicating systems with time and probabilities. In: Proceedings of the RTSS 1990, pp. 278–287. IEEE (1990)
23. Hennessy, M.: Exploring probabilistic bisimulations, part I. Formal Aspects Comput. **24**, 749–768 (2012)
24. Hennessy, M., Milner, R.: On observing nondeterminism and concurrency. In: de Bakker, J., van Leeuwen, J. (eds.) ICALP 1980. LNCS, vol. 85, pp. 299–309. Springer, Heidelberg (1980). https://doi.org/10.1007/3-540-10003-2_79
25. Lee, M.D., de Vink, E.P.: Logical characterization of bisimulation for transition relations over probability distributions with internal actions. In: Faliszewski, P., Muscholl, A., Niedermeier, R. (eds.) Proceedings of the MFCS 2016. LIPIcs, vol. 58, pp. 29:1–29:14. Schloss Dagstuhl - Leibniz-Zentrum für Informatik (2016)
26. Milner, R.: Communication and Concurrency. Prentice Hall, Upper Saddle River (1989)
27. Milner, R.: A complete axiomatization for observational congruence of finite-state behaviours. Inf. Comput. **81**(2), 227–247 (1989)

28. Segala, R.: Modeling and verification of randomized distributed real-time systems. PhD thesis, MIT (1995). Technical report MIT/LCS/TR-676
29. Segala, R., Lynch, N.: Probabilistic simulations for probabilistic processes. In: Jonsson, B., Parrow, J. (eds.) CONCUR 1994. LNCS, vol. 836, pp. 481–496. Springer, Heidelberg (1994). https://doi.org/10.1007/978-3-540-48654-1_35
30. Stoelinga, M.: Alea Jacta est: verification of probabilistic, real-time and parametric systems. Ph.D. thesis, Radboud Universiteit (2002)
31. Turrini, A., Hermanns, H.: Polynomial time decision algorithms for probabilistic automata. Inf. Comput. **244**, 134–171 (2015)
32. Valmari, A.: Simple bisimilarity minimization in $O(m \log n)$ time. Fundam. Inf. **105**(3), 319–339 (2010)
33. Valmari, A., Franceschinis, G.: Simple $O(m \log n)$ time Markov chain lumping. In: Esparza, J., Majumdar, R. (eds.) TACAS 2010. LNCS, vol. 6015, pp. 38–52. Springer, Heidelberg (2010). https://doi.org/10.1007/978-3-642-12002-2_4

Walking Through the Semantics of Exclusive and Event-Based Gateways in BPMN Choreographies

Flavio Corradini, Andrea Morichetta, Barbara Re[(✉)], and Francesco Tiezzi

School of Science and Technology, University of Camerino, Camerino, Italy
{flavio.corradini,andrea.morichetta,barbara.re,francesco.tiezzi}@unicam.it

Abstract. With the evolution of distributed systems, nowadays BPMN choreography diagrams have acquired more and more importance for modelling systems interaction. However, one of the drawbacks of this model is the lack of formal semantics, which leads to different interpretations, and hence implementations, of some of its features. Among the BPMN choreography elements, particularly ambiguous is the semantics of the exclusive and event-based gateways, used to represent different forms of choices. Formalisations of these elements have been proposed in the literature, but none of them is derived from a direct and faithful modelling of the description provided by the BPMN standard. In this work, instead, we provide a direct formalisation, in terms of an operational semantics, that aims at shedding light on the intricacies of the behaviour of the exclusive and event-based gateways. The effectiveness of the approach is shown by illustrating how our semantics can disambiguate tricky behaviours in choreography models.

Keywords: BPMN 2.0 · Choreographies · Exclusive and event-based gateways · Operational semantics

1 Introduction

The BPMN 2.0 OMG standard [23] (in the following just BPMN) is more and more adopted by academia and industry as a modelling language for distributed systems. Its diffusion is mainly due to its capability to describe the behaviour of components by means of an appealing graphical notation. In particular, a BPMN *choreography* diagram provides a global specification focusing on component interactions, while a BPMN *collaboration* diagram describes the implementation of every single component, possibly deployed and managed by different organisations, in terms of exchanged messages and internal behaviour. In such a setting, organisations that are willing to cooperate to achieve a specific objective can refer to choreography specifications for describing the interactions between different parties. On the other hand, interested organisations can put in place the cooperation by relying on collaboration models able to describe both the communication patterns and the internal activities of a component.

M. S. Alvim et al. (Eds.): Palamidessi Festschrift, LNCS 11760, pp. 163–181, 2019.
https://doi.org/10.1007/978-3-030-31175-9_10

The community widely accepts the BPMN standard for its expressiveness and adaptability to many fields. Despite this advantage, its complex semantics can generate possible flaws during the design phase [4,7]. This phenomenon is accentuated by the lack of rigour of the standard in the element descriptions, due to the use of natural language for the semantic definitions. This problem is witnessed by the differences that can be observed between the available business process management systems: from one implementation to another different semantics are used for the same element, disorienting the expectation of the designer [11]. This lack of transparency in the standard, joint with the inability of being executable, has reduced the adoption of BPMN choreography diagrams in favour of collaboration diagrams that played for years a more relevant role [2]. Despite this, BPMN choreography is considered the most appropriate notation to model the coordination of the interactions between different participants when the process cannot be controlled in a centralised manner [3]. Moreover, the increasing diffusion of blockchain technology, natively implementing decentralised trusted scenarios, is asking for adequate modelling languages [13]. Choreographies are promising modelling abstractions for contracts between two or more organisations, being both communication-centric and human understandable by business analysts and IT specialists. The whole area of choreographies may be revitalised by such technological evolution [2]. This raises the need of making even clearer the semantics of choreography diagrams before their wider adoption.

In a distributed scenario, without a clear semantics for choreographies, the difficulty on producing high-quality models for the global perspective may represent a barrier in the development of each single components. The problem is further compounded by the fact that ambiguities on the semantics also regard largely used elements, such as the *exclusive* and the *event-based gateways*, exploited to express in BPMN choreographies different forms of choice among alternative execution paths. To fill this gap, in this paper we first present a detailed analysis of the natural language descriptions provided by the BPMN standard about the semantics of key choreography elements. We specifically focus on the two gateway elements mentioned above. We then propose an informal characterisation, by resorting to a graphical notation, of the semantics of the two gateways, which combines all requirements stated in the standard. Finally, to provide a clear understanding of the ambiguous points of the standard, we provide a formalisation of the BPMN choreographies semantics, given in terms of an operational semantics defined on top of a textual representation of the models. The proposed work is motivated and validated by relying on an example coming from the literature and presenting an evident issue caused by a misunderstanding of the standard.

The rest of the paper is organised as follows. Section 2 provides background notions on BPMN choreographies and collaborations. Section 3 motivates our work by detailing issues resulting from the standard. Section 4 informally describes the semantics of exclusive and event-based gateways. Section 5 proposes the formalisation of choreography syntax and semantics. Section 6 discusses related works. Finally, Sect. 7 concludes the paper and discusses future work.

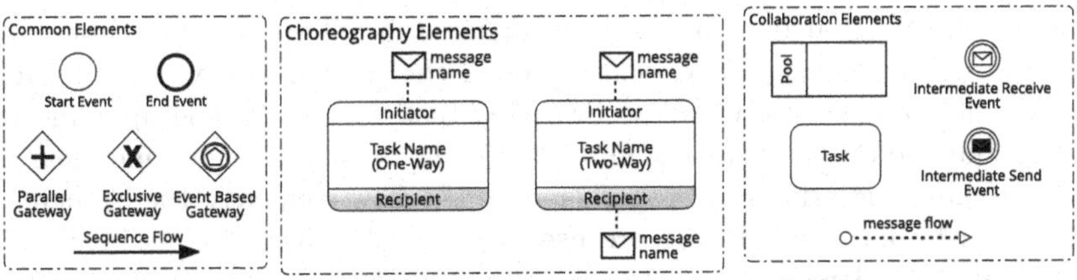

a. Flow elements b. Choreography tasks c. Collaboration elements

Fig. 1. BPMN 2.0 elements.

2 Background

This section presents the relevant elements of choreography and collaboration diagrams we use in the paper.

Figure 1(a) depicts the most used modelling elements that can be included in both diagrams. **Events** are used to represent something that can happen. An event can be a *start event*, representing the point in which the choreography/collaboration starts, while an *end event* is raised when the choreography/collaboration terminates. **Gateways** are used to manage the flow of a choreography/collaboration both for parallel activities and choices. Gateways act as either join nodes (merging incoming sequence edges) or split nodes (forking into outgoing sequence edges). Different types of gateways are available. A *parallel gateway (AND)* in join mode has to wait to be reached by all its incoming edges to start, and respectively all the outgoing edges are started simultaneously in the split case. An *exclusive gateway (XOR)* describes choices; it is activated each time the gateway is reached in join mode and, in split mode, it activates exactly one outgoing edge. An *event-based gateway* is similar to the XOR-split gateway, but its outgoing branches activation depends on the occurrence of a catching event in the collaboration and on the reception of a message in the choreography; these events/messages are in a race condition, where the first one that is triggered wins and disables the other ones. **Sequence Flows** are used to connect collaboration/choreography elements to specify the execution flow.

Focusing on the choreography diagram, we underline its ability to specify the message exchanges between two or more participants. This is done by means of **Choreography Tasks** in Fig. 1(b). They are drawn as rectangles divided in three bands: the central one refers to the name of the task, while the others refer to the involved participants (the white one is the initiator, while the gray one is the recipient). Messages can be sent either by one participant (One-Way tasks) or by both participants (Two-Way tasks). Concerning choreography tasks, we rely on the follow design choices. In relation to the *Two-Way choreography task*, the OMG standard states that it is "an atomic activity in a choreography process" execution [23, p. 323]. However, this does not mean that the task blocks the whole execution of the choreography. In fact, participants are usually distributed, and we assume that other choreography tasks involved in different parallel paths of the choreography can be executed. Thus, here we intend atomicity to mean that both messages exchanged in a Two-Way task have to be received before triggering the execution along the sequence flow outgoing from the task. Therefore, even if we allow Two-Way tasks in the choreography models, we safely manage them as pairs of One-Way tasks preserving the same meaning.

In a collaboration diagram, together with the flow elements, the elements in Fig. 1(c) can be included. **Pools** are used to represent participants involved in the collaboration. **Tasks** are used to represent specific works to perform within a collaboration by a participant. **Intermediate Events** represent something that happens during the flow of the process, such as sending or receiving of a message. **Message Edges** are used to visualize communication flows between participants, by connecting communication elements within different pools.

3 Motivations

A choreography model represents a guideline for driving the communication interactions between organisations and represents a reference point for the implementation of each single component. For this reason a shared and clear understanding of the meaning of BPMN elements is needed in order to improve the quality of the designed model.

Unfortunately, such a common understanding cannot be taken for granted. Even if we deal with BPMN, which being a standard language should guarantee a certain level of rigorousness, the practice gives evidences of various issues. Looking in the literature, only few choreography models are available. Most of them are partial specifications, which miss to specify, e.g., messages and conditions. Moreover, when these are included, they are often incorrectly used, due to misunderstandings resulting by inaccuracies and inconsistencies in the standard. A typical example of such problems concerns the exclusive and event-based gateways, two elements that despite their tricky semantics are largely used in choreography diagrams. As a reference example, we report in Fig. 2 an erroneous

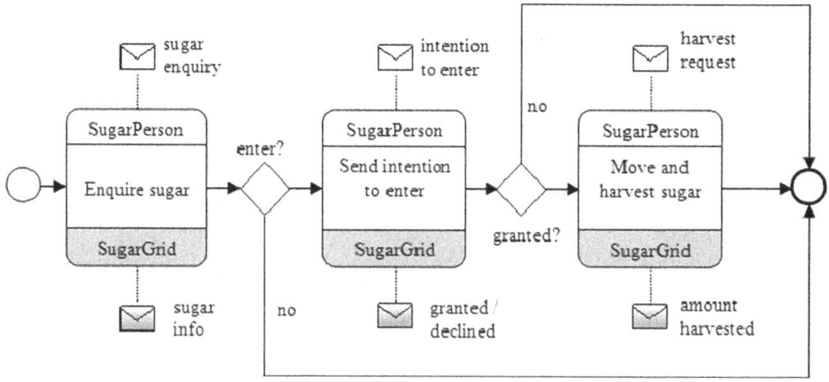

Fig. 2. An erroneous choreography model (source [24]).

- The data used for **Gateway** *Conditions* MUST have been in a **Message** that was sent prior to (upstream from) the **Gateway**.
    - More specifically, all *Participants* that are directly affected by the **Gateway** MUST have either sent or received the **Message(s)** that contained the data used in the *Conditions*.
        - Furthermore, all these *Participants* MUST have the same understanding of the data. That is, the actual values of the data cannot selectively change after a *Participant* has seen a **Message**. Changes to data during the course of the **Choreography** MUST be visible to all the *Participants* affected by the **Gateway**.

Fig. 3. Exclusive gateway [23, p. 345].

choreography model drawn from the literature [24]. It shows the interactions between two participants: the *SugarPerson*, asking for sugar, and *SugarGrid*, looking for sugar. The model under consideration contains a mistake on the specification of the first exclusive gateway. In particular, the enforcement of the *"enter?"* condition is not correct. In fact, the information about the intention of the *SugarPerson* to enter in the sugar grid is sent by the *SugarPerson* to the *SugarGrid* in the *"Send intention to enter"* task, which is executed after the execution of the gateway. The exclusive gateway, as explained in the next section, requires all participants involved in the subsequent tasks to be able to take the same decision expressed by the gateway condition. But, in this example, the decision cannot be properly taken by the *SugarGrid* participant, since he does not have yet the information about the *SugarPerson*'s intention. This model indeed violates a prescription of the standard [23, p. 345], requesting that the value used by the gateway in the condition evaluation has to be included in a message sent before the gateway execution.

Let us now focus more deeply on the description of choreography diagrams provided by the BPMN standard. We comment on observed contradictions that can be misleading for designers intending to use it. Here in the following, we report some excerpt focusing on the use of exclusive and event-based gateways.

Figure 3 is an extract of the standard [23, p. 348] describing the general behaviour of the exclusive gateway. The three bullet points stress the importance to share control information between all participants before to be used in decisions. The text is supported in the standard by two examples representing a choreography interaction between three participants and the corresponding implementation through collaboration, reported in Figs. 4 and 5, respectively. Few pages later, the standard [23, p. 348] also states that a choreography configuration can be valid only if all the participants have a shared information upon which the decision is made. Figure 6 shows an excerpt. At this point a first contradiction in the standard has been detected. It is between the text in Figs. 3 and 6, and the example in Fig. 5. The text stresses the need to share knowledge through messages between senders and receivers, while in Fig. 5 the participant C does not receive any message by participant A and only the message *M1* is exchanged between participants A and B.

Reading more about the description in Fig. 7 coming from [23, p. 349] we observe some constraints on the business process implementing a choreography.

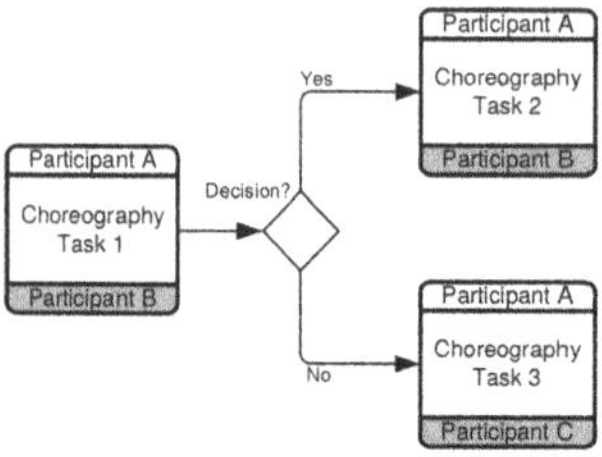

Fig. 4. An example of the exclusive gateway [23, p. 348].

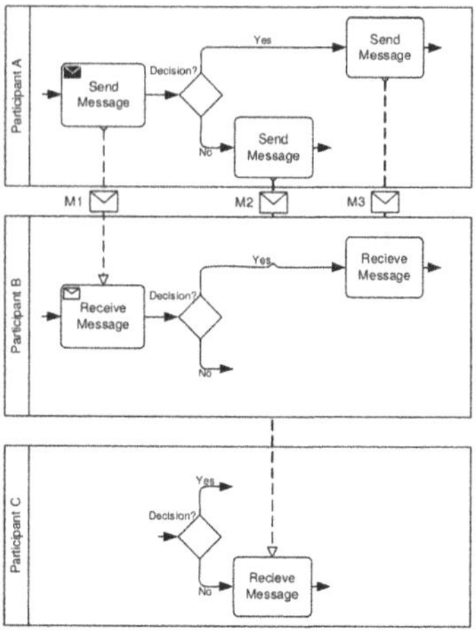

Fig. 5. The corresponding collaboration view of the choreography in Fig. 4 [23, p. 349].

This configuration can only be valid if <u>all</u> the *Participants* in the **Choreography Activities** that follow the **Gateway** have seen the data upon which the decision is made. If either "Participant B" or "Participant C" had not sent or received a **Message** with the appropriate data, then that *Participant* would not be able to know if they are suppose to receive a **Message** at that point in the **Choreography**. There is also the assumption that the value of the data is consistent from the point of view of all *Participants*.

Fig. 6. Valid choreography configuration [23, p. 348].

If we check the alignment between the text and the model in Fig. 5 we observe some issues. In particular, the text confirms the usage of the exclusive gateway in the process of the participant that is the initiator but erroneously state about the usage of the event-based gateway for the receiver participants (see point 3). The text in Fig. 7 is also misaligned with respect to the general description of exclusive gateway in Fig. 3.

Another confirmation of the inaccuracy of the text in Fig. 7 can be obtained by reasoning on the definition of the event-based gateway reported in Fig. 8. It seems that there is no difference in the implementation of the exclusive gateway and event-based gateway when passing from a choreography to a collaboration model. This is obviously not possible, and the recognised overlap is a clear contradiction of the standard.

Summing up, in some parts the standard seems to be written with few attention. Moreover, the used examples are not complete and in most of the cases are inaccurate to substantiate the text. Finally, we believe that a major stumbling block is the language in which the standard is defined: natural language is inadequate when used to define the semantics of choreography elements.

The REQUIRED execution behavior of the **Gateway** and associated **Choreography Activities** are enforced through the **Business Processes** of the *Participants* as follows:

- Each **Choreography Activity** and the **Sequence Flow** connections are reflected in each *Participant* **Process**.
- The **Gateway** is reflected in the **Process** of each *Participant* **Process** that is an initiator of **Choreography Activities** that follow the **Gateway**.
- For the receivers in **Choreography Activities** that follow the **Gateway**, an **Event-Based Gateway** is used to consume the associated **Message** (sent as an outcome of the **Gateway**). When a *Participant* is the receiver of more than one of the alternative **Messages**, the corresponding receives follow the **Event-Based Gateway**. If the *Participant* is the receiver of only one such **Message**, that is also consumed through a receive following the **Event-Based Gateway**. This is because the *Participant* **Process** does not know whether it will receive a **Message** (since the **Gateway** entails a choice of outcomes).

Fig. 7. Text describing collaboration in Fig. 5 [23, p. 349].

The REQUIRED execution behavior of the **Event-Based Gateway** and associated **Choreography Activities** are enforced through the **Business Processes** of the *Participants* as follows:

- Each **Choreography Activity** and the **Sequence Flow** connections is reflected in each *Participant* **Process**.
- If the senders following the **Gateway** are the same, the **Event-Based Gateway** is reflected as an **Exclusive Gateway** in that *Participant's* **Process**. This is because the choice of which **Message** to send is determined by the same Participant. If the senders are different, sending occurs through different **Processes**.
- If the receivers are the same, the senders can be the same or different. In this case, the **Event-Based Gateway** is reflected in the receiver's **Process**, with the different **Message** receives following the **Gateway**.
- If the receivers are different, the senders need to be the same. The **Event-Based Gateway** is reflected for different receiver **Processes** such that the respective receive follows the **Gateway**. A time-out can be used to ensure that the **Gateway** does not wait indefinitely.

Fig. 8. Event-based description [23, p. 351]

4 Discussion on Exclusive and Event-Based Gateways

In this section, we informally clarify the semantics of the choreography diagram discussing the correct implementation of the exclusive gateway and event-based gateway through examples mapping choreographies to collaborations.

To make clear the principle at the basis of the choreography design, we first discuss the *data management* and *realizability*. Data in choreography does not have any control mechanism, so they are not maintained in any central source. The only way to share information is the exchange of messages. On the other hand, to be realizable [5] the choreography should respect a general rule where the initiator of a choreography activity must have been involved as initiator or receiver in the previous choreography activity. This restriction limits the combination of participants in the choreography task and in particular when the tasks sequence includes a gateway in the middle.

Focusing on the admitted combinations of participants, moving from the left-hand-side of a gateway to the right-hand side, we can have five different possible configurations: (i) same sender and same receiver; (ii) same sender and different receivers (iii) different senders and same receiver; (iv) same sender and same receiver swapped; (v) different senders and different receivers.

Tables 1 shows on the left column of each sub-table the possible choreography configurations of the exclusive gateway, and on the right column the respective

Table 1. Exclusive Gateway Implementation

	Choreographies	Collaborations
a)	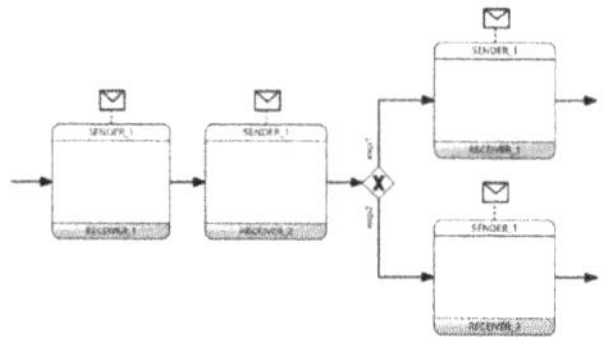	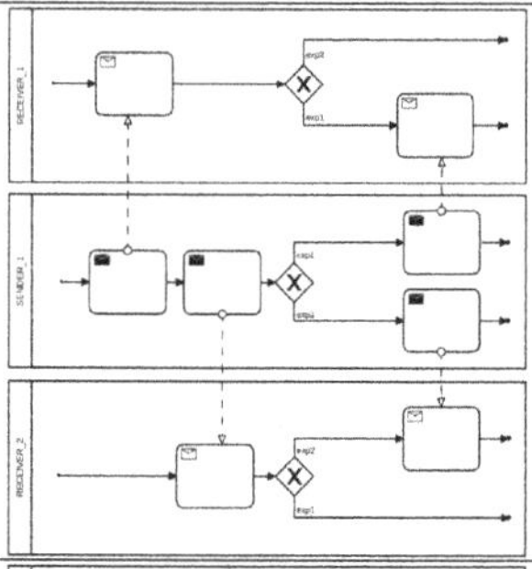
b)	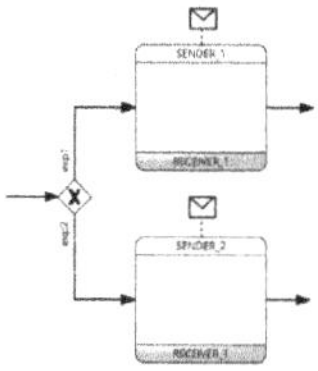	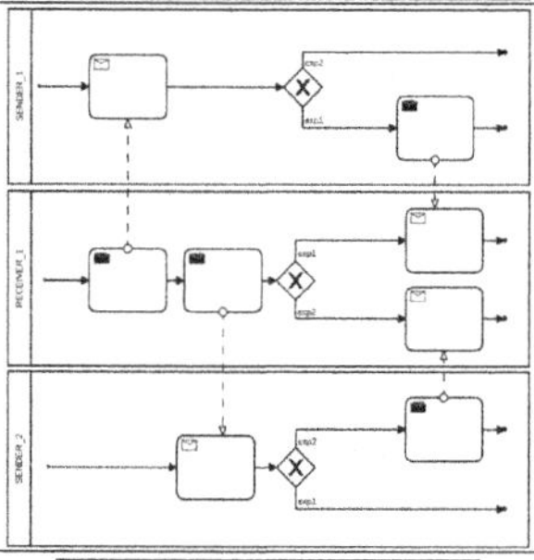
c)	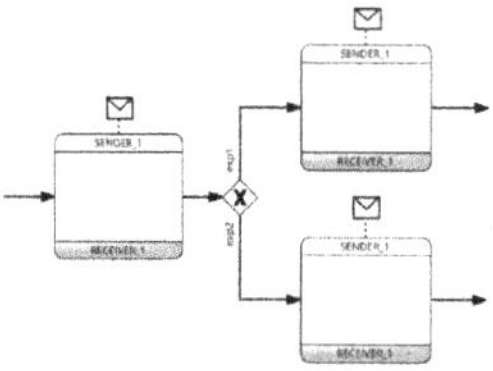	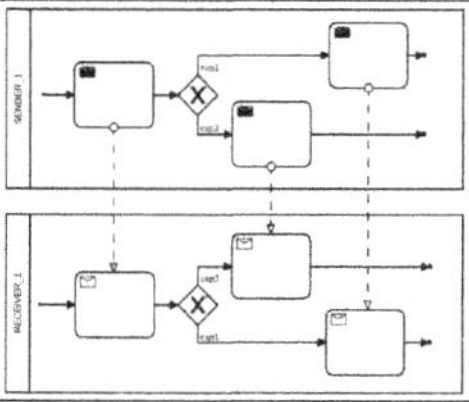
d)	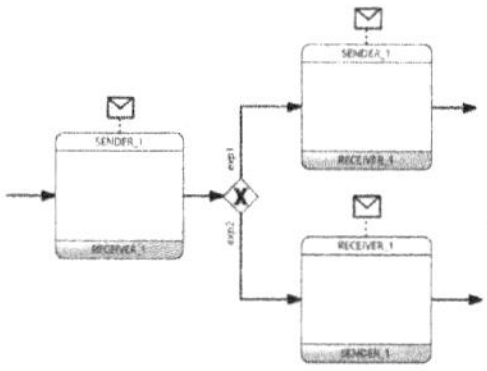	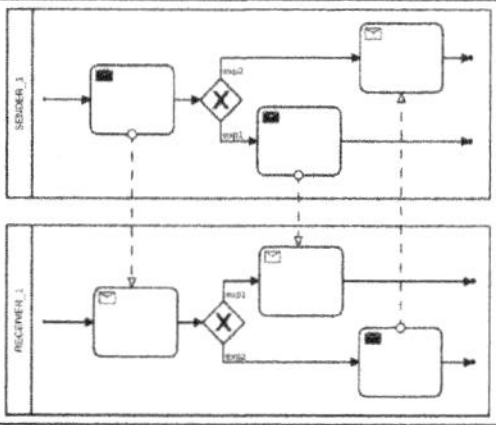
e)	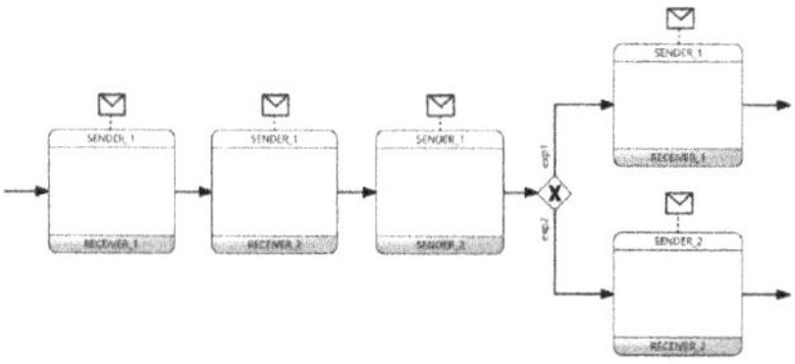	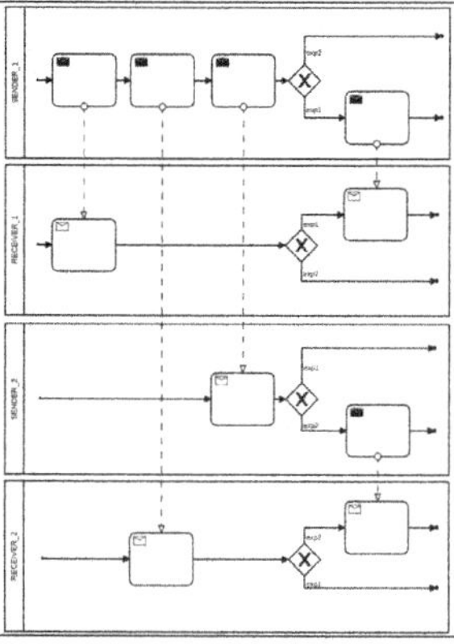

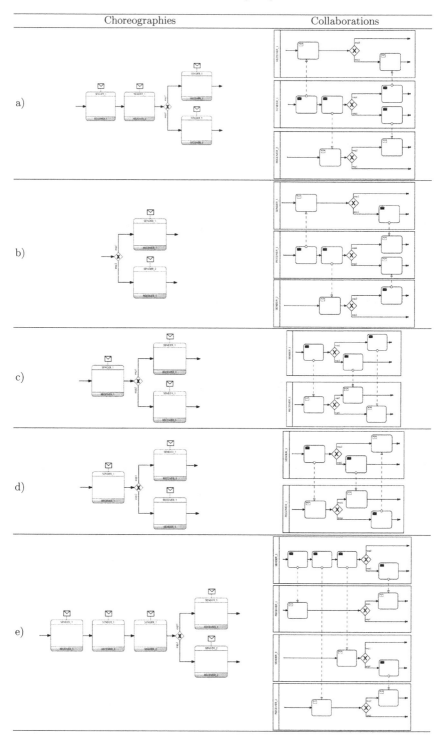

collaborations implementation. In particular, Table 1(a) depicts the design of a choreography containing an exclusive gateway in the configuration of "same senders and different receivers". The corresponding collaboration should contain a first sharing of information from the sender S1 to the receivers R1 and R2, and successively all participants will perform the same choice using the exclusive gateways with the same condition. Table 1(b) represent the configuration with different senders and same receiver. In this case, the task before the gateway was omitted just for the sake of simplicity, but we have always to respect the general rule defining the continuity between participants involved in consecutive tasks. In this configuration, both senders will deliver a message towards the receiver, and it will select the desired message ignoring the others, admitting message loss. The third configuration in Table 1(c) is with same sender and same receiver. The messages exchanged before and after the gateway involve the same participants. The sender will communicate to the receiver its will for the choice, and in the implementation both parties will follow the same path according to the shared information.

The configuration in Table 1(d) is similar to the previous one except for the sender and receiver that are swapped on the right-hand side of the gateway. Again, once the information at the basis of the decision is communicated, the two participants can continue according to either sending a message or waiting for another one. The last case in Table 1(e) represents the configuration with different senders and different receivers. Of course, this can be possible only respecting the right sequence of messages between the involved participants. Despite two separated communications, the participants will move according to the shared information in a coordinated way. Notably, these last two cases are not always possible, since we should consider the structure of the sequence task.

Table 2 shows on the left column the three choreography configurations that are admitted for the event-based gateway, and on the right column the respective collaboration implementations. Differently from the exclusive gateway, here it is not required the sharing of information before a choice. In Table 2(a) it is depicted the case with same sender and different receivers. The sender is the unique participant that takes the decision, and the receivers react consequently. In the implementation, a timer is used for avoiding the deadlock of the receiver not considered. Table 2(b) depicts the configuration with different senders and same receiver. Here the event-based gateway is used in the receiver implementation, and the logic of senders is simply the delivery of the message. Here we are in front of a race condition, where the late message is lost. The configuration with same sender and same receiver depicted in Table 2(c) is the standard communication where the sender takes the decision using an exclusive gateway and the receiver acts according to the choice taken by the counterpart.

5 Formal Semantics

This section presents our formalisation of BPMN choreographies, which in particular deals with the peculiarities of the event-based and exclusive gateways.

Table 2. Event-based gateway implementation

	Choreographies	Collaborations

a)

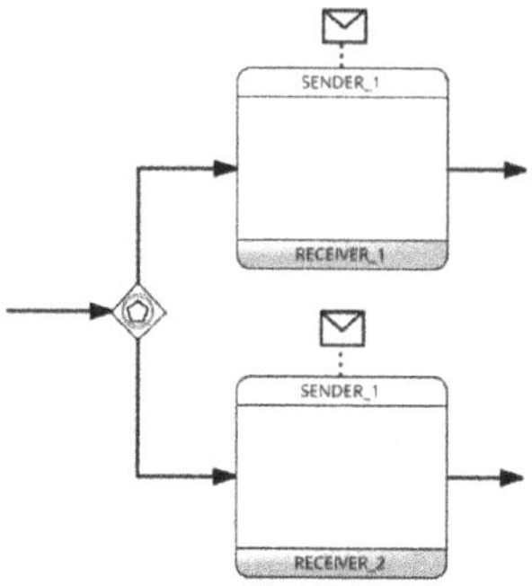

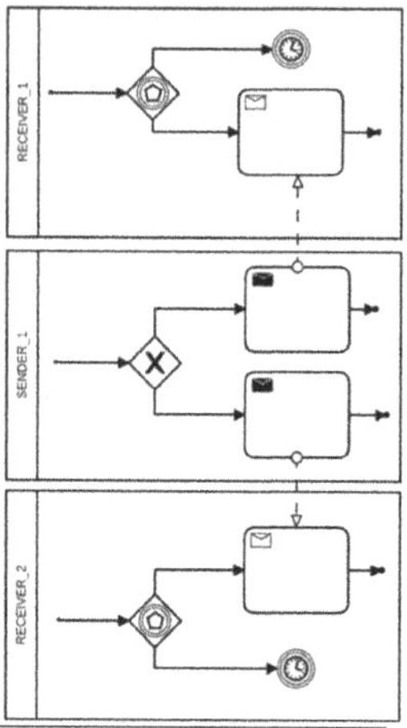

b)

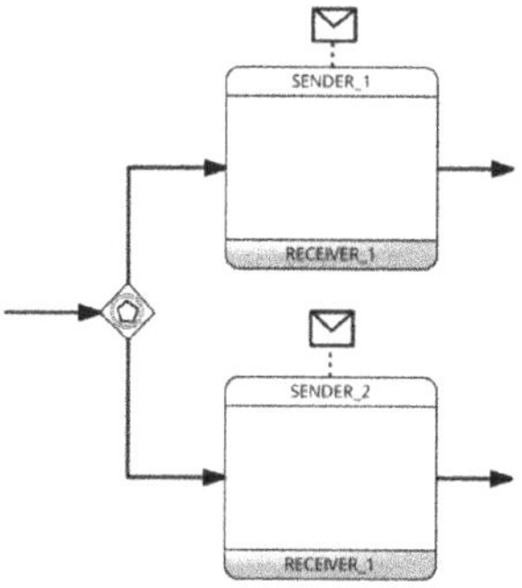

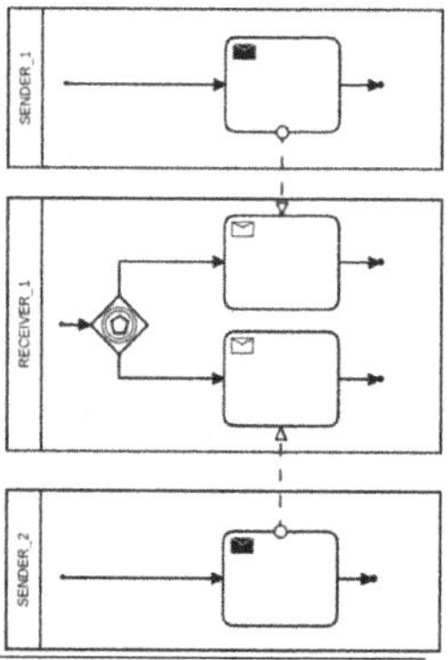

c)

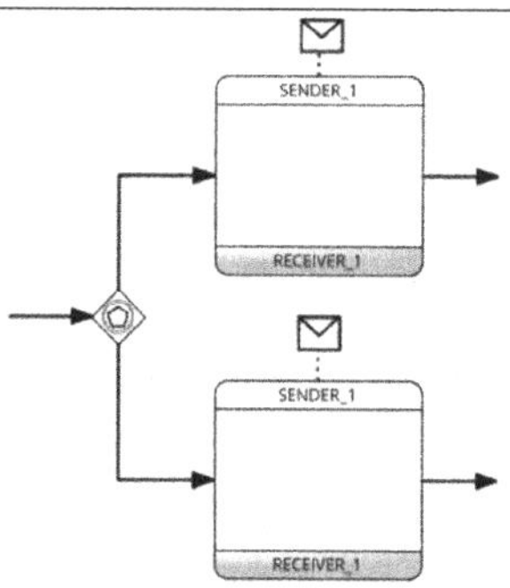

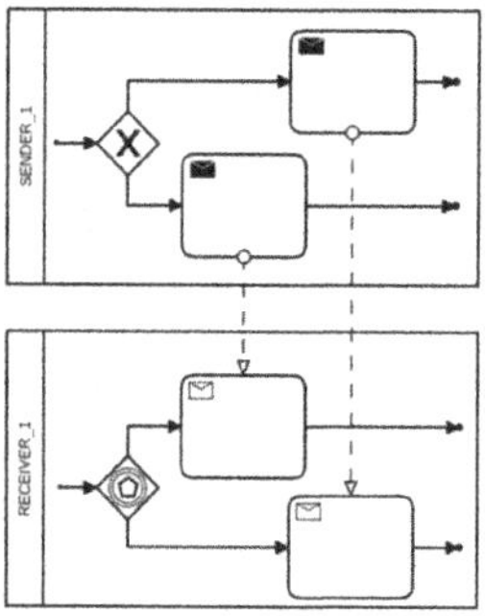

The aim is to shed light on the semantics of these elements, in a way as much faithful as possible with the informal semantics provided by the BPMN standard discussed in the previous section. To enable a formal treatment, we defined a Backus Normal Form (BNF) syntax providing a textual representation of the structure of BPMN choreographies, on top of which we have defined the operational semantics of the BPMN choreography language.

$$
\begin{aligned}
C ::= \; & \mathsf{start}(\mathsf{id},\mathsf{e}) \mid \mathsf{end}(\mathsf{id},\mathsf{e}) \mid \mathsf{andSplit}(\mathsf{id},\mathsf{e},E) \mid \mathsf{andJoin}(\mathsf{id},E,\mathsf{e}) \\
\mid \; & \mathsf{xorSplit}(\mathsf{id},\mathsf{e},(\mathsf{e}_1,\mathsf{exp}_1),\ldots,(\mathsf{e}_k,\mathsf{exp}_k)) \mid \mathsf{xorJoin}(\mathsf{id},E,\mathsf{e}) \mid \mathsf{task}(\mathsf{id},\mathsf{e}_1,\mathsf{e}_2,\mathsf{p},\mathsf{p}',\mathsf{m}) \\
\mid \; & \mathsf{eventBased}(\mathsf{id},\mathsf{e},(\mathsf{e}_1,\mathsf{p}_1,\mathsf{p}'_1,\mathsf{m}_1),\ldots,(\mathsf{e}_k,\mathsf{p}_k,\mathsf{p}'_k,\mathsf{m}_k)) \mid C_1|C_2
\end{aligned}
$$

Fig. 9. Syntax of BPMN choreography structures

The BNF syntax of the choreography models structure is given in Fig. 9. In the grammar, the non-terminal symbol C represents *Choreography Structures*, while the terminal symbols, denoted by the sans serif font, are the considered elements of a BPMN model, i.e. events, gateways and tasks. Notably, we are not proposing a new modeling formalism, but we are only using a textual notation for the BPMN elements. With respect to the graphical notation, the textual one is more manageable for supporting the formal definition of the semantics.

As a matter of notation, $\mathsf{e} \in \mathbb{E}$ denotes a *sequence edge*, $E \in 2^{\mathbb{E}}$ a set of edges, and $\mathsf{m} \in \mathbb{M}$ a message. We also use a set $\mathbb{EXP}$ of *expressions* (ranged over by exp), whose precise syntax is deliberately not specified; we just assume that expressions contain *values* $\mathsf{v} \in \mathbb{V}$. Such design choice has been influenced by the fact that the expression language operating on data is left unspecified even by the BPMN standard. Moreover, id and p denote names uniquely identifying a model element and a participant, respectively.

The correspondence between the syntax used here and the graphical notation of BPMN is as follows.

- $\mathsf{start}(\mathsf{id},\mathsf{e})$ represents a start event identified by id with outgoing edge e.
- $\mathsf{end}(\mathsf{id},\mathsf{e})$ represents an end event identified by id with incoming edge e.
- $\mathsf{andSplit}(\mathsf{id},\mathsf{e},E)$ represents an AND split gateway identified by id with incoming edge e and outgoing edges E (with $|E| > 1$).
- $\mathsf{andJoin}(\mathsf{id},E,\mathsf{e})$ represents an AND join gateway identified by id with incoming edges E (with $|E| > 1$) and outgoing edge e.
- $\mathsf{xorSplit}(\mathsf{id},\mathsf{e},(\mathsf{e}_1,\mathsf{exp}_1),\ldots,(\mathsf{e}_k,\mathsf{exp}_k))$ represents an XOR split gateway identified by id with incoming edge e and outgoing edges $\mathsf{e}_1,\ldots\mathsf{e}_k$ (with $k > 1$) associated to boolean expressions $\mathsf{exp}_1,\ldots,\mathsf{exp}_k$, each expression defining if the corresponding branch can be activated or not.
- $\mathsf{xorJoin}(\mathsf{id},E,\mathsf{e})$ represents an XOR join gateway identified by id with incoming edges E (with $|E| > 1$) and outgoing edge e.
- $\mathsf{task}(\mathsf{id},\mathsf{e}_1,\mathsf{e}_2,\mathsf{p},\mathsf{p}',\mathsf{m})$ represents a one-way task identified by id with incoming edge e_1, outgoing edge e_2, and sending a message m from p_1 to p_2. Two-way tasks are rendered in our formal framework as pairs of one-way tasks, hence they are not explicitly included in the syntax.
- $\mathsf{eventBased}(\mathsf{id},\mathsf{e},(\mathsf{e}_1,\mathsf{p}_1,\mathsf{p}'_1,\mathsf{m}_1),\ldots,(\mathsf{e}_k,\mathsf{p}_k,\mathsf{p}'_k,\mathsf{m}_k))$ represents an event-based gateway identified by id with incoming edge e, and a list (with $k > 1$) of tasks to be processed. It is worth noticing that the definition of the task

list is given by quadruples of the form $(\mathsf{e}, \mathsf{p}, \mathsf{p}', \mathsf{m})$, where e is the outgoing edge of the task, p and p' are interacting participants (sender and receiver, respectively) and m is the exchanged message.
- $C_1 \mid C_2$ represents the composition of elements, which permits to render a choreography structure in terms of a collection of elements.

To achieve a compositional definition, each sequence edge of the BPMN model is split in two parts: the part outgoing from the source element and the part incoming into the target element. The two parts are correlated by means of unique sequence edge names in the BPMN model. Notably, we only consider terms of the syntax that are derived from BPMN models.

The operational semantics we propose is given in terms of configurations of the form $\langle C, \sigma, \gamma \rangle$, where C is a choreography structure, $\sigma : \mathbb{E} \rightarrow \mathbb{N}$ is an *execution state function* mapping sequence edges to numbers of tokens, and $\gamma : \mathbb{M} \rightarrow \mathbb{V}$ is a *message state function* mapping messages to values. The execution state obtained by updating in the state σ the number of tokens of the edge e to n, written as $\sigma \cdot \{\mathsf{e} \mapsto \mathsf{n}\}$, is defined as follows: $(\sigma \cdot \{\mathsf{e} \mapsto \mathsf{n}\})(\mathsf{e}')$ returns n if $\mathsf{e}' = \mathsf{e}$, otherwise it returns $\sigma(\mathsf{e}')$. The message state obtained by updating in the state γ the value of a message m to a value v, written as $\gamma \cdot \{\mathsf{m} \mapsto \mathsf{v}\}$, is defined similarly. The *initial* states of a choreography, where all sequence edges are unmarked and all message values are undefined, denoted respectively by σ_0 and γ_0, are formally defined as: $\sigma_0(\mathsf{e}) = 0 \ \ \forall \mathsf{e} \in \mathbb{E}$, and $\gamma_0(\mathsf{m}) = \mathit{undef} \ \ \forall \mathsf{m} \in \mathbb{M}$.

The operational semantics is defined by means of a labelled transition system (LTS) on choreography configurations, formalising the execution of a choreography in terms of marking evolution and message exchanges. The LTS is a triple $\langle \mathcal{C}, \mathcal{A}, \rightarrow \rangle$ where: $\mathcal{C}$ is the set of choreography configurations; $\mathcal{L}$, ranged over by l, is the set of *labels* (of transitions that choreography configurations can perform); and $\rightarrow \subseteq \mathcal{C} \times \mathcal{L} \times \mathcal{C}$ is the *transition relation*. As usual, we write $\langle C, \sigma, \gamma \rangle \xrightarrow{l} \langle C, \sigma', \gamma' \rangle$ to indicate that $(\langle C, \sigma, \gamma \rangle, l, \langle C, \sigma', \gamma' \rangle) \in \rightarrow$ and say that the choreography configuration $\langle C, \sigma, \gamma \rangle$ performs a transition labelled by l and becomes the configuration $\langle C, \sigma', \gamma' \rangle$. Since choreography execution only affects the current states, and not the choreography structure, for the sake of presentation, we omit the structure from the target configuration of the transition. Thus, a transition $\langle C, \sigma, \gamma \rangle \xrightarrow{l} \langle C, \sigma', \gamma' \rangle$ is written as $\langle C, \sigma, \gamma \rangle \xrightarrow{l} \langle \sigma', \gamma' \rangle$. A label l represents a computational step and is defined by the graphical representation of the executed BPMN element together with further information: the element id and, possibly, the set of participants involved in a decision and/or an exchange of message. Notably, despite the presence of labels, this has to be thought of as a reduction semantics, because labels are not used for synchronisation (as instead it usually happens in labelled semantics), but only for keeping track of the executed elements.

The transition relation over choreography configurations is defined by the rules in Fig. 10. Before commenting on the rules, we introduce the auxiliary functions they exploit. Specifically, function $\mathit{inc} : \mathbb{S}_e \times \mathbb{E} \rightarrow \mathbb{S}_e$ (resp. $\mathit{dec} : \mathbb{S}_e \times \mathbb{E} \rightarrow \mathbb{S}_e$), where $\mathbb{S}_e$ is the set of execution states, allows updating a state by incrementing (resp. decrementing) by one the number of tokens marking an edge

(*Start*)

$$\langle \mathsf{start}(\mathsf{id},\mathsf{e}),\sigma_0,\gamma_0\rangle \xrightarrow{\bigcirc_{\mathsf{id}}} \langle inc(\sigma_0,\mathsf{e}),\gamma_0\rangle$$

(*End*)

$$\langle \mathsf{end}(\mathsf{id},\mathsf{e}),\sigma,\gamma\rangle \xrightarrow{\boldsymbol{\bigcirc}_{\mathsf{id}}} \langle dec(\sigma,\mathsf{e}),\gamma\rangle \qquad \sigma(\mathsf{e})>0$$

(*AndSplit*)

$$\langle \mathsf{andSplit}(\mathsf{id},\mathsf{e},E),\sigma,\gamma\rangle \xrightarrow{\diamond^{+}_{\mathsf{id}}} \langle inc(dec(\sigma,\mathsf{e}),E),\gamma\rangle \qquad \sigma(\mathsf{e})>0$$

(*AndJoin*)

$$\langle \mathsf{andJoin}(\mathsf{id},E,\mathsf{e}),\sigma,\gamma\rangle \xrightarrow{\diamond^{+}_{\mathsf{id}}} \langle inc(dec(\sigma,E),\mathsf{e}),\gamma\rangle \qquad \forall \mathsf{e}\in E_i\,.\,\sigma(\mathsf{e})>0$$

(*XorSplit*)

$$\langle \mathsf{xorSplit}(\mathsf{id},\mathsf{e},(\mathsf{e}_1,\mathsf{exp}_1),\ldots,(\mathsf{e}_k,\mathsf{exp}_k)),\sigma,\gamma\rangle \xrightarrow{\diamond^{\times}_{\mathsf{id}}\{\mathsf{p}_1,\ldots,\mathsf{p}_n\}} \langle inc(dec(\sigma,\mathsf{e}),\mathsf{e}_i),\gamma\rangle$$

$$\sigma(\mathsf{e})>0 \quad 1\le i\le k \quad eval(\mathsf{exp}_i,\gamma)=true \quad p(\{\mathsf{e}_1,\ldots,\mathsf{e}_k\})=\{\mathsf{p}_1,\ldots,\mathsf{p}_n\}$$

(*XorJoin*)

$$\langle \mathsf{xorJoin}(\mathsf{id},\{\mathsf{e}_1\}\cup E,\mathsf{e}_2),\sigma,\gamma\rangle \xrightarrow{\diamond^{\times}_{\mathsf{id}}} \langle inc(dec(\sigma,\mathsf{e}_1),\mathsf{e}_2),\gamma\rangle \qquad \sigma(\mathsf{e}_1)>0$$

(*Task*)

$$\langle \mathsf{task}(\mathsf{id},\mathsf{e}_1,\mathsf{e}_2,\mathsf{p}_1,\mathsf{p}_2,\mathsf{m}),\sigma,\gamma\rangle \xrightarrow{\square_{\mathsf{id}}\{\}\mathsf{p}_1\to\mathsf{p}_2\,:\,\mathsf{m}} \langle inc(dec(\sigma,\mathsf{e}_1),\mathsf{e}_2),\gamma'\rangle$$

$$\sigma(\mathsf{e}_i)>0 \quad update(\gamma,\mathsf{m})=\gamma'$$

(*EventBased₁*)

$$\langle \mathsf{eventBased}(\mathsf{id},\mathsf{e},(\mathsf{e}_1,\mathsf{p},\mathsf{p}_1,\mathsf{m}_1),\ldots,(\mathsf{e}_k,\mathsf{p},\mathsf{p}_k,\mathsf{m}_k)),\sigma,\gamma\rangle \xrightarrow{\diamond^{\circledcirc}_{\mathsf{id}}\{\mathsf{p}\}\mathsf{p}\to\mathsf{p}_i\,:\,\mathsf{m}_i} \langle inc(dec(\sigma,\mathsf{e}),\mathsf{e}_i),\gamma'\rangle$$

$$\sigma(\mathsf{e})>0 \quad 1\le i\le k, \quad update(\gamma,\mathsf{m})=\gamma'$$

(*EventBased₂*)

$$\langle \mathsf{eventBased}(\mathsf{id},\mathsf{e},(\mathsf{e}_1,\mathsf{p}_1,\mathsf{p},\mathsf{m}_1),\ldots,(\mathsf{e}_k,\mathsf{p}_k,\mathsf{p},\mathsf{m}_k)),\sigma,\gamma\rangle \xrightarrow{\diamond^{\circledcirc}_{\mathsf{id}}\{\}\mathsf{p}_i\to\mathsf{p}\,:\,\mathsf{m}_i} \langle inc(dec(\sigma,\mathsf{e}),\mathsf{e}_i),\gamma'\rangle$$

$$\sigma(\mathsf{e})>0\,,\ 1\le i\le k \quad \exists j,h\,.\,1\le j,h\le k,\ \mathsf{p}_j\ne\mathsf{p}_h \quad update(\gamma,\mathsf{m})=\gamma'$$

$$\frac{\langle C_1,\sigma,\gamma\rangle \xrightarrow{l} \langle\sigma',\gamma'\rangle}{\langle C_1|C_2,\sigma,\gamma\rangle \xrightarrow{l} \langle\sigma',\gamma'\rangle}\,(Int_1) \qquad \frac{\langle C_2,\sigma,\gamma\rangle \xrightarrow{l} \langle\sigma',\gamma'\rangle}{\langle C_1|C_2,\sigma,\gamma\rangle \xrightarrow{l} \langle\sigma',\gamma'\rangle}\,(Int_2)$$

Fig. 10. Choreography semantics.

in the state. Formally, they are defined as follows: $inc(\sigma, \mathsf{e}) = \sigma \cdot \{\mathsf{e} \mapsto \sigma(\mathsf{e}) + 1\}$ and $dec(\sigma, \mathsf{e}) = \sigma \cdot \{\mathsf{e} \mapsto \sigma(\mathsf{e}) - 1\}$. These functions extend in a natural ways to sets of edges as follows: $inc(\sigma, \emptyset) = \sigma$ and $inc(\sigma, \{\mathsf{e}\} \cup E)) = inc(inc(\sigma, \mathsf{e}), E)$; the cases for *dec* are similar. The function $eval : \mathbb{EXP} \times \mathbb{S}_m \to \mathbb{V}$, where $\mathbb{S}_m$ is the set of message states, evaluates an expression with respect to a given message state; since the expression language is left unspecified, the definition of *eval* is unspecified as well. We also use the function $update : \mathbb{S}_m \times \mathbb{M} \to \mathbb{S}_m$ that updates a message state by assigning a value to a given message; formally, the function is defined as follows: $update(\gamma, \mathsf{m}) = \gamma \cdot \{\mathsf{m} \mapsto \mathsf{v}\}$ for $\mathsf{v} \in \mathbb{V}$. Finally, function $p : 2^{\mathbb{E}} \to \mathbb{P}$, where $\mathbb{P}$ is the set of participants, returns all participants involved in the first tasks encountered along the edges in the set passed as input parameter; this function can be simply defined on the syntax of the choreography structure of the model under consideration.

We now briefly comment on the operational rules. Rule *Start* starts the execution of a choreography when it is in its initial state. The effect of the rule is to increment the number of tokens in the edge outgoing from the start event. Rule *AndJoin* decrements the tokens in each incoming edge and increments the number of tokens of the outgoing edge, when each incoming edge has at least one token. Rule *XorSplit* is applied when a token is available in the incoming edge of an XOR split gateway; the rule decrements the token in the incoming edge and increments the tokens in one of the outgoing edges corresponding to a positive evaluation of the associated expression. The produced label reports the set of all participants involved in the first tasks reachable from the gateway; indeed, in this case all such participants, with both sending and receiving roles, internally take the same decision in order to ensure the global behaviour prescribed by the choreography (see Table 1). Rule *XorJoin* is activated every time there is a token in one of the incoming edges, which is then moved to the outgoing edge.

Rule *Task* is activated when there is a token in the incoming edge of a choreography task, and moves the token from the incoming edge to the outgoing one. The rule produces a label describing the message exchange and indicating that no decision is taken by the participants involved in the communication. Moreover, the message state is updated with a new value for the involved message name; we abstract from the computation of the message value, which is indeed non-deterministically selected via the *update* function.

Notably, as prescribed by the BPMN standard [23, p. 315], the communication model is synchronous; indeed, according to the standard, a choreography task completes when the receiver participant reads the message. The rules for the event-based gateway are enabled each time there is a token in the incoming edge, which is moved to the activated outgoing edge. Moreover, the value of the exchanged message is updated in the message state. According to the involved participants we distinguish two cases: rule $EventBased_1$ is used in case the tasks following the gateway have the same sender (see Table 2(a) and (c)), while rule $EventBased_2$ is used in case the tasks have the same receiver and at least two distinguished senders (see Table 2(b)); notably, the case with the same sender and the same receiver is dealt with by the former rule.

Rule *EventBased*$_1$ produces a label recording, besides the message exchange, the fact that the sender participant has undertaken an internal decision, according to which the message to be sent has been selected. Instead, Rule *EventBased*$_2$ produces a label reporting just the information about the message exchange, as the decision about the message to be exchanged is not taken by a participant but it is the result of a race-condition. Finally, rules *Int*$_1$ and *Int*$_2$ deal with interleaving.

The proposed operational semantics has been conceived to clarify the behaviour of the exclusive and event-based gateways in BPMN choreographies. The formalisation of these elements is therefore intentionally articulate, since it aims at faithfully capturing their behaviour as informally described in the standard. Although the two gateways represent choice constructs, our operational rules significantly differ from the conditional and non-deterministic choice operators typically used in process algebras. Indeed, the semantics of the gateways depends on the form of interaction that can be carried out by the participants of the subsequent tasks, while this information does not affect the behaviour of the process algebraic constructs. Hence, the latter cannot be used to directly model the choices on BPMN choreographies. On the other hand, they provide a clearer semantics that facilitates the modelling activity to the choreography designers.

We conclude the section by showing our formalisation at work on the sugar harvesting example introduced in Sect. 3.

Example 1. The BPMN choreography model in Fig. 2 represents a scenario where a *SugarPerson* participant interacts with a *SugarGrid* participant in order to enter in a sugar grid and harvest the sugar in the grid. The model is rendered in our textual notation as shown in Fig. 11 (we have specified unique element identifiers and edge names, which are omitted in the graphical representation).

$$
\begin{array}{l}
\mathsf{start}(\mathsf{id}_1, \mathsf{e}_1) \\
\mid\ \mathsf{task}(\mathsf{id}_2, \mathsf{e}_1, \mathsf{e}_1', \mathsf{SugarPerson}, \mathsf{SugarGrid}, \mathsf{sugarEnquiry}) \\
\mid\ \mathsf{task}(\mathsf{id}_2', \mathsf{e}_1', \mathsf{e}_2, \mathsf{SugarGrid}, \mathsf{SugarPerson}, \mathsf{sugarInfo}) \\
\mid\ \mathsf{xorSplit}(\mathsf{id}_3, \mathsf{e}_2, (\mathsf{e}_3, \mathsf{exp}_{\mathsf{yes}}), (\mathsf{e}_4, \mathsf{exp}_{\mathsf{no}})) \\
\mid\ \mathsf{task}(\mathsf{id}_4, \mathsf{e}_3, \mathsf{e}_3', \mathsf{SugarPerson}, \mathsf{SugarGrid}, \mathsf{intentionToEnter}) \\
\mid\ \mathsf{task}(\mathsf{id}_4', \mathsf{e}_3', \mathsf{e}_5, \mathsf{SugarGrid}, \mathsf{SugarPerson}, \mathsf{response}) \\
\mid\ \mathsf{xorSplit}(\mathsf{id}_5, \mathsf{e}_5, (\mathsf{e}_6, \mathsf{granted}(\mathsf{response}) == \mathsf{yes}), (\mathsf{e}_7, \mathsf{granted}(\mathsf{response}) == \mathsf{no})) \\
\mid\ \mathsf{task}(\mathsf{id}_6, \mathsf{e}_6, \mathsf{e}_6', \mathsf{SugarPerson}, \mathsf{SugarGrid}, \mathsf{harvestRequest}) \\
\mid\ \mathsf{task}(\mathsf{id}_6', \mathsf{e}_6', \mathsf{e}_8, \mathsf{SugarGrid}, \mathsf{SugarPerson}, \mathsf{amountHarvested}) \\
\mid\ \mathsf{xorJoin}(\mathsf{id}_7, \{\mathsf{e}_4, \mathsf{e}_7, \mathsf{e}_8\}, \mathsf{e}_9) \\
\mid\ \mathsf{end}(\mathsf{id}_8, \mathsf{e}_9)
\end{array}
$$

Fig. 11. Textual representation of the sugar harvesting choreography model.

In the graphical representation of the BPMN model, the decision of the first XOR gateway is abstracted by means of the *enter*? condition. According to the BPMN standard, XOR gateway decisions have to be based on the message data. In our case, focussing on the previously exchanged messages, we have to consider $\mathsf{sugarEnquiry}$ and/or $\mathsf{sugarInfo}$. Let us consider only the latter message (the following reasoning does not change if we consider the former message or both of them). In our textual notation we would instantiate the expressions specified in the first XOR gateway (i.e., the element with identifier id_3) as follows:

$$\mathsf{exp_{yes}} = (\mathsf{enter(sugarInfo)} == \mathsf{yes}) \qquad \mathsf{exp_{no}} = (\mathsf{enter(sugarInfo)} == \mathsf{no})$$

where function $\mathsf{enter()}$ abstracts the decision to enter or not taken by the *SugarPerson*. According to our semantics, from the initial choreography configuration $\langle C, \sigma_0, \gamma_0 \rangle$, where C is the choreography structure in Fig. 11, by applying rule *Start* and then twice rule *Task* (and by properly applying, each time, the interleaving rules), we can reach the configuration $\langle C, \sigma, \gamma \rangle$, with $\sigma = \sigma_0 \cdot \{\mathsf{e}_2 \mapsto 1\}$ and $\gamma = \gamma_0 \cdot \{\mathsf{sugarEnquiry} \mapsto \mathsf{v}_1\} \cdot \{\mathsf{sugarinfo} \mapsto \mathsf{v}_2\}$. Now, according to rule *XorSplit*, depending on the evaluation of the guard expressions, we can observe the following transitions:

$$\langle C, \sigma, \gamma \rangle \xrightarrow{\diamondsuit_{\mathsf{id}_3}\{\mathsf{SugarPerson}, \mathsf{SugarGrid}\}} \langle \sigma', \gamma \rangle$$

$$\langle C, \sigma, \gamma \rangle \xrightarrow{\diamondsuit_{\mathsf{id}_3}\{\mathsf{SugarPerson}, \mathsf{SugarGrid}\}} \langle \sigma'', \gamma \rangle$$

with $\sigma' = \sigma_0 \cdot \{\mathsf{e}_3 \mapsto 1\}$ and $\sigma'' = \sigma_0 \cdot \{\mathsf{e}_4 \mapsto 1\}$. Both transitions produce the same label, indicating that the entering decision is taken by both the *SugarPerson* and the *SugarGrid*. However, in this specific scenario, while the *sugarInfo* message data can allow the *SugarPerson* to take the decision, since this participant represents the person that actually decides whether to enter or not in the grid, this data cannot be enough for the *SugarGrid* to decide whether the *SugarPerson* intends to enter or not. In fact, the intention to enter will be reported in the message *intentionToEnter*, which is sent by *SugarPerson* to the *SugarGrid* after the execution of the XOR gateway. Anyway, by using this message as argument of the expressions $\mathsf{exp_{yes}}$ and $\mathsf{exp_{no}}$, we will have that the configuration $\langle C, \sigma, \gamma \rangle$ will be deadlocked, because the evaluation of the two expressions will be undefined as $\gamma(\mathsf{intentionToEnter}) = \mathit{undef}$.

Our semantics has hence spotted a semantic inconsistency in the considered BPMN choreography model. Indeed, in this specific example, the designer has accidentally used the wrong element for the modelling the choice about entering in the sugar grid. The decision to use the event-based vs. the exclusive gateway in fact is strictly correlated to the availability of information for all involved participants, that in this case is absent for the *SugarGrid*. A possible fix to make the model compliant with the standard is to replace the XOR gateway with an event-based one. Since the *SugarGrid* has no knowledge about the intention of the *SugarPerson*, at this stage of the protocol it should be able to accept any decision coming from the counterpart. The suggested replacement falls in the specific case of Table 2(c), and is formally represented by the operational rule *EventBased*$_1$, which indeed produces a label indicating that the decision is taken only by the *SugarPerson*. □

6 Related Works

In this section, we first discuss observed issues of the BPMN language, then we discuss other choreography formalizations available in the literature.

BPMN 2.0 Observed Ambiguities. Due to its complexity, BPMN 2.0 standard has been studied to make more explicit the meaning of the elements and their semantics [1,8,9]. The investigation was also done through an analytic work highlighting the problems of the standard according to supported workflow patterns [7]. In particular, the author states that the standard is ambiguous due to the numerous underspecified descriptions of semantics for relevant concepts (e.g., data conditions, and data dependencies between processes). The highlighted problems are justified by the gap between conceptual and executable BPMN models and the fact that on average the designers use less than the 20% of the available elements. The same problem was discussed some years before in [22]. Other relevant studies investigate the standard and its support for systems implementation. In [14,16,17] the authors suggested a series of implementation clarifications useful to designers for a more accurate tool selection.

Choreography Formalisations. Formalisations of choreography syntax and semantics are proposed in different works [12,15,18,25]. The authors of [12] present an efficient algorithm for extracting concrete choreographic programs with asynchronous messaging. They use the core choreography language, where the semantics is given in terms of labelled reductions. In [18] the authors propose a framework able to synthesize local code for processes starting from global choreographies containing the component ports, the model composition and the coordination elements. In [25] the authors propose two abstract semantics of choreographies formalised as pomsets of communication events and as hypergraphs of events. In [15] the authors demonstrate that it is possible to perform reversible computation monitoring choreography models. Other formal semantics rely on types [19], programs [21], graphs [6,20]. It is worth noticing that many proposed works provide the semantics by translation in other formalisms, while in our work we have preferred to develop a direct semantics. We believe indeed that extending available translations, like those into Petri Nets, may result in the generation of convoluted and large models, which may undermine the understanding of the formal meaning of the BPMN execution semantics, and their verification. In [10] we propose a deeper study based on direct formalizations for choreographies and collaborations, and considering their conformance via behavioural equivalences.

7 Concluding Remarks

In this paper, we discuss in detail issues raised walking through the BPMN standard, with a specific focus on the exclusive and event-based gateways, which have a tricky semantics in case of choreography diagrams. These issues arose from the use in the standard of natural language for providing informal descriptions of the elements, and are emphasised by the lack of a clear semantics for choreography models. We discuss good practices in the use of the notation, and we provide a direct formalisation of the choreography elements in order to remove any ambiguity. As a future work, we intend to exploit the proposed formalisation to develop a modelling tool to support the designers during the specification of choreography models.

References

1. van der Aalst, W.M.: Business process management: a comprehensive survey. ISRN Softw. Eng. **2678**, 1–12 (2013)
2. Mendling, J., et al.: Blockchains for business process management - challenges and opportunities. ACM Trans. Manag. Inf. Syst. **9**(1), 4:1–4:16 (2018)
3. Breu, R., et al.: Towards living inter-organizational processes. In: Business Informatics, pp. 363–366. IEEE Computer Society (2013)
4. Suchenia, A., Potempa, T., Ligęza, A., Jobczyk, K., Kluza, K.: Selected approaches towards taxonomy of business process anomalies. In: Pełech-Pilichowski, T., Mach-Król, M., Olszak, C.M. (eds.) Advances in Business ICT: New Ideas from Ongoing Research. SCI, vol. 658, pp. 65–85. Springer, Cham (2017). https://doi.org/10.1007/978-3-319-47208-9_5
5. Basu, S., Bultan, T., Ouederni, M.: Deciding choreography realizability. In: POPL, pp. 191–202. ACM (2012)
6. Bertolino, A., Marchetti, E., Morichetta, A.: Adequate monitoring of service compositions. In: 9th Joint Meeting of the European Software Engineering Conference and the ACM SIGSOFT Symposium on the Foundations of Software Engineering, pp. 59–69 (2013)
7. Börger, E.: Approaches to modeling business processes. Soft. Syst. Model. **11**(3), 305–318 (2012)
8. Chinosi, M., Trombetta, A.: BPMN: an introduction to the standard. Comput. Standards Interfaces **34**(1), 124–134 (2012)
9. Corradini, F., Fornari, F., Polini, A., Re, B., Tiezzi, F.: A formal approach to modeling and verification of business process collaborations. Sci. Comput. Program. **166**, 35–70 (2018)
10. Corradini, F., Morichetta, A., Polini, A., Re, B., Tiezzi, F.: Collaboration vs. choreography conformance in BPMN 2.0: from theory to practice. In: EDOC, pp. 95–104. IEEE (2018)
11. Corradini, F., Muzi, C., Re, B., Rossi, L., Tiezzi, F.: Global vs. local semantics of BPMN 2.0 OR-join. In: Tjoa, A.M., Bellatreche, L., Biffl, S., van Leeuwen, J., Wiedermann, J. (eds.) SOFSEM 2018. LNCS, vol. 10706, pp. 321–336. Springer, Cham (2018). https://doi.org/10.1007/978-3-319-73117-9_23
12. Cruz-Filipe, L., Larsen, K.S., Montesi, F.: The paths to choreography extraction. In: Esparza, J., Murawski, A.S. (eds.) FoSSaCS 2017. LNCS, vol. 10203, pp. 424–440. Springer, Heidelberg (2017). https://doi.org/10.1007/978-3-662-54458-7_25
13. Dumas, M., Hull, R., Mendling, J., Weber, I.: Blockchain technology for collaborative information systems. Dagstuhl Rep. **8**(8), 67–129 (2018)
14. Evéquoz, F., Sterren, C.: Waiting for the miracle: comparative analysis of twelve business process management systems regarding the support of BPMN 2.0 palette and export. Technical report, HES-SO (2011)

15. Francalanza, A., Mezzina, C.A., Tuosto, E.: Reversible choreographies via monitoring in erlang. In: Bonomi, S., Rivière, E. (eds.) DAIS 2018. LNCS, vol. 10853, pp. 75–92. Springer, Cham (2018). https://doi.org/10.1007/978-3-319-93767-0_6
16. Geiger, M., Wirtz, G.: BPMN 2.0 serialization-standard compliance issues and evaluation of modeling tools. In: Enterprise Modelling and Information Systems Architectures (2013)
17. Gutschier, C., Hoch, R., Kaindl, H., Popp, R.: A pitfall with BPMN execution. In: WEB, pp. 7–13 (2014)
18. Hallal, R., Jaber, M., Abdallah, R.: From global choreography to efficient distributed implementation. In: HPCS, pp. 756–763. IEEE (2018)
19. Honda, K., Yoshida, N., Carbone, M.: Multiparty asynchronous session types. J. ACM **63**(1), 9:1–9:67 (2016)
20. Lange, J., Tuosto, E., Yoshida, N.: From communicating machines to graphical choreographies. In: POPL, pp. 221–232. ACM (2015)
21. Dalla Preda, M., Gabbrielli, M., Giallorenzo, S., Lanese, I., Mauro, J.: Dynamic choreographies. In: Holvoet, T., Viroli, M. (eds.) COORDINATION 2015. LNCS, vol. 9037, pp. 67–82. Springer, Cham (2015). https://doi.org/10.1007/978-3-319-19282-6_5
22. zur Muehlen, M., Recker, J.: How much language is enough? Theoretical and practical use of the business process modeling notation. Seminal Contributions to Information Systems Engineering, pp. 429–443. Springer, Heidelberg (2013). https://doi.org/10.1007/978-3-642-36926-1_35
23. OMG: Business Process Model and Notation (BPMN V 2.0) (2011)
24. Onggo, B.S.: Agent-based simulation model representation using BPMN. In: Formal Languages for Computer Simulation, pp. 378–400. IGI Global (2014)
25. Tuosto, E., Guanciale, R.: Semantics of global view of choreographies. JLAMP **95**, 17–40 (2018)

Stronger Validity Criteria for Encoding Synchrony

Rob van Glabbeek[1,2(✉)], Ursula Goltz[3], Christopher Lippert[3], and Stephan Mennicke[3]

[1] Data61, CSIRO, Sydney, Australia
rvg@cs.stanford.edu
[2] Computer Science and Engineering, University of New South Wales, Sydney, Australia
[3] Institute for Programming and Reactive Systems, TU Braunschweig, Braunschweig, Germany

Abstract. We analyse two translations from the synchronous into the asynchronous π-calculus, both without choice, that are often quoted as standard examples of valid encodings, showing that the asynchronous π-calculus is just as expressive as the synchronous one. We examine which of the quality criteria for encodings from the literature support the validity of these translations. Moreover, we prove their validity according to much stronger criteria than considered previously in the literature.

Keywords: Process calculi · Expressiveness · Translations · Quality criteria for encodings · Valid encoding · Compositionality · Operational correspondence · Semantic equivalences · Asynchronous π-calculus

This paper is dedicated to Catuscia Palamidessi, on the occasion of her birthday. It has always been a big pleasure and inspiration to discuss with her.

1 Introduction

In the literature, many definitions are proposed of what it means for one system description language to encode another one. Each concept C of a valid encoding yields an ordering of system description languages with respect to expressive power: language $\mathscr{L}'$ is *at least as expressive as* language $\mathscr{L}$ (according to C), notation $\mathscr{L} \preceq_C \mathscr{L}'$, iff a valid encoding from $\mathscr{L}$ to $\mathscr{L}'$ exists. The concepts of a valid encoding themselves, the *validity criteria*, also can be ordered: criterion C is *stronger* than criterion D iff for each two system description languages $\mathscr{L}$ and $\mathscr{L}'$ one has

$$\mathscr{L} \preceq_C \mathscr{L}' \Rightarrow \mathscr{L} \preceq_D \mathscr{L}' .$$

This work was partially supported by the DFG (German Research Foundation).

M. S. Alvim et al. (Eds.): Palamidessi Festschrift, LNCS 11760, pp. 182–205, 2019.
https://doi.org/10.1007/978-3-030-31175-9_11

Naturally, employing a stronger validity criterion constitutes a stronger claim that the target language is at least as expressive as the source language.

In this paper, we analyse two well-known translations from the synchronous into the asynchronous π-calculus, one by Boudol and one by Honda & Tokoro. Both are often quoted as standard examples of valid encodings. We examine which of the validity criteria from the literature support the validity of these encodings. Moreover, we prove the validity of these encodings according to much stronger criteria than considered previously in the literature.

A *translation* $\mathscr{T}$ from (or *encoding* of) a language $\mathscr{L}$ into a language $\mathscr{L}'$ is a function from the $\mathscr{L}$-expressions to the $\mathscr{L}'$-expressions. The first formal definition of a *valid* encoding of one system description language into another stems from Boudol [4]. It is parametrised by the choice of a semantic equivalence $\sim$ that is meaningful for the source as well as the target language of the translation—and is required to be a congruence for both. Boudol in particular considers languages whose semantics are given in terms of labelled transition systems. Any semantic equivalence defined on labelled transition systems, such as strong bisimilarity, induces an equivalence on the expressions of such languages, and thus allows comparison of expressions from different languages of this kind. Boudol formulates two requirements for valid translations: (1) they should be *compositional*, and (2) for each source language expression P, its translation $\mathscr{T}(P)$—an expression in the target language—is semantically equivalent to P.

Successive generalisations of the definition of a valid encoding from Boudol [4] appear in [16,17,20]. These generalisations chiefly deal with languages that feature process variables, and that are interpreted in a semantic domain (such as labelled transition systems) where not every semantic value need be denotable by a closed term. The present paper, following [4] and most of the expressiveness literature, deals solely with *closed-term* languages, in which the distinction between syntax and semantic is effectively dropped by taking the domain of semantic values, in which the language is interpreted, to consist of the closed terms of the language. In this setting the only generalisation of the notion of a valid encoding from [16,17,20] over [4] is that Boudol's congruence requirement on the semantic equivalence up to which languages are compared is dropped. In [20] it is also shown that the requirement of compositionality can be dropped, as in the presence of process variables it is effectively implied by the requirement that semantic equivalence is preserved upon translation. But when dealing with languages without process variables, as in the present paper, it remains necessary to require compositionality separately.

A variant of the validity criterion from Boudol is the notion of *full abstraction*, employed in [2,11,29,30,38,40,41]. In this setting, instead of a single semantic equivalence $\sim$ that is meaningful for the source as well as the target language of the translation, two semantic equivalences $\sim_{\mathrm{S}}$ and $\sim_{\mathrm{T}}$ are used as parameters of the criterion, one on the source and one on the target language. Full abstraction requires, for source expressions P and Q, that $P \sim_{\mathrm{S}} Q \Leftrightarrow \mathscr{T}(P) \sim_{\mathrm{T}} \mathscr{T}(Q)$. Full abstraction has been criticised as a validity criteria for encodings in [3,23,32]; a historical treatment of the concept can be found in [19, Sect. 18].

An alternative for the equivalence-based validity criteria reviewed above are the ones employing *operational correspondence*, introduced by Nestmann and Pierce in [30]. Here valid encodings are required to satisfy various criteria, differing subtly from paper to paper; often these criteria are chosen to conveniently establish that a given language is or is not as least as expressive as another. Normally some form of operational correspondence is one of these criteria, and as a consequence of this these approaches are suitable for comparing the expressiveness of process calculi with a reduction semantics, rather than system description languages in general. Gorla [22] has selected five of these criteria as a unified approach to encodability and separation results for process calculi—*compositionality*, *name invariance*, *operational correspondence*, *divergence reflection* and *success sensitiveness*—and since then these criteria have been widely accepted as constituting a standard definition of a valid encoding.

In [31] Catuscia Palamidessi employs four requirements for valid encodings between languages that both contain a parallel composition operator $|$: compositionality, preservation of semantics, a form of name invariance, and the requirement that parallel composition is translated homomorphically, i.e., $\mathscr{T}(P|Q) = \mathscr{T}(P)|\mathscr{T}(Q)$. The latter is not implied by any of the requirements considered above. The justification for this requirement is that it ensures that the translation maintains the degree of distribution of the system. However, Peters, Nestmann and Goltz [35] argue that it is possible to maintain the degree of distribution of a system upon translation without requiring a homomorphic translation of $|$; in fact they introduce the criterion *preservation of distributability* that is weaker then the homomorphic translation of $|$.

This paper analyses the encodings $\mathscr{T}_{\mathrm{B}}$ and $\mathscr{T}_{\mathrm{HT}}$ of Boudol and Honda & Tokoro of the synchronous into the asynchronous π-calculus, both without the choice operator $+$. Our aim is to evaluate the validity of these encodings with respect to all criteria for valid encodings summarised above.

Section 2 recalls the encodings $\mathscr{T}_{\mathrm{B}}$ and $\mathscr{T}_{\mathrm{HT}}$. Section 3 reviews the validity criteria from Gorla [22], and recalls the result from [18] that the encodings $\mathscr{T}_{\mathrm{B}}$ and $\mathscr{T}_{\mathrm{HT}}$ meet all those criteria. Trivially, $\mathscr{T}_{\mathrm{B}}$ and $\mathscr{T}_{\mathrm{HT}}$ also meet Palamidessi's criterion that parallel composition is translated homomorphically, and thus also the criterion on preservation of distributability from [35].

Section 4 focuses on the criterion of compositionality. Gorla's proposal involves a weaker form of this requirement, exactly because encodings like $\mathscr{T}_{\mathrm{B}}$ and $\mathscr{T}_{\mathrm{HT}}$ do not satisfy the default form of compositionality. However, we show that these encodings also satisfy a form of compositionality due to [17] that significantly strengthens the one from [22]. Moreover, depending on how the definition of valid encodings between concrete languages generalises to one between parametrised languages, one may even conclude that $\mathscr{T}_{\mathrm{B}}$ and $\mathscr{T}_{\mathrm{HT}}$ satisfy the default notion of compositionality.

Section 5 focuses on the criterion of operational correspondence. In [30] two forms of this criterion were proposed, one for *prompt* and one for *nonprompt* encodings. Gorla's form of operational correspondence [22] is the natural common weakening of the forms from Nestmann and Pierce [30], and thus applies to prompt as well as nonprompt encodings. As the encodings $\mathscr{T}_{\rm B}$ and $\mathscr{T}_{\rm HT}$ are nonprompt, they certainly do not meet the prompt form of operational correspondence from [30]. In [18] it was shown that they not only satisfy the form of [22], but even the nonprompt form from [30].

Gorla's form of operational correspondence, as well as the nonprompt form of [30], weakens the prompt form in two ways. In [18] a natural intermediate form was contemplated that weakens the prompt form in only one of these ways, and the open question was raised whether $\mathscr{T}_{\rm B}$ and $\mathscr{T}_{\rm HT}$ satisfy this intermediate form of operational correspondence. The present paper answers that question affirmatively.

Gorla's criterion of success sensitiveness is a more abstract form of *barb sensitiveness*. The original barbs were predicates telling whether a process could input or output data over a certain channel. In Sect. 6 we show that whereas $\mathscr{T}_{\rm B}$ is barb sensitive, $\mathscr{T}_{\rm HT}$ is not. The encoding $\mathscr{T}_{\rm HT}$ becomes barb sensitive if we use a weaker form of barb, abstracting from the difference between input and output. This, however, is against the spirit of the asynchronous π-calculus, where instead one abstracts from input barbs altogether. Gorla's criterion of success sensitiveness thus appears to be an improvement over barb sensitiveness.

Section 7 evaluates $\mathscr{T}_{\rm B}$ and $\mathscr{T}_{\rm HT}$ under the original validity criterion of Boudol [4], as generalised in [17]; we call a compositional encoding $\mathscr{T}$ *valid up to* a semantic equivalence $\sim$ iff $\mathscr{T}(P) \sim P$ for all source language expressions P. We observe that the encodings $\mathscr{T}_{\rm B}$ and $\mathscr{T}_{\rm HT}$ are not valid under equivalences that match transition labels, such as early weak bisimilarity, nor under asynchronous weak bisimilarity. Then we show that $\mathscr{T}_{\rm B}$, but not $\mathscr{T}_{\rm HT}$, is valid under weak barbed bisimilarity. This is our main result. Finally, we introduce a new equivalence under which $\mathscr{T}_{\rm HT}$ is valid: a version of weak barbed bisimilarity that drops the distinction between input and output barbs.

Section 8 starts with the result that $\mathscr{T}_{\rm B}$ and $\mathscr{T}_{\rm HT}$ are both valid under a version of weak barbed bisimilarity where an abstract success predicate takes over the role of barbs. That statement turns out to be equivalent to the statement that these encodings are success sensitive and satisfy a form of operational correspondence that is stronger then Gorla's. One can also incorporate Gorla's requirement of divergence reflection into the definition of form of barbed bisimilarity. Finally, we remark that $\mathscr{T}_{\rm B}$ and $\mathscr{T}_{\rm HT}$ remain valid when upgrading weak to branching bisimilarity.

Section 9 applies a theorem from [19] to infer from the validity of $\mathscr{T}_{\rm B}$ and $\mathscr{T}_{\rm HT}$ up to a form of weak barbed bisimilarity, that these encodings are also fully abstract, when taking as source language equivalence weak barbed congruence, and as target language equivalence the congruence closure of that form of weak barbed bisimilarity for the image of the source language within the target language.

2 Encoding Synchrony into Asynchrony

Consider the π-calculus as presented by Milner in [27], i.e., the one of Sangiorgi and Walker [39] without matching, τ-prefixing and choice.

Given a set of *names* $\mathcal{N}$, the set $\mathcal{P}_\pi$ of *processes* or *terms* P of the calculus is given by

$$P ::= \mathbf{0} \mid \bar{x}z.P \mid x(y).P \mid P|Q \mid (y)P \mid !P$$

with x, y, z, u, v, w ranging over $\mathcal{N}$.

$\mathbf{0}$ denotes the empty process. $\bar{x}z$ stands for an output guard that sends the name z along the channel x. $x(y)$ denotes an input guard that waits for a name to be transmitted along the channel named x. Upon receipt, the name is substituted for y in the subsequent process. $P|Q$ ($P, Q \in \pi$) denotes a parallel composition between P and Q. $!P$ is the replication construct and $(y)P$ restricts the scope of name y to P.

Definition 1. An occurrence of a name y in π-calculus process $P \in \mathcal{P}_\pi$ is *bound* if it lies within a subexpression $x(y).Q$ or $(y)Q$ of P; otherwise it is *free*. Let $n(P)$ be the set of names occurring in $P \in \mathcal{P}_\pi$, and $fn(P)$ (resp. $bn(P)$) be the set of names occurring free (resp. bound) in P.

Structural congruence, $\equiv$, is the smallest congruence relation on processes satisfying

$$\begin{array}{llrlr}
(1) & P|(Q|R) \equiv (P|Q)|R & (y)\mathbf{0} & \equiv \mathbf{0} & (5)\\
(2) & P|Q \equiv Q|P & (y)(u)P & \equiv (u)(y)P & (6)\\
(3) & P|\mathbf{0} \equiv P & (w)(P|Q) & \equiv P|(w)Q & (7)\\
 & & (y)P & \equiv (w)P\{^w\!/\!_y\} & (8)\\
(4) & !P \equiv P|!P & x(y).P & \equiv x(w).P\{^w\!/\!_y\}\,. & (9)
\end{array}$$

Here $w \notin n(P)$, and $P\{^w\!/\!_y\}$ denotes the process obtained by replacing each free occurrence of y in P by w. Rules (8) and (9) constitute *α-conversion* (renaming of bound names). In case $w \in n(P)$, $P\{^w\!/\!_z\}$ denotes $Q\{^w\!/\!_z\}$ for some process Q obtained from P by means of α-conversion, such that z does not occur within subexpressions $x(w).Q'$ or $(w)Q'$ of Q.

Definition 2. The *reduction relation*, $\longmapsto \subseteq \mathcal{P}_\pi \times \mathcal{P}_\pi$, is generated by the following rules:

$$\frac{}{\bar{x}z.P|x(y).Q \longmapsto P|Q\{^z\!/\!_y\}} \qquad \frac{P \longmapsto P'}{P|Q \longmapsto P'|Q}$$

$$\frac{P \longmapsto P'}{(y)P \longmapsto (y)P'} \qquad \frac{Q \equiv P \quad P \longmapsto P' \quad P' \equiv Q'}{Q \longmapsto Q'}\,.$$

The asynchronous π-calculus, as introduced by Honda and Tokoro in [24] and by Boudol in [5], is the sublanguage aπ of the fragment π of the π-calculus presented above where all subexpressions $\bar{x}z.P$ have the form $\bar{x}z.\mathbf{0}$, and are written $\bar{x}z$. A characteristic of synchronous communication, as used in π, is that sending a message synchronises with receiving it, so that a process sending a message can only proceed after another party has received it. In the asynchronous π-calculus this feature is dropped, as it is not possible to specify any behaviour scheduled after a send action.

Boudol [5] defines an encoding $\mathscr{T}_B$ from π to aπ inductively as follows:

$$\begin{array}{rcll}
\mathscr{T}_B(\mathbf{0}) & := & \mathbf{0} & \\
\mathscr{T}_B(\bar{x}z.P) & := & (u)(\bar{x}u|u(v).(\bar{v}z|\mathscr{T}_B(P))) & \text{with } u,v \notin fn(P)\cup\{x,z\} \\
\mathscr{T}_B(x(y).P) & := & x(u).(v)(\bar{u}v|v(y).\mathscr{T}_B(P)) & \text{with } u,v \notin fn(P)\cup\{x\} \\
\mathscr{T}_B(P|Q) & := & (\mathscr{T}_B(P)|\mathscr{T}_B(Q)) & \\
\mathscr{T}_B(!P) & := & !\mathscr{T}_B(P) & \\
\mathscr{T}_B((x)P) & := & (x)\mathscr{T}_B(P) &
\end{array}$$

always choosing $u \neq v$. To sketch the underlying idea, suppose a π-process is able to perform a communication, for example $\bar{x}z.P|x(y).Q$. In the asynchronous variant of the π-calculus, there is no continuation process after an output operation. Hence, a translation into the asynchronous π-calculus has to reflect the communication on channel x as well as the guarding role of $\bar{x}z$ for P in the synchronous π-calculus. The idea of Boudol's encoding is to assign a guard to P such that this process must receive an *acknowledgement message* confirming the receipt of z.[1] We write the sender as $P' = (\bar{x}z|u(v).P)$ where $u, v \notin fn(P)$. Symmetrically, the receiver must send the acknowledgement, i. e. $Q' = x(y).(\bar{u}v|Q)$. Unfortunately, this simple transformation is not applicable in every case, because the protocol does not protect the channel u. u should be known to sender and receiver only, otherwise the communication may be interrupted by the environment. Therefore, we restrict the scope of u, and start by sending this private channel to the receiver. The actual message z is now sent in a second stage, over a channel v, which is also made into a private channel between the two processes. The crucial observation is that in $(u)(\bar{x}u|u(v).P^*)$, the subprocess $P^* = \bar{v}z|P$ may only continue after $\bar{x}u$ was accepted by some receiver, and this receiver has acknowledged this by transmitting another channel name v on the private channel u.

The encoding $\mathscr{T}_{HT}$ of Honda and Tokoro [24] differs only in the clauses for the input and output prefix:

$$\begin{array}{rcll}
\mathscr{T}_{HT}(\bar{x}z.P) & := & x(u).(\bar{u}z|\mathscr{T}_{HT}(P)) & u \notin fn(P)\cup\{x,z\} \\
\mathscr{T}_{HT}(x(y).P) & := & (u)(\bar{x}u|u(y).\mathscr{T}_{HT}(P)) & u \notin fn(P)\cup\{x\}.
\end{array}$$

Unlike Boudol's translation, communication takes place directly after synchronising along the private channel u. The synchronisation occurs in the reverse direction, because sending and receiving messages alternate, meaning that the sending process $\bar{x}z.Q$ is translated into a process that receives a message on channel x and the receiving process $x(y).R$ is translated into a process passing a message on x.

3 Valid Encodings According to Gorla

In [22] a *process calculus* is given as a triple $\mathscr{L} = (\mathcal{P}, \longmapsto, \asymp)$, where

[1] As observed by a referee, the encodings $\mathscr{T}_B$ and $\mathscr{T}_{HT}$ do not satisfy this constraint: the continuation process P can proceed before z is received. This issue could be alleviated by enriching the protocol with another communication from Q' to P'.

- $\mathcal{P}$ is the set of language terms (called *processes*), built up from k-ary composition operators op,
- $\longmapsto$ is a binary *reduction* relation between processes,
- $\asymp$ is a semantic equivalence on processes.

The operators themselves may be constructed from a set $\mathcal{N}$ of names. In the π-calculus, for instance, there is a unary operator $\bar{x}y.\_$ for each pair of names $x, y \in \mathcal{N}$. This way names occur in processes; the occurrences of names in processes are distinguished in *free* and *bound* ones; $fn(\boldsymbol{P})$ denotes the set of names occurring free in the k-tuple of processes $\boldsymbol{P} = (P_1, \ldots, P_k) \in \mathcal{P}^k$. A *renaming* is a function $\sigma : \mathcal{N} \to \mathcal{N}$; it extends componentwise to k-tuples of names. If $P \in \mathcal{P}$ and σ is a renaming, then $P\sigma$ denotes the term P in which each free occurrence of a name x is replaced by $\sigma(x)$, while renaming bound names to avoid name capture.

A k-ary $\mathscr{L}$-*context* $C[\_1, \ldots, \_k]$ is a term build by the composition operators of $\mathscr{L}$ from *holes* $\_1, \ldots, \_k$; the context is called *univariate* if each of these holes occurs exactly once in it. If $C[\_1, \ldots, \_k]$ is a k-ary $\mathscr{L}$-*context* and $P_1, \ldots, P_k \in \mathcal{P}$ then $C[P_1, \ldots, P_k]$ denotes the result of substituting P_i for $\_i$ for each $i{=}1, \ldots, k$.

Let $\Longmapsto$ denote the reflexive-transitive closure of $\longmapsto$. One writes $P \longmapsto^{\omega}$ if P *diverges*, that is, if there are P_i for $i \in \mathbb{N}$ such that $P = P_0$ and $P_i \longmapsto P_{i+1}$ for all $i \in \mathbb{N}$. Finally, write $P \longmapsto$ if $P \longmapsto Q$ for some term Q.

For the purpose of comparing the expressiveness of languages, a constant $\surd$ is added to each of them [22]. A term P in the upgraded language is said to *report success*, written $P{\downarrow}$, if it has a *top-level unguarded* occurrence of $\surd$.[2] Write $P{\Downarrow}$ if $P \Longmapsto P'$ for a process P' with $P'{\downarrow}$.

Definition 3 ([22]). An *encoding* of $\mathscr{L}_\mathrm{s} = (\mathcal{P}_\mathrm{s}, \longmapsto_\mathrm{s}, \asymp_\mathrm{s})$ into $\mathscr{L}_\mathrm{t} = (\mathcal{P}_\mathrm{t}, \longmapsto_\mathrm{t}, \asymp_\mathrm{t})$ is a pair $(\mathscr{T}, \varphi_{\mathscr{T}})$ where $\mathscr{T} : \mathcal{P}_\mathrm{s} \to \mathcal{P}_\mathrm{t}$ is called *translation* and $\varphi_{\mathscr{T}} : \mathcal{N} \to \mathcal{N}^k$ for some $k \in \mathbb{N}$ is called *renaming policy* and is such that for $u \neq v$ the k-tuples $\varphi_{\mathscr{T}}(u)$ and $\varphi_{\mathscr{T}}(v)$ have no name in common.

The terms of the source and target languages $\mathscr{L}_\mathrm{s}$ and $\mathscr{L}_\mathrm{t}$ are often called S and T, respectively.

Definition 4 ([22]). An encoding is *valid* if it satisfies the following five criteria.

1. *Compositionality:* for every k-ary operator op of $\mathscr{L}_\mathrm{s}$ and for every set of names $N \subseteq \mathcal{N}$, there exists a univariate k-ary context $C^N_{\mathsf{op}}[\_1, \ldots, \_k]$ such that

$$\mathscr{T}(\mathsf{op}(S_1, \ldots, S_k)) = C^N_{\mathsf{op}}[\mathscr{T}(S_1), \ldots, \mathscr{T}(S_k)]$$

for all $S_1, \ldots, S_k \in \mathcal{P}_\mathrm{s}$ with $fn(S_1, \ldots, S_n) = N$.

[2] Gorla defines the latter concept only for languages that are equipped with a notion of *structural congruence* $\equiv$ as well as a parallel composition $|$. In that case P has a top-level unguarded occurrence of $\surd$ iff $P \equiv Q|\surd$, for some Q [22]. Specialised to the π-calculus, a *(top-level) unguarded* occurrence is one that not lies strictly within a subterm $\alpha.Q$, where α is τ, $\bar{x}y$ or $x(z)$. For De Simone languages [42], even when not equipped with $\equiv$ and $|$, a suitable notion of an unguarded occurrence is defined in [43].

2. *Name invariance:* for every $S \in \mathcal{P}_s$ and $\sigma : \mathcal{N} \to \mathcal{N}$

$$\begin{array}{rcll} \mathscr{T}(S\sigma) & = & \mathscr{T}(S)\sigma' & \text{if } \sigma \text{ is injective} \\ \mathscr{T}(S\sigma) & \asymp_t & \mathscr{T}(S)\sigma' & \text{otherwise} \end{array}$$

   with σ' such that $\varphi_{\mathscr{T}}(\sigma(a)) = \sigma'(\varphi_{\mathscr{T}}(a))$ for all $a \in \mathcal{N}$.
3. *Operational correspondence:*
   Completeness if $S \Longmapsto_s S'$ then $\mathscr{T}(S) \Longmapsto_t \asymp_t \mathscr{T}(S')$
   Soundness if $\mathscr{T}(S) \Longmapsto_t T$ then $\exists S' : S \Longmapsto_s S'$ and $T \Longmapsto_t \asymp_t \mathscr{T}(S')$.
4. *Divergence reflection:* if $\mathscr{T}(S) \longmapsto_t^\omega$ then $S \longmapsto_s^\omega$.
5. *Success sensitiveness:* $S\Downarrow$ iff $\mathscr{T}(S)\Downarrow$.
   For this purpose $\mathscr{T}(\cdot)$ is extended to deal with the added constant $\surd$ by taking $\mathscr{T}(\surd) = \surd$.

The above treatment of success sensitiveness differs slightly from the one of Gorla [22]. Gorla requires $\surd$ to be a constant of any two languages whose expressiveness is compared. Strictly speaking, this does not allow his framework to be applied to the encodings $\mathscr{T}_B$ and $\mathscr{T}_{HT}$, as these deal with languages not featuring $\surd$. Here, following [18], we simply allow $\surd$ to be added, which is in line with the way Gorla's framework has been used [12–15,21,25,34–36]. A consequence of this decision is that one has to specify how $\surd$ is translated—see the last sentence of Definition 4—as the addition of $\surd$ to both languages happens after a translation is proposed. This differs from [22], where it is explicitly allowed to take $\mathscr{T}(\surd) \neq \surd$.

In [18] it is established that the encodings $\mathscr{T}_B$ and $\mathscr{T}_{HT}$, reviewed in Sect. 2, are valid according to Gorla [22]; that is, both encodings enjoy the five correctness criteria above. Here, the semantic equivalences $\asymp_s$ and $\asymp_t$ that Gorla assumes to exist on the source and target languages, but were not specified in Sect. 2, can chosen to be the identity, thus obtaining the strongest possible instantiation of Gorla's criteria. Moreover, the renaming policy required by Gorla as part of an encoding can be chosen to be the identity, taking $k = 1$ in Definition 3. Trivially, $\mathscr{T}_B$ and $\mathscr{T}_{HT}$ also meet Palamidessi's criterion that parallel composition is translated homomorphically, and thus also the criterion on preservation of distributability from [35].

4 Compositionality

Compositionality demands that for every k-ary operator op of the source language there is a k-ary context $C_{\mathsf{op}}[\_1, \ldots, \_k]$ in the target such that

$$\mathscr{T}(\mathsf{op}(S_1, \ldots, S_k)) = C_{\mathsf{op}}[\mathscr{T}(S_1), \ldots, \mathscr{T}(S_k)]$$

for all $S_1, \ldots, S_k \in \mathcal{P}_s$ [4]. Gorla [22] strengthens this requirement by the additional requirement that the context C_{op} should be univariate; at the same time he weakens the requirement by allowing the required context C_{op} to depend on the set of names N that occur free in the arguments $S_1, \ldots, S_k$. The application to the encodings $\mathscr{T}_{\mathrm{B}}$ and $\mathscr{T}_{\mathrm{HT}}$ shows that we cannot simply strengthen the criterion of compositionality by dropping the dependence on N. For then the present encodings would fail to be compositional. Namely, the context $C_{\bar{x}z.\_}$ depends on the choice of two names u and v, and the choice of these names depends on $N = fn(S_1)$, where S_1 is the only argument of output prefixing. That the choice of $C_{\bar{x}z.\_}$ also depends on x and z is unproblematic.

In [20] a form of compositionality is proposed where C_{op} does not depend on N, but the main requirement is weakened to

$$\mathscr{T}(\mathsf{op}(S_1,\ldots,S_k)) \stackrel{\alpha}{=} C_{\mathsf{op}}[\mathscr{T}(S_1),\ldots,\mathscr{T}(S_k)].$$

Here $\stackrel{\alpha}{=}$ denotes equivalence up to *α-conversion*, renaming of bound names and variables, for the π-calculus corresponding with rules (8) and (9) of structural congruence. This suffices to rescue the current encodings, for up to α-conversion u and v can always be chosen outside N. It is an open question whether there are examples of intuitively valid encodings that essentially need the dependence of N allowed by [22], i.e., where $C_{\mathsf{op}}^{N_{\mathrm{s}}}$ and $C_{\mathsf{op}}^{N_{\mathrm{t}}}$ differ by more than α-conversion.

Another method of dealing with the fresh names u and v that are used in the encodings $\mathscr{T}_{\mathrm{B}}$ and $\mathscr{T}_{\mathrm{HT}}$, proposed in [17], is to equip the target language with two fresh names that do not occur in the set of names available for the source language. Making the dependence on the choice of set $\mathcal{N}$ of names explicit, this method calls π expressible into aπ if for each $\mathcal{N}$ there exists an $\mathcal{N}'$ such that there is a valid encoding of $\pi(\mathcal{N})$ into $\mathrm{a}\pi(\mathcal{N}')$. By this definition, the encodings $\mathscr{T}_{\mathrm{B}}$ and $\mathscr{T}_{\mathrm{HT}}$ even satisfy the default definition of compositionality, and its strengthening obtained by insisting the contexts C_{op} to be univariate.

5 Operational Correspondence

Operational completeness (one half of operational correspondence) was formulated by Nestmann and Pierce [30] as

$$S \longmapsto_{\mathrm{s}} S' \text{ then } \mathscr{T}(S) \Longmapsto_{\mathrm{t}} \mathscr{T}(S'). \qquad (\mathfrak{C})$$

It makes no difference whether the antecedent of this implication is rephrased as $S \Longmapsto_{\mathrm{s}} S'$, as done by Gorla. Gorla moreover weakens the criterion to

$$S \Longmapsto_{\mathrm{s}} S' \text{ then } \mathscr{T}(S) \Longmapsto_{\mathrm{t}}\asymp_{\mathrm{t}} \mathscr{T}(S'). \qquad (\mathfrak{C}')$$

This makes the criterion applicable to many more encodings. In the case of $\mathscr{T}_{\mathrm{B}}$ and $\mathscr{T}_{\mathrm{HT}}$, [18] shows that these encodings not only satisfy ($\mathfrak{C}'$), but even ($\mathfrak{C}$).

Operational soundness also stems from Nestmann and Pierce [30], who proposed two forms of it:

$$\text{if } \mathscr{T}(S) \longmapsto_{\mathrm{t}} T \text{ then } \exists S' \colon S \longmapsto_{\mathrm{s}} S' \text{ and } T \asymp_{\mathrm{t}} \mathscr{T}(S'). \qquad (\mathfrak{I})$$

$$\text{if } \mathscr{T}(S) \Longmapsto_{\mathrm{t}} T \text{ then } \exists S' \colon S \Longmapsto_{\mathrm{s}} S' \text{ and } T \Longmapsto_{\mathrm{t}} \mathscr{T}(S'). \qquad (\mathfrak{S})$$

The former is meant for "*prompt* encodings, i.e., those where initial steps of literal translations are committing" [30], whereas the latter apply to "nonprompt encodings", that "allow administrative (or book-keeping) steps to precede a committing step". The version of Gorla is the common weakening of ($\mathfrak{I}$) and ($\mathfrak{S}$):

$$\text{if } \mathscr{T}(S) \Longmapsto_{\mathrm{t}} T \text{ then } \exists S' : S \Longmapsto_{\mathrm{s}} S' \text{ and } T \Longmapsto_{\mathrm{t}} \asymp_{\mathrm{t}} \mathscr{T}(S'). \qquad (\mathfrak{G})$$

It thus applies to prompt as well as nonprompt encodings. The encodings $\mathscr{T}_{\mathrm{B}}$ and $\mathscr{T}_{\mathrm{HT}}$ are nonprompt, and accordingly do not meet ($\mathfrak{I}$). In [18] it was shown that they not only satisfy ($\mathfrak{G}$), but even ($\mathfrak{S}$).

An interesting intermediate form between $\mathfrak{I}$ and $\mathfrak{G}$ is

$$\text{if } \mathscr{T}(S) \Longmapsto_{\mathrm{t}} T \text{ then } \exists S' : S \Longmapsto_{\mathrm{s}} S' \text{ and } T \asymp_{\mathrm{t}} \mathscr{T}(S'). \qquad (\mathfrak{W})$$

whereas ($\mathfrak{G}$) weakens ($\mathfrak{I}$) in two ways, ($\mathfrak{W}$) weakens ($\mathfrak{I}$) in only one of these ways. Moreover, ($\mathfrak{W}$) is the natural counterpart of ($\mathfrak{C}'$). In [18] the open question was raised whether $\mathscr{T}_{\mathrm{B}}$ and $\mathscr{T}_{\mathrm{HT}}$ satisfy ($\mathfrak{W}$), for a reasonable choice of $\asymp_{\mathrm{t}}$. (An unreasonable choice, such as the universal relation, tells us nothing.) As pointed out in [18], they do not when taking $\asymp_{\mathrm{t}}$ to be the identity relation, or structural congruence.

The present paper answers this question affirmatively, taking $\asymp_{\mathrm{t}}$ to be weak barbed bisimilarity. A proof will follow in Sect. 8.

6 Barb Sensitiveness

Gorla's success predicate is one of the possible ways to provide source and target languages with a set of *barbs* Ω, each being a unary predicate on processes. For $\omega \in \Omega$, write $P{\downarrow_\omega}$ if process P has the barb ω, and $P{\Downarrow_\omega}$ if $P \Longmapsto P'$ for a process P' with $P'{\downarrow_\omega}$. In Gorla's case, $\Omega = \{\surd\}$, and P has the barb $\surd$ iff P has a top-level unguarded occurrence of $\surd$. The standard criterion of barb sensitiveness is then $S{\Downarrow_\omega} \Leftrightarrow \mathscr{T}(S){\Downarrow_\omega}$ for all $\omega \in \Omega$.

A traditional choice of barb in the π-calculus is to take $\Omega = \{x, \bar{x} \mid x \in \mathcal{N}\}$, writing $P{\downarrow_x}$, resp. $P{\downarrow_{\bar{x}}}$, when P has an unguarded occurrence of a subterm $x(z).R$, resp. $\bar{x}y.R$, that lies not in the scope of a restriction operator (x) [26,39]. This makes a barb a predicate that tells weather a process can read or write over a given channel. Boudol's encoding keeps the original channel names of a sending or receiving process invariant. Hence, a translated term does exhibit the same barbs as the source term.

Lemma 1. Let $P \in \mathcal{P}_\pi$ and $a \in \{x, \bar{x} \mid x \in \mathcal{N}\}$. Then $P{\downarrow_a}$ iff $\mathscr{T}_{\mathrm{B}}(P){\downarrow_a}$.

Proof. With structural induction on P.

- $\mathbf{0}$ and $\mathscr{T}_{\mathrm{B}}(\mathbf{0})$ have the same strong barbs, namely none.
- $\bar{x}z.P$ and $\mathscr{T}_{\mathrm{B}}(\bar{x}z.P)$ both have only the strong barb $\bar{x}$.
- $x(y).P$ and $\mathscr{T}_{\mathrm{B}}(x(y).P)$ both have only strong barb x.

- The strong barbs of $P|Q$ are the union of the ones of P and Q. Using this, the case $P|Q$ follows by induction.
- The strong barbs of $!P$ are the ones of P. Using this, the case $!P$ follows by induction.
- The strong barbs of $(x)P$ are ones of P except x and $\bar{x}$. Using this, the case $(x)P$ follows by induction. □

It follows that $\mathscr{T}_{\mathrm{B}}$ meets the validity criterion of barb sensitiveness.

The philosophy behind the asynchronous π-calculus entails that input actions $x(z)$ are not directly observable (while output actions can be observed by means of a matching input of the observer). This leads to semantic identifications like $\mathbf{0} = x(y).\bar{x}y$, for in both cases the environment may observe $\bar{x}z$ only if it supplied $\bar{x}z$ itself first. Yet, these processes differ on their input barbs $(\downarrow_x)$. For this reason, in aπ normally only output barbs $\downarrow_{\bar{x}}$ are considered [39]. Boudol's encoding satisfies the criterion of output barb sensitiveness (and in fact also input barb sensitiveness). However, the encoding of Honda & Tokoro does not, as it swaps input and output barbs. As such, it is an excellent example of the benefit of the external barb $\surd$ employed in Gorla's notion of success sensitiveness.

To obtain a weaker form of barb sensitiveness such that also $\mathscr{T}_{\mathrm{HT}}$ becomes barb sensitive, we introduce *channel barbs* $x \in \mathcal{N}$. A process is said to have the channel barb x iff it either has the barb $\bar{x}$ or x. We write $P\downarrow_x^{\mathrm{c}}$ when P has the channel barb x, and $P \Downarrow_x^{\mathrm{c}}$ when a P' exists with $P \Longmapsto P'$ and $P' \downarrow_x^{\mathrm{c}}$.

Definition 5. An encoding $\mathscr{T}$ is *channel barb sensitive* if $S{\Downarrow_\omega} \Leftrightarrow \mathscr{T}(S){\Downarrow_\omega}$ for all $\omega \in \Omega$.

This is a weaker criterion than barb sensitiveness, so $\mathscr{T}_{\mathrm{B}}$ is surely channel barb sensitive. It is easy to see that Honda & Tokoro's encoding $\mathscr{T}_{\mathrm{HT}}$, although not barb sensitive, is channel barb sensitive.

7 Validity up to a Semantic Equivalence

This section deals with the original validity criterion from Boudol [4], as generalised in [17]. Following [17] we call a compositional encoding $\mathscr{T}$ *valid up to* a semantic equivalence $\sim \subseteq \mathcal{P} \times \mathcal{P}$, where $\mathcal{P} \supseteq \mathcal{P}_{\mathrm{s}} \cup \mathcal{P}_{\mathrm{t}}$, iff $\mathscr{T}(P) \sim P$ for all $P \in \mathcal{P}_{\mathrm{s}}$. A given encoding may be valid up to a coarse equivalence, and invalid up to a finer one. The equivalence for which it is valid is then a measure of the quality of the encoding.

Below, we will evaluate the encodings $\mathscr{T}_{\mathrm{B}}$ and $\mathscr{T}_{\mathrm{HT}}$ under a number of semantic equivalences found in the literature. Since these encodings translate a single transition in the source language by a small protocol involving two or three transitions in the target language, they surely will not be valid under *strong* equivalences, demanding step-for-step matching of source transitions by target transitions. Hence we only look at *weak* equivalences.

First we consider equivalences that match transition labels, such as early weak bisimilarity. The encodings $\mathscr{T}_{\mathrm{B}}$ and $\mathscr{T}_{\mathrm{HT}}$ are not valid under such equivalences. Then we show that Boudol's encoding $\mathscr{T}_{\mathrm{B}}$ is valid under weak barbed bisimilarity, and thus certainly under its asynchronous version; however, it is not valid under asynchronous weak bisimilarity. The encoding $\mathscr{T}_{\mathrm{HT}}$ of Honda & Tokoro is not valid under any of these equivalences, but we introduce a new equivalence under which it is valid: a version of weak barbed bisimilarity that drops the distinction between input and output barbs.

7.1 A Labelled Transition Semantics of π

We first present a labelled transition semantics of the (a)synchronous π-calculus, to facilitate the definition of semantic equivalences on these languages. Its labels are drawn from a set of actions $\mathrm{Act} := \{\bar{x}y, x(y), \bar{x}(y) \mid x, y \in \mathcal{N}\} \cup \{\tau\}$. We define free and bound names on transition labels:

$$
\begin{aligned}
fn(\tau) &= \emptyset & bn(\tau) &= \emptyset \\
fn(\bar{x}z) &= \{x, z\} & bn(\bar{x}z) &= \emptyset \\
fn(x(y)) &= \{x\} & bn(x(y)) &= \{y\} \\
fn(\bar{x}(y)) &= \{x\} & bn(\bar{x}(y)) &= \{y\}\ .
\end{aligned}
$$

For $\alpha \in \mathrm{Act}$ we define $n(\alpha) := bn(\alpha) \cup fn(\alpha)$.

Definition 6. The *labelled transition relation of* π is the smallest relation $\longrightarrow \subseteq \mathcal{P}_\pi \times \mathrm{Act} \times \mathcal{P}_\pi$, satisfying the rules of Table 1.

The τ-transitions in the labelled transition semantics play the same role as the reductions in the reduction semantics: they present actual behaviour of the represented system. The transitions with a label different from τ merely represent potential behaviour: a transition $x(y)$ for instance represents the potential of the system to receive a value on channel x, but this potential will only be realised in the presence of a parallel component that sends a value on channel x. Likewise, an output action $\bar{x}z$ or $\bar{x}(y)$ can be realised only in communication with an input action $x(y)$.

The following results show (1) that the labelled transition relations are invariant under structural congruence ($\equiv$), and (2) that the closure under structural congruence of the labelled transition relation restricted to τ-steps coincides with the reduction relation—(2) stems from Milner [27].

Lemma 2 (Harmony Lemma [39, Lemma 1.4.15]).

1. If $P \xrightarrow{\alpha} P'$ and $P \equiv Q$ then $\exists Q'. Q \xrightarrow{\alpha} Q' \equiv P'$
2. $P \longmapsto P'$ iff $\exists P''. P \xrightarrow{\tau} P'' \equiv P'$.

The barbs defined in Sect. 6 can be characterised in terms of the labelled transition relation as follows:

Table 1. SOS rules for the synchronous mini-π-calculus. PAR, COM and CLOSE also have symmetric rules.

$$\text{(OUTPUT-ACT)}\ \frac{}{\bar{x}z.P \xrightarrow{\bar{x}z} P} \qquad \text{(INPUT-ACT)}\ \frac{w \notin fn((y)P)}{x(y).P \xrightarrow{x(w)} P\{w/y\}}$$

$$\text{(PAR)}\ \frac{P \xrightarrow{\alpha} P' \qquad bn(a) \cap fn(Q) = \emptyset}{P|Q \xrightarrow{\alpha} P'|Q} \qquad \text{(COM)}\ \frac{P \xrightarrow{\bar{x}z} P' \qquad Q \xrightarrow{x(y)} Q'}{P|Q \xrightarrow{\tau} P'|Q'\{z/y\}}$$

$$\text{(CLOSE)}\ \frac{P \xrightarrow{\bar{x}(w)} P' \qquad Q \xrightarrow{x(w)} Q'}{P|Q \xrightarrow{\tau} (w)(P'|Q')} \qquad \text{(RES)}\ \frac{P \xrightarrow{\alpha} P' \qquad y \notin n(a)}{(y)P \xrightarrow{\alpha} (y)P'}$$

$$\text{(OPEN)}\ \frac{P \xrightarrow{\bar{x}y} P' \qquad y \neq x \qquad w \notin fn((y)P')}{(y)P \xrightarrow{\bar{x}(w)} P'\{w/y\}} \qquad \text{(REP-ACT)}\ \frac{P \xrightarrow{\alpha} P'}{!P \xrightarrow{\alpha} P'|!P}$$

$$\text{(REP-COMM)}\ \frac{P \xrightarrow{\bar{x}z} P' \qquad P \xrightarrow{x(y)} P''}{!P \xrightarrow{\tau} (P'|P''\{z/y\})|!P} \qquad \text{(REP-CLOSE)}\ \frac{P \xrightarrow{\bar{x}(w)} P' \qquad P \xrightarrow{x(w)} P''}{!P \xrightarrow{\tau} ((w)(P'|P''))|!P}$$

Remark 1. A process P has a *strong barb on* $x \in \mathcal{N}$, $P{\downarrow_x}$, iff there is a P' with $P \xrightarrow{x(y)} P'$ for some $y \in \mathcal{N}$. It has a *strong barb on* $\bar{x}$, $P{\downarrow_{\bar{x}}}$, iff there is a P' with $P \xrightarrow{\bar{x}z} P'$ or $P \xrightarrow{\bar{x}(z)} P'$ for some $z \in \mathcal{N}$. A process P has a *weak barb on* a ($a \in \{x, \bar{x} \mid x \in \mathcal{N}\}$), $P{\Downarrow_a}$, iff there is a P' such that $P \xrightarrow{\tau}{}^{*} P'$ and $P'{\downarrow_a}$.

A process P has a *channel barb on* x, $P \downarrow^{c}_{x}$, iff it can perform an action on channel x, i. e. iff $P \xrightarrow{\alpha} P'$, for some P', where α has the form $\bar{x}y$, $\bar{x}(y)$ or $x(y)$. Moreover, $P \Downarrow^{c}_{x}$ iff a P' exists with $P \xrightarrow{\tau}{}^{*} P'$ and $P' \downarrow^{c}_{x}$.

7.2 Comparing Transition Labels: Early and Late Weak Bisimilarity

As they make use of intermediate steps (namely the acknowledgement protocol), we must fail proving the validity of the encodings $\mathscr{T}_{\mathrm{B}}$ or $\mathscr{T}_{\mathrm{HT}}$ up to semantics based on transition labels, e. g. *early weak bisimilarity* [39].

Definition 7. A symmetric binary relation $\mathcal{R}$ on π-processes P, Q is a *early weak bisimulation* iff $P \,\mathcal{R}\, Q$ implies

1. if $P \xrightarrow{\tau} P'$ then a Q' exists with $Q \xrightarrow{\tau}{}^{*} Q'$ and $P' \,\mathcal{R}\, Q'$,
2. if $P \xrightarrow{\alpha} P'$ where $\alpha = \bar{x}z$ or $\bar{x}(y)$ with $y \notin n(P) \cup n(Q)$, then a Q' exists with $Q \xrightarrow{\tau}{}^{*} \xrightarrow{\alpha} \xrightarrow{\tau}{}^{*} Q'$ and $P' \,\mathcal{R}\, Q'$,
3. if $P \xrightarrow{x(y)} P'$ with $y \notin n(P) \cup n(Q)$ then for all w a Q' exists satisfying $Q \xrightarrow{\tau}{}^{*} \xrightarrow{x(y)} \xrightarrow{\tau}{}^{*} Q'$ and $P'\{w/y\} \,\mathcal{R}\, Q'\{w/y\}$.

We denote the largest early weak bisimulation by $\approx_{\mathrm{EWB}}$.

Here $y \notin n(P) \cup n(Q)$ merely ensures the usage of fresh names. A *late* weak bisimulation is obtained by requiring in Clause 3 above that the choice of Q' is independent of w; this gives rise to a slightly finer equivalence relation.

Observation 1. $\mathscr{T}_{\mathrm{B}}$ is not valid up to $\approx_{\mathrm{EWB}}$.

Proof. Let $P = \bar{x}z.\mathbf{0}$ and $\mathscr{T}_{\mathrm{B}}(P) = (u)(\bar{x}u|u(v).(\bar{v}z|\mathbf{0}))$. We present the relevant parts of the labelled transition semantics:

$$\begin{array}{ccc} & & (u)(\bar{x}u|u(v).(\bar{v}z|\mathbf{0})) \\ & & \big\downarrow \bar{x}(c) \\ \bar{x}z.\mathbf{0} & & (u)(\mathbf{0}|c(v).(\bar{v}z|\mathbf{0})) \\ \big\downarrow \bar{x}z & & \big\downarrow c(d) \\ \mathbf{0} & & (u)(\mathbf{0}|(\bar{d}z|\mathbf{0})) \\ & & \big\downarrow \bar{d}z \\ & & (u)(\mathbf{0}|(\mathbf{0}|\mathbf{0})) \end{array}$$

Here, the translated term may perform an input transition $\xrightarrow{c(d)}$ the source term is not capable of. Hence, the processes are not equivalent up to $\approx_{\mathrm{EWB}}$. □

Since late weak bisimilarity is even finer (more discriminating) than $\approx_{\mathrm{EWB}}$, the encoding $\mathscr{T}_{\mathrm{B}}$ is certainly not valid up to late weak bisimilarity. A similar argument shows that neither $\mathscr{T}_{\mathrm{HT}}$ is valid up to early or late weak bisimilarity.

7.3 Weak Barbed Bisimilarity

A weaker approach does not compare all the transitions with visible labels, for these are merely *potential* transitions, that can occur only in certain contexts. Instead it just compares internal transitions, together with the information whether a state has the potential to perform an input or output over a certain channel: the barbs of Sect. 6. Combining the notion of barbs with the transfer property of classical bisimulation for internal actions only yields *weak barbed bisimilarity* [26]. Here, two related processes simulate each other's internal transitions and furthermore have the same weak barbs.

Definition 8. A symmetric relation $\mathcal{R}$ on $\mathcal{P}_\pi$ is a *weak barbed bisimulation* iff $P \mathrel{\mathcal{R}} Q$ implies

1. if $P{\downarrow_a}$ with $a \in \{x, \bar{x} \mid x \in \mathcal{N}\}$ then $Q{\Downarrow_a}$ and
2. if $P \xrightarrow{\tau} P'$ then a Q' exists with $Q \xrightarrow{\tau}{}^{*} Q'$ and $P' \mathrel{\mathcal{R}} Q'$.

The largest weak barbed bisimulation is denoted by $\stackrel{\bullet}{\approx}$, or $\approx_{\mathrm{WBB}}$.

By Lemma 2 this definition can equivalently be stated with $\longmapsto$ in the role of $\xrightarrow{\tau}$. One of the main results of this paper is that Boudol's encoding is valid up to $\stackrel{\bullet}{\approx}$. The proof of this result is given in the Appendix A.

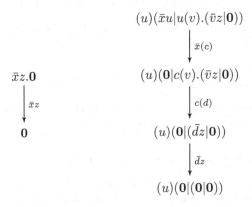

7.4 Asynchronous Weak Barbed Bisimilarity

In *asynchronous weak barbed bisimulation* [1], only the names of output channels are observed. Input barbs are ignored here, as it is assumed that an environment is able to observe output messages, but not (missing) inputs.

Definition 9. A symmetric relation S on $\mathcal{P}_\pi$ is an *asynchronous weak barbed bisimulation* iff $P \mathcal{R} Q$ implies

1. if $P{\downarrow_{\bar{x}}}$, then $Q{\Downarrow_{\bar{x}}}$, and
2. if $P \xrightarrow{\tau} P'$ then a Q' exists with $Q \xrightarrow{\tau}^{*} Q'$ and $P' \mathcal{R} Q'$.

The largest asynchronous weak barbed bisimulation is denoted by $\approx_{\mathrm{AWBB}}$.

Since $\approx_{\mathrm{AWBB}}$ is a coarser equivalence than $\stackrel{\bullet}{\approx}$, we obtain:

Corollary 1. Boudol's encoding is valid up to $\approx_{\mathrm{AWBB}}$.

In [37], a polyadic version of Boudol's encoding was assumed to be valid up to $\approx_{\mathrm{AWBB}}$; see Lemma 17. However, no proof was provided.

7.5 Weak Asynchronous Bisimilarity

We now know that Boudol's translation is valid up to $\approx_{\mathrm{AWBB}}$, but not up to $\approx_{\mathrm{EWB}}$. A natural step is to narrow down this gap by considering equivalences in between. The most prominent semantic equivalence for the asynchronous π-calculus is weak asynchronous bisimilarity, proposed by Amadio et al. [1].

A first strengthening of the requirements for $\approx_{\mathrm{AWBB}}$ is obtained by considering not only output channels but also the messages sent along them.

Definition 10 ([1])**.** A symmetric relation $\mathcal{R}$ on $\mathcal{P}_\pi$ is a *weak $o\tau$-bisimulation* if $\mathcal{R}$ meets Clauses 1 and 2 (but not necessarily 3) from Definition 7. The largest weak $o\tau$-bisimulation is denoted by $\approx_{\mathrm{W}o\tau}$.

Amadio et al. strengthen this equivalence by adding a further constraint for input transitions.

Definition 11 ([1])**.** A relation $\mathcal{R}$ is a *weak asynchronous bisimulation* iff $\mathcal{R}$ is a weak $o\tau$-bisimulation such that $P \mathcal{R} Q$ and $P \xrightarrow{\tau}^{*} \xrightarrow{x(y)} \xrightarrow{\tau}^{*} P'$ implies

- either a Q' exists satisfying a condition akin to Clause 3 of Definition 7,
- or a Q' exists such that $Q \xrightarrow{\tau}^{*} Q'$ and $P' \mathcal{R} (Q' | \bar{x}y)$.

The largest weak asynchronous bisimulation is denoted by $\approx_{\mathrm{WAB}}$.

Observation 2. Boudol's translation $\mathscr{T}_{\mathrm{B}} : \mathcal{P}_\pi \to \mathcal{P}_{\mathrm{a}\pi}$ is not valid up to $\approx_{\mathrm{W}o\tau}$, and thus not up to $\approx_{\mathrm{WAB}}$.

Proof. Consider the proof of Observation 1. $\bar{x}z.\mathbf{0}$ sends a free name along x while $(u)(\bar{x}u | u(v).(\bar{v}z|\mathbf{0}))$ sends a bound name along the same channel. Since $\approx_{\mathrm{W}o\tau}$ differentiates between free and bound names, the transition systems of $\bar{x}z$ and its translation are not $\approx_{\mathrm{W}o\tau}$-equivalent. □

7.6 Weak Channel Bisimilarity

From the equivalences considered, weak barbed bisimilarity, $\stackrel{\bullet}{\approx}$, is the finest one that supports the validity of Boudol's translation. However, it does not validate Honda and Tokoro's translation.

Observation 3. Honda and Tokoro's translation $\mathscr{T}_{\mathrm{HT}}$ is not valid up to $\approx_{\mathrm{AWBB}}$, and thus not up to $\stackrel{\bullet}{\approx}$, $\approx_{\mathrm{W}o\mathcal{T}}$, or $\approx_{\mathrm{WAB}}$.

Proof. Let $P = \bar{x}z.\mathbf{0}$. Then $P{\downarrow_{\bar{x}}}$. The translation is $\mathscr{T}_{\mathrm{HT}}(P) = x(u).(\bar{u}z|\mathbf{0})$ and $\mathscr{T}_{\mathrm{HT}}(P)\not\Downarrow_{\bar{x}}$. □

To address this problem we introduce an equivalence even weaker than $\stackrel{\bullet}{\approx}$, which does not distinguish between input and output channels.

Definition 12. A symmetric relation $\mathcal{R}$ on $\mathcal{P}_\pi$ is a *weak channel bisimulation* if $P\,\mathcal{R}\,Q$ implies

1. if $P \downarrow_x^{\mathrm{c}}$ then $Q \Downarrow_x^{\mathrm{c}}$ and
2. if $P \xrightarrow{\tau} P'$ then a Q' exists with $Q \xrightarrow{\tau}{}^{*} Q'$ and $P'\,\mathcal{R}\,Q'$.

The largest weak channel bisimulation is denoted $\approx_{\mathrm{WCB}}$.

Theorem 1. Honda and Tokoro's encoding $\mathscr{T}_{\mathrm{HT}}$ is valid up to $\approx_{\mathrm{WCB}}$.

The proof is similar to the one of Theorem 4. Here we use that Lemmas 6 and 7 also apply to $\mathscr{T}_{\mathrm{HT}}$ [18] and Lemma 1 now holds with $\downarrow_x^{\mathrm{c}}$ in the role of $\downarrow_a$.

Since $\approx_{\mathrm{WCB}}$ is a coarser equivalence than $\stackrel{\bullet}{\approx}$, we also obtain that Boudol's translation is valid up to $\approx_{\mathrm{WCB}}$.

7.7 Overview

We thus obtain the following hierarchy of equivalence relations on π-calculus processes (cf. Fig. 1), with the vertical lines indicating the realm of validity of $\mathscr{T}_{\mathrm{HT}}$ and $\mathscr{T}_{\mathrm{B}}$, respectively.

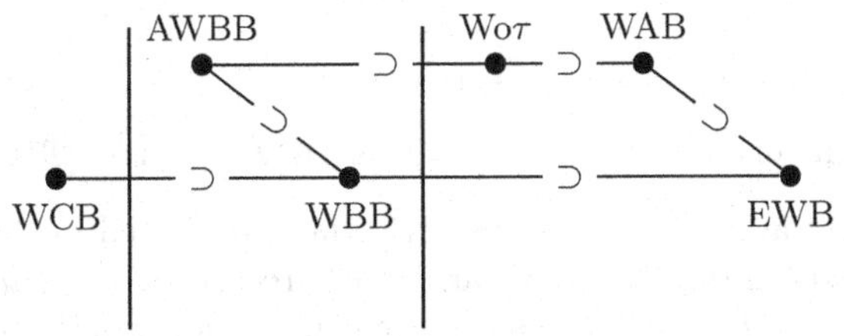

Fig. 1. A hierarchy on semantic equivalence relations for π-calculus processes, with separation lines indicating where the encodings discussed in this paper pass and fail validity.

8 Validity up to an Equivalence Versus Validity à la Gorla

The idea of introducing a success predicate $\surd$ to the source and target language of an encoding, as implicit in Gorla's criterion of success sensitiveness, can be applied to the equivalence based approach as well.

Definition 13. Let $\mathcal{P}_{\rm s}$, $\mathcal{P}_{\rm t}$ be languages equipped with a reduction relation $\longmapsto$, and $\mathcal{P}_{\rm s}^{\surd}$, $\mathcal{P}_{\rm t}^{\surd}$ their extensions with a success predicate $\surd$. A symmetric relation $\mathcal{R}$ on $\mathcal{P} := \mathcal{P}_{\rm s}^{\surd} \uplus \mathcal{P}_{\rm t}^{\surd}$ is a *success respecting weak reduction bisimulation* if $P \mathcal{R} Q$ implies

1. if $P{\downarrow_{\surd}}$ then $Q{\Downarrow_{\surd}}$ and
2. if $P \longmapsto P'$ then a Q' exists with $Q \Longmapsto Q'$ and $P' \mathcal{R} Q'$.

The largest success respecting weak reduction bisimulation is denoted $\stackrel{\bullet}{\approx}^{\surd}$.

An compositional encoding $\mathscr{T} : \mathcal{P}_{\rm s} \rightarrow \mathcal{P}_{\rm t}$ is *valid up to* $\stackrel{\bullet}{\approx}^{\surd}$ if its extension $\mathscr{T}^{\surd} : \mathcal{P}_{\rm s}^{\surd} \rightarrow \mathcal{P}_{\rm t}^{\surd}$, defined by $\mathscr{T}^{\surd}(\surd) := \surd$, satisfies $\mathscr{T}^{\surd}(P) \stackrel{\bullet}{\approx}^{\surd} P$ for all $P \in \mathcal{P}_{\rm s}^{\surd}$.

Trivially, a variant of Lemma 1 with $\surd$ in the role of a holds for $\mathscr{T}_{\rm B}$ as well as $\mathscr{T}_{\rm HT}$: we have $P{\downarrow_{\surd}}$ iff $\mathscr{T}_{\rm B}(P){\downarrow_{\surd}}$ iff $\mathscr{T}_{\rm HT}(P){\downarrow_{\surd}}$. Using this, the material in the Appendix A implies that:

Theorem 2. The encodings $\mathscr{T}_{\rm B}$ and $\mathscr{T}_{\rm HT}$ are valid up to $\stackrel{\bullet}{\approx}^{\surd}$. □

This approach has the distinct advantage over dealing with input and output barbs that both encodings are seen to be valid without worrying on what kinds of barbs to use exactly.

The following correspondence between operational correspondence, success sensitivity and validity up to $\stackrel{\bullet}{\approx}^{\surd}$ was observed in [33], and not hard to infer from the definitions.

Theorem 3. An encoding $\mathscr{T}$ is success sensitive and satisfies operational correspondence criteria ($\mathfrak{C}'$) and ($\mathfrak{W}$), taking $\asymp$ to be $\stackrel{\bullet}{\approx}^{\surd}$, iff it is valid up to $\stackrel{\bullet}{\approx}^{\surd}$. □

This yields the result promised in Sect. 5:

Corollary 2. The encodings $\mathscr{T}_{\rm B}$ and $\mathscr{T}_{\rm HT}$ satisfy criterion ($\mathfrak{W}$).

The validity of $\mathscr{T}_{\rm B}$ and $\mathscr{T}_{\rm HT}$ by Gorla's criteria, established in [18], by the analysis of [33], already implied that $\mathscr{T}_{\rm B}$ and $\mathscr{T}_{\rm HT}$ are valid up to *success respecting coupled reduction similarity* [33], a semantic equivalence strictly coarser than $\stackrel{\bullet}{\approx}^{\surd}$. Theorem 2 yields a nontrivial strengthening of that result.

Gorla's criterion of divergence reflection can be strengthened to *divergence preservation* by requiring

$$\mathscr{T}(S) \longmapsto_{\rm t}^{\omega} \;\Leftrightarrow\; S \longmapsto_{\rm s}^{\omega} ;$$

by [18, Remark 1] this criterion is satisfied by $\mathscr{T}_{\rm B}$ and $\mathscr{T}_{\rm HT}$ as well. A bisimulation $\mathcal{R}$ is said to preserve divergence iff $P \mathrel{\mathcal{R}} Q$ implies $P \longmapsto_{\rm t}^{\omega} \Leftrightarrow Q \longmapsto_{\rm s}^{\omega}$; the largest divergence preserving, success respecting weak reduction bisimulation is denoted $\stackrel{\bullet}{\approx}^{\surd\Delta}$. As observed in [33], Theorem 3 can be extended as follows with divergence preservation:

Observation 4. An encoding $\mathscr{T}$ is success sensitive, divergence preserving, and satisfies operational correspondence criteria ($\mathfrak{C}'$) and ($\mathfrak{W}$), taking $\asymp$ to be $\stackrel{\bullet}{\approx}^{\surd\Delta}$, iff it is valid up to $\stackrel{\bullet}{\approx}^{\surd\Delta}$. □

Hence, $\mathscr{T}_{\rm B}$ and $\mathscr{T}_{\rm HT}$ are valid up to $\stackrel{\bullet}{\approx}^{\surd\Delta}$. This statement implies all criteria of Gorla, except for name invariance.

In [19, Definition 26] the notion of *divergence preserving branching barbed bisimilarity* is defined. This definition is parametrised by the choice of barbs; when taking the success predicate $\surd$ as only barb, it could be called *divergence preserving, success respecting branching reduction bisimilarity.* It is strictly finer then $\stackrel{\bullet}{\approx}^{\surd\Delta}$. It is not hard to adapt the proof of Theorem 4 in the Appendix A to show that $\mathscr{T}_{\rm B}$ and $\mathscr{T}_{\rm HT}$ are even valid up to this equivalence.

9 Full Abstraction

The criterion of full abstraction is parametrised by the choice of two semantic equivalences $\sim_{\rm S}$ and $\sim_{\rm T}$, one on the source and one on the target language. It requires, for source expressions P and Q, that $P \sim_{\rm S} Q \Leftrightarrow \mathscr{T}(P) \sim_{\rm T} \mathscr{T}(Q)$.

It is well known that the encodings $\mathscr{T}_{\rm B}$ and $\mathscr{T}_{\rm HT}$ fail to be fully abstract w.r.t. $\cong^c$ and $\cong^c_a$. Here $\cong^c$ is *weak barbed congruence*, the congruence closure of $\stackrel{\bullet}{\approx}^{\surd}$ (or $\approx_{\rm AWBB}$) on the source language, and $\cong^c_a$ is *asynchronous weak barbed congruence*, the congruence closure of $\stackrel{\bullet}{\approx}^{\surd}$ (or $\approx_{\rm AWBB}$) on the target language. These are often deemed to be the most natural semantic equivalences on π and aπ. The well-known counterexample is given by the π processes $\bar{x}z|\bar{x}z$ and $\bar{x}z.\bar{x}z$. Although related by $\cong^c$, their translations are not related by $\cong^c_a$.

In [10] this problem is addressed by proposing a strict subcalculus $SA\pi$ of the target language that contains the image of the source language under of a version Honda & Tokoro's encoding, such that this encoding is fully abstract w.r.t. $\cong^c$ and the congruence closure of $\stackrel{\bullet}{\approx}^{\surd}$ (or $\approx_{\rm AWBB}$) w.r.t. $SA\pi$. In [37] a similar solution to the same problem was found earlier, but for a variant of Boudol's encoding from the *polyadic* π-calculus to the (monadic) asynchronous π-calculus. They define a class of *well-typed* expressions in the asynchronous π-calculus, such that the well-typed expressions constitute a subcalculus of the target language that contains the image of the source language under the encoding. Again, the encoding is fully abstract w.r.t. $\cong^c$ and the congruence closure of $\stackrel{\bullet}{\approx}^{\surd}$ (or $\approx_{\rm AWBB}$) w.r.t. that sublanguage.

By [20, Theorem 4] such results can always be achieved, namely by taking as target language exactly the image of the source language under the encoding. In this sense a full abstraction result is a direct consequence of the validity of the encodings up to $\dot{\approx}^{\surd}$, taking for $\sim_S$ the congruence closure of $\dot{\approx}^{\surd}$ w.r.t. the source language, and for $\sim_T$ the congruence closure of $\dot{\approx}^{\surd}$ w.r.t. the image of the source language within the target language. What the results of [10,37] add is that the sublanguage may be strictly larger than the image of the source language, and that its definition is not phrased in terms of the encoding.

10 Conclusion

We examined which of the quality criteria for encodings from the literature support the validity of the well-known encodings $\mathscr{T}_B$ and $\mathscr{T}_{HT}$ of the asynchronous into the synchronous π-calculus. It was already known [18] that these encodings are valid à la Gorla [22]; this implies that they are valid up to success respecting coupled reduction similarity [33]. We strengthened this result by showing that they are even valid up to divergence preserving, success respecting weak reduction bisimilarity. That statement implies all criteria of Gorla, except for name invariance. Moreover, it implies a stronger form of operation soundness then considered by Gorla, namely

$$\text{if } \mathscr{T}(S) \Longmapsto_t T \text{ then } \exists S' \colon S \Longmapsto_s S' \text{ and } T \asymp_t \mathscr{T}(S'). \qquad (\mathfrak{W})$$

Crucial for all these results is that we employ Gorla's external barb $\surd$, a success predicate on processes. When reverting to the internal barns x and $\bar{x}$ commonly used in the π-calculus, we see a potential difference in quality between the encodings $\mathscr{T}_B$ and $\mathscr{T}_{HT}$. Boudol's translation $\mathscr{T}_B$ is valid up to weak barbed bisimilarity, regardless whether all barbs are used, or only output barbs $\bar{x}$. However, Honda and Tokoro's translation $\mathscr{T}_{HT}$ is not valid under either of these forms of weak barbed bisimilarity. In order to prove the validity of $\mathscr{T}_{HT}$, we had to use the novel weak channel bisimilarity that does not distinguish between input and output channels. Conversely, we conjecture that there is no natural equivalence for which $\mathscr{T}_{HT}$ is valid, but $\mathscr{T}_B$ is not. Hence, Honda and Tokoro's encoding can be regarded as weaker than the one of Boudol. Whether $\mathscr{T}_B$ is to be preferred, because it meets stronger requirements/equivalences, is a decision that should be driven by the requirements of an application the encoding is used for.

The validity of $\mathscr{T}_B$ under semantic equivalences has earlier been investigated in [7,8], In [7] it is established that $\mathscr{T}_B$ is valid up to *may testing* [9] and *fair testing equivalence* [6,28]. Both results now follow from Theorem 2, since may and fair testing equivalence are coarser then $\dot{\approx}^{\surd}$. On the other hand, [7] also shows that $\mathscr{T}_B$ is not valid up to a form of *must testing*; in [8] this result is strengthened to pertain to any encoding of π into aπ. It follows that this form of must testing equivalence is not implied by $\dot{\approx}^{\surd}$, and not even by $\dot{\approx}^{\surd\Delta}$.

A Appendix: Boudol's Translation is Valid up to $\dot{\approx}$

Before we prove validity of Boudol's translation up to weak barbed bisimulation, we further investigate the protocol steps established by Boudol's encoding. Let

$P' = \bar{x}z.P$ and $Q' = x(y).Q$. Pick u, v not free in P and Q, with $u \neq v$. Write $P^* := \bar{v}z|\mathscr{T}_{\mathrm{B}}(P)$ and $Q^* := v(y).\mathscr{T}_{\mathrm{B}}(Q)$. Then

$$\begin{aligned}\mathscr{T}_{\mathrm{B}}(P'|Q') &= (u)(\bar{x}u|u(v).P^*) \mid x(u).(v)(\bar{u}v|Q^*) \\ &\longmapsto (u)\big(u(v).P^* \mid (v)(\bar{u}v|Q^*)\big) \\ &\longmapsto (v)(P^*|Q^*) \\ &\longmapsto \mathscr{T}_{\mathrm{B}}(P)|(\mathscr{T}_{\mathrm{B}}(Q)\{^z\!/\!_y\}) \ .\end{aligned}$$

Here structural congruence is applied in omitting parallel components **0** and empty binders (u) and (v). Now the crucial idea in our proof is that the last two reductions are *inert*, in that set of the potential behaviours of a process is not diminished by doing (internal) steps of this kind. The first reduction above in general is not inert, as it creates a commitment between a sender and a receiver to communicate, and this commitment goes at the expense of the potential of one of the two parties to do this communication with another partner. We employ a relation that captures these inert reductions in a context.

Definition 14 ([18])**.** Let $\Rrightarrow$ be the smallest relation on $\mathcal{P}_{\mathrm{a}\pi}$ such that

1. $(v)(\bar{v}y|P|v(z).Q) \Rrightarrow P|(Q\{^y\!/\!_z\})$,
2. if $P \Rrightarrow Q$ then $P|C \Rrightarrow Q|C$,
3. if $P \Rrightarrow Q$ then $(w)P \Rrightarrow (w)Q$,
4. if $P \equiv P' \Rrightarrow Q' \equiv Q$ then $P \Rrightarrow Q$,

where $v \notin fn(P) \cup fn(Q\{^y\!/\!_z\})$.

First of all observe that whenever two processes are related by $\Rrightarrow$, an actual reduction takes place.

Lemma 3 ([18])**.** If $P \Rrightarrow Q$ then $P \longmapsto Q$.

The next two lemmas confirm that inert reductions do not diminish the potential behaviour of a process.

Lemma 4 ([18])**.** If $P \Rrightarrow Q$ and $P \longmapsto P'$ with $P' \not\equiv Q$ then there is a Q' with $Q \longmapsto Q'$ and $P' \Rrightarrow Q'$.

Corollary 3. If $P \Rrightarrow^* Q$ and $P \longmapsto P'$ then either $P' \Rrightarrow^* Q$ or there is a Q' with $Q \longmapsto Q'$ and $P' \Rrightarrow^* Q'$.

Proof. By repeated application of Lemma 4. □

Lemma 5. If $P \Rrightarrow Q$ and $P{\downarrow_a}$ for $a \in \{x, \bar{x} \mid x \in \mathcal{N}\}$ then $Q{\downarrow_a}$.

Proof. Let $(\tilde{w})P$ for $\tilde{w} = \{w_1, \ldots, w_n\} \subseteq \mathcal{N}$ with $n \in \mathbb{N}$ denote $(w_1)\cdots(w_n)P$ for some arbitrary order of the (w_i). Using a trivial variant of Lemma 1.2.20 in [39], there are $\tilde{w} \subseteq \mathcal{N}$, $x, y, z \in \mathcal{N}$ and $R, C \in \mathcal{P}_{\mathrm{a}\pi}$, such that $x \in \tilde{w}$ and $P \equiv (\tilde{w})((\bar{x}y|x(z).R)|C) \longmapsto (\tilde{w})((\mathbf{0}|R\{^y\!/\!_z\})|C) \equiv Q$. Since $P{\downarrow_a}$, it must be that $a{=}u$ or $\bar{u}$ with $u \notin \tilde{w}$, and $C{\downarrow_a}$. Hence $Q{\downarrow_a}$. □

The following lemma states, in terms of Gorla's framework, *operational completeness* [22]: if a source term is able to make a step, then its translation is able to simulate that step by protocol steps.

Lemma 6 ([18]). Let $P, P' \in \mathcal{P}_\pi$. If $P \longmapsto P'$ then $\mathscr{T}_{\mathrm{B}}(P) \longmapsto^* \mathscr{T}_{\mathrm{B}}(P')$.

Finally, the next lemma was a crucial step in establishing *operational soundness* [22].

Lemma 7 ([18]). Let $P \in \mathcal{P}_\pi$ and $Q \in \mathcal{P}_{\mathrm{a}\pi}$. If $\mathscr{T}_{\mathrm{B}}(P) \longmapsto Q$ then there is a P' with $P \longmapsto P'$ and $Q \Longrightarrow^* \mathscr{T}_{\mathrm{B}}(P')$.

Using these lemmas, we prove the validity of Boudol's encoding up to weak barbed bisimilarity.

Theorem 4. Boudol's encoding is valid up to $\stackrel{\bullet}{\approx}$.

Proof. Define the relation $\mathcal{R}$ by $P \mathrel{\mathcal{R}} Q$ iff $Q \Longrightarrow^* \mathscr{T}_{\mathrm{B}}(P)$. It suffices to show that the symmetric closure of $\mathcal{R}$ is a weak barbed bisimulation.

To show that $\mathcal{R}$ satisfies Clause 1 of Definition 8, suppose $P \mathrel{\mathcal{R}} Q$ and $P{\downarrow_a}$ for $a \in \{x, \bar{x} \mid x \in \mathcal{N}\}$. Then $\mathscr{T}_{\mathrm{B}}(P){\downarrow_a}$ by Lemma 1. Since $Q \Longrightarrow^* \mathscr{T}_{\mathrm{B}}(P)$, we obtain $Q \longmapsto^* \mathscr{T}_{\mathrm{B}}(P)$ by Lemma 3, and thus $Q{\Downarrow_a}$.

To show that $\mathcal{R}$ also satisfies Clause 2, suppose $P \mathrel{\mathcal{R}} Q$ and $P \longmapsto P'$. Since $Q \Longrightarrow^* \mathscr{T}_{\mathrm{B}}(P)$, by Lemmas 3 and 6 we have $Q \longmapsto^* \mathscr{T}_{\mathrm{B}}(P) \longmapsto^* \mathscr{T}_{\mathrm{B}}(P')$, and also $P' \mathrel{\mathcal{R}} \mathscr{T}_{\mathrm{B}}(P')$.

To show that $\mathcal{R}^{-1}$ satisfies Clause 1, suppose $P \mathrel{\mathcal{R}} Q$ and $Q{\downarrow_a}$. Since $Q \Longrightarrow^* \mathscr{T}_{\mathrm{B}}(P)$, Lemma 5 yields $\mathscr{T}_{\mathrm{B}}(P){\downarrow_a}$, and Lemma 1 gives $P{\downarrow_a}$, which implies $P{\Downarrow_a}$.

To show that $\mathcal{R}^{-1}$ satisfies Clause 2, suppose $P \mathrel{\mathcal{R}} Q$ and $Q \longmapsto Q'$. Since $Q \Longrightarrow^* \mathscr{T}_{\mathrm{B}}(P)$, by Corollary 3 either $Q' \Longrightarrow^* \mathscr{T}_{\mathrm{B}}(P)$ or there is a Q'' with $\mathscr{T}_{\mathrm{B}}(P) \longmapsto Q''$ and $Q' \Longrightarrow^* Q''$. In the first case $P \mathrel{\mathcal{R}} Q'$, so taking $P' := P$ we are done. In the second case, by Lemma 7 there is a P' with $P \longmapsto P'$ and $Q'' \Longrightarrow^* \mathscr{T}_{\mathrm{B}}(P')$. We thus have $P' \mathrel{\mathcal{R}} Q'$. □

References

1. Amadio, R.M., Castellani, I., Sangiorgi, D.: On bisimulations for the asynchronous pi-calculus. Theoret. Comput. Sci. **195**(2), 291–324 (1998). https://doi.org/10.1016/S0304-3975(97)00223-5
2. Baldamus, M., Parrow, J., Victor, B.: A fully abstract encoding of the π-calculus with data terms. In: Caires, L., Italiano, G.F., Monteiro, L., Palamidessi, C., Yung, M. (eds.) ICALP 2005. LNCS, vol. 3580, pp. 1202–1213. Springer, Heidelberg (2005). https://doi.org/10.1007/11523468_97
3. Beauxis, R., Palamidessi, C., Valencia, F.D.: On the asynchronous nature of the asynchronous π-calculus. In: Degano, P., De Nicola, R., Meseguer, J. (eds.) Concurrency, Graphs and Models. LNCS, vol. 5065, pp. 473–492. Springer, Heidelberg (2008). https://doi.org/10.1007/978-3-540-68679-8_29

4. Boudol, G.: Notes on algebraic calculi of processes. In: Apt, K.R. (eds) Logics and Models of Concurrent Systems. NATO ASI Series (Series F: Computer and Systems Sciences), vol. 13, pp. 261–303. Springer, Heidelberg (1985). https://doi.org/10.1007/978-3-642-82453-1_9
5. Boudol, G.: Asynchrony and the π-calculus (Note). Technical Report, 1702, INRIA (1992)
6. Brinksma, E., Rensink, A., Vogler, W.: Fair testing. In: Lee, I., Smolka, S.A. (eds.) CONCUR 1995. LNCS, vol. 962, pp. 313–327. Springer, Heidelberg (1995). https://doi.org/10.1007/3-540-60218-6_23
7. Cacciagrano, D., Corradini, F.: On synchronous and asynchronous communication paradigms. ICTCS 2001. LNCS, vol. 2202, pp. 256–268. Springer, Heidelberg (2001). https://doi.org/10.1007/3-540-45446-2_16
8. Cacciagrano, D., Corradini, F., Palamidessi, C.: Separation of synchronous and asynchronous communication via testing. Theoret. Comput. Sci. **386**(3), 218–235 (2007). https://doi.org/10.1016/j.tcs.2007.07.009
9. De Nicola, R., Hennessy, M.: Testing equivalences for processes. Theoret. Comput. Sci. **34**, 83–133 (1984). https://doi.org/10.1016/0304-3975(84)90113-0
10. Du, W., Yang, Z., Zhu, H.: A fully abstract encoding for sub asynchronous Pi calculus. In: Pang, J., Zhang, C., He, J., Weng, J. (eds.) Proceedings of TASE 2018, pp. 17–27. IEEE Computer Society Press (2018). https://doi.org/10.1109/TASE.2018.00011
11. Fu, Y.: Theory of interaction. Theoret. Comput. Sci. **611**, 1–49 (2016). https://doi.org/10.1016/j.tcs.2015.07.043
12. Given-Wilson, T.: Expressiveness via intensionality and concurrency. In: Ciobanu, G., Méry, D. (eds.) ICTAC 2014. LNCS, vol. 8687, pp. 206–223. Springer, Cham (2014). https://doi.org/10.1007/978-3-319-10882-7_13
13. Given-Wilson, T.: On the expressiveness of intensional communication. In: Borgström, J., Crafa, S. (eds.) Proceedings EXPRESS/SOS 2014. EPTCS, vol. 160, pp. 30–46 (2014). https://doi.org/10.4204/EPTCS.160.4
14. Given-Wilson, T., Legay, A.: On the expressiveness of joining. In: Knight, S., Lanese, I., Lluch Lafuente, A., Torres Vieira, H. (eds.) Proceedings of ICE2015. EPTCS, vol. 189, pp. 99–113 (2015). https://doi.org/10.4204/EPTCS.189.9
15. Given-Wilson, T., Legay, A.: On the expressiveness of symmetric communication. In: Sampaio, A., Wang, F. (eds.) ICTAC 2016. LNCS, vol. 9965, pp. 139–157. Springer, Cham (2016). https://doi.org/10.1007/978-3-319-46750-4_9
16. van Glabbeek, R.J.: On the expressiveness of ACP. In: Ponse, A., Verhoef, C., van Vlijmen, S.F.M. (eds.) Algebra of Communicating Processes. Workshops in Computing, pp. 188–217. Springer, London (1995), https://doi.org/10.1007/978-1-4471-2120-6_8
17. van Glabbeek, R.J.: Musings on encodings and expressiveness. In: Luttik, B., Reniers, M.A. (eds.) Proceedings EXPRESS/SOS 2012. EPTCS, vol. 89, pp. 81–98 (2012). https://doi.org/10.4204/EPTCS.89.7
18. van Glabbeek, R.J.: On the validity of encodings of the synchronous in the asynchronous π-calculus. Inf. Process. Lett. **137**, 17–25 (2018). https://doi.org/10.1016/j.ipl.2018.04.015. https://arxiv.org/abs/1802.09182
19. van Glabbeek, R.J.: A Theory of Encodings and Expressiveness. Technical Report, Data61, CSIRO (2018). https://arxiv.org/abs/1805.10415. Full version of [20]
20. van Glabbeek, R.J.: A theory of encodings and expressiveness (extended abstract). In: Baier, C., Dal Lago, U. (eds.) FoSSaCS 2018. LNCS, vol. 10803, pp. 183–202. Springer, Cham (2018). https://doi.org/10.1007/978-3-319-89366-2_10

21. Gorla, D.: A taxonomy of process calculi for distribution and mobility. Distrib. Comput. **23**(4), 273–299 (2010). https://doi.org/10.1007/s00446-010-0120-6
22. Gorla, D.: Towards a unified approach to encodability and separation results for process calculi. Inf. Comput. **208**(9), 1031–1053 (2010). https://doi.org/10.1016/j.ic.2010.05.002
23. Gorla, D., Nestmann, U.: Full abstraction for expressiveness: history, myths and facts. Math. Struct. Comput. Sci. **26**(4), 639–654 (2016). https://doi.org/10.1017/S0960129514000279
24. Honda, K., Tokoro, M.: An object calculus for asynchronous communication. In: America, P. (ed.) ECOOP 1991. LNCS, vol. 512, pp. 133–147. Springer, Heidelberg (1991). https://doi.org/10.1007/BFb0057019
25. Lanese, I., Pérez, J.A., Sangiorgi, D., Schmitt, A.: On the expressiveness of polyadic and synchronous communication in higher-order process calculi. In: Abramsky, S., Gavoille, C., Kirchner, C., Meyer auf der Heide, F., Spirakis, P.G. (eds.) ICALP 2010. LNCS, vol. 6199, pp. 442–453. Springer, Heidelberg (2010). https://doi.org/10.1007/978-3-642-14162-1_37
26. Milner, R.: The Polyadic π-Calculus: A Tutorial. Technical Report ECS-LFCS-91-180, The University of Edinburgh. Informatics Report Series (1991)
27. Milner, R.: Functions as processes. Math. Struct. Comput. Sci. **2**(2), 119–141 (1992). https://doi.org/10.1017/S0960129500001407
28. Natarajan, V., Cleaveland, R.: Divergence and fair testing. In: Fülöp, Z., Gécseg, F. (eds.) ICALP 1995. LNCS, vol. 944, pp. 648–659. Springer, Heidelberg (1995). https://doi.org/10.1007/3-540-60084-1_112
29. Nestmann, U.: What is a "Good" encoding of guarded choice? Inf. Comput. **156**(1–2), 287–319 (2000). https://doi.org/10.1006/inco.1999.2822
30. Nestmann, U., Pierce, B.C.: Decoding choice encodings. Inf. Comput. **163**(1), 1–59 (2000). https://doi.org/10.1006/inco.2000.2868
31. Palamidessi, C.: Comparing the expressive power of the synchronous and asynchronous pi-calculi. Math. Struct. Comput. Sci. **13**(5), 685–719 (2003). https://doi.org/10.1017/S0960129503004043
32. Parrow, J.: General conditions for full abstraction. Math. Struct. Comput. Sci. **26**(4), 655–657 (2016). https://doi.org/10.1017/S0960129514000280
33. Peters, K., van Glabbeek, R.J.: Analysing and comparing encodability criteria. In: Crafa, S., Gebler, D.E. (eds.) EXPRESS/SOS 2015. EPTCS, vol. 190, pp. 46–60 (2015). https://doi.org/10.4204/EPTCS.190.4
34. Peters, K., Nestmann, U.: Is it a "Good" encoding of mixed choice? In: Birkedal, L. (ed.) FoSSaCS 2012. LNCS, vol. 7213, pp. 210–224. Springer, Heidelberg (2012). https://doi.org/10.1007/978-3-642-28729-9_14
35. Peters, K., Nestmann, U., Goltz, U.: On distributability in process calculi. In: Felleisen, M., Gardner, P. (eds.) ESOP 2013. LNCS, vol. 7792, pp. 310–329. Springer, Heidelberg (2013). https://doi.org/10.1007/978-3-642-37036-6_18
36. Peters, K., Schicke, J.-W., Nestmann, U.: Synchrony vs causality in the asynchronous pi-calculus. In: Luttik, B., Valencia, F. (eds.) Proceedings EXPRESS 2011, EPTCS, vol. 64, pp. 89–103 (2011). https://doi.org/10.4204/EPTCS.64.7
37. Quaglia, P., Walker, D.: On synchronous and asynchronous mobile processes. In: Tiuryn, J. (ed.) FoSSaCS 2000. LNCS, vol. 1784, pp. 283–296. Springer, Heidelberg (2000). https://doi.org/10.1007/3-540-46432-8_19
38. Riecke, J.G.: Fully abstract translations between functional languages. In: Wise, D.S. (ed.) Proceedings POPL 1991, pp. 245–254. ACM Press (1991). https://doi.org/10.1145/99583.99617

39. Sangiorgi, D., Walker, D.: The π-Calculus: A Theory of Mobile Processes. Cambridge University Press, Cambridge (2001)
40. Shapiro, E.Y.: Separating concurrent languages with categories of language embeddings. In: Koutsougeras, C., Vitter, J.S. (eds.) STOC 1991, pp. 198–208. ACM (1991). https://doi.org/10.1145/103418.103423
41. Shapiro, E.: Embeddings among concurrent programming languages. In: Cleaveland, W.R. (ed.) CONCUR 1992. LNCS, vol. 630, pp. 486–503. Springer, Heidelberg (1992). https://doi.org/10.1007/BFb0084811
42. de Simone, R.: Higher-level synchronising devices in Meije-SCCS. Theoret. Comput. Sci. **37**, 245–267 (1985). https://doi.org/10.1016/0304-3975(85)90093-3
43. Vaandrager, F.W.: Expressiveness results for process algebras. In: de Bakker, J.W., de Roever, W.-P., Rozenberg, G. (eds.) REX 1992. LNCS, vol. 666, pp. 609–638. Springer, Heidelberg (1993). https://doi.org/10.1007/3-540-56596-5_49

Confluence of the Chinese Monoid

Jörg Endrullis[1(✉)] and Jan Willem Klop[1,2]

[1] Vrije Universiteit Amsterdam, Amsterdam, The Netherlands
j.endrullis@vu.nl
[2] Centrum Wiskunde en Informatica, Amserdam, The Netherlands
JWKlop1945@kpnmail.nl

Abstract. The Chinese monoid, related to Knuth's Plactic monoid, is of great interest in algebraic combinatorics. Both are ternary monoids, generated by relations between words of three symbols. The relations are, for a totally ordered alphabet, $cba = cab = bca$ if $a \leq b \leq c$. In this note we establish confluence by tiling for the Chinese monoid, with the consequence that every two words u, v have extensions to a common word: $\forall u, v.\ \exists x, y.\ ux = vy$.

Our proof is given using decreasing diagrams, a method for obtaining confluence that is central in abstract rewriting theory. Decreasing diagrams may also be applicable to various related monoid presentations.

We conclude with some open questions for the monoids considered.

Dedication. *Our paper is dedicated in friendship to Catuscia Palamidessi for her 60th anniversary, with fond memories of the second author of cooperations during her stays around 1990 at CWI Amsterdam; with admiration for her work and accomplishments.*

1 Introduction

This paper is concerned with the Chinese monoid which is the quotient of the free monoid over a totally ordered alphabet with respect to the congruence generated by

$$cba = cab = bca \quad \text{for every } a \leq b \leq c.$$

These relations are equivalent to

$$\begin{aligned} aba = baa,\ bba = bab \quad & \text{for every } a < b, \\ cab = cba = bca \quad & \text{for every } a < b < c. \end{aligned} \tag{1}$$

The Chinese monoid plays an important role in algebra and combinatorics. It is closely related to the Plactic monoid of Knuth [16] generated by

$$\begin{aligned} aba = baa,\ bba = bab \quad & \text{for every } a < b, \\ cab = acb,\ bca = bac \quad & \text{for every } a < b < c. \end{aligned} \tag{2}$$

M. S. Alvim et al. (Eds.): Palamidessi Festschrift, LNCS 11760, pp. 206–220, 2019.
https://doi.org/10.1007/978-3-030-31175-9_12

For the case of two generators, the Chinese monoid coincides with the Plactic monoid. Knuth devised the Eq. (2) in 1970 to analyse Schensted's algorithm [21] for finding the longest increasing subsequence of a sequence of integers. The term 'Plactic monoid' has been coined by Lascoux & Schützenberger [17,18]. Their theory of the Plactic monoid became an important tool in various combinatorial contexts (see further [3,4,12]). For an analysis of monoids using advanced rewriting techniques, see [11,19]. A quantum perspective on the Plactic monoid has been discovered by Date, Jimbo and Miwa [5], 20 years after the equations have been suggested by Knuth. In 1995, Leclerc and Thibon [20] have obtained a quantum characterisation of the Plactic monoid, showing that the Plactic monoid can be interpreted as a maximal torus for the quantum group $U_q(gl(n, C))$. It is likely that the Chinese monoid plays a similar role for another quantum group (see further [5]).

Our Contribution

In this note we use decreasing diagrams, a technique from abstract rewriting theory, to establish confluence by tiling for the Chinese monoid. So for all words u, v there exist extensions x, y such that $ux = vy$.

Abstract rewriting theory is an initial part of term rewriting (see [8]) where the structure of the objects is disregarded. An abstract reduction system is just a set equipped with binary 'reduction' relations. A seminal result in abstract rewriting is the classical Newman's lemma that yields confluence (CR) as a consequence of termination (SN) together with local confluence (WCR or weak Church-Rosser).

Decreasing diagrams [6,8,10,22,23] is a method to prove CR that vastly improves Newman's Lemma, which is one its many corollaries. The technique employs a labelling of the steps with labels from a well-founded partial order in order to conclude confluence of the underlying unlabelled relation. Decreasing diagrams are complete for proving confluence of countable systems.[1] The challenge typically is finding a suitable (decreasing) labelling of the steps.

Somewhat surprisingly, for the Chinese monoid the natural labelling of the steps turns out to be immediately suitable for the application of decreasing diagrams. This is in contrast to the Plactic monoid and the Braid monoid. For $n > 2$ the usual monoid presentation for the Plactic monoid does not admit a straightforward confluence by tiling proof via decreasing diagrams. Yet, confluence by tiling may also be valid there, analogous to the case of braids. The well-known braid presentation does also not admit for its 'canonical' tiles an application of decreasing diagrams. Nevertheless confluence by tiling does hold there for its canonical tiles, and it can be proven using decreasing diagrams employing a more complex labelling [9].

[1] Already two labels suffice for proving confluence using decreasing diagrams of every countable abstract reduction system [10]. The completeness for uncountable systems remains a long-standing open problem [23]. For proving commutation (a generalisation of confluence involving two relations) decreasing diagrams are not complete, as established in [8].

Related Work

Cassaigne, Espie, Krob, Novelli and Hivert [4] study combinatorial properties of the Chinese monoid. In particular, they determine the size and structure of the convertibility classes. Karpuz [14] gives a complete rewriting system for convertibility, obtained by critical pair completion.

2 Reduction Diagrams for Monoids

In the theory of abstract rewriting systems (ARSs) we often apply the technique of constructing reduction diagrams by gluing together 'elementary diagrams' (e.d.'s) or 'tiles' to obtain a finite, completed reduction diagram of which the convergent sides yield the desired confluent reductions. Also in the theory of braids and Garside monoids this is an important tool, there called 'word reversing', see Dehornoy [7].

As a preparation for the main section of this note where we prove confluence for the Chinese monoid C_n on n generators, we introduce this method, somewhat informally, guided by a few examples of monoids. More complete expositions of reduction diagram construction can be found in [1,9,22]. In the latter paper the exposition is for positive braid words.

Example 1. Consider the monoid with two generators 1, 2 and relations

$$121 = 211 \qquad\qquad 221 = 212$$

This is actually the Chinese monoid C_2 on two generators, which coincides with the Plactic monoid P_2 on two generators. In Sects. 4 and 5 we will consider these monoids C_n and P_n in the general case with n generators.

Suppose we are interested in the *confluence* question for this monoid:

$$\forall u, v.\ \exists x, y.\ \ ux = vy\ \ ?$$

where u, v, x, y are elements of C_2, and '=' is the monoid equality. Actually, we work with u, v, x, y as words in $\{1, 2\}^*$, subject to the equality generated by the two relations above. In this way we are dealing with *string rewriting*, see Book & Otto [2]. To address the confluence question, we now invoke the technique of constructing reduction diagrams.

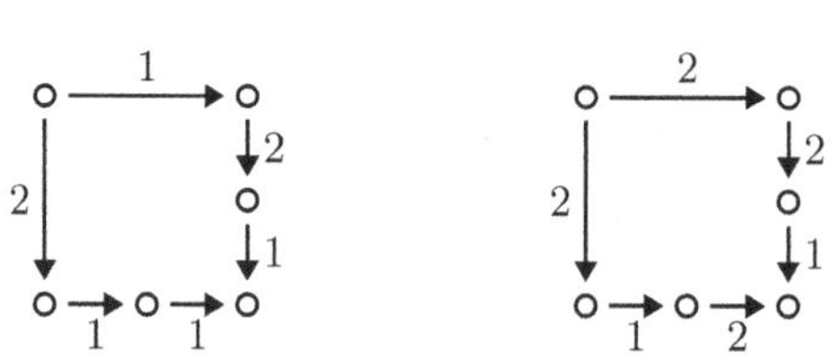

Fig. 1. Tiles for C_2 and P_2.

The relations 121 = 211 and 221 = 212 can be rendered as the elementary diagrams (e.d.'s) shown in Fig. 1. Of these two basic tiles we have an infinite supply of copies, which moreover are scalable, horizontally and vertically.

We can now address the question for common extensions x, y for e.g. $u = 12$ and $v = 222$ by constructing the diagram D_0 in Fig. 2.

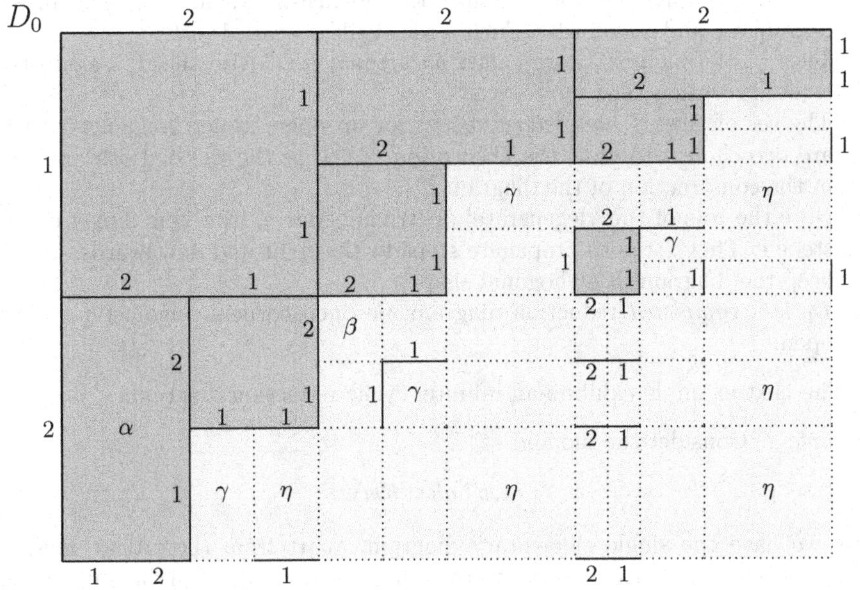

Fig. 2. Completed reduction diagram in C_2. Proper tiles are blue, trivial tiles involving empty steps are grey. (Color figure online)

The result of this diagram construction is

$$12\ 12121 = 222\ 1111$$

In fact, the completed diagram contains also the actual conversion between 1212121 and 2221111:

$$\begin{aligned}
&22\underline{211}11 =\\
&221\underline{211}1 =\\
&2211\underline{211} =\\
&2\underline{211}121 =\\
&21\underline{211}21 =\\
&\underline{211}2121 =\\
&1212121
\end{aligned}$$

This can be seen by traversing the diagram from the right-upper corner and flipping repeatedly an elementary tile. We suppressed the conversion steps corresponding to the trivial tiles with their empty sides; these steps would involve the unit element ϵ of the monoid and equations such as $1\epsilon = \epsilon 1$ and $\epsilon\epsilon = \epsilon\epsilon$.

This example diagram evokes some clarifying comments:

(i) Note that for every 'open corner' or 'peak' in a stage of the construction we have a matching tile, for 1-against-1, 1-against-2, 2-against-2, 1-against-ϵ, 2-against-ϵ and ϵ-against-ϵ. Such a set of tiles is called *full.*

(ii) Tiles γ, of 1-against-1, are called *absorption tiles.* Also tile β, 2-against-2, is an absorption tile.

(iii) The set of tiles is *non-deterministic*: for an open corner 2-against-2 there are two choices to glue, the absorption tile β, or the tile α. Both are used in the construction of the diagram D_0.

(iv) Note the role of the 'degenerate' or 'trivial' tiles η, involving 2 or 4 empty steps ϵ. They serve to propagate steps to the right and downwards, and to keep the diagram in orthogonal shape.

(v) D_0 is a *completed* reduction diagram; no open corners i-against-j are left open.

The next example exhibits an infinite, cyclic reduction diagram.

Example 2. Consider the monoid

$$\langle a, b \mid ab = bba \rangle$$

Here we have the single elementary diagram, apart from the trivial ones, as follows (see Fig. 3). This gives rise to the infinite cyclic reduction diagram, which is a well-known counterexample in abstract rewriting. It shows that in Newman's Lemma, mentioned above, see also [13], the condition SN cannot be missed.

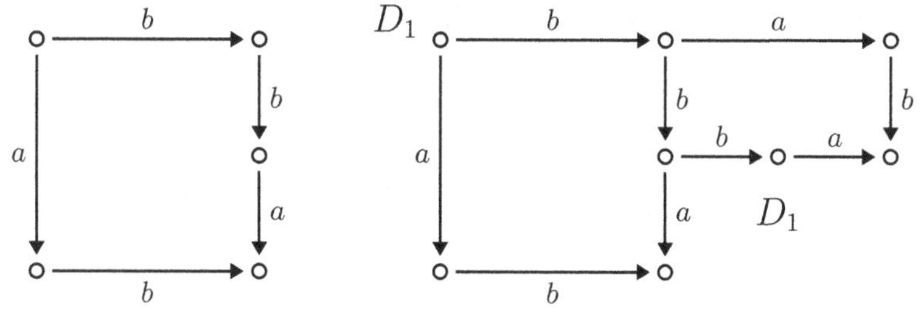

Fig. 3. Cyclic reduction diagram.

Example 3. A similar cyclic diagram construction (see Fig. 4) is exhibited by the monoid

$$\langle a, b \mid aba = bba \rangle$$

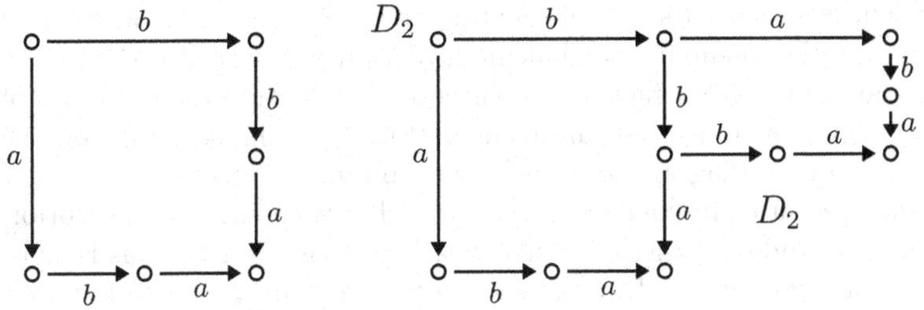

Fig. 4. Another cyclic reduction diagram.

Example 4. An interesting cyclic diagram construction (see Fig. 5) is found in the Artin-Tits monoid, as mentioned in [7]:

$$\langle 1,2,3 \mid 121 = 212,\ 232 = 323,\ 131, 313\rangle$$

In fact, this monoid describes braids on three strands that are placed on a cylinder.

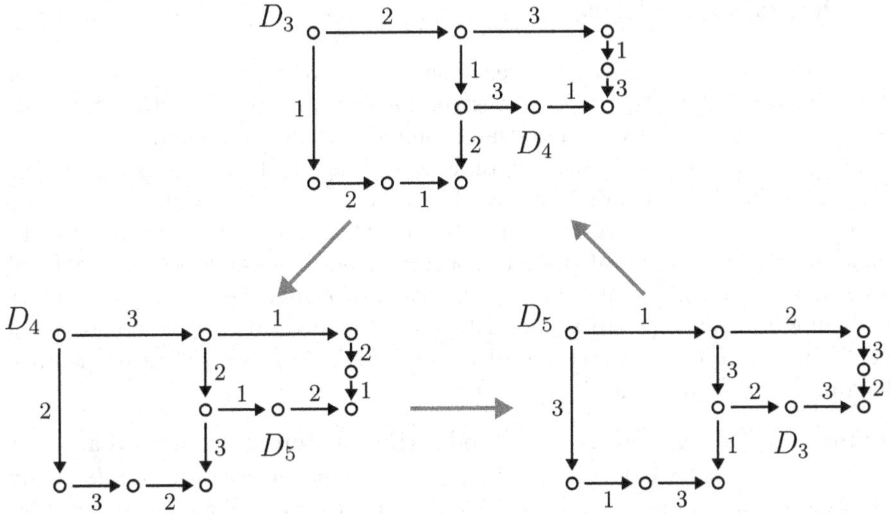

Fig. 5. Cyclic diagram in the Artin-Tits monoid.

Example 5. Consider the monoid

$$\langle 1,2 \mid 121 = 211,\ 121 = 221\rangle$$

This presentation with its two elementary diagrams (in addition to the trivial ones) exhibits the phenomenon that already a proper part of the set of elementary diagrams may be sufficient for successful reduction diagram completion, while the whole set may admit an unsuccessful, diverging diagram construction, by yielding a cyclic diagram or an otherwise infinite diagram.

For the present example the left e.d., together with the two absorption e.d.'s for a-against-a and b-against-b is sufficient; as we will see later this is so because it is a decreasing diagram. But the e.d. on the right may lead to infinite, cyclic diagrams as witnessed by Example 3.

This leads us to define:

Defination 6. Let $\mathcal{T}$ be a set of elementary diagrams:

1. $\mathcal{T}$ is called *sufficient for confluence by tiling*, for short *sufficient*, if for every pair of finite reductions σ set against τ, *some* gluing sequence leads to a completed reduction diagram $D(\sigma, \tau)$ using tiles from $\mathcal{T}$.
2. $\mathcal{T}$ is called *strongly sufficient for confluence by tiling*, for short *strongly sufficient*, if for every pair of finite reductions σ set against τ, *every* glueing sequence leads eventually to a completed reduction diagram $D(\sigma, \tau)$ using tiles from $\mathcal{T}$.

3 Decreasing Diagrams

In the preceding section we have seen that, when we are lucky, confluence can be obtained by tiling, that is, the construction of completed reduction diagrams by repeatedly gluing a tile to a partially completed reduction diagram. However, sometimes this procedure fails. The process of tiling can be infinite without ever completing the reduction diagram, as the cyclic diagrams showed.

The decreasing diagrams technique [6,8,10,22,23] is one of the strongest techniques for guaranteeing that tiling will succeed (terminate and yield a completed reduction diagram). To guarantee termination of tiling, the technique employs a labelling on the steps (together with a well-founded ordering $<$ on the label set I), and it restricts the choice of tiles for gluing to 'decreasing elementary diagrams' as shown in Fig. 6.

Defination 7 (Indexed abstract reduction system). *An* indexed abstract reduction system (ARS) $\mathcal{A} = \langle A, \{\to_i\}_{i \in I}\rangle$ *consists of a set of objects A, an index set I, and a binary relation $\to_i \subseteq A \times A$ for every $i \in I$. We write $\to$ for the union $\bigcup_{i \in I} \to_i$ of all the reduction relations.*

Notation 8. *For $< \subseteq I \times I$, we define $\to_{<\alpha} = \bigcup_{\beta<\alpha} \to_\beta$ and moreover $\to_{<\alpha \cup <\beta} = \to_{<\alpha} \cup \to_{<\beta}$.*

Defination 9 (Decreasing elementary diagrams). *Let $\mathcal{A} = \langle A, \{\to_i\}_{i \in I}\rangle$ be an ARS and $< \subseteq I \times I$ a well-founded partial order. A reduction diagram of the form shown in Fig. 6 is called a* decreasing elementary diagram *for the peak $c \leftarrow_\beta a \to_\alpha b$.*

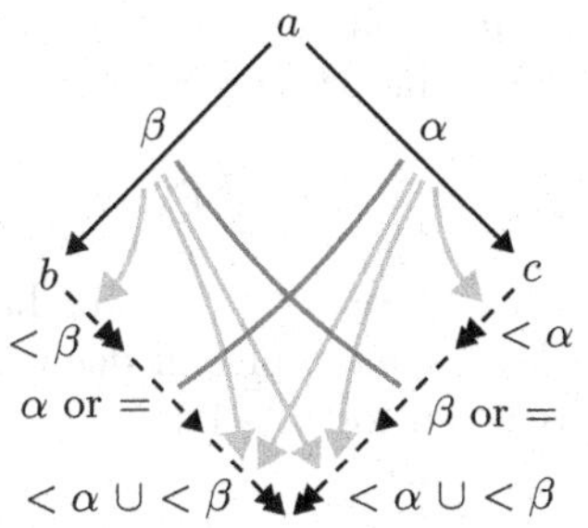

Fig. 6. The red arrows and green lines inside the diagram are intended as a visual aid. The red arrows stand for a strict decrease of the labels (multiple incoming arrows signify choice) while the green lines indicate a label carrying over unchanged. The double-headed arrows $\twoheadrightarrow$ are the transitive-reflexive closure of the one-step reduction relation $\rightarrow$. (Color figure online)

Theorem 10 (Decreasing diagrams [6,23]**).** *Let $\mathcal{A} = \langle A, \{\rightarrow_i\}_{i \in I}\rangle$ be an ARS and $< \subseteq I \times I$ a well-founded partial order. Let T be a set of decreasing elementary diagrams that contains at least one elementary diagram for each peak in $\mathcal{A}$. Then T is strongly sufficient for confluence by tiling of $\mathcal{A}$.*

Example 11. The following is a list of the relations between ternary words over the alphabet $\{1, 2\}$ whose corresponding elementary diagrams are decreasing, with respect to the ordering $1 < 2$:

$$211 = 111 \qquad 121 = 211 \qquad 212 = 212$$
$$212 = 211 \qquad 221 = 211 \qquad 221 = 221$$
$$221 = 212 \qquad 212 = 112$$

Note the last two symmetrical equations with identical sides; in a presentation they would be useless, but they do give rise to decreasing tiles.

Also note the sensitivity with respect to the chosen ordering: for the ordering $2 < 1$ none of the tiles corresponding to the equations as listed below would be decreasing.

Finally, note that equations $211 = 112$ and $121 = 221$, obtained from the listed ones by transitivity, are not decreasing (for $1 < 2$).

Example 12. Earlier we considered the presentation of the monoid

$$\langle a, b \mid ab = bba\rangle$$

called in [7, p. 73, Example 4.28], the Baumslag-Solitar monoid, having a cyclic diverging diagram for its canonical tiles, which is in different notation also included there. The corresponding tile, left in Fig. 4, is therefore not decreasing, for no order on $\{a, b\}$.

A *caveat* may be in order here: at other places in the literature Baumslag-Solitar monoids are said to have the presentation

$$\langle a, b \mid ab = ba^k \rangle$$

Note that the latter presentations for all k, do *not* have a diverging reduction diagram, as the canonical tiles are examples of decreasing diagrams, for $b > a$. (For $a > b$ they are not decreasing, for $k > 1$.)

4 Confluence of the Chinese Monoid

Let Σ be an alphabet equipped with a total order $<$. Let $= \subseteq \Sigma^* \times \Sigma^*$ be the congruence generated by

$$aba = baa, \; bba = bab$$

for every $a < b$, and

$$cab = cba = bca$$

for every $a < b < c$.

So, '=' is an equivalence relation (reflexive, symmetric and transitive) and $lxr = lyr$ whenever $x = y$ is one of the equations above and $l, r \in \Sigma^*$.

Notation 13. For a word $x \in \Sigma^*$, we write $x^=$ for $\{ y \in \Sigma^* \mid x = y \}$, the equivalence class of x. For a set $X \subseteq \Sigma^*$ of words, let $X^= = \{ x^= \mid x \in X \}$.

The (right) word extension in the Chinese monoid can be viewed as an abstract reduction system as follows.

Defination 14. *The* (right) word extension ARS C *for the Chinese monoid over* $\langle \Sigma, < \rangle$ *is*

$$C = \langle \Sigma^{*=}, \{\to_i\}_{i \in \Sigma} \rangle$$

where, for every $i \in \Sigma$, *the relation* $\to_i$ *is defined by*

$$w^= \to_i (wi)^=$$

for every $w \in \Sigma^*$.

The ARS C has elementary diagrams as given in Definition 15. These diagrams arise naturally from the defining equations above of the Chinese monoid.

Defination 15 (Elementary diagrams for the Chinese monoid). *For the Chinese monoid over a totally ordered alphabet, we have the following elementary diagrams*

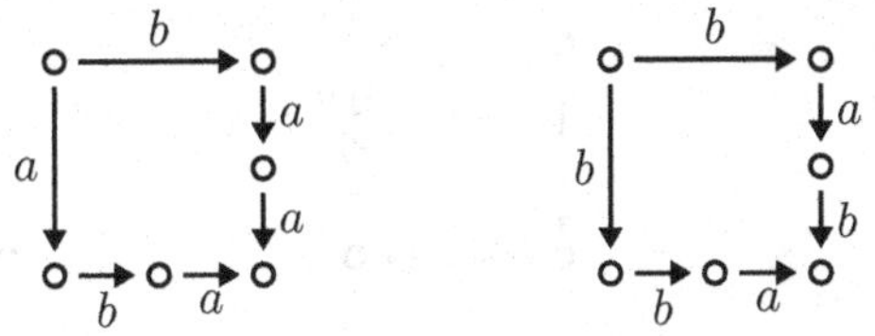

for every $a < b$, and

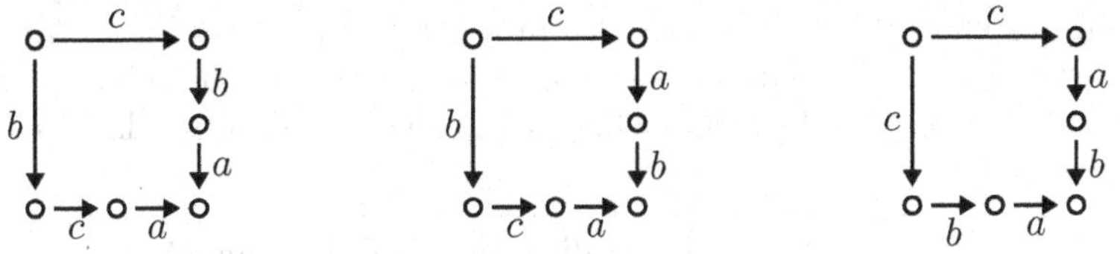

for every $a < b < c$, and trivial elementary diagrams

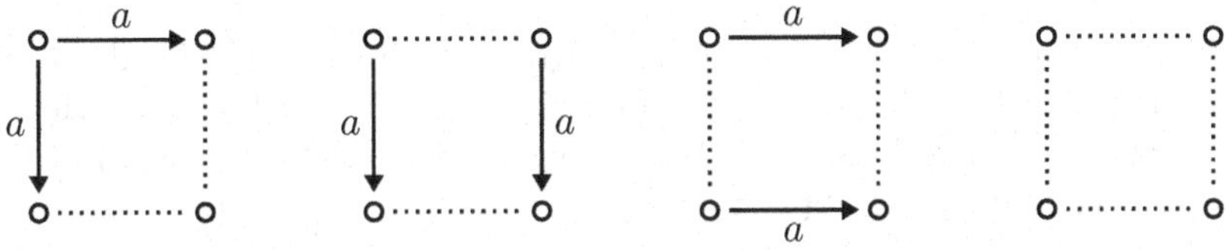

for every a. Here the dotted lines without arrowhead stand for empty steps.

The application of decreasing diagrams for proving confluence typically requires

(a) a careful choice of the labelling of the steps, and
(b) a careful choice of the elementary diagrams (if there are multiple ways to join a peak).

Somewhat surprisingly, for the Chinese monoid, neither of these steps is necessary. It turns out that the natural labelling of the steps with letters from Σ suffices to make all elementary diagrams in Definition 15 decreasing.

Proposition 16. *All elementary diagrams in Definition 15 are decreasing elementary diagrams with respect to the order $<$ on Σ.*

Proof. The trivial elementary diagrams are decreasing for every ordering on the labels. So it suffices to consider the non-trivial elementary diagrams. These are:

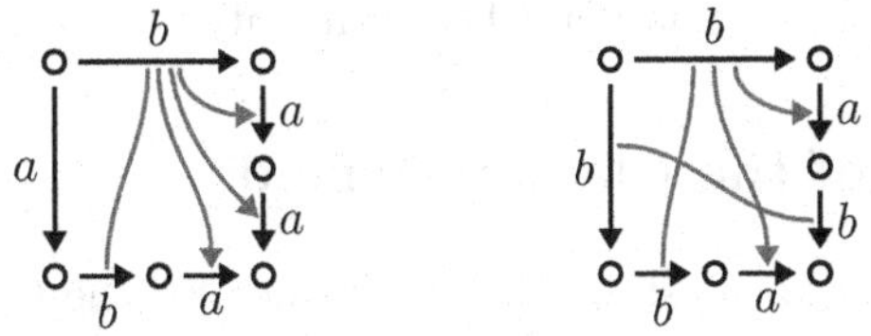

for every $a < b$, and

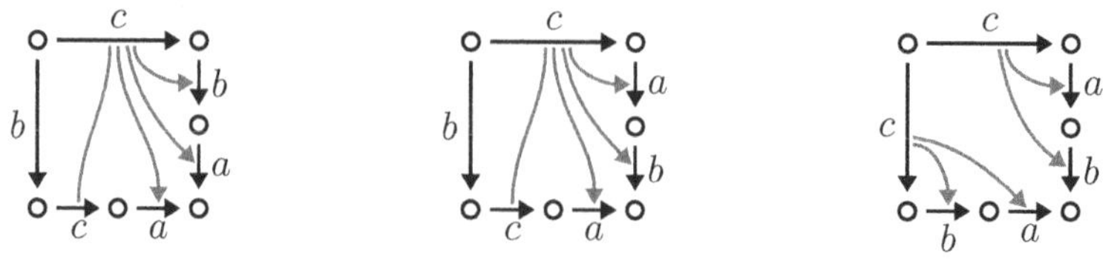

for every $a < b < c$.

All five of these configurations are allowed by the shape of decreasing elementary diagrams shown in Fig. 6. Just like in Fig. 6 we have used red arrows and green lines inside the elementary diagrams as a visual aid. The red arrows indicate a strict decrease of the label, while the green lines signify a label carrying over unchanged.

Theorem 17. *The set of elementary diagrams in Definition 15 is strongly sufficient for confluence by tiling for C (i.e., for confluence of right word extension in the Chinese monoid).*

Proof. By Proposition 16 all elementary diagrams in Definition 15 are decreasing. Moreover, this set of elementary diagrams is exhaustive in the sense that, for every peak in C, there exists a matching elementary diagram: it is a full set of tiles. Thus Theorem 10 is applicable.

Actually, the two diagrams for $a < b$ in Fig. 7 would suffice for establishing confluence. However, the two corresponding equations are not suitable for the intended monoid presentation, as they do not generate the whole equality.

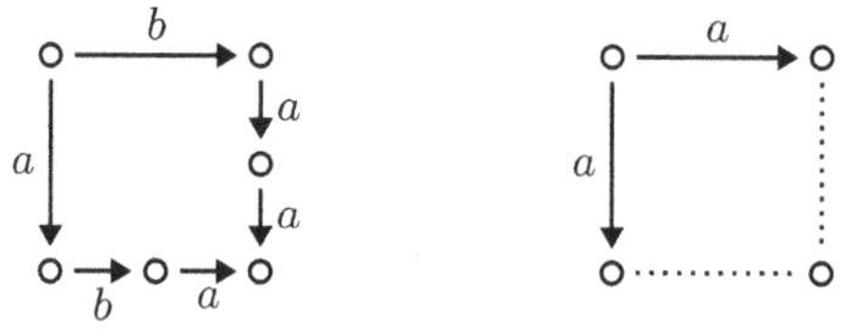

Fig. 7. Subset of tiles sufficient for confluence by tiling for C_n.

The point of Theorem 17 is not merely the confluence property. The crucial observation is that tiling of reduction diagrams always succeeds with the tiles from Definition 15 independent of the gluing strategy. As all these diagrams are decreasing, tiling is always guaranteed to terminate.

5 Confluence of the Plactic Monoid

In the last section we have established confluence for the Chinese monoid C_n. We have as a corollary also the confluence of the Plactic monoid P_n, somewhat trivially. We show this for the case $n = 3$; the general case then follows easily.

Consider the table in Fig. 8. For the Chinese monoid C_3 we have the 8 relations in the left column which gives (together with the trivial tiles involving 2 or 4 empty sides) a nondeterministic, full set of tiles, all of which are decreasing (blue). The grouping shows the nondeterministic choices that are possible when gluing the tiles together towards a completed diagram.

$211 = 121$	$211 = 121$	$(\star)$
	$231 = 213$	
$221 = 212$	$221 = 212$	
$321 = 231$		
$312 = 231$		
$322 = 232$	$322 = 232$	$(\star)$
	$233 = 323$	
$311 = 131$	$311 = 131$	$(\star)$
	$132 = 312$	
$332 = 323$	$332 = 323$	
$331 = 313$	$331 = 313$	

Fig. 8. Relations of C_3 (left) and P_3 (right). Blue is decreasing for the natural ordering, red non-decreasing. Common relations are underlined. The three starred ones are already sufficient for confluence of P_3, they correspond to the two tiles in Fig. 7, but do not generate the whole equality of P_3. (Color figure online)

For the Plactic monoid P_3 we have the 9 relations as in the second column of Fig. 8, where blue is decreasing and red non-decreasing.

The 6 underlined equations in both columns are the intersection between the ones of C_3 and P_3. This intersection corresponds to a full set of decreasing tiles, also nondeterministic. It follows that this set is also for P_3 a sufficient set of tiles. Hence also P_3 is confluent.

The same remark as above for C_n applies: already the three $(\star)$ equations shared with C_n together with the two absorption tiles for 1-against-1 and 2-against-2 are sufficient for confluence. But note that this trio of equations does not generate the whole equality in C_3, nor in P_3.

An interesting question is whether the set of tiles for P_3 is also *strongly* sufficient. We have not been able to find a diverging reduction diagram for P_3; we conjecture that it does not exist, so that the set of tiles is strongly sufficient.

6 Conclusions and Further Questions

1. We have applied the technique of constructing reduction diagrams, confluence by tiling, and decreasing diagrams to the important Chinese and Plactic monoid, obtaining confluence in the sense of having common extensions of elements of these monoids, in analogy to the braids monoid. For braids there is moreover an equivalence between convertibility of braid words and Lévy's *projection equivalence*. For C_n and P_n this is more complicated because their sets of tiles are nondeterministic, admitting choices in the diagram completion. Therefore the notion of 'projection' is not defined unequivocally.
2. A question is whether the full set of tiles for the Plactic monoid is also strongly sufficient, as is the case for the Chinese monoid.
3. An important question is how the various notions such as confluence, confluence by tiling are dependent on the actual presentation of monoids. These presentations can be varied by applying Tietze moves. Some notions are known to be 'absolute' in this respect, they hold for every presentation, and thus are properties of the monoid and not merely of the monoid presentation. Having *finite derivation type* is such an important absolute property (see [15].) We expect that confluence is also an absolute property. For confluence by tiling the absoluteness is also an interesting question.
4. For the braids monoid an interesting fact is that the usual presentation can be via Tietze moves transformed to consist of *short* relations, sometimes called *Ore-conditions*. These are of the form $a = bc$ or $ab = cd$. The corresponding tiles are then *non-splitting*, i.e. have converging sides consisting of a single step or an empty step ϵ. Confluence by tiling is then trivial, as all tiles are simple squares.
   We wonder whether also the Chinese and Plactic monoid possess such a presentation with only short relations.
5. Another question is whether the confluence property for monoid presentations is *decidable*; and the same for the stronger property of confluence by tiling.
6. The infinite diagrams arising from some monoid presentations such as the Baumslag-Solitar monoid in Fig. 3 and the Artin-Tits monoid in Fig. 5 are intriguing objects themselves. The examples of infinite reduction diagrams seen above are all *cyclic*, involving a proper copy of themselves. Two questions arise: are there also *non-cyclic infinite reduction diagrams* arising from the tiles of monoids?
   A second question concerns the *finite and infinite traces* that arise in infinite diagrams starting at the root of the diagram, the left-upper corner. Are the sets of infinite traces arising from infinite diagrams for monoid presentations, as seems to be the case in the examples considered above, always *ω-regular languages*? We offer this puzzle happily to Catuscia!

Applications of confluence by tiling, possibly combined with an application of decreasing diagrams, certainly does not stop with the Chinese monoid. Various other monoids can be subjected to an analysis with these two tools, confluence by tiling and decreasing diagrams. We hope to demonstrate this in subsequent work.

References

1. Bezem, M., Klop, J.W., van Oostrom, V.: Diagram techniques for confluence. Inform. Comput. **141**(2), 172–204 (1998)
2. Book, R.V., Otto, F.: String-Rewriting Systems. Springer, Heidelberg (1993)
3. Cain, A.J., Gray, R.D., Malheiro, A.: Finite Gröbner-Shirshov bases for Plactic algebras and biautomatic structures for Plactic monoids. J. Algebra **423**, 37–53 (2015)
4. Cassaigne, J., Espie, M., Krob, D., Novelli, J.C., Hivert, F.: The Chinese monoid. IJAC **11**(3), 301–334 (2001)
5. Date, E., Jimbo, M., Miwa, T.: Representations of $U_q(gl(n, C))$ at $q = 0$ and the Robinson-Schensted correspondence. World Sci. 185–211 (1990)
6. de Bruijn, N.G.: A note on weak diamond properties. Memorandum 78–08, Eindhoven Uninversity of Technology (1978)
7. Dehornoy, P., Digne, F., Godelle, E., Krammer, D., Michel, J.: Foundations of Garside Theory. Tracts in Mathematics, vol. 22. European Mathematical Society, Zürich (2015)
8. Endrullis, J., Klop, J.W.: De Bruijn's weak diamond property revisited. Indagat. Math. **24**(4), 1050–1072 (2013). In memory of N.G. (Dick) de Bruijn (1918–2012)
9. Endrullis, J., Klop, J.W.: Braids via term rewriting. Theor. Comp. Sci. **777**, 260–295 (2019)
10. Endrullis, J., Klop, J.W., Overbeek, R.: Decreasing diagrams with two labels are complete for confluence of countable systems. In: Proceedings of the Conference on Formal Structures for Computation and Deduction (FSCD 2018), Leibniz International Proceedings in Informatics (LIPIcs), vol. 108, pp. 14:1–14:15. Schloss Dagstuhl-Leibniz-Zentrum fuer Informatik (2018)
11. Guiraud, Y., Malbos, P., Mimram, S.: A homotopical completion procedure with applications to coherence of monoids. In: Proceedings of Conference on Rewriting Techniques and Applications (RTA 2013), LIPIcs, vol. 21, pp. 223–238. Schloss Dagstuhl - Leibniz-Zentrum fuer Informatik (2013)
12. Hage, N.: Finite convergent presentation of plactic monoid for type C. Int. J. Algebra Comput. **25**(8), 1239–1264 (2015)
13. Huet, G.: Confluent reductions: abstract properties and applications to term rewriting systems. J. ACM (JACM) **27**(4), 797–821 (1980)
14. Karpuz, E.G.: Complete rewriting system for the Chinese monoid. Appl. Math. Sci. **4**(22), 1081–1087 (2010)
15. Karpuz, E.G.: Finite derivation type property on the Chinese monoid. Appl. Math. Sci. **4**, 1073–1080 (2010)
16. Knuth, D.E.: Permutations, matrices, and generalized young tableaux. Pacific J. Math. **34**(3), 709–727 (1970)
17. Lascoux, A., Leclerc, B., Thibon, J.: The plactic monoid. In: Algebraic Combinatorics on Words . Cambridge University Press (2001)
18. Lascoux, A., Schützenberger, M.P.: Le monoïde plaxique. Non-Commutative Struct. Algebra Geom. Combin. **109**, 129–156 (1981)
19. Lebed, V.: Plactic monoids: a braided approach. CoRR, abs/1612.05768 (2016)
20. Leclerc, B., Thibon, J.Y.: The robinson-schensted correspondence as the quantum straightening at q = 0. Electron. J. Combin. **3**(2), R11 (1996)
21. Schensted, C.: Longest increasing and decreasing subsequences. Can. J. Math. **13**, 179–191 (1961)

22. Terese: Term Rewriting Systems. Cambridge Tracts in Theoretical Computer Science, vol. 55. Cambridge University Press, Cambridge (2003)
23. van Oostrom, V.: Confluence by decreasing diagrams. Theoret. Comput. Sci. **126**(2), 259–280 (1994)

Logic and Constraint Programming

A Coalgebraic Approach to Unification Semantics of Logic Programming

Roberto Bruni[1(✉)], Ugo Montanari[1], and Giorgio Mossa[2]

[1] Dipartimento di Informatica, Università di Pisa, Pisa, Italy
bruni@di.unipi.it

[2] Dipartimento di Matematica, Università di Pisa, Pisa, Italy

Abstract. In the version of logic programming (LP) based on interpretations where variables occur in atoms, a goal reduction via unification can be seen as a transition labelled by the most general unifier. Categorically, it is thus natural to model a logic program as a coalgebra. In the paper we represent: (i) goals as the substitutive monoid freely generated by the predicate symbols; (ii) the LTS as the structured coalgebra defined by the SOS rules implicit in the LP semantics; (iii) the bisimulation semantics of a logic program as its image on the final coalgebra.

1 Introduction

Logic programming is a paradigm based on first order Horn clauses and SLD resolution. Its fundamental ingredient is to be equipped both with an operational semantics, based on goal reduction via resolution, and with a declarative semantics. The second is defined both in terms of satisfaction (à la Tarski) of the clauses on (standard) interpretations of the Herbrand base (minimal model), and in terms of the least fix point of a transformation on interpretations.

The three approaches are proved equivalent, thus allowing for the famous paradigm Algorithm = Logic + Control. We refer to [18] for the programming motivation and to [22] for the underlying theory.

In the classical version, all the semantics are given as *ground* interpretations, i.e. as sets of *true* atoms without variables. However, it is possible to extend the basic approach to interpretations and refutations containing also atoms with variables. Then also the meaning of a goal changes: it considers also refutations of non-ground atoms. It is defined as the set of all the answer substitutions (possibly with variables) computed by its refutations. It is easy to see that the corresponding operational semantics is more informative, namely finer.

For instance, the goal $P(x)$ is assigned the same semantics in the case of ground interpretations for the two programs

$$\{P(x)\,\text{:-}\,\Box\} \text{ and } \{P(x)\,\text{:-}\,\Box\ ,\ P(a)\,\text{:-}\,\Box\}$$

Research supported by the MIUR PRINs 201784YSZ5 *ASPRA: Analysis of program analyses* and by University of Pisa PRA_2018_66 *DECLWARE: Metodologie dichiarative per la progettazione e il deployment di applicazioni.*

M. S. Alvim et al. (Eds.): Palamidessi Festschrift, LNCS 11760, pp. 223–240, 2019.
https://doi.org/10.1007/978-3-030-31175-9_13

namely $\{(t/x) \mid t \text{ is a term}\}$ (typically an infinite set), while the semantics are different when atoms with variables are considered, namely $\{(y/x)\}$ (a singleton) and $\{(y/x), (a/x)\}$ (a two-element set) respectively. Here answer substitution (y, x) is defined up to variable renaming $\simeq$, e.g. $(y/x) \simeq (z/x)$.

While from a logical point of view the classical semantics is satisfactory, since it includes all the (ground) logical consequences of the clauses, the extended semantics is more convenient from a programming point of view: for instance, for the query $P(x)$ the former program cannot select any specific answer, while the latter does. Thus the two programs should be regarded as different. As for the classical version, also for the extended theory there are declarative semantics based on suitable notions of S-interpretation, S-minimal model and S-transformation which correspond to the extended operational semantics [12].

It is easy to see that the extended operational semantics can be seen as a labelled transition system (LTS), where states are goals, transitions are goal reductions and labels are the substitutions computed by unification in the corresponding reductions. The computed answer substitution corresponding to a path is obtained by composing the substitutions of its transitions, while the semantics of a goal G is the set of substitutions, projected on the variables of G, computed by all the paths from G to the empty goal. In addition, both states and transitions of the LTS have a simple algebraic structure.

The clean structure we outlined here has inspired several authors to take advantage of the universal constructions of category theory to embed operational non-ground logic programming into other well-studied mathematical structures. A general approach for equipping transition systems with an algebraic semantics has been applied to logic programming in the seminal papers [8,10]. Generalizations to richer structures have been studied in [2,13]. Coalgebras turn out to be particularly fit [4,5,15–17], as it should be expected since they have been particularly successful in modelling LTS with all kinds of structures. The meta model of tile systems has been adopted in [7]. An important design point is to choose the category where the coalgebra lives. If the coalgebra can be lifted from category **Set** to a category of algebras (obtaining a structured coalgebra/bialgebra) then arrows become at the same time bisimulations and homomorphisms, thus guaranteeing that bisimilarity is a congruence.

A critical point about categorical LTS for logic programming is about correctly modelling unification, in particular with the co/bialgebraic approach. The universal property of most general unifiers (mgu's) is essential: while restriction to mgu's is not necessary, since any unifier will derive correct answers, the number of unifiers is often infinite, and unification matches completeness and efficiency. Most general unifiers are perfectly representable in category theory: they arise from the pushout construction in the category of substitutions. However to match unification with the homomorphism requirements of coalgebras is difficult. To the authors' knowledge, only [7] and [4] succeed in this task.

In this paper, we define a structured coalgebra for operational non-ground logic programming, which correctly considers only reductions with mgu's. Our construction works as follows. First, we define goals as the algebra of substitutive monoids (SM) freely generated by the set of predicate symbols. The algebraic specification of SM is obtained as the tensor product of the specifications of monoids and of substitutions. The latter specification has natural numbers as sorts (the number of variables) and contains all the substitutions as unary operations, with axioms for substitution composition and unity. The tensor product construction automatically inserts all the exchange axioms, e.g. between the monoidal operation (i.e. the logical conjunction) and substitutions. Second, as base category for the coalgebra we choose the functor category of SM-algebras, but without axioms. Third, we model the LTS as the structured coalgebra defined by the SOS rules implicit in the logic programming semantics. Finally, given a logic program, its image in the final coalgebra yields its bisimulation semantics.

We emphasize the simplicity of our construction for assigning a natural algebraic structure to goals: the ordinary logic representation as conjunction of atomic goals becomes the standard form of the terms of our initial algebra, equipped "automatically" by exchange axioms.

The above construction makes sure that the structural coalgebra we defined fully corresponds to the operational semantics of non-ground logic programming. However the situation is different when we consider abstract semantics. In fact, for logic programming, as mentioned above, the semantics of a goal G is the set of substitutions, projected on the variables of G, computed by all the paths from G to the empty goal. This situation corresponds to a language semantics of LTS (the set of labels of the paths to final states), with the additional difference that the monoid of the labels (the subsitutions) is not free. In the case of coalgebras, the abstract semantics of a LTS is represented by its image on the terminal coalgebra, where all *bisimilar* states are identified. It is easy to see that the categorical, bisimilar abstract semantics is finer than the non-ground semantics.

We are not aware of coalgebraic approaches which yield the classical non-ground semantics. On the other hand, bisimilarity semantics may be natural and convenient when considering LP-based process description languages. In this context, we are not interested in logic computations as refutations of goals for problem solving or artificial intelligence, but we consider LP as a goal rewriting mechanism. We consider logic subgoals as concurrent communicating processes that evolve according to the rules defined by the clauses and that use unification as the fundamental interaction primitive. A presentation of this kind of use of logic programming can be found in [7,19].

Structure of the Paper. In Sect. 2 we fix the notation and recall the operational semantics of logic programming (SLD derivation) together with some preliminaries about coalgebraic representations of labelled transition systems. In Sect. 3 we characterize the algebra of goals as the initial model for the theory of substitutive monoids. In Sect. 4 we define the coalgebra for SLD derivation and prove that it admits a terminal object. Some final remarks are in Sect. 5.

2 Background

In the next subsections we will recall the basic notions of logic programming and structured coalgebras that we will need.

2.1 Operational Semantics with Non-ground atoms à la Levi-Palamidessi-Falaschi-Martelli

Definition 1 (Signatures). *A logic signature is a pair of* sets of symbols (Σ, Π) *and an* arity function $\mathrm{ar}\colon \Sigma + \Pi \to \mathbb{N}$ *associating with each* symbol *in* Σ *and* Π *a natural number, called* its arity. *The symbols in* Σ *are the operation symbols and the symbols in* Π *the predicates.*

The set of operation symbols Σ *with the relative restriction of the arity function form what is called an algebraic signature.*

In general, given an algebraic signature Σ and a set of variables X, we will denote by $T_\Sigma(X)$ the set of terms with variables in X and by $T_\Sigma = T_\Sigma(\emptyset)$ the set of ground terms.

Definition 2 (Atoms, goals). Atomic formulas, *or* atoms *for short, are expressions of the form* $P(t_1, \ldots, t_n)$ *where* $P \in \Pi$ *is a predicate symbol with arity* n*, and the* t_i*'s are just terms in the set* $T_\Sigma(X)$*, for some set variables* X*. A* goal G *is a finite conjunction of atomic formulas over the same set of variables.*

In what follows we will identify goals with finite lists of atoms, so we will write $G \equiv A_1 \ldots A_k$ for the goal made of the conjunction of the atoms A_i's. We will use the symbol $\Box$ to denote the *empty conjunction* of atoms, the *empty goal.*

Definition 3 (Substitutions). *Given two sets* X *and* Y *a substitution from (terms over variables in)* X *to (terms over variables in)* Y *is a function of the form* $\sigma\colon X \to T_\Sigma(Y)$.

Remark 1. These functions have a natural action on terms: for every $\sigma\colon X \to T_\Sigma(Y)$ and every term $t \in T_\Sigma(X)$ we have the term $\sigma(t) \in T_\Sigma(Y)$ obtained by replacing the variable occurrences in t with their images via σ. Similarly, for each atom A and each goal G in the set of variables X we have the atom and goal $\sigma(A)$ and $\sigma(G)$, in the set of variables Y. These are obtained respectively by A and G replacing the occurrences of the variables with their images via σ.

Substitutions form the morphisms of a category whose objects are the sets of variables. Given a pair of substitutions $\sigma\colon X \to T_\Sigma(Y)$ and $\tau\colon Y \to T_\Sigma(Z)$ their composition is the substitution $\tau \circ \sigma\colon X \to T_\Sigma(Z)$ defined as the function sending every variable $x \in X$ into the term $\tau(\sigma(x))$, the result of applying τ to the term $\sigma(x)$. Identity substitutions are given by the functions $\mathrm{id}_X : X \to T_\Sigma(X)$ sending each variable in itself, seen as an element of $T_\Sigma(X)$.

Definition 4 (Substitutions preorder). *Given two substitutions* σ *and* σ' *we say that* σ is more general than σ' *if there is a substitution* τ *such that* $\sigma' = \tau \circ \sigma$.

Table 1. Inference rules for SLD-resolution.

$$\frac{H :\!-\, B \in \mathbb{P} \quad \sigma = \mathrm{mgu}(A, \rho(H))}{\mathbb{P} \models A \Rightarrow_\sigma \sigma(\rho(B))} \quad \text{Atomic goal}$$

where ρ is a variable renaming such that $\rho(H)$ and $\rho(B)$ have no variable in common with A

$$\frac{\mathbb{P} \models G \Rightarrow_\sigma F}{\mathbb{P} \models G', G \Rightarrow_\sigma \sigma(G'), F} \qquad \frac{\mathbb{P} \models G \Rightarrow_\sigma F}{\mathbb{P} \models G, G' \Rightarrow_\sigma F, \sigma(G')} \quad \text{Conjunctive goal}$$

where the goal G' and F have no variable in common

This relation induces a preorder on the substitutions.

Definition 5 (Most general unifier I). *Given two terms/atoms/goals t_1 and t_2 over the same set of variables,* a unifier for them *is a substitution σ such that $\sigma(t_1) = \sigma(t_2)$. A* most general unifier for t_1 and t_2 *is a unifier that is most general in the sense of Definition 4.*

Remark 2. The *mgu*'s are unique up isomorphism, meaning that if σ and σ' are two mgu's for the same terms there is a unique substitution γ such that $\sigma' = \gamma \circ \sigma$ and this γ is an isomorphism in the above mentioned category of substitutions.

Next we introduce logic programs and their operational semantics.

Definition 6 (Horn clauses, logic programs). *A* Horn clause *is an expression of the form $H :\!-\, B$ where H is an atom and B is a goal called respectively the* head *and the* body *of the clause. Without loss of generality we assume the head and body are formulas over the same set of variables. A finite set of Horn clauses is called a* logic program.

A Horn clause is basically a formula stating that the head is a logical consequence of the body, i.e., that every valuation satisfying the body satisfies the head too. In particular any clause whose body is the empty goal $\square$ is a formula stating that the head is satisfied by all valuations.

Logic programs are able to express computations; they do so via the (operational) SLD-reduction semantics. This semantics is described in Table 1 via inference rules that describe the transitions $G \Rightarrow_\sigma F$ of a *labelled transition system* whose states are goals and whose labels are substitutions.

Each sequent of the form $\mathbb{P} \models G \Rightarrow_\sigma F$ asserts that the program $\mathbb{P}$ produces a computation step from a state G to a state F with an observation σ. These computation steps represent the classical behavior of logic program systems, in which at each point of a computation the system selects an atom from the current goal, it tries to unify the selected atom with the head of a clause in the program (up to a renaming with fresh names), and then it updates the current goal, replacing the atom with the body of the clause and then applying the computed unifier.

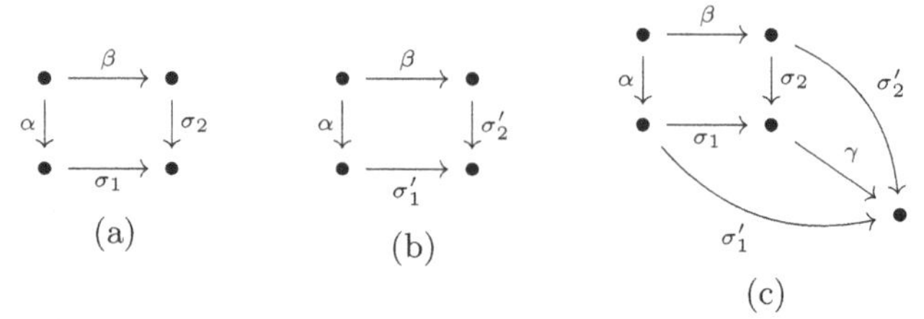

Fig. 1. Mgu's via pushout

From a logic point of view the sequent $\mathbb{P} \models G \Rightarrow_\sigma F$ states that every substitution γ which makes the goal $\gamma(F)$ a consequence of the axioms in $\mathbb{P}$ makes also $\gamma(\sigma(G))$ a consequence of $\mathbb{P}$.

We stress the fact that in logic programming we are always interested in unifying goals and clauses with distinct names, which is the reason for the renaming in the *atomic goal rule* in Table 1. It is possible to avoid this variable renaming changing the definitions of unifiers and mgu's.

Definition 7 (Most general unifier II). *Given a pair of terms/atoms/goals t_1 and t_2, a* unifier *for them is a pair of substitutions (σ_1, σ_2) such that $\sigma_1(t_1) = \sigma_2(t_2)$. Given two unifiers $\sigma = (\sigma_1, \sigma_2)$ and $\sigma' = (\sigma'_1, \sigma'_2)$ we say that σ is* more general *than σ' if there is a substitution γ such that $\sigma'_i = \gamma \circ \sigma_i$ for $i = 1, 2$. As before, the relation of being more general induces a preorder on the unifiers. A* most general unifier *(mgu) for the terms t_1 and t_2 is a unifier which is most general among all the possible unifiers.*

The notion of unifier and mgu admit a nice categorical description.

Remark 3. Every atom $P(t_1, \ldots, t_n)$ can be written in the form $\sigma_t(P(x_1, \ldots, x_n))$ where $\{x_1, \ldots, x_n\}$ is a canonical set of variables, $\sigma_t \colon \{x_1, \ldots, x_n\} \to T_\Sigma(X)$ is the unique substitution such that $\sigma_t(x_i) = t_i$ for each i (where X is the set of variables over which $P(t_1, \ldots, t_n)$ is defined).

Remark 3 provides a canonical decomposition of an atom as the result of a substitution to a *canonical predicate*, namely $P(x_1, \ldots, x_n)$. As a matter of notation, in what follows we write just P for $P(x_1, \ldots, x_n)$.

Remark 4 (Mgu's via pushout). Two atoms $\alpha(P)$ and $\beta(Q)$ unify only when $P = Q$, and in this case a unifier for them is a pair of substitutions $\sigma = (\sigma_1, \sigma_2)$ such that $\sigma_1 \circ \alpha = \sigma_2 \circ \beta$, i.e. such that the square in Fig. 1a commutes in the category of substitutions. Such a square is an mgu if for any other commuting square of the form in Fig. 1b there is a, necessarily unique, γ which makes commute the diagram in Fig. 1c. This is equivalent to say that $(\alpha, \beta, \sigma_1, \sigma_2)$ is a pushout square. So, unifiers are commutative squares in the form above and mgu's are those squares that are pushouts.

Using this characterization of unification via pushouts we can modify the *Atomic goal* rule of Table 1 with the following rule:

$$\frac{\alpha(P) :\text{-} B \in \mathbb{P} \quad (\alpha, \beta, \sigma_1, \sigma_2) \text{ is a pushout}}{\mathbb{P} \models \beta(P) \Rightarrow_{\sigma_2} \sigma_1(B)}.$$

In this way the LTS is not changed, but there is no need of creating new variables.

2.2 LTSs as Coalgebras

Definition 8 (F-coalgebra). *Given a category* $\mathbf{C}$ *and an endofunctor* $F \colon \mathbf{C} \to \mathbf{C}$ *a coalgebra for the functor* F, *or* F-*coalgebra for short, is a pair* $\langle X, \alpha \rangle$ *with* X *an object of* $\mathbf{C}$ *and* $\alpha \colon X \to F(X)$. *Given two coalgebras* $\langle X, \alpha \rangle$ *and* $\langle Y, \beta \rangle$ *a cohomomorphism* h *from* $\langle X, \alpha \rangle$ *to* $\langle Y, \beta \rangle$, *written as* $h \colon \langle X, \alpha \rangle \to \langle Y, \beta \rangle$, *is given by a morphism* $h \colon X \to Y$ *such that the following diagram commutes.*

$$
\begin{array}{ccc}
X & \xrightarrow{\ \ h\ \ } & Y \\
{\scriptstyle \alpha}\big\downarrow & & \big\downarrow{\scriptstyle \beta} \\
F(X) & \xrightarrow[F(h)]{} & F(Y)
\end{array}
$$

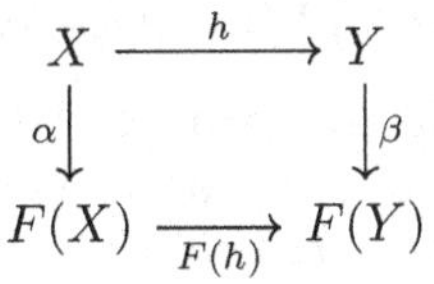

F-coalgebras and relative cohomomorphisms form the category $\mathbf{Coalg}(F)$.

Coalgebras provide an elegant way to encode different notions of dynamic systems in a categorical framework. As an example we can consider the case of LTSs. We recall that a labeled transition system is a triple $\langle S, L, \longrightarrow \rangle$ where S is a set of states, L is a set of labels, and $\longrightarrow \subset S \times L \times S$ is a transition relation. Every LTS $\langle S, L, \to \rangle$ can be regarded as a coalgebra for the endofunctor $\mathcal{P}_L \colon \mathbf{Set} \to \mathbf{Set}$ defined on objects as $\mathcal{P}_L(X) = \mathcal{P}(L \times X)$ where $\mathcal{P}$ is the *powerset functor*. If $\langle S, L, \longrightarrow \rangle$ is an LTS, the relation $\longrightarrow \subseteq S \times L \times S$ gives the function $p \colon S \to \mathcal{P}_L(S)$ that sends every $s \in S$ into the set $p(s) = \left\{ (l,t) \mid s \xrightarrow{\ l\ } t \right\}$.

One of the most interesting thing about coalgebras is that they give an abstract semantics for dynamic systems in term of final objects: if $\langle X, h \rangle$ is a coalgebra, we say that two states, i.e. two elements of X, are *bisimilar* if they have the same image via the unique cohomomorphism $t \colon \langle X, h \rangle \to \langle T, \tau \rangle$ into a final coalgebra $\langle T, \tau \rangle$. This definition of *bisimilarity* generalizes the classical one for transition systems, meaning that two states are bisimilar in the classical sense if and only if they are bisimilar in the coalgebraic sense.

The advantage of coalgebras over classical LTS is that the states' space now can be an object of a generic category, not just a set. Hence coalgebras allow to work with states that have an additional structure. This justifies the name *structured coalgebras* for coalgebras over categories of structures, in particular we will be interested in coalgebras over categories of algebras.

3 Algebraic Structures

In this section we introduce an algebraic structure over the goals of logic programming. The main result is to provide a characterization of this *algebra of goals* as the initial model for *the theory of substitutive monoids*. This theory is many-sorted and its sorts represents the number of variables that can occur in a goal. Each sort is equipped with a binary operation and a constant, representing the conjunction and the empty goal respectively, and there are unary typed operations that correspond to substitution operations of logic programming.

Instead of presenting the theory of substitutive monoids directly, we first recall the *theory of monoids*, then, we introduce the *theory of substitutions*, and finally we take the tensor product of these theories and add some constants. This way the different operations of substitutive monoids are presented to the reader in a gradual way and the distributive axioms between the two algebras are introduced automatically by the tensor product construction.

3.1 Monoids, Substitutions and Substitutive Monoids

Definition 9 (The theory of monoids). *We let* $\Gamma_{\mathbf{Mon}} = (S_{\mathbf{Mon}}, \Sigma_{\mathbf{Mon}}, E_{\mathbf{Mon}})$ *be the* algebraic theory of monoids *having a unique sort* M*, a binary operation* $\cdot\colon M \times M \to M$ *and a constant* $\square$*. The axioms of* $E_{\mathbf{Mon}}$ *are the following*

$$x \cdot (y \cdot z) = (x \cdot y) \cdot z \textit{ for all } x, y, z \textit{ (associativity);}$$
$$x \cdot \square = \square \cdot x = x \textit{ for all } x \textit{ (unit).}$$

Algebras of this theory are monoids, i.e. algebraic structures with a binary associative operation $\cdot$ whose unit is $\square$.

The monoidal structure captures the algebraic operation of goal conjunction. Indeed, for every set of variables X, the goals having free variables in X form a monoid with conjunction and the unit of the monoid is the empty goal.

Definition 10 (The theory of substitutions). *Given a signature* Σ*, we let* $\Gamma_{\mathbf{Sub}} = (S_{\mathbf{Sub}}, \Sigma_{\mathbf{Sub}}, E_{\mathbf{Sub}})$ *be the* algebraic theory of substitutions over Σ. *The set of sorts is the countable set* $S_{\mathbf{Sub}} = \{\underline{n}\colon n \in \mathbb{N}\}$*. For every substitution* $\sigma\colon \{x_1, \ldots, x_n\} \to T_\Sigma(\{x_1, \ldots, x_m\})$ *we have an operation symbol* $\underline{\sigma} \in \Sigma_{\mathbf{Sub}}$ *with arity* $\underline{\sigma}\colon \underline{n} \to \underline{m}$ *and we have the following axioms*

$$\underline{\tau}(\underline{\sigma}(x)) = \underline{\tau \circ \sigma}(x) \textit{ for all } \sigma\colon \underline{n} \to \underline{m} \textit{ and } \tau\colon \underline{m} \to \underline{k} \textit{ and any } x;$$
$$\underline{\mathrm{id}_n}(x) = x \textit{ for any identity substitution } \mathrm{id}_n\colon \underline{n} \to \underline{n} \textit{ and any } x.$$

The algebras for this theory are rather simple, they are basically **Set**-valued functors (not necessarily cartesian ones) from the *(opposite of the) Lawvere theory over the signature* Σ, that is functors from the category of finite sets of variables and substitutions between them, in the sense of logic.

While the monoidal operation captures goal conjunction, the substitution operations capture the operation of variable instantiation, which is fundamental for logic programming. Again the goals provide examples of this substitutive structure in which the operations of $\Gamma_{\mathbf{Sub}}$ are interpreted with their corresponding substitutions.

As seen above, goals mix together monoidal and substitutive structures, but these structures interact with each other in a specific way. Their interaction gives rise to a different algebraic structure which is captured by the so called *tensor product of the two theories.*

Using the machinery of algebraic theories we can build a new theory as the tensor product of $\Gamma_{\mathbf{Mon}}$ and $\Gamma_{\mathbf{Sub}}$, which we call the *theory of substitutive monoids* and denote by $\Gamma_{\mathbf{SM}}$. We will not describe the construction in details (see e.g. [14]), instead we describe the resulting theory.

Definition 11 (The theory of substitutive monoids). *The* theory of substitutive monoids $\Gamma_{\mathbf{SM}} = (S_{\mathbf{SM}}, \Sigma_{\mathbf{SM}}, E_{\mathbf{SM}})$ *has sorts those of* $\Gamma_{\mathbf{Sub}}$*, that is* $S_{\mathbf{SM}} = S_{\mathbf{Sub}}$*, each operation* $\underline{\sigma}\colon \underline{n} \to \underline{m}$ *in* $\Sigma_{\mathbf{Sub}}$ *is also an operation of* $\Sigma_{\mathbf{SM}}$ *with the same arity and for every sort* $\underline{n} \in S_{\mathbf{SM}}$ *we have a binary typed operation* $\cdot_{\underline{n}}\colon \underline{n} \times \underline{n} \to \underline{n}$ *and a constant* $\square_{\underline{n}}\colon \underline{n}$*. The axioms in* $E_{\mathbf{SM}}$ *contain those in* $E_{\mathbf{Sub}}$*, with the addition, for every sort* $\underline{n}$*, of the following monoid axioms*

$$x \cdot_{\underline{n}} (y \cdot_{\underline{n}} z) = (x \cdot_{\underline{n}} y) \cdot_{\underline{n}} z \qquad\qquad x \cdot_{\underline{n}} \square_{\underline{n}} = \square_{\underline{n}} \cdot_{\underline{n}} x = x$$

and of axioms of the form

$$\underline{\sigma}(x \cdot_{\underline{n}} y) = \underline{\sigma}(x) \cdot_{\underline{m}} \underline{\sigma}(y) \qquad\qquad \underline{\sigma}(\square_{\underline{n}}) = \square_{\underline{m}}$$

for every substitution operation $\underline{\sigma}\colon \underline{n} \to \underline{m}$ *in* $\Sigma_{\mathbf{SM}} \cap \Sigma_{\mathbf{Sub}}$.

We call the $\Gamma_{\mathbf{SM}}$-algebras substitutive monoids or also **SM**-algebras. An algebra for $\Gamma_{\mathbf{SM}}$ is basically a countable family of monoids (parametrized by the sorts) with a family of monoid homomorphisms between them (parametrized by the substitutions), whence the name substitutive monoids. Another way to see these algebras is as functors from the above mentioned category of substitutions into the category of monoids. Goals provide algebras for this algebraic theory, the sorts are interpreted by sets of goals with fixed-finite sets of variables, the monoidal operations are given by conjunction operations and finally the substitution operations are given by the variable substitutions.

Definition 12 (The theory of Π-substitutive monoids). *The* theory of Π-substitutive monoids $\Gamma_{\Pi\text{-}\mathbf{SM}} = (S_{\Pi\text{-}\mathbf{SM}}, \Sigma_{\Pi\text{-}\mathbf{SM}}, E_{\Pi\text{-}\mathbf{SM}})$ *has sorts and equations as those of* $\Gamma_{\mathbf{SM}}$ *but the operation symbols are given by* $\Sigma_{\Pi\text{-}\mathbf{SM}} = \Sigma_{\mathbf{SM}} \cup \Pi$*, that is they are those of* $\Sigma_{\mathbf{SM}}$ *plus the predicates of the logic signature. Each predicate* $P \in \Pi$ *with arity* n *is interpreted as a constant symbol of type* $P\colon \underline{n}$*. The other operators, inherited from* $\Sigma_{\mathbf{SM}}$*, keep the signature they have in* $\Gamma_{\mathbf{SM}}$*.*

The algebras of $\Gamma_{\Pi\text{-}\mathbf{SM}}$ are substitutive monoids with selected constants parametrized by predicates. In the next section we focus on the *most important* Π - **SM**-*algebra*, the algebra of goals: this will show how $\Gamma_{\Pi\text{-}\mathbf{SM}}$ is actually *the theory of goals*, meaning that it characterizes the algebraic structure of goals.

3.2 The Goal Algebra as the Initial Π - SM-algebra

Definition 13 (The goal algebra). *Goals form a Π - **SM**-algebra $\mathbb{G}$ such that*

- *$\mathbb{G}$ interprets sort $\underline{n} \in S_{\Pi\text{-}\mathbf{SM}}$ into the set $\mathbb{G}_{\underline{n}}$ of the goals with free variables in the canonical set $\{x_1, \ldots, x_n\}$;*
- *each predicate symbol $P \colon \underline{n}$ is interpreted in the atomic formula*

$$P^{\mathbb{G}} = P(x_1, \ldots, x_n);$$

- *each substitution symbol $\underline{\sigma} \colon \underline{n} \to \underline{m}$ is interpreted in the corresponding substitution operation $\underline{\sigma}^{\mathbb{G}} = \sigma \colon \mathbb{G}_{\underline{n}} \to \mathbb{G}_{\underline{m}}$;*
- *each monoid operation $\cdot_{\underline{n}}$ is interpreted into goal conjunction, i.e. $\cdot_{\underline{n}}^{\mathbb{G}} = \wedge \colon \mathbb{G}_{\underline{n}} \times \mathbb{G}_{\underline{n}} \to \mathbb{G}_{\underline{n}}$;*
- *each $\square_{\underline{n}}$ is interpreted into the empty goal in $\mathbb{G}_n$.*

It is easy to prove that these data satisfy the axiom of $\Gamma_{\Pi\text{-}\mathbf{SM}}$ hence that $\mathbb{G}$ is indeed as Π - **SM**-algebra.

Remark 5. It is not hard to see that every atomic formula $P(t_1, \ldots, t_n)$ can be uniquely represented as the element $\underline{\sigma_t}^{\mathbb{G}}(P^{\mathbb{G}})$, where σ_t is the unique substitution sending the variable x_i into the term t_i (see Remark 3).

In the same way, if we have a goal G in the form $G = A_1 \wedge \cdots \wedge A_k$ with $A_i = P_i(t^i_1, \ldots, t^i_{n_i})$ we have the representation

$$G = A_1^{\mathbb{G}} \cdot_{\underline{n}}^{\mathbb{G}} (\ldots (A_{k-1}^{\mathbb{G}} \cdot_{\underline{n}}^{\mathbb{G}} A_k^{\mathbb{G}}) \ldots) ,$$

where we put $A_i^{\mathbb{G}} = \underline{\sigma_{t^i}}^{\mathbb{G}}(P_i^{\mathbb{G}})$.

This representation allows us to express any goal as a canonical term in the language of $\Gamma_{\Pi\text{-}\mathbf{SM}}$, more specifically as a term in the form

$$\underline{\sigma_1}(P_1) \cdot_{\underline{n}} (\cdots \cdot_{\underline{n}} (\underline{\sigma_{k-1}}(P_{k-1}) \cdot_{\underline{n}} \underline{\sigma_k}(P_k)) \ldots) .$$

In particular, the atomic formulas of the form $P(x_1, \ldots, x_n)$ can be expressed via the term $\underline{\mathrm{id}_n}(P)$. We will exploit this representation in what follows.

The *algebra of goals* has an important characterization.

Theorem 1 (Initiality of $\mathbb{G}$). *The algebra $\mathbb{G}$ is the* initial Π - **SM**-algebra.

Proof. We prove that for any other Π - **SM**-algebra A there is a unique homomorphism $f_A \colon \mathbb{G} \to A$.

We start observing that for every goal $G \in \mathbb{G}_{\underline{n}}$ in the form

$$G = \underline{\sigma_1}(P_1) \cdot_{\underline{n}} \cdots \cdot_{\underline{n}} \underline{\sigma_k}(P_k)$$

every homomorphism $f \colon \mathbb{G} \to A$ should satisfy the following equation

$$f(G) = \underline{\sigma_1}^A(P_1^A) \cdot_{\underline{n}}^A \cdots \cdot_{\underline{n}}^A \underline{\sigma_k}^A(P_k^A) .$$

So the only possible homomorphism is given by the family of maps $f_{\underline{n}}^A \colon \mathbb{G}_{\underline{n}} \to A_{\underline{n}}$ defined by the equation

$$f_{\underline{n}}^A(\underline{\sigma_1}(P_1) \cdot_{\underline{n}} \cdots \cdot_{\underline{n}} \underline{\sigma_k}(P_k)) = \underline{\sigma_1}^A(P_1^A) \cdot_{\underline{n}}^A \cdots \cdot_{\underline{n}}^A \underline{\sigma_k}^A(P_k^A) .$$

By calculations it is easy to prove that this indeed is an homomorphism. $\square$

Table 2. SOS-rules for $\Sigma_{\Pi\text{-SM}}$ algebras.

$$\frac{\sigma(P) :\text{-}\, B \in \mathbb{P} \quad \gamma \in \mathbf{Th}(\Sigma)[m,m] \text{ is an isomorphism}}{P \xrightarrow{\gamma\circ\sigma} \underline{\gamma}(B)} \quad \text{(constant-rule)}$$

where $\sigma \in \mathbf{Th}(\Sigma)[n,m]$ is a substitution, $P \in \Pi$ has arity n and $B \in (T_{\Sigma_{\Pi\text{-SM}}})_{\underline{m}}$

$$\frac{G \xrightarrow{\sigma} B \quad (\tau,\sigma,\sigma',\tau') \text{ is a pushout}}{\underline{\tau}(G) \xrightarrow{\sigma'} \underline{\tau'}(B)} \quad \text{(substitution-rule)}$$

$$\frac{G \xrightarrow{\sigma} B}{G \cdot G' \xrightarrow{\sigma} B \cdot \sigma(G')} \qquad \frac{G \xrightarrow{\sigma} B}{G' \cdot G \xrightarrow{\sigma} \sigma(G') \cdot B} \quad \text{(monoid-rule)}$$

where G and G' are terms of the same type

4 Coalgebraic Semantics

In this section we introduce a structured coalgebra that provides the operational semantics for logic programs. To this aim we proceed as follows: First we provide a set of SOS rules which describe how to generate the transitions of the semantics and then using the abstract machinery of structured coalgebras, developed by Plotkin and Turi [25] (see also [9]), we show how the transitions form a structured coalgebra over the algebra of goals $\mathbb{G}$. In order to do that we need to prove that the axioms of $\Gamma_{\Sigma_{\Pi\text{-SM}}}$ bisimulate; this is an important property that we have to prove, because our state-space is an algebra satisfying some axioms and not just a syntactic (term) algebra. Next we prove that our SOS-rules generate the transitions of the classical operational semantics for logic programs. Finally we prove the existence of a terminal coalgebra: this result allows us to use the coalgebraic-bisimulation semantics described in Sect. 2.

4.1 SOS Rules and Coalgebras

In what follows we work in the category $\mathbf{Alg}(\Sigma_{\Pi\text{-SM}})$ of $\Sigma_{\Pi\text{-SM}}$-algebra, that is those algebras for the algebraic theory obtained by $\Gamma_{\Pi\text{-SM}}$ dropping the axioms. Working in this larger category allows us to reuse the results of Plotkin-Turi for automatically generate endofunctors and coalgebras from an SOS specification. This machinery does not generalize well to categories of algebras satisfying axioms, hence the choice to drop the axioms.

Table 2 provides a set of SOS rules. These rules are in *De Simone format* [11], whence the following result.

Proposition 1. *The SOS-rules in Table 2 induce a functor* $\mathbf{B}^{\mathbb{P}}\colon \mathbf{Alg}(\Sigma_{\Pi\text{-SM}}) \to \mathbf{Alg}(\Sigma_{\Pi\text{-SM}})$ *and a coalgebra* $p\colon T_{\Sigma_{\Pi\text{-SM}}} \to \mathbf{B}^{\mathbb{P}}(T_{\Sigma_{\Pi\text{-SM}}})$.

The functor $\mathbf{B}^{\mathbb{P}}$ associates with each $\Sigma_{\Pi\text{-}\mathbf{SM}}$-algebra A the algebra $\mathbf{B}^{\mathbb{P}}(A)$ that interprets each sort $\underline{n}$ with the set

$$\mathbf{B}^{\mathbb{P}}(A)_{\underline{n}} = \mathcal{P}_f\left(\left(\coprod_m \mathbf{Th}(\Sigma)[n,m] \times A_{\underline{m}}\right) \amalg A_{\underline{n}}\right)$$

whose elements are sets of substitutions' labeled transitions (i.e. pairs of the form $(\sigma, a) \in \coprod_m \mathbf{Th}(\Sigma)[n,m] \times A_{\underline{m}}$) and unlabeled (idle) transitions.

The coalgebra p associates with each term t in $T_{\Sigma_{\Pi\text{-}\mathbf{SM}}}$, the initial $\Sigma_{\Pi\text{-}\mathbf{SM}}$-algebra, the set

$$p(t) = \left\{(\sigma, t') : t \xrightarrow{\sigma} t' \text{ is a derivable sequent}\right\} \cup \{t\}\ .$$

The goal algebra $\mathbb{G}$ is a $\Sigma_{\Pi\text{-}\mathbf{SM}}$-algebra and so it has a natural (unique) homomorphism $\pi\colon T_{\Sigma_{\Pi\text{-}\mathbf{SM}}} \to \mathbb{G}$ from the initial algebra $T_{\Sigma_{\Pi\text{-}\mathbf{SM}}}$. Since $\mathbb{G}$ is the initial $\Gamma_{\Sigma_{\Pi\text{-}\mathbf{SM}}}$-algebra (as shown in Theorem 1), $\mathbb{G}$ is obtained from $T_{\Sigma_{\Pi\text{-}\mathbf{SM}}}$ quotienting for the axioms in $\Gamma_{\Sigma_{\Pi\text{-}\mathbf{SM}}}$ and π is a surjective homomorphism.

Since $\pi\colon T_{\Sigma_{\Pi\text{-}\mathbf{SM}}} \to \mathbb{G}$ is a surjective homomorphism, there can be at most only one $p'\colon \mathbb{G} \to \mathbf{B}^{\mathbb{P}}(\mathbb{G})$ such that the diagram below commutes

$$\begin{array}{ccc} T_{\Sigma_{\Pi\text{-}\mathbf{SM}}} & \xrightarrow{\pi} & \mathbb{G} \\ {\scriptstyle p}\downarrow & & \downarrow{\scriptstyle p'} \\ \mathbf{B}^{\mathbb{P}}(T_{\Sigma_{\Pi\text{-}\mathbf{SM}}}) & \xrightarrow[\mathbf{B}^{\mathbb{P}}(\pi)]{} & \mathbf{B}^{\mathbb{P}}(\mathbb{G}) \end{array}$$

and such p' exists if and only if the morphism $\mathbf{B}^{\mathbb{P}}(\pi) \circ p$ respects the axioms of $\Gamma_{\Sigma_{\Pi\text{-}\mathbf{SM}}}$, that is if and only if for every equation $t = t'$ derivable from the axioms of $\Gamma_{\Sigma_{\Pi\text{-}\mathbf{SM}}}$ the sets $\mathbf{B}^{\mathbb{P}}(p(t))$ and $\mathbf{B}^{\mathbb{P}}(p'(t'))$ are equal.

By the definitions of the functor $\mathbf{B}^{\mathbb{P}}$ and the homomorphism p, this amounts to prove that for every $\Gamma_{\Sigma_{\Pi\text{-}\mathbf{SM}}}$-derivable equation $t = t'$ and every $s \in T_{\Sigma_{\Pi\text{-}\mathbf{SM}}}$ such that $t \xrightarrow{\sigma} s$ is derivable from the rules in Table 2, there is a s' such that $t' \xrightarrow{\sigma} s'$ is also derivable and $s = s'$ is an equation provable from the axioms in $\Gamma_{\Sigma_{\Pi\text{-}\mathbf{SM}}}$, and the symmetric property holds for every transition $t' \xrightarrow{\gamma} s'$. We say that an equation $t = t'$ *bisimulates* if it satisfies this property.

Theorem 2. *Let Γ be an algebraic theory and let R be a set of SOS rules defined over the signature of Γ. If every closed instance of the axioms in Γ bisimulates, then every Γ-derivable equation bisimulates as well.*

Proof. Since every Γ-derivable closed equation can be derived by reflexivity, symmetry, transitivity, and congruence by the axioms, we can prove the thesis by induction on these inference rules.

The proof is straightforward for all the rules with the exception of *congruence*, which is the rule requiring that SOS-rules are in *De Simone Format* [11], that is why we will focus only on this case.

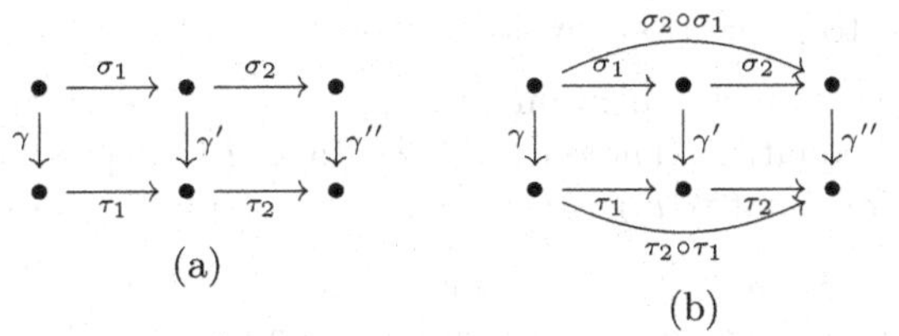

Fig. 2. Pushout (de)composition

We recall that a rule is in *De Simone Format* if it is in the format

$$\frac{\{x_i \xrightarrow{\gamma_i} y_i\}_{i\in I}}{o(x_1,\dots,x_n) \xrightarrow{\gamma} g(y_1,\dots,y_n)} \tag{1}$$

where o is an operation symbol of the signature, $g(y_1,\dots,y_n)$ is a term containing only the variables y_i's, and the variables x_i's and y_i's are all distinct with the exception of the pairs (x_i, y_i) for the $i \notin I$, in this case we have $x_i \equiv y_i$.

Let assume we have a family of derivable equations $t_i = s_i$ that bisimulate. We want to prove that the equation $o(t_1,\dots,t_n) = o(s_1,\dots,s_n)$, derived by congruence-rule, bisimulates too.

Let $o(t_1,\dots,t_n) \xrightarrow{\gamma} u$ be a transition derived by the SOS-rules. We may assume that such transition is derived by a rule as the one in formula (1), so there must exist a family of terms u_i such that we have derivable transitions of the form $t_i \xrightarrow{\gamma_i} u_i$, for the $i \in I$, $t_i \equiv u_i$ for the $i \notin I$, and with $u \equiv g(u_1,\dots,u_n)$.

By the hypothesis that the $t_i = s_i$ bisimulate we can conclude that for every index $i \in I$ there must be a $\bar{u}_i$ such that $s_i \xrightarrow{\gamma_i} \bar{u}_i$, and by letting $\bar{u}_i \equiv s_i$ for $i \notin I$, applying the De Simone rule, we get a transition $o(s_1,\dots,s_n) \xrightarrow{\gamma} g(\bar{u}_1,\dots,\bar{u}_n)$.

By theorems of equational logic, since the equalities $u_i = \bar{u}_i$ are derivable, it follows that the equality

$$u \equiv g(u_1,\dots,u_n) = g(\bar{u}_1,\dots,\bar{u}_n)$$

is derivable. So by letting $\bar{u} \equiv g(\bar{u}_1,\dots,\bar{u}_n)$ we have found that for the transition $o(t_1,\dots,t_n) \xrightarrow{\gamma} u$ there is a transition $o(s_1,\dots,s_n) \xrightarrow{\gamma} \bar{u}$ with $u = \bar{u}$ derivable.

By a symmetric argument we can prove the bisimulation property for transitions of the form $o(s_1,\dots,s_n) \xrightarrow{\gamma} \bar{u}$.

As stated in the beginning of the proof, by induction on the inference rules of equational logic it follows that every derivable equation bisimulates. □

Proposition 2 (Axioms in $\Gamma_{\Sigma_\Pi\text{-SM}}$ bisimulate). *The axioms of $\Gamma_{\Sigma_\Pi\text{-SM}}$ bisimulate with respect to the SOS rules of Table 2.*

Proof. For each instance of the axioms in $\Gamma_{\Sigma_\Pi\text{-SM}}$ one can prove by induction on the inference rules (of the SOS specification in Table 2) the thesis. As an interesting case we consider the substitution axiom $(\underline{\sigma_2}(\underline{\sigma_1}(t)) = (\underline{\sigma_2 \circ \sigma_1})(t))$.

First, we want to prove that for every closed term t and for every transition $\underline{\sigma_2}(\underline{\sigma_1}(t)) \xrightarrow{\gamma''} s''$ there is a transition $(\underline{\sigma_2 \circ \sigma_1})(t) \xrightarrow{\gamma''} \bar{s}''$ such that $s'' = \bar{s}''$ is a $\Gamma_{\Sigma_{\Pi\text{-}\mathbf{SM}}}$-derivable equation. The only rule in Table 2 that produces a transition for terms of the form $\underline{\sigma_2}(\underline{\sigma_1}(t))$ is the substitution-rule, hence we may assume that there is a transition $\underline{\sigma_1}(t) \xrightarrow{\gamma'} s'$ and a pushout square $(\gamma', \sigma_2, \tau_2, \gamma'')$ such that $s'' \equiv \underline{\tau_2}(s')$. For the same reason we may assume that there is a transition $t \xrightarrow{\gamma} s$ and a pushout square $(\gamma, \sigma_1, \tau_1, \gamma')$ such that $s' \equiv \underline{\tau_1}(s)$. We have that, for the diagram in Fig. 2a the two internal squares are pushouts, hence the external rectangle $(\gamma, \sigma_2 \circ \sigma_1, \tau_2 \circ \tau_1, \gamma'')$ is a pushout as well. By the substitution rule we have the transition $(\underline{\sigma_2 \circ \sigma_1})(t) \xrightarrow{\gamma''} (\underline{\tau_2 \circ \tau_1})(s)$, and by the substitution axiom $(\underline{\tau_2 \circ \tau_1})(s) = \underline{\tau_2}(\underline{\tau_1}(s))$, so letting $\bar{s}'' \equiv (\underline{\tau_2 \circ \tau_1})(s)$ we get our claim.

Second, for any transition $(\underline{\sigma_2 \circ \sigma_1})(t) \xrightarrow{\gamma''} s''$ we want to prove that there is a transition $\underline{\sigma_2}(\underline{\sigma_1}(t)) \xrightarrow{\gamma''} \bar{s}''$ such that $s'' = \bar{s}''$ is a derivable equation. Indeed every transition for a term of the form $(\underline{\sigma_2 \circ \sigma_1})(t)$ can be only derived by the substitution rule, hence we may assume that there is a transition $t \xrightarrow{\gamma} s$ and a pushout square $(\gamma, \sigma_2 \circ \sigma_1, \gamma'', \tau)$ such that $s'' \equiv \tau(s)$. It is well known that the pushout square can be decomposed in two pushouts as shown in Fig. 2b, where $\tau \equiv \tau_2 \circ \tau_1$. It follows that $\underline{\sigma_1}(t) \xrightarrow{\gamma'} \underline{\tau_1}(s)$ and $\underline{\sigma_2}(\sigma_1(t)) \xrightarrow{\gamma''} \underline{\tau_2}(\underline{\tau_1}(s))$. By the substitution axiom we also have $\underline{\tau_2}(\underline{\tau_1}(s)) = (\underline{\tau_2 \circ \tau_1})(s)$. □

Combining Theorem 2 with Proposition 2 we get:

Corollary 1 (Equalities in $\Gamma_{\Sigma_{\Pi\text{-}\mathbf{SM}}}$ bisimulate). *All $\Gamma_{\Sigma_{\Pi\text{-}\mathbf{SM}}}$-derivable equations bisimulate.*

From the above discussion it follows that:

Theorem 3 (Coalgebraic semantics of $\mathbb{G}$). *There is a unique $\mathbf{B}^{\mathbb{P}}$-coalgebra $p' \colon \mathbb{G} \to \mathbf{B}^{\mathbb{P}}(\mathbb{G})$ that makes the diagram below commute:*

$$
\begin{array}{ccc}
T_{\Sigma_{\Pi\text{-}\mathbf{SM}}} & \xrightarrow{\quad \pi \quad} & \mathbb{G} \\
{\scriptstyle p}\big\downarrow & & \big\downarrow{\scriptstyle p'} \\
\mathbf{B}^{\mathbb{P}}(T_{\Sigma_{\Pi\text{-}\mathbf{SM}}}) & \xrightarrow[\mathbf{B}^{\mathbb{P}}(\pi)]{} & \mathbf{B}^{\mathbb{P}}(\mathbb{G})
\end{array}
$$

Remark 6. The coalgebra $p' \colon \mathbb{G} \to \mathbf{B}^{\mathbb{P}}(\mathbb{G})$ associates with every goal g those transitions (σ, g') such that $g \xrightarrow{\sigma} g'$ is an instance of a derivable sequent of our SOS rules in the algebra $\mathbb{G}$.

The following theorem establishes the relation between our coalgebra and the operational semantics of logic programs, the SLD-reduction.

Theorem 4 (SLD-reduction as a coalgebra). *The LTS underlying the coalgebra $p' \colon \mathbb{G} \to \mathbf{B}^{\mathbb{P}}(\mathbb{G})$ is the same generated by the rules for the SLD derivation.*

Proof. The LTS generated by the SLD-reduction is the one generated by the SOS-rules obtained via the application of the syntactic transformation that turns every sequent of the form $\mathbb{P} \models G \Rightarrow_\sigma G'$ into a sequent of the form $G \xrightarrow{\sigma} G'$.

It can be shown that these SOS-rules can be derived by the SOS-rules of Table 2 and vice versa. □

4.2 Final $\mathbf{B}^{\mathbb{P}}$-coalgebra

We conclude this section with a proof of the existence of the terminal $\mathbf{B}^{\mathbb{P}}$-coalgebra. This allows to use the bisimulation semantics via terminal-coalgebra as described in Sect. 2. We need the following result.

Theorem 5. *For every locally presentable category* $\mathbf{C}$ *and every accessible endofunctor* $B \colon \mathbf{C} \to \mathbf{C}$ *the forgetful functor* $V_B \colon \mathbf{Coalg}(B) \to \mathbf{C}$ *is a left adjoint.*

Proof. The proof is basically the same as the one presented in [3, Theorem 1.2]. It follows from the *Special Adjoint Functor Theorem* using the fact that $\mathbf{C}$ and $\mathbf{Coalg}(B)$ are locally small and locally presentable, hence cocomplete and with a generating set, that $\mathbf{Coalg}(B)$ is cowellpowered, for [1, Theorem 1.58], and that the forgetful functor V_B preserves colimits. □

Corollary 2. *With the hypothesis of Theorem 5, if* $\mathbf{C}$ *has a terminal object,* $\mathbf{Coalg}(B)$ *has a terminal object as well.*

Proof. If $F_B \colon \mathbf{C} \to \mathbf{Coalg}(B)$ is the right adjoint of V_B, which exists for Theorem 5, it preserves limits and so, letting $T \in \mathbf{C}$ be a terminal object, $F_B(T)$ is a terminal object too. □

Theorem 6 ($\mathbf{B}^{\mathbb{P}}$ is accessible). *The functor* $\mathbf{B}^{\mathbb{P}}$ *is accessible.*

Proof. Since $\mathbf{Alg}(\Sigma_{\Pi\text{-}\mathbf{SM}})$ is an algebraic category $\mathbf{B}^{\mathbb{P}}$ is accessible if and only if for every sort $\underline{n} \in S_{\Sigma_{\Pi\text{-}\mathbf{SM}}}$ the functor $\mathbf{B}^{\mathbb{P}}_{\underline{n}} \colon \mathbf{Alg}(\Sigma_{\Pi\text{-}\mathbf{SM}}) \longrightarrow \mathbf{Set}$ such that $\mathbf{B}^{\mathbb{P}}_{\underline{n}}(A) = \mathbf{B}^{\mathbb{P}}(A)_{\underline{n}}$ is accessible. Remember that

$$\mathbf{B}^{\mathbb{P}}_{\underline{n}}(A) = \mathcal{P}_f\left(\left(\coprod_m \mathbf{Th}(\Sigma)[n,m] \times A_{\underline{m}}\right) \amalg A_{\underline{n}}\right) .$$

This equation shows that $\mathbf{B}^{\mathbb{P}}_{\underline{n}}$ is obtained combining the valuation functors (those the sends algebras in their carriers), that are accessible, with accessible functors, namely the finite powerset $\mathcal{P}_f$ and the multiplication functors $\mathbf{Th}(\Sigma)[n,m] \times -$. Since these functors are composed via coproduct and functor composition, that preserve accessibility, it follows that $\mathbf{B}^{\mathbb{P}}_{\underline{n}}$ is accessible as well.

Since this holds for every $\underline{n} \in S_{\Sigma_{\Pi\text{-}\mathbf{SM}}}$, by the above mentioned property of algebraic categories, it follows that $\mathbf{B}^{\mathbb{P}}$ is accessible. □

By Theorem 6 with Corollary 2 it follows that

Proposition 3 (Finality). *The category* $\mathbf{Coalg}(\mathbf{B}^{\mathbb{P}})$ *has a terminal object.*

4.3 Examples

We conclude this section by showing some examples of (in)equivalence.

Example 1. Let us consider the logic program:

$$\begin{array}{lll} P(x,y,z) \mathrel{:-} Q(x,y), R(y,z) & S(x,y,z) \mathrel{:-} T(x,y,z) & \\ Q(a,c) \mathrel{:-} \square & T(a,c,z) \mathrel{:-} V(z) & V(b) \mathrel{:-} \square \\ R(c,b) \mathrel{:-} \square & T(x,c,b) \mathrel{:-} U(x) & U(a) \mathrel{:-} \square \end{array} \qquad (2)$$

The goals $P(x, y, z)$ and $S(x, y, z)$ are bisimilar: they yield isomorphic LTSs.

Example 2. Let us consider the logic program:

$$P(f(x)) \mathrel{:-} P(x) \qquad Q(f(x)) \mathrel{:-} R(x) \qquad R(f(x)) \mathrel{:-} Q(x) \qquad (3)$$

The goals $P(x)$, $Q(x)$ and $R(x)$ are all bisimilar. They are logic *perpetual* processes [21]: even with the impossibility to terminate, at each transition new substitutions of the form $f(y)/z$ are computed to approximate the result.

Example 3. Let us consider the logic program:

$$\begin{array}{llll} P(a,y) \mathrel{:-} Q(y) & Q(b) \mathrel{:-} \square & S(a,y) \mathrel{:-} T(y) & T(b) \mathrel{:-} \square \\ P(a,y) \mathrel{:-} R(y) & R(c) \mathrel{:-} \square & & T(c) \mathrel{:-} \square \end{array} \qquad (4)$$

The goals $P(x, y)$ and $S(x, y)$ have the same answer substitutions but are not bisimilar, because the choice to substitute b or c for y is done by P implicitly at the first step while it is postponed to the second transition by S.

5 Conclusion

When exploiting LP as a process description language defined by an LTS, it is natural to look for a structured coalgebraic semantics.

In the paper, states, i.e. goals, are represented as the substitutive monoids freely generated by predicate symbols, and transitions are goal reductions via unification. The construction guarantees the existence of a final coalgebra yielding the abstract semantics. More precisely, coalgebras live in the category of algebras equipped with the operations of substitutive monoids, but without their axioms. Thus goal bisimulations respect monoidal and substitution operations.

In [4] the authors introduced a structured coalgebra that models the operational semantics of logic programming in order to apply the theory of reactive systems [20]. Their coalgebra uses a presheaf to model the state-space, which can be viewed as a multisorted-algebra having only unary-typed operations indexed by a family of substitutions. We have chosen a different algebraic structure that seems closer to the natural structure of goals of logic programming. For instance, to define the coalgebraic endofunctor we refer to an original, suggestive SOS specification quite close to the rules for SLD derivation (see Table 1). Nevertheless the two approaches are related. It is possible to obtain two forgetful functors from the two categories of algebras in the category $\mathbb{N}$ - **Set**, of families of sets indexed by natural numbers, and an endofunctor B over $\mathbb{N}$ - **Set** such that the forgetful functors send the coalgebras into the same B-coalgebra in $\mathbb{N}$ - **Set**, up to some technicalities due to differences into the behavioural endofunctors used for the structured coalgebras. Thus the two constructions can be considered as consisting of two different enrichments applied to the same coalgebra in **Set**.

Our construction could be adapted to simulate other models of computation which are structurally similar to LP, importing the well known properties of coalgebras regarding congruence, logic semantics and higher order. For instance, in [19] two process calculi, Fusion Calculus (a variant of pi-calculus) and Synchronized Hyperedge Replacement with Hoare Synchronization (a graph rewriting calculus with synchronization and mobility) are mapped into LP. Implementation efficiency can also get involved: constraint problems may be described by networks of constraints with an algebraic specification similar to our substitutive monoids; interestingly, an additional operation of restriction [24] allows to equip the networks with a hierarchical structure, which allows often to decompose the constraint problem and to solve it by means of an efficient dynamic programming algorithm. Moreover, in a Datalog-style setting [23], the decomposition process is automatically suggested by the goal reductions themselves.

We plan to study the relations between these models of computation taking advantage of the well known expressive power of the categorical approach.

In the present work we modeled the coalgebra in the category of algebras for a given signature—without axioms—to use the machinery of [25]. In [6] the authors provide some tools to generate behavioural endofunctors over categories of algebras *with* axioms. It could be interesting to investigate how to use these tools to turn our bi-algebraic semantics in a coalgebra over the category of substitutive monoids, instead of the wider category of algebras that are not required to respect the axioms of $\Gamma_{\Sigma_{\Pi}\text{-}\mathbf{SM}}$. However, one of the advantages of the current presentation is to be self-contained and more concrete than the framework in [6], so to be possibly accessible to a wider audience.

Acknowledgement. We thank Andrea Corradini who read a preliminary version of this paper and helped us to improve the presentation. We also thank the anonymous referees for their useful remarks and pointers to the literature.

References

1. Adamek, J., Rosicky, J.: Locally Presentable and Accessible Categories. London Mathematical Society Lecture Note Series. Cambridge University Press, Cambridge (1994)
2. Amato, G., Lipton, J., McGrail, R.: On the algebraic structure of declarative programming languages. Theor. Comp. Sci. **410**(46), 4626–4671 (2009)
3. Barr, M.: Terminal coalgebras in well-founded set theory. Theor. Comp. Sci. **114**(2), 299–315 (1993)
4. Bonchi, F., Montanari, U.: Reactive systems, (semi-)saturated semantics and coalgebras on presheaves. Theor. Comput. Sci. **410**(41), 4044–4066 (2009). Festschrift for Mogens Nielsen's 60th birthday

5. Bonchi, F., Zanasi, F.: Bialgebraic semantics for logic programming. Logical Methods Comput. Sci. **11**(1), 1–47 (2015)
6. Bonsangue, M.M., Hansen, H.H., Kurz, A., Rot, J.: Presenting distributive laws. In: Heckel, R., Milius, S. (eds.) CALCO 2013. LNCS, vol. 8089, pp. 95–109. Springer, Heidelberg (2013). https://doi.org/10.1007/978-3-642-40206-7_9
7. Bruni, R., Montanari, U., Rossi, F.: An interactive semantics of logic programming. Theory Pract. Logic Program. **1**(6), 647–690 (2001)
8. Corradini, A., Asperti, A.: A categorical model for logic programs: indexed monoidal categories. In: de Bakker, J.W., de Roever, W.-P., Rozenberg, G. (eds.) REX 1992. LNCS, vol. 666, pp. 110–137. Springer, Heidelberg (1993). https://doi.org/10.1007/3-540-56596-5_31
9. Corradini, A., Heckel, R., Montanari, U.: From SOS specifications to structured coalgebras: how to make bisimulation a congruence. ENTCS **19**, 118–141 (1999)
10. Corradini, A., Montanari, U.: An algebraic semantics for structured transition systems and its applications to logic programs. Theor. Comput. Sci. **103**(1), 51–106 (1992)
11. de Simone, R.: Higher-level synchronising devices in meije-sccs. Theor. Comput. Sci. **37**, 245–267 (1985)
12. Falaschi, M., Levi, G., Palamidessi, C., Martelli, M.: Declarative modeling of the operational behavior of logic languages. Theor. Comput. Sci. **69**(3), 289–318 (1989)
13. Finkelstein, S.E., Freyd, P., Lipton, J.: Logic programming in tau categories. In: Pacholski, L., Tiuryn, J. (eds.) CSL 1994. LNCS, vol. 933, pp. 249–263. Springer, Heidelberg (1995). https://doi.org/10.1007/BFb0022261
14. Gray, J.: The category of sketches as a model for algebraic semantics. In: Categories in Computer Science and Logic. Contemporary Mathematics, vol. 92. AMS (1989)
15. Komendantskaya, E., Power, J.: Coalgebraic semantics for derivations in logic programming. In: Corradini, A., Klin, B., Cîrstea, C. (eds.) CALCO 2011. LNCS, vol. 6859, pp. 268–282. Springer, Heidelberg (2011). https://doi.org/10.1007/978-3-642-22944-2_19
16. Komendantskaya, E., Power, J.: Logic programming: laxness and saturation. J. Log. Algebr. Meth. Program. **101**, 1–21 (2018)
17. Komendantskaya, E., Power, J., Schmidt, M.: Coalgebraic logic programming: from semantics to implementation. J. Log. Comput. **26**(2), 745–783 (2016)
18. Kowalski, R.A.: Algorithm = logic + control. Comm. ACM **22**(7), 424–436 (1979)
19. Lanese, I., Montanari, U.: Mapping fusion and synchronized hyperedge replacement into logic programming. Theory Pract. Logic Program. **7**(1–2), 123–151 (2007)
20. Leifer, J.J., Milner, R.: Deriving bisimulation congruences for reactive systems. In: Palamidessi, C. (ed.) CONCUR 2000. LNCS, vol. 1877, pp. 243–258. Springer, Heidelberg (2000). https://doi.org/10.1007/3-540-44618-4_19
21. Levi, G., Palamidessi, C.: Contributions to the semantics of logic perpetual processes. Acta Inf. **25**(6), 691–711 (1988)
22. Lloyd, J.W.: Foundations of Logic Programming, 2nd edn. Springer, Heidelberg (1987). https://doi.org/10.1007/978-3-642-83189-8
23. Montanari, U., Rossi, F.: Perfect relaxation in constraint logic programming. In: ICLP 1991, pp. 223–237. MIT Press (1991)
24. Montanari, U., Sammartino, M., Tcheukam, A.: Decomposition structures for soft constraint evaluation problems: an algebraic approach. In: Heckel, R., Taentzer, G. (eds.) Graph Transformation, Specifications, and Nets. LNCS, vol. 10800, pp. 179–200. Springer, Cham (2018). https://doi.org/10.1007/978-3-319-75396-6_10
25. Turi, D., Plotkin, G.D.: Towards a mathematical operational semantics. In: LICS 1997, pp. 280–291. IEEE Computer Society (1997)

Polyadic Soft Constraints

Filippo Bonchi[1], Laura Bussi[1], Fabio Gadducci[1(✉)], and Francesco Santini[2]

[1] Dipartimento di Informatica, University of Pisa, Pisa, Italy
{filippo.bonchi,laura.bussi,fabio.gadducci}@unipi.it

[2] Dipartimento di Matematica e Informatica, University of Perugia, Perugia, Italy
francesco.santini@unipg.it

Abstract. We propose a formalism for manipulating soft constraints based on polyadic algebras. The choice of such algebras in place of classical cylindric ones simplifies the structure of the partial order of preference values by removing diagonals, a family of constants used for modelling parameter passing and variable substitution, whose presence require completeness. Removing diagonals also allows for an easy representation of preference/cost functions in terms of polynomials, thus streamlining their manipulation on languages based on (stores of) constraints. Besides presenting the main features of the new formalism, the paper investigates how the operators of polyadic algebras interact with the residuated monoid structure that is used for representing the set of preference values.

Keywords: Soft constraints · Polyadic algebras · Residuated monoids

1 Introduction

Computer scientists often face combinatorial problems, as e.g. in operational research, artificial intelligence, or circuit design, and constraints are a tool for naturally modelling such problems. In their simplest form, constraints are just sets of inequations: given a set of variables V and a domain of values D, a constraint over a subset of variables limits the combinations of values that such variables can take. Combinatorial problems can thus be easily modelled by constraints: *constraint satisfaction problems* (CSPs) are defined over a set V, a domain D and a set of constraints C. Solving a CSP simply means to find an assignment of the variables such that all its constraints are satisfied.

Consider, for instance, a problem where we want to respect a given relation among two measures, e.g. $y = 2x$, for x representing the width and y the length of an object. Of course, both x and y have to be positive reals and might have to respect some upper bounds, let us say 5 and 9. The problem can be formalised as

Research partially supported by the MIUR PRIN 2017FTXR7S "IT-MaTTerS".

M. S. Alvim et al. (Eds.): Palamidessi Festschrift, LNCS 11760, pp. 241–257, 2019.
https://doi.org/10.1007/978-3-030-31175-9_14

- the set of variables V is $\{x, y\}$;
- the domain D is the set of positive reals $\mathbb{R}^+$;
- the constraints in C are given by polynomials: $\{2x - y = 0,\ x - 5 \leq 0,\ y - 9 \leq 0\}$.

where clearly the assignment $\{x \leftarrow 4,\ y \leftarrow 8\}$ is a solution for the problem.

Constraint programming (CP) is a computational framework that allows to implement algorithms for solving CSPs: it has been intensively studied and a set of algorithms for solving CSPs have been implemented since the Nineties [14]. *Concurrent constraint programming* (CCP) extends CP by allowing parallel programs to interact by means of a shared store representing a constraint. As the basic operations in imperative programming are *read* and *write*, in CP and CCP we have operations to manipulate the store

- *tell(c)* adds a constraint c to the store;
- *ask(c)* checks if a constraint c is satisfied, i.e. if the store *entails* such a constraint;

and depending on the domain we may additionally require subtraction (as for example it has been implemented in [9], where constraints are formulas in linear logic)

- *remove(c)* removes c from the store, thus possibly lifting some of the requirements that had been enforced by the constraint.

In order to model e.g. procedure calls, further operators are needed on constraints. The standard solution [19] is via cylindric algebras: a family of variable-indexed unary operations, for representing existential quantifiers, and a family of constants called *diagonals*, for modelling parameters passing and variable substitution.

Despite CSP expressiveness, in many real-life situations we need a more flexible way to model problems, since some constraints could be "less important" than others and might not have to be necessarily satisfied. Soft constraints have been introduced to model this kind of situation. In informal terms, they are classical constraints where a value from a partially ordered set is associated to each instantiation of the variables of a constraint. It is thus possible to state that a constraint is either more or less significant than others in order to find a good approximation of the solution, even if not all constraints are satisfied at the same time. Then, a *soft constraint* is defined as a function of type $(V \to D) \to A$, where functions $V \to D$ are assignments of variables in V to a domain D and the values in A represent either cost or preferability, depending on whether representing negative or positive preferences. Combining positive and negative preferences results in what is called a bipolar approach [5].

To evaluate preferences, the set A is equipped with a semi-lattice structure and, to make possible the combination of soft constraints, a monoidal operator. Therefore, A turns out to be an idempotent semiring (also called tropical semiring or dioid). For instance, *Tropical* semirings associate a constraint to its cost and the aim is to minimise the sum, while *Probabilistic* semirings associate a constraint to its probability and the aim is to maximise the joint probability. A further operation is needed over the set A, in order to model the removal of constraints, resulting in a residuated monoid: the basics of the formalism within the bipolar approach have been presented in [10].

Some extensions of CCP for allowing soft constraints have been already proposed, as for instance in [4], and they required a reworking of the underlying theory. The starting point has been again the operators of cylindric algebras. However, the use of diagonals for soft constraints puts some strong requirements on the structure of the set of values A, which are not always met in actual case studies. Moreover, diagonals also present some serious issues in finding a compact representation.

The proposal developed in this paper is to consider polyadic algebras instead of cylindric ones. This means to replace diagonals with a family of polyadic operators (see e.g. [18]) that precisely axiomatise variable substitution, and to investigate how these operators interact with the residuated monoid structure of the set of values A.

The benefits are twofold. On the one hand, we relax some of the requirements necessary for A: the join semi-lattice on A must be complete for cylindric algebras, while it is not so for polyadic ones. Such relaxation is relevant also for more recent developments on CCP: while in standard denotational models the semantics universe is supposed to have a complete semi-lattice structure, this is not needed with the bisimulation semantics introduced in [1]. On the other hand, replacing diagonals with polyadic operators allows for a compact – *polynomial* – representation of soft constraints.

In Sect. 2 we rephrase some basic definitions for residuated monoids. In Sect. 3 we offer a presentation of polyadic algebras tailored for constraints and we present some results concerning the combination of these two structures. These results are used in Sect. 4, where we provide a novel formalisation of soft constraints in terms of polyadic algebras whose carrier is a residuated monoid. Finally, in Sect. 5 we introduce a set of constraints that enjoy a polynomials representation and we discuss the algebra of linear polynomials with coefficients in $\mathbb{N}$ as a case study.

The (Soft) Constraints of Catuscia. The word "constraint" occurs in the titles of 40 publications co-authored by Catuscia. Between 1991 and 1997, most of these papers were devoted to the semantics [6] and analysis [8] of Constraint Logic Programming and Concurrent Constraint Programming. In 2001, the latter language was extended to Temporal Concurrent Constraint Programming [16,17]. Since then, Catuscia has exploited constraints in several applications domains like security [15], biological systems [7], and the modelling of knowledge in social networks [13].

Despite her scientific interest in hard constraints, Catuscia seems to prefer a rather soft approach in her personal and professional relationships: although always surrounded by students, collaborators, and visitors, Catuscia is always available for chatting and joking, and for helping out, both in technical and in personal matters. Also well-known among friends is her soft spot for the wee hours of the morning. In particular, the first author recalls that, while working on the afore-mentioned bisimulation semantics for CCP [1], after many weeks trying hard to prove the main theorem with the other collaborators, he spent one afternoon in Catuscia's office looking together for a proof. When he woke up the morning after, in his mail-box there was a message from her: a beautiful and crystal-clear proof of the long awaited result. This message was sent at about 3 am.

2 An Introduction to Residuated Monoids

This section recalls some results on residuated monoids, which are our chosen algebraic structure for modelling soft constraints. They are mostly drawn from [11], and they are presented without proofs. Note however that, to the best of our knowledge, the material from Proposition 1 up to Example 2 appears to be original in the literature on constraints.

2.1 Preliminaries on Ordered Monoids

The first step is to define an algebraic structure for modelling preferences, where it is possible to compare values and combine them. Our choice falls into the range of *bipolar* approaches, in order to represent both positive and negative preferences: we refer to [10] for a detailed introduction and a comparison with other proposals.

Definition 1 (partial order). *A partial order (PO) is a pair $\langle A, \leq \rangle$ such that A is a set and $\leq \subseteq A \times A$ is a reflexive, transitive, and anti-symmetric relation. A (join) semi-lattice (SL) is a PO such that any non-empty finite subset of A has a least upper bound (LUB).*

The LUB of a (possibly infinite or empty) subset $X \subseteq A$ is denoted $\bigvee X$, and it is clearly unique. Should they exist, $\bigvee A$ and $\bigvee \emptyset$ correspond respectively to the top, denoted as $\top$, and to the bottom, denoted as $\bot$, of the PO.

Definition 2 (monoid). *A (commutative) monoid is a triple $\langle A, \otimes, \mathbf{1} \rangle$ such that A is a set, $\otimes : A \times A \to A$ is a commutative and associative function, and $\mathbf{1} \in A$ is the* identity *element, namely, $\forall a \in A. a \otimes \mathbf{1} = a$.*

A partially ordered (semi-lattice) monoid is a 4-tuple $\langle A, \leq, \otimes, \mathbf{1} \rangle$ such that $\langle A, \leq \rangle$ is a PO (SL) and $\langle A, \otimes, \mathbf{1} \rangle$ a monoid.

As usual, we use the infix notation: $a \otimes b$ stands for $\otimes(a, b)$.

Definition 3 (distributivity). *Let $\langle A, \leq, \otimes, \mathbf{1} \rangle$ be a semi-lattice monoid. It is distributive if for any non-empty finite $X \subseteq A$*

- $\forall a \in A.\, a \otimes \bigvee X = \bigvee \{a \otimes x \mid x \in X\}$.

Note that distributivity implies that $\otimes$ is monotone with respect to $\leq$.

Remark 1. It is almost straightforward to show that our proposal encompasses many other formalisms in the literature. Indeed, distributive semi-lattice monoids are *tropical* semirings (also known as dioids), namely, semirings with an idempotent sum operator $a \oplus b$, which in our formalism is obtained as $\bigvee\{a, b\}$. If $\mathbf{1}$ is the top of the SL we end up in *absorptive* semirings [12], which are known as *c*-semirings in the soft constraint jargon [2] (see e.g. [3] for a brief survey on residuation for such semirings). Note that requiring the monotonicity of $\otimes$ and imposing $\mathbf{1}$ to be the top of the partial order means that preferences are negative, i.e., that it holds $\forall a, b \in A. a \otimes b \leq a$.

Example 1. Given a (possibly infinite) set V of variables, two semi-lattice monoids are going to play a key role in the following sections.

The first one is the semi-lattice monoid $\mathbb{M}(V) = \langle 2^V_{fin}, \subseteq, \cup, \emptyset \rangle$ of finite subsets of V, with the usual order given by sub-set inclusion.

For the second one, we start by defining the support of an endofunction $f \colon V \to V$ as the set $sv(f) = \{x \in V \mid f(x) \neq x\}$ and $F(V)$ as the set of functions $f \colon V \to V$ with finite support. The semi-lattice monoid of interest is $\mathbb{F}(V) = \langle F(V), id, \circ, \iota \rangle$ where ι is the identity function, $\circ$ is function composition and id is the discrete ordering on $F(V)$.

2.2 Remarks on Residuation

It is often needed to be able to "remove" part of a preference, due e.g. to the non-monotone nature of the language at hand for manipulating constraints. The structure of our choice is given by residuated monoids [12]. They introduce a new operator $\ominus$, which represents a "weak" (due to the presence of partial orders) inverse of $\otimes$.

Definition 4 (residuation). *A residuated monoid (RePO) is a 5-tuple* $\langle A, \leq, \otimes, \ominus, \mathbf{1} \rangle$ *such that* $\langle A, \leq, \otimes, \mathbf{1} \rangle$ *is a partially ordered monoid and* $\ominus : A \times A \to A$ *is a function satisfying*

- $\forall a, b, c \in A. b \otimes c \leq a \iff c \leq a \ominus b.$

An ReSL is an RePO such that the underlying PO is a SL.

In order to confirm the intuition about weak inverses, Lemma 1 below precisely states that residuation conveys the meaning of an approximated form of subtraction.

Lemma 1. *Let* $\langle A, \leq, \otimes, \ominus, \mathbf{1} \rangle$ *be an RePO. Then*

- $\forall a, b \in A. a \ominus b = \bigvee\{c \mid b \otimes c \leq a\},$

In order to ease the verification of the algebraic structure, it is often needed a characterisation of residuation via simpler properties, as the ones given below.

Lemma 2. *Let* $\langle A, \leq, \otimes, \mathbf{1} \rangle$ *be a partially ordered monoid and* $\ominus : A \times A \to A$ *a function. Then* $\langle A, \leq, \otimes, \ominus, \mathbf{1} \rangle$ *is an RePO if and only if*

- $\forall a, b \in A. b \otimes (a \oplus b) \leq a \leq (b \otimes a) \oplus b$,
- $\forall a, b, c \in A.\, a \leq b \implies a \otimes c \leq b \otimes c$ *and* $a \oplus c \leq b \oplus c$.

It is easy to show that in any RePO the $\oplus$ operator is also anti-monotone on the second argument, i.e., $\forall a, b, c \in A.\, a \leq b \implies c \oplus b \leq c \oplus a$. Other properties are also straightforward, such as $\forall a \in A. \mathbf{1} \leq a \oplus a$, which in turn implies that $\forall a \in A. a \otimes (a \oplus a) = a$ and $\forall a, b \in A. a < b \implies \mathbf{1} \not\leq a \oplus b$, where $a < b$ means $a \leq b$ and $a \neq b$. The latter fact suggests the definition below, which identifies sub-classes of residuated monoids that are suitable for an easier manipulation of constraints (see e.g. [3]).

Definition 5 (localisation/invertibility). *An RePO* $\langle A, \leq, \otimes, \oplus, \mathbf{1} \rangle$ *is*

- localised *if* $\forall a, b \in A. a \leq b \implies a \oplus b \leq \mathbf{1}$*;*
- invertible *if* $\forall a, b \in A. a \leq b \implies b \otimes (a \oplus b) = a$.

Note that if a RePO is localised then $\forall a \in A. a \oplus a = \mathbf{1}$.

Remark 2. Some well-known structures used for soft constraints are the *Fuzzy* ($\langle [0,1], \leq, \min, 1 \rangle$), *Probabilistic* ($\langle [0,1], \leq, \times, 1 \rangle$), and *Tropical* ($\langle \mathbb{R}^+, \geq, +, 0 \rangle$) semirings, for $\geq$ the inverse of the standard order (thus 0 the top of the SL). In all these cases the underlying monoids are both invertible and localised, thus the $\oplus$ operator can be also used to (partially) relax constraints (see again [3]).

Moving to ReSLs, next lemma ensures that residuation implies distributivity.

Lemma 3. *Let* $\langle A, \leq, \otimes, \oplus, \mathbf{1} \rangle$ *be an ReSL. Then the underlying SL is distributive.*

Distributivity holds also for the empty set and for infinite sets, if the necessary LUBs exist. Instead, it holds only partially for $\oplus$: this follows directly from the monotonicity of $\oplus$ on the first argument, since it implies that $x \oplus a \leq \bigvee X \oplus a$ for all $x \in X$.

Lemma 4. *Let* $\langle A, \leq, \otimes, \oplus, \mathbf{1} \rangle$ *be an ReSL and* $X \subseteq A$ *a finite non-empty set. Then*

- $\forall a \in A.\, \bigvee \{x \oplus a \mid x \in X\} \leq \bigvee X \oplus a$

Also this inequation holds for the empty set and for infinite sets, if the necessary LUBs exist. Moreover, it also holds that $\bigvee \{a \oplus x \mid x \in X\} \geq a \oplus \bigvee X$, since $\oplus$ is anti-monotone on the second argument.

Proposition 1. *Let* $\langle A, \leq, \otimes, \oplus, \mathbf{1} \rangle$ *be an ReSL. The following are equivalent*

1. $\forall a \in A.\, a \leq \mathbf{1}$
2. $\forall a \in A.\, \mathbf{1} \oplus a = \mathbf{1}$
3. $\forall a, b \in A.\, a \leq b \implies b \oplus a = \mathbf{1}$

Remark 3. The proposition above provides an important characterisation for all absorptive ReSLs, including all those mentioned in Remark 2.

There are some important classes of ReSLs such that $\ominus$ is easily proved to be distributive in the first argument, while it is not so with respect to the second argument, not even in the absorptive case.

Lemma 5. *Let $\langle A, \leq, \otimes, \ominus, \mathbf{1}\rangle$ be an ReSL such that $\langle A, \leq\rangle$ is a total order and $X \subseteq A$ a finite non-empty set. Then*

- $\forall a \in A.\ \bigvee\{x \ominus a \mid x \in X\} = \bigvee X \ominus a$

Example 2. Let n be a positive integer and $[n] = \{0, \ldots, n\}$ the segment of integers from 0 to n. We can now define the (bounded) monoid $\mathbb{M}_n$ as the tuple $\langle [n], \geq, \oplus, \ominus, 0\rangle$, where $\oplus$ and $\ominus$ are the bounded sum and subtraction, which are given as $m \oplus p = min\{n, m + p\}$ and $m \ominus p = max\{0, m - p\}$.

Now, it can be shown that $\mathbb{M}_n$ is an absorptive ReSL, and since it is a total order, $\ominus$ is distributive on the first argument. However, in general it is not distributive on the second one. Consider an integer m such that $m \neq n$ and the set $\{m, m+1\}$: we then have that $(m+1) \ominus \bigvee\{m, m+1\} = 1$, while instead $\bigvee\{(m+1) \ominus m, (m+1) \ominus (m+1)\} = 0$.

3 An Alternative Proposal for Costraint Manipulation

This section presents our personal take on polyadic algebras for ordered monoids: the standard axiomatisation of e.g. [18] has been completely reworked, in order to be adapted to the constraints formalism. We close the section by offering some preliminary insights on the laws for polyadic operators in residuated monoids.

3.1 Cylindric and Polyadic Operators for Ordered Monoids

We now introduce two families of operators that will be used for modelling variables hiding and substitution, which represent key features in languages for manipulating constraints. One is a well-known abstraction for existential quantifiers, the other an axiomatisation of the notion of substitution, and it is proposed as a weaker alternative to diagonals [19], the standard tool for modelling equivalence in constraint programming.[1]

Our first step is the introduction of a technical notion that allows for factorising the common properties in the definition of the two families of operators.

Definition 6 (pomonoid action). *Let $\mathbb{M} = \langle A, \leq, \otimes, \mathbf{1}\rangle$ be a partially ordered monoid and $\mathbb{P} = \langle S, \leq\rangle$ a partial order. A pomonoid action of $\mathbb{M}$ on $\mathbb{P}$ is a function $\phi : A \times S \to S$ such that*

- $\forall s \in S.\ \phi(\mathbf{1}, s) = s,$
- $\forall a, b \in A,\ s \in S.\ \phi(a, \phi(b, s)) = \phi(a \otimes b, s),$
- $\forall a, b \in A,\ s, t \in S.\ a \leq b \wedge s \leq t \implies \phi(a, s) \leq \phi(b, t).$

The first two requirements just state that ϕ is a monoid action of $\mathbb{M}$ on S, while the latter states that ϕ is monotone. Sometimes, we say that $\mathbb{P}$ is an $\mathbb{M}$-PO.

[1] "Weaker alternative" here means that diagonals allow for axiomatising substitutions at the expenses of working with complete partial orders: see e.g. [11, Definition 11].

Cylindric Operators. We fix a partially ordered monoid $\mathbb{S} = \langle A, \leq, \otimes, \mathbf{1}\rangle$ and a set V of variables, and we then define a family of operators axiomatising existential quantifiers.

Definition 7 (cylindrification). *A cylindric operator $\exists$ over $\mathbb{S}$ and V is a pomonoid action $\exists : 2^V_{fin} \times A \to A$ such that for all $X \in 2^V_{fin}$*

1. $\exists(X, \mathbf{1}) = \mathbf{1}$,
2. $\forall a, b \in A.\ \exists(X, a \otimes \exists(X, b)) = \exists(X, a) \otimes \exists(X, b)$.

Let $a \in A$. The support *of a is the set of variables $sv(a) = \{x \mid \exists(\{x\}, a) \neq a\}$.*

Note that, since by Definition 6 we have $\exists(\emptyset, a) = a$, the requirements of Definition 7 trivially hold whenever X is the empty set. Note also that $\exists(X, \mathbf{1}) = \mathbf{1}$ would be a consequence of monotonicity, should $\mathbf{1}$ be the top element. Also, the support is not necessarily finite. Finally, and importantly, note that $X \cap sv(\exists(X, a)) = \emptyset$.

In the following, we often use $\exists_X a$ for $\exists(X, a)$, and $\exists_x a$ whenever $X = \{x\}$.

Polyadic Operators. We now move to define a family of operators axiomatising substitutions. They interact with quantifiers, thus, beside a partially ordered monoid $\mathbb{S}$ and a set V of variables, we fix a cylindric operator $\exists$ over $\mathbb{S}$ and V.

As for notation, for a function $\sigma : V \to V$ and a set $X \subseteq V$, we denote by $\sigma\mid_X : X \to V$ the obvious restriction, and by $\sigma^c(X) \subseteq V$ the counter-image of X along σ.

Definition 8 (polyadification). *A polyadic operator s for $\exists$ is a pomonoid action $s : F(V) \times A \to A$ such that for all $X \in 2^V_{fin}$ and $\sigma, \tau \in F(V)$*

1. $\forall a, b \in A.\ s(\sigma, a \otimes b) = s(\sigma, a) \otimes s(\sigma, b)$,
2. $\forall a \in A.\ \sigma\mid_{sv(a)} = \tau\mid_{sv(a)} \implies s(\sigma, a) = s(\tau, a)$,
3. $\sigma\mid_{\sigma^c(X)}$ *injective* $\implies \forall a \in A.\ \exists(X, s(\sigma, a)) = s(\sigma, \exists(\sigma^c(X), a))$.

Clearly item 3 always holds for an empty X. We usually denote $s(\tau, a)$ as $s_\tau a$. Now, being an action implies that $s_\iota a = a$, and together with item 2 it implies that $s_\tau \mathbf{1} = \mathbf{1}$. A polyadic operator offers enough structure for modelling variable substitution. In the following, we fix a polyadic operator s for $\exists$.

Remark 4. The laws are directly adapted from [18], with the exception of 2, which is stated as for a finite non-empty $X \subseteq V$ and $a \in A$

2'. $\sigma\mid_{V\setminus X} = \tau\mid_{V\setminus X} \implies \forall a \in A.\ s(\sigma, \exists(X, a)) = s(\tau, \exists(X, a))$.

However, the two formulations are equivalent. Indeed, note that $\sigma\mid_{V\setminus X} = \tau\mid_{V\setminus X}$ implies $\sigma\mid_{sv(a)\setminus X} = \tau\mid_{sv(a)\setminus X}$, which in turn implies that $\sigma\mid_{sv(\exists(X,a))} = \tau\mid_{sv(\exists(X,a))}$, and assuming item 2 the result follows. For the vice-versa, first of all note that $\sigma\mid_{V\setminus X} = \tau\mid_{V\setminus X}$ coincides with $\sigma\mid_{Y\setminus X} = \tau\mid_{Y\setminus X}$ for $Y = sv(\sigma) \cup sv(\tau) \subseteq V$, and that Y is finite since both σ and τ are finitely supported. Now, $\sigma\mid_{sv(a)} = \tau\mid_{sv(a)}$ implies that $\sigma\mid_{Y\setminus(Y\setminus sv(a))} = \tau\mid_{Y\setminus(Y\setminus sv(a))}$, thus by 2' we have $s(\sigma, \exists(Y \setminus sv(a), a)) = s(\tau, \exists(Y \setminus sv(a), a))$. Since by definition we have $\exists(Y \setminus sv(a), a)) = a$, the result follows.

Remark 5. Note also that $\sigma(\sigma^c(X)) \subseteq X$, so, when restricted to singleton, we have that item 3 in Definition 8 is equivalent to

3'. $\forall a \in A.\ \sigma^c(x) = \{y\} \implies \exists_x s_\sigma a = s_\sigma \exists_y a$,
3". $\forall a \in A.\ \sigma^c(x) = \emptyset \implies \exists_x s_\sigma a = s_\sigma a$.

3.2 Properties of Polyadic Operators

In this section we just show some facts concerning polyadic operators: they ensure that indeed these operators suitably axiomatise substitutions. More precisely, we consider a few simple properties that mimic those holding for substitutions modelled via diagonals, as considered e.g. in [11, Lemma 2 and Lemma 3].

Definition 9 (inverse functions). *Let $\sigma \in F(V)$ be* invertible, *i.e., such that $\sigma \mid_{sv(\sigma)}$ is injective. Its* inverse *is defined as*

$$\sigma^{-1}(y) = \begin{cases} x \ \textit{if}\ y = \sigma(x) \ \textit{and}\ x \in sv(\sigma) \\ y \ \textit{otherwise} \end{cases}$$

Its injective lifting *is defined as*

$$\sigma_l(y) = \begin{cases} \sigma(y) & \textit{if}\ y \in sv(\sigma) \\ \sigma^{-1}(y) & \textit{otherwise} \end{cases}$$

In other terms, note that an element $y \in \sigma(sv(\sigma))$ can be the image along σ of at most two elements: should this be the case, one of them is the element itself and the other belongs to $sv(\sigma)$. When defining the inverse, we give precedence to the element in $sv(\sigma)$. Instead, the injective lifting is indeed an injective substitution.

Lemma 6. *Let $\sigma \in F(V)$ be invertible. Then it holds*

- $sv(\sigma_l) = \sigma(sv(\sigma)) \cup sv(\sigma) = sv(\sigma_l^{-1})$,
- $sv(\sigma^{-1}) = \sigma(sv(\sigma))$,
- $sv(\sigma^{-1} \circ \sigma) = \sigma(sv(\sigma)) \setminus sv(\sigma)$,
- $sv(\sigma \circ \sigma^{-1}) = sv(\sigma) \setminus \sigma(sv(\sigma))$.

The set-theoretical proofs are immediate and are omitted. Lemma 7 helps to ensure that polyadic substitution behaves correctly with respect to $\exists$. We recall that the properties below mirror those obtained for substitutions via diagonal operators [11].

Lemma 7. *Let $\sigma \in F(V)$ be invertible and $W \subseteq V$ finite. Then it holds*

1. $\forall a \in A.\ s_\sigma \exists_{sv(\sigma)} a = \exists_{sv(\sigma)} a$,
2. $\forall a \in A.\ \sigma(sv(\sigma)) \cap sv(a) = \emptyset \implies \exists_{sv(\sigma)} a = \exists_{\sigma(sv(\sigma))} s_\sigma a$,
3. $W \cap (sv(\sigma) \cup \sigma(sv(\sigma))) = \emptyset \implies \forall a \in A.\ s_\sigma \exists_W a = \exists_W s_\sigma a$.

Proof. Proofs are immediate. As for item 1, it is a consequence of $\sigma\mid_{V\setminus sv(\sigma)} = id\mid_{V\setminus sv(\sigma)}$.

Concerning now item 3, since W satisfies $W \cap sv(\sigma) = \emptyset$ then $\sigma(W) = W$, and since additionally $W \cap \sigma(sv(\sigma)) = \emptyset$ it also follows that $\sigma^c(W) = W$. Thus we have that $\exists_W s_\sigma a = \exists_{\sigma(W)} s_\sigma a = s_\sigma \exists_{\sigma^c(W)} a = s_\sigma \exists_W a$, where the intermediate equality holds by item 3 of Lemma 6.

Let us finally move to item 2, and let us consider the lifting σ_l. This can be less parsimoniously described as

$$\sigma_l(y) = \begin{cases} \sigma(y) & \textit{if } y \in sv(\sigma) \\ \sigma^{-1}(y) & \textit{if } y \in \sigma(sv(\sigma)) \setminus sv(\sigma) \\ y & \textit{otherwise} \end{cases}$$

By the hypothesis $\sigma(sv(\sigma)) \cap sv(a) = \emptyset$ we have that $\sigma\mid_{sv(a)} = \sigma_l\mid_{sv(a)}$, hence it holds $\exists_{sv(\sigma)} a = s_\sigma \exists_{sv(\sigma)} a = s_{\sigma_l} \exists_{sv(\sigma)} a$. Now note that $(\sigma_l)^c(\sigma(sv(\sigma))) = sv(\sigma)$, for $(\sigma_l)^c$ the set-theoretical inverse of σ_l, thus again by item 3 of Definition 8 we have that $s_{\sigma_l} \exists_{sv(\sigma)} a = \exists_{\sigma(sv(\sigma))} s_{\sigma_l} a = \exists_{\sigma(sv(\sigma))} s_\sigma a$. □

Finally, we rephrase some further laws of the crisp case (see [1, p. 140]).

Lemma 8. *Let $\sigma \in F(V)$ be invertible. Then it holds*

1. $\forall a \in A.\ (sv(\sigma) \setminus \sigma(sv(\sigma))) \cap sv(s_\sigma a) = \emptyset$.
2. $\forall a \in A.\ (\sigma(sv(\sigma)) \setminus sv(\sigma)) \cap sv(a) = \emptyset \implies s_{\sigma^{-1}} s_\sigma a = a$,

Proof. As for item 1, note that $sv(s_\sigma a) = \sigma(sv(a))$. Let $Z = sv(\sigma) \cap sv(a)$: then $\sigma(Z) \subseteq \sigma(sv(\sigma))$ and $\sigma\mid_{sv(a)\setminus Z} = \iota\mid_{sv(a)\setminus Z}$, which implies $(sv(a)\setminus Z) \cap sv(\sigma) = \emptyset$. Since $\sigma(Z) \subseteq \sigma(sv(\sigma))$, it holds $(sv(\sigma)\setminus\sigma(sv(\sigma)) \cap \sigma(sv(a)) = (sv(\sigma)\setminus\sigma(sv(\sigma)) \cap sv(s_\sigma a) = \emptyset$.

As for item 2, by definition it holds $s_{\sigma^{-1}} s_\sigma a = s_{\sigma^{-1}\circ\sigma} a$. Now, item 3 of Lemma 6 implies that $sv(\sigma^{-1} \circ \sigma) = \sigma(sv(\sigma)) \setminus sv(\sigma)$, and since $\sigma(sv(\sigma)) \cap sv(a) = \emptyset$ it holds $(\sigma^{-1} \circ \sigma)\mid_{sv(a)} = \iota\mid_{sv(a)}$. The result then follows by item 2 of Definition 8. □

3.3 Cylindric and Polyadic Operators for Residuated Monoids

Both algebraic structures introduced in the previous section are quite standard, even if polyadic operators are less-known in the soft-constraints literature: we tailored their presentation to our needs, and indeed the properties presented in Sect. 3.2 appear to be original. It is now time to consider the interaction of such structures with residuation.

To this end, in the following we assume that $\mathbb{S}$ is a RePO (see Definition 4).

Lemma 9. *Let $X \subseteq V$ be finite. Then it holds*

- $\forall a, b \in A.\ \exists_X (a \ominus \exists_X b) \leq \exists_X a \ominus \exists_X b$.

Proof.

$$\exists_X b \otimes (a \oplus \exists_X b) \leq a \implies \exists_X(\exists_X b \otimes (a \oplus \exists_X b)) \leq \exists_X a \implies$$

$$\exists_X b \otimes \exists_X(a \oplus \exists_X b)) \leq \exists_X a \implies \exists_X(a \oplus \exists_X b) \leq \exists_X a \oplus \exists_X b$$

□

Remark 6. Looking at the proof above, it is clear that $\exists_x(a \oplus \exists_x b) \leq \exists_x a \oplus \exists_x b$ is actually equivalent to state that $\exists_x(a \otimes \exists_x b) \geq \exists_x a \otimes \exists_x b$.

Similarly, it is easy to show that it holds $\forall a, b \in A$. $\exists_X(\exists_X a \oplus b) \leq \exists_X a \oplus \exists_X b$.

A similar result relates residuation and polyadic operators.

Lemma 10. *Let $\sigma, \tau \in F(V)$. Then it holds*

- $\forall a, b \in A.\ s_\sigma(a \oplus b) \leq s_\sigma a \oplus s_\sigma b$.

4 Polyadic Soft Constraints

In the past sections we mentioned a few ReSLs such as the Fuzzy and the Tropical semiring. Building on such examples, in this section we give a main case study: a ReSL where the notion of cylindric and polyadic operators can be easily given. It exploits the notion of soft constraint: indeed, our proposal follows yet generalises [4], whose underlying algebraic structure is the one of absorptive semirings.

Definition 10. ((soft) constraints). *Let V be a set of variables, D a finite domain of interpretation and $\mathbb{S} = \langle A, \leq, \otimes, \oplus, \mathbf{1} \rangle$ a ReSL. A* (soft) constraint *$c : (V \to D) \to A$ is a function associating a value in A with each assignment $\eta : V \to D$ of the variables.*

In this section and in the following one, we denote by $\mathcal{C}$ the set of constraints that can be built starting from chosen $\mathbb{S}$, V, and D. The application of a constraint function $c : (V \to D) \to A$ to a variable assignment $\eta : V \to D$ is denoted $c\eta$.

Even if a constraint involves all the variables in V, it may depend on the assignment of a finite subset of them, called its support. For instance, a binary constraint c with $supp(c) = \{x, y\}$ is a function $c : (V \to D) \to A$ that depends only on the assignment of variables $\{x, y\} \subseteq V$, meaning that two assignments $\eta_1, \eta_2 : V \to D$ differing only for the image of variables $z \notin \{x, y\}$ coincide (i.e., $c\eta_1 = c\eta_2$). The support corresponds to the classical notion of scope of a constraint. We often refer to a constraint with support X as c_X. Moreover, an assignment over a support X of cardinality k is concisely represented by a tuple t in D^k, and we often write $c_X(t)$ instead of $c_X\eta$.

The set of constraints forms a ReSL, with the structure lifted from $\mathbb{S}$.

Lemma 11 (the ReSL of constraints). *The ReSL of constraints* $\mathbb{C}$ *is defined as the tuple* $\langle \mathscr{C}, \leq, \otimes, \oplus, \mathbf{1} \rangle$ *such that*

- $c_1 \leq c_2$ *if* $c_1\eta \leq c_2\eta$ *for all* $\eta : V \to D$,
- $(c_1 \otimes c_2)\eta = c_1\eta \otimes c_2\eta$,
- $(c_1 \oplus c_2)\eta = c_1\eta \oplus c_2\eta$,
- $\mathbf{1}\eta = \mathbf{1}$.

Combining constraints by the $\otimes$ operator means building a new constraint whose support involves at most the variables of the original ones. The resulting constraint associates with each tuple of domain values for such variables the element that is obtained by multiplying those associated by the original constraints to the appropriate sub-tuples.

Lemma 12 (cylindric and polyadic operators for (soft) constraints). *The ReSL of constraints* $\mathbb{C}$ *admits cylindric and polyadic operators, defined as*

- $(\exists_X c)\eta = \bigvee_\rho \{c\rho \mid \eta \mid_{V \setminus X} = \rho \mid_{V \setminus X}\}$ *for all* $c \in \mathscr{C}, X \subseteq V$,
- $(s_\sigma c)\eta = c(\eta \circ \sigma)$ *for all* $c \in \mathscr{C}, \sigma \in F(V)$.

Hiding means eliminating variables from the support: $supp(\exists_X c) \subseteq supp(c) \setminus X$.[2]

Proof. The properties of the pomonoid action (see Definition 6) are easily shown to hold for both operators. As for the cylindric laws (see Definition 7), first note that the set of functions ρ such that $\eta \mid_{V \setminus X} = \rho \mid_{V \setminus X}$ is actually finite. Thus, we have that

$$\begin{aligned}
(\exists_X (c \otimes \exists_X d))\eta &= \bigvee_\rho \{(c \otimes \exists_X d)\rho \mid \eta \mid_{V \setminus X} = \rho \mid_{V \setminus X}\} \\
&= \bigvee_\rho \{c\rho \otimes (\exists_X d)\rho \mid \eta \mid_{V \setminus X} = \rho \mid_{V \setminus X}\} \\
&= \bigvee_\rho \{c\rho \otimes (\bigvee_\xi \{d\xi \mid \rho \mid_{V \setminus X} = \xi \mid_{V \setminus X}\}) \mid \eta \mid_{V \setminus X} = \rho \mid_{V \setminus X}\} \\
&= \bigvee_\rho \{c\rho \mid \eta \mid_{V \setminus X} = \rho \mid_{V \setminus X}\} \otimes \bigvee_\xi \{d\xi \mid \eta \mid_{V \setminus X} = \xi \mid_{V \setminus X}\} \\
&= (\exists_X c)\eta \otimes (\exists_X d)\eta
\end{aligned}$$

Let us now move to the polyadic laws (see Definition 8).We just consider the third item, and we assume that $\sigma \mid_{\sigma^c(X)}$ is injective, thus

$$\begin{aligned}
(\exists_X s_\sigma c)\eta &= \bigvee_\rho \{(s_\sigma c)\rho \mid \eta \mid_{V \setminus X} = \rho \mid_{V \setminus X}\} = \bigvee_\rho \{c(\rho \circ \sigma) \mid \eta \mid_{V \setminus X} = \rho \mid_{V \setminus X}\} \\
&= \bigvee_\xi \{c\xi \mid (\eta \circ \sigma) \mid_{V \setminus \sigma^c(X)} = \xi \mid_{V \setminus \sigma^c(X)}\} \\
&= (\exists_{\sigma^c(X)} c)(\eta \circ \sigma) = (s_\sigma \exists_{\sigma^c(X)} c)\eta
\end{aligned}$$

[2] The operator is called *projection* in the soft framework, and $\exists_X c$ is denoted $c \Downarrow_{V \setminus X}$.

where it always holds that $\eta \mid_{V\setminus X} = \rho \mid_{V\setminus X}$ implies $(\eta \circ \sigma) \mid_{V\setminus\sigma^c(X)} = (\rho \circ \sigma) \mid_{V\setminus\sigma^c(X)}$, while since $\sigma \mid_{\sigma^c(X)}$ is injective we have that a ξ satisfying $(\eta \circ \sigma) \mid_{V\setminus\sigma^c(X)} = \xi \mid_{V\setminus\sigma^c(X)}$ can be decomposed as $\rho \circ \sigma$ for a ρ such that $\eta \mid_{V\setminus X} = \rho \mid_{V\setminus X}$ (otherwise, it could happen that for some $\{x, y\} \subseteq \sigma^c(X)$ we have that $\sigma(x) = \sigma(y)$ and $\xi(x) \neq \xi(y)$). □

5 Towards a Polynomial Presentation

Soft constraints are an expressive specification formalism, yet they suffer from requiring a potentially complex representation. This section considers a sub-class of constraints that allows for a simple presentation by exploiting the notion of polyadic operators.

Let us start by defining the ReSL of natural numbers $\langle \mathbb{N}, \geq, +, 0 \rangle$, which is the sub-ReSL of the tropical semiring of positive reals (see Remark 2). Also, let $\mathbb{N}(V)$ be the ReSL of functions $p : V \to \mathbb{N}$ with finite support (i.e., such that $supp(p) = \{x \in V \mid p(x) \neq 0\}$ is finite), whose monoidal structure is lifted from $\mathbb{N}$.

Finally, let us now assume for the sake of simplicity that $D \subseteq A$ (even if the definitions below could be parametric with respect to a function $D \to A$), and given $d \in D$ let d^n be just $d \otimes \ldots \otimes d$ for n times, with $\mathbf{1}$ for $n = 0$.

Definition 11 (affine constraints). *A constraint c is* affine *if there exist $p \in \mathbb{N}(V)$ and $i \in A$ such that $\forall \eta.\ c\eta = \bigotimes_{x \in V} \eta(x)^{p(x)} \otimes i$.*

Any such pair $\langle p, i \rangle$ is called a *presentation* of the constraint c. Note that it holds $supp(c) \subseteq supp(p)$: clearly, no $x \notin supp(p)$ belongs to $supp(c)$, while for any $x \in supp(p)$ it also depends on the element i, which might e.g. be the top of the monoid.

A presentation is not necessarily unique, even if a canonical one may exist, as it is going to be shown later on in this section.

Example 3 (A polynomial presentation). Affine constraints can be represented as linear polynomials with coefficients in $\mathbb{N}$ and constant term in $\mathbb{S}$. Consider, for instance, as $\mathbb{S}$ the ReSL of non-negative reals, and as D a finite subset of such reals. An affine constraint with variables in $V = \{x, y\}$ is interpreted as the soft constraint associating with a function $\eta : V \to D$ the real obtained as $(a \times \eta(x)) + (b \times \eta(y)) + i$ (where $a = p(x)$ and $b = p(y)$ are coefficients in $\mathbb{N}$, and i a constant in $\mathbb{S}$). What is relevant is that the composition of such constraints is precisely the addition of polynomials.

As expected, the ordering depends both on the coefficients and the constant term. For example, let us consider the polynomials $2x + 1$ and $x + 5$ and let us assume that $D = \{1, 2, 3\}$: It holds that $(2x + 1) \otimes (x + 5) = (3x + 6)$ and $2x + 1 \leq x + 5$.

Let $\mathscr{A}$ be the set of affine constraints. It is clearly closed with respect to $\otimes$, hence it forms a sub-monoid of $\mathscr{C}$. Moreover, it admits cylindric and polyadic operators.

Proposition 2 (cylindric and polyadic operators for affine constraints). *Let $c \in \mathscr{A}$ be an affine constraint with presentation $\langle p, i \rangle$. Then*

- *$\exists_X c$ admits $\langle p[^0/_X], i \otimes \bigvee_\rho \bigotimes_{x \in X} \rho(x)^{p(x)} \rangle$ as a presentation for all $X \subseteq V$,*
- *$s_\sigma c$ admits $\langle \sum_{y \in \sigma^c(-)} p(y), i \rangle$ as a presentation for all $\sigma \in F(V)$.*

for $p[^0/_X](x) = 0$ if $x \in X$, and $p(x)$ otherwise; and $(\sum_{y \in \sigma^c(-)} p(y))(x) = \sum_{y \in \sigma^c(\{x\})} p(y)$.

In other terms, $\exists_X c$ is obtained by first removing the coefficients of the variables in X in order to get them out of the support of p, and then by adding to i the sup with respect to the mappings $\rho : X \to D$ of the values $\bigotimes_{x \in X} \rho(x)^{p(x)}$.

Proof. First of all, note that $\bigvee_\rho \bigotimes_{x \in X} \rho(x)^{p(x)}$ surely exists, since only the restrictions $\rho|_X : X \to D$ actually matter.

$$\begin{aligned}
(\exists_X c)\eta &= \bigotimes_{y \in V} \eta(y)^{p[^0/_X](y)} \otimes i \otimes \bigvee_\rho \bigotimes_{x \in X} \rho(x)^{p(x)} \\
&= \bigotimes_{y \in V \setminus X} \eta(y)^{p(y)} \otimes i \otimes \bigvee_\rho \bigotimes_{x \in X} \rho(x)^{p(x)} \\
&= \bigvee_\rho \{ \bigotimes_{y \in V \setminus X} \eta(y)^{p(y)} \otimes i \otimes \bigotimes_{x \in X} \rho(x)^{p(x)} \} \\
&= \bigvee_\rho \{ i \otimes \bigotimes_{w \in V} \rho(w)^{p(w)} \mid \eta_{|V \setminus X} = \rho_{|V \setminus X} \} \\
&= \bigvee_\rho \{ c\rho \mid \eta_{|V \setminus X} = \rho_{|V \setminus X} \}
\end{aligned}$$

Let us now consider the polyadic operator and let $\sigma \in F(V)$. Then

$$(s_\sigma c)\eta = c(\eta \circ \sigma) = \bigotimes_{x \in V} (\eta \circ \sigma)(x)^{p(x)} \otimes i$$

$$= \bigotimes_{x \in V} \eta(\sigma(x))^{p(x)} \otimes i = \bigotimes_{y \in \sigma(V)} \eta(y)^{\sum_{x \in \sigma^c(\{y\})} p(x)} \otimes i$$

The summation takes care of the fact that y could be the image along σ of more than one variable. Noting that the value of the summation is 0 for $z \in V \setminus \sigma(V)$, we conclude

$$\bigotimes_{y \in \sigma(V)} \eta(y)^{\sum_{x \in \sigma^c(\{y\})} p(x)} \otimes i = \bigotimes_{y \in V} \eta(y)^{\sum_{x \in \sigma^c(\{y\})} p(x)} \otimes i$$

□

Example 4 (Continued ...). The cylindric operator for the polynomial representation is easily described: e.g. $\exists_x(2x + 1) = \bigvee_{d \in D} 2d + 1 = 3$, that is, the maximum obtained by the evaluation of the constraint with respect to the elements in D. Even simpler is s_σ, which is just the application of substitution σ.

So far, we exploited only the distributive semi-lattice structure of $\mathbb{S}$, precisely in the characterisation of $\exists$. However, the partially ordered monoid $\langle \mathscr{A}, \leq, \otimes, \mathbf{1} \rangle$ is not easily equipped with either a semi-lattice or a residuated structure. Given e.g. the polynomials x and y, there is little chance that $x \vee y$ can be expressed in a linear fashion, since the meaning of $x + y$ is based on the monoidal operator. Similarly for $x \ominus y$, which would require that $\mathbb{S}$ be a group, and possibly the choice of integers $\mathbb{Z}$ as coefficients.

We leave as future work the required investigations on residuation for affine constraints, and we close with a simple characterisation of canonical presentations, as announced at the beginning of the section.

Lemma 13. *Let* $\langle p_1, i_1 \rangle$ *and* $\langle p_2, i_2 \rangle$ *be presentations of an affine constraint* c. *If* $\mathbf{1} \in D$ *then* $i_1 = i_2$ *and* $\langle p_1 \vee p_2, i_1 \rangle$ *is also a presentation of* c.

Proof. Since $\langle p_1, i_1 \rangle$ and $\langle p_2, i_2 \rangle$ are both presentations of c, it holds

$$\forall \eta. \bigotimes_{x \in V} \eta(x)^{p_1(x)} \otimes i_1 = \bigotimes_{x \in V} \eta(x)^{p_2(x)} \otimes i_2$$

Since $\mathbf{1} \in D$, by considering the constant function $\mathbf{1}$ we have

$$\bigotimes_{x \in V} \mathbf{1}^{p_1(x)} \otimes i_1 = \bigotimes_{x \in V} \mathbf{1}^{p_2(x)} \otimes i_2 \implies i_1 = i_2$$

It is now easy to check that $\langle p_1 \vee p_2, i_1 \rangle$ is a presentation of c, since both $\langle p_1, i_1 \rangle$ and $\langle p_2, i_1 \rangle$ are so. □

Finally, note that the ReSL $\mathbb{N}(V)$ of functions $p : V \to \mathbb{N}$ with finite support admits a maximum for possibly infinite subsets, since 0 is the top element of the chosen ReSL of naturals. Thus, an affine constraint c admits a maximal presentation, denoted $\langle p_c, i_c \rangle$. The constant can be additionally restrained, further simplifying the presentation.

Lemma 14. *Let* c_1, c_2 *be affine constraints such that* $c_1 \leq c_2$. *If* $\mathbf{1} \in D$, *then* $i_{c_1} \leq i_{c_2}$.

6 Concluding Remarks

Polynomial constraints are used in many areas of system analysis and verification. For instance, when synthesising program invariants or analysing reachability of hybrid systems. A further application is represented by the generation of measures for proving termination of symbolic programs, as well as rewrite systems. This work strives in the direction of bringing polynomial constraints in the soft constraints framework.

Our proposal moved from previous works, such as [10,11], to build a new and more general framework for representing soft constraints. In this paper, we have observed properties of residuation in monoids equipped with partial orders or semi-lattices; moreover, we have introduced cylindric and polyadic operators in terms of pomonoid actions. This allowed us to exploit properties of 2^V_{fin} and $F(V)$ monoids and to generalise the notion of substitution, by defining any possible function as a substitution and building a new axiomatisation, which mirrors those obtained for previous definitions. This provides a more elegant formalisation, as well as more general and compact laws. Such results can be useful in giving an easier representation of soft constraints in SCSPs.

Furthermore, the absence of diagonals allowed the use of more general structures, which are not necessarily complete lattices, making our formalisation applicable to a larger class of case studies. Also, polyadic algebras with constraints can be used to model many problems via a friendlier polynomial representation.

Polynomials can easily fit real-life problems and can be used in SCCP paradigm to describe a knowledge basis shared among agents: indeed, a possible development of our works is a refitting of observational and behavioural equivalences as described in [11], which needed complete lattices, that could lead to the definition of a denotational semantics for SCCP. As future work, we can also think of defining the class of Polynomial *Soft* Constraint Satisfaction Problems (PSCSPs), as accomplished in [20] with crisp constraints, in order to achieve a similar generalisation with respect to CSPs.

References

1. Aristizábal, A., Bonchi, F., Palamidessi, C., Pino, L., Valencia, F.: Deriving labels and bisimilarity for concurrent constraint programming. In: Hofmann, M. (ed.) FoSSaCS 2011. LNCS, vol. 6604, pp. 138–152. Springer, Heidelberg (2011). https://doi.org/10.1007/978-3-642-19805-2_10
2. Bistarelli, S., Montanari, U., Rossi, F.: Semiring-based constraint satisfaction and optimization. J. ACM **44**(2), 201–236 (1997)
3. Bistarelli, S., Gadducci, F.: Enhancing constraints manipulation in semiring-based formalisms. In: Brewka, G., Coradeschi, S., Perini, A., Traverso, P. (eds.) ECAI 2006. FAIA, vol. 141, pp. 63–67. IOS Press (2006)
4. Bistarelli, S., Montanari, U., Rossi, F.: Soft concurrent constraint programming. ACM Trans. Comput. Logic **7**(3), 563–589 (2006)
5. Bistarelli, S., Pini, M.S., Rossi, F., Venable, K.B.: From soft constraints to bipolar preferences: modelling framework and solving issues. Exp. Theor. Artif. Intell. **22**(2), 135–158 (2010)
6. de Boer, F.S., Palamidessi, C.: A fully abstract model for concurrent constraint programming. In: Abramsky, S., Maibaum, T.S.E. (eds.) CAAP 1991. LNCS, vol. 493, pp. 296–319. Springer, Heidelberg (1991). https://doi.org/10.1007/3-540-53982-4_17
7. Chiarugi, D., Falaschi, M., Olarte, C., Palamidessi, C.: A declarative view of signaling pathways. In: Bodei, C., Ferrari, G.-L., Priami, C. (eds.) Programming Languages with Applications to Biology and Security. LNCS, vol. 9465, pp. 183–201. Springer, Cham (2015). https://doi.org/10.1007/978-3-319-25527-9_13
8. De Boer, F.S., Gabbrielli, M., Marchiori, E., Palamidessi, C.: Proving concurrent constraint programs correct. ACM Trans. Program. Lang. Syst. **19**(5), 685–725 (1997)

9. Fages, F., Ruet, P., Soliman, S.: Linear concurrent constraint programming: operational and phase semantics. Inf. Comput. **165**(1), 14–41 (2001)
10. Gadducci, F., Santini, F.: Residuation for bipolar preferences in soft constraints. Inf. Process. Lett. **118**, 69–74 (2017)
11. Gadducci, F., Santini, F., Pino, L.F., Valencia, F.D.: Observational and behavioural equivalences for soft concurrent constraint programming. Logical Algebraic Methods Program. **92**, 45–63 (2017)
12. Golan, J.: Semirings and Affine Equations over Them. Kluwer (2003)
13. Knight, S., Palamidessi, C., Panangaden, P., Valencia, F.D.: Spatial and epistemic modalities in constraint-based process calculi. In: Koutny, M., Ulidowski, I. (eds.) CONCUR 2012. LNCS, vol. 7454, pp. 317–332. Springer, Heidelberg (2012). https://doi.org/10.1007/978-3-642-32940-1_23
14. Kumar, V.: Algorithms for constraint-satisfaction problems: a survey. AI Mag. **13**(1), 32–44 (1992)
15. López, H.A., Palamidessi, C., Pérez, J.A., Rueda, C., Valencia, F.D.: A declarative framework for security: secure concurrent constraint programming. In: Etalle, S., Truszczyński, M. (eds.) ICLP 2006. LNCS, vol. 4079, pp. 449–450. Springer, Heidelberg (2006). https://doi.org/10.1007/11799573_43
16. Nielsen, M., Palamidessi, C., Valencia, F.D.: Temporal concurrent constraint programming: denotation, logic and applications. Nordic J. Comput. **9**(1), 145–188 (2002)
17. Palamidessi, C., Valencia, F.D.: A temporal concurrent constraint programming calculus. In: Walsh, T. (ed.) CP 2001. LNCS, vol. 2239, pp. 302–316. Springer, Heidelberg (2001). https://doi.org/10.1007/3-540-45578-7_21
18. Sági, G.: Polyadic algebras. In: Andréka, H., Ferenczi, M., Németi, I. (eds.) Cylindric-like Algebras and Algebraic Logic. BSMS, vol. 22, pp. 367–389. Springer, Heidelberg (2013). https://doi.org/10.1007/978-3-642-35025-2_18
19. Saraswat, V.A., Rinard, M.C., Panangaden, P.: Semantic foundations of concurrent constraint programming. In: Wise, D.S. (ed.) POPL 1991, pp. 333–352. ACM Press (1991)
20. Scott, A.D., Sorkin, G.B.: Polynomial constraint satisfaction problems, graph bisection, and the Ising partition function. ACM Trans. Algorithms **5**(4), 45:1–45:27 (2009)

Security and Privacy

Core-concavity, Gain Functions and Axioms for Information Leakage

Arthur Américo, M. H. R. Khouzani, and Pasquale Malacaria(✉)

School of Electronic Engineering and Computer Science,
Queen Mary University of London, Mile End Road, London, UK
{a.passosderezende,arman.khouzani,p.malacaria}@qmul.ac.uk

Abstract. This work explores connections between core-concavity and gain functions, two alternative approaches that emerged in the quantitative information flow community to provide a general framework to study information leakage. In particular (1) we revisit "Axioms for Information Leakage" by replacing averaging with η-averaging and convexity with core-concavity. An interesting consequence of these changes is that the revised axioms capture all Rényi entropies, including the ones not captured by the original formulation of the axioms. (2) We provide an alternative proof for the Coriaceous Theorem based on core-concavity. The general approach of this work is more information theoretical in nature than the work based on gain functions and provides an alternative foundational view of quantitative information flow, rooted on the essential properties of entropy as a measure of uncertainty.

1 Introduction

In their seminal CSF 2016 paper [1] Mario S. Alvim, Konstantinos Chatzikokolakis, Annabelle McIver, Carroll Morgan, Catuscia Palamidessi and Geoffrey Smith studied quantitative information flow (QIF) axiomatically, proving important interactions among different axioms related to information leakage, like non-interference, data processing and monotonicity (conditioning reduces entropy). The work had two important contributions: (1) to present a meaningful axiomatization for QIF and (2) to show the completeness of g-vulnerabilities – an information measure based on gain functions [2] – in terms of these axioms. Hence, g-vulnerabilities, by corresponding to convex functions, constitute one of the fundamental axioms for QIF. However, it did not recover all known entropy measures, for instance, the Rényi family.

At the same conference, another contribution was presented [9], which also introduced a general notion of leakage and studied its properties. The motivation in [9] was to study "universal optimality" i.e. to design systems that leak as little information as possible no matter how leakage is measured. To study universal optimality, a generalization of entropy was introduced, under the name of *core-concavity*. As the name suggests, a core-concave function is in its essence a concave function; in fact, it is defined as an increasing function composed with a

M. S. Alvim et al. (Eds.): Palamidessi Festschrift, LNCS 11760, pp. 261–275, 2019.
https://doi.org/10.1007/978-3-030-31175-9_15

concave function. This definition generalizes all definitions of entropy that we are aware of, including families known in the information theoretical community like Sharma-Mittal entropies (which include Rényi's entropies and thus Shannon and min-entropy as limit cases). Core-concave functions encompass other well-known and meaningful leakage measures like guessing entropy that fall outside of this family. Notably, the core-concave formulation can recover any g-vulnerability based entropy as well.

A natural question is how core-concavity and gain functions relate in the context of information leakage? In particular, since core-concave entropies generalise g-vulnerabilities, can they be obtained from a more general set of axioms? This paper approaches this question by revisiting the axioms for information leakage from [1] and replacing convexity with core-concavity. For the axioms to work, a new notion of averaging is introduced, called η-average. Similar results as in [1] are then established.

A second contribution of this work is to provide an alternative proof of the Coriaceous Theorem. The proof uses the elegant work by Dahl [5] in 1999 on matrix majorization. Matrix majorization is a more general order than channel refinement, in that it is defined on all matrices over the real numbers, not just channels; the class of functions reflecting this matrix order turns out to be positively homogeneous concave functions, which are shown to be in correspondence with our posterior core-concave entropies.

2 Preliminaries

In this section we introduce some notations and briefly overview some basic concepts of QIF.

2.1 Notational Conventions

Throughout the paper, we will use $X, Y, Z, \ldots$ for random variables. Each random variable (r.v., for short) takes values over a subset $\{1, \ldots, n\}$ of natural numbers—thus, all random variables are assumed to have finite range. Probability distributions will be noted by p, with $p(x)$ or p_x denoting elements of the "categorical" distribution. We may specify the r.v. with a subscript, e.g., write $p_X(x)$, if it is not clear from the context.

Let $\Delta_n \subset \mathbb{R}^n$ be the $(n-1)$-dimensional probability simplex. Given a probability distribution p over $\{1, \ldots, n\}$, we overload the notation and use p to refer to its probability vector $(p_1, \ldots, p_n) \in \Delta_n$. Hence, given a function F over Δ_n and a random variable X with distribution $p = (p_1, \ldots, p_n)$, we may use $F(X)$, $F(p_1, \ldots, p_n)$ and $F(p)$ interchangeably.

2.2 Secrets and Channels

The main goal in QIF is to quantify leakage of information in systems. A *secret* is represented by a r.v. X, which takes values in a nonempty finite set $\mathcal{X} = \{1, \ldots, n\}$, according to a given distribution $p \in \Delta_n$.

A system takes as *input* a secret value $x \in \mathcal{X}$ and produces an *observable* y in a nonempty finite set $\mathcal{Y}$. It is represented by a channel C, which is a row stochastic matrix with rows indexed by $\mathcal{X}$ and columns indexed by $\mathcal{Y}$. The value $C(x, y)$ is the conditional probability that y is produced when x is the secret value. The notation $C : \mathcal{X} \to \mathcal{Y}$ means that the channel C has $\mathcal{X}$ as input set and $\mathcal{Y}$ as output set.

The framework postulates the existence of an *adversary* interested in the value of the secret, who knows both the prior distribution p and the channel C. This knowledge yields a joint distribution $p(x, y) = p(x)C(x, y)$, and the marginal $p(y) = \sum_x p(x, y)$. By observing the observable produced by the system, they are able to update their knowledge to the distribution $p_{X|y}$, given by Bayes' rule as $p_{X|y}(x) = {p(x,y)}/{p(y)}$.

Using an *entropy measure*, which is a map from probability distributions to the real numbers, it is possible to quantify the information of the adversary about the secret given distributions p and $p_{X|y}$, and therefore, infer the amount of information leaked by the system.

2.3 Entropy Measures

In this paper, we define an entropy measure as a real valued function over Δ_n, for a fixed $n = |\mathcal{X}|$. It reflects how much *uncertainty* a probability distribution represents. It is also possible to measure the dual notion, that of *vulnerability* or Bayes' risk, which in QIF measures how vulnerable the secret value is by quantifying the quality of the best estimator of the adversary.

The most well-known of such functions is perhaps Shannon Entropy [14]. Given a random variable X with probability distribution p, it is defined as

$$H_1(X) = -\sum_{x \in \mathcal{X}} p(x) \log(p(x))$$

with the convention that $p_i \log(p_i)$ is zero if $p_i = 0$. Shannon entropy captures uncertainty as it reflects the expected number of subset membership questions ("is the value in a given subset") in order to identify the correct realization of X. Given a random variable Y, the *conditional* or *posterior* Shannon entropy is given by

$$H_1(X|Y) = \sum_{y \in \mathcal{Y}^+} p(y) H_1(p_{X|y})$$

where $\mathcal{Y}^+$ is the support of p_Y, i.e., all the observables that can be produced with non-zero probability. $H_1(X|Y)$ therefore represents the average Shannon Entropy of the secret after the execution of the system.

Many other such measures have been introduced in the literature, such as *guessing entropy* [11], which is the expected number of guesses to get to the correct realisation, and generalizations of Shannon entropy to the Rényi family [13]. The Rényi entropy of order α is given by

$$H_\alpha(X) = \frac{\alpha}{1-\alpha} \log \|p\|_\alpha$$

for $\alpha \in (0,1) \cup (1,\infty)$. Shannon entropy can then be recovered by taking $\alpha \to 1$ and min-entropy, i.e., $-\log \max_x p(x)$, by taking $\alpha \to \infty$.

2.4 The g-Leakage Framework

While being the preferred choice for most of early work in QIF, Smith [16] argued that Shannon Entropy might be unsuitable for some scenarios, such as when an attacker can make a single guess about the secret. In these settings, he suggested the use of min-entropy, defined as $H_\infty(X) = -\log \max_x p(x)$, as it depends only on the best guess of the adversary. Following a similar reasoning, one could think about a myriad of attack scenarios that would warrant different uncertainty measures, such as an attacker being able to make a finite number of guesses, or just being interested in a property of the secret or an approximation of its value.

This motivated the introduction of the g-leakage framework [2], which is predicated in a gain function $g : \mathcal{W} \times \mathcal{X} \to \mathbb{R}$. The set $\mathcal{W}$ is the set of *actions* the adversary can take, and $\mathcal{X}$ is the possible values of the secret. The value of $g(w,x)$ reflects the gain of the adversary when choosing $w \in \mathcal{W}$ when the secret value is $x \in \mathcal{X}$. In this framework, one measures how *vulnerable* the secret is by the g-vulnerability:

$$V_g(X) = \sup_w \sum_{x \in \mathcal{X}} p(x) g(w,x).$$

The *posterior* g-vulnerability is then computed as

$$V_g(X|Y) = \sum_{y \in \mathcal{Y}^+} p(y) V_g(X|y)$$

where $\mathcal{Y}^+$ is the support of $p(y)$. In [1], the set of g-vulnerabilities was shown to coincide with that of nonnegative, continuous convex functions over the probability simplex.

3 Core-concavity and Basic Properties

The essence of entropy is to be a measure of "uncertainty". Typically this translates into a real valued function over probability distributions, such that the more uncertainty is associated with a distribution, the higher the value of the function. For example, if the function is symmetric, it should attain its minimum on a *point distribution* $(1,0,\dots,0)$ and its maximum on the uniform distribution $(1/n,\dots,1/n)$.

Consider the following setting as argued in [1]: Suppose we have a coin with "head" and "tail" probabilities of α and $1-\alpha$ and a random variable X. If the outcome of the coin flip is "head", the distribution that X is drawn from is p^1, and if it is "tails", it is p^2. If the outcome of the coin flip is not observed, the distribution of X will be $p^3 = \alpha p^1 + (1-\alpha)p^2$. One would expect that a measure of uncertainty F to be lower in the scenario that the outcome of the coin-flip is observed compared to when it is not; that is:

$$\alpha F(p^1) + (1-\alpha)F(p^2) \leq F(\alpha p^1 + (1-\alpha)p^2) \tag{1}$$

In other words, F should be concave. However, we can make the observation that applying an increasing function η to both sides of the inequality in (1) preserves it, i.e.:

$$\eta(\alpha F(p^1) + (1-\alpha)F(p^2)) \leq \eta(F(\alpha p^1 + (1-\alpha)p^2))$$

Hence the above inequality, using the pair (η, F), can be justified as a measure of uncertainty in the same way as F is; moreover the pair (η, F) includes strictly more than concave functions. The pair (η, F) defines what we named a core-concave function:

Definition 1. *A* core-concave entropy $H = (\eta, F)$ *is a pair such that:*

1. *F is a real-valued function over Δ_n that is concave and continuous.*
2. *$\eta : image(F) \to \mathbb{R}$ is a continuous and strictly increasing real valued function defined over $image(F)$, i.e., the image of F.*

Let $\mathcal{H}$ denote the set of core-concave entropies.

Given $H = (\eta, F) \in \mathcal{H}$, we define $H(p) = \eta(F(p))$. We call the function $\eta \circ F$ the prior form, *or simply* prior, *of H.*

Definition 2. *Given a core-concave $H = (\eta, F)$, its* conditional *or* posterior *form is given by, where $\mathcal{Y}^+$ is the support of $p(y)$,*

$$H(X|Y) = \eta\left(\sum_{y\in\mathcal{Y}^+} p(y)F(X|y)\right).$$

For instance, Shannon entropy (H_1) can be readily captured by the pair (η, F) where $\eta(t) = t$ and $F(p) = -\sum_i p_i \log p_i$, min-entropy ($H_\infty$) by $\eta(t) = -\log(-t)$ and $F(p) = -\max p_i$, and guessing entropy (H_G) by $\eta(t) = (t)$ and $F(p) = \sum i p_{[i]}$ (where $p_{[1]}, \ldots, p_{[n]}$ is a non-increasing rearrangement of p).

Core-concave entropies also encompass a more general family of entropies referred to as Sharma-Mittal [15] defined as follows:

$$H_{\alpha,\beta}(p) = \frac{1}{\beta-1}\left(1 - (\|p\|_\alpha^\alpha)^{\frac{1-\beta}{1-\alpha}}\right), \quad \alpha \geq 0,\ \alpha, \beta \neq 1. \tag{2}$$

Up to a scaling factor of $1/\log(e)$, this family generalises Rényi $H_{\alpha,\beta\to1}(p)$, Shannon $H_{\alpha\to1,\beta\to1}(p)$, and Havrda-Tsallis entropies [6,17]: $H_{\alpha,\alpha}(p) = \frac{1}{1-\alpha}(1 - \|p\|_\alpha^\alpha)$. Core-concavity of Sharma-Mittal family $H_{\alpha,\beta}(p)$ can be seen by taking, if $\alpha > 1$,

$$\eta(t) = \frac{1}{\beta-1}\left(1 - (-t)^{\frac{1-\beta}{1-\alpha}}\right), \quad F(p) = -\|p\|_\alpha^\alpha,$$

and, if $0 < \alpha < 1$,

$$\eta(t) = \frac{1}{\beta-1}\left(1 - t^{\frac{1-\beta}{1-\alpha}}\right), \quad F(p) = \|p\|_\alpha^\alpha.$$

3.1 Core-concavity, Quasi-concavity and Schur-concavity

Besides being a generalizing framework for entropy measures, core-concavity can be seen as a property of real valued functions over Δ_n. Given $f : \Delta_n \to \mathbb{R}$, we say that f is a *core-concave function* if there is $(\eta, F) \in \mathcal{H}$ such that $f(p) = \eta(F(p))$. The set of core-concave functions includes, in particular, the prior forms of all entropy measures defined above.

In the following, we shed some light on the relation of the core-concave property with quasi-concavity and Schur-concavity, which are other generalisations of concavity. By definition, a real valued function ϕ over some convex subset of $\mathbb{R}^n$ is *quasi-concave* if for all $\alpha \in \mathbb{R}$, the set $\{x | \phi(x) \leq \alpha\}$ is a convex set. Equivalently, it must be that for all $\lambda \in [0, 1]$, and all x, y in the domain of ϕ we have: $\phi(\lambda x + (1 - \lambda)y) \geq \min\{\phi(x), \phi(y)\}$.

Proposition 1. *Any core-concave function is also quasi-concave.*

Proof. The statement of the proposition follows from the following two facts:

1. All concave functions are quasi-concave; and
2. If F is quasi-concave and η is increasing then $\eta \circ F$ is quasi-concave.

□

Given $p, q \in \Delta_n$, we say that p *majorizes* q if for all $k \leq n$, $\sum p_{[i]} \geq \sum q_{[i]}$, where $(p_{[1]}, \ldots, p_{[n]})$, $(q_{[1]}, \ldots, q_{[n]})$ are nondecreasing rearrangements of p, q. A function ϕ is said to be *Schur-concave* if $\phi(p) \leq \phi(q)$ whenever p majorizes q. As any symmetric quasi-concave function is Schur-concave [10, Chapter 3.C], an immediate consequence of Proposition 1 is the following.

Corollary 2. *Any symmetric core-concave function is Schur-concave.*

Note, however, that we are not assuming symmetry in this paper.

3.2 Choices of η and F

When describing an entropy with our framework the choices of η and F are, in general, not unique. For example, one could have equally define Shannon entropy in this framework by choosing $\eta(x) = 2x$ and $F(p) = -\frac{1}{2}\sum_i p_i \log p_i$. While the choices of η and F are immaterial to the values of the prior form $H(p)$, they can radically change its posterior form. Consider, for instance, the Rényi entropies:

$$H_\alpha(p) = \frac{\alpha}{1-\alpha} \log \|p\|_\alpha,$$

which, for $\alpha > 1$ can be recovered by choosing either $\eta(x) = ({}^{\alpha}/_{1-\alpha}) \log(-x)$ and $F(p) = -\|p\|_\alpha$ or $\eta'(x) = ({}^{1}/_{1-\alpha}) \log(-x)$ and $F'(p) = -\|p\|_\alpha^\alpha$. These choices induce, respectively, the following two posterior forms:

$$H_\alpha(X|Y) = \frac{\alpha}{1-\alpha} \log \sum_{y \in \mathcal{Y}^+} p(y) \|p_{X|y}\|_\alpha, \tag{3a}$$

$$H'_\alpha(X|Y) = \frac{1}{1-\alpha}\log \sum_{y\in\mathcal{Y}^+} p(y)\|p_{X|y}\|_\alpha^\alpha. \tag{3b}$$

Indeed, both forms have been proposed in the literature [3,7], and they do not coincide. In particular, only the first form coincides with posterior min-entropy as $\alpha \to \infty$—that is, $\lim_{\alpha\to\infty} H_\alpha(X|Y) = H_\infty(X|Y)$ [8].

However, the posterior form of an entropy uniquely identifies, up to a linear transformation, the choice of η and F.

Theorem 3. *Let $H^1{=}(\eta_1, F_1)$ and $H^2{=}(\eta_2, F_2)$ be core-concave entropies. If $H^1(X|Y) = H^2(X|Y)$ for all r.v. X, Y, then $\eta_2(x) = \eta_1(ax+b)$ and $F_2(p) = ({}^1\!/\!{}_a)F_1(p) - {}^b\!/\!{}_a$ for some $a, b \in \mathbb{R}$.*

Proof. Let $H^1 = (\eta_1, F_1), H^2 = (\eta_2, F_2) \in \mathcal{H}$ such that for all random variables X, Y, $\eta_1(F_1(X|Y)) = \eta_2(F_2(X|Y))$. As the image of η_1 and η_2 coincide, we can define $\phi = \eta_1^{-1} \circ \eta_2$. Now, for all probability vectors $p \in \Delta_n$, we have:

$$\eta_1(F_1(p)) = \eta_2(F_2(p)) \implies F_1(p) = \phi(F_2(p))$$

Given any $p^1, p^2 \in \Delta_n$, there is always a choice of X, Y that yields a posterior entropy of the form $\eta_1(tF(p^1) + (1-t)F(p^2))$. To see that, consider as prior $p = tp^1 + (1-t)p^2$ and channel C as in the proof of Proposition 6. Then, $H^1(X|Y) = \eta_1(tF(p^1) + (1-t)F(p^2))$.

As the two descriptions coincide on the posterior entropy, we obtain, for any $t \in [0,1]$:

$$\phi(tF_2(p^1){+}(1{-}t)F_2(p^2)) = tF_1(p^1){+}(1{-}t)F_1(p^2) = t\phi(F_2(p^1)){+}(1{-}t)\phi(F_2(p^2))$$

Since F is a continuous function over the compact metric space Δ_n, it attains its maximum and minimum. Let $w = \min_p F_2(p)$ and $z = \max_p F_2(p)$ over this space. Thus, the range of F_2 is $[w, z]$. Let $r \in [w, z]$. Substituting $t = \frac{r-w}{z-w}$ in the above equation, we obtain:

$$\phi(r) = \frac{r-w}{z-w}\phi(z) + \left(1 - \frac{r-w}{z-w}\right)\phi(w) = \left(\frac{\phi(z)-\phi(w)}{z-w}\right)r + \frac{z\phi(w) - w\phi(z)}{z-w}$$

Thus, $\phi(r) = ar + b$, for $a = \frac{\phi(z)-\phi(w)}{z-w}$ and $b = \frac{z\phi(w)-w\phi(z)}{z-w}$. Therefore, $\eta_2(r) = \eta_1(\phi(r)) = \eta_1(ar+b)$, and $F_1(p) = \phi(F_2(p)) = aF_2(p) + b$. □

3.3 Relation Between Core-concavity and g-Leakage

There is a clear intuitive connection between core-concavity and g-vulnerability, in that core-concave functions are pairs (η, F) where F is concave and g-vulnerabilities are equivalent to convex functions [1]. Indeed, there is a simple injection from g-vulnerabilities to core-concave entropies, which maps the convex g-vulnerability V_g to the core-concave function $(I, -V_g)$ where I is the identity function: $I(t) = t$.

This embedding is non-surjective because there are core-concave functions which cannot be written as $(I, -V_g)$ for any convex function V_g. Moreover consider the mapping from posterior vulnerability to posterior core-concave:

$$\sum_{y \in \mathcal{Y}^+} p(y) V_g(X|y) \quad \mapsto \quad I(\sum_{y \in \mathcal{Y}^+} -p(y) V_g(X|y)),$$

this is not only non-surjective but is the "wrong" mapping for most entropies, e.g., either forms of the posterior Rényi entropies in (3) (recall that the η's for Rényi entropies were not the identity). This is the main motivation why in revisiting the axioms for information leakage, we will not use the average axiom but the η-averaging.

4 Axioms

In this section, we expand the axiomatic approach in [1] to core-concave entropies. Let $H = (\eta, F)$ be a pair such that η is a strictly increasing real-valued function over a real interval and $F : \Delta_n \to \mathbb{R}$, and define $H(X) = \eta(F(X))$. Notice that every real-valued function $H : \Delta_n \to \mathbb{R}$ can be constructed in this way, by taking $\eta(x) = x$ and $F = H$.

Given the pair $H = (\eta, F)$, we associate with it a posterior form $H(X|Y)$, which is the extension of H to a distribution over distributions. In particular, $(X|Y)$ represents the family of random variables $(X|y)_{y \in Y}$ where $(X|y)$ is the random variable with distribution $p_{X|y}$. Given a channel $C : \mathcal{X} \to \mathcal{Y}$ and a distribution over $\mathcal{X}$, the family of distributions $(X|Y)$ can be computed using Bayes' rule. This construction is similar to *hyper-distributions*, as defined in [1,12].

Definition 3. *A pair $H = (\eta, F)$ defined as above respects:*

`CCV` ***(Core-concavity)****: if H is core-concave (as in Definition 1).*
`EAVG` ***(η-Averaging)****: if, given r.v. X, Y, the posterior form $H(X|Y)$ is defined as:*

$$H(X|Y) = \eta\left(\sum_{y \in \mathcal{Y}^+} p(y) F(X|y)\right).$$

`NI` ***(Non-Interference)****: if given r.v. X, Y such that Y independent from X,*

$$H(X|Y) = H(X).$$

`MONO` ***(Monotonicity)****:*[1] *if for all r.v. X, Y, $H(X|Y) \leq H(X)$.*
`DPI` ***(Data-Processing Inequality)****: if, for all r.v. X, Y, Z such that $X \to Y \to Z$,*

$$H(X|Y) \leq H(X|Z).$$

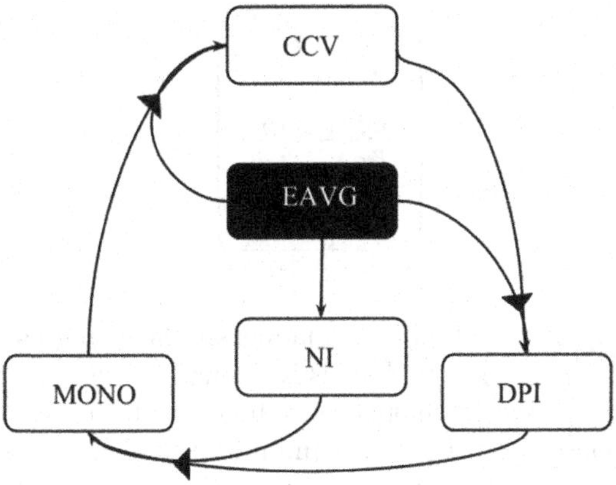

Fig. 1. The implications graph among the axioms.

4.1 Relations Between the Axioms

This section proves the results illustrated in Fig. 1 (cf. Fig. 2 of [1]).

Proposition 4. EAVG $\Rightarrow$ NI.

Proof. Assume X and Y are independent random variables. Then:

$$H(X|Y) = \eta\left(\sum_{y\in\mathcal{Y}^+} p(y)F(X|y)\right) = \eta\left(\sum_{y\in\mathcal{Y}^+} p(y)F(X)\right) = \eta \circ F(X) = H(X).$$

□

Proposition 5. NI + DPI $\Rightarrow$ MONO.

Proof. Let Z be a random variable over the singleton set $\mathcal{Z} = z_1$, so that $p(z_1) = 1$. Consider the composition $X \to Y \to Z$. Then:

$$H(X|Y) \underset{\texttt{DPI}}{\leq} H(X|Z) \underset{\texttt{NI}}{=} H(X).$$

NI equality follows by noting that Z is independent of X. □

Proposition 6. EAVG + MONO $\Rightarrow$ CCV.

Proof. Let X_1, X_2 be random variables with arbitrary distributions $p^1, p^2 \in \Delta_n$, and let X be random variable with distribution $p = \alpha p^1 + (1-\alpha)p^2$ for an $\alpha \in [0,1]$. Suppose that p is full support and consider the following channel, which takes in X and produces Y from a set of two observables:

[1] This is also known in the literature as "conditioning reduces entropy" (CRE).

$$\begin{pmatrix} \frac{\alpha p_1^1}{p_1} & \frac{(1-\alpha)p_1^2}{p_1} \\ \vdots & \vdots \\ \frac{\alpha p_i^1}{p_i} & \frac{(1-\alpha)p_i^2}{p_i} \\ \vdots & \vdots \\ \frac{\alpha p_n^1}{p_n} & \frac{(1-\alpha)p_n^2}{p_n} \end{pmatrix}$$

If p is not full support, we build the channel similarly, but choosing any values for the kth row whenever $p_k = 0$. This is incidentally the same channel C^* used in Proposition 8 of [1]. The channel is constructed such that the posterior for the observable corresponding to the first column (call it y_1), i.e., $p(X|y_1)$ is p^1 and the posterior for the second column y_2, i.e., $p(X|y_2)$ is p^2; moreover, $p(y_1) = \alpha$ and $p(y_2) = 1 - \alpha$. Hence, EAVG implies

$$H(X|Y) \;\; = \eta\left(\alpha F(p^1) + (1-\alpha)F(p^2)\right)$$

On the other hand, $H(X)$ can be written as $\eta(F(p))$. From MONO, we have $H(X|Y) \le H(X)$. Hence:

$$\eta(\alpha F(p^1) + (1-\alpha)F(p^2)) \le \eta\left(F\left(\alpha p^1 + (1-\alpha)p^2\right)\right)$$

Since η is strictly increasing, the above yields concavity of F:

$$\alpha F(p^1) + (1-\alpha)F(p^2) \le F(\alpha p^1 + (1-\alpha)p^2).$$

Hence $H = (\eta, F)$ is core-concave. □

Proposition 7. EAVG + CCV ⇒ DPI.

Proof. Assume $X \to Y \to Z$. Since $p(y) = \sum_{z\in\mathcal{Z}^+} p(z)p(y|z)$, we can write:

$$\begin{aligned}
&\sum_{y\in\mathcal{Y}^+} p(y)F\left(X|y\right) \\
&= \sum_{y\in\mathcal{Y}^+}\left(\sum_{z\in\mathcal{Z}^+} p(z)p(y|z)\right) F\left(X|y\right) \\
&=_{(a)} \sum_{y\in\mathcal{Y}^+, z\in\mathcal{Z}^+} p(z)p(y|z)F\left(X|y,z\right) \\
&\le \sum_{z\in\mathcal{Z}^+} p(z)F\left(\sum_{y\in\mathcal{Y}^+} p(y|z)p_{X|y,z}\right) \\
&=_{(b)} \sum_{z\in\mathcal{Z}^+} p(z)F\left(X|z\right)
\end{aligned}$$

Equality (a) follows because $X \rightarrow Y \rightarrow Z$ implies $p_{X|y} = p_{X|y,z}$. Next, Jensen's inequality is applied for concave F, noting that $p(y|z)$ for $y \in \mathcal{Y}^+$ constitute convex coefficients. Equality (b) uses:

$$\sum_{y \in \mathcal{Y}^+} p(y|z) p_{X|y,z} = p_{X|z}.$$

The proposition follows by applying $\eta(\cdot)$ to the steps, noting that η preserves the inequality since it is increasing. □

This completes all the relations between axioms illustrated in Fig. 1.

5 Recovering MAX as a Limit Case of CCV+EAVG

In Section V-B of [1], the authors define an alternative definition of posterior vulnerability based on the "worst case" scenario, by replacing the averaging rule with maximum. In the language of posterior entropy, that would be defining:

$$H_{\max}(X|Y) = -\max_{y \in \mathcal{Y}^+} V_g(X|y) \tag{4}$$

where $\mathcal{Y}^+$ designates the support of the probability distribution over the output (or the "outer" distribution as referred to in [1]).

Here, we show how such definition can be retrieved as a limit case of our framework of CCV+EAVG. Following [1] V_g is always non-negative. We have then:

Proposition 8

$$\lim_{\alpha \to \infty} \left(\sum_{y \in \mathcal{Y}^+} p(y) \left(V_g(X|y)\right)^{\alpha} \right)^{\frac{1}{\alpha}} = \max_{y \in \mathcal{Y}^+} V_g(X|y)$$

Proof. Because $\sum_{y \in \mathcal{Y}^+} p(y) = 1$, we have:

$$p(y^*) \left(V_g(X|y^*)\right)^{\alpha} \leq \sum_{y \in \mathcal{Y}^+} p(y) \left(V_g(X|y)\right)^{\alpha} \leq \max_{y \in \mathcal{Y}^+} \left(V_g(X|y)\right)^{\alpha}$$

where $y^* \in \arg\max_{y \in \mathcal{Y}^+} V_g^{\alpha}(X|y)$. Since, $f(t) = t^{1/\alpha}$ for $\alpha > 0$ is an increasing function over $\mathbb{R}^+$, the above inequalities yield:

$$\left(p(y^*) \left(V_g(X|y^*)\right)^{\alpha}\right)^{1/\alpha} \leq \left(\sum_{y \in \mathcal{Y}^+} p(y) \left(V_g(X|y)\right)^{\alpha} \right)^{1/\alpha} \leq \left(\max_{y \in \mathcal{Y}^+} \left(V_g(X|y)\right)^{\alpha} \right)^{1/\alpha}$$

The first and second inequalities respectively imply:

$$\liminf_{\alpha \to \infty} \left(\sum_{y \in \mathcal{Y}^+} p(y) \left(V_g(X|y)\right)^{\alpha} \right)^{1/\alpha} \geq \left(V_g(X|y^*)\right)^{\alpha} = \max_{y \in \mathcal{Y}^+} \left(V_g(X|y)\right)^{\alpha}$$

$$\limsup_{\alpha \to \infty} \left(\sum_{y \in \mathcal{Y}^+} p(y) \left(V_g(X|y)\right)^{\alpha} \right)^{1/\alpha} \leq \max_{y \in \mathcal{Y}^+} \left(V_g(X|y)\right)^{\alpha}$$

which establish the claim. □

Corollary 9. *The worst case posterior entropy as defined in* (4) *for any concave* V_g *can be recovered as a limit case of a family of* CCV+EAVG.

Proof. The corollary follows from Proposition 8 by taking $\eta(x) = -(-x)^{1/\alpha}$ and $F(p) = -(V_g(p))^{\alpha}$, noting that for $\alpha > 0$, $-(-x)^{1/\alpha}$ is increasing for $x \in \mathbb{R}_{\leq 0}$, and F is concave over the probability simplex, since it is a composition of the function $f(t) = -t^{\alpha}$, which is decreasing and concave over $\mathbb{R}_{\geq 0}$, and V_g, which is convex over the probability simplex. □

6 The Coriaceous Theorem for Core-Concave Functions

Given channels C_1, C_2 the refinement ordering [12] is defined as:[2]

$$C_1 \sqsubseteq_\circ C_2 \Leftrightarrow \text{ there exists a channel } R. \ s.t. \ C_2 = C_1 R.$$

The Coriaceous Theorem [12] establishes an equivalence between this ordering and the ordering of g-vulnerabilities [2]. In particular it states that $C_1 \sqsubseteq_\circ C_2$ if and only if, for all priors p and all g-vulnerabilities V_g, the posterior vulnerability of C_1 with prior p is never less than the posterior vulnerability of C_2 with prior p.

An alternative proof for the Coriaceous theorem can be derived directly from core-concavity, without appealing to the relation between them and the g-leakage framework. We first need some auxiliary results and definitions:

Definition 4. *Given a continuous and concave function* $F : \Delta_n \to \mathbb{R}$, *we define* G_F, *for all* $\mathbf{q} = \{q_1, \dots, q_n\} \in \mathbb{R}^n_{\geq 0}$, *as*

$$G_F(\mathbf{q}) = \left(\sum_{i=1}^{n} q_i\right) F\left(\frac{q_1}{\sum_{j=1}^{n} q_j}, \dots, \frac{q_n}{\sum_{j=1}^{n} q_j}\right)$$

if $\mathbf{q}$ *is not the null vector, and* $G_F(0, \dots, 0) = 0$[3]. *In short, using the 1-norm notation, we have* $G_F(\mathbf{q}) = \|\mathbf{q}\|_1 F\left(\frac{\mathbf{q}}{\|\mathbf{q}\|_1}\right)$.

Given $H = (\eta, F) \in \mathcal{H}$, it is possible to define the posterior entropy $H(X|Y)$ in terms of the functions G_F and of the joint probability distribution $p(x, y)$ in the following way

$$H(X|Y) = \eta\left(\sum_{y \in \mathcal{Y}^+} G_F(p(x_1, y), \dots p(x_n, y))\right) \tag{5}$$

Definition 5. *A* cone $V \subset \mathbb{R}^n$ *is a set such that for all* $\lambda \in \mathbb{R}_{\geq 0}$, $\mathbf{q} \in V \implies \lambda\mathbf{q} \in V$. *A function* $\phi : V \to \mathbb{R}$ *over a cone* $V \subset \mathbb{R}^n$ *is* positively homogeneous *if, for all* $\lambda \in \mathbb{R}_{\geq 0}$ *and* $\mathbf{q} \in V$, $\phi(\lambda\mathbf{q}) = \lambda\phi(\mathbf{q})$.

[2] Note that the symbol $\sqsubseteq_\circ$ has its direction reversed in older literature [2,12].

[3] Notice that, as F is continuous over a compact set (and thus, bounded), $\lim_{\epsilon \to 0^+} G_F(\epsilon\mathbf{u}) = 0 = G_F(0, \dots, 0)$, for all $\mathbf{u} \in \mathbb{R}^n_{\geq 0}$. Hence, G_F is continuous.

Proposition 10. *A function $\phi : \mathbb{R}^n_{\geq 0} \to \mathbb{R}$ is continuous, positively homogeneous and concave if, and only if, it coincides with G_F for some continuous and concave $F : \Delta_n \to \mathbb{R}$.*

Proof. First, we prove that G_F is continuous, positively homogeneous and concave. It is immediate to see that G_F is continuous and positively homogeneous. To see that G_F is concave, consider the affine map

$$g_1(q_1, \dots, q_n) = (q_1, \dots, q_n, \sum_i q_i)$$

and the perspective function $g_2(\mathbf{q}, t) = tF(\mathbf{q}/t)$, which is concave since F is concave [4, Chap. 3.2.6]. Then, for all non zero $\mathbf{q} \in \mathbb{R}^n_{\geq 0}$,

$$G_F(\mathbf{q}) = g_2(g_1(\mathbf{q})).$$

Therefore, being the composition of an affine map and a concave function, G_F is concave [4, Chap. 3.2.2].

Conversely, suppose $\phi : \mathbb{R}^n_{\geq 0} \to \mathbb{R}$ is continuous, positively homogeneous and concave. It is then possible to write ϕ as a G_F by taking $F = \phi|_{\Delta_n}$, where $\phi|_{\Delta_n}$ the restriction of ϕ to Δ_n. To see that, notice that $\phi(\mathbf{q}/\|\mathbf{q}\|_1) = \phi(\mathbf{q})/\|\mathbf{q}\|_1$, and hence:

$$G_{\phi|_{\Delta_n}}(\mathbf{q}) = \|\mathbf{q}\|_1 \phi\left(\frac{\mathbf{q}}{\|\mathbf{q}\|_1}\right) = \phi(\mathbf{q}).$$

□

The proof of the Coriaceous Theorem for core-concave entropies now follows directly from the following result from [5]:

Theorem 11. *Let A and B be real-valued matrices with m rows, and denote by A^i, B^i their i-th column. There is a row stochastic matrix R such that $AR = B$ if and only if for all positively homogeneous convex functions $\phi : \mathbb{R}^m \to \mathbb{R}$,*

$$\sum_i \phi(A^i) \geq \sum_j \phi(B^j).$$

Theorem 12 (Coriaceous Theorem for core-concave). *Given channels $C : \mathcal{X} \to \mathcal{Y}$ and $D : \mathcal{X} \to \mathcal{Z}$, $D = CR$ for some channel $R : \mathcal{Y} \to \mathcal{Z}$ if, and only if, for all distributions p over X, $H(X|Y) \leq H(X|Z)$ for all core-concave functions H.*

Proof. The forward implication is the "data processing" inequality and is proven in Proposition 7.

For the reverse implication, suppose that $H(X|Y) \leq H(X|Z)$ for a full support p and all core-concave H. Let diag(p) be the matrix with the values of the prior in the diagonal and 0 otherwise, and let $A = \text{diag}(p)C$, $B = \text{diag}(p)D$—that is, A and B are the matrices of the joint distributions obtained from p and

C, D. Then (5) yields, for all $(\eta, F) \in \mathcal{H}$ (and therefore, for all concave functions F),

$$\sum_{y \in \mathcal{Y}^+} G_F(A^y) \leq \sum_{z \in \mathcal{Z}^+} G_F(B^z).$$

Notice that all positively homogeneous functions over $\mathbb{R}^n$ are also positively homogeneous over $\mathbb{R}^n_{\geq 0}$, and that a convex function over $\mathbb{R}^n$ is continuous. Therefore, from Theorem 11 and Proposition 10, there is a row stochastic (channel) matrix R such that $AR = B$. As p has full support, $\text{diag}(p)$ is non-singular, and therefore $CR = \text{diag}(p)^{-1}AR = \text{diag}(p)^{-1}B = D$. □

7 Conclusion

This paper provided a more general axiomatic foundations for quantitative information flow based on the notion of core-concavity. Compared to the g-vulnerability approach, core-concavity is more information theoretical in nature, and the alternative axioms here presented includes all entropy based leakage. Also the axioms based on `MAX` are shown to be a limit case of the η-averaging axioms. The work presents also an alternative proof, based on core-concavity and matrix majorization, for the Coriaceous theorem.

Based on these results it will be interesting to study characterizations and further properties of core-concavity and its relation to other families of functions.

References

1. Alvim, M.S., Chatzikokolakis, K., McIver, A., Morgan, C., Palamidessi, C., Smith, G.: Axioms for information leakage. In: Proceedings of CSF, pp. 77–92 (2016). https://doi.org/10.1109/CSF.2016.13
2. Alvim, M.S., Chatzikokolakis, K., Palamidessi, C., Smith, G.: Measuring information leakage using generalized gain functions. In: 2012 IEEE 25th Computer Security Foundations Symposium, pp. 265–279 (2012). https://doi.org/10.1109/CSF.2012.26
3. Arimoto, S.: Information measures and capacity of order α for discrete memoryless channels. Top. Inf. Theory (1977)
4. Boyd, S., Vandenberghe, L.: Convex Optimization. Cambridge University Press, Cambridge (2004)
5. Dahl, G.: Matrix majorization. Linear Algebra Appl. **288**, 53–73 (1999). https://doi.org/10.1016/S0024-3795(98)10175-1. http://www.sciencedirect.com/science/article/pii/S0024379598101751
6. Havrda, J., Charvát, F.: Quantification method of classification processes: concept of structural α-entropy. Kybernetika **3**(1), 30–35 (1967)
7. Hayashi, M.: Exponential decreasing rate of leaked information in universal random privacy amplification. IEEE Trans. Inf. Theory **57**(6), 3989–4001 (2011)
8. Iwamoto, M., Shikata, J.: Information theoretic security for encryption based on conditional Rényi entropies. In: Padró, C. (ed.) Information Theoretic Security, pp. 103–121. Springer, Cham (2014)

9. Khouzani, M.H.R., Malacaria, P.: Relative perfect secrecy: universally optimal strategies and channel design. In: IEEE 29th Computer Security Foundations Symposium, CSF 2016, Lisbon, Portugal, 27 June–1 July 2016, pp. 61–76 (2016). https://doi.org/10.1109/CSF.2016.12
10. Marshall, A.W., Olkin, I., Arnold, B.C.: Inequalities: Theory of Majorization and Its Applications. Mathematics in Science and Engineering, vol. 143. Academic Press, Cambridge (1979)
11. Massey: Guessing and entropy. In: Proceedings of the IEEE International Symposium on Information Theory, p. 204. IEEE (1994)
12. McIver, A., Morgan, C., Smith, G., Espinoza, B., Meinicke, L.: Abstract channels and their robust information-leakage ordering. In: Abadi, M., Kremer, S. (eds.) POST 2014. LNCS, vol. 8414, pp. 83–102. Springer, Heidelberg (2014). https://doi.org/10.1007/978-3-642-54792-8_5
13. Rényi, A.: On measures of entropy and information. In: Proceedings of the 4th Berkeley Symposium on Mathematics, Statistics, and Probability, pp. 547–561 (1961)
14. Shannon, C.E.: A mathematical theory of communication. Bell Syst. Tech. J. **27**(379–423), 625–656 (1948)
15. Sharma, B.D., Mittal, D.P.: New non-additive measures of entropy for discrete probability distributions. J. Math. Sci. (Soc. Math. Sci. Delhi India) **10**, 28–40 (1975)
16. Smith, G.: On the foundations of quantitative information flow. In: de Alfaro, L. (ed.) FoSSaCS 2009. LNCS, vol. 5504, pp. 288–302. Springer, Heidelberg (2009). https://doi.org/10.1007/978-3-642-00596-1_21
17. Tsallis, C.: Possible generalization of Boltzmann-Gibbs statistics. J. Stat. Phys. **52**(1–2), 479–487 (1988)

Formalisation of Probabilistic Testing Semantics in Coq

Yuxin Deng[1(✉)] and Jean-Francois Monin[2]

[1] Shanghai Key Laboratory of Trustworthy Computing, MOE International Joint Lab of Trustworthy Software, East China Normal University, Shanghai, China
yxdeng@sei.ecnu.edu.cn

[2] Univ. Grenoble Alpes, CNRS, Grenoble INP, VERIMAG, 38000 Grenoble, France

Abstract. Van Breugel et al. [*Theor. Comput. Sci.* 333(1–2):171–197, 2005] have given an elegant testing framework that can characterise probabilistic bisimulation, but its completeness proof is highly involved. Deng and Feng [*Inf. Comput.* 257:58–64, 2017] have simplified that result for finite-state processes. The crucial part in the latter work is an algorithm that can construct enhanced tests. We formalise the algorithm and prove its correctness by maintaining a number of subtle invariants in Coq. To support the formalisation, we develop a reusable library for manipulating finite sets. This sets an early example of formalising probabilistic concurrency theory or quantitative aspects of concurrency theory at large, which is a rich field to be pursued.

1 Introduction

One of the central concepts in concurrency theory is bisimulation [24,25]. Its generalisation in probabilistic concurrency theory is put forward by Larsen and Skou in [22]. Various characterisations of the largest probabilistic bisimulation (aka bisimilarity) by probabilistic extensions of Hennessy-Milner logic [17] have appeared in the literature [6,8,10–12,16,18,22,26]. For example, it is shown in [22] that probabilistic bisimilarity can be characterised by a very simple testing framework for reactive probabilistic processes [22,28] with a minimal probability assumption. In [27] van Breugel et al. generalise the testing characterisation of [22] to labelled Markov processes, i.e., reactive probabilistic processes [22,28] with continuous state spaces, and surprisingly, with an even simpler testing language. Generally speaking, the simpler the logical or testing characterisation, the more difficult the completeness proof of the chararacterisation. The reason is that this kind of proofs usually involve constructing distinguishing formulae or tests for non-bisimilar states, which is more challenging if there are fewer modalities to use. Van Breugel et al. succeeded in proving such an elegant result by

Supported by the National Natural Science Foundation of China (61672229, 61832015), the French national research organization ANR (grant ANR-15-CE25-0008), and the Inria-CAS joint project Quasar.

M. S. Alvim et al. (Eds.): Palamidessi Festschrift, LNCS 11760, pp. 276–292, 2019.
https://doi.org/10.1007/978-3-030-31175-9_16

making use of advanced machinery such as the Lawson topology on probabilistic powerdomains [19] and Banach algebras. In [7], Deng and Feng consider finite-state reactive probabilistic processes and give an extremely elementary proof of the coincidence of bisimilarity with the aforementioned testing equivalence while avoiding all the advanced machinery used in [27]. The core of that proof is to construct a sort of enhanced tests from basic tests by a tricky algorithm. Therefore, the correctness of the algorithm is crucial for the validity of the testing characterisation of probabilistic bisimilarity. A manual proof is given in [7], but to increase our confidence, a machine-checkable proof would be preferable. This is worthwhile because, as far as we know, among all the modal or testing characterisations of probabilistic bisimilarity, the one in [27] is the simplest, and the completeness proof presented in [7] is the most elementary and thus accessible.

In the current work, we formalise the algorithm of [7] in Coq [5] and prove its termination and correctness. We choose Coq because it is one of the mainstream proof assistants that has a large number of users in both industry and research communities. It has been successfully used for formal specifications of the X86 and LLVM instruction sets and programming languages such as C [20,21,29]. It has also been used to build CompCert [23], a fully-verified optimizing C compiler, and CertiKos [15], a fully verified hypervisor, for proving the correctness of many algorithms. Moreover, some important results in mathematics, such as the four-color theorem [13] and the Feit-Thompson theorem [14] are formally proved with Coq. We also claim that some features of the type system of Coq are very important in our formalisation in the presence of nested loops (cf. Sect. 4.5).

The algorithm that we formalise in the current work involves two nested loops that render termination and correctness proofs highly non-trivial. We carefully design a number of invariants and show that they are preserved by appropriate loops. Sometimes, the invariants are not mutually independent. As we will see in Sect. 5.3, there are scenarios where we have two invariants (a) and (b), and after some steps of execution, invariant (b) holds as a post-condition because in the precondition both (a) and (b) are required to hold. What is more, as invariants are predicates on program states, they rely on another kind of invariants on program states. This happens because we heavily use finite sets. On one hand, if a set contains elements of the form (i, j) where i and j are natural numbers with $i < j$, we must make sure that no matter how the set expands or shrinks, its elements always keep that form. This is a particular invariant originated from program states. On the other hand, since we represent finite sets as lists, a general invariant to be maintained is that there is no duplicated elements in the lists no matter how the lists evolve under different set operations. The outcome of our development is not only a formal proof of the correctness of the non-trivial algorithm for constructing enhanced tests, but also a convenient library for manipulating finite sets.

Although there are efforts on the mechanisation of reasoning about randomised algorithms [3] and on applications to cryptography [4] there exists little work on formalising probabilistic concurrency theory, or quantitative aspects of concurrency theory at large. The current work sets an example towards this direction. Undoubtedly, much more can be done in the future.

The rest of the paper is structured as follows. In Sect. 2 we recall the background on probabilistic testing semantics and the algorithm that we will consider in Sect. 4. In Sect. 3 we outline our use of Coq. In Sect. 4 we formalise the algorithm in Coq. In Sect. 5 we introduce the main invariants used in proving the correctness of the algorithm. Finally, we conclude in Sect. 6.

The Coq scripts are available at the following link http://www-verimag.imag.fr/~monin/Proof/ProbaTesting/.

2 Preliminaries

In this section, we recall the probabilistic testing semantics and the algorithm for computing enhanced tests introduced in [7].

Let S be a finite set. A *(discrete) probability distribution* over S is a function $\Delta : S \to [0,1]$ with $\sum_{s \in S} \Delta(s) = 1$. Its *support* is the set $\{s \in S \mid \Delta(s) > 0\}$. Let $\mathcal{D}(S)$ denote the set of all distributions over S. We write $\overline{s}$ for the point distribution satisfying $\overline{s}(t) = 1$ if $t = s$, and 0 otherwise. If $p_i \geq 0$ and Δ_i is a distribution for each i in some finite index set I, then $\sum_{i \in I} p_i \cdot \Delta_i$ is the function given by $(\sum_{i \in I} p_i \cdot \Delta_i)(s) = \sum_{i \in I} p_i \cdot \Delta_i(s)$. If $\sum_{i \in I} p_i = 1$ then this is easily seen to be a distribution in $\mathcal{D}(S)$.

Definition 1. *A* reactive probabilistic labelled transition system *(rpLTS) is a triple $(S, A, \to)$, where S is a finite set of states, A is a finite set of actions, and the transition relation $\to$ is a partial function from $S \times A$ to $\mathcal{D}(S)$.*

We write $s \xrightarrow{a} \Delta$ for $\to (s, a) = \Delta$. In the probabilistic setting, we often need to compare distributions. There is a way of lifting relations on states to relations on distributions [9].

Definition 2. *Given two sets S, T and a relation $\mathcal{R} \subseteq S \times T$, the lifted relation $\mathcal{R}^\dagger \subseteq \mathcal{D}(S) \times \mathcal{D}(T)$ is the smallest relation that satisfies:*

(i) $s \mathrel{\mathcal{R}} t$ implies $\overline{s} \mathrel{\mathcal{R}^\dagger} \overline{t}$;
(ii) $\Delta_i \mathrel{\mathcal{R}^\dagger} \Theta_i$ for all $i \in I$ implies $(\sum_{i \in I} p_i \cdot \Delta_i) \mathrel{\mathcal{R}^\dagger} (\sum_{i \in I} p_i \cdot \Theta_i)$, where I is a finite index set and $\sum_{i \in I} p_i = 1$.

The above lifting operation is used to define probabilistic bisimulation.

Definition 3. *A binary relation $\mathcal{R} \subseteq S \times S$ is a* probabilistic simulation *if $s \mathrel{\mathcal{R}} t$ and $s \xrightarrow{a} \Delta$ implies the existence of some transition $t \xrightarrow{a} \Theta$ with $\Delta \mathrel{\mathcal{R}^\dagger} \Theta$.*

If both $\mathcal{R}$ and $\mathcal{R}^{-1}$ are probabilistic simulations, then $\mathcal{R}$ is a probabilistic bisimulation. *The largest probabilistic bisimulation is* probabilistic bisimilarity.

Let us fix a rpLTS $(S, A, \to)$ and recall the testing framework given in [27].

Definition 4. *Let $\mathcal{T}$ be a testing language given by the grammar*

$$t \ ::= \ \omega \mid a \cdot t \mid \langle t, t \rangle$$

where a ranges over the set of labels A of our rpLTS. The function $Pr : S \times \mathcal{T} \mapsto [0, 1]$ *prescribes the probability of applying a test to a state as follows:*

$$
\begin{aligned}
Pr(s, \omega) &= 1 \\
Pr(s, a \cdot t) &= \begin{cases} \sum_{s' \in S} \Delta(s') \cdot Pr(s', t) & \text{if } s \xrightarrow{a} \Delta \\ 0 & \text{otherwise.} \end{cases} \\
Pr(s, \langle t_1, t_2 \rangle) &= Pr(s, t_1) \cdot Pr(s, t_2)
\end{aligned}
$$

We call $\langle t_1, t_2 \rangle$ a *conjunction* of two tests, which models the copying capacity of probabilistic testing. Here, conjunction is given the arithmetic interpretation as multiplication.

Definition 5. *The testing language* $\mathcal{T}$ *induces a testing equivalence relation, written* $=_{\mathcal{T}}$*, by letting* $s_1 =_{\mathcal{T}} s_2$ *if* $Pr(s_1, t) = Pr(s_2, t)$ *for any* $t \in \mathcal{T}$*.*

It is shown in [7] that $=_{\mathcal{T}}$ is a probabilistic bisimulation [22]. The key ingredient is to introduce a notion of enhanced tests. By making use of Algorithm 1, we can construct enhanced tests that satisfy the conditions in Lemma 1. From that lemma, it is not difficult to prove that $=_{\mathcal{T}}$ is a probabilistic bisimulation. Let us first set up the scenario where Algorithm 1 applies. Observe that $=_{\mathcal{T}}$ is an equivalence relation. Hence, we can partition the state space S according to $=_{\mathcal{T}}$. Let $C_1, \cdots, C_n$ be the equivalence classes induced by $=_{\mathcal{T}}$, where $n \geq 1$. Within each equivalence class C_i, the states are testing equivalent. So we can write $Pr(C_i, t)$ for $Pr(s_{ij}, t)$, where s_{ij} is any state in C_i and t is any test. Nevertheless, for any two states in different equivalence classes, there exist some tests that can tell them apart. For any i, j with $1 \leq i < j \leq n$, let t_{ij} be a test that distinguishes C_i from C_j; that is, $Pr(C_i, t_{ij}) \neq Pr(C_j, t_{ij})$. Notice that here t_{ij} is only a distinguishing test for C_i and C_j, and in general it says nothing about a third equivalence class C_k when $k \neq i, j$. For example, applying t_{ij} to C_i and then to C_k might yield the same outcome. This is normal because t_{ij} is not necessarily a distinguishing test for C_i and C_k. The surprising fact discovered in [7] is that it is possible to construct an *enhanced test* that sharpens testing outcomes to distinguish many equivalence classes. More precisely, applying the enhanced test to some equivalence classes gives either 0 or distinct positive values.

Lemma 1. *For any* $I \subseteq \{1, \cdots, n\}$ *with* $I \neq \emptyset$*, there exist a nonempty* $I' \subseteq I$ *and an enhanced test t such that*

(i) for all $k \in I$*,* $Pr(C_k, t) > 0$ *iff* $k \in I'$*;*
(ii) for any $i \neq j \in I'$*,* $Pr(C_i, t) \neq Pr(C_j, t)$*.*

The lemma is valid because we can use Algorithm 1 to construct such an enhanced test. The current work proposes (and proves the correctness of) a functional version of this algorithm. However, its design was driven by the original imperative version given in [7]. Proving that the functional version fits in with the imperative one could be performed using standard transformation techniques, but this is not needed: what matters is the existence of an algorithm satisfying the required specification. The algorithm initially sets I' to be I and t to be ω,

Algorithm 1. Compute an enhanced test

input : A nonempty $I = \{1, \cdots, n\}$ with the distinguishing tests t_{ij}, $i \neq j$.
output: A nonempty $I' \subseteq I$ and an enhanced test t satisfying Lemma 1.

begin
  $\mathcal{I}_{pass} \leftarrow \emptyset$;
  $\mathcal{I}_{rem} \leftarrow \{(i,j) \in I \times I : i < j\}$;
  $I' \leftarrow I$;
  $t \leftarrow \omega$;
  while $\mathcal{I}_{rem} \neq \emptyset$ **do**
    Choose arbitrarily $(i,j) \in \mathcal{I}_{rem}$;
    $I' \leftarrow \{k \in I' : Pr(C_k, t_{ij}) > 0\}$;
    $\mathcal{I}_{dis} \leftarrow \{(k,l) \in \mathcal{I}_{rem} \cap I' \times I' : Pr(C_k, t_{ij}) \neq Pr(C_l, t_{ij})\}$;
    $\mathcal{I}_{rem} \leftarrow (\mathcal{I}_{rem} \cap I' \times I') \backslash \mathcal{I}_{dis}$;
    $\mathcal{I}_{pass} \leftarrow (\mathcal{I}_{pass} \cap I' \times I') \cup \mathcal{I}_{dis}$;
    $t \leftarrow \langle t, t_{ij} \rangle$;
    $\mathcal{I}_{tem} \leftarrow \emptyset$;
    $\mathcal{I} \leftarrow \mathcal{I}_{pass}$;
    while $\mathcal{I} \neq \emptyset$ **do**
      $\mathcal{I} \leftarrow \{(k,l) \in \mathcal{I}_{pass} \backslash \mathcal{I}_{tem} : Pr(C_k, t) = Pr(C_l, t)\}$;
      if $\mathcal{I} \neq \emptyset$ **then**
        $t \leftarrow \langle t, t_{ij} \rangle$;
        $\mathcal{I}_{tem} \leftarrow \mathcal{I}_{tem} \cup \mathcal{I}$;
      end
    end
  end
  return I', t;
end

then it gradually updates the test t to equipe it with more and more discriminating power. The construction of the new tests involves removing the indices k with $Pr(C_k, t) = 0$ from I' and keeping the indices i, j such that $Pr(C_i, t)$ and $Pr(C_j, t)$ are distinct and both positive. The outer loop uses four auxiliary sets: $\mathcal{I}_{pass}$, $\mathcal{I}_{rem}$, $\mathcal{I}_{dis}$, and $\mathcal{I}_{tem}$. Among them, $\mathcal{I}_{pass}$ and $\mathcal{I}_{rem}$ form a partition of the set $\{(i,j) \in I' \times I' : i < j\}$. The subset $\mathcal{I}_{pass}$ contains the pairs (i,j) such that the current test t can distinguish C_i from C_j, while $\mathcal{I}_{rem}$ contains the remaining pairs to be processed. Each iteration of the outer loop picks up any pair (i,j) in $\mathcal{I}_{rem}$, uses the distinguishing test t_{ij} to form $\mathcal{I}_{dis}$, which is a subset of $\mathcal{I}_{rem}$, and moves it from $\mathcal{I}_{rem}$ to $\mathcal{I}_{pass}$. Each pair being moved, e.g., $(k,l) \in \mathcal{I}_{dis}$ indicates that C_k and C_l can be differed by t_{ij}. However, just expanding $\mathcal{I}_{pass}$ with $\mathcal{I}_{dis}$ is insufficient. A newly added index k might conflict with another index l, which occurs either already in the old $\mathcal{I}_{pass}$ or in the set $\mathcal{I}_{dis}$, in the sense that C_k and C_l cannot be distinguished by t. To solve this problem, the inner loop tries to update t by padding it with enough copies of t_{ij} until it can distinguish all the equivalence classes indicated by the pairs in $\mathcal{I}_{pass}$. Interestingly, the padding procedure only involves conjunctions of tests and this suffices for our purpose!

The auxiliary set $\mathcal{I}_{tem}$ is introduced to facilitate this procedure and contains all the pairs indicating the equivalence classes distinguishable by the final enhanced test. When all the pairs in $\mathcal{I}_{rem}$ are explored, the whole procedure terminates.

Two main properties that interest us are the termination and the correctness of the algorithm. Let us give a more detailed analysis of the termination property. We first look at the inner **while** loop. In each iteration, $\mathcal{I}$ is assigned a new value, which is a subset of $\mathcal{I}_{pass}\backslash\mathcal{I}_{term}$. If it becomes empty, the loop terminates immediately. Otherwise, the set $\mathcal{I}_{tem}$ is enlarged to include $\mathcal{I}$. Since the set $\mathcal{I}_{pass}$ does not change in the inner loop, in the next iteration the set $\mathcal{I}_{pass}\backslash\mathcal{I}_{term}$ becomes smaller and so does $\mathcal{I}$. Eventually, $\mathcal{I}$ must become empty and the loop terminates. For the outer **while** loop, in each iteration we choose a pair, say (i, j), from $\mathcal{I}_{rem}$, and then update $\mathcal{I}_{dis}$ and $\mathcal{I}_{rem}$. Since t_{ij} is a distinguishing test for (i, j), the two values $Pr(C_i, t_{ij})$ and $Pr(C_j, t_{ij})$ cannot be 0 at the same time. If both of them are positive, then $\mathcal{I}_{dis}$ contains at least the pair (i, j) and is not empty. If exactly one of them, say $Pr(C_i, t_{ij})$, is 0, then the corresponding index i is removed from I', which causes I' to shrink. In both cases, the assignment of $\mathcal{I}_{rem}$ by $(\mathcal{I}_{rem} \cap I' \times I')\backslash\mathcal{I}_{dis}$ makes $\mathcal{I}_{rem}$ strictly smaller. Eventually, $\mathcal{I}_{rem}$ becomes empty and the outer loop terminates. The number of iterations for the inner loop depends on the size of the set $\mathcal{I}_{pass}$, and for the outer loop depends on the size of $\mathcal{I}_{rem}$.

The correctness of the algorithm relies on the following four invariants, namely, at the beginning of each run of the outer **while** loop,

(a) $\mathcal{I}_{pass} \cup \mathcal{I}_{rem} = \{(i, j) \in I' \times I' : i < j\}$;
(b) $I' \neq \emptyset$;
(c) for all $k \in I$, $Pr(C_k, t) > 0$ iff $k \in I'$;
(d) for any $(i, j) \in \mathcal{I}_{pass}$, $Pr(C_i, t) \neq Pr(C_j, t)$.

A manual proof of these statements is provided in [7]. It is involved as it mixes logical reasoning with quantitative computation. Therefore, a machine-checkable proof is needed.

The above analysis of the algorithm serves as a guide for our formal proof of the correctness in Coq. However, we do not faithfully follow the manual proofs in [7]. In particular, for statement (d) we provide a proof that deviates a lot from the original one and is easier to implement inductively.

3 Our Use of Coq

In order to complete our case study, several issues need to be addressed:

1. Representing an (originally imperative) algorithm in Coq;
2. Dealing with finite sets;
3. Formalising and proving assertions and invariants;
4. Proving the termination property.

The previous issues are not independent from each other. For instance, in the strongly normalising purely functional setting of Coq, only terminating functions can be defined. In practice, it means that termination has to be taken into account at the definition time of a function, making issues 1 and 4 interact. Moreover, part of the invariants can be relevant to termination. Finally, design choices related to issue 2 interact with all other issues.

About issue 1, an important design decision has to be taken from the very beginning. If we stick to imperative programs, there are at least two distinct techniques for representing them: by a function from states to states – typically, a while loop will be encoded as a tail-recursive function – or, by an inductive relation between input and output states. An extension of the last option is even to consider an IMP-like toy imperative language (including ad-hoc operations on sets), with a big-step or small-step operational semantics [1].

For our purpose here, we only need to prove that there exists an effective terminating algorithm satisfying the expected input-output relation. The imperative or functional nature of the presentation of the algorithm does not matter. Therefore we just provide a Coq function returning the desired final state from the initial state and the above considerations about imperative programming just vanish. Nevertheless, the manual proof given before in an imperative setting is an important guide in the Coq development. In particular, the given invariants remain very important and provide the structure of the proof. Some amount of dependent typing is needed, in particular in order to deal with the issue of termination. For clarity, at the end, a simple executable functional program can be extracted and it is easy (but not crucial) to manually check that this program corresponds to the imperative one given in the previous section.

Issue 2 is not fundamental by itself: models and basic results for finite sets are available in the standard library of Coq [2]. However, for the management of our proofs, the introduction of concise and familiar set notations on top of it turned out to be very important: our very first formalization, based only on available libraries, resulted in statements very hard to reread and follow, and completing the last steps of the proofs became a discouraging hassle. Another choice had to be made: the representation of finite sets. Several options are possible and available in existing libraries: lists, with or without duplicates, ordered or unordered; binary search trees; AVL, to quote a few. Here we only need to prove the correctness of an abstract algorithm, which does not depend on a specific representation of finite sets. Efficiency of computation of the various operations is an orthogonal issue which is not relevant here. In summary, we choose a very simple representation of finite sets based on lists without duplicates, already available in the Coq standard library, completed with convenient notations and suitable basic lemmas.

Most of our efforts are spent on issue (3) because we are interested in the correctness of the algorithm, which follows from a non-trivial combination of invariants. In the presence of nested loops, the design of invariants is subtle. We should pay attention to not only (i) the internal logic of the algorithm but also (ii) the program states. Both (i) and (ii) give rise to different invariants and furthermore one invariant may depend on another.

4 Formalisation of the Algorithm

4.1 Preliminaries

We first formalise the testing semantics. According to the algorithm, an enhanced test t is always in the form $\langle \omega, t_{(i_1,j_1)}, t_{(i_2,j_2)}, ..., t_{(i_m,j_m)} \rangle$ for some $m > 0$ with $t_{(i_l,j_l)}$ being a distinguishing test and i_l, j_l the indices for two different equivalence classes. In other words, as far as the correctness of Algorithm 1 is concerned, we do not need to consider tests of the form $a \cdot t$, where a is a label and t is a test, as previously given in Definition 4. Since the input of Algorithm 1 has distinguishing tests and the final enhanced test is constructed out of them, we make a distinction between the two types of tests. We formalise a distinguishing test as a *basic test* and the enhanced test is constructed as a list of basic tests.

```
Inductive basic_test : Set :=
  | mk_bt : ℕ -> ℕ -> basic_test.

Definition test : Set := list basic_test.
```

Now that we have two types of tests, it is natural to model the function Pr given in Definition 4 in two steps. We first define a variable Prb that prescribes the probability of applying a basic test to the states in an equivalence class. We use the hypothesis that all probabilities are non-negative. Then we define the function Pr to calculate the probability of applying a general test to a state by multiplying the probabilities of applying Prb to basic tests and that state.

```
Variable Prb : ℕ -> basic_test -> Q.
Hypothesis not_neg_bt : ∀ i t, 0 <= Prb i t.

Fixpoint Pr (i : ℕ) (t : test) : Q :=
  match t with
  | nil => 1
  | bt :: t' => Qmult (Prb i bt) (Pr i t')
  end.
```

We see from the first case of the above definition that `Pr i nil = 1`, which corresponds nicely to the fact that $Pr(C_i, \omega) = 1$. Therefore, the special test ω is modelled by the empty list and there is no need to define a special basic test.

4.2 States

Different parts of the algorithm involve different variables. States are represented by records constrained by a structural invariant. Transitions are described by functions between such records. In particular, loops are modelled by tail-recursive functions. Instead of using a notion of global states containing all variables, we use *inner states* and *outer states* to model the data flow of the inner loop and the outer loop of the algorithm, respectively.

The inner states are determined by the three sets $\mathcal{I}$, $\mathcal{I}_{tem}$, and $\mathcal{I}_{pass}$, together with the test t. However, $\mathcal{I}_{pass}$ is not modified in the inner loop. An advantage of the functional style is that we can consider this set as a constant parameter of the corresponding function. This way, we get the invariance of $\mathcal{I}_{pass}$ for free, without additional proof obligation. In Coq, we can conveniently use the scoping mechanism of sections as follows.

```
Section sec_with_Y_pass.
Variable Y_pass : set ℕ².
```

The other sets are then modelled by the three components `Y`, `Y_term` and `tt` of the following record type.

```
Record in_iter_data : Type := mk_in_iter_data {
  Y : set ℕ²;
  Y_term : set ℕ²;
  tt : test; }.
```

The outer states are determined by the three sets `I'`, $\mathcal{I}_{rem}$, and $\mathcal{I}_{pass}$, together with the test t. They are modelled by the four components `I'`, `Y_rem`, `Y_pass` and `et` of the following record type.

```
Record out_iter_data : Type := mk_out_iter_data {
  I' : set ℕ;
  Y_rem : set ℕ²;
  Y_pass : set ℕ²;
  et : test; }.
```

An invariant maintained by the outer states is the following. It says that each pair (i, j) in $\mathcal{I}_{rem}$ consists of two different indices, and there are no duplicates in the three sets I', $\mathcal{I}_{rem}$, and $\mathcal{I}_{pass}$.

```
Inductive out_data_inv0 (r: out_iter_data) : Prop :=
| out_data_inv0_intro :
    (∀ i j, (i,j) ∈ (Y_rem r) -> i <> j) ->
    NoDup (I' r) -> NoDup (Y_rem r) -> NoDup (Y_pass r) ->
    out_data_inv0 r.
```

Lines 2–5 in Algorithm 1 are the initialisation step. The effect is to create the following initial state.

```
Definition init_data (m : ℕ) : out_iter_data :=
  {| I' := createI m;
     Y_rem := createY m;
     Y_pass := ∅;
     et := ∅ |}.
```

The two auxiliary functions `createI` and `createY` are introduced in order to create the sets $I = \{1, ..., n\}$ and $\{(i, j) \in I \times I : i < j\}$, respectively.

```
Fixpoint createI (n : ℕ) : set ℕ :=
  match n with
  | O => ∅
  | S n' => n :: createI n'
  end.

Definition createY n : set ℕ² :=
  filter_lt (nodup (createI n × createI n)).
```

Here we see an example of using filters: from the set $I \times I$ we filter out all pairs (i, j) not satisfying the condition $i < j$. This is a convenient means of creating subsets from a given set that we often use in our Coq development.

The final output data of the algorithm is the set I' and the enhanced test t, as described by the type `final_data`.

```
Record final_data : Type := mk_final_data {
  I'f : set ℕ;
  etf : test; }.
```

4.3 Inner Loop

To formalise loops, our general strategy is to first define one round of iteration, which is a function from a state possibly with additional parameters to a new state, and then repeatedly apply the iteration until the state meets the termination condition.

The inner loop (lines 15–21 of Algorithm 1) makes use of an iteration step described below, where `next_Y_inner_loop` computes the new value to be assigned to $\mathcal{I}$, as stated in line 16.

```
Definition next_Y_inner_loop r := filter_eq (tt r) (Y_pass \ Y_term r).
Definition inner_loop_iter
  (r : in_iter_data) (tij : basic_test) : in_iter_data :=
  match Y r with
  | ∅ => r
  | _ :: _ => let Y' := next_Y_inner_loop r in
              match Y' with
              | ∅ => {| Y := Y';
                        Y_term := Y_term r;
                        tt := tt r |}
              | _ :: _ => {| Y := Y';
                            Y_term := Y_term r ∪ Y';
                            tt := tij :: tt r |}
              end
  end.
```

The structural invariant of the record maintained by the inner loop is:

```
Inductive in_data_inv (d: in_iter_data) : Prop :=
| in_data_inv_intro :
   Y d ⊆ Y_pass -> Y_term d ⊆ Y_pass ->
   NoDup (Y d) -> NoDup (Y_term d) -> in_data_inv d.
```

It requires that for an inner state $(\mathcal{I}, \mathcal{I}_{term}, t)$ to be legal, we must have that $\mathcal{I} \subseteq \mathcal{I}_{pass}$, $\mathcal{I}_{term} \subseteq \mathcal{I}_{pass}$, and there is no duplicated element in $\mathcal{I}$, $\mathcal{I}_{term}$ and $\mathcal{I}_{pass}$.

The inner loop itself is formulated as a recursive function that repeatedly applies the iteration step `inner_loop_iter`. Note that each iteration step either makes $\mathcal{I}$ empty, or decreases the size of the set $\mathcal{I}_{pass} \backslash \mathcal{I}_{tem}$. So we define the measure `size_of_data` for inner states, which yields a strictly decreasing argument for the function `in_loop`.

```
Definition size_of_data r := size (Y r) (Y_pass \ Y_term r).
Function in_loop (r : in_iter_data) (di : in_data_inv r) tij
  {measure size_of_data r} : in_iter_data := match Y r with
   | ∅ => r
   | _ :: _ => in_loop (inner_loop_iter r tij)
                       (inner_loop_iter_inv di tij) tij
  end.
```

The previous definitions are written inside Section `sec_with_Y_pass` mentioned in Sect. 4.2. When refering to them outside of this section, namely when they are used in functions and proofs modelling the outer loop, an actual parameter for `Y_pass` has to be provided. Moreover, `in_loop` needs a parameter stating that `Y_pass` has no duplicate in order to ensure that `di` is maintained invariant. This assumption is stated at the beginning of Section `sec_with_Y_pass`.

4.4 Interface Between Inner and Outer Loops

Before entering the inner loop, we need to pass the information stored in the outer state to the inner state. The interface between the two loops is modelled by the function mk_in_from_out that creates an inner state from an outer state. It contains the formalisation of lines 13–14 in Algorithm 1.

```
Definition mk_in_from_out
  (r : out_iter_data) (bt : basic_test) : in_iter_data :=
  let rnew := outer_loop_iter1 r bt in
  {| Y := Y_pass rnew;
     Y_term := ∅;
     tt := et rnew |}.
```

```
Lemma out_data_in_data_inv r (dio : out_data_inv0 r) bt :
  let rnew := outer_loop_iter1 r bt in
  in_data_inv (Y_pass rnew) (mk_in_from_out r bt).
```

Note that the actual parameter given for Y_pass to in_data_inv is Y_pass rnew.

4.5 Outer Loop

As we did for the inner loop, we first specify one iteration step of the outer loop. The sequence of assignments before entering the inner loop (lines 7–12 of Algorithm 1) is described by the function outer_loop_iter1.

```
Definition outer_loop_iter1
  (r : out_iter_data) (bt : basic_test) : out_iter_data :=
  let nI' := filter_pos bt (I' r) in
  let Ydis := filter_neq bt (Y_rem r ∩ (nI' × nI')) in
  {| I' := nI';
     Y_rem := (Y_rem r ∩ (nI' × nI')) \ Ydis;
     Y_pass := (Y_pass r ∩ (nI' × nI')) ∪ Ydis;
     et := bt :: et r |}.
```

Then the function outer_loop_iter2 formalises one iteration of the outer loop by first calling outer_loop_iter1, then creating an inner state through the interface mk_in_from_out, and finally invoking in_loop to deal with the tasks required by the inner loop.

```
Definition outer_loop_iter2 r (dio : out_data_inv0 r) bt : out_iter_data :=
  let rnew := outer_loop_iter1 r bt in
  let rin := mk_in_from_out r bt in
  let ndYp := nd_Y_pass r dio bt in
  let rin' := in_loop (Y_pass rnew) ndYp rin
                      (out_data_in_data_inv r dio bt) bt in
  {| I' := I' rnew;
     Y_rem := Y_rem rnew;
     Y_pass := Y_pass rnew;
     et := tt rin' |}.
```

By repeatedly applying the iteration step `outer_loop_iter2`, we obtain a formalisation of the outer loop. The decreasing argument for the recursive function `out_loop` is the size of the set $\mathcal{I}_{rem}$.

```
Function out_loop r (di : out_data_inv0 r)
                    {measure size_of_out_data r} : out_iter_data :=
  match (Y_rem r) with
  | nil => r
  | (i , j) :: _ => let (bt, _) := oracle i j in
    out_loop (outer_loop_iter2 r di bt) (outer_loop_iter_inv0 r di bt)
  end.
```

The definition of the outer loop uses the variable `oracle` with (`oracle i j`) representing the distinguishing test t_{ij} for all equivalence class indices $i \neq j$. The distinguishing tests are part of the input of the algorithm. A technical difference from the inner loop is that the definition of the iteration step is defined only on states satisfying the loop invariant, entailing a more subtle use of dependent typing. More specifically, if we compare the definitions of `in_loop` and `out_loop`, both of them contain a tail recursive call with the first argument for the state reached after one iteration step then the second argument embedding the invariant satisfied by this state. This kind of dependency, where properties depending on data are kept separated from them, is rather common in Coq developments: for instance, it avoids complications related to proof irrelevance that arise with dependent records (and any inductive type made of one constructor with dependencies between its arguments). However, in contrast to the definition of `in_loop`, where the state argument of the recursive call refers to data only, the corresponding argument for `out_loop`, namely `outer_loop_iter2 r di bt`, makes a crucial use of invariant `di`. We believe that this would be hard to express in a concise and accurate way without dependent types. In other words, we benefit from the rich type system of Coq which includes dependent types, while avoiding issues related to dependent records.

Interestingly, the exact reason for this situation is that invariant `di` is needed in the body of `outer_loop_iter2` in order to allow for a call to `in_loop`. A similar situation can reasonably be expected with the formalisation of many algorithms containing nested loops, at least when non-trivial interactions occur between these loops.

4.6 The Whole Algorithm

Once the outer loop is formalised, the whole algorithm easily follows. We first initialise an outer state and then call `out_loop` before obtaining the final set I' and the enhanced test t.

```
Definition algo_compt_enhanced_test (m : ℕ) : final_data :=
  let r := init_data m in
  let final_r := out_loop r (init_data_inv0 m) in
  {| I'f := I' final_r;
     etf := et final_r |}.
```

5 Formal Proofs

In this section, we take a close look at some important invariants that finally entail the correctness of the algorithm.

5.1 Invariants of the Inner Loop

As mentioned earlier, the definition of `in_loop` embeds the structural invariant. The following invariants are used for invariant (c) of the outer loop.

```
Lemma inner_loop_iter_inv_c {k r bt} : 0 < Pr k (tt r) ->
  (∃ t, tt r = bt :: t) -> Pr k (tt (inner_loop_iter r bt)) > 0.
Lemma inner_loop_iter_inv_c2 {k r} :
  Pr k (tt r) == 0 -> ∀ bt, Pr k (tt (inner_loop_iter r bt)) == 0.
```

The first one says that if $Pr(C_k, t) > 0$ and t is made from basic test bt then after running one iteration of the inner loop with bt, the test may evolve into some t', but we still have $Pr(C_k, t') > 0$. The second one states that if $Pr(C_k, t) = 0$ and t is changed into some t' then we always have $Pr(C_k, t') = 0$. The next two lemmas tell us that executing the whole inner loop preserves similar properties.

```
Lemma in_loop_inv_c {k r bt di} : 0 < Pr k (tt r) ->
  (∃ t, (tt r) = bt :: t) -> 0 < Pr k (tt (in_loop r di bt)).
Lemma in_loop_inv_c2 {k r di} :
  Pr k (tt r) == 0 -> ∀ bt, Pr k (tt (in_loop r di bt)) == 0.
```

Many additional invariants and post-conditions are used for invariant (d) of the outer loop. For example, the lemma `in_loop_nextI` says that if the component $\mathcal{I}$ in an inner state is empty, then it remains empty after the execution of the inner loop.

```
Definition loop_body bt r := inner_loop_iter r bt.

Inductive ufp (iid: in_iter_data) r bt : Prop :=
  ufp_intro :
  ∀ n, iid = rditer _ n (loop_body bt) r ->
    (Y r <> ∅ -> (0 < n)%nat /\
            Y (rditer _ (pred n) (loop_body bt) r) <> ∅ /\
            next_Y_inner_loop (rditer _ (pred n) (loop_body bt) r) = ∅) ->
    (Y r = ∅ -> (n = 0)%nat)
    -> ufp iid r bt.

Lemma unfold_fixed_point r (di : in_data_inv r) bt :
  ufp (in_loop r di bt) r bt.

Lemma in_loop_nextI r di bt : Y r <> ∅ ->
  next_Y_inner_loop (in_loop r di bt) = ∅.
```

5.2 Invariants of the Outer Loop

The outer loop maintains four invariants as given in page 6. They are formalised as the predicates on outer states: out_data_inv_a, out_data_inv_b, out_data_inv_c, and finally out_data_inv_d. Altogether, they are used to form the global invariant out_data_global_inv.

```
Inductive out_data_inv_a (d: out_iter_data) : Prop :=
| out_data_inv_a_intro :
    (Y_rem d ∪ Y_pass d) ⊆ (filter_lt (I' d × I' d)) ->
    (filter_lt (I' d × I' d)) ⊆ (Y_rem d ∪ Y_pass d) ->
    out_data_inv_a d.

Inductive out_data_inv_b (d: out_iter_data) : Prop :=
| out_data_inv_b_intro : I' d <> ∅ -> out_data_inv_b d.

Inductive out_data_inv_c (d: out_iter_data) I : Prop :=
| out_data_inv_c_intro :
    (∀ k, k ∈ I' d -> 0 < Pr k (et d)) ->
    (∀ k, k ∈ I \ I' d  -> Pr k (et d) == 0) ->
    out_data_inv_c d I.

Inductive out_data_inv_d (d: out_iter_data) : Prop :=
| out_data_inv_d_intro :
    all_false (Y_pass d) (eq_prob (et d)) -> out_data_inv_d d.

Inductive out_data_global_inv m I (r: out_iter_data) : Prop :=
  odgi_intro :
    out_data_inv_a r ->
    ((m > 0)%nat -> out_data_inv_b r) ->
    out_data_inv_c r I ->
    out_data_inv_d r ->
    out_data_global_inv m I r.
```

5.3 Preservation Lemmas for the Outer Loop

In order to show that the global invariant is preserved by the outer loop, we consider invariants (a)–(d) separately. Each of them is preserved after one iteration of the outer loop, and it is the same case for out_data_inv0 given in Sect. 4.2. Note that those invariants are not completely independent. For example, the preservation of invariant (b) depends on invariant (a), and the preservation of invariant (d) relies on invariant (c).

```
Lemma outer_loop_iter_inv0 : ∀ r (dio : out_data_inv0 r) bt,
  out_data_inv0 (outer_loop_iter2 r dio bt).

Lemma outer_loop_iter_inv_a {r dio bt} : out_data_inv_a r ->
  out_data_inv_a (outer_loop_iter2 r dio bt).

Lemma outer_loop_iter_inv_b {r i j dio bt} : out_data_inv_a r ->
  (i,j) ∈ (Y_rem r) -> (i <> j -> Qeq_bool (Prb i bt) (Prb j bt) = false) ->
  out_data_inv_b (outer_loop_iter2 r dio bt).

Lemma outer_loop_iter_inv_c {r bt dio I} : out_data_inv_c r I ->
  out_data_inv_c (outer_loop_iter2 r dio bt) I.
```

```
Lemma outer_loop_iter_inv_d {r bt dio I} :
  out_data_inv_c r I -> out_data_inv_d r ->
  out_data_inv_d (outer_loop_iter2 r dio bt).
Lemma out_loop_inv m I r (di : out_data_inv0 r) :
  out_data_global_inv m I r -> out_data_global_inv m I (out_loop r di).
```

In the proofs of the lemmas `outer_loop_iter_inv_c` and `outer_loop_iter_inv_d`, we need to make an in-depth analysis of the inner loop and use a number of invariants discussed in Sect. 5.1. Moreover, the emptyness of $\mathcal{I}\_rem$ is (trivially) ensured after running the outer loop, which implies the termination of the outer loop.

```
Lemma out_loop_ensures_empty_Yrem r di : Y_rem (out_loop r di) = ∅.
```

5.4 Main Theorem

The desired post-condition is ensured after running the algorithm. It says that the final set I' and the enhanced test t satisfy the conditions required by Lemma 1. In other words, we have formally proved the correctness of the algorithm.

```
Theorem correctness : ∀ m,
  let (fI, ft) := algo_compt_enhanced_test m in
  ((0 < m)%nat -> fI <> ;) /\
  (∀ k, k ∈ fI -> 0 < Pr k ft) /\
  (∀ k, k ∈ (createI m \ fI)  -> Pr k ft == 0) /\
  (∀ i j, i ∈ fI -> j ∈ fI -> i <> j -> Qeq_bool (Pr i ft) (Pr j ft) = false).
```

6 Conclusion

We have demonstrated a mechanisation of proofs in probabilistic testing semantics with Coq. Proving properties in this setting requires subtle reasoning both on algorithmic and quantitative aspects of program states. Our development includes more than 500 lines of specifications and definitions, and more than 900 lines of proof scripts. Along with a machine-checkable proof of the correctness of the non-trivial algorithm with nested loops for constructing enhanced tests, we obtain a convenient library for manipulating finite sets, which we believe will benefit future formalisation efforts such as formal reasoning with probabilistic bisimulations and testing equivalences.

Acknowledgment. We would like to thank Yves Bertot for helpful discussion.

References

1. https://softwarefoundations.cis.upenn.edu/lf-current/index.html
2. https://coq.inria.fr
3. Audebaud, P., Paulin-Mohring, C.: Proofs of randomized algorithms in Coq. Sci. Comput. Program. **74**(8), 568–589 (2009)

4. Barthe, G., Crespo, J.M., Grégoire, B., Kunz, C., Zanella Béguelin, S.: Computer-aided cryptographic proofs. In: Beringer, L., Felty, A. (eds.) ITP 2012. LNCS, vol. 7406, pp. 11–27. Springer, Heidelberg (2012). https://doi.org/10.1007/978-3-642-32347-8_2
5. Bertot, Y., Castéran, P.: Interactive Theorem Proving and Program Development. Coq'Art: The Calculus of Inductive Constructions. Springer, Heidelberg (2004). https://doi.org/10.1007/978-3-662-07964-5
6. Deng, Y.: Semantics of Probabilistic Processes: An Operational Approach. Springer, Heidelberg (2015). https://doi.org/10.1007/978-3-662-45198-4
7. Deng, Y., Feng, Y.: Probabilistic bisimilarity as testing equivalence. Inf. Comput. **257**, 58–64 (2017)
8. Deng, Y., van Glabbeek, R.: Characterising probabilistic processes logically. In: Fermüller, C.G., Voronkov, A. (eds.) LPAR 2010. LNCS, vol. 6397, pp. 278–293. Springer, Heidelberg (2010). https://doi.org/10.1007/978-3-642-16242-8_20
9. Deng, Y., van Glabbeek, R., Hennessy, M., Morgan, C.: Testing finitary probabilistic processes. In: Bravetti, M., Zavattaro, G. (eds.) CONCUR 2009. LNCS, vol. 5710, pp. 274–288. Springer, Heidelberg (2009). https://doi.org/10.1007/978-3-642-04081-8_19
10. Deng, Y., Wu, H.: Modal characterisations of probabilistic and fuzzy bisimulations. In: Merz, S., Pang, J. (eds.) ICFEM 2014. LNCS, vol. 8829, pp. 123–138. Springer, Cham (2014). https://doi.org/10.1007/978-3-319-11737-9_9
11. Desharnais, J., Edalat, A., Panangaden, P.: Bisimulation for labelled Markov processes. Inf. Comput. **179**(2), 163–193 (2002)
12. Desharnais, J., Gupta, V., Jagadeesan, R., Panangaden, P.: Approximating labelled Markov processes. Inf. Comput. **184**(1), 160–200 (2003)
13. Gonthier, G.: Formal proof – the four-color theorem. Not. Am. Math. Soc. **55**(11), 1382–1393 (2008)
14. Gonthier, G., et al.: A machine-checked proof of the odd order theorem. In: Blazy, S., Paulin-Mohring, C., Pichardie, D. (eds.) ITP 2013. LNCS, vol. 7998, pp. 163–179. Springer, Heidelberg (2013). https://doi.org/10.1007/978-3-642-39634-2_14
15. Gu, R., Shao, Z., Chen, H., Wu, X.N., Kim, J., Sjöberg, V., Costanzo, D.: CertiKOS: an extensible architecture for building certified concurrent OS kernels. In: Proceedings of OSDI 2016, pp. 653–669. USENIX Association (2016)
16. Hennessy, M.: Exploring probabilistic bisimulations, Part I. Formal Aspects Comput. **24**(4–6), 749–768 (2012)
17. Hennessy, M., Milner, R.: Algebraic laws for nondeterminism and concurrency. J. ACM **32**(1), 137–161 (1985)
18. Hermanns, H., Parma, A., Segala, R., Wachter, B., Zhang, L.: Probabilistic logical characterization. Inf. Comput. **209**(2), 154–172 (2011)
19. Jones, C.: Probabilistic nondeterminism. Ph.d. thesis, University of Edinburgh (1990)
20. Kennedy, A., Benton, N., Jensen, J.B., Dagand, P.: Coq: the world's best macro assembler? In: Proceedings of PPDP 2013, pp. 13–24. ACM (2013)
21. Krebbers, R.: The C standard formalized in Coq. Ph.d. thesis, Radboud University Nijmegen (2015)
22. Larsen, K.G., Skou, A.: Bisimulation through probabilistic testing. Inf. Comput. **94**, 1–28 (1991)
23. Leroy, X., Blazy, S., Kästner, D., Schommer, B., Pister, M., Ferdinand, C.: CompCert - a formally verified optimizing compiler. In: Proceedings of the 8th European Congress on Embedded Real Time Software and Systems. SEE (2016). https://hal.inria.fr/hal-01238879

24. Milner, R.: Communication and Concurrency. Prentice Hall, Upper Saddle River (1989)
25. Park, D.: Concurrency and automata on infinite sequences. In: Deussen, P. (ed.) GI-TCS 1981. LNCS, vol. 104, pp. 167–183. Springer, Heidelberg (1981). https://doi.org/10.1007/BFb0017309
26. Parma, A., Segala, R.: Logical characterizations of bisimulations for discrete probabilistic systems. In: Seidl, H. (ed.) FoSSaCS 2007. LNCS, vol. 4423, pp. 287–301. Springer, Heidelberg (2007). https://doi.org/10.1007/978-3-540-71389-0_21
27. van Breugel, F., Mislove, M.W., Ouaknine, J., Worrell, J.: Domain theory, testing and simulation for labelled Markov processes. Theoret. Comput. Sci. **333**(1–2), 171–197 (2005)
28. van Glabbeek, R.J., Smolka, S.A., Steffen, B., Tofts, C.M.N.: Reactive, generative, and stratified models of probabilistic processes. In: Proceedings of LICS 1990, pp. 130–141. IEEE Computer Society (1990)
29. Zhao, J., Nagarakatte, S., Martin, M.M.K., Zdancewic, S.: Formalizing the LLVM intermediate representation for verified program transformations. In: Proceedings of POPL 2012, pp. 427–440. ACM (2012)

Fully Syntactic Uniform Continuity Formats for Bisimulation Metrics

Valentina Castiglioni[1,2(✉)], Ruggero Lanotte[3], and Simone Tini[3]

[1] INRIA Saclay - Ile de France, Palaiseau, France
vale.castiglioni@gmail.com
[2] Reykjavik University, Reykjavik, Iceland
[3] University of Insubria, Como, Italy

Abstract. Behavioral metrics play a fundamental role in the analysis of probabilistic systems. They allow for a robust comparison of the behavior of processes and provide a formal tool to study their performance, privacy and security properties. Gebler, Larsen and Tini showed that the *bisimilarity metric* is also suitable for *compositional reasoning*, expressed in terms of *continuity properties* of the metric. Moreover, Gebler and Tini provided *semantic formats* guaranteeing, respectively, the *non-extensiveness*, *non-expansiveness* and *Lipschitz continuity* of this metric. In this paper, starting from their work, we define three *specification formats* for the bisimilarity metric, one for each continuity property, namely sets of *syntactic constraints* over the SOS rules defining process operators that guarantee the desired continuity property of the metric.

1 Introduction

With the ever-increasing interest in probabilistic processes, robust notions and formal tools are needed to analyze their behavior. It has been argued many times by now that *behavioral metrics* [2,6,7,13,14,20,23,34] are preferable to behavioral relations in this setting in that they overcome the high sensitivity of equivalences and preorders with respect to tiny variations in the values of probabilities. Instead of stating whether the behavior of two processes is exactly the same or not, behavioral metrics *measure* the disparities in their behavior.

A fundamental aspect for process specification and verification is *compositional reasoning*, namely to prove the compatibility of the language operators with the chosen behavioral semantics. In [16] the authors proposed to express the compositional properties of the *bisimilarity metric* [6,13,14], i.e., the quantitative analogue to bisimulation equivalence, in terms of its *continuity properties*. In detail, a notion of *uniform continuity* was proposed subsuming the properties of *non-extensiveness* [4], *non-expansiveness* [14] and *Lipschitz continuity* of the metric. Informally, an operator is non-extensive wrt. a metric if the distance of two systems defined via that operator is not greater than the *maximum* of the pairwise distances of their components. The notion of non-expansiveness relaxes that of non-extensiveness by allowing the distance of the two systems to be

M. S. Alvim et al. (Eds.): Palamidessi Festschrift, LNCS 11760, pp. 293–312, 2019.
https://doi.org/10.1007/978-3-030-31175-9_17

bounded by the *sum* of the pairwise distances of their components. Lipschitz continuity extends non-expansiveness by considering as bound a *multiplicative factor* of the sum of pairwise distances of the components of the two systems.

Since [33], a *specification format* is a set of *syntactic constraints* on the form of the SOS inference rules [29,30] used to define the operational semantics of processes, ensuring by construction a given compositionality property (see [1,28] for surveys). A first step in the definition of specification formats for uniform continuity properties of bisimilarity metric can be found in [19]. However, the formats in [19] are not fully syntactic since they do not simply require an inspection of the pattern of the rules, but they require to compute how many copies of a process argument can be spawned by each operator. Such computation is necessarily recursive since the SOS rules define operators in terms of operators.

Our Contribution. In this paper we provide a classic syntactic format for uniform continuity properties for the bisimilarity metric. In detail, we propose three formats: the *non-extensiveness*, the *non-expansiveness*, and the *Lipschitz continuity* format. These formats build on the PGSOS format [11,12] and on *syntactic constraints* regulating the form of the *targets of rules*. We recall that the PGSOS format guarantees that probabilistic bisimilarity, i.e., the kernel of the bisimilarity metric, is a congruence [12]. Hence, to obtain the desired continuity properties it is fundamental to start from the PGSOS format, since it ensures that by composing processes which are at distance 0, i.e., bisimilar, we obtain processes which are at distance 0. We will see that less demanding continuity properties require less demanding rule constraints.

2 Background

In the process algebra setting, *processes* are constructed inductively as *closed terms* over a suitable *signature*, namely a countable set Σ of *operators*. We let f range over Σ and $\mathfrak{n}$ range over the *rank* of $f \in \Sigma$. Operators with rank 0 are called *constants*. We let Σ^0 denote the subset of constants in the signature Σ.

Assume a set of (*state*, or process) *variables* $\mathbf{V}_s$ disjoint from Σ and ranged over by $x, y, \ldots$. The set $\mathsf{T}(\Sigma, V)$ of *terms* over Σ and a set of variables $V \subseteq \mathbf{V}_s$ is the least set satisfying: (i) $V \subseteq \mathsf{T}(\Sigma, V)$, and (ii) $f(t_1, \ldots, t_\mathfrak{n}) \in \mathsf{T}(\Sigma, V)$ whenever $f \in \Sigma$ and $t_1, \ldots, t_\mathfrak{n} \in \mathsf{T}(\Sigma, V)$. By $\mathbf{T}(\Sigma)$ we denote the set of the *closed terms* $\mathsf{T}(\Sigma, \emptyset)$, or *processes*. By $\mathbb{T}(\Sigma)$ we denote the set of all *open terms* $\mathsf{T}(\Sigma, \mathbf{V}_s)$. By $\mathrm{var}(t)$ we denote the set of the variables occurring in the term t.

To equip processes with a semantics we rely on *nondeterministic probabilistic labeled transition systems* (PTSs) [31], which allow for modeling reactive behavior, nondeterminism and probability. The state space in a PTS is the set of processes $\mathbf{T}(\Sigma)$. Transitions take processes to discrete probability distributions over processes, i.e., mappings $\pi \colon \mathbf{T}(\Sigma) \to [0,1]$ with $\sum_{t \in \mathbf{T}(\Sigma)} \pi(t) = 1$. The *support* of a distribution π is $\mathsf{supp}(\pi) = \{t \in \mathbf{T}(\Sigma) \mid \pi(t) > 0\}$. By $\Delta(\mathbf{T}(\Sigma))$ we denote the set of all *finitely supported* distributions over $\mathbf{T}(\Sigma)$.

Definition 1 (PTS, [31]). *A PTS is a triple* $(\mathbf{T}(\Sigma), \mathcal{A}, \rightarrow)$*, where* Σ *is a signature,* $\mathcal{A}$ *a countable set of* action labels, *and* $\rightarrow \subseteq \mathbf{T}(\Sigma) \times \mathcal{A} \times \Delta(\mathbf{T}(\Sigma))$ *a* transition relation.

As usual, a *transition* $(t, \alpha, \pi) \in \rightarrow$ is denoted $t \xrightarrow{\alpha} \pi$. Then, $t \stackrel{\alpha}{\not\rightarrow}$ denotes that there is no distribution π with $t \xrightarrow{\alpha} \pi$ and is called a *negative transition*.

Distributions and Distribution Terms. For a process $t \in \mathbf{T}(\Sigma)$, δ_t is the *Dirac distribution* with $\delta_t(t) = 1$ and $\delta_t(s) = 0$ for all $s \neq t$. For $f \in \Sigma$ and $\pi_i \in \Delta(\mathbf{T}(\Sigma))$, $f(\pi_1, \ldots, \pi_n)$ is the distribution defined by $f(\pi_1, \ldots \pi_n)(f(t_1, \ldots t_n)) = \prod_{i=1}^{n} \pi_i(t_i)$ and $f(\pi_1, \ldots, \pi_n)(t) = 0$ for all t not in the form $f(t_1, \ldots, t_n)$. The convex combination $\sum_{i \in I} p_i \pi_i$ of a family of distributions $\{\pi_i\}_{i \in I} \subseteq \Delta(\mathbf{T}(\Sigma))$ with $p_i \in (0, 1]$ and $\sum_{i \in I} p_i = 1$ is defined by $(\sum_{i \in I} p_i \pi_i)(t) = \sum_{i \in I}(p_i \pi_i(t))$ for all $t \in \mathbf{T}(\Sigma)$.

Then, we need the notion of *distributions term*. Assume a countable set of *distribution variables* $\mathbf{V}_d$ disjoint from Σ and $\mathbf{V}_s$ and ranged over by $\mu, \nu, \ldots$. Intuitively, any $\mu \in \mathbf{V}_d$ may instantiate to a distribution in $\Delta(\mathbf{T}(\Sigma))$. The set of *distribution terms* over Σ and the sets of variables $V_s \subseteq \mathbf{V}_s$ and $V_d \subseteq \mathbf{V}_d$, notation $\mathsf{DT}(\Sigma, V_s, V_d)$, is the least set satisfying: (i) $\{\delta_t \mid t \in \mathsf{T}(\Sigma, V_s)\} \subseteq \mathsf{DT}(\Sigma, V_s, V_d)$, (ii) $V_d \subseteq \mathsf{DT}(\Sigma, V_s, V_d)$, (iii) $f(\Theta_1, \ldots, \Theta_n) \in \mathsf{DT}(\Sigma, V_s, V_d)$ whenever $f \in \Sigma$ and $\Theta_i \in \mathsf{DT}(\Sigma, V_s, V_d)$, (iv) $\sum_{j \in J} p_j \Theta_j \in \mathsf{DT}(\Sigma, V_s, V_d)$ whenever $\Theta_j \in \mathsf{DT}(\Sigma, V_s, V_d)$ and $p_j \in (0, 1]$ with $\sum_{j \in J} p_j = 1$. We write $\mathbb{DT}(\Sigma)$ for $\mathsf{DT}(\Sigma, \mathbf{V}_s, \mathbf{V}_d)$, i.e. the set of all *open distribution terms*, and $\mathbf{DT}(\Sigma)$ for $\mathsf{DT}(\Sigma, \emptyset, \emptyset)$, i.e. the set of the *closed distribution terms*. Notice that closed distribution terms denote distributions. We denote by $\mathrm{var}(\Theta)$ the set of the variables occurring in distribution term Θ. We let $\delta_{\mathbf{V}_s}$ denote the set $\{\delta_x \mid x \in \mathbf{V}_s\}$.

A distribution term is in *normal form* if the Dirac operator is applied only to single variables δ_x and not to non-variable terms δ_t with $t \notin \mathbf{V}_s$ and, then, either there is no convex combination or there is only one convex combination as outermost operation.

Definition 2 (Normal form, [19]). *A distribution term* $\Theta \in \mathbb{DT}(\Sigma)$ *is in* normal form *iff: either (i)* $\Theta = \mu \in \mathbf{V}_d$*, or (ii)* $\Theta = \delta_x$ *for* $x \in \mathbf{V}_s$*, or (iii)* $\Theta = f(\Theta_1, \ldots, \Theta_n)$ *where all the* Θ_i *are in normal form and no convex combination occurs in any of the* Θ_i*, or (iv)* $\Theta = \sum_{j \in J} p_j \Theta_j$ *where all the* Θ_j *are in normal form and no convex combination occurs in any of the* Θ_j*.*

A *substitution* is a mapping $\sigma \colon \mathbf{V}_s \cup \mathbf{V}_d \rightarrow \mathbb{T}(\Sigma) \cup \mathbb{DT}(\Sigma)$ with $\sigma(x) \in \mathbb{T}(\Sigma)$ for all $x \in \mathbf{V}_s$ and $\sigma(\mu) \in \mathbb{DT}(\Sigma)$ for all $\mu \in \mathbf{V}_d$. A substitution σ is *closed* if it maps terms to closed terms in $\mathbf{T}(\Sigma) \cup \mathbf{DT}(\Sigma)$.

We say that two distribution terms Θ_1, Θ_2 are *equivalent*, written $\Theta_1 \equiv \Theta_2$ if under all closed substitutions σ, the closed distribution terms $\sigma(\Theta_1)$ and $\sigma(\Theta_2)$ express the same probability distribution. Each distribution term is equivalent to a normal form.

Proposition 1 ([19]). *For each distribution term* $\Theta \in \mathbb{DT}(\Sigma)$ *there is a distribution term* $\Theta' \in \mathbb{DT}(\Sigma)$ *in normal form such that* $\Theta \equiv \Theta'$.

PGSOS Specifications. PTSs are defined by means of SOS rules [30], which are syntax-driven inference rules allowing us to infer the behavior of processes inductively with respect to their structure. They are based on expressions of the form $t \xrightarrow{\alpha} \Theta$ and $t \stackrel{\alpha}{\not\longrightarrow}$, called, resp., *positive* and *negative literals* with t a term and Θ a distribution term, which will instantiate to transitions through substitutions. The literals $t \xrightarrow{\alpha} \Theta$ and $t \stackrel{\alpha}{\not\longrightarrow}$ with the same term t in the left hand side and the same label α, are said to *deny* each other.

We assume SOS rules in the *probabilistic GSOS* (PGSOS) format [12], which allows for specifying most of probabilistic process algebras operators [16]. As examples, all SOS rules in Tables 1, 2 and 3 are PGSOS rules according to the following definition.

Definition 3 (PGSOS rules, [12]). *A* PGSOS rule r *has the form:*

$$\frac{\{x_h \xrightarrow{\alpha_{h,m}} \mu_{h,m} \mid h \in H, m \in M_h\} \ \{x_h \stackrel{\alpha_{h,n}}{\not\longrightarrow} \mid h \in H, n \in N_h\}}{f(x_1, \ldots, x_{\mathtt{n}}) \xrightarrow{\alpha} \Theta} \tag{1}$$

where $f \in \Sigma$ is an operator, $H \subseteq \{1, \ldots, \mathtt{n}\}$, M_h and N_h are finite sets of indexes, $\alpha_{h,m}, \alpha_{h,n}, \alpha \in \mathcal{A}$ are action labels, $x_h \in \mathbf{V}_s$ are state variables, $\mu_{h,m} \in \mathbf{V}_d$ are distribution variables and $\Theta \in \mathbb{DT}(\Sigma)$ is a distribution term. Furthermore: (i) all distribution variables $\mu_{h,m}$ with $h \in H$ and $m \in M_h$ are distinct, (ii) all variables $x_1, \ldots x_{\mathtt{n}}$ are distinct, (iii) $\mathrm{var}(\Theta) \subseteq \{\mu_{h,m} \mid h \in H, m \in M_h\} \cup \{x_1, \ldots, x_{\mathtt{n}}\}$.

Constraints (i)–(iii) are inherited by the classical GSOS rules in [5] and are necessary to ensure that probabilistic bisimulation is a congruence [8,9,12]. For a PGSOS rule r, the positive (resp. negative) literals above the line are called the *positive* (resp. *negative*) *premises*, notation $\mathrm{pprem}(r)$ (resp. $\mathrm{nprem}(r)$). We denote by $\mathrm{der}(x, r) = \{\mu \mid x \xrightarrow{a} \mu \in \mathrm{pprem}(r)\}$ the set of *derivatives* of x in r. The literal $f(x_1, \ldots, x_{\mathtt{n}}) \xrightarrow{\alpha} \Theta$ is called the *conclusion*, notation $\mathrm{conc}(r)$, the term $f(x_1, \ldots, x_{\mathtt{n}})$ is called the *source*, notation $\mathrm{src}(r)$, and the distribution term Θ is called the *target*, notation $\mathrm{trg}(r)$.

Definition 4 (PTSS, [12]). *A* PGSOS-transition system specification (PTSS) *is a tuple $P = (\Sigma, \mathcal{A}, R)$, with Σ a signature, $\mathcal{A}$ a set of actions and R a set of PGSOS rules.*

We conclude with the notion of *disjoint extension* for a PTSS which allows us to introduce new operators without affecting the behavior of those already specified.

Definition 5 (Disjoint extension). *A PTSS $P' = (\Sigma', \mathcal{A}, R')$ is a* disjoint extension *of a PTSS $P = (\Sigma, \mathcal{A}, R)$, written $P \sqsubseteq P'$, if $\Sigma \subseteq \Sigma'$, $R \subseteq R'$ and R' introduces no new rule for any operator in R.*

3 Behavioral Metrics and Their Compositional Properties

Behavioral equivalences, such as *probabilistic bisimulation* [27,32], answer the question of whether two processes behave precisely the same way or not with respect to the *observations* we can make on them. *Behavioral metrics* [2,6,13,14, 34] answer the more general question of measuring the differences in processes behavior. They are usually defined as 1-bounded *pseudometrics* expressing the behavioral distance on processes, namely they quantify the disparities in the observations that we can make on them.

A 1-bounded *pseudometric* on a set X is a function $d\colon X \times X \to [0,1]$ such that: (i) $d(x,x) = 0$, (ii) $d(x,y) = d(y,x)$, and (iii) $d(x,y) \leq d(x,z) + d(z,y)$, for all $x, y, z \in X$. The *kernel* of a pseudometric d on X consists in the set of the pairs of elements in X that are at distance 0, namely $ker(d) = \{(x,y) \in X \times X \mid d(x,y) = 0\}$. For simplicity, we denote by $\mathcal{D}(X)$ the set of 1-bounded pseudometrics on the set X.

As elsewhere in the literature, we will sometimes use the term *metric* in place of pseudometric. Below we recall the notion of *bisimilarity metric*, which is one of the most studied behavioral metrics in the literature, and we discuss its compositionality.

The notion of *bisimulation metric* [6,13,14] bases on the quantitative analogous to the *bisimulation game*: two processes can be at some given distance $\varepsilon < 1$ only if they can mimic each other's transitions and evolve to distributions that are, in turn, at a distance $\leq \varepsilon$. To formalize this intuition, we need to lift pseudometrics on processes to pseudometrics on distributions. To this purpose, we rely on the notions of *matching* (also known as *coupling* or *weight function*) and *Kantorovich lifting*.

Definition 6 (Matching). *Assume two sets X and Y. A* matching *for distributions $\pi \in \Delta(X)$ and $\pi' \in \Delta(Y)$ is a distribution over the product space $\mathfrak{w} \in \Delta(X \times Y)$ with π and π' as left and right marginal, namely: (i) $\sum_{y \in Y} \mathfrak{w}(x,y) = \pi(x)$, for all $x \in X$, and (ii) $\sum_{x \in X} \mathfrak{w}(x,y) = \pi'(y)$, for all $y \in Y$. We let $\mathfrak{W}(\pi,\pi')$ denote the set of all matchings for π and π'.*

Definition 7 (Kantorovich metric, [21]). *Given $d \in \mathcal{D}(X)$, the* Kantorovich lifting *of d is the pseudometric $\mathbf{K}(d)\colon \Delta(X) \times \Delta(X) \to [0,1]$ defined for all $\pi, \pi' \in \Delta(X)$ by*

$$\mathbf{K}(d)(\pi,\pi') = \min_{\mathfrak{w} \in \mathfrak{W}(\pi,\pi')} \sum_{x,y \in X} \mathfrak{w}(x,y) \cdot d(x,y).$$

Bisimulation metrics are normally parametric with respect to a *discount factor* allowing us to specify how much the distance of future transitions is mitigated [3,14]. Informally, any difference that can be observed only after a long sequence of computation steps does not have the same impact of the differences that can be witnessed at the beginning of the computation. In our context, the discount factor is a value $\lambda \in (0,1]$ such that the distance arising at step n is mitigated by λ^n ($\lambda = 1$ means no discount).

Definition 8 (Bisimulation metric, [13]**).** *Let* $\lambda \in (0,1]$. *We say that* $d \in \mathcal{D}(\mathbf{T}(\Sigma))$ *is a* bisimulation metric *if for all* $s, t \in \mathbf{T}(\Sigma)$ *with* $d(s,t) < 1$, *we have that:*

$$\forall s \xrightarrow{a} \pi_s \; \exists t \xrightarrow{a} \pi_t \text{ such that } \lambda \cdot \mathbf{K}(d)(\pi_s, \pi_t) \leq d(s,t).$$

Notice that any bisimulation metric is a pseudometric by definition. In [13] it was proved that the *smallest* bisimulation metric exists and it is called *bisimilarity metric*, denoted by $\mathbf{b}^\lambda$. Moreover, its kernel induces an equivalence relation that coincides with probabilistic bisimilarity, namely $\mathbf{b}^\lambda(s,t) = 0$ if and only if s and t are bisimilar [6,13].

3.1 Compositionality

In order to specify and verify systems in a compositional manner, it is necessary that the behavioral semantics be compatible with all operators of the language that describe these systems. In the behavioral metric approach, we must guarantee that the distance between two composed systems $f(s_1, \ldots, s_\mathfrak{n})$ and $f(t_1, \ldots, t_\mathfrak{n})$ depends on the distance between all pairs of arguments s_i, t_i, so that the closer all pairs s_i and t_i the closer the composed systems $f(s_1, \ldots, s_\mathfrak{n})$ and $f(t_1, \ldots, t_\mathfrak{n})$.

Following [18,19], we define compositionality properties for bisimulation metric relying on the notion of modulus of continuity for an operator with respect to a distance.

Definition 9 (Modulus of continuity). *Let* $P = (\Sigma, \mathcal{A}, R)$ *be a PTSS,* $f \in \Sigma$ *some* $\mathfrak{n}$*-ary operator and* $d \in \mathcal{D}(\mathbf{T}(\Sigma))$. *Then, a mapping* $\mathfrak{m} \colon [0,1]^\mathfrak{n} \to [0,1]$ *is a* modulus of continuity *for* f *with respect to* d *iff:*

- *for all processes* $s_i, t_i \in T(\Sigma)$ *we have:*

$$d(f(s_1, \ldots, s_\mathfrak{n}), f(t_1, \ldots, t_\mathfrak{n})) \leq \mathfrak{m}(d(s_1, t_1), \ldots, d(s_\mathfrak{n}, t_\mathfrak{n})),$$

- $\mathfrak{m}$ *is continuous at* $(0, \ldots, 0)$, *i.e.* $\lim_{(\varepsilon_1 \ldots \varepsilon_\mathfrak{n}) \to (0 \ldots 0)} \mathfrak{m}(\varepsilon_1, \ldots, \varepsilon_\mathfrak{n}) = \mathfrak{m}(0, \ldots, 0)$,
- $\mathfrak{m}(0, \ldots, 0) = 0$.

In [16,18,19] an operator is called *uniformly continuous* if this operator admits any modulus of continuity with respect to $\mathbf{b}^\lambda$. Intuitively, a uniformly continuous $\mathfrak{n}$-ary operator f ensures that for any non-zero bisimulation distance ε (understood as the admissible tolerance from the operational behavior of the composed processes $f(s_1, \ldots, s_\mathfrak{n})$ and $f(t_1, \ldots, t_\mathfrak{n})$) there are non-zero bisimulation distances ε_i such that the distance between the composed processes $f(s_1, \ldots, s_\mathfrak{n})$ and $f(t_1, \ldots, t_\mathfrak{n})$ is at most ε whenever the component t_i is in distance of at most ε_i from s_i.

Definition 10 (Uniformly continuous operator). *Let* $P = (\Sigma, \mathcal{A}, R)$ *be a PTSS. We say that an* $\mathfrak{n}$*-ary operator* $f \in \Sigma$ *is*

- non extensive *if* $\mathfrak{m}(\varepsilon_1, \dots, \varepsilon_n) = \max_{i=1}^{n} \varepsilon_i$ *is a modulus of continuity for* f;
- non expansive *if* $\mathfrak{m}(\varepsilon_1, \dots, \varepsilon_n) = \sum_{i=1}^{n} \varepsilon_i$ *is a modulus of continuity for* f;
- Lipschitz continuous *if* $\mathfrak{m}(\varepsilon_1, \dots, \varepsilon_n) = L \cdot \sum_{i=1}^{n} \varepsilon_i$ *is a modulus of continuity for* f *for some* $L > 1$.

Non-extensiveness and non-expansiveness are the most studied notion of compositionality (see, e.g., [4,10,13–15,17,22,34,35]). However, they do not allow for reasoning on recursive processes [16,19,24]. Essentially, recursive operators are Lipschitz continuous but not non-expansive [16].

We conclude this section by showing that the PGSOS format is necessary to obtain uniform continuity properties of operators. In the following Example we show an operator f defined by a rule which does not respect the PGSOS format, and for which no modulus of continuity among those in Definition 10 can be established wrt. $\mathbf{b}^\lambda$.

Example 1 (Counterexample for PGSOS format). For simplicity of notation, in all the Examples we will let $a.t$ stand for $a.\delta_t$. Assume the prefix operator $a.\bigoplus_{i \in I}[p_i]\_$ in Table 1 and a PTSS containing the following rules

$$\frac{x \parallel_{\mathcal{A}} x \xrightarrow{a} \mu}{f(x) \xrightarrow{a} f(\mu)} \qquad \frac{}{\eta \xrightarrow{\surd} \delta_{\mathrm{nil}}} \qquad \frac{x \xrightarrow{a} \mu \quad y \xrightarrow{a} \nu \quad a \neq \surd}{x \parallel_{\mathcal{A}} y \xrightarrow{a} \mu \parallel_{\mathcal{A}} \nu} \qquad \frac{x \xrightarrow{a} \surd \quad y \xrightarrow{a} \surd}{x \parallel_{\mathcal{A}} y \xrightarrow{\surd} \delta_{\mathrm{nil}}}$$

in which $\parallel_{\mathcal{A}}$ denotes the fully synchronous parallel composition operator, η denotes the *skip* process, nil is the processes that cannot perform any action and $\surd$ is a special action in $\mathcal{A}$ denoting successful termination. Clearly, the rule for f falls out of the PGSOS format since an arbitrary term different from a single process variable occurs as left hand side of the premise. Given any process $u \in \mathbf{T}(\Sigma)$ s.t. $\mathbf{b}^\lambda(u, \eta) = 1$, consider processes $s_1 = a.u$ and $t_1 = a.([1-\varepsilon]u \oplus [\varepsilon]\eta)$ and for all $k > 1$ define $s_k = a.s_{k-1}$ and $t_k = a.t_{k-1}$. Clearly, for all $k \geq 1$ we have $\mathbf{b}^\lambda(s_k, t_k) = \varepsilon \cdot \lambda^k$. We now aim at evaluating $\mathbf{b}^\lambda(f(s_k), f(t_k))$, for all $k \geq 1$. Since the target of the rule for f is of the form $f(\mu)$, where μ is derived from the synchronous parallel composition of the source variable with itself, we can infer that $f(s_k)$ via a sequence of k a-labeled transitions reaches a distribution that assigns probability 1 to the parallel composition of 2^k copies of process u. Similarly, after k a-labeled transitions $f(t_k)$ reaches a distribution assigning probability $(1-\varepsilon)^{2^k}$ to the process corresponding to the parallel composition of 2^k copies of u; probability ε^{2^k} to the parallel composition of 2^k copies of η, and the remaining probability $1-(1-\varepsilon)^{2^k} - \varepsilon^{2^k}$ to processes that cannot execute any action (being the parallel composition of copies of u and η). Therefore we get $\mathbf{b}^\lambda(f(s_k), f(t_k)) = \lambda^k(1-(1-\varepsilon)^{2^k})$. This implies that for any k with $2^k > L$, $\mathbf{b}^\lambda(f(s), f(t))/\mathbf{b}^\lambda(s,t) = (1-(1-\varepsilon)^{2^k})/\varepsilon > L$ holds for $s = s_k$, $t = t_k$ and $\varepsilon \in (0, (2^k - L)/(2^{k-1}(2^k - 1)))$. We infer that no Lipschitz constant can be found for f and thus none of the continuity properties in Definition 10 can be established for f.

4 The Non-extensiveness Format

In this section we provide a specification format ensuring that all operators in PTSSs respecting the syntactic constraints imposed by the format are non-extensive. Briefly, the *non-extensiveness format* admits only convex combinations of variables and constants as targets of the rules. In the following, we rely on Proposition 1 and without loss of generality we always assume that the targets of all PGSOS-rules are in normal form.

Definition 11 (Specification format for non-extensiveness). *A PTSS $P = (\Sigma, \mathcal{A}, R)$ is in* non-extensiveness format *if all rules $r \in R$ are PGSOS rules as in Definition 3 and $\Theta = \mathrm{trg}(r)$ is of the form $\Theta = \sum_{j \in J} p_j \Theta_j$, with $\Theta_j \in \mathbf{V}_d \cup \delta_{\mathbf{V}_s} \cup \Sigma^0$ for all $j \in J$.*

*Then, all operators in Σ are called P-*non-extensive.

Essentially, in each element Θ_j we have at most one occurrence of one process argument x_h or of one of its derivatives $\mu_{h,m}$, thus implying that Θ_j contributes to the distance $\mathbf{b}^\lambda(f(s_1,\dots,s_n), f(t_1,\dots,t_n))$ by $\mathbf{b}^\lambda(s_h,t_h)$. As a consequence, we have that $\mathbf{b}^\lambda(f(s_1,\dots,s_n), f(t_1,\dots,t_n)) \le \sum_{j \in J} p_j \max_{i=1}^n \mathbf{b}^\lambda(s_i,t_i) = \max_{i=1}^n \mathbf{b}^\lambda(s_i,t_i)$.

We notice that all rules in Table 1 respect Definition 11. Conversely, for each operator defined in Tables 2 and 3 there is at least one rule violating the constraints in Definition 11. For instance, the target of the first rule for operator of sequentialization $x;y$ contains both the derivative μ of the process argument x and the process argument y. If we consider the first rule for the operator of parallel composition $x \parallel_B y$, we note that the target contains one derivative for each of the two process arguments.

To obtain the non-extensiveness property of bisimilarity metric from our format, we need to apply a quantitative analogous to the standard technique used to prove the congruence property of behavioural equivalences. Informally, we introduce the notion of *non-extensive closure* of a pseudometric, seen as the quantitative analogous to the congruence closure of behavioral relations. Briefly, the non-extensive closure allows us to lift a pseudometric to a pseudometric that satisfies the non-extensiveness property.

Definition 12 (Non-extensive closure). *Assume a PTSS $(\Sigma, \mathcal{A}, R)$. The* non-extensive closure $\mathrm{Ext}\colon \mathcal{D}(\mathbf{T}(\Sigma)) \to \mathcal{D}(\mathbf{T}(\Sigma))$ *is defined for $d \in \mathcal{D}(\mathbf{T}(\Sigma))$ and $s,t \in \mathbf{T}(\Sigma)$ by*

$$\mathrm{Ext}(d)(s,t) = \begin{cases} \min\{d(s,t),\ \max\limits_{i=1\dots,n} \mathrm{Ext}(d)(s_i,t_i)\} & \text{if } s = f(s_1,\dots,s_n), \\ & t = f(t_1,\dots,t_n) \text{ and } f \in \Sigma \\ d(s,t) & \text{otherwise.} \end{cases}$$

In particular we denote by $\mathbf{Ext}^\lambda$ the non-extensive closure of $\mathbf{b}^\lambda$.

Clearly, focusing on the non-extensive closure of bisimilarity metric $\mathbf{Ext}^\lambda$, we observe that $\mathbf{Ext}^\lambda(s,t) \leq \mathbf{b}^\lambda(s,t)$ for all $s,t \in \mathbf{T}(\Sigma)$. Moreover, notice that each operator f admits its modulus of continuity for non-extensiveness wrt. $\mathbf{Ext}^\lambda$. Therefore, the proof of the *non-extensiveness theorem* (Theorem 1 below), that states the correctness of the non-extensiveness format (Definition 11), consists in proving that under the constraints of our format the metric $\mathbf{Ext}^\lambda$ is itself a bisimulation metric. As a consequence, from $\mathbf{Ext}^\lambda(s,t) \leq \mathbf{b}^\lambda(s,t)$ and $\mathbf{b}^\lambda$ being the least bisimulation metric, we can infer that $\mathbf{Ext}^\lambda = \mathbf{b}^\lambda$ and thus that $\mathbf{b}^\lambda$ is non-extensive wrt. all the operators satisfying our non-extensiveness format.

Theorem 1 (Non-extensiveness). *Let $\lambda \in (0,1]$ and $P = (\Sigma, \mathcal{A}, R)$ be a PTSS in non-extensiveness format. Then the operators in Σ are non-extensive wrt.* $\mathbf{b}^\lambda$.

Proof sketch. We present only a sketch of the proof. The proofs of Theorems 2 and 3 follow the same reasoning schema.

As $\mathbf{Ext}^\lambda \leq \mathbf{b}^\lambda$ by definition and $\mathbf{b}^\lambda$ is the least bisimulation metric, the thesis is equivalent to prove that $\mathbf{Ext}^\lambda$ is a bisimulation metric. We need to show that

$$\begin{aligned} &\text{whenever } \mathbf{Ext}^\lambda(\sigma(t),\sigma'(t)) < 1 \text{ and } P \vdash \sigma(t) \xrightarrow{a} \pi \\ &\text{then } P \vdash \sigma'(t) \xrightarrow{a} \pi' \text{ and } \lambda \cdot \mathbf{K}(\mathbf{Ext}^\lambda)(\pi,\pi') \leq \mathbf{Ext}^\lambda(\sigma(t),\sigma'(t)). \end{aligned} \tag{2}$$

This is proved by induction over the structure of t.

For the non-trivial inductive step with $t = f(t_1,\ldots,t_\mathtt{n})$ and $\mathbf{Ext}^\lambda(\sigma(t),\sigma'(t)) = \max_{i=1,\ldots,\mathtt{n}} \mathbf{Ext}^\lambda(\sigma(t_i),\sigma'(t_i))$ the proof can be sketched as follows. Assume that $P \vdash \sigma(t) \xrightarrow{a} \pi$, so that there are a non-extensive rule

$$r = \frac{\{x_h \xrightarrow{a_{h,m}} \mu_{h,m} \mid h \in H, m \in M_h\} \;\; \{x_h \stackrel{a_{h,n}}{\not\longrightarrow} \mid h \in H, n \in N_h\}}{f(x_1,\ldots,x_\mathtt{n}) \xrightarrow{a} \sum_{j \in J} p_j \Theta_j}$$

and a closed substitution σ_1, such that $\sigma_1(x_i) = \sigma(x_i)$ for all $i = 1,\ldots,\mathtt{n}$ and $\sigma_1(\sum_{j\in J} p_j\Theta_j) = \pi$, from which we can derive such a transition.

For each $h \in H, m \in M_h$, the inductive hypothesis gives that $\sigma'(t_h) \xrightarrow{a_{h,m}} \pi_{h,m}$ and $\lambda \cdot \mathbf{K}(\mathbf{Ext}^\lambda)(\sigma_1(\mu_{h,m}),\pi_{h,m}) \leq \mathbf{Ext}^\lambda(\sigma(t_h),\sigma'(t_h))$. For the closed substitution σ_2 with $\sigma_2(x_i) = \sigma'(t_i)$ for all $i = 1,\ldots,\mathtt{n}$ and $\sigma_2(\mu_{h,m}) = \pi_{h,m}$ for all $h \in H, m \in M_h$, we get that $\{\sigma_2(x_h) \xrightarrow{a_{h,m}} \sigma_2(\mu_{h,m}) \mid h \in H, m \in M_h\}$ is provable from P. For each $h \in H, n \in N_h$, by the inductive hypothesis we can infer that $\sigma'(t_h) \stackrel{a_{h,n}}{\not\longrightarrow}$ for all $h \in H, n \in N_h$. For σ_2 defined as above, we get that the set of negative premises $\{\sigma_2(x_h) \stackrel{a_{h,n}}{\not\longrightarrow} \mid h \in H, n \in N_h\}$ is provable from P. Hence, all the premises of the closed instance of r with respect to σ_2 are provable from P, thus implying that $\sigma'(t) \xrightarrow{a} \sigma'(\sum_{j\in J} p_j\Theta_j)$ is provable from P.

Then, $\lambda \cdot \mathbf{K}(\mathbf{Ext}^\lambda)(\sum_{j\in J} p_j\sigma_1(\Theta_j), \sum_{j\in J} p_j\sigma_2(\Theta_j)) \leq \mathbf{Ext}^\lambda(\sigma(t),\sigma'(t))$ follows by: 1. the convexity properties of the Kantorovich metric, 2. the form of the Θ_j in the target of a non-extensive rule, and 3. the inductive hypothesis. $\square$

Table 1. Non-extensive process algebra operators. $\mu \oplus_p \nu$ stays for $p \cdot \mu + (1-p) \cdot \nu$.

$$\frac{}{a.\bigoplus_{i=1}^{n}[p_i]x_i \xrightarrow{a} \sum_{i=1}^{n} p_i \delta_{x_i}} \qquad \frac{x \xrightarrow{a} \mu}{x + y \xrightarrow{a} \mu} \qquad \frac{y \xrightarrow{a} \nu}{x + y \xrightarrow{a} \nu}$$

$$\frac{x \xrightarrow{a} \mu \quad y \stackrel{a}{\not\rightarrow}}{x +_p y \xrightarrow{a} \mu} \qquad \frac{x \stackrel{a}{\not\rightarrow} \quad y \xrightarrow{a} \nu}{x +_p y \xrightarrow{a} \nu} \qquad \frac{x \xrightarrow{a} \mu \quad y \xrightarrow{a} \nu}{x +_p y \xrightarrow{a} \mu \oplus_p \nu}$$

By applying Theorem 1, we get the non-extensiveness of all operators in Table 1.

Corollary 1. *The operators defined by the PGSOS rules in Table 1 are non-extensive.*

Since we already know from [16] that none of the operators in Tables 2 and 3 is non-extensive, we can conclude that our format is not too restrictive, meaning that no non-extensive operator among those used in standard probabilistic processes algebras is out of the format. Moreover, we can also infer that the constraints of the format cannot be relaxed in any immediate way.

5 The Non-expansiveness Format

In this section we focus on the non-expansiveness property. We show that by relaxing the constraints of the non-extensiveness format, we obtain a specification *format for non-expansiveness*. In detail, we allow the target of each rule to be a convex combination of distribution terms in which also non-constant operators can appear and can be applied to source variables and to their derivatives, provided that for each source variables at most one occurrence of that variable or of its derivatives occurs.

Definition 13 (Specification format for non-expansiveness). *Assume a PTSS $P_1 = (\Sigma_1, \mathcal{A}, R_1)$ in non-extensiveness format. A PTSS $P_2 = (\Sigma_2, \mathcal{A}, R_2)$ with $P_1 \sqsubseteq P_2$ is in* non-expansiveness format *if all rules $r \in R_2 \setminus R_1$ are of the form*

$$\frac{\{x_h \xrightarrow{\alpha_{h,m}} \mu_{h,m} \mid h \in H, m \in M_h\} \ \ \{x_h \stackrel{\alpha_{h,n}}{\not\longrightarrow} \mid h \in H, n \in N_h\}}{f(x_1, \ldots, x_\mathfrak{n}) \xrightarrow{\alpha} \Theta}$$

where:

1. *the target Θ is of the form $\Theta = \sum_{j \in J} p_j \Theta_j$,*
2. *for all $j \in J$ and $h \in H$, at most one among the variables in $(\{x_h\} \cup \{\mu_{h,m} \mid m \in M_h\})$ occurs in Θ_j and there must be at most one such occurrence, and*
3. *for all $j \in J$ and $i \in \{1, \ldots, \mathfrak{n}\} \setminus H$, the variable x_i can occur at most once in Θ_j.*

Moreover, all operators defined by rules in $R_2 \setminus R_1$ *are called* P_2-non-expansive.

Since in each Θ_j we can have at most one occurrence of a source variable x_h or of one of its derivatives $\{\mu_{h,m} \mid m \in M_h\}$, we can infer that Θ_j contributes to $\mathbf{b}^\lambda(f(s_1,\ldots,s_n), f(t_1,\ldots,t_n))$ by, at most, $\sum_{i=1}^n \mathbf{b}^\lambda(s_i,t_i)$. Consequently, we obtain $\mathbf{b}^\lambda(f(s_1,\ldots,s_n), f(t_1,\ldots,t_n)) \leq \sum_{j\in J} p_j \sum_{i=1}^n \mathbf{b}^\lambda(s_i,t_i) = \sum_{i=1}^n \mathbf{b}^\lambda(s_i,t_i)$.

Notice that all non-extensive operators are also non-expansive. Further, all operators defined by rules in Table 2, as well as those in Table 1, respect the non-expansiveness format. Conversely, the operators in Table 3, which are known to be expansive [16], have at least one rule falling out of the format. For instance, the rule for the bang operator $!x$, which stands for the parallel composition of infinitely many copies of the process argument, violates Definition 13 since both the source variable x and its derivative μ occur in the same distribution term in the target.

As in the case of non-extensiveness, the formal proof of the correctness of the non-expansiveness format bases on a proper notion of closure of a pseudometric. We define the *non-expansiveness closure* of a pseudometric as its lifting to a pseudometric which is non-extensive on operators satisfying Definition 11, and non-expansive on the others.

Definition 14 (Non-expansive closure). *Assume the PTSSs* $P_1 = (\Sigma_1, \mathcal{A}, R_1)$ *in non-extensiveness format and* $P_2 = (\Sigma_2, \mathcal{A}, R_2)$ *with* $P_1 \sqsubseteq P_2$. *The* non-expansive closure $\mathrm{Exp}\colon \mathcal{D}(\mathbf{T}(\Sigma_2)) \to \mathcal{D}(\mathbf{T}(\Sigma_2))$ *is defined for* $d \in \mathcal{D}(\mathbf{T}(\Sigma_2))$ *and* $s,t \in \mathbf{T}(\Sigma_2)$ *by*

$$\mathrm{Exp}(d)(s,t) = \begin{cases} \min\{d(s,t), \max\limits_{i=1\ldots,n} \mathrm{Exp}(d)(s_i,t_i)\} & \text{if } s = f(s_1,\ldots,s_n),\\ & t = f(t_1,\ldots,t_n), f \in \Sigma_1 \\ \min\{d(s,t), \sum\limits_{i=1}^{n} \mathrm{Exp}(d)(s_i,t_i)\} & \text{if } s = f(s_1,\ldots,s_n),\\ & t = f(t_1,\ldots t_n), f \in \Sigma_2 \setminus \Sigma_1 \\ d(s,t) & \text{otherwise.} \end{cases}$$

In particular we denote by $\mathbf{Exp}^\lambda$ *the non-expansive closure of* $\mathbf{b}^\lambda$.

We can then show that under the constraints of Definition 13, the non-expansive closure $\mathbf{Exp}^\lambda$ of $\mathbf{b}^\lambda$ is indeed a bisimulation metric. Since by definition $\mathbf{Exp}^\lambda(s,t) \leq \mathbf{b}^\lambda(s,t)$ for all $s,t \in \mathbf{T}(\Sigma)$ and $\mathbf{b}^\lambda$ is the least bisimulation metric, we can conclude that $\mathbf{Exp}^\lambda = \mathbf{b}^\lambda$ and the operators in a PTSS in non-expansiveness format are non-expansive with respect to $\mathbf{b}^\lambda$, as stated in Theorem 2 below.

Theorem 2 (Non-expansiveness). *Let* $\lambda \in (0,1]$ *and* $(\Sigma, \mathcal{A}, R)$ *be a PTSS in non-expansiveness format. Then the operators is* Σ *are non-expansive with respect to* $\mathbf{b}^\lambda$.

By applying Theorem 2 we get the non-expansiveness of all operators in Table 2.

Corollary 2. *The operators defined by the PGSOS rules in Tables 1 and 2 are non-expansive with respect to* $\mathbf{b}^\lambda$.

Since we already know from [16] that none of the operators in Table 3 is non-expansive, we can conclude that our format is not too restrictive, meaning that no non-expansive operator among those used in standard probabilistic processes algebras is out of the format. Moreover, the following examples show that the constraints in Definition 13 cannot be relaxed in any trivial way.

Example 2 (Counterexample for Definition 13.2, *I).* Consider (i) $s_1 = a.([1/2]u_1 \oplus [1/2]u_2)$; (ii) $t_1 = a.([1/3]u_1 \oplus [1/3]u_2 \oplus [1/3]u_3)$; (iii) $u_1 = b.\eta$; (iv) $u_2 = c.\eta$; (v) $u_3 = u_1 + u_2$. Clearly we have $\mathbf{b}^\lambda(s_1, t_1) = 1/3 \cdot \lambda$, due to u_3. Let $\|$ denote $\|_\emptyset$. Consider the operator f defined by the following PGSOS rule

$$\frac{x \xrightarrow{a} \mu}{f(x) \xrightarrow{a} \delta_x \parallel \mu}$$

which violates the non-expansiveness format of Definition 13 since both x and μ occur in the target. Let us evaluate $\mathbf{b}^\lambda(f(s_1), f(t_1))$. Let $\pi_{s_1} = 1/2\delta(s_1 \parallel u_1) + 1/2\delta(s_1 \parallel u_2)$ and $\pi_{t_1} = 1/3\delta(t_1 \parallel u_1) + 1/3\delta(t_1 \parallel u_2) + 1/3\delta(t_1 \parallel u_3)$. We have $f(s_1) \xrightarrow{a} \pi_{s_1}$, $f(t_1) \xrightarrow{a} \pi_{t_1}$ and $\mathbf{b}^\lambda(f(s_1), f(t_1)) \geq \lambda \cdot \mathbf{K}(\mathbf{b}^\lambda)(\pi_{s_1}, \pi_{t_1})$. Then $\mathbf{b}^\lambda(s_1 \parallel u_1, t_1 \parallel u_3) = 1 = \mathbf{b}^\lambda(s_1 \parallel u_2, t_1 \parallel u_3)$, whereas $\mathbf{b}^\lambda(s_1 \parallel u_1, t_1 \parallel u_1) = \mathbf{b}^\lambda(s_1 \parallel u_2, t_1 \parallel u_2) = \mathbf{b}^\lambda(s_1, t_1)$. Hence, $\mathbf{b}^\lambda(f(s_1), f(t_1)) = 1/3 \cdot \lambda + 2/9 \cdot \lambda^2 > \mathbf{b}^\lambda(s_1, t_1)$.

Example 3 (Counterexample for Definition 13.2, *II).* Consider process u_1 from Example 2 and processes $s_2 = a.([1/2]u_1 \oplus [1/2]\eta)$ and $t_2 = a.u_1$. Clearly, $\mathbf{b}^\lambda(s_2, t_2) = 1/2 \cdot \lambda$. Consider the following PGSOS rule defining the behavior of operator g

$$\frac{x \xrightarrow{a} \mu}{g(x) \xrightarrow{a} \mu \parallel_{\mathcal{A}} \mu}$$

which violates the non-expansiveness format as variable μ occurs in the right hand side of premise and twice in the target. Let us evaluate $\mathbf{b}^\lambda(g(s_2), g(t_2))$. We have that $g(s_2) \xrightarrow{a} 1/4\delta(u_1 \parallel_{\mathcal{A}} u_1) + 1/4\delta(\eta \parallel_{\mathcal{A}} \eta) + 1/4\delta(u_1 \parallel_{\mathcal{A}} \eta) + 1/4\delta(\eta \parallel_{\mathcal{A}} u_1)$. Conversely, $g(t_2) \xrightarrow{a} \delta(u_1 \parallel_{\mathcal{A}} u_1)$. Therefore, since $\parallel_{\mathcal{A}}$ corresponds to the fully synchronous parallel composition, we can directly infer that $\mathbf{b}^\lambda(g(s_2), g(t_2)) \geq 3/4 \cdot \lambda > \mathbf{b}^\lambda(s_2, t_2)$.

Example 4 (Counterexample for Definition 13.3*).* Consider process u_1 from Example 2 and processes s_2, t_2 from Example 3. Consider the following PGSOS rule defining the behavior of operator h

$$\frac{x \xrightarrow{b} \mu}{h(x, y) \xrightarrow{b} \delta_y \parallel_{\mathcal{A}} \delta_y}$$

Table 2. Non-expansive process algebra operators. The symmetric version of the first rule for $\|_B$ and the one for $|||_p$ have been omitted.

$$\frac{}{\eta \xrightarrow{\surd} \delta_{\text{nil}}} \qquad \frac{x \xrightarrow{a} \mu \quad a \neq \surd}{x;y \xrightarrow{a} \mu;\delta_y} \qquad \frac{x \xrightarrow{\surd} \mu \quad y \xrightarrow{a} \nu}{x;y \xrightarrow{a} \nu}$$

$$\frac{x \xrightarrow{a} \mu \quad a \notin B \cup \{\surd\}}{x \|_B\, y \xrightarrow{a} \mu \|_B\, \delta_y} \qquad \frac{x \xrightarrow{a} \mu \quad y \xrightarrow{a} \nu \quad a \in B \setminus \{\surd\}}{x \|_B\, y \xrightarrow{a} \mu \|_B\, \nu} \qquad \frac{x \xrightarrow{\surd} \mu \quad y \xrightarrow{\surd} \nu}{x \|_B\, y \xrightarrow{\surd} \delta_{\text{nil}}}$$

$$\frac{x \xrightarrow{a} \mu \quad y \not\xrightarrow{a} \quad a \neq \surd}{x\, |||_p\, y \xrightarrow{a} \mu\, |||_p\, \delta_y} \qquad \frac{x \xrightarrow{a} \mu \quad y \xrightarrow{a} \nu \quad a \neq \surd}{x\, |||_p\, y \xrightarrow{a} \mu\, |||_p\, \delta_y \oplus_p \delta_x\, |||_p\, \nu} \qquad \frac{x \xrightarrow{\surd} \mu \quad y \xrightarrow{\surd} \nu}{x\, |||_p\, y \xrightarrow{\surd} \delta_{\text{nil}}}$$

which violates the non-expansiveness format in Definition 13 as variable y occurs twice in the target. From similar calculations to those applied in Example 3, we can infer that $\mathbf{b}^\lambda(h(u_1, s_2), h(u_1, t_2)) \geq 3/4 \cdot \lambda^2$, which is greater than $\mathbf{b}^\lambda(s_2, t_2)$ for $\lambda > 2/3$.

6 The Lipschitz Continuity Format

We dedicate this section to the definition of a specification *format for Lipschitz continuity*. This is obtained by relaxing the constraints of the non-expansiveness format in Definition 13 as follows: firstly, for each operator f we fix $n_f, m_f \in \mathbb{N}$. Then, we allow the targets of the rules for f to be convex combinations of distribution terms in which (i) the number of occurrences of source variables and their derivatives is bounded by n_f; (ii) the number of nested operators in which a source variable can occur is bounded by m_f; (iii) derivatives can only occur in the scope of non-expansive operators.

Definition 15 (Specification format for Lipschitz continuity). *Assume a PTSS $P_1 = (\Sigma_1, \mathcal{A}, R_1)$ in non-expansiveness format. A PTSS $P_2 = (\Sigma_2, \mathcal{A}, R_2)$ with $P_1 \sqsubseteq P_2$ is in* Lipschitz continuity format *if for all operators $f \in \Sigma_2 \setminus \Sigma_1$ there are naturals n_f, m_f and all rules $r \in R_2 \setminus R_1$ are of the form*

$$\frac{\{x_h \xrightarrow{\alpha_{h,m}} \mu_{h,m} \mid h \in H, m \in M_h\} \;\; \{x_h \stackrel{\alpha_{h,n}}{\not\longrightarrow} \mid h \in H, n \in N_h\}}{f(x_1, \ldots, x_{\mathtt{n}}) \xrightarrow{\alpha} \Theta}$$

where:

1. *the target Θ is of the form $\Theta = \sum_{j \in J} p_j \Theta_j$,*
2. *for all $j \in J$ and $i \in \{1, \ldots, \mathtt{n}\}$, x_i occurs at most n_f times in Θ_j and such occurrences appear in the scope of at most m_f operators, and*
3. *for all $j \in J$ and $h \in H$, there are at most n_f occurrences of variables in $\{\mu_{h,m} \mid m \in M_h\}$ in Θ_j, which appear only in the scope of non-expansive operators in Σ_1.*

Moreover, all operators defined by rules in $R_2 \setminus R_1$ *are called* P_2-Lipschitz.

Informally, for each $h \in H$, bounding to n_f the number of occurrences of derivatives $\{\mu_{h,m} \mid m \in M_h\}$ that may occur in Θ_j and requiring them in the scope of non-expansive operators, guarantees that process argument x_h and Θ_j contribute to $\mathbf{b}^\lambda(f(s_1,\dots,s_n), f(t_1,\dots,t_n))$ by, at most, $n_f \cdot \mathbf{b}^\lambda(s_h,t_h)$. Considering instead source variables, constraint 2 of Definition 15 guarantees that for each $i = 1,\dots,n$ there is a constant $L_i \geq 1$ such that process argument x_i and Θ_j contribute to distance $\mathbf{b}^\lambda(f(s_1,\dots,s_n), f(t_1,\dots,t_n))$ by, at most, $\lambda \cdot n_f \cdot L_i^{m_f} \cdot \mathbf{b}^\lambda(s_i,t_i)$, where L_i is the maximal Lipschitz constant among those of operators applied on top of x_i, and factor λ arises since x_i is delayed by one computation step. Therefore, we can infer that for each $j \in J$ there is a constant $L_j \geq n_f$ such that Θ_j contributes to $\mathbf{b}^\lambda(f(s_1,\dots,s_n), f(t_1,\dots,t_n))$ by, at most, $L_j \sum_{i=1}^{n} \mathbf{b}^\lambda(s_i,t_i)$. Finally, we obtain $\mathbf{b}^\lambda(f(s_1,\dots,s_n), f(t_1,\dots,t_n)) \leq \sum_{j\in J} p_j L_j \cdot \sum_{i=1}^{n} \mathbf{b}^\lambda(s_i,t_i) = L \cdot \sum_{i=1}^{n} \mathbf{b}^\lambda(s_i,t_i)$ for some constant $L \geq 1$.

Notice that the rules in Table 3, as those in Tables 1 and 2, respect the constraints in Definition 15. We observe that Table 3 contains the specification of standard recursive operators x^ω, $!x$ and x^*y which were ruled out by Definitions 11 and 13.

Next, we formalize the definition of a *Lipschitz factor* for an operator f, namely the constant L_f wrt. which f is Lipschitz continuous, and its construction over the rules for f and their targets. Since this construction is recursive for recursive operators, in order to have well-defined Lipschitz factors for these operators we need a discount factor strictly less than one.

Definition 16 (Lipschitz factor). *Assume* $\lambda \in (0,1)$, *a PTSS* $P_1 = (\Sigma_1, \mathcal{A}, R_1)$ *in non-extensiveness format and a PTSS* $P_2 = (\Sigma_2, \mathcal{A}, R_2)$ *with* $P_1 \sqsubseteq P_2$. *For each operator* $f \in \Sigma_2 \setminus \Sigma_1$ *let* $R_f \subseteq R_2$ *denote the set of rules defining the behavior. Then, the set of* Lipschitz factors *of operators in* $\Sigma_2 \setminus \Sigma_1$ *is defined as the set of minimum reals* $L_f \geq 1$ *satisfying the system of inequalities of the form*

$$L_f \geq \max_{r\in R_f} \max_{x\in \mathrm{src}(r)} \Big\{ \ell_{\mathrm{trg}(r)}(x) + \sum_{\mu \in \mathrm{der}(x,r)} \ell_{\mathrm{trg}(r)}(\mu) \Big\}$$

for each $f \in \Sigma_2 \setminus \Sigma_1$ *where, for any distribution term* $\Theta \in \mathbb{DT}(\Sigma)$ *in normal form, the factor* ℓ_Θ *is defined inductively as follows*

$$\ell_\Theta(x) = \begin{cases} 0 & \text{if } x \notin \mathrm{var}(\Theta) \\ \lambda & \text{if } \Theta = \delta_x \\ \max_{i=1,\dots,n} \ell_{\Theta_i}(x) & \text{if } \Theta = f(\Theta_1,\dots,\Theta_n) \text{ and } f \in \Sigma_1 \\ L_f \sum_{i=1}^{n} \ell_{\Theta_i}(x) & \text{if } \Theta = f(\Theta_1,\dots,\Theta_n) \text{ and } f \in \Sigma_2 \setminus \Sigma_1 \\ \sum_{j\in J} p_j \ell_{\Theta_j}(x) & \text{if } \Theta = \sum_{j\in J} p_j \Theta_j \end{cases}$$

$$\ell_\Theta(\mu) = \begin{cases} 0 & \textit{if } \mu \notin \mathrm{var}(\Theta) \\ 1 & \textit{if } \Theta = \mu \\ \max_{i=1,\dots,\mathrm{n}} \ell_{\Theta_i}(\mu) & \textit{if } \Theta = f(\Theta_1,\dots,\Theta_\mathrm{n}) \textit{ and } f \in \Sigma_1 \\ L_f \sum_{i=1}^{\mathrm{n}} \ell_{\Theta_i}(\mu) & \textit{if } \Theta = f(\Theta_1,\dots,\Theta_\mathrm{n}) \textit{ and } f \in \Sigma_2 \setminus \Sigma_1 \\ \sum_{j\in J} p_j \ell_{\Theta_j}(\mu) & \textit{if } \Theta = \sum_{j \in J} p_j \Theta_j \end{cases}$$

The following proposition states that Lipschitz factors are well-defined if the PTSS is in Lipschitz format and $\lambda \in (0,1)$. Moreover, $L_f = 1$ if f is non-expansive.

Proposition 2. *Let $\lambda \in (0,1)$. Assume a PTSS $P_1 = (\Sigma_1, \mathcal{A}, R_1)$ in non-extensiveness format and an extension $P_1 \sqsubseteq P_2 = (\Sigma_2, \mathcal{A}, R_2)$ in Lipschitz format. Then:*

- *for each operator $f \in \Sigma_2$, the Lipschitz factor $L_f < +\infty$ is well-defined.*
- *moreover, if $f \in \Sigma_2$ is non-expansive then $L_f = 1$.*

The following two examples show that if P_2 falls out of the Lipschitz continuity format, then Proposition 2 does not hold.

Example 5. Consider operators f and g specified by the following rules for all $k \in \mathbb{N}$:

$$\frac{x \xrightarrow{a_k} \mu}{f(x) \xrightarrow{a_k} \underbrace{\mu \parallel_\mathcal{A} \dots \parallel_\mathcal{A} \mu}_{k-\text{times}}} \qquad \overline{g(x) \xrightarrow{a_k} \delta(\underbrace{h(\dots h(}_{k-\text{times}} x)))} \qquad \frac{x \xrightarrow{a_k} \mu}{h(x) \xrightarrow{a_k} \mu \parallel_\mathcal{A} \mu}$$

Notice that rules for both f and g are out of the Lipschitz format since we cannot define any n_f and m_g accordingly to Definition 15. Then, the inductive definition of Lipschitz factors in Definition 16 gives that $L_{\parallel_\mathcal{A}} = 1$, $L_h = 2$ and, then, $L_f \geq k$ and $L_g \geq \lambda \cdot 2^k$ for all $k \in \mathbb{N}$, thus implying that $L_f = L_g = +\infty$.

Example 6. Consider the *copy* operator [5] defined by the PGSOS rules below

$$\frac{x \xrightarrow{a} \mu \quad a \notin \{\mathrm{l}, \mathrm{r}\}}{\mathrm{cp}(x) \xrightarrow{a} \mu} \qquad \frac{x \xrightarrow{\mathrm{l}} \mu \quad x \xrightarrow{\mathrm{r}} \nu}{\mathrm{cp}(x) \xrightarrow{\mathrm{s}} \mathrm{cp}(\mu) \parallel_\mathcal{A} \mathrm{cp}(\nu)}.$$

Firstly, notice that the second rule violates the non-expansiveness format since we have two occurrences of derivatives of x in the same distribution term (Definition 13.2). Consequently, it also violates the Lipschitz continuity format in that μ and ν occur in the scope of an expansive operator (Definition 15.3). By Definition 16 we would have $L_{\mathrm{cp}} \geq L_{\parallel_\mathcal{A}} \cdot L_{\mathrm{cp}} + L_{\parallel_\mathcal{A}} \cdot L_{\mathrm{cp}} = 2 \cdot L_{\mathrm{cp}}$ which is satisfied only by $L_{\mathrm{cp}} = +\infty$.

Then, the following example shows that, due to the presence of recursive operators, we must require λ to be strictly lesser than one to guarantee the validity of Proposition 2.

Example 7. Consider the rule for the bang operator $!\_$ in Table 3. By Definition 16 we obtain that $L_! \geq \lambda \cdot L_! + 1$, thus giving $L_! \geq \frac{1}{1-\lambda}$, which does not hold for $\lambda = 1$.

We remark that by restricting to non-recursive operators, such as x^n and $!^n$ in Table 3, Proposition 2 would hold also for $\lambda = 1$.

As one can expect, the proof of the correctness of our Lipschitz continuity format bases on the notion of *Lipschitz closure* of a pseudometric that lifts it to a pseudometric which is non-extensive on non-extensive operators, non-expansive on non-expansive operators and Lipschitz continuous on Lipschitz operators.

Definition 17 (Lipschitz closure). *Let $\lambda \in (0,1)$. Assume a PTSS $P_1 = (\Sigma_1, \mathcal{A}, R_1)$ in non-extensiveness format and a PTSS $P_2 = (\Sigma_2, \mathcal{A}, R_2)$ with $P_1 \sqsubseteq P_2$. Then, the* Lipschitz closure Lip: $\mathcal{D}(\mathbf{T}(\Sigma_2)) \to \mathcal{D}(\mathbf{T}(\Sigma_2))$ *is defined for all $d \in \mathcal{D}(\mathbf{T}(\Sigma_2))$ and $s, t \in \mathbf{T}(\Sigma_2)$ by*

$$\mathrm{Lip}(d)(s,t) = \begin{cases} \min\{d(s,t),\ \max\limits_{i=1\ldots,n} \mathrm{Lip}(d)(s_i,t_i)\} & \text{if } s = f(s_1,\ldots,s_n),\\ & t = f(t_1,\ldots,t_n), f \in \Sigma_1 \\ \min\{d(s,t),\ L_f \cdot \sum\limits_{i=1}^{n} \mathrm{Lip}(d)(s_i,t_i)\} & \text{if } s = f(s_1,\ldots,s_n),\\ & t = f(t_1,\ldots t_n), f \in \Sigma_2 \setminus \Sigma_1 \\ d(s,t) & \text{otherwise.} \end{cases}$$

In particular we denote by $\mathbf{Lip}^\lambda$ *the Lipschitz closure of* $\mathbf{b}^\lambda$.

We can show that on a PTSS respecting Definition 15 the Lipschitz closure $\mathbf{Lip}^\lambda$ of $\mathbf{b}^\lambda$ is a bisimulation metric and thus the operators in the PTSS are Lipschitz continuous wrt. $\mathbf{b}^\lambda$. In particular, the constraints of the Lipschitz continuity format combined with the definition of the Lipschitz factor L_f of operator f (Definition 16) guarantee that f is L_f-Lipschitz wrt. $\mathbf{b}^\lambda$.

Theorem 3 (Lipschitz continuity). *Let $(\Sigma, \mathcal{A}, R)$ be a PTSS in Lipschitz continuity format and let $\lambda \in (0,1)$. Then the operators is Σ are Lipschitz continuous wrt.* $\mathbf{b}^\lambda$. *In particular, for a Lipschitz operator f, we have that f is L_f-Lipschitz wrt.* $\mathbf{b}^\lambda$.

As an application of Theorem 3, we obtain that $\mathbf{b}^\lambda$ is compositional wrt. all the operators defined in Table 3 in the sense of Lipschitz continuity.

Corollary 3. *Let $\lambda \in (0,1)$. The operators defined by the PGSOS rules in Tables 1, 2 and 3 are Lipschitz continuous with respect to* $\mathbf{b}^\lambda$.

Again, we remark that in Corollary 3 the condition λ strictly lesser than one is required by the operators x^ω, $!x$ and x^*y.

Since all operators that have been proved in [16] to be Lipschitz continuous are in Table 3, we can infer that our format is not too restrictive as it does not exclude any know Lipschitz continuous operator. Moreover, the following two examples show that the constraint of Definition 15 cannot be trivially relaxed.

Table 3. Lipschitz continuous process algebra operators. We recall that $\|$ denotes $\|_\emptyset$.

$$\frac{x \xrightarrow{a} \mu \quad a \neq \surd}{x^{n+1} \xrightarrow{a} \mu; \delta_{x^n}} \qquad \frac{x \xrightarrow{\surd} \mu}{x^{n+1} \xrightarrow{\surd} \mu} \qquad \frac{}{x^0 \xrightarrow{\surd} \delta_{\mathrm{nil}}} \qquad \frac{x \xrightarrow{\surd} \mu \quad x \xrightarrow{a} \nu \quad a \neq \surd \quad n > m}{x^n \xrightarrow{a} \nu; \delta_{x^m}}$$

$$\frac{x \xrightarrow{a} \mu \quad a \neq \surd}{x^\omega \xrightarrow{a} \mu; \delta_{x^\omega}} \qquad \frac{x \xrightarrow{a} \mu \quad a \neq \surd}{x^* y \xrightarrow{a} \mu; \delta_{x^* y}} \qquad \frac{y \xrightarrow{a} \nu}{x^* y \xrightarrow{a} \nu}$$

$$\frac{x \xrightarrow{a} \mu \; y \xrightarrow{a} \nu \; a \neq \surd}{x^{*_p} y \xrightarrow{a} \nu \oplus_p (\mu; \delta_{x^{*_p} y})} \quad \frac{x \xrightarrow{a} \mu \; y \not\xrightarrow{a} \; a \neq \surd}{x^{*_p} y \xrightarrow{a} \mu; \delta_{x^{*_p} y}} \quad \frac{x \not\xrightarrow{a} \; y \xrightarrow{a} \nu \; a \neq \surd}{x^{*_p} y \xrightarrow{a} \nu} \quad \frac{y \xrightarrow{\surd} \nu}{x^{*_p} y \xrightarrow{\surd} \nu}$$

$$\frac{x \xrightarrow{a} \mu \quad a \neq \surd}{!^{n+1} x \xrightarrow{a} \mu \parallel \delta_{!^n x}} \qquad \frac{x \xrightarrow{\surd} \mu}{!^{n+1} x \xrightarrow{\surd} \mu} \qquad \frac{}{!^0 x \xrightarrow{\surd} \delta_{\mathrm{nil}}}$$

$$\frac{x \xrightarrow{a} \mu \quad a \neq \surd}{!x \xrightarrow{a} \mu \parallel \delta_{!x}} \qquad \frac{x \xrightarrow{a} \mu \quad a \neq \surd}{!_p x \xrightarrow{a} \mu \oplus_p (\mu \parallel \delta_{!_p x})}$$

Example 8 (Counterexample for bounded number of occurrences). Assume $\mathcal{A} = \{a_k \mid k \in \mathbb{N}\}$. Consider the operators f and g as in Example 5, where we already noticed that the rules defining them violate Definition 15. Let $\varepsilon \in (0,1)$. Following [19], consider processes $u_k = a_k.a_k.\mathrm{nil}$ and $v_k = a_k.([1-\varepsilon]a_k.\mathrm{nil} \oplus [\varepsilon]\mathrm{nil})$ with $k \in \mathbb{N}$. Clearly, $\mathbf{b}^\lambda(u_k, v_k) = \lambda\varepsilon$ for all $k \in \mathbb{N}$.

Consider first operator f. We have $f(u_k) \xrightarrow{a_k} \pi_{u,k}$ and $f(v_k) \xrightarrow{a_k} \pi_{v,k}$, with $\pi_{u,k}(a_k.\mathrm{nil} \parallel_\mathcal{A} \ldots \parallel_\mathcal{A} a_k.\mathrm{nil}) = 1$, $\pi_{v,k}(a_k.\mathrm{nil} \parallel_\mathcal{A} \ldots \parallel_\mathcal{A} a_k.\mathrm{nil}) = (1-\varepsilon)^k$ and $\pi_{v,k}$ giving the remaining probability $1-(1-\varepsilon)^k$ to processes that cannot move. We get $\mathbf{b}^\lambda(f(u_k), f(v_k)) = \lambda(1-(1-\varepsilon)^k)$. Hence, $\sup_{k \in \mathbb{N}} \mathbf{b}^\lambda(f(u_k), f(v_k)) = \sup_{k \in \mathbb{N}} \lambda(1-(1-\varepsilon)^k) = \lambda$, thus giving that the distance between f-composed processes is bounded by $\mathfrak{m}(\varepsilon) = \lambda$ if $\varepsilon > 0$ and $\mathfrak{m}(0) = 0$, which is not a modulus of continuity since it is not continuous at 0. Hence, f is not uniformly continuous accordingly to Definition 10.

Consider now operator g. We have $g(u_k) \xrightarrow{a_k} \delta(h(\ldots h(u_k)\ldots))$ (with k nested occurrences of h) and $g(v_k) \xrightarrow{a_k} \delta(h(\ldots h(v_k)\ldots))$. Then, $h(\ldots h(u_k)\ldots) \xrightarrow{a_k} \pi'_{u,k}$ and $h(\ldots h(v_k)\ldots) \xrightarrow{a_k} \pi'_{v,k}$, with $\pi'_{u,k}(a_k.\mathrm{nil} \parallel_\mathcal{A} \ldots \parallel_\mathcal{A} a_k.\mathrm{nil}) = 1$ (parallel composition of 2^k copies of $a_k.\mathrm{nil}$), $\pi'_{v,k}(a_k.\mathrm{nil} \parallel_\mathcal{A} \ldots \parallel_\mathcal{A} a_k.\mathrm{nil}) = (1-\varepsilon)^{2^k}$ and $\pi'_{v,k}$ giving probability $1-(1-\varepsilon)^{2^k}$ to processes that cannot move. We get $\mathbf{b}^\lambda(g(u_k), g(v_k)) = \lambda^2(1-(1-\varepsilon)^{2^k})$ and, then, $\sup_{k \in \mathbb{N}} \mathbf{b}^\lambda(g(u_k), g(v_k)) = \sup_{k \in \mathbb{N}} \lambda^2(1-(1-\varepsilon)^{2^k}) = \lambda^2$. As for operator f, we conclude that g is not uniformly continuous.

Example 9 Counterexample for occurrences in non-expansive operators). In Example 6 we noticed that the rules for the copy operator cp violate Definition 15. Let $\varepsilon \in (0,1)$. Following [19], consider processes $s_1 = \mathrm{l}.([1-\varepsilon]a.\mathrm{nil} \oplus [\varepsilon]\mathrm{nil}) + \mathrm{r}.([1-\varepsilon]a.\mathrm{nil} \oplus [\varepsilon]\mathrm{nil})$ and $t_1 = \mathrm{l}.a.\mathrm{nil} + \mathrm{r}.a.\mathrm{nil}$ for some $a \neq \mathrm{l}, \mathrm{r}$. Then, for $k > 1$, let $s_k = \mathrm{l}.s_{k-1} + \mathrm{r}.s_{k-1}$ and $t_k = \mathrm{l}.t_{k-1} + \mathrm{r}.t_{k-1}$.

Clearly, $\mathbf{b}^\lambda(s_k, t_k) = \lambda^k \cdot \varepsilon$. Similar calculations to those in Example 1 give $\mathbf{b}^\lambda(\mathrm{cp}(s_k), \mathrm{cp}(t_k)) = \lambda^k(1 - (1 - \varepsilon)^{2^k})$. Thus, for any k with $2^k > L$, $\mathbf{b}^\lambda(\mathrm{cp}(s), \mathrm{cp}(t))/\mathbf{b}^\lambda(s,t) = (1 - (1 - \varepsilon)^{2^k})/\varepsilon > L$ holds for $s = s_k$, $t = t_k$ and $\varepsilon \in (0, (2^k - L)/(2^{k-1}(2^k - 1)))$. We can conclude that cp is not Lipschitz continuous.

7 Conclusions

We have provided three specification formats for the bisimilarity metric. These formats guarantee, respectively, the *non-extensiveness*, *non-expansiveness* and *Lipschitz continuity* of the operators satisfying the related format wrt. the bisimilarity metric.

To the best of our knowledge, the only other formats for the bisimilarity metric proposed in the literature are those in [19]. As outlined in the Introduction, the difference between the two proposals is in that our formats are purely syntactic, whereas the ones in [19] are *semantic*, in the sense that they require a recursive computation of how many copies of a process argument can be spawned by each operator. However, we remark that we have applied the same technique of [19] to the evaluation of Lipschitz factors of operators. This is due to the fact that from the syntactic constraints on PGSOS rules for an operator f we can only guarantee that f is Lipschitz continuous *for some* L_f and that $L_f \geq n_f$, but no further information can be inferred. This is reasonable since the validity of the Lipschitz continuity property does not depend on the value of the Lipschitz constant, but only on its existence. The actual value of L_f can be established, in general, only from the semantics of f.

As future work, we plan to provide specification formats for *weak* bisimulation metrics [15], whose applicability has been demonstrated in [24–26]. Intuitively, we need to consider TSSs in a format guaranteeing that the *kernel* of the chosen metric is a *congruence* and enrich our three formats with constraints allowing us to deal with silent moves.

Acknowledgements. We thank the anonymous reviewers for their detailed comments and feedback. V. Castiglioni has been partially supported by the project 'Open Problems in the Equational Logic of Processes' (OPEL) of the Icelandic Research Fund (grant nr. 196050-051). We thank Carlos, Kostas, Mario and Frank for inviting us to submit this paper. Last but not least, we extend our heartiest wishes to Catuscia. Valentina is indebted to Catuscia for her support, mentoring and friendship during the period spent at INRIA, in her team COMETE. We wish Catuscia a very happy birthday!

References

1. Aceto, L., Fokkink, W.J., Verhoef, C.: Structural operational semantics. In: Handbook of Process Algebra, pp. 197–292. Elsevier (2001)
2. de Alfaro, L., Faella, M., Stoelinga, M.: Linear and branching system metrics. IEEE Trans. Softw. Eng. **35**(2), 258–273 (2009)

3. de Alfaro, L., Henzinger, T.A., Majumdar, R.: Discounting the future in systems theory. In: Baeten, J.C.M., Lenstra, J.K., Parrow, J., Woeginger, G.J. (eds.) ICALP 2003. LNCS, vol. 2719, pp. 1022–1037. Springer, Heidelberg (2003). https://doi.org/10.1007/3-540-45061-0_79
4. Bacci, G., Bacci, G., Larsen, K.G., Mardare, R.: Computing behavioral distances, compositionally. In: Chatterjee, K., Sgall, J. (eds.) MFCS 2013. LNCS, vol. 8087, pp. 74–85. Springer, Heidelberg (2013). https://doi.org/10.1007/978-3-642-40313-2_9
5. Bloom, B., Istrail, S., Meyer, A.R.: Bisimulation can't be traced. J. ACM **42**(1), 232–268 (1995)
6. van Breugel, F., Worrell, J.: A behavioural pseudometric for probabilistic transition systems. Theor. Comput. Sci. **331**(1), 115–142 (2005)
7. Castiglioni, V.: Trace and testing metrics on nondeterministic probabilistic processes. In: Proceedings of the EXPRESS/SOS 2018. EPTCS, vol. 276, pp. 19–36 (2018)
8. Castiglioni, V., Gebler, D., Tini, S.: Modal decomposition on nondeterministic probabilistic processes. In: Proceedings of the CONCUR 2016, pp. 36:1–36:15 (2016)
9. Castiglioni, V., Gebler, D., Tini, S.: SOS-based modal decomposition onnondeterministic probabilistic processes. Logical Methods Comput. Sci. **14**(2) (2018)
10. Chatzikokolakis, K., Gebler, D., Palamidessi, C., Xu, L.: Generalized bisimulation metrics. In: Baldan, P., Gorla, D. (eds.) CONCUR 2014. LNCS, vol. 8704, pp. 32–46. Springer, Heidelberg (2014). https://doi.org/10.1007/978-3-662-44584-6_4
11. D'Argenio, P.R., Gebler, D., Lee, M.D.: A general SOS theory for the specification of probabilistic transition systems. Inf. Comput. **249**, 76–109 (2016)
12. D'Argenio, P.R., Gebler, D., Lee, M.D.: Axiomatizing bisimulation equivalences and metrics from probabilistic SOS rules. In: Muscholl, A. (ed.) FoSSaCS 2014. LNCS, vol. 8412, pp. 289–303. Springer, Heidelberg (2014). https://doi.org/10.1007/978-3-642-54830-7_19
13. Deng, Y., Chothia, T., Palamidessi, C., Pang, J.: Metrics for action-labelled quantitative transition systems. Electr. Notes Theor. Comput. Sci. **153**(2), 79–96 (2006)
14. Desharnais, J., Gupta, V., Jagadeesan, R., Panangaden, P.: Metrics for labelled markov processes. Theor. Comput. Sci. **318**(3), 323–354 (2004)
15. Desharnais, J., Jagadeesan, R., Gupta, V., Panangaden, P.: The metric analogue of weak bisimulation for probabilistic processes. In: Proceedings of the LICS 2002, pp. 413–422 (2002)
16. Gebler, D., Larsen, K.G., Tini, S.: Compositional bisimulation metric reasoningwith probabilistic process calculi. Logical Methods Comput. Sci. **12**(4) (2016)
17. Gebler, D., Tini, S.: Compositionality of approximate bisimulation for probabilistic systems. In: Proceedings of the EXPRESS/SOS 2013, pp. 32–46 (2013)
18. Gebler, D., Tini, S.: Fixed-point characterization of compositionality properties of probabilistic processes combinators. In: Proceedings of the EXPRESS/SOS 2014, pp. 63–78 (2014)
19. Gebler, D., Tini, S.: SOS specifications for uniformly continuous operators. J. Comput. Syst. Sci. **92**, 113–151 (2018)
20. Giacalone, A., Jou, C.C., Smolka, S.A.: Algebraic reasoning for probabilistic concurrent systems. In: Proceedings of the IFIP Work, Conference on Programming, Concepts and Methods, pp. 443–458 (1990)
21. Kantorovich, L.V.: On the transfer of masses. Dokl. Akad. Nauk SSSR **37**(2), 227–229 (1942)

22. Katoen, J.P., Baier, C., Latella, D.: Metric semantics for true concurrent real time. TCS **254**(1–2), 501–542 (2001)
23. Kwiatkowska, M., Norman, G.: Probabilistic metric semantics for a simple language with recursion. In: Penczek, W., Szałas, A. (eds.) MFCS 1996. LNCS, vol. 1113, pp. 419–430. Springer, Heidelberg (1996). https://doi.org/10.1007/3-540-61550-4_167
24. Lanotte, R., Merro, M., Tini, S.: Compositional weak metrics for group key update. In: MFCS 2017. LIPIcs, vol. 42, pp. 72:1–72:16 (2017)
25. Lanotte, R., Merro, M., Tini, S.: Equational reasonings in wireless networkgossip protocols. Logical Methods Comput. Sci. **14**(3) (2018)
26. Lanotte, R., Merro, M., Tini, S.: Towards a formal notion of impact metric for cyber-physical attacks. In: Furia, C.A., Winter, K. (eds.) IFM 2018. LNCS, vol. 11023, pp. 296–315. Springer, Cham (2018). https://doi.org/10.1007/978-3-319-98938-9_17
27. Larsen, K.G., Skou, A.: Bisimulation through probabilistic testing. Inf. Comput. **94**(1), 1–28 (1991)
28. Mousavi, M.R., Reniers, M.A., Groote, J.F.: Sos formats and meta-theory: 20 years after. Theoret. Comput. Sci. **373**(3), 238–272 (2007)
29. Plotkin, G.: A structural approach to operational semantics. Report DAIMI FN-19, Aarhus University (1981). Reprinted in JLAP, 60–61:17–139 (2004)
30. Plotkin, G.D.: A structural approach to operational semantics. Report DAIMI FN-19, Aarhus University (1981)
31. Segala, R.: Modeling and verification of randomized distributed real-time systems. Ph.D. thesis, MIT (1995)
32. Segala, R., Lynch, N.A.: Probabilistic simulations for probabilistic processes. Nord. J. Comput. **2**(2), 250–273 (1995)
33. de Simone, R.: Higher-level synchronising devices in MEIJE-SCCS. Theor. Comput. Sci. **37**, 245–267 (1985)
34. Song, L., Deng, Y., Cai, X.: Towards automatic measurement of probabilistic processes. In: Proceedings of the QSIC 2007, pp. 50–59 (2007)
35. Tracol, M., Desharnais, J., Zhioua, A.: Computing distances between probabilistic automata. In: Proceedings of the QAPL 2011, pp. 148–162 (2011)

Fooling the Parallel or Tester with Probability 8/27

Jean Goubault-Larrecq(✉)

LSV, ENS Paris-Saclay, CNRS, Université Paris-Saclay, Cachan, France
goubault@lsv.fr

Abstract. It is well-known that the higher-order language PCF is not fully abstract: there is a program—the so-called parallel or tester, meant to test whether its input behaves as a parallel or—which never terminates on any input, operationally, but is denotationally non-trivial. We explore a probabilistic variant of PCF, and ask whether the parallel or tester exhibits a similar behavior there. The answer is no: operationally, one can feed the parallel or tester an input that will fool it into thinking it is a parallel or. We show that the largest probability of success of such would-be parallel ors is exactly 8/27. The bound is reached by a very simple probabilistic program. The difficult part is to show that that bound cannot be exceeded.

1 Capsule

Catuscia and I have collaborated for a few years, but that was quite some time ago [10]. We were both working in computer security then, and I have diverged since. I still felt I had to contribute to this Festschrift, but anything I could contribute in computer security would be outdated, the few things I have worked on in process algebra have more of an algebraic topological flavor, and so the most relevant of Catuscia's research interests that I could contribute to is probabilistic transition systems. Here is a small result related to that theme.

2 Introduction

There is a recurring theme in security: to defeat a strong adversary, you need to rely on random choice. This paper will be a somewhat devious illustration of that principle, in the field of programming language semantics.

The higher-order, functional language PCF [15] forms the core of actual programming languages such as Haskell [3]. Plotkin [15], and independently Sazonov [16], had shown that PCF, while being adequate (i.e., its operational and denotational semantics match, in a precise sense), is not fully abstract: there are

This research was partially supported by Labex DigiCosme (project ANR-11-LABEX-0045-DIGICOSME) operated by ANR as part of the program "Investissement d'Avenir" Idex Paris-Saclay (ANR-11-IDEX-0003-02).

M. S. Alvim et al. (Eds.): Palamidessi Festschrift, LNCS 11760, pp. 313–328, 2019.
https://doi.org/10.1007/978-3-030-31175-9_18

programs that are contextually equivalent (a notion arising from the operational semantics), but have different denotational semantics. (One should note that, conversely, two programs with the same denotational semantics are always contextually equivalent.)

The argument is as follows. In the denotational model, there is a function of type $\texttt{int} \rightarrow \texttt{int} \rightarrow \texttt{int}$ called *parallel or*, which maps the pair $1, 1$ to 1, and both $0, N$ and $N, 0$ to 0, for whatever program N (including non-terminating programs). One can show that parallel or is undefinable in PCF. More is true. One can define a PCF program, the *parallel or tester*, which takes an argument $f \colon \texttt{int} \rightarrow \texttt{int} \rightarrow \texttt{int}$, and tests whether f is a parallel or, by testing whether $f11 = 1$, $f0\Omega = 0$, and $f\Omega 0 = 0$, where Ω is a canonical non-terminating program. The parallel or tester is contextually equivalent to the always non-terminating program $\lambda f.\Omega$, meaning that applying it to any PCF program (for f) will never terminate. However, the denotational semantics of the parallel or tester and of $\lambda f.\Omega$ differ: applied to any given parallel or map (which exists in the denotational model), one returns and the other one does not.

We introduce a probabilistic variant of PCF which we call PCF_P, and we define a suitable parallel or tester `portest`. A PCF_P program M *fools* the parallel or tester if `portest` applied to M terminates. In PCF, there is no way of fooling the parallel or tester. Our purpose is to show that one can fool the parallel or tester of PCF_P with probability at most $8/27$, and that this bound is attained. The optimal fooler is easy to define. The hard part is to show that one cannot do better.

A final word before we start. Even though we started by motivating it from matters related to full abstraction, which involves both operational and denotational semantics, the question we are addressing is purely *operational* in nature: it is only concerned with the behavior of `portest` under its operational semantics, under arbitrary PCF_P contexts. Nonetheless, denotational semantics will be essential in our proof.

Outline. We define the syntax of $\textit{PCF}_\text{P}$ in Sect. 3, its operational semantics in Sect. 4, and—once we have stated the required basic facts we need from domain theory in Sect. 5—its denotational semantics in Sect. 6. We state the adequacy theorem at the end of the latter section. This says that the operational and denotational probabilities that a term M of type `int` terminates on any given value $n \in \mathbb{Z}$ are the same. We define the parallel tester, and show that it can be fooled with probability $8/27$ at most, in Sect. 7. We conclude by citing some recent related work in Sect. 8.

3 The Syntax of PCF_P

PCF_P is a typed language. The *types* are given by the grammar:

$$\begin{aligned} \sigma, \tau, \cdots ::= {} & \texttt{int} && \text{basic types} \\ \mid {} & D\tau && \text{type of (subprobability) distributions on } \tau \\ \mid {} & \sigma \rightarrow \tau && \text{function types.} \end{aligned}$$

Mathematically, $D\tau$ will be the type of subprobability valuations of elements of type τ. Operationally, an element of type $D\tau$ is just a random value of type τ. There is only one basic type, `int`, but one could envision a more expressive algebra of datatypes.

A *computation type* is a type of the form $D\tau$ or $\sigma \to \tau$ where τ is a computation type. The computation types are the types where one can do computation, in particular whose objets can be defined by recursion.

Our language will have functions, and a function mapping inputs of type σ to outputs of type τ will have type $\sigma \to \tau$. We write $\sigma_1 \to \sigma_2 \to \cdots \to \sigma_n \to \tau$ for $\sigma_1 \to (\sigma_2 \to (\cdots \to (\sigma_n \to \tau)\cdots))$, and this is a type of functions taking n inputs, of respective types $\sigma_1, \sigma_2, \ldots, \sigma_n$ and returning outputs of type τ.

We fix a countably infinite set of *variables* x_τ, y_τ, z_τ, ..., for each type τ. Each variable has a unique type, which we read off from its subscript. We will occasionally omit the type subscript when it is clear from context, or irrelevant.

$$\frac{}{x_\tau : \tau} \qquad \frac{}{n : \mathtt{int}}\ (n \in \mathbb{Z}) \qquad \frac{M : \mathtt{int}}{\mathtt{s}M : \mathtt{int}} \qquad \frac{M : \mathtt{int}}{\mathtt{p}M : \mathtt{int}}$$

$$\frac{M : \mathtt{int} \quad N : \tau \quad P : \tau}{\mathtt{if}\ M = 0\ \mathtt{then}\ N\ \mathtt{else}\ P : \tau} \qquad \frac{M : \tau \to \tau}{\mathtt{rec}_\tau\, M}\ (\tau \text{ computation type})$$

$$\frac{M : \sigma \to \tau \quad N : \sigma}{MN : \tau} \qquad \frac{M : \tau}{\lambda x_\sigma . M : \sigma \to \tau}$$

$$\frac{M : D\tau \quad N : D\tau}{M \oplus N : D\tau} \qquad \frac{M : \sigma}{\mathtt{ret}_\sigma\, M : D\sigma} \qquad \frac{M : D\sigma \quad N : \sigma \to D\tau}{\mathtt{bind}_{\sigma,\tau}\, MN : D\tau}$$

Fig. 1. The syntax of PCF_P

The *terms* M, N, ..., of our language are defined inductively, together with their types, in Fig. 1. We agree to write $M : \tau$ to mean "M is a term, of type τ". We shall write $MN_1N_2\cdots N_n$ for $(\cdots((MN_1)N_2)\cdots)N_n$, and $\lambda x_1, \cdots, x_n.M$ for $\lambda x_1.\lambda x_2.\cdots.\lambda x_n.M$. We shall also use the abbreviations $\mathtt{let}\ x_\sigma = M\ \mathtt{in}\ N$ for $(\lambda x_\sigma.N)M$ and $\mathtt{letrec}\ f_\tau = M\ \mathtt{in}\ N$, where $M : \tau$, for $\mathtt{let}\ f_\tau = \mathtt{rec}\ (\lambda f_\tau.M)\ \mathtt{in}\ N$. Finally, we shall write $\mathtt{do}\, x_\sigma \leftarrow M; N$ for $\mathtt{bind}_{\sigma,\tau}\, M(\lambda x_\sigma.N)$, of type $D\tau$ (draw x_σ at random along distribution M, then run N). $M \oplus N$ is meant to execute either M or N with probability 1/2.

The free variables and the bound variables of a term M are defined as usual. A term with no free variable is *ground*. For a substitution $\theta \stackrel{\text{def}}{=} [x_1 := N_1, \cdots, x_k := N_k]$ (where each N_i has the same type as x_i, and the variables x_i are pairwise distinct), we write $M\theta$ for the parallel substitution of each N_i for each x_i, and $\mathrm{dom}\,\theta$ for $\{x_1, \cdots, x_k\}$. We say that θ is *ground* if N_1, ..., N_k are all ground.

Example 1. The term $\texttt{rand\_int} \stackrel{\text{def}}{=} \texttt{rec}_{\texttt{int}\to D\texttt{int}}(\lambda r.\lambda m_{\texttt{int}}.r(\texttt{s}m) \oplus \texttt{ret}_{\texttt{int}}\, m)0$ is of type $D\texttt{int}$. As we will see, this draws a natural number n at random, with probability $1/2^{n+1}$.

Example 2. Rejection sampling is a process by which one draws an element of a subset A of a space X, as follows: we draw an element of X at random, and we return it if it lies in A, otherwise we start all over again. Here is a simple example of rejection sampling, meant to draw a number uniformly among $\{0,1,2\}$. The idea is to draw two independent bits at random, representing a number in $X \stackrel{\text{def}}{=} \{0,1,2,3\}$, and to use rejection sampling on $A \stackrel{\text{def}}{=} \{0,1,2\}$. Formally, we define the PCF_{P} term $\texttt{rand3} \stackrel{\text{def}}{=} \texttt{rec}_{D\texttt{int}}(\lambda p_{D\texttt{int}}.((\texttt{ret}_{\texttt{int}}\, 0 \oplus \texttt{ret}_{\texttt{int}}\, 1) \oplus (\texttt{ret}_{\texttt{int}}\, 2 \oplus p_{D\texttt{int}})))$. Note that this uses recursion to define a distribution, not a function.

4 Operational Semantics

The *elementary contexts* E, with their types $\sigma \vdash \tau$, are defined as:

- $[\_N]$ of type $(\sigma \to \tau) \vdash \tau$, for every $N : \sigma$, and for every type τ;
- $[\texttt{s}\_]$ and $[\texttt{p}\_]$, of type $\texttt{int} \vdash \texttt{int}$;
- $[\texttt{if}\ \_ = 0\ \texttt{then}\ N\ \texttt{else}\ P]$, of type $\texttt{int} \vdash \tau$, for all $N, P : \tau$;
- $[\texttt{bind}_{\sigma,\tau}\ \_N]$, of type $D\sigma \vdash D\tau$, for every $N : \sigma \to D\tau$.

The *initial contexts* are $[\_]$ (of type $\sigma \vdash \sigma$ for any σ) and $[\texttt{ret}_{\texttt{int}}\ \_]$ (of type $\texttt{int} \vdash D\texttt{int}$). The (evaluation) *contexts* C are the finite sequences $E_0E_1 \cdots E_n$, $n \in \mathbb{N}$, where E_0 is an initial context of type $\sigma_1 \vdash \sigma_0$, each E_i $(1 \leq i \leq n)$ is an elementary context of type $\sigma_{i+1} \vdash \sigma_i$. Then we say that C has type $\sigma_{n+1} \vdash \sigma_0$.

The notation $C[M]$ makes sense for every context $C \stackrel{\text{def}}{=} E_0E_1 \cdots E_n$ of type $\sigma \vdash \tau$ and every $M : \sigma$, and is defined as $E_0[E_1[\cdots[E_n[M]]]]$, where $E[M]$ is defined by removing the square brackets in E and replacing the hole $\_$ by M. E.g., if $C = [\texttt{ret}_{\texttt{int}}\ \_][\texttt{p}\_]$, then $C[M] = \texttt{ret}_{\texttt{int}}(\texttt{p}M)$.

Exploration rules	
$C \cdot E[M] \xrightarrow{1} CE \cdot M$ (E elem. context)	$[\_] \cdot \texttt{ret}_{\texttt{int}}\, M \xrightarrow{1} [\texttt{ret}_{\texttt{int}}\ \_] \cdot M$
Computation rules	
$C[\_N] \cdot \lambda x_\sigma.M \xrightarrow{1} C \cdot M[x_\sigma := N]$	$C \cdot \texttt{rec}_\tau\, M \xrightarrow{1} C \cdot M(\texttt{rec}_\tau\, M)$
$C \cdot M \oplus N \xrightarrow{1/2} C \cdot M$	$C \cdot M \oplus N \xrightarrow{1/2} C \cdot N$
$C[\texttt{bind}_{\sigma,\tau}\ \_N] \cdot \texttt{ret}_\sigma\, M \xrightarrow{1} C \cdot NM$	$C[\texttt{p}\_] \cdot n \xrightarrow{1} C \cdot n-1$ $C[\texttt{s}\_] \cdot n \xrightarrow{1} C \cdot n+1$

Fig. 2. Operational semantics

A *configuration* (of type τ) is a pair $C \cdot M$, where C is a context of type $\sigma \vdash \tau$ and $M : \sigma$.

The operational semantics of $\mathrm{PCF_P}$—an abstract interpreter that runs $\mathrm{PCF_P}$ programs—is a probabilistic transition system on configurations, defined by the rules of Fig. 2. We write $s \xrightarrow{\alpha} s'$ to say that one can go from configuration s to configuration s' in one step, with probability α.

A *trace* is a sequence $s_0 \xrightarrow{\alpha_1} s_1 \xrightarrow{\alpha_2} \cdots \xrightarrow{\alpha_m} s_m$, where $m \in \mathbb{N}$, and where each $s_{i-1} \xrightarrow{\alpha_i} s_i$ is an instance of a rule of Fig. 2. The trace starts at s_0, ends at s_m, its *length* is m and its *weight* is the product $\alpha \stackrel{\text{def}}{=} \alpha_1 \cdots \alpha_2 \cdots \alpha_m$. In that case, we also write $s_0 \xrightarrow{\alpha}{}^{*} s_m$.

The *run* starting at s_0 is the tree of all traces starting at s_0. Its root is s_0 itself, and for each vertex s in the tree, for each instance of a rule of the form $s \xrightarrow{\alpha} t$, t is a successor of s, and the edge from s to t is labeled α.

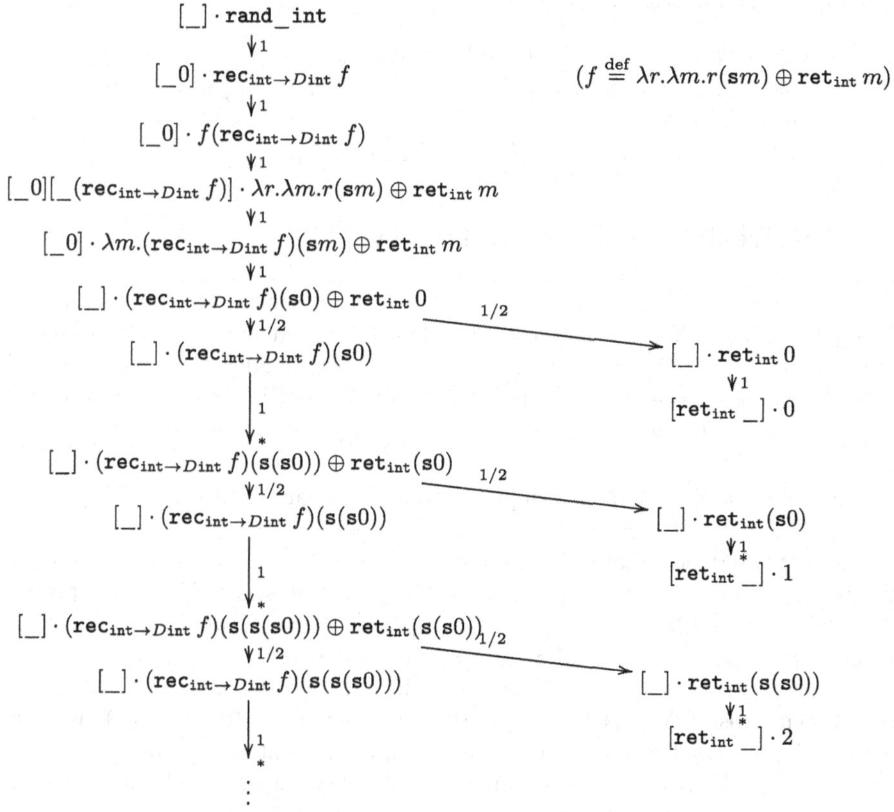

Fig. 3. An example run in $\mathrm{PCF_P}$

For every configuration s of type $D\mathtt{int}$, and every $n \in \mathbb{Z}$, we define $\Pr[s \downarrow n]$ as the sum of the weights of all traces that start at s and end at $[\mathtt{ret}_{\mathtt{int}}\, \_] \cdot n$. This is the *subprobability that s eventually computes n*. We also write $\Pr[M \downarrow n]$ for $\Pr[[\_] \cdot M \downarrow n]$, where $M \colon D\mathtt{int}$.

Example 3. The run starting at `rand_int` (see Example 1) is shown in Fig. 3. We have abbreviated some sequences of $\xrightarrow{1}$ steps as $\xrightarrow{1}{}^*$. One sees that $\Pr[\texttt{rand\_int}\downarrow n] = 1/2^{n+1}$ for every $n \in \mathbb{N}$, and is zero for every $n < 0$. Notice the infinite branch on the left, whose weight is 0.

Example 4. We let the reader draw the run starting at `rand3` (see Example 2), and check that $\Pr[\texttt{rand3}\downarrow n]$ is equal to $1/3$ if $n \in \{0, 1, 2\}$, 0 otherwise. Explicitly, if $n \in \{0, 1, 2\}$, show that the traces that start at `rand3` and end at $[\texttt{ret}_{\texttt{int}}\ \_] \cdot n$ have respective weights $1/4$, $1/4 \cdot 1/4$, ..., $(1/4)^n \cdot 1/4$, ..., and that the sum of those weights is $1/3$.

The following is immediate.

Lemma 1. *The following hold:*

1. *For every rule $s \xrightarrow{\alpha} t$, t and s have the same type.*
2. *For every rule of the form $s \xrightarrow{1} t$ of type* `Dint`*, for every $n \in \mathbb{Z}$, $Pr[t \downarrow n] = Pr[s \downarrow n]$.*
3. *$Pr[C \cdot M \oplus N \downarrow n] = \frac{1}{2} Pr[C \cdot M \downarrow n] + \frac{1}{2} Pr[C \cdot N \downarrow n]$.* □

5 A Refresher on Domain Theory

We will require some elementary domain theory, for which we refer the reader to [1,5,6]. A *poset* X is a set with a partial ordering, which we will always write as $\leq$. A *directed family* $D \subseteq X$ is a non-empty family such that every pair of points of D has an upper bound in D. A *dcpo* is a poset in which every directed family D has a supremum $\sup^{\uparrow} D$. If $D = (x_i)_{i \in I}$, we also write $\sup^{\uparrow}_{i \in I} x_i$ for $\sup^{\uparrow} D$.

The *product* $X \times Y$ of two dcpos is the set of pairs (x, y), $x \in X$, $y \in Y$, ordered by $(x, y) \leq (x', y')$ if and only if $x \leq x'$ and $y \leq y'$.

For any two dcpos X and Y, a map $f : X \to Y$ is *Scott-continuous* if and only if it is monotonic ($x \leq x'$ implies $f(x) \leq f(x')$) and preserves directed suprema (for every directed family $(x_i)_{i \in I}$ in X, $\sup^{\uparrow}_{i \in I} f(x_i) = f(\sup^{\uparrow}_{i \in I} x_i)$). There is a category **Dcpo** of dcpos and Scott-continuous maps.

We order maps from X to Y by $f \leq g$ if and only if $f(x) \leq g(x)$ for every $x \in X$. The poset $[X \to Y]$ of all Scott-continuous maps from X to Y is then again a dcpo, and directed suprema are computed pointwise: $(\sup^{\uparrow}_{i \in I} f_i)(x) = \sup^{\uparrow}_{i \in I}(f_i(x))$. **Dcpo** is a Cartesian-closed category—a model of simply-typed λ-calculus—and that can be said more concretely as follows:

- for all dcpos X, Y, there is a Scott-continuous map $\mathsf{App} : [X \to Y] \times X \to Y$ defined by $\mathsf{App}(f, x) \stackrel{\text{def}}{=} f(x)$;
- for all dcpos X, Y, Z, for every Scott-continuous map $f : Z \times X \to Y$, the map $\Lambda_X(f) : Z \to [X \to Y]$ defined by $\Lambda_X(f)(z)(x) \stackrel{\text{def}}{=} f(z, x)$ is Scott-continuous;
- those satisfy certain equations which we will not require.

If the dcpo X is *pointed*, namely if it has a least element $\bot$, then every Scott-continuous map $f\colon X \to X$ has a least fixed point $\mathrm{lfp}_X(f) \stackrel{\text{def}}{=} \sup^{\uparrow}_{n\in\mathbb{N}} f^n(\bot)$. This is used to interpret recursion. Additionally, the map $\mathrm{lfp}_X\colon [X \to X] \to X$ is itself Scott-continuous.

The set $\overline{\mathbb{R}}_+ \stackrel{\text{def}}{=} \mathbb{R}_+ \cup \{\infty\}$ of extended non-negative real numbers is a dcpo under the usual ordering. We write $\mathcal{L}X$ for $[X \to \overline{\mathbb{R}}_+]$. Its elements are called the *lower semicontinuous* functions in analysis.

A *Scott-open* subset U of a dcpo X is an upwards-closed subset ($x \in U$ and $x \leq y$ imply $y \in U$) that is inaccessible from below (every directed family D such that $\sup^{\uparrow} D \in U$ intersects U). The lattice of Scott-open subsets is written $\mathcal{O}X$, and forms a topology, the *Scott topology* on X. Note that $\mathcal{O}X$ is itself a dcpo under inclusion, and directed suprema are computed as unions.

The *Scott-closed* sets are the complements of Scott-open sets, i.e., the downwards-closed subsets C such that for every directed family $D \subseteq C$, $\sup^{\uparrow} D \in C$.

In order to give a denotational semantics to probabilistic choice, we will follow Jones [11,12]. A *continuous valuation* on X is a map $\nu\colon \mathcal{O}X \to \overline{\mathbb{R}}_+$ that is *strict* ($\nu(\emptyset) = 0$), *monotone* ($U \subseteq V$ implies $\nu(U) \leq \nu(V)$), *modular* ($\nu(U) + \nu(V) = \nu(U \cup V) + \nu(U \cap V)$), and Scott-continuous ($\nu(\bigcup^{\uparrow}_{i\in I} U_i) = \sup^{\uparrow}_{i\in I} \nu(U_i)$). A *subprobability valuation* additionally satisfies $\nu(X) \leq 1$. Continuous valuations and measures are very close concepts: see [14] for details.

Among subprobability valuations, one finds the *Dirac valuation* δ_x, for each $x \in X$, defined by $\delta_x(U) \stackrel{\text{def}}{=} 1$ if $x \in U$, 0 otherwise. One can integrate any Scott-continuous map $f\colon X \to \overline{\mathbb{R}}_+$, and the integral $\int_{x\in X} f(x) d\nu$ is Scott-continuous and linear (i.e., commutes with sums and scalar products by elements of $\mathbb{R}_+$) both in f and in ν.

We write $\mathbf{V}_{\leq 1}X$ for the poset of subprobability valuations on X. This is a dcpo under the pointwise ordering ($\mu \leq \nu$ if and only if $\mu(U) \leq \nu(U)$ for every $U \in \mathcal{O}X$), and directed suprema are computed pointwise ($(\sup^{\uparrow}_{i\in I} \nu_i)(U) = \sup^{\uparrow}_{i\in I}(\nu_i(U))$). Additionally, $\mathbf{V}_{\leq 1}$ defines a *monad* on **Dcpo**. Concretely:

- there is a *unit* $\eta\colon X \to \mathbf{V}_{\leq 1}X$, which is the continuous map $x \mapsto \delta_x$;
- every Scott-continuous map $f\colon X \to \mathbf{V}_{\leq 1}Y$ has an *extension* $f^\dagger\colon \mathbf{V}_{\leq 1}X \to \mathbf{V}_{\leq 1}Y$, defined by $f^\dagger(\nu)(V) \stackrel{\text{def}}{=} \int_{x\in X} f(x)(V) d\nu$;
- those satisfy a certain number of equations, of which we will need the following:

$$f^\dagger(\eta(x)) = f(x) \tag{1}$$

$$\int_{y\in Y} h(y) df^\dagger(\nu) = \int_{x\in X} \left(\int_{y\in Y} h(y) df(x)\right) d\nu, \tag{2}$$

for all Scott-continuous maps $f\colon X \to Y$, $h\colon Y \to \overline{\mathbb{R}}_+$, and every $\nu \in \mathbf{V}_{\leq 1}X$.

Note that the map $f \mapsto f^\dagger$ is itself Scott-continuous.

6 Denotational Semantics

The types τ are interpreted as dcpos $[\![\tau]\!]$, as follows: $[\![\texttt{int}]\!] \stackrel{\text{def}}{=} \mathbb{Z}$, with equality as ordering; $[\![D\tau]\!] \stackrel{\text{def}}{=} \mathbf{V}_{\leq 1}[\![\tau]\!]$; and $[\![\sigma \to \tau]\!] \stackrel{\text{def}}{=} [[\![\sigma]\!] \to [\![\tau]\!]]$. Note that $[\![\tau]\!]$ is pointed for every computation type τ, so $\text{lfp}_{[\![\tau]\!]}$ makes sense in those cases.

An *environment* is a map ρ sending each variable x_τ to an element $\rho(x_\tau)$ of $[\![\tau]\!]$. The dcpo *Env* of environments is the product $\prod_{x_\tau \text{ variable}}[\![\tau]\!]$, with the usual componentwise ordering. When $V \in [\![\sigma]\!]$, we write $\rho[x_\sigma := V]$ for the environment that maps x_σ to V, and all other variables y to $\rho(y)$.

$$[\![x_\tau]\!]\,\rho \stackrel{\text{def}}{=} \rho(x_\tau) \quad [\![n]\!]\,\rho \stackrel{\text{def}}{=} n\ (n \in \mathbb{Z}) \quad [\![\mathtt{s}M]\!]\,\rho \stackrel{\text{def}}{=} [\![M]\!]\,\rho + 1 \quad [\![\mathtt{p}M]\!]\,\rho \stackrel{\text{def}}{=} [\![M]\!]\,\rho - 1$$

$$[\![\mathtt{if}\ M = 0\ \mathtt{then}\ N\ \mathtt{else}\ P]\!]\,\rho \stackrel{\text{def}}{=} \begin{cases} [\![N]\!]\,\rho \text{ if } [\![M]\!]\,\rho = 0 \\ [\![P]\!]\,\rho \text{ otherwise} \end{cases}$$

$$[\![MN]\!]\,\rho \stackrel{\text{def}}{=} \mathsf{App}([\![M]\!]\,\rho, [\![N]\!]\,\rho) \qquad [\![\lambda x_\sigma.M]\!]\,\rho \stackrel{\text{def}}{=} (V \in [\![\sigma]\!] \mapsto [\![M]\!]\,\rho[x_\sigma := V])$$

$$[\![\mathtt{rec}_\tau\,M]\!]\,\rho \stackrel{\text{def}}{=} \text{lfp}_{[\![\tau]\!]}([\![M]\!]\,\rho) \qquad [\![M \oplus N]\!]\,\rho \stackrel{\text{def}}{=} \tfrac{1}{2}([\![M]\!]\,\rho + [\![N]\!]\,\rho)$$

$$[\![\mathtt{ret}_\sigma\,M]\!]\,\rho \stackrel{\text{def}}{=} \eta([\![M]\!]\,\rho) = \delta_{[\![M]\!]\rho} \qquad [\![\mathtt{bind}_{\sigma,\tau}\,MN]\!]\,\rho \stackrel{\text{def}}{=} ([\![N]\!]\,\rho)^\dagger([\![M]\!]\,\rho).$$

Fig. 4. Denotational semantics

Let us write $V \in X \mapsto f(V)$ for the function that maps every $V \in X$ to the value $f(V)$. We can now define the value $[\![M]\!]$ of terms $M : \tau$, as Scott-continuous maps $\rho \in Env \mapsto [\![M]\!]\rho$, by induction on M, see Fig. 4.

The operational semantics and the denotational semantics match, namely:

Theorem 1 (Adequacy). *For every ground term M:* $\texttt{D}$*int, for every $n \in \mathbb{Z}$, $[\![M]\!](\{n\}) = Pr[M \downarrow n]$.*

The proof is relatively standard, using appropriate logical relations, and is given in the full version of this paper [8, Appendices].

Example 5. We retrieve the result of Example 3 using adequacy as follows. $[\![\lambda r_{\mathtt{int}\to D\mathtt{int}}.\lambda m_{\mathtt{int}}.r(\mathtt{s}m) \oplus \mathtt{ret}_{\mathtt{int}}\,m]\!]$ is the function F that maps every $\varphi \in [\![\mathtt{int} \to D\mathtt{int}]\!]$ (the value of r) and every $m \in [\![\mathtt{int}]\!] = \mathbb{Z}$ to $1/2\varphi(m+1) + 1/2\delta_m$. Let $\varphi_k \stackrel{\text{def}}{=} F^k(\bot)$, for every $k \in \mathbb{N}$. Then $\varphi_0 = \bot$ maps every $m \in \mathbb{N}$ to the zero valuation 0, $\varphi_1(m) = 1/2\delta_m$ for every $m \in \mathbb{N}$, $\varphi_2(m) = 1/4\delta_{m+1} + 1/2\delta_m$ for every $m \in \mathbb{N}$, etc. By induction on k, $\varphi_k(m) = \sum_{i=0}^{k-1} 1/2^{i+1}\delta_{m+i}$. Taking suprema over k, we obtain that $\text{lfp}_{[\![\mathtt{int}\to D\mathtt{int}]\!]}(F)$ maps every $m \in \mathbb{N}$ to $\sum_{i=0}^{\infty} 1/2^{i+1}\delta_{m+i}$. Then $[\![\mathtt{rand\_nat}]\!] = \text{lfp}_{[\![\mathtt{int}\to D\mathtt{int}]\!]}(F)(0) = \sum_{n\in\mathbb{N}} \frac{1}{2^{n+1}}\delta_n$.

Example 6. We retrieve the result of Example 2, using adequacy, as follows. The semantics of $\lambda p_{D\mathtt{int}}.((\mathtt{ret}_{\mathtt{int}}\,0 \oplus \mathtt{ret}_{\mathtt{int}}\,1) \oplus (\mathtt{ret}_{\mathtt{int}}\,2 \oplus p_{D\mathtt{int}}))$ is the function f that maps every $\nu \in [\![D\mathtt{int}]\!]$ to $\frac{1}{4}\delta_0 + \frac{1}{4}\delta_1 + \frac{1}{4}\delta_2 + \frac{1}{4}\nu$. For every $n \in \mathbb{N}$, $f^n(0) = a_n\delta_0 + a_n\delta_1 + a_n\delta_2$ where $a_n = 1/4 + (1/4)^2 + \cdots + (1/4)^n = 1/4(1 - (1/4)^n)/(1 - 1/4)$. Since $[\![\mathtt{int}]\!]$ has equality as ordering, the ordering on $[\![D\mathtt{int}]\!]$ is given by comparing the coefficients of each δ_N, $N \in [\![\mathtt{int}]\!]$. In particular, the least fixed point of f is obtained as $a\delta_0 + a\delta_1 + a\delta_2$, where $a \stackrel{\text{def}}{=} \sup^{\uparrow}_{n\in\mathbb{N}} a_n = 1/3$.

Example 7. Here is a lengthier example, which we will leave to the reader. While lengthy, working denotationally is doable. Proving the same argument operational would be next to impossible, even in the special case $\tau = \mathtt{int}$.

We define a more general form of rejection sampling, as follows. Let τ be any type. We consider the PCF$_\mathrm{P}$ term:

$$\begin{aligned}\mathtt{sample} \stackrel{\mathrm{def}}{=} \lambda p_{D\tau}.\lambda sel_{\tau\to\mathtt{Dint}}.\\ \mathtt{rec}_{D\tau}(\lambda r_{D\tau}.\,\mathtt{do}\, x_\tau \leftarrow p_{D\tau};\\ \mathtt{do}\, b_{\mathtt{int}} \leftarrow sel_{\tau\to\mathtt{Dint}} x_\tau;\\ \mathtt{if}\, b_{\mathtt{bool}} = 0\,\mathtt{then}\,\mathtt{ret}_\tau\, x_\tau\;\;\mathtt{else}\, r_{D\tau}).\end{aligned}$$

The idea is that we draw x according to distribution p, then we call *sel* as a predicate on x. If the result, b, is true (zero) then we return x, otherwise we start all over. Note that *sel* can itself return a *random* b.

For every $g \in \mathcal{L}[\![\tau]\!]$, and every $\nu[\![D\tau]\!]$, we let $g \cdot \nu$ (sometimes written $g\, d\nu$) be the continuous valuation defined from ν by using g as a density, namely $(g \cdot \nu)(U) \stackrel{\mathrm{def}}{=} \int_{x\in[\![\tau]\!]} \chi_U(x) g(x) d\nu$ for every open subset U of $[\![\tau]\!]$, where χ_U is the characteristic map of U. One can check that $g \cdot \nu = (x \mapsto g(x)\delta_x)^\dagger(\nu)$, using the equality $\chi_U(x) = \delta_x(U)$, and, using (2), that for every $h \in \mathcal{L}[\![\tau]\!]$, $\int_{x\in X} h(x) d(g \cdot \nu) = \int_{x\in X} h(x)g(x)d\nu$.

For every $s \in [\![\tau \to \mathtt{Dint}]\!]$, for every $x \in [\![\tau]\!]$, let $s_0(x) \stackrel{\mathrm{def}}{=} s(x)(\{0\})$, $s_1(x) \stackrel{\mathrm{def}}{=} s(x)(\mathbb{Z} \smallsetminus \{0\})$. We let the reader check that, for every environment ρ, $[\![\mathtt{sample}]\!]\rho$ maps every subprobability valuation ν on $[\![\tau]\!]$ and every $s \in [\![\tau \to \mathtt{Dint}]\!]$ to the subprobability valuation $\frac{1}{1-(s_1\cdot\nu)([\![\tau]\!])}(s_0 \cdot \nu)$ if $(s_1 \cdot \nu)([\![\tau]\!]) \neq 1$, to the zero valuation otherwise.

In particular, if s is a predicate, implemented as a function that maps every $x \in U \subseteq [\![\tau]\!]$ to δ_0 and every $x \in V \subseteq [\![\tau]\!]$ (for some disjoint open sets U and V) to δ_1, so that $s_0 = \chi_U$ and $s_1 = \chi_V$, then $[\![\mathtt{sample}]\!]\rho(\nu)(s)$ is the subprobability valuation $\frac{1}{1-\nu(V)}\nu_{|U}$ if $\nu(V) \neq 1$, the zero valuation otherwise. ($\nu_{|U}$ denotes the restriction of ν to U, defined by $\nu_{|U}(V) \stackrel{\mathrm{def}}{=} \nu(U \cap V)$.)

In the special case where V is the complement of U, it follows that `sample` implements *conditional probabilities*: $[\![\mathtt{sample}]\!]\rho(\nu)(s)(W)$ is the probability that a ν-random element lies in W, conditioned on the fact that it is in U.

7 The Parallel or Tester

In PCF$_\mathrm{P}$, computation happens at type $D\mathtt{int}$, not $\mathtt{int}$, hence let us call *parallel or* function any $f \in [\![D\mathtt{int} \to D\mathtt{int} \to D\mathtt{int}]\!]$ such that $f(\delta_1)(\delta_1) = \delta_1$ and $f(\delta_0)(\nu) = f(\nu)(\delta_0) = \delta_0$ for every $\nu \in [\![D\mathtt{int}]\!]$. Realizing that every element of $[\![D\mathtt{int}]\!]$ is of the form $a\delta_0 + b\delta_1$, with $a, b \in \mathbb{R}_+$ such that $a + b \leq 1$, the function *por* defined by $por(a\delta_0 + b\delta_1)(a'\delta_0 + b'\delta_1) \stackrel{\mathrm{def}}{=} (a + a' - aa')\delta_0 + bb'\delta_1$ is such a parallel or function.

Note how parallel ors differ from the usual *left-to-right sequential or* used in most programming languages:

$$\begin{aligned}\texttt{lror} &\stackrel{\text{def}}{=} \lambda p_{D\texttt{int}}.\lambda q_{D\texttt{int}}.\\ &\quad \texttt{do}\, x_{\texttt{int}} \leftarrow p_{D\texttt{int}}; \texttt{if}\, x_{\texttt{int}} = 0\, \texttt{then}\, \texttt{ret}_{\texttt{int}}\, 0\ \texttt{else}\, q_{D\texttt{int}}\end{aligned}$$

whose semantics is given by $[\![\texttt{lror}]\!](a\delta_0 + b\delta_1)(a'\delta_0 + b'\delta_1) = (a + ba')\delta_0 + bb'\delta_1$—so $[\![\texttt{lror}]\!]$ maps δ_1, δ_1 to δ_1, and δ_0, ν to δ_0, but maps $a\delta_0 + b\delta_1, \delta_0$ to $(a + b)\delta_0$, not δ_0. Symmetrically, there is a *right-to-left sequential or*:

$$\begin{aligned}\texttt{rlor} &\stackrel{\text{def}}{=} \lambda p_{D\texttt{int}}.\lambda q_{D\texttt{int}}.\\ &\quad \texttt{do}\, x_{\texttt{int}} \leftarrow q_{D\texttt{int}}; \texttt{if}\, x_{\texttt{int}}\, \texttt{then}\, \texttt{ret}_{\texttt{int}}\, 0\ \texttt{else}\, p_{D\texttt{int}}.\end{aligned}$$

We define a *parallel or tester* as follows:

$$\begin{aligned}\texttt{portest} \stackrel{\text{def}}{=} \lambda f_{D\texttt{int}\to D\texttt{int}\to D\texttt{int}}.&\\ \texttt{do}\, x_{\texttt{int}} \leftarrow f(\texttt{ret}_{\texttt{int}}\, 1)(\texttt{ret}_{\texttt{int}}\, 1);&\\ \texttt{if}\, x_{\texttt{int}} = 0\, \texttt{then}\, \Omega&\\ \texttt{else}\, (\texttt{do}\, y_{\texttt{int}} \leftarrow f(\texttt{ret}_{\texttt{int}}\, 0)(\Omega);&\\ \quad \texttt{if}\, y_{\texttt{int}} = 0\, \texttt{then}\, (\texttt{do}\, z_{\texttt{int}} \leftarrow f(\Omega)(\texttt{ret}_{\texttt{int}}\, 0);&\\ \qquad\qquad \texttt{if}\, z_{\texttt{int}} = 0\, \texttt{then}\, \texttt{ret}_{\texttt{unit}}\, 0\ \texttt{else}\, \Omega)&\\ \quad \texttt{else}\, \Omega),&\end{aligned}$$

where $\Omega \stackrel{\text{def}}{=} \texttt{rec}\ (\lambda a_{D\texttt{int}}.a_{D\texttt{int}})$. One can check that $[\![\texttt{portest}]\!](por) = \delta_0$, and that would hold for any other parallel or function instead of *por*. If things worked in PCF_{P} as in PCF, we would be able to show that `portest` is contextually equivalent to the constant map that loops on every input $f_{D\texttt{int}\to D\texttt{int}\to D\texttt{int}}$.

However, that is not the case. As we will now see, there is a PCF_{P} term, the *poor man's parallel or* `pmpor`, such that `portest pmpor` terminates with non-zero probability. That term takes its two arguments of type $D\texttt{int}$, then decides to do one of the following three actions with equal probability 1/3: (1) call `lror` on the two arguments; (2) call `rlor` on the two arguments; or (3) return true (0), regardless of its arguments.

In order to define `pmpor`, we need to draw an element out of three with equal probability. We do that by rejection sampling, imitating `rand3` (Examples 2, 4 and 6): we draw one element among four with equal probability, and we repeat until it falls in a specified subset of three. Hence we define:

$$\begin{aligned}\texttt{pmpor} &\stackrel{\text{def}}{=} \lambda p_{D\texttt{int}}.\lambda q_{D\texttt{int}}.\, \texttt{rec}_{D\texttt{int}}(\lambda r.\\ &\quad ((\texttt{lror}\, p\, q) \oplus (\texttt{rlor}\, p\, q)) \oplus (\texttt{ret}_{\texttt{int}}\, 0 \oplus r))\end{aligned}$$

One can show that $[\![\texttt{pmpor}]\!]$ maps every pair of subprobability distributions μ, ν on $[\![\texttt{int}]\!]$ to $\frac{1}{3}[\![\texttt{rlor}]\!](\mu)(\nu) + \frac{1}{3}[\![\texttt{lror}]\!](\mu)(\nu) + \frac{1}{3}\delta_0$. Intuitively, `portest pmpor` will terminate with probability $(2/3)^3 = 8/27 \approx 0.296296\ldots$: with $f = $ `pmpor`, the first test $f(\delta_1)(\delta_1) = \delta_1$ will succeed whether f acts as `lror` or as `rlor` (but not as the constant map returning δ_0), which happens with probability 2/3; the second test $f(\delta_0)(0) = \delta_0$ will succeed whether f acts as `lror` or as the constant map returning δ_0 (but not as `rlor`), again with probability 2/3; and the final test $f(0)(\delta_0) = \delta_0$ will symmetrically succeed with probability 2/3.

We now show that the probability 8/27 is optimal. To this end, we need to use a logical relation $(\rhd_\tau)_{\tau \text{ type}}$, namely a family of relations $\rhd_\tau$, one for each type τ, and related by certain constraints to be described below. Each $\rhd_\tau$ will be an I-ary relation on values in $[\![\tau]\!]$, for some non-empty set I, namely $\rhd_\tau \subseteq ([\![\tau]\!])^I$. In practice, we will take $I \stackrel{\text{def}}{=} \{1, 2, 3\}$, but the proofs are easier if we keep I arbitrary for now.

Our construction will be parameterized by an I-ary relation $\rhd \subseteq \overline{\mathbb{R}}_+^I$. We will also define an auxiliary family of relations $\rhd_\tau^\perp$, as certain subsets of $(\mathcal{L}[\![\tau]\!])^I$. We require $\rhd$ to contains the all zero tuple $\mathbf{0} \stackrel{\text{def}}{=} (0)_{i \in I}$, to be closed under directed suprema, and to be convex. (By *convex*, we mean that for all $\boldsymbol{x}, \boldsymbol{y} \in \rhd$ and $a \in [0, 1]$, $a\boldsymbol{x} + (1 - a)\boldsymbol{y}$ is in $\rhd$ as well.)

We define:

- $(n_i)_{i \in I} \in \rhd_{\texttt{int}}$ if and only if all n_i are equal;
- $(f_i)_{i \in I} \in \rhd_{\sigma \to \tau}$ if and only if for all $(V_i)_{i \in I} \in \rhd_\sigma$, $(f_i(V_i))_{i \in I} \in \rhd_\tau$;
- $(\nu_i)_{i \in I} \in \rhd_{D\tau}$ if and only if for all $(h_i)_{i \in I} \in \rhd_\tau^\perp$, $\left(\int_{V \in [\![\tau]\!]} h_i(V) d\nu_i\right)_{i \in I} \in \rhd$;
- $(h_i)_{i \in I} \in \rhd_\tau^\perp$ if and only if for all $(V_i)_{i \in I} \in \rhd_\tau$, $(h_i(V_i))_{i \in I} \in \rhd$.

We also define $\rhd_* \subseteq Env^I$ by $(\rho_i)_{i \in I} \in \rhd_*$ if and only if for every variable x_σ, $(\rho_i(x_\sigma))_{i \in I} \in \rhd_\sigma$. We prove the following *basic lemma of logical relations*:

Proposition 1. *For all $(\rho_i)_{i \in I} \in \rhd_*$, for every $M : \tau$, $([\![M]\!]\rho_i)_{i \in I}$ is in $\rhd_\tau$.*

Proof. Step 1. We claim that for every type τ, $\rhd_\tau$ is closed under directed suprema taken in $([\![\tau]\!])^I$, and contains the least element $(\perp_\tau)_{i \in I}$ if τ is a computation type. This is by induction on τ. The claim is trivial for `int`, since $[\![\texttt{int}]\!]^I$ is ordered by equality. For every directed family $(\boldsymbol{f}_j)_{j \in J}$ in $\rhd_{\sigma \to \tau}$, with $\boldsymbol{f}_j \stackrel{\text{def}}{=} (f_{ji})_{i \in I}$, we form its supremum $\boldsymbol{f} \stackrel{\text{def}}{=} (f_i)_{i \in I}$ pointwise, namely $f_i \stackrel{\text{def}}{=} \sup^\uparrow_{j \in J} f_{ji}$. For every $(V_i)_{i \in I} \in \rhd_\sigma$, $(f_{ji}(V_i))_{i \in I}$ is in $\rhd_\tau$ for every $j \in J$, so by induction hypothesis $(f_i(V_i))_{i \in I}$ is also in $\rhd_\tau$. It follows that $(f_i)_{i \in I}$ is in $\rhd_{\sigma \to \tau}$. For every directed family $(\boldsymbol{\nu}_j)_{j \in J}$ in $\rhd_{D\tau}$, with $\boldsymbol{\nu}_j \stackrel{\text{def}}{=} (\nu_{ji})_{i \in I}$, we form its supremum $\boldsymbol{\nu} \stackrel{\text{def}}{=} (\nu_i)_{i \in I}$ pointwise, that is $\nu_i \stackrel{\text{def}}{=} \sup^\uparrow_{j \in J} \nu_{ji}$. For all $(h_i)_{i \in I} \in \rhd_\tau^\perp$, $\left(\int_{V \in [\![\tau]\!]} h_i(V) d\nu_{ji}\right)_{i \in I} \in \rhd$ for every $j \in J$, by induction hypothesis. We take suprema over $j \in J$. Since $\rhd$ is closed under directed suprema, and integration is Scott-continuous in the valuation, $\left(\int_{V \in [\![\tau]\!]} h_i(V) d\nu_i\right)_{i \in I}$ is in $\rhd$. Since $(h_i)_{i \in I}$ is arbitrary, $(\nu_i)_{i \in I} \in \rhd_{D\tau}$.

We also show that $(\perp_\tau)_{i \in I} \in \rhd_\tau$ for every computation type τ. For function types, this is immediate. For types of the form $D\tau$, we must check that $\mathbf{0}$ is in $\rhd_{D\tau}$. For all $(h_i)_{i \in I} \in \rhd_\tau^\perp$, we indeed have $\left(\int_{V \in [\![\tau]\!]} h_i(V) d0\right)_{i \in I} \in \rhd$, since $\mathbf{0} \in \rhd$.

Step 2. We claim that for all $(\nu_i)_{i\in I} \in \rhd_{D\sigma}$, for all $(f_i)_{i\in I} \in \rhd_{\sigma\to D\tau}$, $(f_i^\dagger(\nu_i))_{i\in I} \in \rhd_{D\tau}$. We wish to use the definition of $\rhd_{D\tau}$, so we consider an arbitrary tuple $(h_i)_{i\in I} \in \rhd_\tau^\perp$, and we aim to prove that $(\int_{V\in[\![\tau]\!]} h_i(V)df_i^\dagger(\nu_i))_{i\in I}$ is in $\rhd$. For that, we use Eq. (2): $\int_{V\in[\![\tau]\!]} h_i(V)df_i^\dagger(\nu_i) = \int_{x\in[\![\sigma]\!]} \left(\int_{V\in[\![\tau]\!]} h_i(V)df_i(x)\right) d\nu_i$, for every $i \in I$.

Let us define $h_i'(x) \stackrel{\text{def}}{=} \int_{V\in[\![\tau]\!]} h_i(V)df_i(x)$. We claim that $(h_i')_{i\in I} \in \rhd_\sigma^\perp$. Let $(x_i)_{i\in I} \in \rhd_\sigma$. Then $(f_i(x_i))_{i\in I} \in \rhd_{D\tau}$, and since $(h_i)_{i\in I} \in \rhd_\tau^\perp$, $(h_i'(x_i))_{i\in I}$ is in $\rhd$, by definition of $\rhd_{D\tau}$. Since $(x_i)_{i\in I}$ is arbitrary, $(h_i')_{i\in I} \in \rhd_\sigma^\perp$.

Since $(h_i')_{i\in I} \in \rhd_\sigma^\perp$ and $(\nu_i)_{i\in I} \in \rhd_{D\sigma}$, by definition of $\rhd_{D\sigma}$ we obtain that $(\int_{x_i\in[\![D\sigma]\!]} h_i'(x_i)d\nu_i)_{i\in I}$ is in $\rhd$, and this is exactly what we wanted to prove.

We now prove the claim by induction on M. If M is a variable, this is by assumption. If $M = 0$, this is trivial. If M is of the form $\mathtt{s}N$, then all the values $[\![N]\!]\rho_i$ are equal, hence also all the values $[\![M]\!]\rho_i = [\![N]\!]\rho_i + 1$. Similarly for terms of the form $\mathtt{p}N$. The case of applications is by definition of $\rhd_{\sigma\to\tau}$. In the case of abstractions $\lambda x_\sigma.M$ with $M\colon \tau$, we must show that, letting f_i be the map $V \in [\![\sigma]\!] \mapsto [\![M]\!](\rho_i[x_\sigma \mapsto V])$ $(i \in I)$, for all $(V_i)_{i\in I} \in \rhd_\sigma$, $(f_i(V_i))_{i\in I} \in \rhd_\tau$. This boils down to checking that $([\![M]\!](\rho_i[x_\sigma \mapsto V_i]))_{i\in I} \in \rhd_\tau$ for all $(V_i)_{i\in I} \in \rhd_\sigma$, which follows immediately from the induction hypothesis and the easily checked fact that $(\rho_i[x_\sigma \mapsto V_i])_{i\in I}$ is in $\rhd_*$.

The case of terms of the form $\mathtt{rec}_\tau\, M$, where τ is a computation type, is more interesting. Let f_i be the map $[\![M]\!]\rho_i\colon [\![\tau]\!] \to [\![\tau]\!]$. By induction hypothesis $(f_i)_{i\in I}$ is in $\rhd_{\tau\to\tau}$, so for all $(a_i)_{i\in I} \in \rhd_\tau$, $(f_i(a_i))_{i\in I}$ is in $\rhd_\tau$. Iterating this, we have $(f_i^n(a_i))_{i\in I} \in \rhd_\tau$ for every $n \in \mathbb{N}$. By Step 1, $(\bot_\tau)_{i\in I}$ is in $\rhd_\tau$. Hence $(f_i^n(\bot_\tau))_{i\in I} \in \rhd_\tau$ for every $n \in \mathbb{N}$. Since $\rhd_\tau$ is closed under directed suprema by Step 1, $(\mathrm{lfp}_{[\![\tau]\!]}\, f_i)_{i\in I} = ([\![\mathtt{rec}_\tau\, M]\!]\rho_i)_{i\in I}$ is in $\rhd_\tau$.

For terms of the form $M \stackrel{\text{def}}{=} \mathtt{if}\ N = 0\ \mathtt{then}\ P\ \mathtt{else}\, Q$ of type τ, by induction hypothesis $([\![N]\!]\rho_i)_{i\in I} \in \rhd_{\mathtt{int}}$, so all values $[\![N]\!]\rho_i$ are the same integer, say n. (And this term exists because I is non-empty.) If $n = 0$, then for every $i \in I$, $[\![M]\!]\rho_i$ is then equal to $[\![P]\!]\rho_i$, so $([\![M]\!]\rho_i)_{i\in I} = ([\![P]\!]\rho_i)_{i\in I}$ is in $\rhd_\tau$. We reason similarly if $n \neq 0$.

For terms of the form $M \oplus N$, of type $D\tau$, we consider an arbitrary tuple $(h_i)_{i\in I} \in \rhd_\tau^\perp$. By induction hypothesis $([\![M]\!]\rho_i)_{i\in I}$ and $([\![N]\!]\rho_i)_{i\in I}$ are in $\rhd_{D\tau}$, so $(\int_{V\in[\![\tau]\!]} h_i(V)d[\![M]\!]\rho_i)_{i\in I}$ and $(\int_{V\in[\![\tau]\!]} h_i(V)d[\![N]\!]\rho_i)_{i\in I}$ are in $\rhd$. Since $\rhd$ is convex, and integration is linear in the valuation, $(\int_{V\in[\![\tau]\!]} h_i(V)d[\![M \oplus N]\!]\rho_i)_{i\in I}$ is also in $\rhd$. Since $(h_i)_{i\in I}$ is arbitrary, $([\![M \oplus N]\!]\rho_i)_{i\in I}$ is in $\rhd_{D\tau}$.

For terms of the form $\mathtt{ret}_\sigma\, M$, we again consider an arbitrary tuple $(h_i)_{i\in I}$ in $\rhd_\sigma^\perp$. By induction hypothesis, $([\![M]\!]\rho_i)_{i\in I}$ is in $\rhd_\sigma$, so by definition of $\rhd_\sigma^\perp$, $(h_i([\![M]\!]\rho_i))_{i\in I}$ is in $\rhd$. Equivalently, $(\int_{V\in[\![\sigma]\!]} h_i(V)d\delta_{[\![M]\!]\rho_i})_{i\in I}$ is in $\rhd$, and that means that $([\![\mathtt{ret}_\sigma\, M]\!]\rho_i)_{i\in I}$ is in $\rhd_{D\sigma}$.

Finally, for terms $\mathtt{bind}_{\sigma,\tau}\, MN$, we have $([\![M]\!]\rho_i)_{i\in I} \in \rhd_{D\sigma}$ and $([\![N]\!]\rho_i)_{i\in I} \in \rhd_{\sigma\to D\sigma}$ by induction hypothesis, so $([\![\mathtt{bind}_{\sigma,\tau}\, MN]\!]\rho_i)_{i\in I} \in \rhd_{D\tau}$by Step 2. $\square$

Proposition 2. *For every ground $PCF_{\mathbb{P}}$ term P:* `Dint` $\rightarrow$ `Dint` $\rightarrow$ `Dint`, $[\![\texttt{portest}\ P]\!] \leq 8/27 \cdot \delta_0$.

Proof. We specialize the construction of the logical relation $(\rhd_\tau)_{\tau \text{ type}}$ to $I \stackrel{\text{def}}{=} \{1,2,3\}$ and to $\rhd$, defined as the downward closure in $\mathbb{R}^3_+$ of the convex hull $\{a \cdot (1,0,1) + b \cdot (1,1,0) + c \cdot (0,1,1) \mid a,b,c \in \mathbb{R}_+, a+b+c \leq 1\}$ of the three points $\boldsymbol{\alpha}_1 \stackrel{\text{def}}{=} (1,0,1)$, $\boldsymbol{\alpha}_2 \stackrel{\text{def}}{=} (1,1,0)$, and $\boldsymbol{\alpha}_3 \stackrel{\text{def}}{=} (0,1,1)$. The relation $\rhd$ has an alternate description as the set of those points (a,b,c) of $\mathbb{R}^3_+$ such that $a,b,c \leq 1$ and $a+b+c \leq 2$. This is depicted on the right.

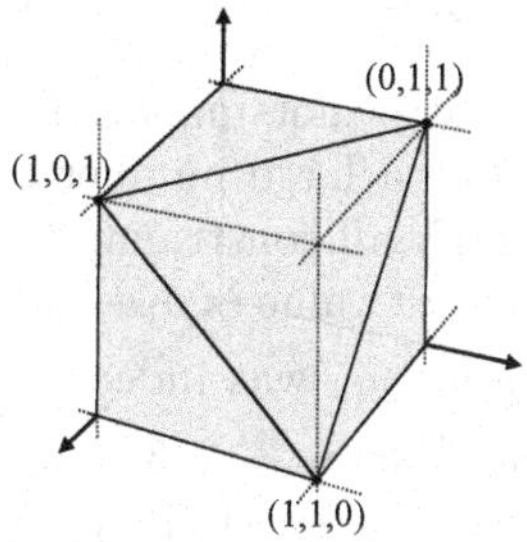

The relations $\rhd$ and $\rhd_\tau$ are ternary to account for the three calls to f in the definition of `portest`, and $\rhd$ is designed so that $\rhd_{\texttt{Dint}}$ is as small a relation as possible that contains the triples $(\delta_1, \delta_0, 0)$ and $(\delta_1, 0, \delta_0)$. Considering the three tests $f(\delta_1)(\delta_1) = \delta_1$, $f(\delta_0)(0) = \delta_0$ and $f(0)(\delta_0) = \delta_0$, the triple $(\delta_1, \delta_0, 0)$ consists of the first arguments to f in those tests, and the triple $(\delta_1, 0, \delta_0)$ consists of the second arguments. Hence, with f bound to P, the triple consisting of the three values of $f(\delta_1)(\delta_1)$, $f(\delta_0)(0)$ and $f(0)(\delta_0)$ respectively will also be contained in $\rhd_{\texttt{Dint}}$, by the basic lemma of logical relations (Proposition 1). We will then show that the largest probability that those values are 1, 0 and 0 respectively is 8/27, and this will complete the proof.

First, let us check that $(\delta_1, \delta_0, 0)$ and $(\delta_1, 0, \delta_0)$ are in $\rhd_{\texttt{Dint}}$. To that end, we simplify the expression of $\rhd_{\texttt{Dint}}$. For all $h_1, h_2, h_3 \in \mathcal{L}[\![\texttt{int}]\!]$, $(h_1, h_2, h_3) \in \rhd^\perp_{\texttt{int}}$ if and only if for every $n \in [\![\texttt{int}]\!]$, $(h_1(n), h_2(n), h_3(n)) \in \rhd$. Next, $(a_1\delta_0 + b_1\delta_1, a_2\delta_0 + b_2\delta_1, a_3\delta_0 + b_3\delta_1)$ is in $\rhd_{\texttt{Dint}}$ if and only if for all $(h_1, h_2, h_3) \in \rhd^\perp_{\texttt{int}}$, $(a_1h_1(0) + b_1h_1(1), a_2h_2(0) + b_2h_2(1), a_3h_3(0) + b_3h_3(1)) \in \rhd$. Since $\rhd$ is convex and downwards-closed, it suffices to check the latter when the triples $(h_1(0), h_2(0), h_3(0))$ and $(h_1(1), h_2(1), h_3(1))$ each range over the three points $\boldsymbol{\alpha}_i$, $1 \leq i \leq 3$ (nine possibilities). Let us write $\boldsymbol{\alpha}_i$ as $(\alpha_{i1}, \alpha_{i2}, \alpha_{i3})$. Hence $(a_1\delta_0 + b_1\delta_1, a_2\delta_0 + b_2\delta_1, a_3\delta_0 + b_3\delta_1)$ is in $\rhd_{\texttt{Dint}}$ if and only if the nine triples $(a_1\alpha_{i1} + b_1\alpha_{j1}, a_2\alpha_{i2} + b_2\alpha_{j2}, a_3\alpha_{i3} + b_3\alpha_{j3})$ $(1 \leq i,j \leq 3)$ are in $\rhd$, namely consist of non-negative numbers ≤ 1 that sum up to a value at most 2. Verifying that this holds for $(\delta_1, \delta_0, 0)$ $(a_1 \stackrel{\text{def}}{=} 0,\ b_1 \stackrel{\text{def}}{=} 1,\ a_2 \stackrel{\text{def}}{=} 1,\ b_2 \stackrel{\text{def}}{=} 0,\ a_3 \stackrel{\text{def}}{=} b_3 \stackrel{\text{def}}{=} 0)$ and $(\delta_1, 0, \delta_0)$ $(a_1 \stackrel{\text{def}}{=} 0,\ b_1 \stackrel{\text{def}}{=} 1,\ a_2 \stackrel{\text{def}}{=} b_2 \stackrel{\text{def}}{=} 0,\ a_3 \stackrel{\text{def}}{=} 1,\ b_3 \stackrel{\text{def}}{=} 0)$ means verifying that for all i, j between 1 and 3, $(\alpha_{j1}, \alpha_{i2}, 0)$ and $(\alpha_{j1}, 0, \alpha_{i3})$ are in $\rhd$, which is obvious since those are triples of numbers equal to 0 or to 1.

Using Proposition 1, $([\![P]\!](\delta_1)(\delta_1), [\![P]\!](\delta_0)(0), [\![P]\!](0)(\delta_0))$ is also in $\rhd_{\texttt{Dint}}$. Let us write that triple as $(a_1\delta_0 + b_1\delta_1, a_2\delta_0 + b_2\delta_1, a_3\delta_0 + b_3\delta_1)$. Then $[\![\texttt{portest}\ P]\!]$ is equal to $b_1a_2a_3 \cdot \delta_0$, as one can check. We wish to maximize $b_1a_2a_3$ subject to the constraint $(a_1\delta_0 + b_1\delta_1, a_2\delta_0 + b_2\delta_1, a_3\delta_0 + b_3\delta_1) \in \rhd_{\texttt{Dint}}$. That constraint rewrites to the following list of twelve inequalities, not mentioning the constraints that say that each a_i and each b_i is non-negative:

- $a_1 + b_1$, $a_2 + b_2$, and $a_3 + b_3$ should be at most 1,
- and the nine values $a_1 + b_1 + a_3 + b_3$, $a_1 + b_1 + b_2 + a_3$, $a_1 + b_2 + a_3 + b_3$, $a_1 + b_1 + a_2 + b_3$, $a_1 + b_1 + a_2 + b_2$, $a_1 + a_2 + b_2 + b_3$, $b_1 + a_2 + a_3 + b_3$, $b_1 + a_2 + b_2 + a_3$ and $a_2 + b_2 + a_3 + b_3$ should be at most 2.

That is not manageable. To help us, we have run a Monte-Carlo simulation: draw a large number of values at random for the variables a_i and b_i so as to verify all constraints (using rejection sampling), and find those that lead to the largest value of $b_1a_2a_3$. That simulation gave us the hint that the maximal value of $b_1a_2a_3$ was indeed $8/27$, attained for $a_1 \stackrel{\text{def}}{=} 0$, $b_1 \stackrel{\text{def}}{=} 2/3$, $a_2 \stackrel{\text{def}}{=} 2/3$, $b_2 \stackrel{\text{def}}{=} 0$, $a_3 \stackrel{\text{def}}{=} 0$, $b_3 \stackrel{\text{def}}{=} 2/3$. We now have to verify that formally. Knowing which values of a_i and b_i maximize $b_1a_2a_3$ allows us to select which constraints are the important ones, and then one can simplify slightly further.

In order to obtain a formal argument, we therefore choose to maximize $b_1a_2a_3$ with respect to the relaxed constraints that $a_1 + b_1 + a_2 + b_2 + a_3 + b_3 \leq 2$ (an inequality implied by all the above constraints), all numbers being non-negative. This will give us an upper bound, which may fail to be optimal (but won't).

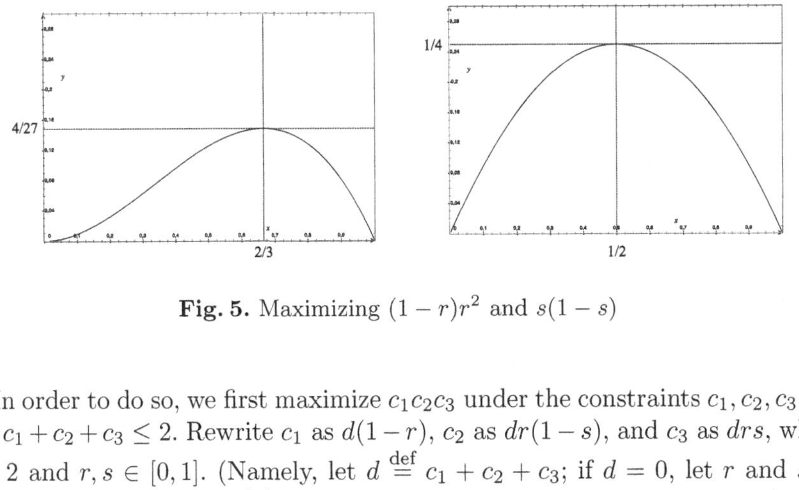

Fig. 5. Maximizing $(1 - r)r^2$ and $s(1 - s)$

In order to do so, we first maximize $c_1c_2c_3$ under the constraints $c_1, c_2, c_3 \geq 0$ and $c_1 + c_2 + c_3 \leq 2$. Rewrite c_1 as $d(1-r)$, c_2 as $dr(1-s)$, and c_3 as drs, where $d \leq 2$ and $r, s \in [0,1]$. (Namely, let $d \stackrel{\text{def}}{=} c_1 + c_2 + c_3$; if $d = 0$, let r and s be arbitrary; otherwise, let $r \stackrel{\text{def}}{=} 1 - c_1/d$; if $r = 0$, then let s be arbitrary; otherwise, let $s \stackrel{\text{def}}{=} c_3/(dr)$.) The maximal value of $c_1c_2c_3 = d^3(1-r)r^2s(1-s)$ is obtained by maximizing:

- d (as 2),
- $(1-r)r^2$ when $r \in [0,1]$ (value $4/27$ obtained at $r \stackrel{\text{def}}{=} 2/3$, see Fig. 5, left),
- and $s(1-s)$ when $s \in [0,1]$ (value $1/4$ obtained at $s \stackrel{\text{def}}{=} 1/2$, see Fig. 5, right),

hence is equal to $2 \cdot (4/27) \cdot (1/4) = 8/27$. It follows that for all $a_1, b_1, a_2, b_2, a_3, b_3 \in [0,1]$ such that $a_1 + b_1 + a_2 + b_2 + a_3 + b_3 \leq 2$, $b_1a_2a_3 \leq (a_1 + b_1)(a_2 + b_2)(a_3 + b_3) \leq 8/27$, by taking $c_i \stackrel{\text{def}}{=} a_i + b_i$ for each i. □

We sum up our results as follows. Note that $\Pr[\texttt{portest}\ P \downarrow n] = 0$, for any P, if $n \neq 0$.

Theorem 2. *For every ground PCF*$_{\mathsf{P}}$ *term* $P\colon \texttt{Dint} \to \texttt{Dint} \to \texttt{Dint}$, *the probability* $Pr[\texttt{portest}\ P \downarrow 0]$ *that* P *fools the parallel or tester never exceeds* 8/27. *That bound is attained by taking* $P \stackrel{def}{=} \texttt{pmpor}$. □

8 Conclusion and Related Work

There is an extensive literature on the semantics of higher-order functional languages, and extensions that include probabilistic choice are now attracting attention more than ever.

Concerning denotational semantics, we should cite the following. *Probabilistic coherence spaces* provide a fully abstract semantics for a version of PCF with probabilistic choice, as shown by Ehrhard, Tasson, and Pagani [4]. *Quasi-Borel spaces* and predomains have recently been used to give adequate semantics to typed and untyped probabilistic programming languages, see e.g. [17]. *QCB spaces* form a convenient category in which various effects, including probabilistic choice, can be modeled [2]. Comparatively, the domain-theoretic semantics we are using in this paper is rather mundane, and I have used similar models for further extensions that also include angelic [7] and demonic [9] non-deterministic choice. In those papers, I obtain full abstraction at the price of adding some extra primitives, but also of considering a richer semantics that also includes forms of non-deterministic choice. The latter allows us to work in categories with nice properties. That is not available in the context of PCF$_{\mathsf{P}}$, because there is no known Cartesian-closed category of continuous dcpos that is closed under $\mathbf{V}_{\leq 1}$ [13].

Let me remind the reader that denotational semantics is only a tool here: the result we have presented concerns the operational semantics, and domain-theory is only used, through adequacy, in order to bound $\Pr[\texttt{portest}\ P \downarrow *]$. One may wonder whether a direct operational approach would work, but I doubt it strongly. Eventually, any operational approach would have to find suitable invariants, and such invariants will be hard to distinguish from an actual denotational semantics.

One may wonder whether such semantical proofs would be useful in the realm of probabilistic process algebras as well. In non-probabilistic process algebras, syntactic reasoning is usually enough, using bisimulations and up-to techniques. The case of probabilistic processes is necessarily more complex, and may benefit from such semantical arguments.

References

1. Abramsky, S., Jung, A.: Domain theory. In: Abramsky, S., Gabbay, D.M., Maibaum, T.S.E. (eds.) Handbook of Logic in Computer Science, vol. III, pp. 1–168. Oxford University Press (1994)

2. Battenfeld, I.: Computational effects in topological domain theory. Electron. Notes Theor. Comput. Sci. **158**, 59–80 (2006)
3. Bird, B.: Introduction to Functional Programming using Haskell. Prentice-Hall Series in Computer Science. Prentice-Hall, Inc., Upper Saddle River, New Jersey (1998)
4. Ehrhard, T., Tasson, C., Pagani, M.: Probabilistic coherence spaces are fully abstract for probabilistic PCF. In: Jagannathan, S., Sewell, P. (eds.) Proceedings of 41st Annual ACM SIGPLAN-SIGACT Symposium on Principles of Programming Languages (POPL 2014), pp. 309–320 (2014)
5. Gierz, G., Hofmann, K.H., Keimel, K., Lawson, J.D., Mislove, M., Scott, D.S.: Continuous Lattices and Domains, Encyclopedia of Mathematics and its Applications, vol. 93. Cambridge University Press, Cambridge (2003)
6. Goubault-Larrecq, J.: Non-Hausdorff Topology and Domain Theory-Selected Topics in Point-Set Topology. New Mathematical Monographs, vol. 22. Cambridge University Press, Cambridge (2013)
7. Goubault-Larrecq, J.: Full abstraction for non-deterministic and probabilistic extensions of PCF I: the angelic cases. J. Logic Algebraic Methods Program. **84**(1), 155–184 (2015). https://doi.org/10.1016/j.jlamp.2014.09.003
8. Goubault-Larrecq, J.: Fooling the parallel or tester with probability 8/27 (2019). full version on arXiv:1903.12653 [cs.LO]
9. Goubault-Larrecq, J.: A probabilistic and non-deterministic call-by-push-value language. In: 34th Annual ACM/IEEE Symposium on Logic in Computer Science (LICS 2019) (2019, to appear). full version on arXiv:1812.11573 [cs.LO]
10. Goubault-Larrecq, J., Palamidessi, C., Troina, A.: A probabilistic applied pi–calculus. In: Shao, Z. (ed.) APLAS 2007. LNCS, vol. 4807, pp. 175–190. Springer, Heidelberg (2007). https://doi.org/10.1007/978-3-540-76637-7_12
11. Jones, C.: Probabilistic non-determinism. Ph.D. thesis, University of Edinburgh (1990). Technical report ECS-LFCS-90-105
12. Jones, C., Plotkin, G.: A probabilistic powerdomain of evaluations. In: Proceedings of the 4th Annual Symposium on Logic in Computer Science, pp. 186–195. IEEE Computer Society (1989)
13. Jung, A., Tix, R.: The troublesome probabilistic powerdomain. In: Edalat, A., Jung, A., Keimel, K., Kwiatkowska, M. (eds.) Proceedings of 3rd Workshop on Computation and Approximation. Electronic Lecture Notes in Computer Science, vol. 13, 23 p. Elsevier (1998)
14. Keimel, K., Lawson, J.: Measure extension theorems for T_0-spaces. Topol. Appl. **149**(1–3), 57–83 (2005)
15. Plotkin, G.D.: LCF considered as a programming language. Theor. Comput. Sci. **5**(1), 223–255 (1977)
16. Sazonov, V.Y.: Expressibility of functions in D. Scott's LCF language. Algebra i Logika **15**(3), 308–330 (1976). Translated from Russian
17. Vákár, M., Kammar, O., Staton, S.: A domain theory for statistical probabilistic programming. In: Proceedings of 46th ACM Symposium on Principles of Programming Languages (POPL 2019) (2019). arXiv:1811.04196 [cs.LO]

Categorical Information Flow

Tahiry Rabehaja[1(✉)], Annabelle McIver[2], Carroll Morgan[3], and Georg Struth[4]

[1] Optus Macquarie University Cyber Security Hub, Sydney, Australia
tahiry.rabehaja@mq.edu.au
[2] Department of Computing, Macquarie University, Sydney, Australia
[3] School of Computer Science and Engineering, UNSW, and Data61, Kensington, Australia
[4] Department of Computer Science, The University of Sheffield, Sheffield, UK

Abstract. We propose a categorical model for information flows of correlated secrets in programs. We show how programs act as transformers of such correlations, and that they can be seen as natural transformations between probabilistic constructors. We also study some basic properties of the construction.

1 Introduction

The foundations for Quantitative Information Flow (*QIF*) [2–5,14,15] has had great successes in explaining the impact of information flows in security systems. A basic assumption in *QIF* is that there is a single secret, it does not change over time, and that one risk of observing the behaviour of the system is that partial information about the secret leaks out. A *QIF* analysis is then able to answer questions such as "can adversaries use the information leaked to their advantage?"

In contexts where the secret does change over time however, the analysis becomes considerably more difficult. Even more challenging are contexts where there are multiple secrets, some correlated with others, some that change, whilst others remain unchanged. Here we are concerned with this most egregious case.

Consider the scenario of A who is burdened by her company's policy of forcing employees to change passwords every month. Moreover, since the company wants its employees to put serious effort into picking a strong password, they measure the time it takes for the new password to be selected, to avoid the situation that the employees will simply add another digit to their current password. A decides to use a workaround to this draconian practice: she writes a short but effective password selection program, which first allows time to pass for some random interval determined by the current value of the password, and then selects independently a completely random password which is ultimately stored automatically in A's keychain. Her code looks something like this, where `pw` stands for her current password.

M. S. Alvim et al. (Eds.): Palamidessi Festschrift, LNCS 11760, pp. 329–343, 2019.
https://doi.org/10.1007/978-3-030-31175-9_19

```
while (pw>0) { // Wait for some time
  pw--;
}
pw:= uniform(0, N) // Select uniformly at random a new password
```

Over lunch, A tells her friend G about her scheme; G is a little concerned because of a known timing attack on the loop which counts down from the initial value of `pw`. However, since it only reveals the *initial* value of `pw`, which is then immediately reset to a completely random value, there seems little to worry about.

A few months later the IT department introduces a new computer system to supplement the existing one. Whilst some aspects of her work environment improve, unfortunately now A has to reset *two* passwords every month—one for the old system, and an additional one for the supplementary system. She decides on an easy extension of her password updating program, setting her new password `npw` to her old password, and then updating `pw` as before:

```
npw:= pw;
while (pw>0) { // Wait for some time
  pw--;
}
pw:= uniform(0, N) // Select uniformly at random a new password
```

Unfortunately, since it has been months since she started using the original scheme, she has forgotten about G's brief analysis concerning the leaking of the initial value of `pw`, and indeed there was nothing to worry about since there was only one secret. But now, in this extended context, there is a serious vulnerability, not for `pw`, but for `npw`—its value is correlated with the initial value of `pw` which is completely revealed through the timing attack during the iteration of the loop. Thus G's original analysis, which assumed a single secret, although effective under that assumption, turns out to be unhelpful in general.[1]

In this paper we consider the general situation illustrated by that story. We find that any general analysis should take into account possible *future correlations* with other secrets. Our general approach suggests that secrets should never be considered in isolation, and that instead the basic currency of *QIF* should be *correlations of secrets*, and that any analysis should study leaks about correlations. It turns out that a categorical approach can be used to give a description of this situation. In particular we show how information flow can be summarised by a natural transformation between two constructors for structuring prior and posterior knowledge.

This paper is organised as follows. In Sect. 2, we revise the use of *HMM*'s in modelling information flow pertaining to static and dynamic secrets. We also introduce the notion of correlated secrets. In Sect. 3, we construct a category within which correlated secrets live. This form the fundamental basis for our

[1] Recall the Ariane disaster, which occurred when the software was executed within an environment for which it was not originally designed.

new compositional semantics where security programs modelled as *HMM*'s are in turn interpreted as natural transformations over this category. We also explore some characteristic properties of this category and the new semantic maps. We conclude in Sect. 4.

2 Channels, *HMM*'s and Secrets that Change

Hidden Markov Models or *HMM*'s are commonly used to solve inference problems such as to infer a sequence of hidden states that could have produced a sequence of observations. The most well known practical application is speech recognition, where a *HMM* is trained to recognise and translate speech waveforms into texts. *HMM*'s have also been applied to cryptanalysis where a cryptographic algorithm can emit a sequence of observations if given a sequence of states. In [7], Karlof and Wagner use *HMM*'s to model noisy side channels. Their work provides a generalised framework through which a large class of side channel attacks can be modelled. More recently, Smith provided a general foundation upon which the notion of information leakage is precisely defined, and then computed [14]. Even though Smith works mainly with static secrets, his notions of channel and their leakages are central to the subsequent advancement of Quantitative Information Flow (*QIF*).

Fundamental to *QIF* is the idea of an information flow channel. It models the process by which a secret with values drawn from a basic type $\mathcal{X}$ can be partially revealed to an adversary. We say that a channel C is an $\mathcal{X}\times\mathcal{Y}$ *stochastic matrix* denoted by $\mathcal{X}{\rightarrow\!\!\!\!\rightarrow}\mathcal{Y}$. Here $\mathcal{Y}$ denotes the set of observables available to a passive but curious adversary.[2] An entry C_{xy} is the (conditional) probability that, given the secret has some value x in $\mathcal{X}$, then the observation is $y \in \mathcal{Y}$. Stochastic means that, for each x in $\mathcal{X}$, we have that $\sum_{y\in\mathcal{Y}} C_{xy} = 1$. What can an adversary do with this information? We assume first that the adversary has some prior knowledge about the possible values of $\mathcal{X}$ and that this is modelled as a probability distribution $\pi \in \mathbb{D}\mathcal{X}$. Once we have a channel C and a prior π we can form the joint distribution $\pi\rangle C$ in $\mathbb{D}(\mathcal{X} \times \mathcal{Y})$ defined

$$(\pi\rangle C)_{xy} \quad := \quad \pi_x C_{xy} \,, \tag{1}$$

which describes the probability that the secret is x and the observation is y. We focus on two interesting distributions that provide some insight into what exactly has been leaked by the channel, and what the adversary can do with that leak.

The first is the marginal distribution over the observation set $\mathcal{Y}$—this is the probability that a specific y occurs when we don't know which particular x occurred. It is

$$(\pi\rangle C)_{-y} \quad := \quad \sum_{x\in\mathcal{X}} (\pi\rangle C)_{xy}.$$

[2] We do not assume that $\mathcal{X}$ and $\mathcal{Y}$ are finite. We do however assume that they are discrete and countable.

Next we look at the adversary's revised knowledge of the secret given that observation y has occurred. This is the *posterior* distribution wrt y and is the conditional distribution Eq. (1) relative to y:

$$(\delta^y)_x \quad := \quad (\pi\rangle C)_{xy}/(\pi\rangle C)_{-y}.$$

It turns out that the marginal and the posterior distributions can be combined to form a hyper-distribution –a distribution over distributions– of type $\mathbb{D}\mathbb{D}\mathcal{X} = \mathbb{D}^2\mathcal{X}$, where the outer distribution corresponds to the marginal (at some y) and the inner distribution corresponds to the posterior. We write $[\pi\rangle C]$ for the hyper-distribution corresponding to the joint distribution $\pi\rangle C$, where for $\delta{:}\,\mathbb{D}\mathcal{X}$ the posterior corresponding to observation y, we have

$$[\pi\rangle C]_\delta \quad := \quad \sum_{y:\mathcal{Y}_\delta} (\pi\rangle C)_{-y}\ ,$$

where $\mathcal{Y}_\delta \subseteq \mathcal{Y}$ is the subset of $\mathcal{Y}$ consisting of observations y such that δ and δ^y are the same distribution, i.e. $\delta_x = (\delta^y)_x$ for all x. Note that the hyper-distribution formulation represents the collection of posteriors and their respective marginal probabilities related to the joint distribution Eq. (1) *except* that the outer probability $[\pi\rangle C]_\delta$ is the *sum* of the marginal probabilities associated with any other posterior that is the same as δ. The reason is that separating those posteriors does not affect the computation of information flow [12]. Using this fact allows us to give an abstract information flow semantics as *abstract channels* of type priors to hyper-distributions, that is $[\![C]\!]{:}\,\mathbb{D}\mathcal{X}{\rightarrow}\mathbb{D}^2\mathcal{X}$ defined by[3]

$$[\![C]\!].\pi \quad := \quad [\pi\rangle C].$$

In our example in Sect. 1 above, we see a more complicated case of information flow, where the original secret `pw` was updated. In other work [9] we showed how a generalisation of abstract channels called abstract Hidden Markov Models (*HMM*'s) can be used to model this situation. *HMM*'s combine the effect of information flow (as defined above) for channels together with "Markov updates" to describe how secrets can be changed. Intuitively, a single *HMM* step describes the effect of first leaking some information about a secret through a channel matrix C followed by an immediate update of the secret via a Markov matrix M. This single step *HMM* is denoted by a matrix $H = (C{:}M){:}\,\mathcal{X}{\rightarrow\!\!\triangleright}\mathcal{Y}\times\mathcal{X}$ such that

$$H_{xyx'} \quad := \quad C_{xy}M_{xx'}.$$

More complicated *HMM*'s can be formed by sequentially composing smaller *HMM* matrices to give a pattern of leaking information about the *current* value of the secret, followed by a possible update. For example the sequential composition of two *HMM*'s $H{:}\,\mathcal{X}{\rightarrow\!\!\triangleright}\mathcal{Y}_1\times\mathcal{X}$ and $K{:}\,\mathcal{X}{\rightarrow\!\!\triangleright}\mathcal{Y}_2\times\mathcal{X}$ gives a *HMM* $H;K{:}\,\mathcal{X}{\rightarrow\!\!\triangleright}(\mathcal{Y}_1\times\mathcal{Y}_2)\times\mathcal{X}$ such that

$$(H;K)_{x(y_1,y_2)x'} \quad := \quad \sum_z H_{xy_1z}K_{zy_2x'}. \tag{2}$$

[3] Here $f.x$ denotes the application of a function f to the argument x.

We note that the "observations" now record a pair of elements from the observable set $\mathcal{Y}_1 \times \mathcal{Y}_2$, and in general the observations for n sequential compositions will be a trace of length n.

Just as for channels, *HMM*'s also transform initial prior information into a hyper-distribution through the construction of a joint distribution. But now consider the situation of executing an *HMM* over basic secret of type $\mathcal{X}$ which is correlated with some other secret of type $\mathcal{Z}$, by which we mean that there is a dependency between $\mathcal{X}$ and $\mathcal{Z}$ described by a joint distribution. This was the situation described in Sect. 1, where the *explicit* leak about `pw` caused a *collateral* leak about a different dependent secret `npw`. To account for this "collateral leak", we start with a correlation between the secret and some arbitrary secret ranging over $\mathcal{Z}$ modelled as a joint distribution $\Pi{:}\,\mathbb{D}(\mathcal{X} \times \mathcal{Z})$. We can now track what happens to the correlation of the two secrets when *HMM* H executes. It results in a joint distribution in $\mathbb{D}(\mathcal{Z} \times \mathcal{X} \times \mathcal{Y})$:

$$(\Pi\rangle H)_{zx'y} \quad := \quad \sum_x \Pi_{xz} H_{xyx'} .$$

Next, as we did for abstract channels, we abstract from the observation name y to obtain a hyper-distribution in $[\Pi\rangle H]{:}\,\mathbb{D}^2(\mathcal{X} \times \mathcal{Z})$ such that $[\Pi\rangle H]_\Delta$ is the probability that $\Delta{:}\,\mathbb{D}^2(\mathcal{X} \times \mathcal{Z})$ occurs as a (posterior) distribution. This implies that the abstract type of our *HMM* H, given some other secret $\mathcal{Z}$ that is not changed by H, but could be correlated with $\mathcal{X}$ is now

$$[\![H]\!]^{\mathcal{Z}}{:}\,\mathbb{D}(\mathcal{X} \times \mathcal{Z}) {\rightarrow} \mathbb{D}^2(\mathcal{X} \times \mathcal{Z}) \;, \tag{3}$$

where correlation $\Pi{:}\,\mathbb{D}(\mathcal{X} \times \mathcal{Z})$ is mapped to hyperdistribution

$$[\![H]\!]^{\mathcal{Z}}.\Pi = [\Pi\rangle H]. \tag{4}$$

Here, we see that this type captures clearly the role of H—that it is a mechanism for leaking and changing secrets, but in a way that it has an effect on other, possibly correlated secrets, which might or might not become relevant in a wider context. In fact the abstract semantics is now a set of transformations (parameterised by $\mathcal{Z}$). We show below in Sec. (3) that in fact H can be seen as a natural transformation.

Before moving further, let us see how the program used by A in the previous section is expressed in our framework. It is made up of 3 parts: the first assignment `npw:= pw`, followed by the while loop, and then the final resetting of `pw` to a random value. The first assignment establishes a correlation Π between `pw` and `npw`, which describes a distribution with the property that the only non-zero entries occur when `pw` and `npw` are the same. The while loop leaks the exact value of `pw` and then sets it to 0; recall that a curious eavesdropper would be able to count how many iterations the loop performs. If we assume that $\mathcal{X} = \{0, 1\}$, the while loop is then represented by the following *HMM* matrix:

$$\begin{array}{c} \\ 0 \\ 1 \end{array} \begin{array}{c} \overbrace{\texttt{0}\quad\texttt{1}}^{\circ\texttt{S}}\ \overbrace{\texttt{0}\quad\texttt{1}}^{\circ\texttt{B}} \\ \left(\begin{array}{cccc} 1 & 0 & 0 & 0 \\ 0 & 0 & 1 & 0 \end{array}\right) \end{array}$$

The labels ∘B and ∘S denote the observations that the body of the loop was executed or not, respectively. The other column labels come from the column labels in the execution of the body of the loop. Thus each column is labelled by a pair in $\{\circ\texttt{B}, \circ\texttt{S}\} \times \mathcal{X}$. The first row (labelled 0) describes the scenario where the initial value of pw is 0 so that the loop terminates immediately (i.e. there is a single 1 in the column $(\circ\texttt{S}, \texttt{0})$).

The last program that randomly sets pw is simply a Markov matrix with 0.5 in all positions. The two programs are composed using the sequential composition defined at Eq. (2) above to model to model the effect of information leakage on the secret stored in npw.

3 A Category of Correlations

In order to study the collateral consequences of information flow of a program operating over a single secret, we must first find a mathematical space that treats correlations as "first class citizens". In this section we describe the space of collateral types, show that it forms a category and we study some of its properties.

We begin by considering a basic relationship between two secrets. In general this relationship is not necessarily symmetric and can be expressed in the form of a channel, which means that observation of the value of one collateral secret may leak information regarding the value of the other collateral secret.

Definition 1. *The category* CORR *of collateral types contains finite sets* $\mathcal{Z}_1, \mathcal{Z}_2$ *as objects and channel matrices* $Z\colon \mathcal{Z}_1 \rightarrow\!\!\!\!\triangleright \mathcal{Z}_2$ *between them as arrows. The composition of channel in this category is the matrix multiplication of channel matrices. That is, given* $Z_1\colon \mathcal{Z}_1 \rightarrow\!\!\!\!\triangleright \mathcal{Z}_2$ *and* $Z_2\colon \mathcal{Z}_2 \rightarrow\!\!\!\!\triangleright \mathcal{Z}_3$*, we have* $Z_1 \cdot Z_2\colon \mathcal{Z}_1 \rightarrow\!\!\!\!\triangleright \mathcal{Z}_3$.

Intuitively, objects of CORR correspond to spaces in which all of A's passwords live. This includes pw and npw.

Lemma 1. CORR *is a category.*

Proof. This is clear since matrix multiplication behaves exactly as a functional composition and identity arrows are given by identity matrices.

Recall that Eq. (3) gives the type of a *HMM* on the secret type $\mathcal{X}$ and taking into account a collateral type $\mathcal{Z}$. The main intuition here is that, instead of simply transforming information regarding the *mutable secret type* $\mathcal{X}$, the program transforms joint information, which is an object in $\mathbb{D}(\mathcal{X} \times \mathcal{Z})$. When the mutable secret type $\mathcal{X}$ is fixed, the set $\mathbb{D}(\mathcal{X} \times \mathcal{Z})$ is parametrised by $\mathcal{Z}$.

More precisely, for a fixed $\mathcal{X}$ we can construct a functor $\mathbb{F}_{\mathcal{X}}$ from CORR to a category containing $\mathbb{D}(\mathcal{X} \times \mathcal{Z})$ as an object instance. Since $\mathcal{X} \times \mathcal{Z}$ is a finite set, the set $\mathbb{D}(\mathcal{X} \times \mathcal{Z})$ is a compact metric space with respect to the Kantorovich metric [9,13][4]. We know that the compact metric spaces and the continuous

[4] In $\mathbb{D}(\mathcal{X} \times \mathcal{Z})$, the Kantorovich metric is equivalent to the standard Euclidian distance.

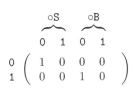

functions between them form a category which we refer to as COMP [1,16]. The functor $\mathbb{F}_{\mathcal{X}}$: CORR→COMP constructs the set of correlations $\mathbb{D}(\mathcal{X} \times \mathcal{Z})$, for each type $\mathcal{X}$.

Definition 2. *Let $\mathcal{X}$ be a fixed mutable type. We define the* collateral functor $\mathbb{F}_{\mathcal{X}}$: CORR→COMP *as follows:*

- *For objects, $\mathbb{F}_{\mathcal{X}}(\mathcal{Z}){:=}\mathbb{D}(\mathcal{X}{\times}\mathcal{Z})$ is the set of all joint distributions over $\mathcal{X}$ and $\mathcal{Z}$ for any collateral type $\mathcal{Z}$.*
- *For an arrow $Z{:}\,\mathcal{Z}_1{\rightarrow\!\!\!\!\rightarrow}\mathcal{Z}_2$, we have $\mathbb{F}_{\mathcal{X}}(Z){:}\,\mathbb{D}(\mathcal{X}{\times}\mathcal{Z}_1){\rightarrow}\mathbb{D}(\mathcal{X}{\times}\mathcal{Z}_2)$ such that, for every Π in $\mathbb{D}(\mathcal{X}{\times}\mathcal{Z}_1)$ we have*

$$\mathbb{F}_{\mathcal{X}}(Z)(\Pi) \quad := \quad \Pi{\cdot}Z\ , \tag{5}$$

which is the matrix multiplication of Π and Z.

The extra line npw := pw in the second program above creates a diagonal matrix in $\mathbb{F}_{\mathcal{X}}(\mathcal{X}')$ where $\mathcal{X}'$ is a copy of type $\mathcal{X}$. This produces a diagonal correlation between the values of these two variables.

Lemma 2. *$\mathbb{F}_{\mathcal{X}}$ is a functor.*

Proof. *Let $\mathcal{Z}$ be a collateral type. It is clear that $\mathbb{F}_{\mathcal{X}}(\mathrm{id}){:}\,\mathbb{D}(\mathcal{X} \times \mathcal{Z}){\rightarrow}\mathbb{D}(\mathcal{X} \times \mathcal{Z})$ is the identity map when* id: $\mathcal{Z}{\rightarrow\!\!\!\!\rightarrow}\mathcal{Z}$ *is the identity channel matrix on $\mathcal{Z}$.*

The functor $\mathbb{F}_{\mathcal{X}}$ preserves composition follows from the associativity of matrix multiplication.

Finally, for well-definedness, as long as we start with finite (or compact) $\mathcal{X}$ and $\mathcal{Z}$, *the set $\mathbb{D}(\mathcal{X}{\times}\mathcal{Z})$ is a compact metric space with the Kantorovich distance and the function $\mathbb{F}_{\mathcal{X}}(Z)$ is continuous* [13].

We can rewrite (3) using the newly defined functors $\mathbb{F}_{\mathcal{X}}$. That is, the denotation of a program H with mutable type $\mathcal{X}$ and collateral type $\mathcal{Z}$ has type

$$[\![H]\!]^{\mathcal{Z}}{:}\,\mathbb{F}_{\mathcal{X}}(\mathcal{Z}) \rightarrow \mathbb{D}\mathbb{F}_{\mathcal{X}}(\mathcal{Z}) \tag{6}$$

Note however that the denotation of H should only depend on $\mathcal{X}$ explicitly. In other words, $[\![H]\!]^{\mathcal{Z}}$ is parametric in $\mathcal{Z}$. This means that $[\![H]\!]^{\mathcal{Z}}$ in (6) is actually a component of the denotation of the *HMM* H given that we know the collateral type is $\mathcal{Z}$. This is an important observation because it implies that the "unparametrised" denotation of H should have type

$$\mathbb{F}_{\mathcal{X}} \Rightarrow \mathbb{D}\mathbb{F}_{\mathcal{X}}\ ,$$

that is, $[\![H]\!]$ should transform the functor $\mathbb{F}_{\mathcal{X}}$ into the composite functor $\mathbb{D}\mathbb{F}_{\mathcal{X}}$. This suggests that the denotation of a program computing with secrets is a natural transformation because the parameter $\mathcal{Z}$ should be manipulated in a syntactic manner. More precisely, $[\![H]\!]$ should apply to the collateral type $\mathcal{Z}$ polymorphically. This is exactly the rationale behind our next definition.

Definition 3. *An* abstract program *with mutable type $\mathcal{X}$ is a natural transformation $h: \mathbb{F}_{\mathcal{X}} \Rightarrow \mathbb{D}\mathbb{F}_{\mathcal{X}}$. That is, for every correlated type $\mathcal{Z}_1, \mathcal{Z}_2$ and channel arrow $Z: \mathcal{Z}_1 \rightarrow \mathcal{Z}_2$, we have that the following diagram commutes:*

$$
\begin{array}{ccc}
\mathbb{D}(\mathcal{X} \times \mathcal{Z}_1) & \xrightarrow{h^{\mathcal{Z}_1}} & \mathbb{D}^2(\mathcal{X} \times \mathcal{Z}_1) \\
{\scriptstyle \mathbb{F}_{\mathcal{X}}(Z)} \downarrow & & \downarrow {\scriptstyle \mathbb{D}\mathbb{F}_{\mathcal{X}}(Z)} \\
\mathbb{D}(\mathcal{X} \times \mathcal{Z}_2) & \xrightarrow[h^{\mathcal{Z}_2}]{} & \mathbb{D}^2(\mathcal{X} \times \mathcal{Z}_2)
\end{array}
$$

This definition implies that an abstract program h operates on the category CORR directly. Its denotation is independent of the collateral types, that is, any relationship (channel) between $\mathcal{Z}_1$ and $\mathcal{Z}_2$ can be accounted for before or after the execution of h while preserving information flow.

The following result shows that if H is a *HMM* matrix then $[\![H]\!]$ is indeed a natural transformation.

Theorem 1. *For any HMM matrix H, $[\![H]\!]$ is a natural transformation.*

Proof. Let $\Pi: \mathbb{D}(\mathcal{X} \times \mathcal{Z}_1)$ and $Z: \mathcal{Z}_1 \rightarrow \mathcal{Z}_2$ be any channel arrow. We need to show that

$$\mathbb{D}\mathbb{F}_{\mathcal{X}}(Z) \circ [\![H]\!]^{\mathcal{Z}_1}.\Pi = [\![H]\!]^{\mathcal{Z}_2}.\mathbb{F}_{\mathcal{X}}(Z)(\Pi). \tag{7}$$

We use Eqs. (4) and (5) to instantiate the inners of the left and right hand side hyper distributions.

On the one hand, an inner γ^y of the left-hand side of Eq. (7), associated to observation y, takes the form

$$\gamma^y_{x'z_2} = \frac{\sum_{z_1} \left(\sum_x \Pi_{xz_1} H_{xyx'}\right) Z_{z_1 z_2}}{\overline{\gamma}^y}$$

where $\overline{\gamma}^y = \sum_{xx'z_1z_2} \Pi_{xz_1} H_{xyx'} Z_{z_1z_2}$.

On the other hand, an inner δ^y of the right hand side, associated with the observation y, takes the form

$$\delta^y_{x'z_2} = \frac{\sum_x \left(\sum_{z_1} \Pi_{xz_1} Z_{z_1z_2}\right) H_{xyx'}}{\overline{\delta}^y}$$

where $\overline{\delta}^y = \sum_{xx'z_1z_2} \Pi_{xz_1} Z_{z_1z_2} H_{xyx'}$. It follows, after some arithmetic, that $\delta^y = \gamma^y$.

This theorem shows that there are indeed natural transformations from $\mathbb{F}_{\mathcal{X}}$ to $\mathbb{D}\mathbb{F}_{\mathcal{X}}$, of which the denotations of *HMM* matrices form a subset. Thus, the first program in the introduction is also a natural transformation where $\mathcal{X}$ is the set of all possible values that A can choose for variable `pw`.

To compose arrows of type $\mathbb{D}(\mathcal{X} \times \mathcal{Z}) \rightarrow \mathbb{D}^2(\mathcal{X} \times \mathcal{Z})$, we need to define the Giry monad $(\mathbb{D}, \eta, \mathsf{avg})$. The natural transformation η is the identity with $\eta: \mathcal{X} \Rightarrow \mathbb{D}$ and

$\eta_{\mathcal{X}}(x)$ is the point distribution at x. The natural transformation avg "squashes" higher level distributions (such as hyper-distributions) by averaging. That is, $\mathsf{avg}\colon \mathbb{D}^2 \Rightarrow \mathbb{D}$ where, for each $\Delta \in \mathbb{D}^2\mathcal{X}$, we intuitively have[5]

$$(\mathsf{avg}_{\mathcal{X}}(\Delta))_x = \sum_{\delta \in \mathbb{D}\mathcal{X}} \delta_x \Delta_\delta .$$

It is shown in [10] that

$$[\![H;K]\!]^{\mathcal{Z}} \quad = \quad [\![H]\!]^{\mathcal{Z}};[\![K]\!]^{\mathcal{Z}} \quad := \quad \mathsf{avg}_{\mathbb{D}(\mathcal{X}\times\mathcal{Z})} \circ \mathbb{D}[\![K]\!]^{\mathcal{Z}} \circ [\![H]\!]^{\mathcal{Z}} \qquad (8)$$

where the right hand side is the definition of sequential composition in the Giry monad [6].

Equation (8) is central for the compositionality result we show in [10]. It assumes that the mutable secrets of H and K are represented by the exact same variables ranging over $\mathcal{X}$. A slightly more general version is proven in the Appendix (Lemma 4).

Next, we show that the sequential composition of *HMM*'s in (2) is defined using the vertical composition of natural transformations $(k \bullet h)^{\mathcal{Z}} := k^{\mathcal{Z}} \circ h^{\mathcal{Z}}$.

Theorem 2. *Let H and K be HMM matrices on the mutable secret type $\mathcal{X}$. Then $[\![H;K]\!]$ is also a natural transformation where*

$$[\![H;K]\!] = \mathsf{avg}_{\mathbb{F}_{\mathcal{X}}} \bullet \mathbb{D}[\![K]\!] \bullet [\![H]\!] \quad ,$$

in which $\bullet$ is the vertical composition of natural transformations.

Proof. Let $\mathcal{Z}$ be a collateral type. By definition of the vertical composition, we have

$$(\mathsf{avg}_{\mathbb{F}_{\mathcal{X}}} \bullet \mathbb{D}[\![K]\!] \bullet [\![H]\!])^{\mathcal{Z}} = \mathsf{avg}_{\mathbb{F}_{\mathcal{X}}(\mathcal{Z})} \circ (\mathbb{D}[\![K]\!])^{\mathcal{Z}} \circ [\![H]\!]^{\mathcal{Z}} = \mathsf{avg}_{\mathbb{D}(\mathcal{X}\times\mathcal{Z})} \circ \mathbb{D}[\![K]\!]^{\mathcal{Z}} \circ [\![H]\!]^{\mathcal{Z}}$$

The second equality follows from the definition of $\mathbb{F}_{\mathcal{X}}(\mathcal{Z})$ and the composition of a functor with natural transformation, namely, $(\mathbb{D}[\![H]\!])^{\mathcal{Z}} = \mathbb{D}[\![H]\!]^{\mathcal{Z}}$.

The result now follows from Eq. (8).

The denotation of a *HMM* is therefore polymorphic with respect to the object of the underlying category CORR. In particular, the mutable type $\mathcal{X}$ of the *HMM* is also an object of CORR and we will show that the arrow $h^{\mathcal{X}}$ completely characterises the natural transformation.

Let $\mathcal{Z}$ be an arbitrary collateral type and $J\colon \mathbb{D}(\mathcal{X} \times \mathcal{Z})$ be any prior joint distribution. We can decompose J into a joint distribution $\Pi\colon \mathbb{D}\mathcal{X}^2$ and channel matrix $Z\colon \mathcal{X} \rightarrowtriangle \mathcal{Z}$ such that $\Pi\rangle Z = J$. This decomposition can be achieved within the context of the collateral functor $\mathbb{F}_{\mathcal{X}}$ and the construction is given in the following lemma.

[5] Rigorously, the sum in right hand side is an integral when Δ is not a discrete distribution. In this case, the left hand side should be applied to a measurable subset rather than the singleton.

Lemma 3. *Let $\mathcal{X}$ be a finite mutable type and $\mathcal{Z}$ be a non-empty object in* Corr. *For every joint-distribution $J{:}\,\mathbb{D}(\mathcal{X}{\times}\mathcal{Z})$, there exists $\Pi{:}\,\mathbb{D}\mathcal{X}^2$ and $Z{:}\,\mathcal{X}{\rightarrow}\mathcal{Z}$ such that*

$$\mathbb{F}_{\mathcal{X}}(Z)(\Pi) \quad = \quad J$$

Proof. One way to decompose $J{:}\,\mathbb{D}(\mathcal{X} \times \mathcal{Z})$ is through the marginal-diagonal-conditional decomposition [10]. *Let $\Pi{:}\,\mathbb{D}\mathcal{X}^2$ be the diagonal matrix such that*

$$\Pi_{xx'} = \begin{cases} \sum_z J_{xz} & if x = x' \\ 0 & otherwise \end{cases},$$

and $Z{:}\,\mathcal{X}{\rightarrow}\mathcal{Z}$ such that

$$Z_{xz} = \begin{cases} J_{xz}/\Pi_{xx} & if \Pi_{xx} > 0 \\ 1/|\mathcal{Z}| & otherwise. \end{cases}$$

Since Π is a diagonal matrix, we have

$$\Pi\rangle Z = \Pi_{xx} Z_{xz} = \sum_{x'} \Pi_{xx'} Z_{x'z} = \mathbb{F}_{\mathcal{X}}(Z)(\Pi).$$

Moreover, for $\Pi_{xx} > 0$, we have $\Pi_{xx} Z_{xz} = J_{xz}$. If $\Pi_{xx} = 0$*for some x,* then $J_{xz} = 0$ *for every z. In either cases, $J = \mathbb{F}_{\mathcal{X}}(Z)(\Pi)$.*

Note that the decomposition in Lemma 3 is not necessarily unique. For instance, the convention $Z_{xz} = 1/|\mathcal{Z}|$ is arbitrary and can be replaced with any collection of numbers such that $\sum_z Z_{xz} = 1$.

The following corollary of Lemma 3 shows that a natural transformation $h{:}\,\mathbb{F}_{\mathcal{X}} \Rightarrow \mathbb{D}\mathbb{F}_{\mathcal{X}}$ is fully characterised by its realisation $h^{\mathcal{X}}{:}\,\mathbb{F}_{\mathcal{X}}(\mathcal{X}) \rightarrow \mathbb{D}\mathbb{F}_{\mathcal{X}}(\mathcal{X})$.

Corollary 1. *$h^{\mathcal{Z}}$ is determined by $h^{\mathcal{X}}$ i.e.*

$$h^{\mathcal{Z}}.J = \mathbb{D}\mathbb{F}_{\mathcal{X}}(Z) \circ h^{\mathcal{X}}(\Pi)$$

for every $J{:}\,\mathbb{D}(\mathcal{X} \times \mathcal{Z})$ where Z, Π is a decomposition of J.

Proof. Lemma 3 *and Definition.* 3.

This corollary has practical importance as it allows us to focus any formal analysis of a program H on the arrow $[\![H]\!]^{\mathcal{X}}$ without any thought about the possibly unknown collateral type $\mathcal{Z}$. Thus, we do not need to exhaust all possible collateral types to know the collateral effect of executing H. We only need to compute the leaked information regarding the initial and final values of the mutable secret with type $\mathcal{X}$ [10].

In particular, if two programs H, K leak the exact same amount of information about the initial and final values of the secret with type $\mathcal{X}$, then $[\![H]\!]^{\mathcal{Z}}$ and $[\![K]\!]^{\mathcal{Z}}$ leak the exact same amount of information about the collateral type $\mathcal{Z}$. For instance, once A understands that she needs to account for *both* the initial and final value of `pw`, she realises that the collateral secret `npw` is not safe anymore well before she creates the new password.

Let us now explore a few special cases which reinforce the fact that our natural transformation semantics fully captures the notion of collateral information leakage.

3.1 Z Is a Null Channel: $\mathcal{Z}_2 = \{*\}$

This first case explains that the closed semantics of a *HMM* in [9] is a particular instance of our natural transformation semantics. We obtain this by replacing $\mathcal{Z}_2$ with the singleton set $\{*\}$.

An arrow $Z: \mathcal{Z}_1 \rightarrow\!\!\!\!\!\triangleright \{*\}$ is equivalent to the **null channel** on $\mathcal{Z}_1$ (i.e. single column filled with ones). Definition 3 reduces to

$$\begin{array}{ccc} \mathbb{D}(\mathcal{X} \times \mathcal{Z}_1) & \xrightarrow{h^{\mathcal{Z}}} & \mathbb{D}^2(\mathcal{X} \times \mathcal{Z}_1) \\ {\scriptstyle \mathbb{F}_{\mathcal{X}}(Z)}\downarrow & & \downarrow{\scriptstyle \mathbb{D}\mathbb{F}_{\mathcal{X}}(Z)} \\ \mathbb{D}\mathcal{X} & \xrightarrow[h^*]{} & \mathbb{D}^2\mathcal{X} \end{array}$$

where $\mathbb{F}_{\mathcal{X}}(Z): \mathbb{D}(\mathcal{X} \times \mathcal{Z}_1) \rightarrow \mathbb{D}\mathcal{X}$ such that, for $J: \mathbb{D}(\mathcal{X} \times \mathcal{Z}_1)$ we have

$$\mathbb{F}_{\mathcal{X}}(Z)(J)_x = \sum_z J_{xz} \quad ,$$

which is the **$\mathcal{X}$-marginal** of J and becomes our prior, which is then fed into the closed semantics h^*. In other words, *h^* is obtained by forgetting any correlated type i.e. working with the $\mathcal{X}$ -marginals of priors and posteriors.*

3.2 Z Selects a Collateral Prior: $\mathcal{Z}_1 = \{*\}$

Our second case shows that the closed semantics can only express independent correlations. Two independent secrets are expressed as a joint distribution in $\mathbb{D}(\mathcal{X} \times \mathcal{Z})$ that is obtained as the product of marginals on $\mathcal{X}$ and $\mathcal{Z}$. We obtain this by setting $\mathcal{Z}_1$ to be the singleton set $\{*\}$.

An arrow $Z: \{*\} \rightarrow\!\!\!\!\!\triangleright \mathcal{Z}_2$ is identified with a distribution in $\mathbb{D}\mathcal{Z}_2$. In this case, Definition 3 reduces to

$$\begin{array}{ccc} \mathbb{D}\mathcal{X} & \xrightarrow{h^*} & \mathbb{D}^2\mathcal{X} \\ {\scriptstyle \mathbb{F}_{\mathcal{X}}(Z)}\downarrow & & \downarrow{\scriptstyle \mathbb{D}\mathbb{F}_{\mathcal{X}}(Z)} \\ \mathbb{D}(\mathcal{X} \times \mathcal{Z}_2) & \xrightarrow[h^{\mathcal{Z}}]{} & \mathbb{D}^2(\mathcal{X} \times \mathcal{Z}_2) \end{array}$$

where $\mathbb{F}_{\mathcal{X}}(Z): \mathbb{D}\mathcal{X} \rightarrow \mathbb{D}(\mathcal{X} \times \mathcal{Z}_2)$ such that, for $\pi: \mathbb{D}\mathcal{X}$ we have

$$\mathbb{F}_{\mathcal{X}}(Z)(\pi)_{xz} = \pi_x Z_z \quad . \tag{9}$$

So what does this mean? Well, we know that the denotation h^* cannot express the information flow of variables correlated to the secret with type $\mathcal{X}$ [10]. However, Eq. (9) says that $\mathbb{F}_{\mathcal{X}}(Z)$ simply constructs the independent product of Z and a prior π, and similarly for $\mathbb{D}\mathbb{F}_{\mathcal{X}}(Z)$. Thus the above commutative diagram says that, when $\mathcal{Z}_2$ is independent of $\mathcal{X}$, we can construct the joint distributions and encapsulating hypers by simply multiplying the marginals pointwise. This is the standard definition of a joint distribution expressing independence of the column and row random variables. This is an expected sanity check.

In other words, the non-compositional denotation h^* *is also a special case of the compositional semantics where every other variable not specified in the underlying program is independent of the secret of type* $\mathcal{X}$*. In this case, and this case only,* $h^{\mathcal{Z}}$ *can be expressed via* h^*.

3.3 Collateral Only: $\mathcal{X} = \{*\}$

In this case, we will always know the value of the mutable secret variable (which is $*$, of course) and the uncertainty about any correlated secret should remain the same pre- and post execution of a program h.

Let us make that intuition precise. Let $Z: \mathcal{Z}_1 \rightarrow\!\!\!\!\!\!\!\circ\;\; \mathcal{Z}_2$. We have $\mathbb{F}_*(Z): \mathbb{D}\mathcal{Z}_1 \rightarrow \mathbb{D}\mathcal{Z}_2$ (up to the isomorphism $\{*\} \times \mathcal{Z} \simeq \mathcal{Z}$) where, for every $\pi: \mathbb{D}\mathcal{Z}_1$

$$\mathbb{F}_*(Z)(\pi) = \sum_z \pi_z Z_{z,z'}.$$

That h is natural gives us

$$\begin{array}{ccc} \mathbb{D}\mathcal{Z}_1 & \xrightarrow{h^{\mathcal{Z}_1}} & \mathbb{D}^2\mathcal{Z}_1 \\ {\scriptstyle \mathbb{F}_*(Z)}\downarrow & & \downarrow{\scriptstyle \mathbb{D}\mathbb{F}_*(Z)} \\ \mathbb{D}\mathcal{Z}_2 & \xrightarrow[h^{\mathcal{Z}_2}]{} & \mathbb{D}^2\mathcal{Z}_2 \end{array}$$

But what is $h^{\mathcal{Z}_1}$? Well, Corollary 1 tells us that $h^{\mathcal{Z}}$ is determined by $h^{\mathcal{X}}$ for every $\mathcal{Z}$. But since $\mathcal{X} = \{*\}$, we deduce that $h^*: \mathbb{D}\{*\} \rightarrow \mathbb{D}^2\{*\}$. The only arrow of this type is $\mathbb{D}\eta$ where η maps the point $*$ to the point distribution η_*. That is, $h^* = \mathbb{D}\eta$ which means that it is the **Skip** program.

Finally, for every $\pi: \mathbb{D}\mathcal{Z}_1$ (actually $\mathbb{D}(\{*\} \times \mathcal{Z}_1)$), the only decomposition given by Lemma 3 is given by the joint distribution matrix $(1) \in \mathbb{D}\{*\}^2$ and channel $Z: \{*\} \rightarrow\!\!\!\!\!\!\!\circ\;\; \mathcal{Z}_1$ with $Z_* = \pi$. So

$$h^{\mathcal{Z}}.\pi = \mathbb{D}\mathbb{F}_*(Z) \circ h^*((1)) = \mathbb{D}\mathbb{F}_*(Z)(\eta_{(1)}) = \eta_{\mathbb{F}_*(Z)((1))} = \eta_\pi.$$

Thus $h^{\mathcal{Z}}$ is also **Skip** as expected.

4 Conclusion

We have shown that programs computing with secrets can be interpreted as natural transformations acting on a category CORR of correlations. This semantic space is compositional. For each collateral type $\mathcal{Z}$, the natural transformation $[\![P]\!]$, for an arbitrary program P, can be instantiated on $\mathcal{Z}$ by working with the arrow $[\![P]\!]^{\mathcal{Z}}$. Moreover, the natural transformation $[\![P]\!]$ is completely characterised by the component $[\![P]\!]^{\mathcal{X}}$. This implies that, instead of trying to capture all possible correlations (known and unknown), one can simply look at programs as constructing correlations between initial and final secret values. Thus a fresh correlation can be attached to the output using the preserved initial secret values. This is captured abstractly by the notion of polymorphism.

Our main goal was to provide a categorical compositional semantics that can be used as a foundation for further frameworks suitable for program with secrets. In particular, we did not elaborate on the explicit construction of a refinement ordering on the space of natural transformation nor did we study any notion of convergence of a sequence of natural transformations. Though both of these items are important to obtain a sound logical framework for programs, we leave them for future work.

A Compositional sequential composition

The sequential composition (8) assumes that the mutable secrets of H and K are represented by the exact same variables ranging in $\mathcal{X}$. A slightly more general version of that result is to assume that H and K may share a hidden variable ranging in $\mathcal{X}$ and that they also have other hidden variables specific to each *HMM*. In this case, if H contains another hidden secret with type $\mathcal{X}_1$ then this secret must be considered as a collateral secret in K—thus K leaks collateral information about it, but does not update that secret. More precisely, we have the following lemma.

Lemma 4. *Let* $H\colon(\mathcal{X}\times\mathcal{X}_1)\twoheadrightarrow\mathcal{Y}_1\times(\mathcal{X}\times\mathcal{X}_1)$ *with* $K\colon(\mathcal{X}\times\mathcal{X}_2)\twoheadrightarrow\mathcal{Y}_2\times(\mathcal{X}\times\mathcal{X}_2)$ *be two HMM matrices. We have*

$$[\![H;K]\!]^{\mathcal{Z}} = [\![H]\!]^{\mathcal{X}_2\times\mathcal{Z}};[\![K]\!]^{\mathcal{X}_1\times\mathcal{Z}}. \tag{10}$$

where the composition $H;K$ is given by Eq. (2) and the composition operator (;) *on the right takes the isomorphism between $\mathcal{X}_1\times\mathcal{X}_2$ and $\mathcal{X}_2\times\mathcal{X}_1$ into account.*

Proof. Firstly, we have the following types

- $[\![H;K]\!]^{\mathcal{Z}}\colon\mathbb{D}(\mathcal{X}\times\mathcal{X}_1\times\mathcal{X}_2\times\mathcal{Z})\to\mathbb{D}^2(\mathcal{X}\times\mathcal{X}_1\times\mathcal{X}_2\times\mathcal{Z})$
- $[\![H]\!]^{\mathcal{X}_2\times\mathcal{Z}}\colon\mathbb{D}(\mathcal{X}\times\mathcal{X}_1\times\mathcal{X}_2\times\mathcal{Z})\to\mathbb{D}^2(\mathcal{X}\times\mathcal{X}_1\times\mathcal{X}_2\times\mathcal{Z})$
- $[\![K]\!]^{\mathcal{X}_1\times\mathcal{Z}}\colon\mathbb{D}(\mathcal{X}\times\mathcal{X}_2\times\mathcal{X}_1\times\mathcal{Z})\to\mathbb{D}^2(\mathcal{X}\times\mathcal{X}_2\times\mathcal{X}_1\times\mathcal{Z})$

Let us assume that we have applied the isomorphism that permutes the positions of the $\mathcal{X}_1$ and $\mathcal{X}_2$ *in the domain of* $[\![K]\!]^{\mathcal{X}_1\times\mathcal{Z}}$ *so that the right hand side composition is well defined and is given by the sequential composition in the Giry monad* [6]. *This type matching function is implicitly embedded into the right hand side composition in* (10).

Let $\Pi\colon\mathbb{D}(\mathcal{X}\times\mathcal{X}_1\times\mathcal{X}_2\times\mathcal{Z})$, *Eqs.* (1) and (2) *give*

$$(\Pi\rangle(H;K))_{y_1y_2,x''x_1'x_2'z} = \sum_{xx_1x_2}\Pi_{xx_1x_2z}\sum_{x'}H_{xx_1y_1x'x_1'}K_{x'x_2y_2x''x_2'} \tag{11}$$

On the other hand, $[\![H]\!]^{\mathcal{X}_2\times\mathcal{Z}};[\![K]\!]^{\mathcal{X}_1\times\mathcal{Z}} = \mathsf{avg}\circ\mathbb{D}[\![K]\!]^{\mathcal{X}_1\times\mathcal{Z}}\circ[\![H]\!]^{\mathcal{X}_2\times\mathcal{Z}}$ *uses the Kleisli lifting. Let us consider an inner* δ of $[\![H]\!]^{\mathcal{X}_2\times\mathcal{Z}};[\![K]\!]^{\mathcal{X}_1\times\mathcal{Z}}.\Pi$. *There exists an inner* α *of* $[\![H]\!]^{\mathcal{X}_2\times\mathcal{Z}}.\Pi$ *such that* δ *is an inner of* $[\![K]\!]^{\mathcal{X}_1\times\mathcal{Z}}.\alpha$. *That is, there exist two observations* y and y' *such that*

$$\alpha^{y_1}_{x'x'_1x_2z} = \frac{\sum_{xx_1} \Pi_{xx_1x_2z} H_{xx_1y_1x'x'_1}}{\overline{\alpha}^{y_1}}$$

where $\overline{\alpha}^{y_1} = \sum_{x'x'_1x_2z}\left(\sum_{xx_1} \Pi_{xx_1x_2z}H_{xx_1y_1x'x'_1}\right)$, *and*

$$\delta^{y_1y_2}_{x''x'_1x'_2z} = \frac{\sum_{x'x_2} \alpha^{y_1}_{x'x'_1x_2z} K_{x'x_2y_2x''x'_2}}{\overline{\delta}^{y_1y_2}}$$

where

$$\overline{\delta}^{y_1y_2} = \sum_{x''x'_1x'_2z}\left(\sum_{x'x_2} \alpha^{y_1}_{x'x'_1x_2z} K_{x'x_2y_2x''x'_2}\right)$$

By substituting α *into the expression of* δ *and simplifying* $\overline{\alpha}^{y_1}$, *we have*

$$\begin{aligned}\delta^{y_1y_2}_{x''x'_1x'_2z} &= \frac{\sum_{x'x_2}\left(\sum_{xx_1}\Pi_{xx_1x_2z}H_{xx_1y_1x'x'_1}\right)K_{x'x_2y_2x''x'_2}}{\sum_{x''x'_1x'_2z}\left(\sum_{x'x_2}\left(\sum_{xx_1}\Pi_{xx_1x_2z}H_{xx_1y_1x'x'_1}\right)K_{x'x_2y_2x''x'_2}\right)}\\ &= \frac{\sum_{xx_1x_2}\Pi_{xx_1x_2z}\sum_{x'}H_{xx_1y_1x'x'_1}K_{x'x_2y_2x''x'_2}}{\sum_{x''x'_1x'_2z}\left(\sum_{xx_1x_2}\Pi_{xx_1x_2z}\sum_{x'}H_{xx_1y_1x'x'_1}K_{x'x_2y_2x''x'_2}\right)}\end{aligned}$$

The inner $\delta^{y_1y_2}$ *corresponds exactly to a normalized* (y_1, y_2)*-column of* $\Pi\rangle(H;K)$ *as per Eq.* (11).

Lemma 4 is a straightforward generalisation our sequential composition for systems independent of any other collateral type [8,9] as well as our compositional, but single secret, semantics [10,11]. That is, the closed sequential composition is obtained by setting $\mathcal{X}_1 = \mathcal{X}_2 = \mathcal{Z} = \{*\}$ while the context aware version is obtained by setting $\mathcal{X}_1 = \mathcal{X}_2 = \{*\}$. Lemma 4 is slightly more general than the composition we define in [10] because it allows the declaration of new variables "on the go". For instance, if we set $\mathcal{X}_1 = \{*\}$, then a new secret variable of type $\mathcal{X}_2$ that is declared in K does not change in H. The lifted map $[\![H]\!]^{\mathcal{X}_2}$ is aware of this upcoming new secret and accounts for the correlation between the new variable and current secret of type $\mathcal{X}$.

References

1. Adámek, J., Herrlich, H., Strecker, G.: Abstract and Concrete Categories. Wiley, New York (1990)
2. Alvim, M., Andrés, M., Palamidessi, C.: Probabilistic information flow. In Proceedings of the 25th IEEE Symposium on Logic in Computer Science, pp. 314–321 (2010)
3. Alvim, M., Chatzikokolakis, K., McIver, A., Morgan, C., Palamidessi, C., Smith, G.: Additive and multiplicative notions of leakage, and their capacities. In: Proceedings of the IEEE 27th Computer Security Foundations Symposium, pp. 308–322 (2014)

4. Alvim, M., Chatzikokolakis, K., Palamidessi, C., Smith, G.: Measuring information leakage using generalized gain functions. In: Proceedings of the 25th IEEE Computer Security Foundations Symposium, pp. 265–279, June 2012
5. Braun, C., Chatzikokolakis, K., Palamidessi, C.: Quantitative notions of leakage for one-try attacks. In: Proceedings of the 25th International Conference on Mathematical Foundations of Programming Semantics, pp. 75–91 (2009)
6. Giry, M.: A categorical approach to probability theory. In: Banaschewski, B. (ed.) Categorical Aspects of Topology and Analysis. LNM, vol. 915, pp. 68–85. Springer, Heidelberg (1982). https://doi.org/10.1007/BFb0092872
7. Karlof, C., Wagner, D.: Hidden Markov model cryptanalysis. In: Walter, C.D., Koç, Ç.K., Paar, C. (eds.) CHES 2003. LNCS, vol. 2779, pp. 17–34. Springer, Heidelberg (2003). https://doi.org/10.1007/978-3-540-45238-6_3
8. McIver, A., Meinicke, L., Morgan, C.: Compositional closure for Bayes risk in probabilistic noninterference. In: Abramsky, S., Gavoille, C., Kirchner, C., Meyer auf der Heide, F., Spirakis, P.G. (eds.) ICALP 2010. LNCS, vol. 6199, pp. 223–235. Springer, Heidelberg (2010). https://doi.org/10.1007/978-3-642-14162-1_19
9. McIver, A., Morgan, C., Rabehaja, T.: Abstract hidden Markov models: a monadic account of quantitative information flow. In: Proceedings of the 30th Annual ACM/IEEE Symposium on Logic in Computer Science, pp. 597–608 (2015)
10. McIver, A.K., Morgan, C.C., Rabehaja, T.: Algebra for quantitative information flow. In: Höfner, P., Pous, D., Struth, G. (eds.) RAMICS 2017. LNCS, vol. 10226, pp. 3–23. Springer, Cham (2017). https://doi.org/10.1007/978-3-319-57418-9_1
11. Bordenabe, N., McIver, A., Morgan, C., Rabehaja, T.: Reasoning about distributed secrets. In: Bouajjani, A., Silva, A. (eds.) FORTE 2017. LNCS, vol. 10321, pp. 156–170. Springer, Cham (2017). https://doi.org/10.1007/978-3-319-60225-7_11
12. McIver, A., Morgan, C., Smith, G., Espinoza, B., Meinicke, L.: Abstract channels and their robust information-leakage ordering. In: Abadi, M., Kremer, S. (eds.) POST 2014. LNCS, vol. 8414, pp. 83–102. Springer, Heidelberg (2014). https://doi.org/10.1007/978-3-642-54792-8_5
13. Parthasarathy, K.R.: Probability Measures on Metric Spaces. Academic Press, Cambridge (1967)
14. Smith, G.: On the foundations of quantitative information flow. In: de Alfaro, L. (ed.) FoSSaCS 2009. LNCS, vol. 5504, pp. 288–302. Springer, Heidelberg (2009). https://doi.org/10.1007/978-3-642-00596-1_21
15. Smith, G.: Quantifying information flow using min-entropy. In: Proceedings of the 8th International Conference on Quantitative Evaluation of SysTems, pp. 159–167 (2011)
16. van Breugel, F.: The metric monad for probabilistic nondeterminism (2005). Draft available at http://www.cse.yorku.ca/~franck/research/drafts/monad.pdf

Statistical Epistemic Logic

Yusuke Kawamoto(✉)

National Institute of Advanced Industrial Science and Technology (AIST), Tsukuba, Japan
yusuke.kawamoto.aist@gmail.com

Abstract. We introduce a modal logic for describing statistical knowledge, which we call *statistical epistemic logic*. We propose a Kripke model dealing with probability distributions and stochastic assignments, and show a stochastic semantics for the logic. To our knowledge, this is the first semantics for modal logic that can express the statistical knowledge dependent on non-deterministic inputs and the statistical significance of observed results. By using statistical epistemic logic, we express a notion of statistical secrecy with a confidence level. We also show that this logic is useful to formalize statistical hypothesis testing and differential privacy in a simple and abstract manner.

Keywords: Epistemic logic · Possible world semantics · Divergence · Statistical hypothesis testing · Differential privacy

1 Introduction

Knowledge representation and reasoning have been studied in two research areas: *logic* and *statistics*. Broadly speaking, logic describes our knowledge using formal languages and reasons about it using symbolic techniques, while statistics interprets collected data having random variation and infers properties of their underlying probability models. As research advances demonstrate, logical and statistical approaches are respectively successful in many applications, including artificial intelligence, software engineering, and information security.

The techniques of these two approaches are basically orthogonal and could be integrated to get the best of both worlds. For example, in a large system with artificial intelligence (e.g., an autonomous car), both rule-based knowledge and statistical machine learning models may be used, and the way of combining them would be crucial to the performance and security of the whole system. However, even in theoretical research on knowledge models, there still remains much to be done to integrate techniques from the two approaches. For a very basic example, *epistemic logic* [39], a formal logic for representing and reasoning about knowledge, has not yet been able to model "statistical knowledge" with

This work was supported by JSPS KAKENHI Grant Number JP17K12667, and by Inria under the project LOGIS.

M. S. Alvim et al. (Eds.): Palamidessi Festschrift, LNCS 11760, pp. 344–362, 2019.
https://doi.org/10.1007/978-3-030-31175-9_20

sampling and statistical significance, although a lot of epistemic models [14,20, 21] have been proposed so far.

One of the important challenges in integrating logical and statistical knowledge is to design a logical model for statistical knowledge, which can be updated by a limited number of sampling of probabilistic events and by the non-deterministic inputs from an external environment. Here we note that non-deterministic inputs are essential to model the security of the system, because we usually do not have a prior knowledge of the probability distribution of adversarial inputs and need to reason about the worst scenarios caused by the attack. Nevertheless, to the best of our knowledge, no previous work on epistemic logic has proposed an abstract model for the statistical knowledge that involves non-deterministic inputs and the statistical significance of observed results.

In the present paper, we propose an epistemic logic for describing statistical knowledge. To define its semantics, we introduce a variant of a Kripke model [29] in which each possible world is defined as a probability distribution of states and each variable is probabilistically assigned a value. In this model, the stochastic behaviour of a system is modeled as a distribution of states at each world, and each non-deterministic input to the system corresponds to a distinct possible world. As for applications of this model, we define an accessibility relation as a statistical distance between distributions of observations, and show that our logic is useful to formalize statistical hypothesis testing and differential privacy [11] of statistical data.

Our Contributions. The main contributions of this work are as follows:

- We introduce a modal logic, called *statistical epistemic logic* (StatEL), to describe statistical knowledge.
- We propose a Kripke model incorporating probability distributions and stochastic assignments by regarding each possible world as a distribution of states and by defining an accessibility relation using a metric/divergence between distributions.
- We introduce a stochastic semantics for StatEL based on the above models. As far as we know, this is the first semantics for modal logic that can express the statistical knowledge dependent on non-deterministic inputs and the statistical significance of observed results.
- We present basic properties of the probability quantification and epistemic modality in StatEL. In particular, we show that the transitivity and Euclidean axioms rely on the agent's capability of observation in our model.
- By using StatEL we introduce a notion of statistical secrecy with a significance level α. We also show that StatEL is useful to formalize statistical hypothesis testing and differential privacy in a simple and abstract manner.

Paper Organization. The rest of this paper is organized as follows. Section 2 introduces background and notations used in this paper. Section 3 presents an example of coin flipping to explain the motivation for a logic of statistical knowledge. Section 4 shows the syntax and semantics of the statistical epistemic logic

StatEL. Section 5 presents basic properties of the logic. As for applications, Sects. 6 and 7 respectively model statistical hypothesis testing and statistical data privacy using StatEL. Section 8 presents related work and Sect. 9 concludes.

2 Preliminaries

In this section we recall the definitions of divergence and metrics, which are used in later sections to quantitatively model an agent's capability of distinguishing possible worlds.

2.1 Notations

Let $\mathbb{R}^{\geq 0}$ be the set of non-negative real numbers, and $[0, 1] = \{r \in \mathbb{R}^{\geq 0} \mid r \leq 1\}$. We denote by $\mathbb{D}\mathcal{O}$ the set of all probability distributions over a set $\mathcal{O}$. For a finite set $\mathcal{O}$ and a distribution $\mu \in \mathbb{D}\mathcal{O}$, the probability of sampling a value y from μ is denoted by $\mu[y]$. For a subset $R \subseteq \mathcal{O}$, let $\mu[R] = \sum_{y \in R} \mu[y]$. The *support* of a distribution μ over a finite set $\mathcal{O}$ is $\mathsf{supp}(\mu) = \{v \in \mathcal{O} \colon \mu[v] > 0\}$. For a set $\mathcal{D}$, a randomized algorithm $A : \mathcal{D} \to \mathbb{D}\mathcal{O}$ and a set $R \subseteq \mathcal{O}$ we denote by $A(d)[R]$ the probability that given input $d \in \mathcal{D}$, A outputs one of the elements of R.

2.2 Metric and Divergence

A *metric* over a non-empty set $\mathcal{O}$ is a function $d : \mathcal{O} \times \mathcal{O} \to \mathbb{R}^{\geq 0}$ such that for all $y, y', y'' \in \mathcal{O}$, (i) $d(y, y') \geq 0$; (ii) $d(y, y') = 0$ iff $y = y'$; (iii) $d(y, y') = d(y', y)$; (iv) $d(y, y'') \leq d(y, y') + d(y', y'')$. Recall that (iii) and (iv) are respectively referred to as symmetry and subadditivity.

A *divergence* over a non-empty set $\mathcal{O}$ is a function $D(\cdot \parallel \cdot) : \mathbb{D}\mathcal{O} \times \mathbb{D}\mathcal{O} \to \mathbb{R}^{\geq 0}$ such that for all $\mu, \mu' \in \mathbb{D}\mathcal{O}$, (i) $D(\mu \parallel \mu') \geq 0$ and (ii) $D(\mu \parallel \mu') = 0$ iff $\mu = \mu'$. Note that a divergence may not be symmetric or subadditive.

To describe a statistical hypothesis testing in Sect. 6, we recall the definition of χ^2 divergence due to Pearson [16] as follows:

Definition 1 (Pearson's χ^2divergence). Given two distributions μ, μ' over a finite set $\mathcal{O}$, the χ^2*-divergence* $D_{\chi^2}(\mu \parallel \mu')$ of μ from μ' is defined by:

$$D_{\chi^2}(\mu \parallel \mu') = \sum_{y \in \mathsf{supp}(\mu)} \frac{(\mu'[y] - \mu[y])^2}{\mu[y]}.$$

χ^2 *statistics* is the multiplication of χ^2-divergence with a sample size n.

To introduce a notion of statistical data privacy in Sect. 7, we recall the definition of the max-divergence D_∞ as follows.

Definition 2 (Max divergence). For two distributions μ, μ' over a finite set $\mathcal{O}$, the *max divergence* $D_\infty(\mu \parallel \mu')$ of μ from μ' is defined by:

$$D_\infty(\mu \parallel \mu') = \max_{R \subseteq \mathsf{supp}(\mu)} \ln \frac{\mu[R]}{\mu'[R]}.$$

Note that neither D_{χ^2} nor D_∞ is symmetric.

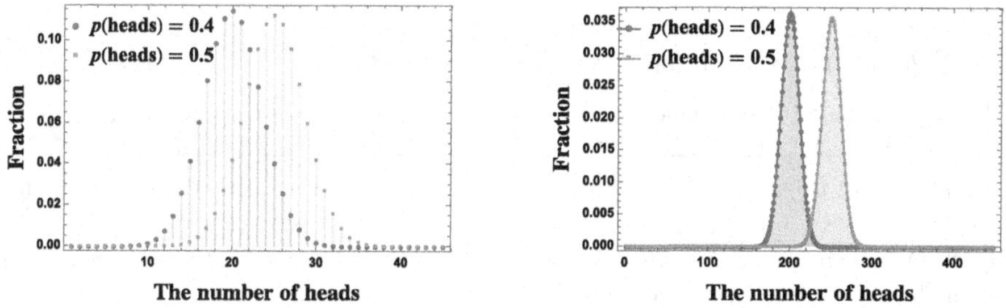

(a) Given 50 coin flips, the two distributions overlap much.

(b) Given 500 coin flips, the two distributions are distinguished more clearly.

Fig. 1. The frequency distributions of the numbers of heads in coin flipping.

3 Motivating Example

In this section we present a motivating example to explain why we need to introduce a new model for epistemic logic to describe statistical knowledge.

Example 1 (Coin flipping). Let us consider a simple running example of flipping a coin in two possible worlds w_0 and w_1 respectively. We assume that in the world w_0 the coin is fair (represented by $p(heads) = 0.5$), whereas in w_1 the probability of getting a heads is 0.4 (represented by $p(heads) = 0.4$). Here we do not have any prior belief on the probabilities of the worlds w_0 and w_1. This does not mean $p(w_0) = p(w_1) = 0.5$, but means we have no idea on the values of $p(w_0)$ and $p(w_1)$ at all, i.e., either w_0 or w_1 is chosen non-deterministically.

When we flip a coin just once and observe its outcome (heads or tails), we do not know whether the coin is fair or biased, that is, we cannot tell whether we are located in the world w_0 or w_1.

As shown in Fig. 1, however, when we increase the number n of coin flips, we can more clearly see the difference between the numbers of getting heads in w_0 and in w_1. If the fraction of observing heads goes to 0.5 (resp. 0.4), then we learn we are located in the world w_0 (resp. w_1) with a stronger confidence, namely, we have a stronger belief that the coin is fair (resp. biased). This implies that a larger number of observing the outcome enables us to distinguish two possible worlds more clearly, hence to obtain a stronger belief.

To model such statistical beliefs, we regard each possible world as a probability distribution of two states *heads* and *tails* as shown in Fig. 2 (e.g., $w_1[heads] = 0.4$ and $w_1[tails] = 0.6$). Then for a divergence D between two distributions, we define an accessibility relation $\mathcal{R}_\varepsilon$ between worlds such that for any worlds w and w', $(w, w') \in \mathcal{R}_\varepsilon$ iff $D(w \parallel w') \leq \varepsilon$. Then $(w_0, w_1) \in \mathcal{R}_\varepsilon$ for a smaller threshold ε represents that a larger number of sampling is required to distinguish w_0 from w_1.

This relation $\mathcal{R}_\varepsilon$ is used to formalize statistical knowledge in a model of epistemic logic in Sect. 4. Intuitively, given a threshold ε determining a confidence

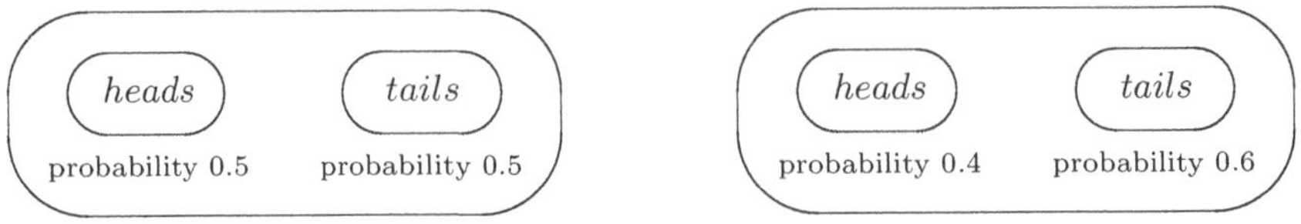

(a) The world w_0 with the fair coin. (b) The world w_1 with the biased coin.

Fig. 2. One of the possible worlds (i.e., w_0 or w_1) is chosen non-deterministically. Then one of the states (i.e., *heads* or *tails*) is chosen probabilistically.

level, we say that we know a proposition φ in a world w if φ is satisfied in all possible worlds that are indistinguishable from w in terms of $\mathcal{R}_\varepsilon$. In Sect. 6 we will revisit the coin flipping example to see how we formalize it using our logic.

To our knowledge, no previous work on epistemic logic has modeled a statistical knowledge that depends on the agent's capability of observing events. In fact, in most of the Kripke models used in previous work, a possible world represents a single state instead of a probability distribution of states, hence the relation between possible worlds does not involve the probability of distinguishing them. Therefore, no prior work on epistemic logic has proposed an abstract model for the statistical knowledge that involves the sample size of observing random variables and the statistical significance of the observed results.

4 Statistical Epistemic Logic (StatEL)

In this section we introduce the syntax and semantics of the *statistical epistemic logic* (StatEL).

4.1 Syntax

We first present the syntax of the statistical epistemic logic as follows. To express both deterministic and probabilistic properties, we introduce two levels of formulas: *static formulas* and *epistemic formulas*. Intuitively, a static formula represents a proposition that can be satisfied at a state with probability 1, while an epistemic formula represents a proposition that can be satisfied at a probability distribution of states with some probability.

Formally, let $\mathtt{Mes}$ be a set of symbols called *measurement variables*, and Γ be a set of atomic formulas of the form $\gamma(x_1, x_2, \ldots, x_n)$ for a predicate symbol γ and $x_1, x_2, \ldots, x_n \in \mathtt{Mes}$ ($n \geq 0$). Let $I \subseteq [0, 1]$ be a finite union of intervals, and $\mathcal{A}$ be a finite set of indices (typically associated with the names of agents and/or statistical tests). Then the static and epistemic formulas are defined by:

Static formulas: $\psi ::= \gamma(x_1, x_2, \ldots, x_n) \mid \neg\psi \mid \psi \wedge \psi$
Epistemic formulas: $\varphi ::= \mathbb{P}_I\, \psi \mid \neg\varphi \mid \varphi \wedge \varphi \mid \psi \supset \varphi \mid \mathsf{K}_a\, \varphi$

where $a \in \mathcal{A}$. Let $\mathcal{F}$ be the set of all epistemic formulas. Note that we have no quantifiers over measurement variables. (See Sect. 4.5.)

The *probability quantification* $\mathbb{P}_I \psi$ represents that a static formula ψ is satisfied with a probability belonging to a set I. For instance, $\mathbb{P}_{(0.5,1]} \psi$ represents that ψ holds with a probability greater than 0.5. The *non-classical implication* $\supset$ is used to represent conditional probabilities. For example, by $\psi_0 \supset \mathbb{P}_I \psi_1$ we represent that the conditional probability of ψ_1 given ψ_0 is included in a set I. The *epistemic knowledge* $\mathsf{K}_a \varphi$ expresses that an agent a knows φ. The formal meaning of these operators will be shown in the definition of semantics.

As syntax sugar, we use *disjunction* $\vee$, *classical implication* $\rightarrow$, and *epistemic possibility operator* P_a, defined by: $\varphi_0 \vee \varphi_1 ::= \neg(\neg\varphi_0 \wedge \neg\varphi_1)$, $\varphi_0 \rightarrow \varphi_1 ::= \neg\varphi_0 \vee \varphi_1$, and $\mathsf{P}_a \varphi ::= \neg \mathsf{K}_a \neg\varphi$. When I is a singleton $\{i\}$, we abbreviate $\mathbb{P}_{[i,i]}$ as $\mathbb{P}_i$.

4.2 Modeling of Systems

In this work we deal with a simple stochastic system with measurement variables. Let $\mathcal{O}$ be the finite set of all data that can be assigned to the measurement variables in $\mathtt{Mes}$. We assume that all possible worlds share the same domain $\mathcal{O}$. We define a *stochastic system* as a pair (S, σ) consisting of:

- a stochastic program S that deals with input and output data through measurement variables in $\mathtt{Mes}$, behaves deterministically or probabilistically (by using some randomly generated data), and terminates with probability 1;
- a *stochastic assignment* $\sigma : \mathtt{Mes} \rightarrow \mathbb{D}\mathcal{O}$ representing that each measurement variable x has an observed value v with probability $\sigma(x)[v]$.

Here we present only a general model and do not specify the data type of those measurement variables, which can be (sequences of) bit strings, floating point numbers, texts, or other types of data. Thanks to the assumption on the program termination and on the finite range of data, the program S can reach finitely many states. For the sake of simplicity, our model does not take timing into account. Extension to time and temporal modality is left for future work.

4.3 Distributional Kripke Model

To define a semantics for StatEL, we recall the notion of a Kripke model [29]:

Definition 3 (Kripke model). Given a set Γ of atomic formulas, a *Kripke model* is defined as a triple $(\mathcal{W}, \mathcal{R}, V)$ consisting of a non-empty set $\mathcal{W}$, a binary relation $\mathcal{R}$ on $\mathcal{W}$, and a function V that maps each atomic formula $\gamma \in \Gamma$ to a subset $V(\gamma)$ of $\mathcal{W}$. The set $\mathcal{W}$ is called a *universe*, its elements are called *possible worlds*, $\mathcal{R}$ is called an *accessibility relation*, and V is called a *valuation*.

Now we introduce a Kripke model called a "distributional" Kripke model where each possible world is a probability distribution of states over $\mathcal{S}$ and each world w is associated with a stochastic assignment σ_w to measurement variables.

Definition 4 (Distributional Kripke model). Let $\mathcal{A}$ be a finite set of indices (typically associated with the names of agents and/or statistical tests), $\mathcal{S}$ be a finite set of states[1], and $\mathcal{O}$ be a finite set of data. A *distributional Kripke model* is a tuple $\mathfrak{M} = (\mathcal{W}, (\mathcal{R}_a)_{a\in\mathcal{A}}, (V_s)_{s\in\mathcal{S}})$ consisting of:

- a non-empty set[2] $\mathcal{W}$ of probability distributions of states over $\mathcal{S}$;
- for each $a \in \mathcal{A}$, an accessibility relation $\mathcal{R}_a \subseteq \mathcal{W} \times \mathcal{W}$;
- for each $s \in \mathcal{S}$, a valuation V_s that maps each k-ary predicate γ to a set $V_s(\gamma) \subseteq \mathcal{O}^k$.

We assume that each $w \in \mathcal{W}$ is associated with a function $\rho_w : \mathtt{Mes} \times \mathcal{S} \rightarrow \mathcal{O}$ that maps each measurement variable x to its value $\rho_w(x, s)$ observed at a state s. We also assume that each state s in a world w is associated with the assignment $\sigma_s : \mathtt{Mes} \rightarrow \mathcal{O}$ defined by $\sigma_s(x) = \rho_w(x, s)$.

Note that this model assumes a constant domain $\mathcal{O}$; i.e., all measurement variables range over the same set $\mathcal{O}$ in every world. Since each world w is a probability distribution of states, we denote by $w[s]$ the probability that a state s is sampled from w. Then the probability that a variable x has a value v in a world w is given by:

$$\sigma_w(x)[v] = \sum_{s\in\mathtt{supp}(w),\, \sigma_s(x)=v} w[s].$$

This means that when a state s is drawn from the distribution w, an input value $\sigma_s(x)$ is sampled from the distribution $\sigma_w(x)$.

4.4 Divergence-Based Accessibility Relation

Next we introduce a family of accessibility relations used in typical statistical inferences. Since many notions of statistical distance are not metrics but divergences, we introduce an accessibility relation based on a divergence as follows.

Suppose that an agent a observes some data through a single measurement variable x. Then the distribution of the observed data at a world w is represented by $\sigma_w(x)$. Assume that the agent a distinguishes distributions in terms of a divergence $D(\cdot \,\|\, \cdot) : \mathbb{D}\mathcal{O} \times \mathbb{D}\mathcal{O} \rightarrow \mathbb{R}^{\geq 0}$. Then given a threshold $\varepsilon \geq 0$, we define a *divergence-based accessibility relation* $\mathcal{R}_{a,\varepsilon}$ by:

$$\mathcal{R}_{a,\varepsilon} \stackrel{\text{def}}{=} \{(w, w') \in \mathcal{W} \times \mathcal{W} \mid D(\sigma_w(x) \,\|\, \sigma_{w'}(x)) \leq \varepsilon\}.$$

For a smaller value of ε, the capability of distinguishing worlds is stronger.

If D is a metric instead, we call $\mathcal{R}_{a,\varepsilon}$ a *metric-based accessibility relation*. We often omit a to write $\mathcal{R}_\varepsilon$ when we do not compare different agents' knowledge.

---

[1] It is left for future work to investigate the case of infinite numbers of states.

[2] Since $\mathcal{W}$ is not a multiset, each world in $\mathcal{W}$ is a different distribution of states. However, this is still expressive enough when we take $\mathcal{S}$ to be sufficiently large.

Intuitively, $(w, w') \in \mathcal{R}_{a,\varepsilon}$ represents that the distribution of the data observed in w is indistinguishable from that in w' in terms of D. By the definition of a divergence/metric D, $D(\sigma_w(x) \parallel \sigma_{w'}(x)) = 0$ implies $\sigma_w(x) = \sigma_{w'}(x)$. Therefore, the relation $\mathcal{R}_{a,0}$ expresses that the agent a has an unlimited capability of observing the distributions $\sigma_w(x)$ and $\sigma_{w'}(x)$. In Sects. 6 and 7 we will show examples of divergence-based accessibility relations.

4.5 Stochastic Semantics

In this section we define the *stochastic semantics* for the StatEL formulas over a distributional Kripke model $\mathfrak{M} = (\mathcal{W}, (\mathcal{R}_a)_{a\in\mathcal{A}}, (V_s)_{s\in\mathcal{S}})$ with $\mathcal{W} = \mathbb{D}\mathcal{S}$.

The interpretation of static formulas ψ at a state s is given by:

$$s \models \gamma(x_1, x_2, \ldots, x_k) \text{ iff } (\sigma_s(x_1), \sigma_s(x_2), \ldots, \sigma_s(x_k)) \in V_s(\gamma)$$
$$s \models \neg\psi \text{ iff } s \not\models \psi$$
$$s \models \psi \wedge \psi' \text{ iff } s \models \psi \text{ and } s \models \psi'.$$

Note that the satisfaction of the static formulas does not involve probability.

To interpret the non-classical implication $\supset$, we define the *restriction* $w|_\psi$ of a world w to a state formula ψ as follows. If there exists a state s such that $w[s] > 0$ and $s \models \psi$, then $w|_\psi$ can be defined as the distribution over the finite set $\mathcal{S}$ of states such that:

$$w|_\psi[s] = \begin{cases} \frac{w[s]}{\sum_{s':s' \models \psi} w[s']} & \text{if } s \models \psi \\ 0 & \text{otherwise.} \end{cases}$$

Then $\sum_s w|_\psi[s] = 1$. Note that $w|_\psi$ is undefined if w does not have a state s that satisfies ψ and has a non-zero probability in w.

Now we define the interpretation of epistemic formulas at a world w in $\mathfrak{M}$ by:

$$\mathfrak{M}, w \models \mathbb{P}_I\, \psi \text{ iff } \Pr\Big[s \xleftarrow{\$} w :\ s \models \psi\Big] \in I$$
$$\mathfrak{M}, w \models \neg\varphi \text{ iff } \mathfrak{M}, w \not\models \varphi$$
$$\mathfrak{M}, w \models \varphi \wedge \varphi' \text{ iff } \mathfrak{M}, w \models \varphi \text{ and } \mathfrak{M}, w \models \varphi'$$
$$\mathfrak{M}, w \models \psi \supset \varphi \text{ iff } w|_\psi \text{ is defined and } \mathfrak{M}, w|_\psi \models \varphi$$
$$\mathfrak{M}, w \models \mathsf{K}_a\, \varphi \text{ iff for every } w' \text{ s.t. } (w, w') \in \mathcal{R}_a, \mathfrak{M}, w' \models \varphi,$$

where $s \xleftarrow{\$} w$ represents that a state s is sampled from the distribution w.

Finally, the interpretation of an epistemic formula φ in $\mathfrak{M}$ is given by:

$$\mathfrak{M} \models \varphi \text{ iff for every world } w \text{ in } \mathfrak{M},\ \mathfrak{M}, w \models \varphi.$$

We remark that in each world w, measurement variables can be interpreted using σ_w, as shown in Sect. 4.3. This allows one to assign different values to distinct occurrences of a variable in a formula; E.g., in $\varphi(x) \to \mathsf{K}_a\, \varphi'(x)$, the measurement variable x occurring in $\varphi(x)$ can be interpreted using σ_w in a world w, while x in $\varphi'(x)$ can be interpreted using $\sigma_{w'}$ in another w' s.t. $(w, w') \in \mathcal{R}_a$.

Note that our semantics for probability quantification is different from that in the previous work. Halpern [19] shows two approaches to defining semantics: giving probabilities (1) on the domain and (2) on possible worlds. However, our semantics is different from both. It defines probabilities on the states belonging to a possible world, while each world is not assigned a probability. Hence, unlike Halpern's approaches, our model can deal with both probabilistic behaviours of systems and non-deterministic inputs from an external environment.

We also remark that StatEL can be used to formalize conditional probabilities. If the conditional probability of satisfying a static formula ψ_1 given another static formula ψ_0 is included in a set I at a world w, then we have $\Pr\left[s \xleftarrow{\$} w|_{\psi_0} : s \models \psi_1\right] \in I$, hence we obtain $\mathfrak{M}, w \models \psi_0 \supset \mathbb{P}_I\, \psi_1$.

5 Basic Properties of StatEL

In this section we present basic properties of StatEL. In particular, we show the transitivity and Euclidean axioms rely on the agent's capability of observation.

5.1 Properties of Probability Quantification

We can define a dual operator of $\mathbb{P}_I$ as follows. Given a finite union $I \subseteq [0,1]$ of disjoint intervals, let $I^c \stackrel{\text{def}}{=} [0,1] \setminus I$ and $\overline{I} \stackrel{\text{def}}{=} \{1 - p \mid p \in I\}$. Then $\overline{I^c} = \overline{I}^c$. Negation with $\mathbb{P}_I$ has the following properties.

Proposition 1 (Negation with probability quantification). *For any world w in a model $\mathfrak{M}$ and any static formula ψ, we have:*

1. $\mathfrak{M}, w \models \neg\, \mathbb{P}_I\, \psi$ *iff* $\mathfrak{M}, w \models \mathbb{P}_{I^c}\, \psi$
2. $\mathfrak{M}, w \models \mathbb{P}_I \neg\psi$ *iff* $\mathfrak{M}, w \models \mathbb{P}_{\overline{I}}\, \psi$.

By Proposition 1, $\neg\, \mathbb{P}_I \neg\psi$ is logically equivalent to $\mathbb{P}_{\overline{I^c}}\, \psi$. For instance, $\neg\, \mathbb{P}_{(0,1]} \neg\psi$ is equivalent to $\mathbb{P}_1\, \psi$, and $\neg\, \mathbb{P}_{[0,1)} \neg\psi$ is equivalent to $\mathbb{P}_0\, \psi$.

5.2 Properties of Epistemic Modality

Next we show some properties of epistemic modality. As with the standard modal logic, StatEL satisfies the necessitation rule and distribution axiom.

Proposition 2 (Minimal properties). *For any distributional Kripke model $\mathfrak{M}$, any $\varphi, \varphi_0, \varphi_1 \in \mathcal{F}$, and any $a \in \mathcal{A}$, we have:*

(**N**) *necessitation:* $\mathfrak{M} \models \varphi$ *implies* $\mathfrak{M} \models \mathsf{K}_a\, \varphi$
(**K**) *distribution:* $\mathfrak{M} \models \mathsf{K}_a(\varphi_0 \rightarrow \varphi_1) \rightarrow (\mathsf{K}_a\, \varphi_0 \rightarrow \mathsf{K}_a\, \varphi_1)$.

The satisfaction of other properties depends on the definition of the accessibility relation. Since many notions of statistical distance are not metrics but divergences, we present some basic properties when $\mathfrak{M}$ has a divergence-based accessibility relation: $\mathcal{R}_{a,\varepsilon} = \{(w, w') \in \mathcal{W} \times \mathcal{W} \mid D(\sigma_w(x) \parallel \sigma_{w'}(x)) \leq \varepsilon\}$.

Proposition 3 (Properties with divergence-based accessibility). *Let $a \in \mathcal{A}$ and $\varepsilon \geq \varepsilon' \geq 0$. For any distributional Kripke model $\mathfrak{M}$ with a divergence-based accessibility relation $\mathcal{R}_{a,\varepsilon}$ and any $\varphi \in \mathcal{F}$, we have:*

(**T**) *reflexivity:* $\mathfrak{M} \models \mathsf{K}_{a,\varepsilon}\,\varphi \rightarrow \varphi$
($\geq$) *comparison of observability:* $\mathfrak{M} \models \mathsf{K}_{a,\varepsilon}\varphi \rightarrow \mathsf{K}_{a,\varepsilon'}\varphi$.

If $\mathcal{R}_{a,\varepsilon}$ is symmetric (e.g., based on the Jensen-Shannon divergence [33]) then:

(**B**) *symmetry:* $\mathfrak{M} \models \varphi \rightarrow \mathsf{K}_{a,\varepsilon}\,\mathsf{P}_{a,\varepsilon}\,\varphi$.

Here the axiom ($\geq$) represents that an agent having a stronger capability of distinguishing worlds may have more beliefs.

Finally, we show some properties when $\mathcal{R}_{a,\varepsilon}$ is based on a metric (e.g. the p-Wasserstein metric [38], including the Earth mover's distance).

Proposition 4 (Properties with metric-based accessibility). *Let $a \in \mathcal{A}$ and $\varepsilon, \varepsilon' \geq 0$. For any distributional Kripke model $\mathfrak{M}$ with a metric-based accessibility relation $\mathcal{R}_{a,\varepsilon}$ and any $\varphi \in \mathcal{F}$, we have* (**T**)*reflexivity,* (**B**)*symmetry, and:*

(**4q**) *quantitative transitivity:* $\mathfrak{M} \models \mathsf{K}_{a,\varepsilon+\varepsilon'}\varphi \rightarrow \mathsf{K}_{a,\varepsilon}\,\mathsf{K}_{a,\varepsilon'}\varphi$
(**5q**) *relaxed Euclidean:* $\mathfrak{M} \models \mathsf{P}_{a,\varepsilon}\varphi \rightarrow \mathsf{K}_{a,\varepsilon'}\mathsf{P}_{a,\varepsilon+\varepsilon'}\varphi$.

If the agent has an unlimited capability of observation (i.e., $\varepsilon = \varepsilon' = 0$), then:

(**4**) *transitivity:* $\mathfrak{M} \models \mathsf{K}_{a,0}\,\varphi \rightarrow \mathsf{K}_{a,0}\,\mathsf{K}_{a,0}\,\varphi$
(**5**) *Euclidean:* $\mathfrak{M} \models \mathsf{P}_{a,0}\,\varphi \rightarrow \mathsf{K}_{a,0}\,\mathsf{P}_{a,0}\,\varphi$.

By this proposition, for $\varepsilon = 0$, StatEL has the axioms of **S5**, hence the epistemic operator $\mathsf{K}_{a,0}$ represents knowledge rather than beleif.

However, if the agent has a limited observability (i.e., $\varepsilon > 0$), then neither transitivity nor Euclidean may hold. This means that, even when he know whether φ holds or not with some confidence, he may not be perfectly confident that he knows it.

6 Modeling Statistical Hypothesis Testing Using StatEL

In this section we formalize statistical hypothesis testing by using StatEL formulas, and introduce a notion of statistical secrecy with a confidence level.

6.1 Statistical Hypothesis Testing

A *statistical hypothesis testing* is a method of statistical inference to check whether given datasets provide sufficient evidence to support some hypothesis. Typically, given two datasets, a *null hypothesis* H_0 is defined to claim that there is no statistical relationship between the two datasets (e.g., no difference between the result of a medical treatment and the placebo effect), while an *alternative hypothesis* H_1 represents that there is some relationship between them (e.g., the result of a medical treatment is better than the placebo effect).

Before performing a hypothesis test, we specify a *significance level* α, i.e., the probability that the test might reject the null hypothesis H_0, given that H_0 is true. Typically, α is 0.05 or 0.01. $1 - \alpha$ is called a *confidence level.*

6.2 Formalization of Statistical Hypothesis Testing

Now we define a distributional Kripke model $\mathfrak{M}$ with a universe $\mathcal{W}$ that includes at least two worlds w_{real} and w_{ideal} corresponding to the two datasets we compare:

- the real world w_{real} where we have a dataset sampled from actual experiments (e.g., from a medical treatment whose effectiveness we want to know);
- the ideal world w_{ideal} where we have a dataset that is synthesized from the null hypothesis setting (e.g., the dataset obtained from the placebo effect).

Note that $\mathcal{W}$ may include other worlds corresponding to different possible datasets.

Let n be the size of the dataset, and x be a measurement variable denoting a single data value chosen from the dataset we have. We assume that each world w has a state s corresponding to each single data value $\sigma_s(x)$ in the dataset. Then $\sigma_{w_{\mathsf{real}}}(x)$ is the empirical distribution (histogram) calculated from the dataset observed in the actual experiments in w_{real}, while $\sigma_{w_{\mathsf{ideal}}}(x)$ is the distribution calculated from the synthetic dataset in w_{ideal}. Then the number of data having a value v in the dataset in a world w is given by $n \cdot \sigma_w(x)[v]$.

Assume that $\mathfrak{M}$ has an accessibility relation $\mathcal{R}_{c_\alpha/n}$ that is specific to the sample size n, the statistical hypothesis test, and the *critical value* c_α for a significance level α we use. For brevity let $\varepsilon_{\alpha,n} = c_\alpha/n$. Intuitively, $(w_{\mathsf{real}}, w_{\mathsf{ideal}}) \in \mathcal{R}_{\varepsilon_{\alpha,n}}$ represents that the hypothesis test cannot distinguish the actual dataset from the synthetic one. For instance, when we use Pearson's χ^2-test as the hypothesis test, then $\mathcal{R}_{\varepsilon_{\alpha,n}}$ is defined by:

$$\mathcal{R}_{\varepsilon_{\alpha,n}} \stackrel{\text{def}}{=} \{(w, w') \in \mathcal{W} \times \mathcal{W} \mid D_{\chi^2}(\sigma_w(x) \parallel \sigma_{w'}(x)) \leq \varepsilon_{\alpha,n}\},$$

where D_{χ^2} is Pearson's χ^2 divergence (Definition 1).

Observe that when the confidence level $1 - \alpha$ increases, then c_α decreases, hence $\varepsilon_{\alpha,n} = c_\alpha/n$ is smaller, i.e., the capability of distinguishing possible worlds is stronger.

Let φ_{syn} be a formula representing that the dataset is synthesized from the null hypothesis setting (e.g., representing the placebo effect). Then $\mathfrak{M}, w_{\mathsf{ideal}} \models$

φ_{syn}. Since each world in $\mathcal{W}$ corresponds to a different dataset, it holds for any $w' \neq w_{\mathsf{ideal}}$ that $\mathfrak{M}, w' \models \neg\varphi_{\mathsf{syn}}$. For instance, $\mathfrak{M}, w_{\mathsf{real}} \models \neg\varphi_{\mathsf{syn}}$, since the actual dataset is used in w_{real} even when it looks indistinguishable from the synthetic dataset by the hypothesis test.

When the null hypothesis is rejected with a confidence level $1-\alpha$, then $(w_{\mathsf{real}}, w_{\mathsf{ideal}}) \notin \mathcal{R}_{\varepsilon_{\alpha,n}}$. Since $\mathfrak{M}, w' \models \neg\varphi_{\mathsf{syn}}$ holds for any $w' \neq w_{\mathsf{ideal}}$, this rejection of the null hypothesis implies:

$$\mathfrak{M}, w_{\mathsf{real}} \models \mathsf{K}_{\varepsilon_{\alpha,n}} \neg\varphi_{\mathsf{syn}},$$

which is logically equivalent to $\mathfrak{M}, w_{\mathsf{real}} \models \neg\,\mathsf{P}_{\varepsilon_{\alpha,n}} \varphi_{\mathsf{syn}}$. This means that with the confidence level $1-\alpha$, we know we are not located in the world w_{ideal}, hence do not have a synthetic dataset.

On the other hand, when the null hypothesis is not rejected with a confidence level $1-\alpha$, then $(w_{\mathsf{real}}, w_{\mathsf{ideal}}) \in \mathcal{R}_{\varepsilon_{\alpha,n}}$. Thus we obtain:

$$\mathfrak{M}, w_{\mathsf{real}} \models \mathsf{P}_{\varepsilon_{\alpha,n}} \varphi_{\mathsf{syn}}. \tag{1}$$

This means that we cannot recognize whether we are located in the world w_{real} or w_{ideal}, i.e., we are not sure which database we have. To see this in details, let φ' be a formula representing that we have a third database (different from those in w_{real} and w_{ideal}). Suppose that another null hypothesis of satisfying φ' is not rejected with a confidence level $1-\alpha$. Then we have $\mathfrak{M}, w_{\mathsf{real}} \models \mathsf{P}_{\varepsilon_{\alpha,n}} \varphi'$. Since each world in $\mathcal{W}$ corresponds to a different database, we obtain $\mathfrak{M}, w_{\mathsf{real}} \models \mathsf{P}_{\varepsilon_{\alpha,n}} \neg\varphi_{\mathsf{syn}}$, which implies $\mathfrak{M}, w_{\mathsf{real}} \models \neg\,\mathsf{K}_{\varepsilon_{\alpha,n}} \varphi_{\mathsf{syn}}$. This represents that, when the null hypothesis is not rejected, we are not sure whether the null hypothesis is true or false.

6.3 Formalization of Statistical Secrecy

Now let us formalize the coin flipping in Example 1 in Sect. 3 by using StatEL as follows. Recall that $p(heads) = 0.5$ in w_0 and $p(heads) = 0.4$ in w_1. Let ψ be a static formula representing that the coin is a heads. Then $\mathfrak{M}, w_0 \models \mathbb{P}_{0.5}\,\psi$ and $\mathfrak{M}, w_1 \models \mathbb{P}_{0.4}\,\psi$. Assume that either $p(heads) = 0.5$ or $p(heads) = 0.4$ holds, i.e., $\mathfrak{M} \models \mathbb{P}_{0.5}\,\psi \vee \mathbb{P}_{0.4}\,\psi$.

When we have a sufficient number n of coin flips (e.g., $n = 500$), we can distinguish $p(heads) = 0.5$ from $p(heads) = 0.4$ (i.e., w_0 from w_1) by a hypothesis test. Hence we learn the probability $p(heads)$ with some confidence level $1-\alpha$, i.e., $\mathfrak{M}, w_0 \models \mathsf{K}_{\varepsilon_{\alpha,n}} \mathbb{P}_{0.5}\,\psi$ and $\mathfrak{M}, w_1 \models \mathsf{K}_{\varepsilon_{\alpha,n}} \mathbb{P}_{0.4}\,\psi$. Therefore we obtain:

$$\mathfrak{M} \models \big(\mathbb{P}_{0.5}\,\psi \rightarrow \mathsf{K}_{\varepsilon_{\alpha,n}} \mathbb{P}_{0.5}\,\psi\big) \wedge \big(\mathbb{P}_{0.4}\,\psi \rightarrow \mathsf{K}_{\varepsilon_{\alpha,n}} \mathbb{P}_{0.4}\,\psi\big).$$

Note that for a larger sample size $n' > n$, we have $\varepsilon_{\alpha,n'} = c_\alpha/n' < c_\alpha/n = \varepsilon_{\alpha,n}$, hence it follows from the axiom ($\geq$) in Proposition 3 that:

$$\mathfrak{M} \models \big(\mathbb{P}_{0.5}\,\psi \rightarrow \mathsf{K}_{\varepsilon_{\alpha,n'}} \mathbb{P}_{0.5}\,\psi\big) \wedge \big(\mathbb{P}_{0.4}\,\psi \rightarrow \mathsf{K}_{\varepsilon_{\alpha,n'}} \mathbb{P}_{0.4}\,\psi\big).$$

This means that if our knowledge derived from a smaller sample is statistically significant, then we derive the same conclusion from a larger sample.

On the other hand, when we have a very small number n'' of coin flips, we cannot distinguish w_0 from w_1. Then we are not sure about $p(heads)$ with a confidence level $1-\alpha$, i.e., $\mathfrak{M}, w_0 \models \mathsf{P}_{\varepsilon_{\alpha,n''}} \mathbb{P}_{0.5}\,\psi$ and $\mathfrak{M}, w_1 \models \mathsf{P}_{\varepsilon_{\alpha,n''}} \mathbb{P}_{0.4}\,\psi$. Hence:

$$\mathfrak{M} \models (\mathbb{P}_{0.5}\,\psi \vee \mathbb{P}_{0.4}\,\psi) \rightarrow (\mathsf{P}_{\varepsilon_{\alpha,n''}} \mathbb{P}_{0.5}\,\psi \wedge \mathsf{P}_{\varepsilon_{\alpha,n''}} \mathbb{P}_{0.4}\,\psi).$$

This expresses a secrecy of $p(heads)$. We generalize this to introduce the following definition of secrecy.

Definition 5 ((α, n)-statistical secrecy). Let Φ be a finite set of formulas, $\alpha \in [0,1]$ be a significance level, and n be a sample size. We say that Φ is (α, n)-*statistically secret* if we have:

$$\mathfrak{M} \models \bigvee_{\varphi \in \Phi} \varphi \rightarrow \bigwedge_{\varphi \in \Phi} \mathsf{P}_{\varepsilon_{\alpha,n}}\, \varphi.$$

In the above coin flipping example, $\{\mathbb{P}_{0.5}\,\psi, \mathbb{P}_{0.4}\,\psi\}$ is (α, n)-statistically secret for some significance level α and sample size n. Syntactically, (α, n)-statistical secrecy resembles the notion of *total anonymity* [21], whereas in our definition, the epistemic operator $\mathsf{P}_{\varepsilon_{\alpha,n}}$ deals with the statistical significance and φ is not limited to a formula representing an agent's action.

7 Modeling Statistical Data Privacy Using StatEL

In this section we formalize a notion of statistical data privacy by using StatEL.

7.1 Differential Privacy

Differential privacy [11,12] is a popular measure of data privacy guaranteeing that by observing a statistics about a database d, we cannot learn whether an individual user's record is included in d or not.

As a toy example, let us assume that the body weight of individuals is sensitive information, and we publish the average weight of all users recorded in a database d. Then we denote by d' the database obtained by adding to d a single record of a new user u's weight. If we also disclose the average weight of all users in d', then you learn u's weight from the difference between these two averages.

To mitigate such privacy leaks, many studies have proposed *obfuscation mechanisms*, i.e., randomized algorithms that add random noise to the statistics calculated from databases. In the above example, an obfuscation mechanism receives a database d and outputs a statistics of average weight to which some random noise is added. Then you cannot learn much information on u's weight from the perturbed statistics of average weight.

The privacy achieved by such obfuscation is often formalized as differential privacy. Intuitively, an ε-differential privacy mechanism makes every two "adjacent" (i.e., close) database d and d' indistinguishable with a degree of ε.

Definition 6 (Differential privacy). Let e be the base of natural logarithm, $\varepsilon \geq 0$, $\mathcal{D}$ be the set of all databases, and $\Psi \subseteq \mathcal{D} \times \mathcal{D}$ be an adjacency relation between two databases. A randomized algorithm $A : \mathcal{D} \to \mathbb{D}\mathcal{O}$ provides *ε-differential privacy* w.r.t. Ψ if for any $(d, d') \in \Psi$ and any $R \subseteq \mathcal{O}$,

$$\Pr[A(d) \in R] \leq e^{\varepsilon} \Pr[A(d') \in R]$$

where the probability is taken over the randomness in A.

For a smaller ε, the protection of differential privacy is stronger. It is known that differential privacy can be defined using the max-divergence D_∞ (Definition 2) as follows [12].

Proposition 5. *An obfuscation mechanism $A : \mathcal{D} \to \mathbb{D}\mathcal{O}$ provides ε-differential privacy w.r.t. $\Psi \subseteq \mathcal{D} \times \mathcal{D}$ iff for any $(d, d') \in \Psi$, $D_\infty(A(d) \parallel A(d')) \leq \varepsilon$ and $D_\infty(A(d') \parallel A(d)) \leq \varepsilon$.*

7.2 Formalization of Differential Privacy

Next we define a distributional Kripke model $\mathfrak{M} = (\mathcal{W}, \mathcal{R}_\varepsilon, (V_s)_{s \in \mathcal{S}})$ where there is a possible world corresponding to each database in $\mathcal{D}$. We assume that each world is a probability distribution of states in each of which an obfuscation mechanism A uses a different value of random seed for providing a probabilistically perturbed output. Let x (resp. y) be a measurement variable denoting the input (resp. output) of the obfuscation mechanism A. In each world w, $\sigma_w(x)$ is the database that A receives as input, and $\sigma_w(y)$ is the distribution of statistics that A outputs. Then the set of all databases is denoted by $\mathcal{D} = \{\sigma_w(x) \mid w \in \mathcal{W}\}$.

Now we define the accessibility relation $\mathcal{R}_\varepsilon$ in $\mathfrak{M}$ by using the max divergence D_∞ as follows[3]:

$$\mathcal{R}_\varepsilon \stackrel{\text{def}}{=} \{(w, w') \in \mathcal{W} \times \mathcal{W} \mid D_\infty(\sigma_w(y) \,\|\, \sigma_{w'}(y)) \leq \varepsilon,\ D_\infty(\sigma_{w'}(y) \,\|\, \sigma_w(y)) \leq \varepsilon\}.$$

Intuitively, $(w, w') \in \mathcal{R}_\varepsilon$ represents that, when we observe an output y of the obfuscation mechanism A, we do not know which of the two worlds w and w' we are located at. Hence we do not see which of the two databases $\sigma_w(x)$ and $\sigma_{w'}(x)$ was the input to A.

For each $d \in \mathcal{D}$, let φ_d be a formula representing that we have a database d. Then the ε-differential privacy of A w.r.t. an adjacency relation Ψ is expressed as:

$$\mathfrak{M} \models \bigwedge_{d \in \mathcal{D}} \Big(\varphi_d \to \bigwedge_{d' \in \Psi(d)} \mathsf{P}_\varepsilon\, \varphi_{d'} \Big).$$

Note that the privacy of user attributes defined as distribution privacy [27] can also be expressed using StatEL, since it is defined as the differential privacy w.r.t. a relation between the probability distributions that represent user attributes. We will elaborate on this in future work.

---

[3] Since the relation $\mathcal{R}_\varepsilon$ is symmetric, the symmetry axiom (**B**) also holds.

8 Related Work

In this section, we overview related work, including the integration of logical and statistical techniques, epistemic logic, and logical formalization of privacy.

Integration of Logical and Statistical Techniques. There have been various studies on integrating logical and statistical techniques in software engineering. Notable examples are *probabilistic programming* [18], which has sampling from distributions and conditioning by observations, and *statistical model checking* [31,36,40], which checks the satisfiability of logical formulas by simulations and statistical hypothesis tests. In research of privacy, a few papers (e.g., [5]) present hybrid methods combining symbolic and statistical analyses to quantify privacy leaks. In future work, our logic may be used to define specifications of these techniques and characterize their properties.

Non-determinism and Probability in Kripke Models. Although many epistemic models have been proposed [14,20,21], they often assume that each possible world is a single deterministic state. To formalize the behaviours of stochastic systems in their model, they assume that every world is assigned a probability (e.g., [28]), which means the non-determinism needs to be resolved in advance.

However, not only probability but also non-deterministic inputs are essential to reason about security and many applications in statistics. In the context of security, we usually do not have a prior knowledge of the probability distribution of adversarial inputs. Also in the statistical hypothesis testing (Sect. 6.1), we do not assume the prior probabilities of the null/alternative hypotheses. The notion of differential privacy (Definition 6) is also independent of the prior distribution on the databases. Therefore, unlike ours, the Kripke models in previous work cannot be used for the purpose of formalizing such statistical knowledge.

Kripke Model for Some Aspects of Statistics. The *random worlds model* [20] is an epistemic model that tries to formalize some aspects of statistics. In that model, they assume that each possible world has an identical probability at the initial time, although this causes problems as mentioned in Chap. 10 of [20]. Unlike our distributional model, their model employs neither distributions of states nor statistical significance. They assume only finite intervals of errors, and analyze only the ideal situation that corresponds to an infinite sample size. Therefore, the random worlds model cannot formalize statistical knowledge in our sense.

In research of philosophical logic, [2,32] formalize the idea that when a random value has various possible probability distributions, those distributions should be represented on different possible worlds. Unlike our work, however, they do not model statistical significance or explore accessibility relations.

Independently of our work, French et al. [15] propose a probability model for a dynamic epistemic logic where each world is associated with a (subjective) probability distribution *over the universe* and may have a different probability for a propositional variable to be true. This is different from our distributional Kripke model in that their model does not associate each world with a probability distribution of observable variables, hence deals with neither non-deterministic inputs, divergence-based accessibility relations, nor statistical significance.

Epistemic Logic for Privacy Properties. Epistemic logic has been used to formalize and reason about privacy properties, including anonymity [4,6,13,17,21, 25,35,37], role-interchangeability [34], receipt-freeness of electronic voting protocols [4,23], and its extension called coercion-resistance [30]. Unlike our formalization in Sect. 7, however, these do not regard possible worlds as probability distributions and cannot formalize privacy properties with a statistical significance.

Logical Approaches to Differential Privacy. There have been studies that formalize differential privacy using logics, such as Hoare logic [3] and HyperPCTL [1]. Compared to StatEL, these formalizations need to explicitly describe inequalities of probabilities without much abstraction, hence the formulas are more complicated. In addition, none of them formalizes the situation with finite sample sizes or statistical significance.

9 Conclusion

We introduced statistical epistemic logic (StatEL) to describe statistical knowledge, and showed its stochastic semantics based on the distributional Kripke model. By using StatEL we introduced (α, n)-statistical secrecy with a significance level α and a sample size n, and showed that StatEL is useful to formalize hypothesis testing and differential privacy in a simple way. As shown in [24], StatEL can also express certain properties of statistical machine learning.

In our ongoing work, we extend StatEL to deal with the security of cryptography based on computational complexity theory. As for future work, we will extend this logic with temporal modality and give its axiomatization. Our future work includes an extension of StatEL to formalize the quantitative notions of anonymity [9] and asymptotic anonymity [26]. We are also interested in clarifying the relationships between our distributional Kripke model and the main stream probabilistic epistemic logic assigning probabilities to worlds. Furthermore, we plan to develop statistical epistemic logic for process calculi in an analogous way to [6,8,10,22], and to investigate the relationships between statistical epistemic logic and bisimulation metrics analogously to [7].

Acknowledgments. When I was a postdoctoral researcher with Catuscia Palamidessi in 2014, I tried to work alone on this research, but could not manage. Through my collaboration with her on quantitative information flow for the last several years, I obtained missing pieces of techniques needed to develop my ideas into this paper. I am grateful to her for our research collaboration and her helpful advice until today.

I would like to thank the reviewers for their helpful and insightful comments. I am also grateful to Ken Mano, Gergei Bana, and Ryuta Arisaka for their useful comments on preliminary manuscripts.

References

1. Ábrahám, E., Bonakdarpour, B.: HyperPCTL: a temporal logic for probabilistic hyperproperties. In: McIver, A., Horvath, A. (eds.) QEST 2018. LNCS, vol. 11024, pp. 20–35. Springer, Cham (2018). https://doi.org/10.1007/978-3-319-99154-2_2
2. Bana, G.: Models of objective chance: an analysis through examples. In: Hofer-Szabó, G., Wroński, L. (eds.) Making it Formally Explicit. ESPS, vol. 6, pp. 43–60. Springer, Cham (2017). https://doi.org/10.1007/978-3-319-55486-0_3
3. Barthe, G., Gaboardi, M., Arias, E.J.G., Hsu, J., Kunz, C., Strub, P.: Proving differential privacy in Hoare logic. In: Proceedings of CSF, pp. 411–424 (2014)
4. Baskar, A., Ramanujam, R., Suresh, S.P.: Knowledge-based modelling of voting protocols. In: Proceedings of TARK, pp. 62–71 (2007)
5. Biondi, F., Kawamoto, Y., Legay, A., Traonouez, L.: Hybrid statisticalestimation of mutual information and its application to information flow. Formal Asp. Comput. **31**(2), 165–206 (2019). https://doi.org/10.1007/s00165-018-0469-z
6. Chadha, R., Delaune, S., Kremer, S.: Epistemic logic for the applied pi calculus. In: Lee, D., Lopes, A., Poetzsch-Heffter, A. (eds.) FMOODS/FORTE -2009. LNCS, vol. 5522, pp. 182–197. Springer, Heidelberg (2009). https://doi.org/10.1007/978-3-642-02138-1_12
7. Chatzikokolakis, K., Gebler, D., Palamidessi, C., Xu, L.: Generalized bisimulation metrics. In: Baldan, P., Gorla, D. (eds.) CONCUR 2014. LNCS, vol. 8704, pp. 32–46. Springer, Heidelberg (2014). https://doi.org/10.1007/978-3-662-44584-6_4
8. Chatzikokolakis, K., Knight, S., Palamidessi, C., Panangaden, P.: Epistemic strategies and games on concurrent processes. ACM Trans. Comput. Logic **13**(4), 28:1–28:35 (2012). https://doi.org/10.1145/2362355.2362356
9. Chatzikokolakis, K., Palamidessi, C., Panangaden, P.: Anonymity protocols as noisy channels. Inf. Comput. **206**(2–4), 378–401 (2008). https://doi.org/10.1016/j.ic.2007.07.003
10. Dechesne, F., Mousavi, M.R., Orzan, S.: Operational and epistemic approaches to protocol analysis: bridging the gap. In: Dershowitz, N., Voronkov, A. (eds.) LPAR 2007. LNCS (LNAI), vol. 4790, pp. 226–241. Springer, Heidelberg (2007). https://doi.org/10.1007/978-3-540-75560-9_18
11. Dwork, C.: Differential privacy. In: Bugliesi, M., Preneel, B., Sassone, V., Wegener, I. (eds.) ICALP 2006. LNCS, vol. 4052, pp. 1–12. Springer, Heidelberg (2006). https://doi.org/10.1007/11787006_1
12. Dwork, C., Roth, A., et al.: The algorithmic foundations of differential privacy. Found. Trends® Theor. Comput. Sci. **9**(3–4), 211–407 (2014)
13. van Eijck, J., Orzan, S.: Epistemic verification of anonymity. Electr. Notes Theor. Comput. Sci. **168**, 159–174 (2007). https://doi.org/10.1016/j.entcs.2006.08.026
14. Fagin, R., Halpern, J., Moses, Y., Vardi, M.: Reasoning About Knowledge. The MIT Press, Cambridge (1995)
15. French, T., Gozzard, A., Reynolds, M.: Dynamic aleatoric reasoning in games of bluffing and chance. In: Proceedings of AAMAS, pp. 1964–1966 (2019)
16. Pearson, K.: On the criterion that a given system of deviations from the probable in the case of a correlated system of variables is such that it can be reasonably supposed to have arisen from random sampling. Lond. Edinb. Dublin Philos. Mag. J. Sci. **50**(302), 157–175 (1900)
17. Garcia, F.D., Hasuo, I., Pieters, W., van Rossum, P.: Provable anonymity. In: Proceedings of FMSE, pp. 63–72 (2005). https://doi.org/10.1145/1103576.1103585

18. Gordon, A.D., Henzinger, T.A., Nori, A.V., Rajamani, S.K.: Probabilistic programming. In: Proceedings of FOSE, pp. 167–181 (2014). https://doi.org/10.1145/2593882.2593900
19. Halpern, J.Y.: An analysis of first-order logics of probability. Artif. Intell. **46**(3), 311–350 (1990). https://doi.org/10.1016/0004-3702(90)90019-V
20. Halpern, J.Y.: Reasoning About Uncertainty. The MIT Press, Cambrdige (2003)
21. Halpern, J.Y., O'Neill, K.R.: Anonymity and information hiding in multiagent systems. In: Proceedings of CSFW, pp. 75–88 (2003)
22. Hughes, D., Shmatikov, V.: Information hiding, anonymity and privacy: a modular approach. J. Comput. Secur. **12**(1), 3–36 (2004)
23. Jonker, H.L., Pieters, W.: Receipt-freeness as a special case of anonymity in epistemic logic. In: Proceedings of Workshop On Trustworthy Elections (WOTE 2006), June 2006
24. Kawamoto, Y.: Towards logical specification of statistical machine learning. In: Proceedings of SEFM (2019, to appear)
25. Kawamoto, Y., Mano, K., Sakurada, H., Hagiya, M.: Partial knowledge of functions and verification of anonymity. Trans. Jpn. Soc. Ind. Appl. Math. **17**(4), 559–576 (2007). https://doi.org/10.11540/jsiamt.17.4_559. (in Japanese)
26. Kawamoto, Y., Murakami, T.: On the anonymization of differentially private location obfuscation. In: Proceedings of ISITA, pp. 159–163 (2018)
27. Kawamoto, Y., Murakami, T.: Local obfuscation mechanisms for hiding probability distributions. In: Proceedings of ESORICS (2019, to appear)
28. Kooi, B.P.: Probabilistic dynamic epistemic logic. J. Log. Lang. Inf. **12**(4), 381–408 (2003). https://doi.org/10.1023/A:1025050800836
29. Kripke, S.A.: Semantical analysis of modal logic i normal modal propositional calculi. Math. Log. Q. **9**(5–6), 67–96 (1963)
30. Küsters, R., Truderung, T.: An epistemic approach to coercion-resistance for electronic voting protocols. In: Proceedings of S&P, pp. 251–266 (2009). https://doi.org/10.1109/SP.2009.13
31. Legay, A., Delahaye, B., Bensalem, S.: Statistical model checking: an overview. In: Barringer, H., et al. (eds.) RV 2010. LNCS, vol. 6418, pp. 122–135. Springer, Heidelberg (2010). https://doi.org/10.1007/978-3-642-16612-9_11
32. Lewis, D.: A subjectivist's guide to objective chance. In: Harper, W.L., Stalnaker, R., Pearce G. (eds)Studies in Inductive Logic and Probability, vol. II, pp. 263–293. University of California Press, Berkeley (1980)
33. Lin, J.: Divergence measures based on the Shannon entropy. IEEE Trans. Inf. Theory **37**(1), 145–151 (1991). https://doi.org/10.1109/18.61115
34. Mano, K., Kawabe, Y., Sakurada, H., Tsukada, Y.: Role interchange for anonymity and privacy of voting. J. Log. Comput. **20**(6), 1251–1288 (2010). https://doi.org/10.1093/logcom/exq013
35. van der Meyden, R., Su, K.: Symbolic model checking the knowledge of the dining cryptographers. In: Proceedings of CSFW, p. 280 (2004). https://doi.org/10.1109/CSFW.2004.19
36. Sen, K., Viswanathan, M., Agha, G.: Statistical model checking of black-box probabilistic systems. In: Alur, R., Peled, D.A. (eds.) CAV 2004. LNCS, vol. 3114, pp. 202–215. Springer, Heidelberg (2004). https://doi.org/10.1007/978-3-540-27813-9_16
37. Syverson, P.F., Stubblebine, S.G.: Group principals and the formalization of anonymity. In: Wing, J.M., Woodcock, J., Davies, J. (eds.) FM 1999. LNCS, vol. 1708, pp. 814–833. Springer, Heidelberg (1999). https://doi.org/10.1007/3-540-48119-2_45

38. Vaserstein, L.: Markovian processes on countable space product describing large systems of automata. Probl. Peredachi Inf. **5**(3), 64–72 (1969)
39. von Wright, G.H.: An Essay in Modal Logic. North-Holland Pub. Co., Amsterdam (1951)
40. Younes, H.L.: Verification and planning for stochastic processes with asynchronous events. Ph.D. thesis, Carnegie Mellon University (2005)

Approximate Model Counting, Sparse XOR Constraints and Minimum Distance

Michele Boreale[1(✉)] and Daniele Gorla[2]

[1] Dip. Statistica, Informatica, Applicazioni (DiSIA), Università di Firenze, Florence, Italy
michele.boreale@unifi.it

[2] Department of Computer Science, Università di Roma "La Sapienza", Rome, Italy

Abstract. The problem of counting the number of models of a given Boolean formula has numerous applications, including computing the leakage of deterministic programs in Quantitative Information Flow. Model counting is a hard, #P-complete problem. For this reason, many approximate counters have been developed in the last decade, offering formal guarantees of confidence and accuracy. A popular approach is based on the idea of using random XOR constraints to, roughly, successively halving the solution set until no model is left: this is checked by invocations to a SAT solver. The effectiveness of this procedure hinges on the ability of the SAT solver to deal with XOR constraints, which in turn crucially depends on the length of such constraints. We study to what extent one can employ sparse, hence short, constraints, keeping guarantees of correctness. We show that the resulting bounds are closely related to the geometry of the set of models, in particular to the minimum Hamming distance between models. We evaluate our theoretical results on a few concrete formulae. Based on our findings, we finally discuss possible directions for improvements of the current state of the art in approximate model counting.

Keywords: Model counting · Approximate counting · XOR sampling

1 Introduction

#SAT (aka *model-counting*) is the problem of counting the number of satisfying assignments of a given Boolean formula and is a #P-complete problem. Indeed, every NP Turing machine can be encoded as a formula whose satisfying assignments correspond to the accepting paths of the machine [24]. Thus, model-counting is harder than satisfiability: #SAT is indeed intractable in cases for which SAT is tractable (e.g., sets of Horn clauses or sets of 2-literal clauses) [25]. Still, there are cases in which model-counting is tractable (e.g., OBDDs and d-DNNFs). For a very good overview of the problem and of some approaches to it see [14].

Our interest in model counting originates from its applications in the field of Quantitative Information Flow (QIF) [8,20]. Indeed, a basic result in QIF is

M. S. Alvim et al. (Eds.): Palamidessi Festschrift, LNCS 11760, pp. 363–378, 2019.
https://doi.org/10.1007/978-3-030-31175-9_21

that the maximum min-entropy leakage of a deterministic program is $\log_2 k$, with k the number of distinct outputs the program can return [20], varying the input. If the program is modeled as a Boolean formula, then computing its leakage reduces to #SAT, specifically to computing the number of models of the formula obtained by existentially projecting out the non-output variables; see [3,17].

Over the years, several *exact* counting algorithms have been put forward and implemented, such as, among others, [12,17,19,23], with applications to QIF [16]. The problem with exact counters is that, although performing reasonably well when certain parameters of the formula – size, number of variables, number of clauses – are relatively small, they rapidly go out of memory as these parameters grow.

For this reason, *approximate* counters have more and more been considered. Indeed, in many applications, the exact count of models is not required: it may suffice to provide an estimate, as long as the method is quick and it is equipped with a formal guarantee of correctness. This is typically the case in QIF, where knowing the exact count within a factor of η is sufficient to estimate leakage within $\log_2 \eta$ bits. For probabilistic counters, correctness is usually expressed in terms of two parameters: *accuracy* – the desired maximum difference between the reported and the true count; and *confidence* – the probability that the reported result is actually within the specified accuracy.

We set ourselves in the line of research pioneered by [25] and followed, e.g., by [1,6,13,17]. The basic idea of a probabilistic model counting algorithm is the following (Sect. 2): given a formula ϕ in the Boolean variables $y_1, \ldots, y_m$, one chooses at random $\langle a_0, \ldots, a_m \rangle \in \{0,1\}^{m+1}$. The resulting *XOR constraint* $a_0 = a_1 y_1 \oplus \cdots \oplus a_m y_m$ splits evenly the set of models of ϕ into two parts: those satisfying the constraint and those not satisfying it. If one independently generates s such constraints $c_1, \ldots, c_s$, the formula $\phi' \stackrel{\triangle}{=} \phi \wedge c_1 \wedge \cdots \wedge c_s$ has an expected $\frac{N}{2^s}$ models, where N is the number of models of ϕ (i.e., the number we aim at estimating). If ϕ' is still satisfiable, then with high probability $N \geq 2^s$, otherwise $N < 2^s$. By repeating this process, one can arrive at a good estimate of N. This procedure can be implemented by relying on any SAT-solver capable of dealing with XOR constraints, e.g CryptoMiniSat [22]; or even converting the XOR constraints into CNF before feeding ϕ' to the SAT solver. In any case, the branching factor associated with searching for models of ϕ' quickly explodes as the *length* (number of variables) of the XOR constraints grows. The random generation outlined above will lead to an expected length of $\frac{m}{2}$ for each constraint, making the procedure not easily scalable as m grows.

In the present paper, we study under what conditions one can employ sparse, hence shorter, constraints, keeping the same guarantees of correctness. We generalize the results of [9] to arrive at an improved understanding of how sparsity is related to minimum distance between models, and how this affects the counting procedure. Based on these results, we also suggest a possible direction for a new counting methodology based on the use of Low Density Parity Check

(LDPC) codes [11,18]; however, we leave a through experimentation with this new methodology for future work.

The main point is to generate the coefficients $a_1, \ldots, a_m$ according to a probability value $\lambda \in \left(0, \frac{1}{2}\right]$, rather than uniformly. This way, the constraints will have an average length of $\lambda m \leq \frac{m}{2}$ each. Basically, the correctness guarantees of the algorithm depend, via the Chebyshev inequality, on keeping the variance of the number of models of ϕ', the formula obtained by joining the constraints, below a certain threshold. A value of the density λ that achieves this is said to be *feasible* for the formula. In our main result (Sect. 3), we provide a bound on the variance that also depends on the minimum Hamming distance d between the formula's models: a larger d yields a smaller variance, hence smaller feasible λ's. Our bound essentially coincides with that of [9] for $d = 1$. Therefore, in principle, a lower bound on the minimum distance can be used to obtain tighter bounds on λ, making the XOR constraints shorter and the counting procedure pragmatically more efficient.

We will show this phenomenon at work (Sect. 4) on some formulae where the value of d is known by construction, comparing our results with the state of the art model counter ApproxMC3 [21]. These considerations also suggest that, if no information on d is available, one can encode the formula's models using an error correcting code with a known minimum distance. We will briefly discuss the use of LDPC codes to this purpose, although at the moment we have no experimental results available in this respect. A comparison with recent related work concludes the paper (Sect. 5). For reasons of space, all proofs have been sketched and are fully available in [4].

2 A General Counting Algorithm

In what follows, we let $\phi(y)$, or just ϕ, denote a generic boolean formula with boolean variables $y = (y_1, ..., y_m)$ and $m \geq 1$.

2.1 A General Scheme

According to a well-known general scheme [13], the building block of a statistical counting procedure is a probabilistic decision algorithm: with high probability, this algorithm correctly decides whether the cardinality of the set is, or is not, below a given threshold 2^s, within some tolerance factors, given by the slack parameters α and β below.

Definition 1 (#SAT decision algorithm). *Let* $0 \leq \delta < frac12$ *(error probability),* $\alpha > 1$ *and* $\beta > 1$ *(two slack parameters) be three reals. An* (α, β, δ)-decision algorithm *(for #SAT) is a probabilistic algorithm* $A(\cdot, \cdot)$*, taking a pair of an integer* s *and a boolean formula* ϕ*, and returning either* 1 *(meaning '*$\#\phi \geq 2^{s-\alpha}$*') or* 0 *(meaning '*$\#\phi \leq 2^{s+\beta}$*') and such that for each integer* $s \geq 0$ *and formula* ϕ*:*

1. $\#\phi > 2^{s+\beta}$ *implies* $\Pr(A(s,\phi) = 0) \leq \delta$*;*
2. $\#\phi < 2^{s-\alpha}$ *implies* $\Pr(A(s,\phi) = 1) \leq \delta$*.*

The use of two different slack parameters in the definition above is justified by the need of stating formal guarantees about the outcome of the algorithm, while keeping the precision of the algorithm as high as possible.

As usual, we can boost the confidence in the reported answer, and get an arbitrarily small error probability, by running $A(s,\phi)$ several times independently. In particular, consider the algorithm $RA_t(s,\phi)$ obtained by running $A(s,\phi)$ an odd $t \geq 1$ number times independently, and then reporting the majority answer. Call Err the event that RA_t reports a wrong answer; then,

$$\begin{aligned}\Pr(Err) &= \Pr(\text{at least } \left\lceil \frac{t}{2} \right\rceil \text{ runs of } A(s,\phi) \text{ report the wrong answer}) \\ &= \sum_{k=\lceil \frac{t}{2} \rceil}^{t} \Pr(\text{exactly } k \text{ runs of } A(s,\phi) \text{ report the wrong answer}) \\ &= \sum_{k=\lceil \frac{t}{2} \rceil}^{t} \binom{t}{k} p^k (1-p)^{t-k} \qquad (1)\end{aligned}$$

where

$$\begin{aligned}p &\triangleq \Pr(A(s,\phi) \text{ reports the wrong answer}) \\ &= \begin{cases} \Pr(A(s,\phi) = 0) & \text{if } \#\phi > 2^{s+\beta} \\ \Pr(A(s,\phi) = 1) & \text{if } \#\phi < 2^{s-\alpha} \end{cases} \\ &\leq \delta \qquad (2)\end{aligned}$$

Now, replacing (2) in (1), we obtain

$$\Pr(Err) \leq \sum_{k=\lceil \frac{t}{2} \rceil}^{t} \binom{t}{k} \delta^k (1-\delta)^{t-k} \qquad (3)$$

Let us call $\Delta(t,\delta)$ the right hand side of (3); then, RA_t is an $(\alpha, \beta, \Delta(t,\delta))$-decision algorithm whenever A is an (α, β, δ)-decision algorithm.

We now show that any (α, β, δ)-decision algorithm A for #SAT can be used as a building block for a counting algorithm, $C_A(\phi)$, that determines an interval $[\ell, u]$ such that $\lfloor 2^\ell \rfloor \leq \#\phi \leq \lceil 2^u \rceil$ with high probability. Informally, starting with an initial interval $[-1, m]$, the algorithm C_A performs a binary search, using A to decide which half of the current interval $\log_2(\#\phi)$ lies in. The search stops when the current interval cannot be further narrowed, taking into account the slack parameters α and β, or when a certain predefined number of iterations is reached. Formally, let $I_0 \triangleq [-1, m]$. Assume $k > 0$ and $I_k = [l_k, u_k]$, then:

(a) if $u_k - l_k \leq 2\max(\alpha, \beta) + 1$ or $k = \lceil \log_2(m) \rceil$, then return I_k;

(b) otherwise, let $s = round(\frac{u_k+l_k}{2})$; if $A(s,\phi) = 0$ then $I_{k+1} \triangleq [l_k, s+\beta]$ otherwise $I_{k+1} \triangleq [s-\alpha, u_k]$.

Theorem 1. *Let A be a (α,β,δ)-decision algorithm. Then:*

1. *$C_A(\phi)$ terminates in $k \leq \lceil \log_2 m \rceil$ iterations returning an interval $I_k = [l,u]$ such that $u - l \leq 2\max(\alpha,\beta)+2$;*
2. *The probability that $\#\phi \notin [\lfloor 2^l \rfloor, \lceil 2^u \rceil]$ is at most $\lceil \log_2 m \rceil \delta$.*

Proof (Sketch). If the algorithm terminates because $u_k - l_k \leq 2\max(\alpha,\beta)+1$, the first claim is trivial. Otherwise, by construction of the algorithm, we have that $|I_0| = m+1$. Furthermore, by passing from I_{k-1} to I_k, we have that $s_k = round\left(\frac{u_{k-1}+l_{k-1}}{2}\right)$ and

$$|I_k| \leq \begin{cases} \frac{u_{k-1}-l_{k-1}}{2} + \alpha + \frac{1}{2} & \text{if } A(s_k, m) = 1 \text{ and } s_k = \left\lfloor \frac{u_{k-1}-l_{k-1}}{2} \right\rfloor \\ \frac{u_{k-1}-l_{k-1}}{2} + \beta & \text{if } A(s_k, m) = 0 \text{ and } s_k = \left\lfloor \frac{u_{k-1}-l_{k-1}}{2} \right\rfloor \\ \frac{u_{k-1}-l_{k-1}}{2} + \alpha & \text{if } A(s_k, m) = 1 \text{ and } s_k = \left\lceil \frac{u_{k-1}-l_{k-1}}{2} \right\rceil \\ \frac{u_{k-1}-l_{k-1}}{2} + \beta + \frac{1}{2} & \text{if } A(s_k, m) = 0 \text{ and } s_k = \left\lceil \frac{u_{k-1}-l_{k-1}}{2} \right\rceil \end{cases}$$

Thus, by letting $M = \max(\alpha,\beta) + \frac{1}{2}$, we have that $|I_k| \leq \frac{|I_{k-1}|}{2} + M$; this suffices to obtain $|I_{\lceil \log_2 m \rceil}| \leq 2\max(\alpha,\beta)+2$.

The error probability is the probability that either $\#\phi < \lfloor 2^l \rfloor$ or $\#\phi > 2^u$; this is the probability that one of the $\lceil \log_2 m \rceil$ calls to A has returned a wrong answer. □

In all the experiments we have run, the algorithm has always returned an interval of width at most $2\max(\alpha,\beta)+1$, sometimes in less than $\lceil \log_2(m) \rceil$ iterations: this consideration pragmatically justifies the exit condition we used in the algorithm. We leave for future work a more elaborated analysis of the algorithm to formally establish the $2\max(\alpha,\beta)+1$ bound.

2.2 XOR-based Decision Algorithms

Recall that a XOR constraint c on the variables $y_1, ..., y_m$ is an equality of the form

$$a_0 = a_1 y_1 \oplus \cdots \oplus a_m y_m$$

where $a_i \in \mathbb{F}_2$ for $i = 0, ..., m$ (here $\mathbb{F}_2 = \{0,1\}$ is the two elements field.) Hence c can be identified with a $(m+1)$-tuple in $\mathbb{F}_2^{m+1}$. Assume that a probability distribution on $\mathbb{F}_2^{m+1}$ is fixed. A simple proposal for a decision algorithm $A(s,\phi)$ is as follows:

1. generate s XOR constraints $c_1, \ldots, c_s$ independently, according to the fixed probability distribution;

2. if $\phi \wedge c_1 \wedge \cdots \wedge c_s$ is unsatisfiable then return 0, else return 1.

Indeed [13], every XOR constraint splits the set of boolean assignments in two parts, according to whether the assignment satisfies the constraint or not. Thus, if ϕ has less than 2^s models (and so less than $2^{s+\beta}$), the formula $\phi \wedge c_1 \wedge \cdots \wedge c_s$ is likely to be unsatisfiable.

In step 2 above, any off-the-shelf SAT solver can be employed: one appealing possibility is using CryptoMiniSat [22], which offers support for specifying XOR constraints (see e.g. [3,6,17]). Similarly to [13], it can be proved that this algorithm yields indeed an (α, β, δ)-decision algorithm, for a suitable $\delta < \frac{1}{2}$, if the constraints c_i at step 1 are chosen uniformly at random. This however will generate 'long' constraints, with an average of $\frac{m}{2}$ variables each, which a SAT solver will not be able to manage as m grows.

3 Counting with Sparse XORs

We want to explore alternative ways of generating constraints, which will make it possible to work with short ('sparse') XOR constraints, while keeping the same guarantees of correctness. In what follows, we assume a probability distribution over the constraints, where each coefficient a_i, for $i > 0$, is chosen independently with probability λ, while a_0 is chosen uniformly (and independently from all the other a_i's). In other words, we assume that the probability distribution over $\mathbb{F}_2^{m+1}$ is of the following form, for $\lambda \in \left(0, \frac{1}{2}\right]$:

$$\begin{aligned} \Pr(a_0, a_1, ..., a_m) &\triangleq p(a_0)p'(a_1) \cdots p'(a_m) \\ \text{where:} \quad & p(1) = p(0) = \frac{1}{2} \qquad p'(1) = \lambda \qquad p'(0) = 1 - \lambda \end{aligned} \tag{4}$$

The expected number of variables appearing in a constraint chosen according to this distribution will therefore be $m\lambda$. Let us still call A the algorithm presented in Sect. 2.2, with this strategy in choosing the constraints. We want to establish conditions on λ under which A can be proved to be a decision algorithm.

Throughout this section, we fix a boolean formula $\phi(y_1, ..., y_m)$ and $s \geq 1$ XOR constraints. Let

$$\chi_s \triangleq \phi \wedge c_1 \wedge \cdots \wedge c_s$$

where the c_i are chosen independently according to (4). For any assignment (model) σ from variables $y_1, ..., y_m$ to $\mathbb{F}_2$, let us denote by Y_σ the Bernoulli r.v. which is 1 iff σ satisfies χ_s.

We now list the steps needed for proving that A is a decision algorithm. This latter result is obtained by Proposition 1(2) (that derives from Lemma 1(1)) and by combining Proposition 1(1), Lemma 2(3) (that derives from Lemma 1(2)) and Lemma 3 in Theorem 2 later on. In what follows, we shall let $\rho \triangleq 1 - 2\lambda$.

Lemma 1.

1. $\Pr(Y_\sigma = 1) = E[Y_\sigma] = 2^{-s}$.
2. *Let σ, σ' be any two assignments and d be their Hamming distance in $\mathbb{F}_2^m$ (i.e., the size of their symmetric difference seen as subsets of $\{1, ..., m\}$). Then* $\Pr(Y_\sigma = 1, Y_{\sigma'} = 1) = E[Y_\sigma \cdot Y_{\sigma'}] = \left(\frac{1+\rho^d}{4}\right)^s$ *(where we let $\rho^d \triangleq 1$ whenever $\rho = d = 0$.)*

Proof (Sketch). The first claim holds by construction. For the second claim, the crucial thing to prove is that, by fixing just one constraint c_1 (so, $s = 1$), we have that $\Pr(Y_\sigma = 1, Y_{\sigma'} = 1) = \frac{1+\rho^d}{4}$. To this aim, call A and B the sets of variables that are assigned value 1 in σ and σ', respectively; then, we let $U = A \setminus B$, $V = B \setminus A$ and $I = A \cap B$. Let C be the set of variables appearing in the constraint c_1; then, U_e and U_o abbreviate $\Pr(|C \cap U|$ is even$)$ and $\Pr(|C \cap U|$ is odd$)$, respectively; similarly for I_e, I_o and V_e, V_o. Then,

$$\Pr(Y_\sigma = 1, Y_{\sigma'} = 1 | a_0 = 0) = U_e I_e V_e + U_o I_o V_o$$
$$\Pr(Y_\sigma = 1, Y_{\sigma'} = 1 | a_0 = 1) = U_o V_o + U_e V_e - U_o I_o V_o - U_e I_e V_e$$

By elementary probability theory, $\Pr(Y_\sigma = 1, Y_{\sigma'} = 1) = \frac{1}{2}(1 - V_e - U_e + 2U_e V_e)$; the result is obtained by noting that $U_e = \frac{1}{2}(1 + \rho^{|U|})$, $V_e = \frac{1}{2}(1 + \rho^{|V|})$ and $d = |U \cup V| = |U| + |V|$. □

Now let T_s be the random variable that counts the number of models of χ_s, when the constraints $c_1, ..., c_s$ are chosen independently according to distribution (4):

$$T_s \triangleq \#\chi_s \,.$$

The event that χ_s is unsatisfiable can be expressed as $T_s = 0$. A first step toward establishing conditions under which A yields a decision algorithm is the following result. It makes it clear that a possible strategy is to keep under control the variance of T_s, which depends in turn on λ. Let us denote by μ_s the expectation of T_s and by $\mathrm{var}(T_s)$ its variance. Note that $\mathrm{var}(T_s) > 0$ if $\#\phi > 0$.

Proposition 1.

1. $\#\phi > 2^{s+\beta}$ *implies* $\Pr(A(s, \phi) = 0) \leq \frac{1}{1+\frac{\mu_s^2}{\mathrm{var}(T_s)}}$;
2. $\#\phi < 2^{s-\alpha}$ *implies* $\Pr(A(s, \phi) = 1) < 2^{-\alpha}$.

Proof (Sketch). The first claim relies on a version of the Cantelli-Chebyshev inequality for integer nonnegative random variables (a.k.a. Alon-Spencer's inequality); the second claim relies on Lemma 1(1) and Markov's inequality. □

By the previous proposition, assuming $\alpha > 1$, we obtain a decision algorithm (Definition 1) provided that $\mathrm{var}(T_s) < \mu_s^2$. This will depend on the value of λ that is chosen, which leads to the following definition.

Definition 2 (feasibility). *Let ϕ, s and β be given. A value $\lambda \in \left(0, \frac{1}{2}\right]$ is said to be (ϕ, s, β)-*feasible *if $\#\phi > 2^{s+\beta}$ implies $\mathrm{var}(T_s) < \mu_s^2$, where the constraints in χ_s are chosen according to (4).*

Our goal is now to give a method to minimize λ while preserving feasibility. Recall that $T_s \triangleq \#\chi_s$. Denote by $\sigma_1, ..., \sigma_N$ the distinct models of ϕ (hence, $T_s \leq N$). Note that $T_s = \sum_{i=1}^{N} Y_{\sigma_i}$. Given any two models σ_i and σ_j, $1 \leq i, j \leq N$, let d_{ij} denote their Hamming distance. The following lemma gives exact formulae for the expected value and variance of T_s.

Lemma 2. *Let $\rho = 1 - 2\lambda$.*

1. $\mu_s = E[T_s] = N2^{-s}$;
2. $\mathrm{var}(T_s) = \mu_s + 4^{-s} \sum_{i=1}^{N} \sum_{j \neq i} (1 + \rho^{d_{ij}})^s - \mu_s^2$;
3. *If $N \neq 0$, then* $\frac{\mathrm{var}(T_s)}{\mu^2} = \mu_s^{-1} + N^{-2} \sum_{i=1}^{N} \sum_{j \neq i} (1 + \rho^{d_{ij}})^s - 1$

Proof (Sketch). The first two items are a direct consequence of Lemma 1 and linearity of expectation; the third item derives from the previous ones. □

Looking at the third item above, we clearly see that the upper bound on the error probability we are after depends much on 'how sparse' the set of ϕ's models is in the Hamming space $\mathbb{F}_2^m$: the sparser, the greater the distance, the lower the value of the double summation, the better. Let us denote by S the double summation in the third item of the above lemma:

$$S \triangleq \sum_{i=1}^{N} \sum_{j \neq i} (1 + \rho^{d_{ij}})^s$$

In what follows, we will give an upper bound on S which is easy to compute and depends on the minimum Hamming distance d among any two models of ϕ. We need some notation about models of a formula. Below, we let $j = d, ..., m$.

$$l_j \triangleq \begin{cases} j - \lceil \frac{d}{2} \rceil + 1 & \text{if } j \leq \frac{m}{2} \\ \max\{0, m - j - \lceil \frac{d}{2} \rceil + 1\} & \text{if } j > \frac{m}{2} \end{cases}$$

$$w^* \triangleq \min \left\{ w : d \leq w \leq m \text{ and } \sum_{j=d}^{w} \binom{m}{l_j} \geq N - 1 \right\} \tag{5}$$

$$N^* \triangleq \sum_{j=d}^{w^*-1} \binom{m}{l_j} \tag{6}$$

where we stipulate that $\min \emptyset = 0$. Note that the definitions of w^* and N^* depend solely on N, m and d.

With the above definitions and results, we have the following upper bound on S.

Lemma 3. *Let the minimal distance between any two models of ϕ be at least d. Then*

$$S \leq N\left(\sum_{j=d}^{w^*-1}\binom{m}{l_j}(1+\rho^j)^s + (N-1-N^*)(1+\rho^{w^*})^s\right)$$

Proof (Sketch). Fix one of the models of ϕ (say σ_i), and consider the subsummation originated by it, $S_i \triangleq \sum_{j\neq i}\left(\frac{1+\rho^{d_{ij}}}{4}\right)^s$. Let us group the remaining $N-1$ models into disjoint families, $\mathcal{F}_d, \mathcal{F}_{d+1}, \ldots$, of models that are at distance $d, d+1, \ldots$, respectively, from σ_i. Note that each of the $N-1$ models gives rise to exactly one term in the summation S_i. Hence, $S_i = \sum_{j=d}^{m}|\mathcal{F}_j|\left(\frac{1+\rho^j}{4}\right)^s$. By the Ray-Chaudhuri-Wilson Lemma [2, Th.4.2], $|\mathcal{F}_j| \leq \binom{m}{l_j}$. Hence, upper-bounding S_i consists, e.g., in choosing a tuple of integers $x_d, \ldots, x_m$ such that $\sum_{j=d}^{m} x_j\left(\frac{1+\rho^j}{4}\right)^s \geq \sum_{j=d}^{m}|\mathcal{F}_j|\left(\frac{1+\rho^j}{4}\right)^s$, under the constraints $0 \leq x_j \leq \binom{m}{l_j}$, for $j = d, \ldots, m$, and $\sum_{j=d}^{m} x_j = N-1$. An optimal solution is

$$x_j = \begin{cases} \binom{m}{l_j} & \text{for } j = d, \ldots, w^*-1 \\ N-1-N^* & \text{for } j = w^* \\ 0 & \text{for } j > w^* \end{cases}$$

The thesis is obtained by summing over all models σ_i. □

Definition 3. *Given $s \geq 1$, $\beta > 0$, $d \geq 1$ and $\lambda \in \left(0, \frac{1}{2}\right]$, let us define*

$$B(s,m,\beta,d,\lambda) \triangleq 2^{-\beta} + 2^{-s-\beta}\left(\sum_{j=d}^{w^*-1}\binom{m}{l_j}(1+\rho^j)^s + (N-1-N^*)(1+\rho^{w^*})^s\right) - 1,$$

where $\rho \triangleq 1-2\lambda$ and $N = \lceil 2^{s+\beta}\rceil$ also in the definition of w^* and N^*.

Using the facts collected so far, the following theorem follows, giving an upper bound on $\frac{\mathrm{var}(T_s)}{\mu_s^2}$.

Theorem 2 (upper bound). *Let the minimal distance between models of ϕ be at least d and $\#\phi > 2^{s+\beta}$. Then, $\frac{\mathrm{var}(T_s)}{\mu_s^2} \leq B(s,m,\beta,d,\lambda)$.*

Proof. First note that we can assume without loss of generality that $\#\phi = N = \lceil 2^{s+\beta}\rceil$. If this was not the case, we can consider in what follows any formula ϕ' whose models are models of ϕ but are exactly $\lceil 2^{s+\beta}\rceil$ (ϕ' can be obtained by adding some conjuncts to ϕ that exclude $\#\phi - \lceil 2^{s+\beta}\rceil$ models). Then, $\Pr(A(s,\phi)=0) \leq \Pr(A(s,\phi')=0)$ and this would suffice, for the purpose of upper-bounding $\Pr(A(s,\phi)=0)$. The result follows from Proposition 1(1), Lemma 2(3), Lemma 3 and by the fact that $N = \lceil 2^{s+\beta}\rceil \geq 2^{s+\beta}$. □

The following notation will be useful in the rest of the paper. For $0 \leq \gamma \leq 1$, define

$$\lambda^*_\gamma(s,m,\beta,d) \triangleq \inf\left\{\lambda \in \left(0, \frac{1}{2}\right] : B(s,m,\beta,d,\lambda) \leq \gamma\right\} \tag{7}$$

where we stipulate $\inf\emptyset = +\infty$.

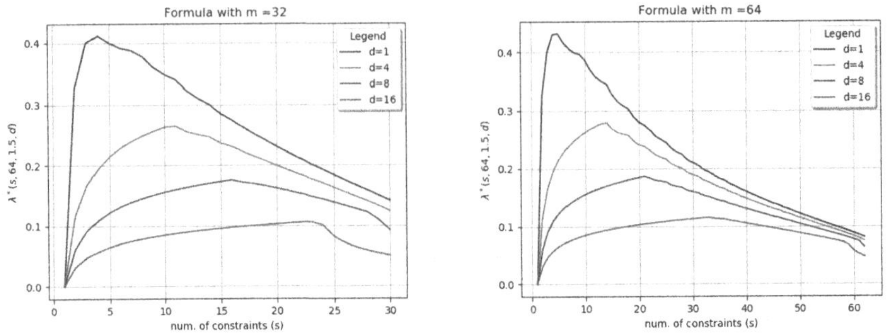

Fig. 1. Plots of λ_1^* as a function of s, for $m = 32$ and $m = 64$, $\beta = 1.5$, and different values of d. For any value of s and d, any value of λ above the curve is feasible.

Corollary 1 (Feasibility). *Assume the minimal distance between any two models of ϕ is at least d. Then every $\lambda \in \left(\lambda_1^*(s, m, \beta, d), \frac{1}{2}\right]$ is (ϕ, s, β)-feasible.*

4 Evaluation

4.1 Theoretical Bounds on Expected Constraint Length

To assess the improvements that our theory introduces on XOR-based approximate counting, we start by considering the impact of minimum Hamming distance on the expected length of the XOR constraints. First, in Fig. 1 we plot λ_1^* as a function of s, for fixed values of $m = 32$ and 64, $\beta = 1.5$, and four different values of d. Note that the difference between different values of d tends to vanish as s gets large – i.e. close to m.

Next, we compare our theoretical bounds with those in [9], where a goal similar to ours is pursued. Interestingly, their bounds coincide with ours when setting $d = 1$ – no assumption on the minimum Hamming distance – showing that our approach generalizes theirs. We report a numerical comparison in Table 1, where several hypothetical values of m (no. of variables) and s (no. of constraints) are considered. Following a similar evaluation conducted in [9, Tab. 1], here we fix the error probability to $\delta = \frac{4}{9}$ and the upper slack parameter to $\beta = 2$, and report the values of $\lambda \times m$ for the minimal value of λ that would guarantee a confidence of at least $1 - \delta$ in case an upper bound is found, computed with their approach and ours. Specifically, in their case λ is obtained via the formulae in [9, Cor.1,Th.3], while in our case $\lambda = \lambda_\gamma^*$ for $\gamma = 0.8$, which entails the wanted confidence according to Proposition 1(2). We see that, under the assumption that lower bounds on d as illustrated are known, in some cases a dramatic reduction of the expected length of the XOR constraints is obtained.

4.2 Execution Times

Although the focus of the present paper is mostly theoretical, it is instructive to look at the results of some simple experiments for a first concrete assessment

Table 1. Comparison with the provable bounds from [9, Tab. 1].

N. Vars (m)	N. Constraints (s)	$\lambda \times m$ from [9]	$\lambda \times m$ present paper $d = 1$	$d = 5$	$d = 20$
50	13	16.85	16.85	11.76	3.88
50	16	15.38	15.38	11.36	4.1
50	20	13.26	13.26	10.26	4.37
50	30	9.57	9.57	8.02	4.75
50	39	7.08	7.08	6.2	4.45
100	11	39.05	39.05	23.65	7.4
100	15	35.44	35.44	25.26	8.07
100	25	27.09	27.09	21.33	9.14
119	7	50.19	50.18	25.07	7.6
136	9	55.63	55.63	30.66	9.46
149	11	60.6	60.6	35.24	11.02
352	10	147.99	147.99	81.42	25.31

of the proposed methodology. To this aim, we have implemented in Python the algorithm C_A with A as described in Sect. 3, relying on CryptoMiniSAT [22] as a SAT solver, and conducted a few experiments[1].

The crucial issue to use Theorem 2 is the knowledge of (a lower bound on) the minimal distance d among the models of the formula we are inputting to our algorithm. In general, this information is unknown and we shall discuss a possible approach to face this problem in the next section. For the moment, we use a few formulae describing the set of codewords of certain error correcting codes, for which the number of models and the minimum distance is known by construction. Each of such formulae derives from a specific BCH code [5,15] (these are very well-known error-correcting codes whose minimal distance among the codewords is lower-bounded by construction). In particular, `Fxx-yy-zz.cnf` is a formula in CNF describing membership to the BCH code for 2^{xx} messages (the number of models), codified via codewords of yy bits and with a distance that is at least zz.

For these formulae, we run 3 kinds of experiments. First, we run the tool for every formula without using the improvements of the bounds given by knowing the minimum distance (i.e., we used Theorem 2 by setting $d = 1$). Second, we run the tool for every formula by using the known minimum distance. Third, we run the state-of-the-art tool for approximate model counting, called ApproxMC3 [21] (an improved version of ApproxMC2 [7]). The results obtained are reported in Table 2.

[1] Run on a MacBook Air, with a 1,7 GHz Intel Core i7 processor, 8 GB of memory (1600 MHz DDR3) and OS X 10.9.5.

Table 2. Results for our tool with $\alpha = \beta = 1.5$, $d = 1$ and $d = d_{min}$, compared to ApproxMC3 with a tolerance $\epsilon = 3$. In all trials, the error probability δ is 0.1.

Formula	*Our tool with* $d = 1$	*Our tool with* $d = d_{min}$	*ApproxMC3*
F21-31-5.cnf	res: $[2^{19.5}, 2^{23.5}]$ time: 8.05 s	res: $[2^{19.5}, 2^{23.5}]$ time: 6.73 s	res: ?? time: > 3 h
F16-31-7.cnf	res: $[2^{14.5}, 2^{18.5}]$ time: 11.75 s	res: $[2^{14.5}, 2^{18.5}]$ time: 9.5 s	res: $[2^{13.86}, 2^{17.85}]$ time: 22 min 43 s
F11-31-9.cnf	res: $[2^{9.5}, 2^{13.5}]$ time: 6.75 s	res: $[2^{9.5}, 2^{13.5}]$ time: 4.32 s	res: $[2^{9.09}, 2^{13.09}]$ time: 15.24 s
F6-31-11.cnf	res: $[2^{4.5}, 2^{8.5}]$ time: 3.01 s	res: $[2^{4.5}, 2^{8.5}]$ time: 2.62 s	res: $[2^4, 2^8]$ time: 1.9 s
F16-63-23.cnf	res: $[2^{14.5}, 2^{18.5}]$ time: 31 min 15 s	res: $[2^{14.5}, 2^{18.5}]$ time: 2 min 36 s	res: $[2^{13.9}, 2^{17.9}]$ time: 54 min 57 s

To compare our results with theirs, we have to consider that, if our tool returns $[l, u]$, then the number of models lies in $[\lfloor 2^l \rfloor, \lceil 2^u \rceil]$ with error probability δ (set to 0.1. in all experiments). By contrast, if ApproxMC3 returns a value M, then the number of models lies in $\left[\frac{M}{1+\epsilon}, M(1+\epsilon)\right]$ with error probability δ (again, set to 0.1 in all experiments). So, we have to choose for ApproxMC3 a tolerance ϵ that produces an interval of possible solutions comparable to what we obtain with our tool. The ratio between the sup and the inf of our intervals is $2^{2\max(\alpha,\beta)+1}$ (indeed, A always returned an interval such that $u - l \leq 2\max(\alpha, \beta) + 1$); when $\alpha = \beta = 1.5$, the value is 16. By contrast, the ratio between the sup and the inf of ApproxMC3's intervals is $(1+\epsilon)^2$; this value is 16 for $\epsilon = 3$.

In all formulae that have "sufficiently many" models – empirically, at least 2^{11} – our approach outperforms ApproxMC3. Moreover, making use of the minimum distance information implies a gain in performance. Of course, the larger the distance, the greater the gain – again, provided that the formula has sufficiently many models: compare, e.g., the first and the last formula.

4.3 Towards a Practical Methodology

To be used in practice, our technique requires a lower bound on the minimum Hamming distance between any two models of ϕ. We discuss below how error-correcting codes might be used, in principle, to obtain such a bound. Generally, speaking an error-correcting code adds redundancy to a string of bits and inflates the minimum distance between the resulting codewords. The idea here is to transform the given formula ϕ into a new formula ϕ' that describes an encoding of the original formula's models: as a result $\#\phi' = \#\phi$, but the models of ϕ' live in a higher dimensional space, where a minimum distance between models is ensured.

Assume that $\phi(y)$ is already in Conjunctive Normal Form. Fix a binary linear $[n, m, d]$ block code $\mathcal{C}$ (i.e. a code with 2^m codewords of n bits and minimum distance d), and let G be its generator matrix, i.e. a $m \times n$ binary matrix such that the codeword associated to $u \in \mathbb{F}_2^m$ is uG (where we use the vector-matrix multiplication in the field $\mathbb{F}_2$). The fact that a $c \in \mathbb{F}_2^n$ is a codeword can be expressed by finding some u that satisfies the conjunction of the n XOR constraints:

$$c_i = \bigoplus_{j=1}^{m} u_j \cdot G_{ij} \qquad \text{for } i = 1, \ldots, n\,.$$

Again, the important condition here is that G be *sparse* (on its columns), so that the above formula effectively corresponds to a conjunction of sparse XOR constraints. That is, we should confine ourselves to low-density parity check (LDPC) codes [11,18]. Now, we consider the formula

$$\phi'(z) \triangleq \exists y(\phi(y) \wedge z = yG)\,.$$

It can be easily proved that ϕ and ϕ' have the same number of models. If we now assume a minimum distance of d when applying Theorem 2, we have a decrease in the feasibility threshold λ^*, as prescribed by (7). This gain must of course be balanced against the increased number of boolean variables in the formula (viz. n). We will have an actual advantage using $\mathcal{C}$ over not using it (and simply assuming $d = 1$) if and only if, by using $\mathcal{C}$, the expected length of the resulting XOR constraints is actually smaller. By letting $\lambda^{*,d} \triangleq \lambda_1^*(s, n, \beta, d)$, the latter fact holds if and only if

$$n\lambda^{*,d} \leq m\lambda^{*,1} \tag{8}$$

or equivalently $\lambda^{*,d} \leq R\lambda^{*,1}$, where $R \triangleq \frac{m}{n}$ is the *rate* of $\mathcal{C}$. This points to codes with high rate and big minimum distance. Despite these two parameters pull one against the other, (8) can be fulfilled, and good expected length bounds obtained, by choosing $\mathcal{C}$ appropriately.

For example, [10] presents a [155, 64, 20]-LDPC code, that is a code with block length of 155, with 2^{64} codewords and with minimum distance between codewords of 20. In Fig. 2 we compare the expected length of the resulting XOR constraints in the two cases – $m \times \lambda^*(s, m, \beta, 1)$ (without the code, for $m = 32$ and $m = 64$) and $155 \times \lambda_1^*(s, 155, \beta, 20)$ (with the code) – as functions of s, for fixed $\beta = 1.5$. As seen from the plots, the use of the code offers a significant advantage in terms of expected length up to $s = 10$ and $s = 26$, respectively.

We have performed a few tests for a preliminary practical assessment of this idea. Unfortunately, in all but a few cases, the use of the said [155, 64, 20]-LDPC code does not offer a significant advantage in terms of execution time. Indeed, embedding a formula in this code implies adding 155 new XOR constraints: the presence of so many constraints, however short, apparently outweights the benefit of a minimum distance $d = 20$. We hope that alternative codes, with a more advantageous block length versus minimum distance tradeoff, would fare

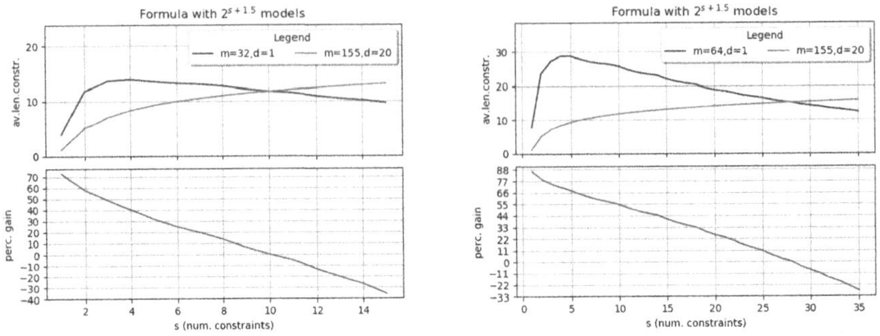

Fig. 2. Plots of the expected length of XOR constraints as a function of s, with and without code, and the relative percentual gain. Here, $m = 32$ (left) and 64 (right), $\beta = 1.5$ and the code is a $[155, 64, 20]$-LDPC.

better. Indeed, as we showed in Table 1, relatively small distances (e.g. $d = 5$) can already give interesting gains, if the number of extra constraints is small. We leave that as subject for future research.

5 Conclusion

We have studied the relation between sparse XOR constraints and minimum Hamming distance in model counting. Our findings suggest that minimum distance plays an important role in making the feasibility threshold for λ (density) lower, thus potentially improving the effectiveness of XOR based model counting procedures. These results also prompt a natural direction for future research: embedding the set of models into a higher dimensional Hamming space, so as to enforce a given minimum distance.

Beside the already mentioned [9], our work also relates to the recent [1]. There, constraints are represented as systems $Ay = b$, for A a random LDPC matrix enjoying certain properties, b a random vector, and y the variable vector. Their results are quite different from ours, but also they take the geometry of the set of models into account, including minimum distance. In particular, they make their bounds depend also on a "boost" parameter which appears quite difficult to compute. This leads to a methodology that is only empirically validated – that is, the model count results are offered with no guarantee.

Acknowledgements. We thank Marco Baldi, Massimo Battaglioni and Franco Chiaraluce for providing us with the generator and parity check matrices of the LDPC code in Subsect. 4.3.

References

1. Achlioptas, D., Theodoropoulos, P.: Probabilistic model counting with short XORs. In: Gaspers, S., Walsh, T. (eds.) SAT 2017. LNCS, vol. 10491, pp. 3–19. Springer, Cham (2017). https://doi.org/10.1007/978-3-319-66263-3_1

2. Babai, L., Frankl, P.: Linear Algebra Methods in Combinatorics. The University of Chicago, Chicago (1992)
3. Biondi, F., Enescu, M.A., Heuser, A., Legay, A., Meel, K.S., Quilbeuf, J.: Scalable approximation of quantitative information flow in programs. Verification, Model Checking, and Abstract Interpretation. LNCS, vol. 10747, pp. 71–93. Springer, Cham (2018). https://doi.org/10.1007/978-3-319-73721-8_4
4. Boreale, M., Gorla, D.: Approximate model counting, sparse XOR constraints and minimum distance. https://arxiv.org/abs/1907.05121 (2019)
5. Bose, R.C., Ray-Chaudhuri, D.K.: On a class of error correcting binary group codes. Inf. Control **3**(1), 68–79 (1960)
6. Chakraborty, S., Meel, K.S., Vardi, M.Y.: A scalable approximate model counter. In: Schulte, C. (ed.) CP 2013. LNCS, vol. 8124, pp. 200–216. Springer, Heidelberg (2013). https://doi.org/10.1007/978-3-642-40627-0_18
7. Chakraborty, S., Meel, K.S., Vardi, M.Y.: Algorithmic improvements in approximate counting for probabilistic inference: from linear to logarithmic SAT calls. In: Proceedings of International Joint Conference on Artificial Intelligence (2016)
8. Chatzikokolakis, K., Palamidessi, C., Panangaden, P.: Anonymity protocols as noisy channels. Inf. Comput. **206**(2–4), 378–401 (2008)
9. Ermon, S., Gomes, C.P., Sabharwal, A., Selman, B.: Low-density parity constraints for hashing-based discrete integration. In: Proceedings of the 31th International Conference on Machine Learning, ICML 2014, Beijing, China, 21–26 June 2014, pp. 271–279 (2014)
10. Fuja, T.E., Sridhara, D., Tanner, R.M.: A class of group-structured LDPC codes. In: International Symposium on Communication Theory and Applications (2001)
11. Gallager, R.G.: Low Density Parity Check Codes. MIT Press, Cambridge (1963)
12. Gebser, M., Kaufmann, B., Neumann, A., Schaub, T.: *clasp*: a conflict-driven answer set solver. In: Baral, C., Brewka, G., Schlipf, J. (eds.) LPNMR 2007. LNCS (LNAI), vol. 4483, pp. 260–265. Springer, Heidelberg (2007). https://doi.org/10.1007/978-3-540-72200-7_23
13. Gomes, C.P., Sabharwal, A., Selman, B.: Model counting: a new strategy for obtaining good bounds. In: Proceedings of AAAI, pp. 54–61 (2006)
14. Gomes, C.P., Sabharwal, A., Selman, B.: Model counting. In: Handbook of Satisfiability, pp. 633–654. IOS Press (2009)
15. Hocquenghem, A.: Codes correcteurs d'erreurs. Chiffres **2**, 147–156 (1959)
16. Klebanov, V., Manthey, N., Muise, C.: SAT-based analysis and quantification of information flow in programs. In: Joshi, K., Siegle, M., Stoelinga, M., D'Argenio, P.R. (eds.) QEST 2013. LNCS, vol. 8054, pp. 177–192. Springer, Heidelberg (2013). https://doi.org/10.1007/978-3-642-40196-1_16
17. Klebanov, V., Weigl, A., Weibarth, J.: Sound probabilistic #SAT with projection. In: Proceedings of QAPL (2016)
18. MacKay, D.J.C.: Good error-correcting codes based on very sparse matrices. IEEE Trans. Inf. Theory **45**(3), 399–432 (1999)
19. Muise, C., McIlraith, S.A., Beck, J.C., Hsu, E.I.: DSHARP: fast d-DNNF compilation with sharpSAT. In: Kosseim, L., Inkpen, D. (eds.) AI 2012. LNCS (LNAI), vol. 7310, pp. 356–361. Springer, Heidelberg (2012). https://doi.org/10.1007/978-3-642-30353-1_36
20. Smith, G.: On the foundations of quantitative information flow. In: de Alfaro, L. (ed.) FoSSaCS 2009. LNCS, vol. 5504, pp. 288–302. Springer, Heidelberg (2009). https://doi.org/10.1007/978-3-642-00596-1_21

21. Soos, M., Meel, K.S.: BIRD: engineering an efficient CNF-XOR SAT solver and its applications to approximate model counting. In: Proceedings of AAAI Conference on Artificial Intelligence (AAAI) (2019)
22. Soos, M., Nohl, K., Castelluccia, C.: Extending SAT solvers to cryptographic problems. In: Kullmann, O. (ed.) SAT 2009. LNCS, vol. 5584, pp. 244–257. Springer, Heidelberg (2009). https://doi.org/10.1007/978-3-642-02777-2_24. http://www.msoos.org/cryptominisat2/
23. Thurley, M.: sharpSAT – counting models with advanced component caching and implicit BCP. In: Biere, A., Gomes, C.P. (eds.) SAT 2006. LNCS, vol. 4121, pp. 424–429. Springer, Heidelberg (2006). https://doi.org/10.1007/11814948_38
24. Valiant, L.G.: The complexity of enumeration and reliability problems. SIAM J. Comput. **8**(3), 410–421 (1979)
25. Valiant, L.G., Vazirani, V.V.: NP is as easy as detecting unique solutions. Theor. Comput. Sci. **47**(3), 85–93 (1986)

Verification and Control of Turn-Based Probabilistic Real-Time Games

Marta Kwiatkowska[1], Gethin Norman[2(✉)], and David Parker[3]

[1] Department of Computing Science, University of Oxford, Oxford, UK
[2] School of Computing Science, University of Glasgow, Glasgow, UK
gethin.norman@glasgow.ac.uk
[3] School of Computer Science, University of Birmingham, Birmingham, UK

Abstract. Quantitative verification techniques have been developed for the formal analysis of a variety of probabilistic models, such as Markov chains, Markov decision process and their variants. They can be used to produce guarantees on quantitative aspects of system behaviour, for example safety, reliability and performance, or to help synthesise controllers that ensure such guarantees are met. We propose the model of turn-based probabilistic timed multi-player games, which incorporates probabilistic choice, real-time clocks and nondeterministic behaviour across multiple players. Building on the digital clocks approach for the simpler model of probabilistic timed automata, we show how to compute the key measures that underlie quantitative verification, namely the probability and expected cumulative price to reach a target. We illustrate this on case studies from computer security and task scheduling.

1 Introduction

Probability is a crucial tool for modelling computerised systems. We can use it to model uncertainty, for example in the operating environment of an autonomous vehicle or a wireless sensor network, and we can reason about systems that use randomisation, from probabilistic routing in anonymity network protocols to symmetry breaking in communication protocols.

Formal verification of such systems can provide us with rigorous guarantees on, for example, the performance and reliability of computer networks [7], the amount of inadvertent information leakage by a security protocol [5], or the safety level of an airbag control system [2]. To do so requires us to model and reason about a range of quantitative aspects of a system's behaviour: probability, time, resource usage, and many others.

Quantitative verification techniques have been developed for a wide range of probabilistic models. The simplest are Markov chains, which model the evolution of a stochastic system over discrete time. Markov decision processes (MDPs) additionally include nondeterminism, which can be used either to model the uncontrollable behaviour of an adversary or to determine an optimal strategy (or policy) for controlling the system. Generalising this model further still, we can add a notion of real time, yielding the model of probabilistic timed automata

M. S. Alvim et al. (Eds.): Palamidessi Festschrift, LNCS 11760, pp. 379–396, 2019.
https://doi.org/10.1007/978-3-030-31175-9_22

(PTAs). This is done in the same way as for the widely used model of timed automata (TAs), adding real-valued variables called clocks to the model. Tools such as PRISM [33] and Storm [21] support verification of a wide range of properties of these different probabilistic models.

Another dimension that we can add to these models and verification techniques is *game-theoretic* aspects. These can be used to represent, for example, the interaction between an attacker and a defender in a computer security protocol [4], between a controller and its environment, or between participants in a communication protocol who have opposing goals [27]. Stochastic multi-player games include choices made by multiple players who can either collaborate or compete to achieve their goals. Tool support for verification of stochastic games, e.g. PRISM-games [41], has also been developed and deployed successfully in a variety of application domains.

In this paper, we consider a modelling formalism that captures all these aspects: probability, nondeterminism, time and multiple players. We define a model called *turn-based probabilistic timed multi-player games* (TPTGs), which can be seen as either an extension of PTAs to incorporate multiple players, or a generalisation of stochastic multi-player games to include time. Building on known techniques for the simpler classes of models, we show how to compute key properties of these models, namely probabilistic reachability (the probability of reaching a set of target states) and expected price reachability (the expected price accumulated before reaching a set of target states).

Existing techniques for PTAs largely fall into two classes: *zone-based* and *digital clocks*, both of which construct and analyse a finite-state abstraction of the model. Zones are symbolic expressions representing sets of clock values. Zone-based approaches for analysing PTAs were first introduced in [32,39,40] and recent work extended them to the analysis of expected time [28] and expected price reachability [34]. The digital clocks approach works by mapping real-valued clocks to integer-valued ones, reducing the problem of solving a PTA to solving a (discrete-time) MDP. This approach was developed for PTAs in [38] and recently extended to the analysis of partially observable PTAs in [46].

In this paper, we show how a similar idea can be used to reduce the verification problem for TPTGs to an equivalent one over (discrete-time) stochastic games. More precisely, for the latter, we use *turn-based stochastic games* (TSGs). We first present the model of TPTGs and give two alternative semantics: one using real-valued clocks and the other using (integer-valued) digital clocks. Then, we prove the correspondence between these two semantics. Next, we demonstrate the application of this approach to two case studies from the domains of computer security and task scheduling. Using a translation from TPTGs to TSGs and the model checking tool PRISM-games, we show that a variety of useful properties can be studied on these case studies. An extended version of this paper, with complete proofs, is available [35].

Related Work. Timed games were introduced and shown to be decidable in [1, 6,43]. These games have since been extensively studied; we mention [19,51], where efficient algorithms are investigated, and [47], which concerns the synthesis of strategies that are robust to stochastic perturbation in clock values. Also related is the tool UPPAAL TIGA [8], which allows the automated analysis of reachability and safety problems for timed games.

Priced (or weighted) timed games were introduced in [3,11,49], which extend timed games by associating integer costs with locations and transitions, and optimal cost reachability was shown to be decidable under certain restrictions. The problem has since been shown to be undecidable for games with three or more clocks [10,17]. Priced timed games have recently been extended to allow partial observability [18] and to energy games [14].

Two-player (concurrent) probabilistic timed games were introduced in [24]. The authors demonstrated that such games are not determined (even when all clock constraints are closed) and investigated the complexity of expected time reachability for such games. Stochastic timed games [13,16] are turn-based games where time delays are exponentially distributed. A similar model, based on interactive Markov chains [26], is considered in [15].

2 Background

We start with some background and notation on *turn-based stochastic games* (TSGs). For a set X, let $Dist(X)$ denote the set of discrete probability distributions over X and $\mathbb{R}$ the set of non-negative real numbers.

Definition 1 (Turn-based stochastic multi-player game). *A turn-based stochastic multi-player game (TSG) is a tuple* $\mathsf{G} = (\Pi, S, \bar{s}, A, \langle S_i \rangle_{i \in \Pi}, \delta, R)$ *where* Π *is a finite set of players,* S *is a (possibly infinite) set of states,* $\bar{s} \in S$ *is an initial state,* A *is a (possibly infinite) set of actions,* $\langle S_i \rangle_{i \in \Pi}$ *is a partition of the state space,* $\delta : S{\times}A \rightarrow Dist(S)$ *is a (partial) transition function and* $R : S \times A \rightarrow \mathbb{R}$ *is a price (or reward) function.*

The transition function is partial in the sense that δ need not be defined for all state-action pairs. For each state s of a TSG G, there is a set of available actions given by $A(s) \stackrel{\text{def}}{=} \{a \in A \mid \delta(s,a) \text{ is defined}\}$. The choice of which available action is taken in state s is under the control of a single player: the player i such that $s \in S_i$. If player i selects action $a \in A(s)$ in s, then the probability of transitioning to state s' equals $\delta(s,a)(s')$ and a price of $R(s,a)$ is accumulated.

Paths and Strategies. A path of a TSG G is a sequence $\pi = s_0 \xrightarrow{a_0} s_1 \xrightarrow{a_1} \cdots$ such that $s_i \in S$, $a_i \in A(s_i)$ and $\delta(s_i,a_i)(s_{i+1}){>}0$ for all $i \geqslant 0$. For a path π, we denote by $\pi(i)$ the $(i{+}1)$th state of the path, $\pi[i]$ the action associated with the $(i{+}1)$th transition and, if π is finite, $last(\pi)$ the final state. The length of a path π, denoted $|\pi|$, equals the number of transitions. For a path π and $k{<}|\pi|$, let $\pi^{(k)}$ be the kth prefix of π. Let $FPaths_{\mathsf{G}}$ and $IPaths_{\mathsf{G}}$ equal the sets of finite and infinite paths starting in the initial state $\bar{s}$.

A strategy for player $i \in \Pi$ is a way of resolving the choice of action in each state under the control of player i, based on the game's execution so far. Formally, a strategy σ for player $i \in \Pi$ is a function $\sigma_i : \{\pi \in FPaths_{\mathsf{G}} \mid$

$last(\pi) \in S_i\} \rightarrow Dist(A)$ such that, if $\sigma_i(\pi)(a)>0$, then $a \in A(last(\pi))$. The set of all strategies of player $i \in \Pi$ is represented by Σ^i_{G} (when clear from the context we will drop the subscript G). A strategy for player i is deterministic if it always selects actions with probability 1, and memoryless if it makes the same choice for any paths that end in the same state.

A strategy profile for G takes the form $\sigma = \langle \sigma_i \rangle_{i \in \Pi}$, listing a strategy for each player. We use $FPaths^\sigma$ and $IPaths^\sigma$ for the sets of finite and infinite paths corresponding to the choices made by the profile σ when starting in the initial state. For a given profile σ, the behaviour of G is fully probabilistic and we can define a probability measure $Prob^\sigma$ over the set of infinite paths $IPaths^\sigma$ [30].

Properties. Two fundamental properties of quantitative models are the probability of reaching a set of target states and the expected price accumulated before doing so. For a strategy profile σ and set of target states F of a TSG G, the probability of reaching F and expected price accumulated before reaching F from the initial state $\bar{s}$ under the profile σ are given by the following (again, when it is clear from the context, we will drop the subscript G):

$$\mathbb{P}^\sigma_{\mathsf{G}}(F) \stackrel{\text{def}}{=} Prob^\sigma(\{\pi \in IPaths^\sigma \mid \pi(i) \in F \text{ for some } i \in \mathbb{N}\})$$
$$\mathbb{E}^\sigma_{\mathsf{G}}(F) \stackrel{\text{def}}{=} \int_{\pi \in IPaths^\sigma} rew(\pi, F) \,\mathrm{d}Prob^\sigma$$

where for any infinite path π:

$$rew(\pi, F) \stackrel{\text{def}}{=} \textstyle\sum_{i=0}^{k_F} R(\pi(i), \pi[i])$$

and $k_F = \min\{k{-}1 \mid \pi(k) \in F\}$ if $\pi(k) \in F$ for some $k \in \mathbb{N}$ and $k_F = \infty$ otherwise.

To quantify the above properties over the strategies of the players, we consider a coalition $C \subseteq \Pi$ who try to maximise the property of interest, while the remaining players $\Pi \backslash C$ try to minimise it. Formally, we have the following definition:

$$\mathbb{P}^C_{\mathsf{G}}(F) \stackrel{\text{def}}{=} \sup\nolimits_{\sigma_1 \in \Sigma^1} \inf\nolimits_{\sigma_2 \in \Sigma^2} \mathbb{P}^{\sigma_1,\sigma_2}_{\mathsf{G}^C}(F)$$
$$\mathbb{E}^C_{\mathsf{G}}(F) \stackrel{\text{def}}{=} \sup\nolimits_{\sigma_1 \in \Sigma^1} \inf\nolimits_{\sigma_2 \in \Sigma^2} \mathbb{E}^{\sigma_1,\sigma_2}_{\mathsf{G}^C}(F)$$

where G^C is the two-player game constructed from G where the states controlled by player 1 equal $\cup_{i \in C} S_i$ and the states controlled by player 2 equal $\cup_{i \in \Pi \backslash C} S_i$.

The above definition yields the *optimal value* of G if it is *determined*, i.e., if the maximum value that the coalition C can ensure equals the minimum value that the coalition $\Pi \setminus C$ can ensure. Formally, the definition of determinacy and optimal strategies for probabilistic reachability properties of TSGs are given below, and the case of expected reachability is analogous (replacing $\mathbb{P}$ with $\mathbb{E}$).

Definition 2. *For a TSG G, target F and coalition of players C, we say the game G^C is* determined *with respect to probabilistic reachability if:*

$$\sup\nolimits_{\sigma_1 \in \Sigma^1} \inf\nolimits_{\sigma_2 \in \Sigma^2} \mathbb{P}^{\sigma_1,\sigma_2}(F) = \inf\nolimits_{\sigma_2 \in \Sigma^2} \sup\nolimits_{\sigma_1 \in \Sigma^1} \mathbb{P}^{\sigma_1,\sigma_2}(F).$$

Furthermore, a strategy $\sigma_1^\star \in \Sigma_1$ is optimal *if $\mathbb{P}^{\sigma_1^\star,\sigma_2}(F) \geqslant \mathbb{P}^C_{\mathsf{G}}(F)$ for all $\sigma_2 \in \Sigma^2$ and strategy $\sigma_2^\star \in \Sigma_2$ is* optimal *if $\mathbb{P}^{\sigma_1,\sigma_2^\star}(F) \leqslant \mathbb{P}^C_{\mathsf{G}}(F)$ for all $\sigma_1 \in \Sigma^1$.*

As we shall demonstrate, the games we consider are determined with respect to probabilistic and expected reachability, and optimal strategies exist. In particular, finite-state and finite-branching TSGs are determined [31] and efficient techniques exist to approximate optimal values and optimal strategies [20,22]. These techniques underlie the model checking algorithms for logics such as rPATL, defined for TSGs and implemented in the tool PRISM-games [41].

3 Turn-Based Probabilistic Timed Multi-Player Games

We now introduce *turn-based probabilistic timed multi-player games* (TPTGs), a framework for modelling systems which allows probabilistic, non-deterministic, real-time and competitive behaviour. Let $\mathbb{T} \in \{\mathbb{R}, \mathbb{N}\}$ be the time domain of either the non-negative reals or natural numbers.

Clocks, Valuations and Clock Constraints. We assume a finite set of *clocks* $\mathcal{X}$. A *clock valuation* is a function $v : \mathcal{X} \to \mathbb{T}$; the set of all clock valuations is denoted $\mathbb{T}^{\mathcal{X}}$. Let $\mathbf{0}$ be the clock valuation that assigns the value 0 to all clocks. For any set of clocks $X \subseteq \mathcal{X}$ and clock valuation $v \in \mathbb{T}^{\mathcal{X}}$, let $v[X{:=}0]$ be the clock valuation such that, for any clock x, we have $v[X{:=}0](x)$ equals 0 if $x \in X$ and $v(x)$ otherwise. Furthermore, for any time instant $t \in \mathbb{T}$, let $v{+}t$ be the clock valuation such that $(v{+}t)(x) = v(x){+}t$ for all $x \in \mathcal{X}$. A closed, diagonal-free clock constraint[1] ζ is a conjunction of inequalities of the form $x \leqslant c$ or $x \geqslant c$, where $x \in \mathcal{X}$ and $c \in \mathbb{N}$. We write $v \models \zeta$ if the clock valuation v satisfies the clock constraint ζ and use $CC(\mathcal{X})$ for the set of all clock constraints over $\mathcal{X}$.

We are now in a position to present the syntax and semantics of TPTGs.

Definition 3 (TPTG syntax). *A* turn-based probabilistic timed multi-player game *(TPTG) is a tuple* $\mathsf{P} = (\Pi, L, \bar{l}, \mathcal{X}, A, \langle L_i \rangle_{i \in \Pi}, inv, enab, prob, r)$ *where:*

- Π *is a finite set of* players;
- L *is a finite set of* locations *and* $\bar{l} \in L$ *is an* initial location;
- $\mathcal{X}$ *is a finite set of* clocks;
- A *is a finite set of* actions;
- $\langle L_i \rangle_{i \in \Pi}$ *is a partition of* L;
- $inv : L \to CC(\mathcal{X})$ *is an* invariant condition;
- $enab : L \times A \to CC(\mathcal{X})$ *is an* enabling condition;
- $prob : L \times A \to Dist(2^{\mathcal{X}} \times L)$ *is a (partial)* probabilistic transition function;
- $r = (r_L, r_A)$ *is a* price structure *where* $r_L : L \to \mathbb{N}$ *is a location price function and* $r_A : L \times A \to \mathbb{N}$ *an action price function.*

As for PTAs [45], a state of a TPTG P is a location-clock valuation pair (l, v) such that the clock valuation satisfies the invariant $inv(l)$. The transition choice in (l, v) is under the control of the player i where $l \in L_i$. A transition is a time-action pair (t, a) which represents letting time t elapse and then performing action a.

[1] A constraint is closed if does not contain strict inequalities and diagonal-free if there are no inequalities of the form $x{-}y \sim c$ for $x, y \in \mathcal{X}$, $\sim \in \{<, \leqslant, \geqslant, >\}$ and $c \in \mathbb{N}$.

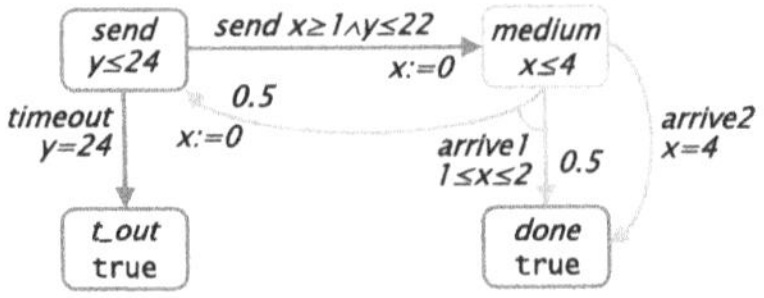

Fig. 1. An example TPTG

Time can elapse if the invariant of the current location remains continuously satisfied and action a can be performed only if the enabling condition is satisfied. If action a is taken in location l, then the probability of moving to location l' and resetting the set of clocks X equals $prob(l,a)(X,l')$. TPTGs have both location prices, which are accumulated at rate $r_L(l)$ when time passes in location l, and action prices, where $r_A(l,a)$ is accumulated when performing action a in location l. Formally, the semantics of a TPTG is a TSG defined as follows.

Definition 4 (TPTG semantics). *For any time domain* $\mathbb{T} \in \{\mathbb{R}, \mathbb{N}\}$ *and TPTG* $\mathsf{P} = (\Pi, L, \bar{l}, \mathcal{X}, A, \langle L_i \rangle_{i \in \Pi}, inv, enab, prob, r)$ *the semantics of* P *with respect to the time domain* $\mathbb{T}$ *is the TSG* $[\![\mathsf{P}]\!]_{\mathbb{T}} = (\Pi, S, \bar{s}, \mathcal{A}, \langle S_i \rangle_{i \in \Pi}, \delta, R)$ *where:*

- $S = \{(l,v) \in L \times \mathbb{T}^{\mathcal{X}} \mid v \models inv(l)\}$ *and* $\bar{s} = (\bar{l}, \mathbf{0})$;
- $\mathcal{A} = \mathbb{T} \times A$;
- $S_i = \{(l,v) \in S \mid l \in L_i\}$ *for* $i \in \Pi$;
- *for any* $(l,v) \in S$ *and* $(t,a) \in \mathcal{A}$ *we have* $\delta((l,v),(t,a)) = \mu$ *if and only if* $v+t' \models inv(l)$ *for all* $0 \leqslant t' \leqslant t$, $v+t \models enab(l,a)$ *and for any* $(l',v') \in S$:

$$\mu(l',v') = \sum_{X \subseteq \mathcal{X} \wedge v' = (v+t)[X:=0]} prob(l,a)(X,l')$$

- $R((l,v),(t,a)) = t \cdot r_L(l) + r_A(l,a)$ *for all* $(l,v) \in S$ *and* $(t,a) \in \mathcal{A}$.

We follow the approach of [24,28,29] and use time-action pairs in the transition function of Definition 4. As explained in [28], this yields a more expressive semantics than having separate time and action transitions.

Example 1. Consider the TPTG in Fig. 1 which represents a simple communication protocol. There are two players: the sender and the medium, with the medium controlling the location *medium* and the sender all other locations. The TPTG has two clocks: x is used to keep track of the time it takes to send a message and y the elapsed time. In the initial location *send*, the sender waits between 1 and 2 time units before sending the message. The message then passes through the medium that can either delay the message for between 1 and 2 time units after which it arrives with probability 0.5, or delay the message for 4 time units after which it arrives with probability 1. If the message does not arrive, then the sender tries to send it again until reaching a timeout after 24 time units.

As for PTAs, in the standard (dense-time) semantics for a TPTG the time domain $\mathbb{T}$ equals $\mathbb{R}$. This yields an infinite state model which is not amenable to verification. One approach that yields a finite state representation used in the case of PTAs is the digital clocks semantics [38]. This is based on replacing the real-valued clocks with clocks taking only values from a bounded set of integers. In order to give the definition for a TPTG P, for any clock x of P we define k_x to be the greatest constant to which x is compared in the clock constraints of P. This allows us to use bounded clock values since, if the value of the clock x exceeds k_x, then the exact value will not affect the satisfaction of the invariants and enabling conditions of P, and therefore does not influence the behaviour.

Definition 5 (Digital clocks semantics). *The digital clocks semantics of a TPTG* P*, denoted* $[\![\mathsf{P}]\!]_{\mathbb{N}}$*, is obtained from Definition 4, by setting* $\mathbb{T}$ *equal to* $\mathbb{N}$ *and for any* $v \in \mathbb{N}^{\mathcal{X}}$*,* $t \in \mathbb{N}$ *and* $x \in \mathcal{X}$ *letting* $(v{+}t)(x) = \min\{v(x){+}t, k_x{+}1\}$*.*

We restrict our attention to time-divergent (also called non-Zeno) behaviour. More precisely, we only consider strategies for the players that do not generate unrealisable executions, i.e., executions in which time does not advance beyond a certain point. We achieve this by restricting to TPTGs that satisfy the syntactic conditions for PTAs given in [45], derived from results on TAs [50,52]. In addition, we require the following assumptions to ensure the correctness of the digital clocks semantics.

Assumption 1 *For any TPTG* P *: (a) all invariants of* P *are bounded; (b) all clock constraints are closed and diagonal free; (c) all probabilities are rational.*

Regarding Assumption 1(a), in fact bounded TAs are as expressive as standard TAs [9], and this result carries over to TPTGs.

To facilitate higher-level modelling, PTAs can be extended with parallel composition, discrete variables, urgent transitions and locations and resetting clocks to integer values [45]. We can extend TPTGs in a similar way, and will use these constructs in Sect. 5.

4 Correctness of the Digital Clocks Semantics

We now show that, under Assumption 1, optimal probabilistic and expected price reachability values agree under the digital and dense-time semantics. As for PTAs [45], by modifying the TPTG under study, we can reduce time-bounded probabilistic reachability properties to probabilistic reachability properties and both expected time-bounded cumulative price properties and expected time-instant price properties to expected reachability properties. In each case the modifications to the TPTG preserve Assumption 1, and therefore the digital clocks semantics can also be used to verify these classes of properties.

For the remainder of this section, we fix a TPTG P, coalition of players C and set of target locations $F \subseteq L$, and let $F_{\mathbb{T}} = \{(l, v) \in F \times \mathbb{T}^{\mathcal{X}} \mid v \models inv(l)\}$ for $\mathbb{T} \in \{\mathbb{R}, \mathbb{N}\}$. We have omitted the proofs that closely follow those for PTAs [38]. The missing proofs can be found in [35].

We first present results relating to the determinacy and existence of optimal strategies for the games $[\![\mathsf{P}]\!]_{\mathbb{R}}^{C}$ and $[\![\mathsf{P}]\!]_{\mathbb{N}}^{C}$ and a correspondence between the strategy profiles of $[\![\mathsf{P}]\!]_{\mathbb{N}}^{C}$ and $[\![\mathsf{P}]\!]_{\mathbb{R}}^{C}$.

Proposition 1. *For any TPTG* P *satisfying Assumption 1, the games* $[\![\mathsf{P}]\!]_{\mathbb{R}}^{C}$ *and* $[\![\mathsf{P}]\!]_{\mathbb{N}}^{C}$ *are determined and have optimal strategies for both probabilistic and expected price reachability properties.*

Proof. In the case of $[\![\mathsf{P}]\!]_{\mathbb{R}}^{C}$, the result follows from [23] and Assumption 1, i.e. since all clock constraints are closed. Considering $[\![\mathsf{P}]\!]_{\mathbb{N}}^{C}$, the result follows from the fact that the game has a finite state space and is finitely branching [31]. □

Proposition 2. *For any strategy profile* σ' *of* $[\![\mathsf{P}]\!]_{\mathbb{N}}^{C}$, *there exists a strategy profile* σ *of* $[\![\mathsf{P}]\!]_{\mathbb{R}}^{C}$ *such that* $\mathbb{P}^{\sigma}(F_{\mathbb{R}}) = \mathbb{P}^{\sigma'}(F_{\mathbb{N}})$ *and* $\mathbb{E}^{\sigma}(F_{\mathbb{R}}) = \mathbb{E}^{\sigma'}(F_{\mathbb{N}})$.

Using the ϵ-digitization approach of TAs [25], which has been extended to PTAs in [38], the following theorem follows, demonstrating the correctness of the digital clocks semantics for probabilistic reachability properties.

Theorem 1. *For any TPTG* P *satisfying Assumption 1, coalition of players* C *and set of locations* $F \subseteq L$: $\mathbb{P}^{C}_{[\![\mathsf{P}]\!]_{\mathbb{R}}}(F_{\mathbb{R}}) = \mathbb{P}^{C}_{[\![\mathsf{P}]\!]_{\mathbb{N}}}(F_{\mathbb{N}})$.

For expected price reachability properties, we extend the approach of [38], by first showing that, for any fixed (dense-time) profile σ of $[\![\mathsf{P}]\!]_{\mathbb{R}}^{C}$ and $n \in \mathbb{N}$, there exist profiles of $[\![\mathsf{P}]\!]_{\mathbb{N}}^{C}$ whose expected price of reaching the target locations F within n transitions that provide lower and upper bounds for that of σ.

For $\mathbb{T} \in \{\mathbb{R}, \mathbb{N}\}$, profile σ of $[\![\mathsf{P}]\!]_{\mathbb{T}}^{C}$, and finite path $\pi \in FPaths^{\sigma}$ we inductively define the values $\langle \mathbb{E}^{\sigma}_{n}(\pi, F_{\mathbb{T}}) \rangle_{n \in \mathbb{N}}$ which equal the expected price, under the profile σ, of reaching the target $F_{\mathbb{T}}$ after initially performing the path π within n steps. To ease presentation we only give the definition for deterministic profiles.

Definition 6. *For* $\mathbb{T} \in \{\mathbb{R}, \mathbb{N}\}$, *strategy profile* $\sigma = (\sigma_1, \sigma_2)$ *of* $[\![\mathsf{P}]\!]_{\mathbb{T}}^{C}$ *and finite path* π *of the profile let* $\mathbb{E}^{\sigma_1,\sigma_2}_{0}(\pi, F_{\mathbb{T}}) = 0$ *and for any* $n \in \mathbb{N}$, *if* $last(\pi) = (l, v) \in S_i$ *for* $1 \leqslant i \leqslant 2$, $\sigma_i(\pi) = (t, a)$ *and* $\mu = P_{[\![\mathsf{P}]\!]_{\mathbb{T}}}((l, v), (t, a))$, *then:*

$$\mathbb{E}^{\sigma}_{n+1}(\pi, F_{\mathbb{T}}) = \begin{cases} 0 & \textit{if } (l, v) \in F_{\mathbb{T}} \\ r_L(l) \cdot t + r_A(l, a) + \sum_{s' \in S} \mu(s') \cdot \mathbb{E}^{\sigma}_{n}(\pi \xrightarrow{t,a} s', F_{\mathbb{T}}) & \textit{otherwise.} \end{cases}$$

We require the following properties of these expected price reachability properties. These then allow us to prove the correctness of the digital clocks semantics for expected price reachability properties (Theorem 2 below).

Lemma 1. *For* $\mathbb{T} \in \{\mathbb{R}, \mathbb{N}\}$ *and profile* σ *of* $[\![\mathsf{P}]\!]_{\mathbb{T}}^{C}$, *the sequence* $\langle \mathbb{E}^{\sigma}_{n}(F_{\mathbb{T}}) \rangle_{n \in \mathbb{N}}$ *is non-decreasing and converges to* $\mathbb{E}^{\sigma}(F_{\mathbb{T}})$, *and, for any player 1 strategy* σ_1 *of* $[\![\mathsf{P}]\!]_{\mathbb{T}}^{C}$, *the sequence of functions* $\mathbb{E}^{\sigma_1,\cdot}_{n}(F_{\mathbb{T}}) : \Sigma^2 \to \mathbb{R}$ *converges uniformly. Furthermore, for any player 1 strategy* σ_1, *the sequence* $\langle \inf_{\sigma_2 \in \Sigma^2} \mathbb{E}^{\sigma_1,\sigma_2}_{n}(F_{\mathbb{T}}) \rangle_{n \in \mathbb{N}}$ *is non-decreasing and converges to* $\inf_{\sigma_2 \in \Sigma^2} \mathbb{E}^{\sigma_1,\sigma_2}(F_{\mathbb{T}})$, *and the sequence of functions* $\inf_{\sigma_2 \in \Sigma^2} \mathbb{E}^{\cdot,\sigma_2}_{n}(F_{\mathbb{T}}) : \Sigma^1 \to \mathbb{R}$ *converges uniformly.*

Proof. In each case, proving that the sequence is non-decreasing and converges follows from Definition 6. Uniform convergence follows from showing the set of strategies for players is compact and using the fact that the sequences are non-decreasing and converge pointwise [48, Theorem 7.13]. In the case when $\mathbb{T} = \mathbb{N}$, compactness follows from the fact the action set is finite, while for $\mathbb{T} = \mathbb{R}$ we must restrict to PTAs for which all invariants are bounded (Assumption 1) to ensure the action set is compact. □

Lemma 2. *For any strategy profile σ of $[\![\mathsf{P}]\!]_{\mathbb{R}}^C$ and $n \in \mathbb{N}$, there exist strategy profiles σ^{lb} and σ^{ub} of $[\![\mathsf{P}]\!]_{\mathbb{N}}^C$ such that: $\mathbb{E}_n^{\sigma^{lb}}(F_{\mathbb{N}}) \leqslant \mathbb{E}_n^{\sigma}(F_{\mathbb{R}}) \leqslant \mathbb{E}_n^{\sigma^{ub}}(F_{\mathbb{N}})$.*

Theorem 2. *For any TPTG P satisfying Assumption 1, coalition of players C and set of locations $F \subseteq L$: $\mathbb{E}_{[\![\mathsf{P}]\!]_{\mathbb{R}}}^C(F_{\mathbb{R}}) = \mathbb{E}_{[\![\mathsf{P}]\!]_{\mathbb{N}}}^C(F_{\mathbb{N}})$.*

Proof. Consider any $n \in \mathbb{N}$. Using Lemma 2 it follows that, for any profile $\sigma = (\sigma_1, \sigma_2)$ of $[\![\mathsf{P}]\!]_{\mathbb{R}}$, there exist profiles $\sigma^{lb} = (\sigma_1^{lb}, \sigma_1^{lb})$ and $\sigma^{ub} = (\sigma_1^{ub}, \sigma_1^{ub})$ of $[\![\mathsf{P}]\!]_{\mathbb{N}}^C$ such that:

$$\mathbb{E}_n^{\sigma_1^{lb},\sigma_2^{lb}}(F_{\mathbb{N}}) \leqslant \mathbb{E}_n^{\sigma_1,\sigma_2}(F_{\mathbb{R}}) \leqslant \mathbb{E}_n^{\sigma_1^{ub},\sigma_2^{ub}}(F_{\mathbb{N}}) .$$

On the other hand, using the construction in the proof of Proposition 2, for any profile $\sigma' = (\sigma_1', \sigma_2')$ of $[\![\mathsf{P}]\!]_{\mathbb{N}}$, there exists a profile $\sigma = (\sigma_1, \sigma_1)$ of $[\![\mathsf{P}]\!]_{\mathbb{R}}^C$ such that:

$$\mathbb{E}_n^{\sigma_1,\sigma_2}(F_{\mathbb{N}}) = \mathbb{E}_n^{\sigma_1',\sigma_2'}(F_{\mathbb{R}}) .$$

Combining these results with Proposition 1 it follows that:

$$\sup\nolimits_{\sigma_1' \in \Sigma^1_{[\![\mathsf{P}]\!]_{\mathbb{N}}^C}} \inf\nolimits_{\sigma_2' \in \Sigma^2_{[\![\mathsf{P}]\!]_{\mathbb{N}}^C}} \mathbb{E}_n^{\sigma_1',\sigma_2'}(F_{\mathbb{N}}) = \sup\nolimits_{\sigma_1 \in \Sigma^1_{[\![\mathsf{P}]\!]_{\mathbb{R}}^C}} \inf\nolimits_{\sigma_2 \in \Sigma^2_{[\![\mathsf{P}]\!]_{\mathbb{R}}^C}} \mathbb{E}_n^{\sigma_1,\sigma_2}(F_{\mathbb{R}}) .$$

Since $n \in \mathbb{N}$ was arbitrary, we have:

$$\lim_{n\to\infty} \sup\nolimits_{\sigma_1' \in \Sigma^1_{[\![\mathsf{P}]\!]_{\mathbb{N}}^C}} \inf\nolimits_{\sigma_2' \in \Sigma^2_{[\![\mathsf{P}]\!]_{\mathbb{N}}^C}} \mathbb{E}_n^{\sigma_1',\sigma_2'}(F_{\mathbb{N}}) = \lim_{n\to\infty} \sup\nolimits_{\sigma_1 \in \Sigma^1_{[\![\mathsf{P}]\!]_{\mathbb{R}}^C}} \inf\nolimits_{\sigma_2 \in \Sigma^2_{[\![\mathsf{P}]\!]_{\mathbb{R}}^C}} \mathbb{E}_n^{\sigma_1,\sigma_2}(F_{\mathbb{R}})$$

and hence using Lemma 1 it follows that:

$$\sup\nolimits_{\sigma_1' \in \Sigma^1_{[\![\mathsf{P}]\!]_{\mathbb{N}}^C}} \inf\nolimits_{\sigma_2' \in \Sigma^2_{[\![\mathsf{P}]\!]_{\mathbb{N}}^C}} \mathbb{E}^{\sigma_1',\sigma_2'}(F_{\mathbb{N}}) = \sup\nolimits_{\sigma_1 \in \Sigma^1_{[\![\mathsf{P}]\!]_{\mathbb{R}}^C}} \inf\nolimits_{\sigma_2 \in \Sigma^2_{[\![\mathsf{P}]\!]_{\mathbb{R}}^C}} \mathbb{E}^{\sigma_1,\sigma_2}(F_{\mathbb{R}}) .$$

The fact that the limit can move inside the sup and inf operators on both sides of the inequality follows from the uniform convergence results of Lemma 1. □

5 Case Studies

In this section, we apply our approach to two case studies, a security protocol and a scheduling problem, both of which have been previously modelled as PTAs [38]. In both case studies, by working with games we are able to give more realistic models that overcome the limitations of the earlier PTA models. We specify the finite-state TSG digital clocks semantic models of the case studies using the PRISM language and employ the PRISM-games tool [41] to perform the

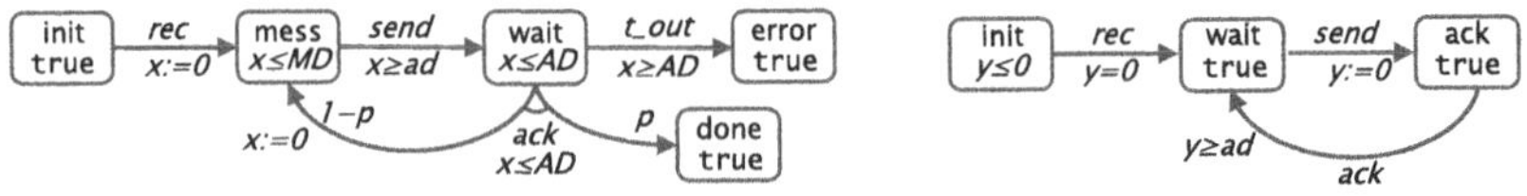

(a) Originator (b) Honest recipient

Fig. 2. PTAs used to model the non-repudiation protocol

analysis. Using PRISM-games we are not only able to find optimal probabilistic and expected reachability values, but also synthesise optimal strategies for the players. PRISM files for the case studies are available from [53].

Non-repudiation Protocol. Markowitch and Roggeman's non-repudiation protocol for information transfer [44] is designed to allow an originator O to transfer information to a recipient R while guaranteeing non-repudiation, that is, neither O nor R can deny that they participated in the transfer.

Randomisation is fundamental to the protocol as, in the initialisation step, O randomly selects a positive integer N that is never revealed to R during execution. Timing is also fundamental as, to prevent R potentially gaining an advantage, if O does not receive an acknowledgement within a specific timeout value (denoted AD), the protocol is stopped and O states R is trying to cheat. In previous PTA models of the protocol [42,45] the originator O had fixed behaviour, while the choices of a (malicious) recipient R included varying the delay between receiving a message from O and sending an acknowledgement. By modelling the protocol as a two-player game we can allow both O and R to make choices which can depend on the history, i.e., the previous behaviour of the parties. The game is naturally turn-based since, in each round, first O sends a message after a delay of their choosing and, after receiving this message, R can respond with an acknowledgement after a delay of their choosing.

We first consider an 'honest' version of the protocol where both O and R can choose delays for their messages but do follow the protocol (i.e., send messages and acknowledgements before timeouts occur). The component PTA models for O and R are presented in Fig. 2. In the PTA for O, the message delay is between $md = 2$ and $MD = 9$ time units, while the acknowledgement delay is at least $ad = 1$ time units and $AD = 5$ is the timeout value. In addition, the probabilistic choice of N is made using a geometric distribution with parameter $p \in (0, 1]$. The parallel composition of these two components then gives the TPTG model of the protocol by assigning control of locations to either O or R, based on which party decides on the delay. There is a complication in the location where O is waiting and R is sending an acknowledgement, as the delay before sending the acknowledgement controlled by R, while if the timeout is reached O should end the protocol. However, since O's behaviour is deterministic in this location, we assign this location to be under the control of R, but add the constraints that an acknowledgement can only be sent before the timeout is reached. If O's behaviour was not deterministic, then a turn-based model would not be sufficient and the protocol would need to be modelled as a concurrent game.

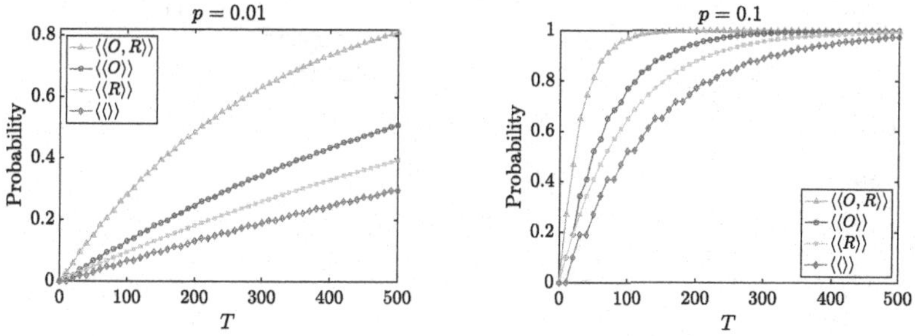

Fig. 3. Max. probability the protocol terminates successfully by time T (honest version)

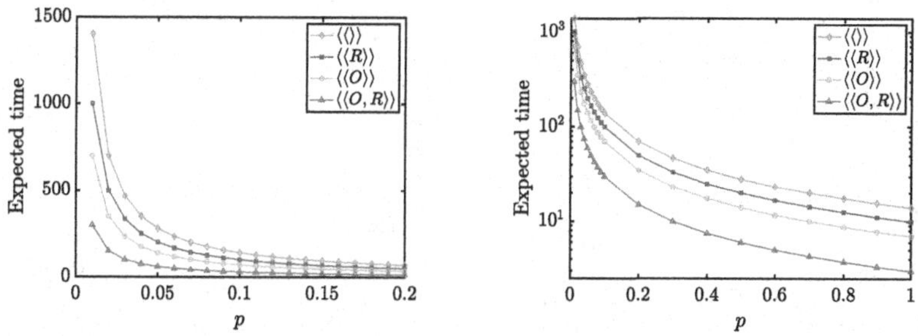

Fig. 4. Min. expected time until the protocol terminates successfully (honest version)

We also consider two 'malicious' versions of the protocol, one in which R is allowed to guess which is the last message (malicious version 1) and a version further extended by giving R additional power through a probabilistic decoder that can decode a message with probability 0.25 before O will timeout (malicious version 2). The TPTG models follow the same structure as that for the 'honest' version, requiring that, in the locations where O is waiting for an acknowledgement, once the timeout has been reached the only possible behaviour is for the protocol to terminate and O states R is trying to cheat.

For the 'honest' version, Figs. 3 and 4 present results when different coalitions try to maximise the probability the protocol terminates successfully by time T when $p = 0.01$ and $p = 0.1$ and minimise expected time for successful termination as the parameter p varies. More precisely, we consider the coalition of both players ($\langle\langle O, R\rangle\rangle$), a single player ($\langle\langle O\rangle\rangle$ or $\langle\langle R\rangle\rangle$) and the empty coalition ($\langle\langle\rangle\rangle$). Using a PTA model, only the first and last cases could be considered. As can be seen, both parties have some control over the time it takes for the protocol to complete and O has greater power as it can delay messages longer than R (if R delays too long then O will terminate the protocol stating R is cheating).

In the case of the versions with a malicious recipient, in Figs. 5 and 6 we have plotted the maximum probability the recipient gains information by time T for versions 1 and 2 respectively. We have included the cases where O works against R ($\langle\langle R\rangle\rangle$) and where they collaborate ($\langle\langle O, R\rangle\rangle$). As we can see, although

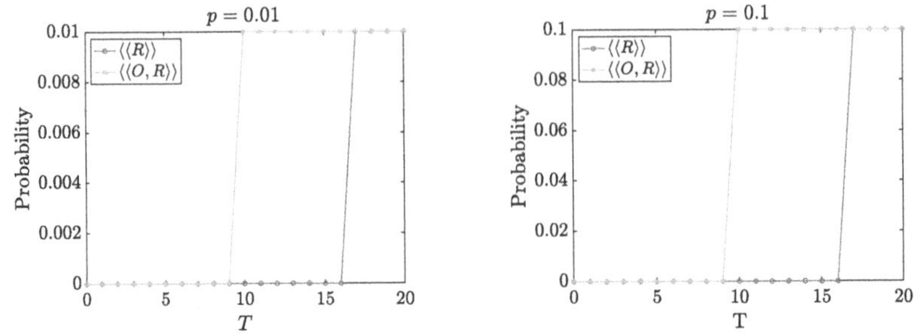

Fig. 5. Maximum probability R gains information by time T (malicious version 1)

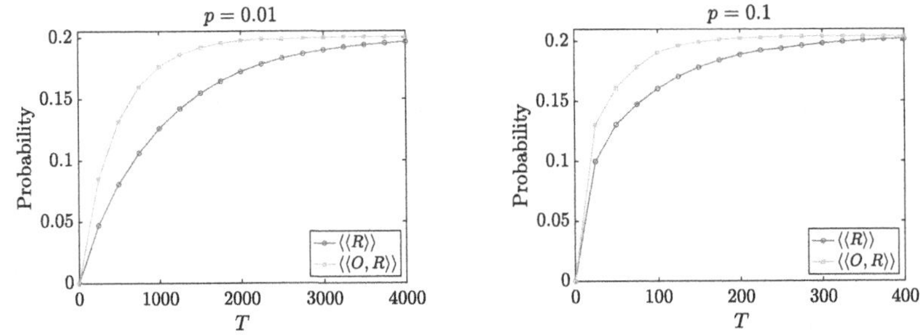

Fig. 6. Maximum probability R gains information by time T (malicious version 2)

O cannot reduce the probability of R obtaining information, it can to some extent increase the time it takes R to obtain this information.

Processor Task Scheduling. This case study is based on the task-graph scheduling problem from [12]. The task-graph is given in Fig. 7 and is for evaluating the expression $D \times (C \times (A{+}B)){+}((A{+}B){+}(C \times D))$ where each multiplication and addition is evaluated on one of two processors, P_1 and P_2. The time and energy required to perform these operations is different, with P_1 being faster than P_2 while consuming more energy as detailed below.

- Time and energy usage of P_1: $[0, 2]$ picoseconds for addition, $[0, 3]$ picoseconds multiplication, 10 Watts when idle and 90 Watts when active.
- Time and energy usage of P_2: $[0, 5]$ picoseconds for addition, $[0, 7]$ picoseconds multiplication, 20 Watts when idle and 30 Watts when active.

A (non-probabilistic) TA model is considered in [12], which is the parallel composition of a TA for each processor and for the scheduler. Previously, in [45], we extended this model by adding probabilistic behaviour to give a PTA. However, the execution time of the processors had to remain fixed since the nondeterminism was under the control of the scheduler, and therefore the optimal scheduler would always choose the minimum execution time for each operation. By moving to a TPTG model, we can allow the execution times to be under the control of a separate player (the environment). We further extend the model by

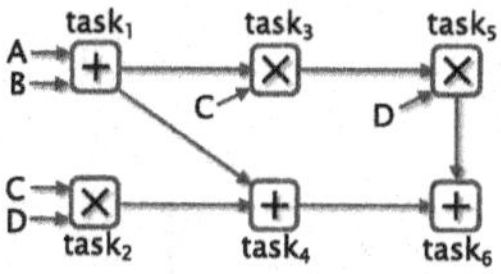

Fig. 7. Task graph for computing $D \times (C \times (A+B))+((A+B)+(C \times D))$

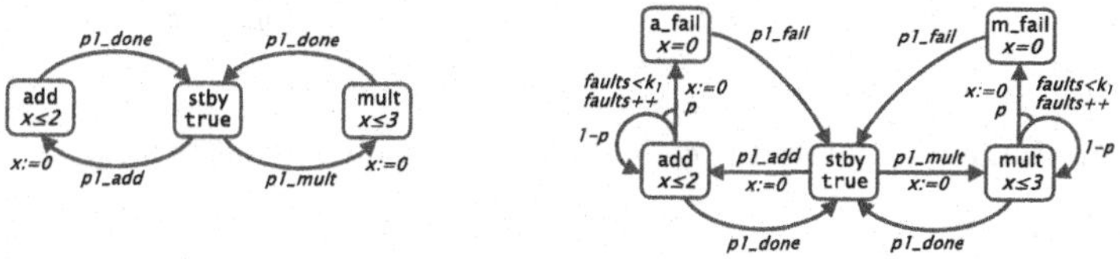

(a) Processor P_1 (b) Faulty version of processor P_1

Fig. 8. PTAs for the task-graph scheduling case study

allowing the processors P_1 and P_2 to have at most k_1 and k_2 faults respectively. We assume that the probability of any fault causing a failure is p and faults can happen at any time a processor is active, i.e., the time the faults occur is under the control of the environment. Again, we could not model this extension with a PTA, since the scheduler would then be in control of when the faults occurred, and therefore could decide that no faults would occur.

As explained in [12], an optimal schedule for a game model in which delays can vary does not yield a simple assignment of tasks to processors at specific times as presented in [45] for PTAs, but instead it is an assignment that also has as input when previous tasks were completed and on which processors.

In Fig. 8 we present both the original TA model for processor P_1, in which the execution time is non-deterministic, and the extended PTA, which allows k_1 faults and where the probability of a fault causing a failure equals p. The PTA includes an integer variable *faults* and the missing enabling conditions equal `true`. To specify the automaton for the scheduler and ensure that we can then build a turn-based game, we restrict the scheduler so that it decides what tasks to schedule initially and immediately after a task ends, then passes control to the environment, which decides the time for the next active task to end.

In Fig. 9 we have plotted both the optimal expected time and energy when there are different number of faults in each processor as the parameter p (the probability of a fault causing a failure) varies. As would be expected, both the optimal expected time and energy consumption increases both as the number of faults increases and the probability that a fault causes a failure increases.

Considering the synthesised optimal schedulers for the expected time case, when $k_1 = k_2 = 1$ and $p = 1$, the optimal approach is to just use the faster processor P_1 and the expected time equals 18.0. The optimal strategy for the environment, i.e., the choices that yield the worst-case expected time, against this scheduler is to delay all tasks as long as possible and cause a fault when

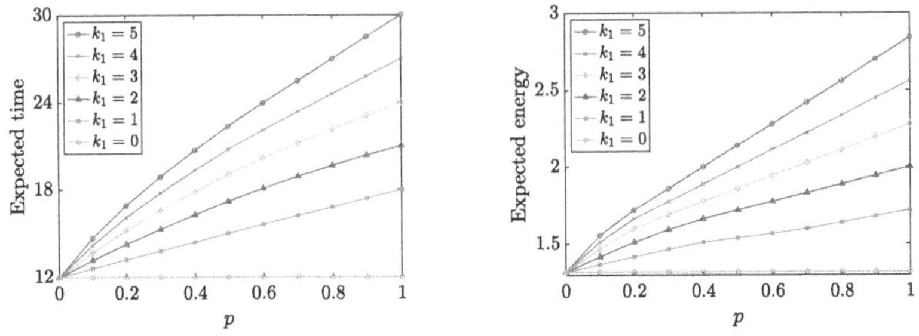

Fig. 9. Minimum expected time and energy to complete all tasks ($k_2 = k_1$)

a multiplication task is just about to complete on P_1 (recall P_2 is never used under the optimal scheduler). A multiplication is chosen as this takes longer (3 picoseconds) than an addition task (2 picoseconds). These choices can be seen through the fact that 18.0 is the time for 4 multiplications and 3 additions to be performed on P_1, while the problem requires 3 multiplications and 3 additions. As soon as the probability of a fault causing a failure is less than 1, the optimal scheduler does use processor P_2 from the beginning by initially scheduling $task_1$ on process P_1 and $task_2$ on processor P_2 (which is also optimal when no faults can occur).

In the case of the expected energy consumption, the optimal scheduler uses both processes unless one has 2 or more faults than the other and there is only a small chance that a fault will cause a failure. For example, if P_1 has 3 faults and P_2 has 1 fault, then P_1 is only used by the optimal scheduler when the probability of a failure causing a fault is approximately 0.25 or less.

6 Conclusions

We have introduced turn-based probabilistic timed games and shown that digital clocks are sufficient for analysing a large class of such games and performance properties. We have demonstrated the feasibility of the approach through two case studies. However, there are limitations of the method since, in particular, as for PTAs [38], the digital clocks semantics does not preserve stopwatch properties or general (nested) temporal logic specifications.

We are investigating extending the approach to concurrent probabilistic timed games. However, since such games are not determined for expected reachability properties [24], this is not straightforward. One direction is to find a class of games which are determined. If we are able to find such a class, then the extension of PRISM-games to concurrent stochastic games [36] could be used to verify this class. Work on finite-state concurrent stochastic games has recently been extended to the case when players have distinct objectives [37] and considering such objectives in the real-time case is also a direction of future research.

Another direction of future research is to formulate a zone-based approach for verifying probabilistic timed games. For the case of probabilistic reachability, this appears possible through the approach of [32] for PTAs. However, it is less clear that the techniques for expected time [28] and expected prices [34], and temporal logic specifications [40] for PTAs, can be extended to TPTGs. Finally, we mention that, although the PRISM language models used in Sect. 5 were built by hand, in future we plan to automate this procedure, extending the one already implemented in PRISM [33] for PTAs.

Acknowledgements. This work is partially supported by the EPSRC Programme Grant on Mobile Autonomy and the PRINCESS project, under the DARPA BRASS programme (contract FA8750-16-C-0045).

References

1. de Alfaro, L., Faella, M., Henzinger, T.A., Majumdar, R., Stoelinga, M.: The element of surprise in timed games. In: Amadio, R., Lugiez, D. (eds.) CONCUR 2003. LNCS, vol. 2761, pp. 144–158. Springer, Heidelberg (2003). https://doi.org/10.1007/978-3-540-45187-7_9
2. Aljazzar, H., Fischer, M., Grunske, L., Kuntz, M., Leitner, F., Leue, S.: Safety analysis of an airbag system using probabilistic FMEA and probabilistic counter examples. In: Proceedings of QEST 2009. IEEE (2009)
3. Alur, R., Bernadsky, M., Madhusudan, P.: Optimal reachability for weighted timed games. In: Díaz, J., Karhumäki, J., Lepistö, A., Sannella, D. (eds.) ICALP 2004. LNCS, vol. 3142, pp. 122–133. Springer, Heidelberg (2004). https://doi.org/10.1007/978-3-540-27836-8_13
4. Alvim, M., Chatzikokolakis, K., Kawamoto, Y., Palamidessi, C.: A game-theoretic approach to information-flow control via protocol composition. Entropy **20**(5), 382 (2018)
5. Alvim, M., Chatzikokolakis, K., Palamidessi, C., Smith, G.: Measuring information leakage using generalized gain functions. In: Proceedings of CSF 2012. IEEE (2012)
6. Asarin, E., Maler, O., Pnueli, A., Sifakis, J.: Controller synthesis for timed automata. In: Proceedings of SSC 1998. Elsevier (1998)
7. Baier, C., Haverkort, B., Hermanns, H., Katoen, J.P.: Performance evaluation and model checking join forces. CACM **53**(9), 76–85 (2010)
8. Behrmann, G., Cougnard, A., David, A., Fleury, E., Larsen, K.G., Lime, D.: UPPAAL-Tiga: time for playing games!. In: Damm, W., Hermanns, H. (eds.) CAV 2007. LNCS, vol. 4590, pp. 121–125. Springer, Heidelberg (2007). https://doi.org/10.1007/978-3-540-73368-3_14
9. Behrmann, G., et al.: Minimum-cost reachability for priced time automata. In: Di Benedetto, M.D., Sangiovanni-Vincentelli, A. (eds.) HSCC 2001. LNCS, vol. 2034, pp. 147–161. Springer, Heidelberg (2001). https://doi.org/10.1007/3-540-45351-2_15
10. Bouyer, P., Brihaye, T., Markey, N.: Improved undecidability results on weighted timed automata. IPL **98**, 188–194 (2006)
11. Bouyer, P., Cassez, F., Fleury, E., Larsen, K.G.: Optimal strategies in priced timed game automata. In: Lodaya, K., Mahajan, M. (eds.) FSTTCS 2004. LNCS, vol. 3328, pp. 148–160. Springer, Heidelberg (2004). https://doi.org/10.1007/978-3-540-30538-5_13
12. Bouyer, P., Fahrenberg, U., Larsen, K., Markey, N.: Quantitative analysis of real-time systems using priced timed automata. Comm. ACM **54**(9), 78–87 (2011)

13. Bouyer, P., Forejt, V.: Reachability in stochastic timed games. In: Albers, S., Marchetti-Spaccamela, A., Matias, Y., Nikoletseas, S., Thomas, W. (eds.) ICALP 2009. LNCS, vol. 5556, pp. 103–114. Springer, Heidelberg (2009). https://doi.org/10.1007/978-3-642-02930-1_9
14. Bouyer, P., Markey, N., Randour, M., Larsen, K., Laursen, S.: Average-energy games. Acta Informatica **55**(2), 91–127 (2018)
15. Brázdil, T., Hermanns, H., Krcál, J., Kretínský, J., Rehák, V.: Verification of open interactive Markov chains. In: Proceedings of FSTTCS 2012, LIPIcs 18 (2012)
16. Brázdil, T., Krčál, J., Křetínský, J., Kučera, A., Řehák, V.: Stochastic real-time games with qualitative timed automata objectives. In: Gastin, P., Laroussinie, F. (eds.) CONCUR 2010. LNCS, vol. 6269, pp. 207–221. Springer, Heidelberg (2010). https://doi.org/10.1007/978-3-642-15375-4_15
17. Brihaye, T., Bruyère, V., Raskin, J.-F.: On optimal timed strategies. In: Pettersson, P., Yi, W. (eds.) FORMATS 2005. LNCS, vol. 3829, pp. 49–64. Springer, Heidelberg (2005). https://doi.org/10.1007/11603009_5
18. Cassez, F., David, A., Larsen, K.G., Lime, D., Raskin, J.-F.: Timed control with observation based and stuttering invariant strategies. In: Namjoshi, K.S., Yoneda, T., Higashino, T., Okamura, Y. (eds.) ATVA 2007. LNCS, vol. 4762, pp. 192–206. Springer, Heidelberg (2007). https://doi.org/10.1007/978-3-540-75596-8_15
19. Cassez, F., David, A., Fleury, E., Larsen, K.G., Lime, D.: Efficient on-the-fly algorithms for the analysis of timed games. In: Abadi, M., de Alfaro, L. (eds.) CONCUR 2005. LNCS, vol. 3653, pp. 66–80. Springer, Heidelberg (2005). https://doi.org/10.1007/11539452_9
20. Condon, A.: On algorithms for simple stochastic games. In: Advances in Computational Complexity Theory, DIMACS Series in DMTCS 13 (1993)
21. Dehnert, C., Junges, S., Katoen, J.-P., Volk, M.: A Storm is coming: a modern probabilistic model checker. In: Majumdar, R., Kunčak, V. (eds.) CAV 2017. LNCS, vol. 10427, pp. 592–600. Springer, Cham (2017). https://doi.org/10.1007/978-3-319-63390-9_31
22. Filar, J., Vrieze, K.: Competitive Markov Decision Processes. Springer, New York (1997)
23. Forejt, V., Kwiatkowska, M., Norman, G., Trivedi, A.: Expected reachability-time games. In: Chatterjee, K., Henzinger, T.A. (eds.) FORMATS 2010. LNCS, vol. 6246, pp. 122–136. Springer, Heidelberg (2010). https://doi.org/10.1007/978-3-642-15297-9_11
24. Forejt, V., Kwiatkowska, M., Norman, G., Trivedi, A.: Expected reachability-time games. TCS **631**, 139–160 (2016)
25. Henzinger, T.A., Manna, Z., Pnueli, A.: What good are digital clocks? In: Kuich, W. (ed.) ICALP 1992. LNCS, vol. 623, pp. 545–558. Springer, Heidelberg (1992). https://doi.org/10.1007/3-540-55719-9_103
26. Hermanns, H.: Interactive Markov Chains and the Quest for Quantified Quality. LNCS, vol. 2428. Springer, Heidelberg (2002). https://doi.org/10.1007/3-540-45804-2
27. van der Hoek, W., Wooldridge, M.: Model checking cooperation, knowledge, and time - a case study. Res. Econ. **57**(3), 235–265 (2003)
28. Jovanovic, A., Kwiatkowska, M., Norman, G., Peyras, Q.: Symbolic optimal expected time reachability computation and controller synthesis for probabilistic timed automata. TCS **669**, 1–21 (2017)
29. Jurdziński, M., Kwiatkowska, M., Norman, G., Trivedi, A.: Concavely-priced probabilistic timed automata. In: Bravetti, M., Zavattaro, G. (eds.) CONCUR 2009.

LNCS, vol. 5710, pp. 415–430. Springer, Heidelberg (2009). https://doi.org/10.1007/978-3-642-04081-8_28
30. Kemeny, J., Snell, J., Knapp, A.: Denumerable Markov Chains. Springer, New York (1976). https://doi.org/10.1007/978-1-4684-9455-6
31. Krčál, J.: Determinacy and optimal strategies in stochastic games. Master's thesis, School of Informatics, Masaryk University, Brno (2009)
32. Kwiatkowska, M., Norman, G., Parker, D.: Stochastic games for verification of probabilistic timed automata. In: Ouaknine, J., Vaandrager, F.W. (eds.) FORMATS 2009. LNCS, vol. 5813, pp. 212–227. Springer, Heidelberg (2009). https://doi.org/10.1007/978-3-642-04368-0_17
33. Kwiatkowska, M., Norman, G., Parker, D.: PRISM 4.0: verification of probabilistic real-time systems. In: Gopalakrishnan, G., Qadeer, S. (eds.) CAV 2011. LNCS, vol. 6806, pp. 585–591. Springer, Heidelberg (2011). https://doi.org/10.1007/978-3-642-22110-1_47
34. Kwiatkowska, M., Norman, G., Parker, D.: Symbolic verification and strategy synthesis for linearly-priced probabilistic timed automata. In: Aceto, L., Bacci, G., Bacci, G., Ingólfsdóttir, A., Legay, A., Mardare, R. (eds.) Models, Algorithms, Logics and Tools. LNCS, vol. 10460, pp. 289–309. Springer, Cham (2017). https://doi.org/10.1007/978-3-319-63121-9_15
35. Kwiatkowska, M., Norman, G., Parker, D.: Verification and control of turn-based probabilistic real-time games. arXiv:1906.09142 (2019)
36. Kwiatkowska, M., Norman, G., Parker, D., Santos, G.: Automated verification of concurrent stochastic games. In: McIver, A., Horvath, A. (eds.) QEST 2018. LNCS, vol. 11024, pp. 223–239. Springer, Cham (2018). https://doi.org/10.1007/978-3-319-99154-2_14
37. Kwiatkowska, M., Norman, G., Parker, D., Santos, G.: Equilibria-based probabilistic model checking for concurrent stochastic games. In: Proceeding of FM 2019, LNCS. Springer, Berlin (2019, to appear)
38. Kwiatkowska, M., Norman, G., Parker, D., Sproston, J.: Performance analysis of probabilistic timed automata using digital clocks. FMSD **29**, 33–78 (2006)
39. Kwiatkowska, M., Norman, G., Segala, R., Sproston, J.: Automatic verification of real-time systems with discrete probability distributions. TCS **282**, 101–150 (2002)
40. Kwiatkowska, M., Norman, G., Sproston, J., Wang, F.: Symbolic model checking for probabilistic timed automata. IC **205**(7), 1027–1077 (2007)
41. Kwiatkowska, M., Parker, D., Wiltsche, C.: PRISM-games: verification and strategy synthesis for stochastic multi-player games with multiple objectives. STTT **20**(2), 195–210 (2018)
42. Lanotte, R., Maggiolo-Schettini, A., Troina, A.: Automatic analysis of a non-repudiation protocol. In: Proceedings of QAPL 2004, ENTCS 112 (2005)
43. Maler, O., Pnueli, A., Sifakis, J.: On the synthesis of discrete controllers for timed systems. In: Mayr, E.W., Puech, C. (eds.) STACS 1995. LNCS, vol. 900, pp. 229–242. Springer, Heidelberg (1995). https://doi.org/10.1007/3-540-59042-0_76
44. Markowitch, O., Roggeman, Y.: Probabilistic non-repudiation without trusted third party. In: Proceedings of Workshop Security in Communication Networks (1999)
45. Norman, G., Parker, D., Sproston, J.: Model checking for probabilistic timed automata. FMSD **43**(2), 164–190 (2013)
46. Norman, G., Parker, D., Zou, X.: Verification and control of partially observable probabilistic systems. RTS **53**(3), 354–402 (2017)

47. Oualhadj, Y., Reynier, P.-A., Sankur, O.: Probabilistic robust timed games. In: Baldan, P., Gorla, D. (eds.) CONCUR 2014. LNCS, vol. 8704, pp. 203–217. Springer, Heidelberg (2014). https://doi.org/10.1007/978-3-662-44584-6_15
48. Rudin, W.: Principles of Mathematical Analysis, 3rd edn. McGraw-Hill, New York (1976)
49. La Torre, S., Mukhopadhyay, S., Murano, A.: Optimal-reachability and control for acyclic weighted timed automata. In: Baeza-Yates, R., Montanari, U., Santoro, N. (eds.) Foundations of Information Technology in the Era of Network and Mobile Computing. ITIFIP, vol. 96, pp. 485–497. Springer, Boston, MA (2002). https://doi.org/10.1007/978-0-387-35608-2_40
50. Tripakis, S.: Verifying progress in timed systems. In: Katoen, J.-P. (ed.) ARTS 1999. LNCS, vol. 1601, pp. 299–314. Springer, Heidelberg (1999). https://doi.org/10.1007/3-540-48778-6_18
51. Tripakis, S., Altisen, K.: On-the-fly controller synthesis for discrete and dense-time systems. In: Wing, J.M., Woodcock, J., Davies, J. (eds.) FM 1999. LNCS, vol. 1708, pp. 233–252. Springer, Heidelberg (1999). https://doi.org/10.1007/3-540-48119-2_15
52. Tripakis, S., Yovine, S., Bouajjan, A.: Checking timed Büchi automata emptiness efficiently. FMSD **26**(3), 267–292 (2005)
53. Supporting material. www.prismmodelchecker.org/files/tptgs/

Refinement Metrics for Quantitative Information Flow

Konstantinos Chatzikokolakis[1(✉)] and Geoffrey Smith[2]

[1] University of Athens, Athens, Greece
kostasc@di.uoa.gr
[2] Florida International University, Miami, USA
smithg@cis.fiu.edu

Abstract. In Quantitative Information Flow, *refinement* ($\sqsubseteq$) expresses the strong property that one channel never leaks more than another. Since two channels are then typically incomparable, here we explore a family of *refinement quasimetrics* offering greater flexibility. We show these quasimetrics let us unify refinement and capacity, we show that some of them can be computed efficiently via linear programming, and we establish upper bounds via the Earth Mover's distance. We illustrate our techniques on the Crowds protocol.

1 Introduction

Completely eliminating the leakage of sensitive information by computer systems is often infeasible, making it attractive to approach the problem quantitatively. *Quantitative information flow* [3] offers a rich family of *g-leakage* measures of the amount of leakage caused by a channel C taking a secret input X to an observable output Y; these measures are parameterized by X's *prior distribution* π and a *gain function* g, which models the adversary's capabilities and goals (See Sect. 2 for a brief review.).

While g-leakage lets us precisely measure information leakage in a rich variety of operational scenarios, we may be unsure about the appropriate prior π or gain function g to use, and hence about the robustness of our conclusions. Two approaches to robustness have proved fruitful: *capacity*, which is the maximum leakage over some sets of gain functions and priors, and *refinement*: channel A is *refined* by channel B, written $\mathsf{A} \sqsubseteq \mathsf{B}$, iff B never leaks more than A, regardless of the prior or gain function. Remarkably, refinement also has a structural characterization: $\mathsf{A} \sqsubseteq \mathsf{B}$ iff there exists a "post-processing" channel matrix R such that $\mathsf{B} = \mathsf{AR}$. Moreover, ($\sqsubseteq$) is a *partial order* on abstract channels.

Unfortunately, refinement is a very *partial* partial order, in that most channels are incomparable. And this "Boolean" nature of refinement is inconsistent with the spirit of quantitative information flow, which is above all motivated by the need to tolerate imperfect security. If $\mathsf{A} \not\sqsubseteq \mathsf{B}$, then we know that B can be worse than A; but we do not know whether B is really terrible, or whether is just slightly worse than A. This is the main issue that we address in this paper.

M. S. Alvim et al. (Eds.): Palamidessi Festschrift, LNCS 11760, pp. 397–416, 2019.
https://doi.org/10.1007/978-3-030-31175-9_23

In mathematics, it is common to use metrics to provide a "finer", quantified generalization of a relation. A relation has a "Boolean" nature: elements are either related or not. In the metric version, related elements have distance 0; but for non-related elements the metric tells you *how much* the relation is violated. For instance, the Euclidean distance can be seen as a quantified generalization of the equality relation ($=$) on $\mathbb{R}$ — the distance $|x - y|$ tells us how much $x = y$ is violated; 0 means that it is not violated at all.[1] This approach is not of course limited to symmetric relations. For instance, the quasimetric

$$\mathbf{q}^{+}_{<}(x, y) \quad := \quad \max\{y - x, 0\},$$

can be seen as a quantified version of ($\geq$); it measures how much ($\geq$) is violated, and is 0 iff $x \geq y$.

In this paper we define a family of *refinement quasimetrics* $\mathrm{ref}^{q}_{\mathcal{D},\mathcal{G}}$ to measure how much refinement is violated. Refinement is then the *kernel* (i.e. elements at distance 0) of these metrics: we have $\mathrm{ref}^{q}_{\mathcal{D},\mathcal{G}}(\mathsf{A}, \mathsf{B}) = 0$ iff $\mathsf{A} \sqsubseteq \mathsf{B}$. Through this approach, we make the following contributions:

- We treat both additive and multiplicative leakage together via additive and multiplicative quasimetrics $\mathbf{q}^{+}_{<}$ and $\mathbf{q}^{\times}_{<}$ (The latter crucially takes the log of the ratio, instead of just the ratio.).
- We observe that the capacity of C can be expressed as $\mathrm{ref}^{q}_{\mathcal{D},\mathcal{G}}(\mathbb{1}, \mathsf{C})$, where $\mathbb{1}$ is the perfect channel leaking nothing, giving us a unified way of looking at both refinement and capacity.
- In Sect. 4 we show that, for a fixed prior π, the additive refinement metric over all gain functions in $\mathbb{G}^{\updownarrow}\mathcal{X}$ can be computed in polynomial time via linear programming.
- In Sect. 5 we prove (again for a fixed π) that the refinement metric is bounded by the Earth Mover's distance between $[\pi \triangleright \mathsf{A}]$ and $[\pi \triangleright \mathsf{B}]$, allowing it to be approximated very efficiently. We also prove a bound on the metric over all priors.
- In Sect. 6 we give a case study of our techniques on the Crowds protocol.

2 Preliminaries

2.1 Quantitative Information Flow

In *quantitative information flow* [6,7,9] an adversary's prior knowledge about a secret input X drawn from a finite set $\mathcal{X}$ is modeled as a probability distribution π, and a probabilistic system taking input X to observable output Y is modeled as an information-theoretic channel matrix C giving the conditional probabilities $p(y|x)$. Assuming that the adversary knows C, then each output y allows the

[1] Another example is bisimulation metrics for probabilistic properties. Because bisimulation is a strong property that can be broken by tiny modifications to transition probabilities, a variety of bisimulation metrics have been proposed, measuring how much bisimulation is violated.

adversary to update her knowledge about X to a posterior distribution $p_{X|y}$, which moreover has probability $p(y)$ of occurring. As a result, the effect of C is to map each prior distribution π to a *distribution on posterior distributions*, called a *hyper-distribution* and denoted $[\pi{\triangleright}\mathsf{C}]$. For example, channel

C	y_1	y_2	y_3	y_4
x_1	$1/2$	$1/2$	0	0
x_2	0	$1/4$	$1/2$	$1/4$
x_3	$1/2$	$1/3$	$1/6$	0

maps $\pi = (1/4, 1/2, 1/4)$ to hyper

$[\pi{\triangleright}\mathsf{C}]$	$1/4$	$1/3$	$7/24$	$1/8$
x_1	$1/2$	$3/8$	0	0
x_2	0	$3/8$	$6/7$	1
x_3	$1/2$	$1/4$	$1/7$	0

.

This mapping is called the *abstract channel* denoted by C.[2]

It is natural to measure the "vulnerability" of X using functions V_g from probability distributions to reals, which are parameterized with a *gain function* $g : \mathcal{W} \times \mathcal{X} \to \mathbb{R}$ that models the operational scenario; here $\mathcal{W}$ is the set of *actions* that the adversary can make, and $g(w, x)$ is the adversary's *gain* for doing w when the secret's actual value is x. We then define the *g-vulnerability* by $V_g(\pi) = \sup_w \sum_x \pi_x g(w, x)$, since that is the maximum expected gain over all possible actions. The choice of g allows us to model a wide variety of operational scenarios; a commonly used one is the *identity* gain function, given by $\mathcal{W} = \mathcal{X}$ and $g_{\mathrm{id}}(w, x) = 1$ iff $w = x$ and 0 otherwise. For this gain function $V_{g_{\mathrm{id}}}(\pi)$ gives the probability of correctly guessing the secret in one try.

Posterior vulnerability is defined as the average g-vulnerability in the hyper distribution: $V_g[\pi{\triangleright}\mathsf{C}] = \sum_y p(y) V_g(p_{X|y})$. And then *g-leakage* is defined as either the *ratio* or the *difference* between the posterior- and prior g-vulnerability; these are respectively *multiplicative leakage* $\mathcal{L}_g^{\times}(\pi, \mathsf{C})$ and *additive leakage* $\mathcal{L}_g^{+}(\pi, \mathsf{C})$.

2.2 Metrics

A *proper metric* on a set $\mathcal{A}$ is a function $d : \mathcal{A} \to [0, \infty]$ satisfying the following axioms for all x, y, z: $\mathcal{A}$:

1. **Reflexivity:** $d(x, x) = 0$.
2. **Symmetry:** $d(x, y) = d(y, x)$.
3. **Anti-symmetry:** $d(x, y) = d(y, x) = 0 \Rightarrow x = y$.[3]
4. **Triangle inequality:** $d(x, z) \le d(x, y) + d(y, z)$.
5. **Finiteness**: $d(x, y) \neq \infty$.

Various prefixes can be added to "metric" to denote that some of the above conditions are dropped. The following are used in this paper:

1. **Quasi:** symmetry is dropped.
2. **Pseudo:** anti-symmetry is dropped.

[2] Because of structural redundancies (e.g. the ordering and labels of columns), distinct channel matrices may denote the same abstract channel.

[3] When d is not symmetric, this is weaker than the axiom $d(x, y) = 0 \Rightarrow x = y$ (which is sometimes used in the literature even in the non-symmetric case).

3. **Extended:** finiteness is dropped.

Any combination of these prefixes is meaningful. For brevity, we use "metric" to generally refer to *any* of these classes, while we use the exact prefixes (or "proper") to refer to a precise one.

Kernel. The *kernel* of a metric d is a relation, containing points at distance 0:

$$(x, y) \in \ker(d) \qquad \text{iff} \qquad d(x, y) = 0 \ .$$

It is easy to see that the metric properties of d imply the corresponding homonymous properties on relations for $\ker(d)$ (reflexivity, symmetry, anti-symmetry), while the triangle inequality of d corresponds to transitivity for $\ker(d)$. Hence the kernel of a proper metric is always the *equality relation* $=$, the kernel of a pseudometric is an *equivalence relation* (reflexive, symmetric and transitive), and the kernel of a quasimetric is a *partial order* (reflexive, anti-symmetric and transitive).

Quasimetrics on $\mathbb{R}, \mathbb{R}_{\geq 0}$. On $\mathbb{R}$ the following quasimetric is of particular interest:

$$\mathbf{q}_<^+(x, y) \quad := \quad \max\{y - x, 0\} \ .$$

Intuitively, $\mathbf{q}_<^+$ measures (additively) "how much smaller" than y is x; 0 means that x is "no smaller" than y. Or we can view $\mathbf{q}_<^+$ as measuring "how much $x \geq y$ is violated"; 0 means that $x \geq y$ holds (is not violated at all).

On $\mathbb{R}_{\geq 0}$ we can define a multiplicative variant of $\mathbf{q}_<^+$ as follows:[4]

$$\mathbf{q}_<^\times(x, y) \quad := \quad \max\{\log_2 \frac{y}{x}, 0\},$$

with the understanding that $\mathbf{q}_<^\times(0, y) = 0$ iff $y = 0$ and ∞ otherwise ($\mathbf{q}_<^\times$ is an extended quasimetric). Again, $\mathbf{q}_<^\times$ measures "how much smaller" than y is x, but this time multiplicatively. Although $\mathbf{q}_<^+$ and $\mathbf{q}_<^\times$ are clearly different, they coincide on their kernel:

$$\ker(\mathbf{q}_<^+) \quad = \quad \ker(\mathbf{q}_<^\times) \quad = \quad \geq \ .$$

Showing the quasimetric properties of $\mathbf{q}_<^+$ is trivial for all but the triangle inequality; for the latter, assuming $x \leq z$ (the case $x > z$ is trivial) we have that

$$\begin{array}{ll} & \mathbf{q}_<^+(x, z) \\ = & z - x \qquad \text{“}x \leq z\text{”} \\ = & z - y + y - x \\ \leq & \max\{z - y, 0\} + \max\{y - x, 0\} \\ = & \mathbf{q}_<^+(x, y) + \mathbf{q}_<^+(y, z) \ . \end{array}$$

[4] The use of logarithm is crucial for reflexivity and the triangle inequality. The choice of the base is arbitrary, but $\log_2$ is interesting due to the connection with min-entropy.

Given a function $f : \mathcal{B} \to \mathcal{A}$ and metric d on $\mathcal{A}$ lets us define a metric $d \circ f$ on $\mathcal{B}$ by $(d \circ f)(x, y) := d(f(x), f(y))$.

Proposition 1. *The composition $d \circ f$ always preserves all metric properties of d except anti-symmetry, which is also preserved if f is injective.*

This construction is useful in that $\mathbf{q}^{\times}_{<} = \mathbf{q}^{+}_{<} \circ \log_2$, from which it follows that $\mathbf{q}^{\times}_{<}$ is an extended quasimetric on $\mathbb{R}_{\geq 0}$.[5]

3 Refinement Metrics

To design metrics for QIF, we start from the simple observation that we can *compare vulnerabilities* using any *(quasi) metric q on $\mathbb{R}_{\geq 0}$*. For instance, *leakage* can be measured by comparing prior and posterior vulnerabilities as follows:

$$\mathcal{L}^q_g(\pi, C) \quad := \quad q(V_g(\pi), V_g[\pi \triangleright C]) \ .$$

The usual *additive* and *multiplicative* variants of leakage can be obtained by instantiating q by $\mathbf{q}^{+}_{<}, \mathbf{q}^{\times}_{<}$ (except that for the latter we get $\log_2$ of the multiplicative leakage[6]):

$$\mathcal{L}^{\mathbf{q}^{+}_{<}}_g(\pi, C) \quad = \quad \mathcal{L}^{+}_g(\pi, \mathsf{C}), \qquad \qquad \mathcal{L}^{\mathbf{q}^{\times}_{<}}_g(\pi, C) \quad = \quad \log_2 \mathcal{L}^{\times}_g(\pi, \mathsf{C}) \ .$$

The conceptual advantage of explicitly using a quasimetric (instead of just the difference or ratio) is twofold: first, we can exploit the metric properties of q and second, we can treat both leakage variants together.

Continuing this line of reasoning, we can obtain a *metric on channels*, by using q to *compare their posterior vulnerabilities.* This brings us to our refinement metric.

Definition 1. *Given classes $\mathcal{D} \subseteq \mathbb{D}\mathcal{X}$ of distributions and $\mathcal{G} \subseteq \mathbb{G}\mathcal{X}$ of gain functions, and metric q on $\mathbb{R}_{\geq 0}$, the refinement metric is defined as*

$$\mathrm{ref}^q_{\mathcal{D},\mathcal{G}}(\mathsf{A}, \mathsf{B}) \quad := \quad \sup_{\pi : \mathcal{D}, g : \mathcal{G}} q(V_g[\pi \triangleright \mathsf{A}], V_g[\pi \triangleright \mathsf{B}]) \ .$$

We write $\mathrm{ref}^q_{\pi,\mathcal{G}}$ instead of $\mathrm{ref}^q_{\{\pi\},\mathcal{G}}$ when $\mathcal{D} = \{\pi\}$, and similarly for $\mathcal{G}$.

Any metric could be meaningful for q (for instance symmetric ones). But here we restrict to quasimetrics such that $\ker(q) = (\geq)$, which we call *order-respecting*. The standard choices of interest are $\mathbf{q}^{+}_{<}$ and $\mathbf{q}^{\times}_{<}$, giving additive and

[5] Technically, to establish this we need to view $\mathbf{q}^{+}_{<}$ as an extended quasimetric on $\mathbb{R} \cup \{-\infty\}$ and $\log_2$ as an injective function $\mathbb{R}_{\geq 0} \to \mathbb{R} \cup \{-\infty\}$.

[6] This is reminiscent of min-entropy; conceptually, however, we did not use $\log_2$ to convert vulnerabilities to entropies, we just used $\mathbf{q}^{\times}_{<}$ to compare vulnerabilities via a metric of multiplicative nature (the log conveniently turns ratios into a metric).

multiplicative comparison, for which we write the refinement metrics as $\mathrm{ref}^{+}_{\mathcal{D},\mathcal{G}}$ and $\mathrm{ref}^{\times}_{\mathcal{D},\mathcal{G}}$ respectively (Note however that many results are independent of the choice of q.).

When $\mathcal{D}$ contains at least one point prior, or when $\mathcal{G}$ contains at least one constant g, then the max in the definition of $\mathbf{q}^{+}_{<}, \mathbf{q}^{\times}_{<}$ can be eliminated, giving

$$\mathrm{ref}^{+}_{\mathcal{D},\mathcal{G}}(\mathsf{A},\mathsf{B}) \quad = \quad \sup_{\pi:\mathcal{D}, g:\mathcal{G}} V_g[\pi\triangleright\mathsf{B}] - V_g[\pi\triangleright\mathsf{A}] \qquad \text{and}$$

$$\mathrm{ref}^{\times}_{\mathcal{D},\mathcal{G}}(\mathsf{A},\mathsf{B}) \quad = \quad \sup_{\pi:\mathcal{D}, g:\mathcal{G}} \log_2 \frac{V_g[\pi\triangleright\mathsf{B}]}{V_g[\pi\triangleright\mathsf{A}]} \;.$$

Intuitively, $\mathrm{ref}^{+}_{\mathcal{D},\mathcal{G}}, \mathrm{ref}^{\times}_{\mathcal{D},\mathcal{G}}$ measure (additively or multiplicatively) *how robust* it is to replace A by B, with respect to the vulnerability of the system. A value of 0 means no risk—B is never worse than A (for $\mathcal{D},\mathcal{G}$)—while a positive value means that there is risk, but it might be small. Note that refinement is *asymmetric* by design: replacing A by B is inherently different than replacing B by A.

For fixed π and g, we can obtain $\mathrm{ref}^{q}_{\pi,g}$ as the composition

$$\mathrm{ref}^{q}_{\pi,g} \quad = \quad q \circ V_g[\pi\triangleright\cdot], \tag{1}$$

and then obtain $\mathrm{ref}^{q}_{\mathcal{D},\mathcal{G}}$ as the sup of a family of quasimetrics

$$\mathrm{ref}^{q}_{\mathcal{D},\mathcal{G}} \quad = \quad \sup_{\pi:\mathcal{D}, g:\mathcal{G}} \mathrm{ref}^{q}_{\pi,g} \;. \tag{2}$$

This brings us to the following proposition.

Proposition 2. $\mathrm{ref}^{q}_{\mathcal{D},\mathcal{G}}$ *always inherits reflexivity, symmetry and the triangle inequality from* q, *but not necessarily anti-symmetry and finiteness.*

Proof. From (1) and Proposition 1 we get that $\mathrm{ref}^{q}_{\pi,g}$ inherits all metric properties of q except anti-symmetry ($V_g[\pi\triangleright\cdot]$ is not injective). Then, from (2), we only need to show that the sup of metrics preserves reflexivity, symmetry and the triangle inequality. The first two are trivial; for the triangle inequality, let $d = \sup_i d_i$, we have:

$$\begin{array}{lll} & d(x,z) & \\ = & \sup_i d_i(x,z) & \\ \leq & \sup_i (d_i(x,y) + d_i(y,z)) & \text{“triangle ineq. for } d_i\text{”} \\ \leq & \sup_i d_i(x,y) + \sup_i d_i(y,z) & \text{“}\sup_i(a_i+b_i) \leq \sup_i a_i + \sup_i b_i\text{”} \\ = & d(x,y) + d(y,z) \;. & \end{array}$$

Clearly, anti-symmetry is not always preserved (take $\mathcal{G}$ to contain only constant gain functions), nor is finiteness (for $\mathcal{G} = \mathbb{G}\mathcal{X}$ the sup can be ∞ even if q itself is finite). □

Hence $\mathrm{ref}^{+}_{\mathcal{D},\mathcal{G}}$ and $\mathrm{ref}^{\times}_{\mathcal{D},\mathcal{G}}$ always satisfy reflexivity and the triangle inequality (i.e. they are extended quasi pseudo metrics); they might satisfy additional properties for specific choices of $\mathcal{D}$ and $\mathcal{G}$.

3.1 Recovering the Refinement Relation

The following result states that if $\mathcal{D}$ and $\mathcal{G}$ are "sufficiently rich", then we can obtain $\sqsubseteq$ as the kernel of the corresponding refinement metric $\mathrm{ref}^q_{\mathcal{D},\mathcal{G}}$.

Theorem 1. *Assume that q is order-respecting, $\mathcal{D}$ contains a full-support prior, and $\mathcal{G}$ contains the non-negative, 1-bounded gain functions with finitely many actions. Then*

$$\ker(\mathrm{ref}^q_{\mathcal{D},\mathcal{G}}) \quad = \quad \sqsubseteq \ .$$

Proof. (Sketch.) If $\mathsf{A} \not\sqsubseteq \mathsf{B}$, then the proof of the Coriaceous Theorem [10, Theorem 9] shows that for the uniform prior π^u, there is a gain function g, restricted as in the statement of the theorem, such that $V_g[\pi^u{\triangleright}\mathsf{A}] < V_g[\pi^u{\triangleright}\mathsf{B}]$; hence if q is order-respecting we have $\mathrm{ref}^q_{\pi^u,\mathcal{G}}(\mathsf{A},\mathsf{B}) > 0$. And by embedding the prior into g, we can replace π^u with any full-support prior. □

Hence the anti-symmetry of $\sqsubseteq$ on abstract channels is transferred to $\mathrm{ref}^q_{\mathcal{D},\mathcal{G}}$:

Corollary 1. *Under the conditions of Theorem 1, $\mathrm{ref}^q_{\mathcal{D},\mathcal{G}}$ is anti-symmetric on abstract channels.*

3.2 Recovering Capacity

Since the noninterfering channel $\mathbb{1}$ always produces a point hyper, capacity is simply given by the corresponding refinement metric between $\mathbb{1}$ and C.

Proposition 3. *For any q and $\mathcal{D},\mathcal{G}$ we have that $\mathcal{ML}^q_{\mathcal{G}}(\mathcal{D},\mathsf{C}) = \mathrm{ref}^q_{\mathcal{D},\mathcal{G}}(\mathbb{1},\mathsf{C})$.*

Proof. Direct consequence of the fact that $V_g[\pi{\triangleright}\mathbb{1}] = V_g(\pi)$. □

Recall that using the quasimetrics $\mathrm{ref}^{+}_{\mathcal{D},\mathcal{G}}$ or $\mathrm{ref}^{\times}_{\mathcal{D},\mathcal{G}}$ for q make $\mathrm{ref}^q_{\mathcal{D},\mathcal{G}}$ itself non-symmetric. In particular $\mathrm{ref}^q_{\mathcal{D},\mathcal{G}}(\mathbb{1},\mathsf{C})$ gives the capacity of C, while $\mathrm{ref}^q_{\mathcal{D},\mathcal{G}}(\mathsf{C},\mathbb{1})$ is always zero since $\mathsf{C} \sqsubseteq \mathbb{1}$.

A corollary of the above proposition is that the refinement metric of A,B can never exceed the capacity of B. This is intuitive since $\mathrm{ref}^q_{\mathcal{D},\mathcal{G}}(\mathsf{A},\mathsf{B})$ measures "how much worse than A can B be", while the capacity measures "how much worse than $\mathbb{1}$ (the best possible channel) can B be".

Corollary 2. *For any order-respecting q and $\mathcal{D},\mathcal{G},\mathsf{A},\mathsf{B}$ we have that*

$$\mathrm{ref}^q_{\mathcal{D},\mathcal{G}}(\mathsf{A},\mathsf{B}) \quad \leq \quad \mathcal{ML}^q_{\mathcal{G}}(\mathcal{D},\mathsf{B}) \ .$$

Proof. We have that:

$$\begin{array}{lll} & \mathrm{ref}^q_{\mathcal{D},\mathcal{G}}(\mathsf{A},\mathsf{B}) & \\ \leq & \mathrm{ref}^q_{\mathcal{D},\mathcal{G}}(\mathsf{A},\mathbb{1}) + \mathrm{ref}^q_{\mathcal{D},\mathcal{G}}(\mathbb{1},\mathsf{B}) & \text{“triangle ineq. for } \mathrm{ref}^q_{\mathcal{D},\mathcal{G}}\text{”} \\ \leq & \mathrm{ref}^q_{\mathcal{D},\mathcal{G}}(\mathbb{1},\mathsf{B}) & \text{“}\mathrm{ref}^q_{\mathcal{D},\mathcal{G}}(\mathsf{A},\mathbb{1}) = 0 \text{ for order-respecting } q\text{”} \\ = & \mathcal{ML}^q_{\mathcal{G}}(\mathcal{D},\mathsf{B}) & \text{“Prop. 3”} \end{array}$$

□

The above corollary in turn implies that $\mathrm{ref}^q_{\mathcal{D},\mathcal{G}}$ is finite whenever the capacity itself is finite. In particular $\mathrm{ref}^{+}_{\mathbb{D},\mathbb{G}\updownarrow}$ and $\mathrm{ref}^{\times}_{\mathbb{D},\mathbb{G}+}$ are both finite.

4 Computing the Additive Refinement Metric

Given channel matrices $\mathsf{A}{:}\,\mathcal{X}{\twoheadrightarrow}\mathcal{Y}$ and $\mathsf{B}{:}\,\mathcal{X}{\twoheadrightarrow}\mathcal{Z}$ where $\mathsf{A} \not\sqsubseteq \mathsf{B}$, we know that sometimes B leaks more than A. To determine how much worse B can be, we wish to compute some version of the refinement metric: here we focus on the additive refinement metric $\mathrm{ref}^{+}_{\pi,\mathbb{G}^{\updownarrow}}(\mathsf{A},\mathsf{B})$, which measures the maximum amount by which the additive leakage of B can exceed that of A on prior π and over gain functions in $\mathbb{G}^{\updownarrow}\mathcal{X}$, the class of 1-bounded gain functions.[7]

It is known that we can do this computation via linear programming [4], where the elements of G (the matrix representation of g) are variables. But here we refine the previously-used approach, enabling it to be done far more efficiently than was previously known. First, to ensure that G is in $\mathbb{G}^{\updownarrow}\mathcal{X}$, we just need to constrain its entries to be at most 1, and (to ensure that its gain values are non-negative) include a special action $\perp$ whose gain values are all 0. A second issue is that the number of possible actions of G is not fixed. But it suffices for G to have $|\mathcal{Y}|+|\mathcal{Z}|+1$ actions, since this allows a different action for each output of A and B, along with the special $\perp$ action. Now we recall the trace-based formulation of posterior vulnerability:

$$V_g[\pi{\triangleright}\mathsf{A}] = \max_{\mathsf{S}^\mathsf{A}} \mathbf{tr}(\mathsf{G}\mathsf{D}^\pi\mathsf{A}\mathsf{S}^\mathsf{A}) \quad \text{and} \quad V_g[\pi{\triangleright}\mathsf{B}] = \max_{\mathsf{S}^\mathsf{B}} \mathbf{tr}(\mathsf{G}\mathsf{D}^\pi\mathsf{B}\mathsf{S}^\mathsf{B})$$

where $\mathsf{S}^\mathsf{A}{:}\,\mathcal{Y}{\twoheadrightarrow}\mathcal{W}$ and $\mathsf{S}^\mathsf{B}{:}\,\mathcal{Z}{\twoheadrightarrow}\mathcal{W}$ are strategies mapping outputs of A and B to actions. While the previously-used approach was to try exponentially-many strategies S^A and S^B, we now argue that this is actually unnecessary. For we can assume without loss of generality that the first $|\mathcal{Y}|$ rows of G contain, in order, optimal actions for the columns of A; the next $|\mathcal{Z}|$ rows of G contain, in order, optimal actions for the columns of B; and finally the last row of G is all 0, for the $\perp$ action. Achieving this just requires reordering the rows of G and possibly duplicating some actions (since the same action might be optimal for more than one column). Once this is done, we can use a fixed S^A that maps column j of A to row j of G, and a fixed S^B that maps column k of B to row $k{+}|\mathcal{Y}|$ of G. Now we can solve a *single* linear programming problem for that S^A and S^B:[8]

> Choose G to maximize $\mathbf{tr}(\mathsf{G}\mathsf{D}^\pi\mathsf{B}\mathsf{S}^\mathsf{B}) - \mathbf{tr}(\mathsf{G}\mathsf{D}^\pi\mathsf{A}\mathsf{S}^\mathsf{A})$ subject to G having elements of at most 1 and a final all-zero row, and $\mathit{opt}(\mathsf{S}^\mathsf{A})$,

where $\mathit{opt}(\mathsf{S}^\mathsf{A})$ constrains S^A to be optimal for G. Remarkably, this means that a maximizing gain function G can be found in *polynomial time*.

[7] These are the gain functions whose gain values are at most 1 and which always produce non-negative vulnerabilities (even though some of their gain values can be negative). We need to restrict to bounded gain functions to get interesting results, since otherwise we could get arbitrarily large leakage differences simply by scaling up the gain values, making the additive refinement metric infinite.

[8] Note that deciding whether $A \sqsubseteq B$ can also be done using a single linear program; but here we achieve more: when $A \not\sqsubseteq B$ we know that B can leak more than A, but we want to also compute *how much* more.

Let us consider some examples. Consider the following channels:

A	y_1	y_2	y_3
x_1	0.1	0.2	0.7
x_2	0.3	0.5	0.2
x_3	0.6	0.1	0.3

B	y_1	y_2	y_3
x_1	0.101	0.200	0.699
x_2	0.298	0.501	0.201
x_3	0.600	0.099	0.301

where B is formed by slightly tweaking the probabilities in A. First, given π we can compute the additive capacity of A, $\mathcal{ML}^+_{\mathbb{G}\updownarrow}(\pi, \mathsf{A}) = \mathrm{ref}^+_{\pi,\mathbb{G}\updownarrow}(\mathbb{1}, \mathsf{A})$. As predicted in [5, Theorem 6], we find that for any full-support π, the capacity is 0.6, which is indeed 1 minus the sum of the column minimums of A; moreover, the gain function that realizes the capacity is indeed the complement of the π-reciprocal gain function. For instance, when $\pi = (1/3, 1/2, 1/6)$ it is

G	x_1	x_2	x_3
w_1	−2	1	1
w_2	1	−1	1
w_3	1	1	−5
$\bot$	0	0	0

Next, because B is so close to A, we expect that its additive refinement metric should be small. On the uniform prior $\pi^u = (1/3, 1/3, 1/3)$, we find indeed that $\mathrm{ref}^+_{\pi^u,\mathbb{G}\updownarrow}(\mathsf{A}, \mathsf{B}) = 1063/255000 \approx 0.00416863$, realized on the gain function

G	x_1	x_2	x_3
w_1	$72/85$	$-88/255$	$8/255$
w_2	1	$-7/3$	1
w_3	1	1	−7
$\bot$	0	0	0

As another example, consider the following channels from [4]:

A	y_1	y_2	y_3	y_4
x_1	0.1	0.4	0.1	0.4
x_2	0.2	0.2	0.3	0.3
x_3	0.5	0.1	0.1	0.3

B	z_1	z_2	z_3
x_1	0.2	0.22	0.58
x_2	0.2	0.4	0.4
x_3	0.35	0.4	0.25

They are interesting in that $\mathsf{A} \not\sqsubseteq \mathsf{B}$, but it was historically difficult to find a gain function that makes B leak more than A. On a uniform prior we find that $\mathrm{ref}^+_{\pi^u,\mathbb{G}\updownarrow}(\mathsf{A}, \mathsf{B}) = 103/13320 \approx 0.00773273$, realized on the gain function

G	x_1	x_2	x_3
w_1	$-107/148$	$409/444$	$19/444$
w_2	$-1/4$	1	$-2/3$
w_3	1	$-7/3$	1
$\bot$	0	0	0

To achieve robustness wrt any prior, we might prefer not to compute $\mathrm{ref}^{+}_{\pi,\mathbb{G}\updownarrow}(\mathsf{A},\mathsf{B})$ for some specific π, but instead to compute $\mathrm{ref}^{+}_{\mathbb{D},\mathbb{G}\updownarrow}(\mathsf{A},\mathsf{B})$. But it is not clear how to do this.

To get some partial insight, we can try an exhaustive search over all priors whose probabilities have a denominator of 1000. We find that the best prior in that set is $\pi = ({}^{624}/{}_{1000}, {}^{29}/{}_{1000}, {}^{347}/{}_{1000})$, which gives $\mathrm{ref}^{+}_{\pi,\mathbb{G}\updownarrow}(\mathsf{A},\mathsf{B}) = {}^{15553}/{}_{1850000} \approx 0.00840703$, realized on gain function

G	x_1	x_2	x_3
w_1	$-5/7696$	$16/1073$	$4/12839$
w_2	$57/208$	1	$-257/347$
w_3	1	$-1179/29$	1
$\perp$	0	0	0

Hence we see that unlike the situation of additive capacity where $\mathcal{ML}^{+}_{\mathbb{G}\updownarrow}(\pi,\mathsf{A})$ is the same for any full support π, the additive refinement metric *does* depend on the prior.[9]

5 Bounding the Additive Refinement Metric

In the previous section we showed that computing the additive refinement metric for a fixed prior can be done by solving a *single* linear program. Although this gives us a solution in time polynomial in the size of C, this solution is still practically feasible only for very modest sizes. In this section, we study efficient techniques for bounding the additive metric based on the Earth Mover's distance, inspired by the use of the same technique for computing capacity [2].

Moreover, we discuss a simple bound on the additive refinement metric when maximizing over both priors and gain functions. The existence of an efficient exact algorithm for this case remains unknown.

5.1 The Earth Mover's Distance

The *Earth Mover's* distance (EMD), also known as *Wasserstein* distance, is a fundamental distance between *probability distributions*, with numerous applications in computer science. EMD gives a metric on the set $\mathbb{D}\mathcal{A}$ of probability distributions over an underlying set $\mathcal{A}$, equipped with its own *"ground" metric* d. As a consequence, it can be thought of as a mapping $\mathbb{W} : \mathbb{M}\mathcal{A} \to \mathbb{M}\mathbb{D}\mathcal{A}$, lifting a metric on $\mathcal{A}$ to a metric on $\mathbb{D}\mathcal{A}$.

The idea behind EMD is that the distance between two distributions $\alpha, \alpha' {:} \mathbb{D}\mathcal{A}$ is the minimum *cost* of *transforming* α into α'. The transformation

[9] There is an interesting phenomenon that we have observed. Given prior π, we can first compute the gain function g that realizes $\mathrm{ref}^{+}_{\pi,\mathbb{G}\updownarrow}(\mathsf{A},\mathsf{B})$, and then we can compute the prior that realizes $\mathrm{ref}^{+}_{\mathbb{D},g}(\mathsf{A},\mathsf{B})$. In every case that we have tried, the prior that is found is exactly the π that we started with, suggesting that π is somehow "encoded" into the gain function g.

is performed by moving probability mass between elements $a, a' \colon \mathcal{A}$, where α_a is seen as the probability mass *available* at a (supply), and $\alpha'_{a'}$ as the probability mass *needed* at a' (demand). The ground metric d determines the *cost* of this transportation: moving a unit of probability from a to a' costs $d(a, a')$. There are many strategies for achieving this transformation, the cost of the best one gives the distance between α and α'.

More precisely, an earth-moving *strategy* is a joint distribution $S \in \mathbb{D}\mathcal{A}^2$ whose two marginals are α and α', i.e. $\sum_{a'} S(a, a') = \alpha_a$ and $\sum_a S_{a,a'} = \alpha'_{a'}$. Intuitively, $S_{a,a'}$ is the probability mass moved from a to a', and the requirements on the marginals capture the fact that both supply and demand should be respected. We write $\mathcal{S}_{\alpha,\alpha'}$ for the set of such strategies. Moreover, we write $\mathcal{E}_S d := \sum_{a,a'} S_{a,a'} d(a, a')$ for the expected value of d wrt the distribution S, in other words the total transportation cost of the strategy S.

Definition 2. *The Earth mover's distance is the mapping* $\mathbb{W} : \mathbb{M}\mathcal{A} \to \mathbb{M}\mathbb{D}\mathcal{A}$ *given by:*

$$\mathbb{W}(d)(\alpha, \alpha') \quad := \quad \inf_{S \colon \mathcal{S}_{\alpha,\alpha'}} \mathcal{E}_S d \ .$$

Note that the properties of $\mathbb{W}(d)$ directly depend on those of d; in particular $\mathbb{W}(d)$ might not be symmetric unless d itself is symmetric.

The EMD can be computed via a linear program with S as variables, since $\mathcal{E}_S d$ is linear on S and respecting supply and demand can be expressed by linear constraints. However, more efficient network flow algorithms also exist (e.g. [11]). Note, however, that some of these algorithms require d to be a proper metric.

5.2 Bounding the Refinement Metric Using the EMD

We now turn our attention to the problem of bounding the additive refinement metric between two channels A and B. To do so, we exploit the view of channels as producing *hyper-distributions*. Recall from Sect. 2 that the effect of a channel A on the prior π can be viewed as a hyper-distribution $[\pi \triangleright \mathsf{A}]$, that is a distribution on the posterior distributions produced by each output of A. In other words, hypers have type $\mathbb{D}\mathbb{D}\mathcal{X}$.

Since hypers are distributions, we can use the EMD to measure the distance between them. But the EMD requires a ground metric on the underlying space, which in our case is $\mathcal{A} = \mathbb{D}\mathcal{X}$. In other words, to measure the distance between hypers we first need to measure the distance between posteriors. It turns out that the appropriate distance for our needs is the following.

Definition 3. *Let* $\pi, \sigma \colon \mathbb{D}\mathcal{X}$. *Define the "convex separation" quasimetric* $\mathbf{q_{cs}} \colon \mathbb{M}\mathbb{D}\mathcal{X}$ *as*

$$\mathbf{q_{cs}}(\sigma, \pi) \quad := \quad \max_{x \colon \lceil \sigma \rceil} \left(1 - \frac{\pi_x}{\sigma_x} \right) \ .$$

The usefulness of this quasimetric lies in the fact that it can be shown [5] to express the maximum that σ, π can be separated by a 1-bounded gain function, that is

$$\mathbf{q_{cs}}(\sigma, \pi) = \sup_{g:\, \mathbb{G}^{\updownarrow}\mathcal{X}} \mathbf{q}^{+}_{<}(V_g(\sigma), V_g(\pi)) \ .$$

This in turn means that V_g is non-expansive wrt $\mathbf{q_{cs}}$, which is important due to the connection between EMD and the Kantorovich metric. The technicalities of this construction are available in the appendix, we only state the most relevant result here.

Theorem 2. *For any prior* π *and channels* A, B *it holds that*

$$\mathrm{ref}^{+}_{\pi,\mathbb{G}^{\updownarrow}}(\mathsf{A}, \mathsf{B}) \quad \leq \quad \mathbb{W}(\mathbf{q_{cs}})([\pi \triangleright \mathsf{A}], [\pi \triangleright \mathsf{B}]) \ .$$

Hence, to obtain a bound on the refinement metric, we can compute the hypers $[\pi \triangleright \mathsf{A}]$ and $[\pi \triangleright \mathsf{B}]$ (i.e. the corresponding output and posterior distributions) and then employ either linear programming or a network flow algorithm to compute the $\mathbb{W}(\mathbf{q_{cs}})$ distance between them. Note that the algorithm for EMD should not require the ground metric to be symmetric, since $\mathbf{q_{cs}}$ is not.

5.3 More Relaxed Bounds

As discussed in the previous sections, the EMD is the cost of the *best* transportation strategy to transform one distribution into another. As a consequence, one can clearly obtain a bound by choosing *any* transportation strategy, not necessarily the best. That is, if S is a strategy for transforming $[\pi \triangleright \mathsf{A}]$ into $[\pi \triangleright \mathsf{B}]$, then the cost (wrt $\mathbf{q_{cs}}$) of S provides a bound to the refinement metric between A and B:

$$\mathrm{ref}^{+}_{\pi,\mathcal{G}}(\mathsf{A}, \mathsf{B}) \quad \leq \quad \mathcal{E}_S \mathbf{q_{cs}} \ .$$

Of course the actual EMD gives the best such bound, but the freedom of choosing *any* strategy can be convenient. In Sect. 6 an analytic bound for the Crowds protocol is given by choosing an arbitrary strategy, without having to show that this is the optimal one.

5.4 Maximizing over All Priors

In view of the apparent difficulty of computing $\mathrm{ref}^{+}_{\mathbb{D},\mathbb{G}^{\updownarrow}}(\mathsf{A}, \mathsf{B})$, we are led to wonder how much the additive refinement bound can be increased by changing from a uniform prior to a non-uniform prior. We find that the possible increase can be bounded:

Theorem 3. *For any channel matrices* $\mathsf{A}\colon \mathcal{X} \rightarrowtriangle \mathcal{Y}$ *and* $\mathsf{B}\colon \mathcal{X} \rightarrowtriangle \mathcal{Z}$*, we have*

$$\mathrm{ref}^{+}_{\mathbb{D},\mathbb{G}^{\updownarrow}}(\mathsf{A}, \mathsf{B}) \quad \leq \quad |\mathcal{X}| \cdot \mathrm{ref}^{+}_{\pi^u,\mathbb{G}^{\updownarrow}}(\mathsf{A}, \mathsf{B}) \ .$$

Proof. Let prior π be arbitrary, and suppose that $\text{ref}^{+}_{\pi,\mathbb{G}^{\updownarrow}}(\mathsf{A},\mathsf{B})$ is realized on gain function g with matrix representation G, which we assume without loss of generality to include an all-zero row. Next recall the trace-based formulation of posterior g-vulnerability: $V_g[\pi{\triangleright}\mathsf{A}] = \max_{\mathsf{S}} \mathbf{tr}(\mathsf{G}\mathsf{D}^{\pi}\mathsf{A}\mathsf{S})$. Observe that we can factor $\mathsf{D}^{\pi} = \mathsf{D}^{\rho}\mathsf{D}^{\pi^u}$, where vector ρ is given by $\rho_x = \pi_x/\pi^u{}_x = |\mathcal{X}| \cdot \pi_x$. Now we can translate from π to the uniform prior π^u:

$$V_g[\pi{\triangleright}\mathsf{A}] \quad = \quad \max_{\mathsf{S}} \mathbf{tr}(\mathsf{G}\mathsf{D}^{\pi}\mathsf{A}\mathsf{S}) \quad = \quad \max_{\mathsf{S}} \mathbf{tr}(\mathsf{G}\mathsf{D}^{\rho}\mathsf{D}^{\pi^u}\mathsf{A}\mathsf{S}) \quad = \quad V_{g'}[\pi^u{\triangleright}\mathsf{A}],$$

where g' is the gain function with matrix representation $\mathsf{G}\mathsf{D}^{\rho}$. The only problem is that g' may not be in $\mathbb{G}^{\updownarrow}\mathcal{X}$. It *does* give non-negative vulnerabilities (since $\mathsf{G}\mathsf{D}^{\rho}$ contains an all-zero row), but it may have gain values bigger than 1. We can deal with this by scaling g' down, dividing its gain values by $\max_x \pi_x/\pi^u{}_x = \max_x \pi_x \cdot |\mathcal{X}|$, which is at most $|\mathcal{X}|$. If we let g'' be the scaled-down version of g', then we have $g'' \in \mathbb{G}^{\updownarrow}\mathcal{X}$ and $|\mathcal{X}| \cdot V_{g''}[\pi^u{\triangleright}\mathsf{A}] = V_g[\pi{\triangleright}\mathsf{A}]$. Since the same equality holds for B, the desired inequality follows. (Note that it is only an inequality, since g'' may not be optimal for π^u.) □

Applying this theorem to the last example from Sect. 4, where the additive refinement metric on the uniform prior was about 0.0077, we see that the metric over *all* priors is at most 3 times as large, or 0.0232. While this bound is considerably larger than 0.0084 (the maximum found in our partial search), it may still be useful.

6 Case Study : The Crowds Protocol

Crowds is a simple protocol for anonymous web surfing. The context is that a user, called the *initiator*, wants to contact a *web server* but does not want to disclose his identity to the server. That is achieved by collaborating with a group of other users, called the *crowd*, who participate in the protocol to facilitate the task. The protocol is essentially a simple probabilistic routing protocol:

- In the first step, the initiator selects a user uniformly (including possibly himself) and forwards the request to him. The user who receives the message now becomes the (new) *forwarder*.
- A forwarder, upon receiving a message, flips a (biased) probabilistic coin: with probability φ he forwards the message to a new user (again chosen uniformly, including himself), but with remaining probability $1-\varphi$ he instead delivers the message directly to the web server.

If $\varphi{<}1$ then with probability 1 the message will eventually arrive at the web server, with the expected number of hops being $1/1{-}\varphi$.

Concerning anonymity, from the point of view of the web server all users seem equally likely to have been the initiator (assuming a uniform prior), due to the first step's always being a forwarding step. The analysis becomes more interesting however if we assume that some users in the crowd are corrupt, reporting to an adversary who is trying to discover messages' initiators.[10]

If user a forwards a message to a corrupted user, we say that a was *detected.* Since a's being detected is more likely when he is the initiator than when he is not, the protocol does leak information: detecting a creates a posterior where a has a greater chance of being the initiator than he had *a priori.* Still, if $\varphi > 0$ then the posterior is *not* a point distribution: user a can "plead not guilty" by claiming the he was merely forwarding for someone else. The greater φ is, the more plausible that claim becomes: the forwarding probability φ can be seen as trading anonymity for utility (expected number of hops).

In their original work, Reiter and Rubin [12] introduced the property of "probable innocence" to analyze Crowds, meaning, roughly speaking, that each user appears more likely to be innocent than guilty. For our purposes, however, we study Crowds in the context of QIF.[11]

6.1 Modeling the Crowds Protocol

To perform an analysis of the anonymity guarantees of Crowds, we start by modeling the protocol as an information-theoretic channel. Letting n, c, m be the number of honest, corrupted, and total users respectively (i.e. $n+c = m$), the secret and observable events are $\mathcal{X}{:=}\,\{x_1, \ldots, x_n\}$, $\mathcal{Y}{:=}\,\{y_1, \ldots, y_n, s\}$, with x_a denoting that (honest) user a is the initiator, y_b denoting that user b was detected (forwarded a message to a corrupted user), and s denoting that the message was delivered to the web server without having been forwarded to a corrupted user.

To construct the channel matrix C, we need for any honest user a to compute the probability of producing each observable event, given the secret event x_a. Due to symmetry, the probability of detecting the initiator, or some non-initiator, is the same for all initiators. In other words, the probabilities $p(y_b|x_a)$ only depend on whether $a{=}b$ or $a \neq b$, and not on the exact value of a, b. Hence the entries of C can only have 3 distinct values, $\mathsf{C}_{x_a,s} = \alpha$, $\mathsf{C}_{x_a,y_a} = \beta$, and $\mathsf{C}_{x_a,y_b} = \gamma$, $a \neq b$, giving the following channel matrix:[12]

$$\mathsf{C} \;=\; \begin{array}{c} \\ x_1 \\ \vdots \\ x_n \end{array}\!\!\begin{array}{c} \begin{array}{ccc} y_1 \cdots y_n & s \end{array} \\ \left[\begin{array}{ccc} \beta & \cdots \; \gamma & \alpha \\ & \ddots & \\ \gamma & \cdots \; \beta & \alpha \end{array}\right] \end{array} \qquad \text{where} \qquad \begin{aligned} \alpha &:= \frac{n - \varphi n}{m - \varphi n}, \\ \beta &:= \frac{c(m - \varphi(n{-}1))}{m(m - \varphi n)}, \\ \gamma &:= \frac{c\varphi}{m(m - \varphi n)} . \end{aligned}$$

Finally, given a prior π on $\mathcal{X}$, modeling how likely each user is to be the initiator a-priori, we can quantify the anonymity of the protocol by considering its

[10] This is a relatively weak adversarial model; a model in which the adversary can view any communication in the network would be too strong for Crowds.

[11] See [3] for a discussion of how probable innocence relates to QIF.

[12] See [3] for the exact derivations.

(additive or multiplicative) g-leakage, for a properly constructed gain function g. For instance, the identity gain function g_{id} is a reasonable choice for this protocol, modeling an adversary that tries to guess the initiator of the message in one try. For this choice of gain function, and the uniform prior π^u, the additive leakage is given by

$$\mathcal{L}^{+}_{g_{\text{id}}}(\pi^u, \mathsf{C}) \quad = \quad \frac{nc-c}{nc+n} \quad . \tag{3}$$

But g_{id} is not the only reasonable choice for this protocol. It could be argued, for instance, that the whole point behind "probable innocence" is that the adversary would not incriminate a user unless there is sufficient evidence—that is, a *wrong* guess should have a *penalty*, compared to *not guessing at all.* Such adversaries can be expressed by suitable gain functions [3].

A crucial question, then, is how can we make *robust* conclusions in our analysis of information leakage, in particular when we tune a parameter of the protocol such as φ?

6.2 Increasing φ : Refinement Relation

A striking fact about Crowds, first observed in [8], is that its Bayes leakage for a uniform prior, given by (3), does not depend on φ! This is remarkable since the fact of forwarding is the only reason why a detected user can plausibly pretend not be the initiator. Intuitively, the more likely a user is to forward the more likely detecting the wrong user becomes, so anonymity is improved. Understanding this paradox is beyond the scope of this paper (an explanation is given in [3]), but we should note that this phenomenon *only* happens in the very specific case of g_{id} with a uniform prior: for most other priors and gain functions, the corresponding leakage does depend on φ, as expected.

This brings us to the natural question: *how* does leakage depend on φ? Answering this question is of great importance to the designer of the system who tries to optimize the parameters, but without compromising its anonymity.

The easier case to answer is when φ is *increased.* Intuitively, this operation should be *safe*, in the sense that the resulting protocol should leak no more than before. But can we be sure that this *always* happens? Could it be the case that for some strange adversary, modeled by some gain function g, increasing φ causes the leakage to increase, leading to a less safe protocol?

The theory of refinement provides a robust answer. Let C^{φ} denote the channel matrix for a particular choice of φ. It can be shown that

$$\mathsf{C}^{\varphi} \sqsubseteq \mathsf{C}^{\widetilde{\varphi}} \qquad \text{iff} \qquad \varphi \leq \widetilde{\varphi},$$

by showing that $\mathsf{C}^{\widetilde{\varphi}} = \mathsf{C}^{\varphi}\mathsf{R}$ for a properly constructed channel R. As a consequence $\mathsf{C}^{\widetilde{\varphi}}$ is a safer protocol; it leaks no more than C^{φ} for *all* priors and gain functions.

6.3 Decreasing φ : Refinement Metric

Although refinement establishes in a very robust way that increasing φ is a safe operation, it can be argued that the interesting case is exactly the opposite: *decreasing* φ. This is because a system designer is typically interested in improving the *utility* of the system, which in this case corresponds to its *latency*. Since the expected number of hops to reach the server is $1/1-\varphi$, decreasing φ lowers the protocol's latency, so such a change is desirable.

However, decreasing φ is *not always* safe. Although Bayes leakage for a uniform prior remains unaffected, we can experimentally verify that for all other priors, and for many gain functions, the leakage indeed increases. We thus have a typical trade-off between anonymity and utility; we might be willing to accept a less safe protocol, and *replace* $\widetilde{\varphi}$ *by a smaller* φ, if we knew that the increase in leakage cannot exceed a certain threshold. Can we answer this question in a *robust* way?

Using the refinement metric. The refinement metric provides a natural answer to our question. For a fixed π, we know that $\mathrm{ref}^{+}_{\pi,\mathbb{G}^{\updownarrow}}(\mathsf{C}^{\widetilde{\varphi}}, \mathsf{C}^{\varphi})$ gives the maximum increase in vulnerability when we replace $\widetilde{\varphi}$ by φ, for any 1-bounded gain function $g{:}\,\mathbb{G}^{\updownarrow}$.

The metric can be computed using the linear programming technique of Sect. 4). As an example, for $n = 110$, $c = 5$, $\varphi = 0.55$ and $\widetilde{\varphi} = 0.6$ we find that the refinement metric is approximately 0.0076. This means that lowering the probability of forwarding from 0.6 to 0.55 cannot have a huge effect on the vulnerability; for instance if the adversary's probability of a correct guess (Bayes vulnerability) was p, it can be at most $p + 0.0076$ after the change. Note that $\varphi = 0.55$ gives 12.5% *lower latency* compared to $\varphi = 0.6$, so we might decide that the benefit outweighs the risk.

If computing the exact metric is not feasible, we might still be able to apply the EMD bound discussed in Sect. 5, given by the EMD (wrt $\mathbf{q_{cs}}$) between the hypers $[\pi^u{\triangleright}\mathsf{C}^{\widetilde{\varphi}}]$ and $[\pi^u{\triangleright}\mathsf{C}^{\varphi}]$. In our example, this is approximately 0.017; it is more than twice as large as the real metric, but it still provides a useful bound on the loss of anonymity.

An Analytic Bound for Crowds. As discussed in Sect. 5, an upper bound to the refinement metric can be obtained by computing, not the actual EMD, but the cost (wrt $\mathbf{q_{cs}}$) of *any transportation strategy* from $[\pi{\triangleright}\mathsf{C}^{\widetilde{\varphi}}]$ to $[\pi{\triangleright}\mathsf{C}^{\varphi}]$. For the uniform prior, $[\pi^u{\triangleright}\mathsf{C}^{\varphi}]$ can be computed analytically; then we can choose any strategy to obtain an analytic bound.

Following the calculations of [3], the hyper distribution $[\pi^u{\triangleright}\mathsf{C}^{\varphi}]$ has

$$
\begin{array}{lcll}
\text{output distribution} & \rho & = & ({}^{1-\alpha}/_{n}, \cdots, {}^{1-\alpha}/_{n}, \alpha), \\
\text{and posteriors} & \delta^{y_b} & = & (\frac{\varphi}{m}, \cdots, \frac{\varphi}{m}, \frac{m-\varphi(n-1)}{m}, \frac{\varphi}{m}, \cdots, \frac{\varphi}{m}), \\
 & \delta^{s} & = & ({}^{1}/_{n}, \cdots, {}^{1}/_{n}) \;.
\end{array}
$$

Note that the detection events y_b produce a posterior δ^{y_b} assigning higher probability to the detected user b, while s provides no information, producing a uniform posterior. Denoting by $\widetilde{\alpha}, \widetilde{\rho}, \ldots$ the corresponding quantities for $\widetilde{\varphi} \geq \varphi$, we can check that $\widetilde{\alpha} \leq \alpha$; hence $\widetilde{\rho}$ has lower probability than ρ on its last element and higher on its first n elements. A reasonable transportation strategy[13] S from $[\pi \triangleright \mathsf{C}^{\widetilde{\varphi}}]$ to $[\pi \triangleright \mathsf{C}^{\varphi}]$ is then to move all probability mass from $\widetilde{\delta^s}$ to δ^s, while splitting the probability of $\widetilde{\delta^{y_b}}$ between δ^{y_b} and δ^s. This strategy produces the following bound.

Theorem 4. *For any* $0 < \varphi \leq \widetilde{\varphi} \leq 1$ *it holds that:*

$$\mathrm{ref}^{+}_{\pi^u, \mathbb{G}\ddagger}(\mathsf{C}^{\widetilde{\varphi}}, \mathsf{C}^{\varphi}) \quad \leq \quad (1-\alpha)(1-\frac{\varphi}{\widetilde{\varphi}}) + (\alpha - \alpha')\left(1 - \frac{m}{mn - \widetilde{\varphi} n(n-1)}\right) .$$

Experimental Evaluation All techniques presented in this paper have been implemented in the libqif library [1]. An experimental evaluation of the additive refinement metric for Crowds is shown in Fig. 1. On the left-hand side, we compute the exact metric as well as the EMD bound between two instances of Crowds with $\varphi = 0.55$ and $\widetilde{\varphi} = 0.6$, for a uniform prior, $c = 5$ corrupted users, while varying the number of honest users n. We see that as n increases $\mathrm{ref}^{+}_{\pi^u, \mathbb{G}\ddagger}(\mathsf{C}^{\widetilde{\varphi}}, \mathsf{C}^{\varphi})$ becomes small, guaranteeing that the decrease of φ has limited effect on the leakage. The EMD bound is not tight, but as n increases it becomes small enough to provide meaningful guarantees.

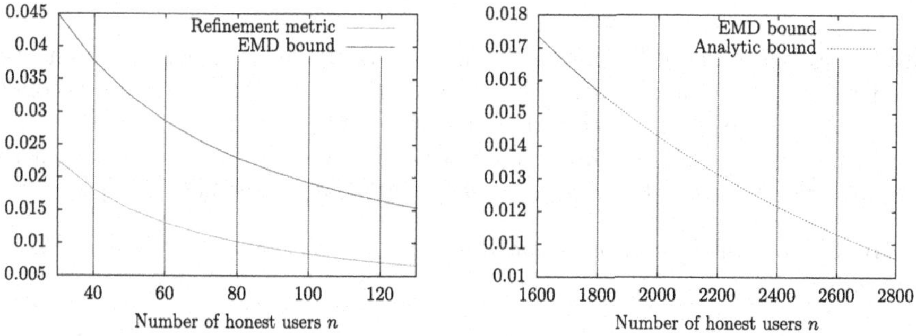

Fig. 1. Refinement metric, EMD and analytic bounds for Crowds

Computing the exact metric was possible for up to $n = 130$ within a computation threshold of 10 min. Computing the EMD bound, on the other hand, was feasible for much larger sizes. On the right-hand side of Fig. 1 we show both the EMD and the analytic bound of the previous section, for the same example, but for larger numbers of users and $c = 70$.

[13] Experiments suggest that S might be optimal, although this is neither required nor proved.

The two bounds coincide, suggesting that the transportation strategy chosen for the analytic bound is likely the optimal one. We were able to compute the EMD bound for up to $n = 1800$, while the analytic bound is applicable to any size. Note that, although the exact refinement metric is unknown, the bounds are sufficiently small to provide meaningful guarantees about the anonymity of the protocol.

A Proofs of Sect. 5

In this section we provide the proof of the main result of Sect. 5.

A.1 The Kantorovich Distance

The Kantorovich metric plays a fundamental role in our bounding technique. Given metrics $d_\mathcal{A} : \mathbb{M}\mathcal{A}, d_\mathcal{B} \in \mathbb{M}\mathcal{B}$, a function $f : \mathcal{A} \to \mathcal{B}$ is called Lipschitz wrt $d_\mathcal{A}, d_\mathcal{B}$ iff

$$d_\mathcal{B}(f(a), f(a')) \quad \leq \quad d_\mathcal{A}(a, a') \qquad \text{for all } a, a' \in \mathcal{A} \,.$$

Let $\mathbb{C}^{d_\mathcal{A}, d_\mathcal{B}}\mathcal{A}$ denote the set of all such functions.

The standard Kantorovich construction transforms metrics on some set $\mathcal{A}$, to metrics on $\mathbb{D}\mathcal{X}$, i.e. metrics measuring the distance between two probability distributions on $\mathcal{A}$. Formally, it is the lifting $\mathbb{K} : \mathbb{M}\mathcal{A} \to \mathbb{M}\mathbb{D}\mathcal{A}$ given by

$$\mathbb{K}(d)(\alpha, \alpha') \quad := \quad \sup_{F\mathbb{C}:\, \mathbb{C}^{d, d_\mathbb{R}}\mathcal{A}} \quad |\mathcal{E}_\alpha F - \mathcal{E}_{\alpha'} F| \;.$$

Note the two uses of the Euclidean distance $d_\mathbb{R}$ in the above definition: first, we consider Lipschitz functions wrt to d (the metric being lifted) and $d_\mathbb{R}$; second, we use $d_\mathbb{R}$ to compare the expected values of these functions.

For the purposes of bounding the refinement metric, it is more convenient to use a *quasimetric variant* of Kantorovich, in which $d_\mathbb{R}$ is replaced by $\mathbf{q}^+_<$.

Definition 4. *The Kantorovich quasimetric is the mapping* $\mathbb{K}^+ : \mathbb{M}\mathcal{A} \to \mathbb{M}\mathbb{D}\mathcal{A}$ *given by:*

$$\mathbb{K}^+(d)(\alpha, \alpha') \quad := \quad \sup_{F\mathbb{C}:\, \mathbb{C}^{d, \mathbf{q}^+_<}\mathcal{A}} \quad \mathbf{q}^+_<(\mathcal{E}_\alpha F, \mathcal{E}_{\alpha'} F) \quad = \quad \sup_{F\mathbb{C}:\, \mathbb{C}^{d, \mathbf{q}^+_<}\mathcal{A}} \quad \mathcal{E}_{\alpha'} F - \mathcal{E}_\alpha F \;.$$

Note that the max in the definition of $\mathbf{q}^+_<$ is removed since the sup is anyway non-negative.

An important property of the Kantorovich metric is its *dual* formulation as the *Earth Mover's* distance $\mathbb{W}(d)$ (Definition 2, also known as Wasserstein metric). The well-known Kantorovich-Rubinstein theorem states that $\mathbb{K}(d) = \mathbb{W}(d)$ (when d is a proper metric and satisfies further assumptions, see below). In our case, however, we use $\mathbb{K}^+$ *instead of* $\mathbb{K}$, and moreover we apply it to distances d that are *not symmetric*. It turns out that the full Kantorovich-Rubinstein theorem still holds in our case (under the same conditions).

Theorem 5. (Kantorovich-Rubinstein, extended). *Let $\mathcal{A}$ be a Polish (i.e. complete and separable) metric space, let d be a lower semi-continuous distance function on $\mathcal{A}$ satisfying reflexivity and the triangle inequality. Then $\mathbb{K}^+(d) = \mathbb{W}(d)$.*

Proof. In [13, Theorem. 1.3, page 19], a "Kantorovich duality" is proven, involving generic (lower semi-continuous) cost functions. Then, Kantorovich-Rubinstein is retrieved as a spacial case when the cost function is a metric [13, Theorem 1.14, page 34]. In the proof of the latter (which is quite short), only the reflexivity and the triangle inequality of d are used. Moreover, the change of $d_{\mathbb{R}}$ to $\mathbf{q}^+_<$ in the definition of $\mathbb{K}^+$ does not affect the proof. □

A surprising consequence of the above result is that, under the assumptions of Kantorovich-Rubinstein, we have:

$$\mathbb{K}^+(d) = \mathbb{K}(d) \quad = \quad \mathbb{W}(d) \; .$$

Note that the classes of Lipschitz functions in $\mathbb{K}^+$ an $\mathbb{K}$ are different, still maximizing over them turns out to give the same result! So, although the formulation of $\mathbb{K}^+$ is essential for computing capacity (see the next section), in fact it gives the standard Kantorovich construction.

A.2 Using Kantorovich/EMD to Bound the Refinement Metric

The connection between Kantorovich and EMD allows us to obtain a bound on the refinement metric. To do so, we use the fact that posterior vulnerability, for any 1-bounded gain function, is non-expansive wrt convex separation quasimetric (Definition 3).

Lemma 1. ([5]). *For any $g \in \mathbb{G}^{\updownarrow}\mathcal{X}$, V_g is Lipschitz wrt $\mathbf{q_{cs}}, d_{\mathbb{R}}$.*

Since Kantorovich distance is given by maximizing over *any* Lipschitz function, it clearly gives an upper bound on the refinement metric. And the connection between Kantorovich and EMD allows us to efficiently compute this bound.
Theorem 2 For any prior π and channels A, B it holds that

$$\mathrm{ref}^+_{\pi,\mathbb{G}^{\updownarrow}}(\mathsf{A},\mathsf{B}) \quad \leq \quad \mathbb{W}(\mathbf{q_{cs}})([\pi{\triangleright}\mathsf{A}],[\pi{\triangleright}\mathsf{B}]) \; .$$

Proof. From Lemma 1 we get that $\mathrm{ref}^+_{\pi,\mathbb{G}^{\updownarrow}}(\mathsf{A},\mathsf{B}) \leq \mathbb{K}^+(\mathbf{q_{cs}})([\pi{\triangleright}\mathsf{A}],[\pi{\triangleright}\mathsf{B}])$, and the result follows from the (extended) Kantorovich-Rubinstein theorem.

References

1. LIBQIF : Quantitative information flow C++ library. https://github.com/chatziko/libqif
2. Alvim, M.S., Chatzikokolakis, K., McIver, A., Morgan, C., Palamidessi, C., Smith, G.: Additive and multiplicative notions of leakage, and their capacities. In: Proceedings of the 27th IEEE Computer Security Foundations Symposium (CSF 2014), pp. 308–322 (2014)
3. Alvim, M.S., Chatzikokolakis, K., McIver, A., Morgan, C., Palamidessi, C., Smith, G.: The Science of Quantitative Information Flow. Springer, Heidelberg (2019)
4. Alvim, M.S., Chatzikokolakis, K., Palamidessi, C., Smith, G.: Measuring information leakage using generalized gain functions. In: Proceedings of 25th IEEE Computer Security Foundations Symposium (CSF 2012), pp. 265–279, June 2012
5. Chatzikokolakis, K.: On the additive capacity problem for quantitative information flow. In: McIver, A., Horvath, A. (eds.) QEST 2018. LNCS, vol. 11024, pp. 1–19. Springer, Cham (2018). https://doi.org/10.1007/978-3-319-99154-2_1
6. Clark, D., Hunt, S., Malacaria, P.: Quantitative analysis of the leakage of confidential data. Electr. Notes Theor. Comput. Sci. **59**(3), 238–251 (2001). (Proceedings of Workshop on Quantitative Aspects of Programming Languages)
7. Clarkson, M., Myers, A., Schneider, F.: Belief in information flow. In: Proceedings of 18th IEEE Computer Security Foundations Workshop (CSFW 2005), pp. 31–45 (2005)
8. Espinoza, B., Smith, G.: Min-entropy as a resource. Inf. Comput. **226**, 57–75 (2013). (Special Issue on Information Security as a Resource)
9. Köpf, B., Basin, D.: An information-theoretic model for adaptive side-channel attacks. In: Proceedings of 14th ACM Conference on Computer and Communications Security (CCS 2007), pp. 286–296 (2007)
10. McIver, A., Morgan, C., Smith, G., Espinoza, B., Meinicke, L.: Abstract channels and their robust information-leakage ordering. In: Abadi, M., Kremer, S. (eds.) POST 2014. LNCS, vol. 8414, pp. 83–102. Springer, Heidelberg (2014). https://doi.org/10.1007/978-3-642-54792-8_5
11. Pele, O., Werman, M.: Fast and robust earth mover's distances. In: ICCV (2009)
12. Reiter, M.K., Rubin, A.D.: Crowds: anonymity for web transactions. ACM Trans. Inf. Syst. Secur. **1**(1), 66–92 (1998)
13. Villani, C.: Topics in optimal transportation, no. 58. American Mathematical Society (2003)

Models and Puzzles

Toward a Formal Model for Group Polarization in Social Networks

Mário S. Alvim[1(✉)], Sophia Knight[2], and Frank Valencia[3,4]

[1] Department of Computer Science, Universidade Federal de Minas Gerais, Belo Horizonte, Brazil
msalvim@dcc.ufmg.br
[2] Department of Computer Science, University of Minnesota Duluth, Duluth, USA
sophia.knight@gmail.com
[3] CNRS & LIX, École Polytechnique, Palaiseau, France
frank.valencia@lix.polytechnique.fr
[4] Pontificia Universidad Javeriana Cali, Cali, USA

Abstract. In this paper we develop a preliminary model for social networks, and a measure of the level of polarization in these social networks, based on Esteban and Ray's classic measure of polarization for economic situations. Our model includes information about each agent's quantitative strength of belief in a proposition of interest and a representation of the strength of each agent's influence on every other agent. We consider how the model changes over time as agents interact and communicate, and include several different options for belief update, including rational belief update and update taking into account irrational responses such as confirmation bias and the backfire effect. Under various scenarios, we consider the evolution of polarization over time, and the implications of these results for real world social networks.

1 Introduction

Social networks may facilitate civil discourse by making possible a prompt exchange of facts, stories, and opinions. They also allow and even suggest users to join groups that share similar interests or ideologies. These are some of the appealing features that have made of social networks one of the most, if not the most, popular media for interaction and communication in today's world.

Nevertheless, social networks can also *filter* and *direct* (or allow others to direct) information to millions of users based on their social media history, social graph connectivity, preferences or group affiliation. While this may increase a healthy exposure to diverse perspectives, which is beneficial to our society, it may also increases ideological segregation [11]. People with opposing views would tend to interpret this information with their own bias, therefore reinforcing their original beliefs rather than questioning them. This may cause their opinions to move

Mário S. Alvim was partially supported by the National Council for Scientific and Technological Development (CNPq), CAPES, and FAPEMIG. Frank Valencia has been partially supported by the ECOS-NORD project FACTS (C19M03).

M. S. Alvim et al. (Eds.): Palamidessi Festschrift, LNCS 11760, pp. 419–441, 2019.
https://doi.org/10.1007/978-3-030-31175-9_24

further apart [36]. Similarly, in groups with uniform views users may become more extreme by reinforcing each other's opinions [1]. In fact, social psychology tells us that groups may strengthen their beliefs by interpreting information in their favor (*confirmation bias*) or by strongly opposing it if it contradicts their views (*backfire effect*). As a result, social networks often make their users become radical and isolated in their own ideological circle causing dangerous splits in society [3] in a phenomenon known as *group polarization* [1].

Indeed, social media platforms have driven polarization in the political process. Referenda such as Brexit and the Colombian peace agreement, as well as Brazil/Colombia/USA presidential elections are compelling examples of this phenomenon [23]. They bear witness to the fact that divisive messages in social networks with elements of extremist ideology, misleading information, or demagoguery may influence fundamental decision-making processes. In fact, we have reached a critical point where the production of malicious information to cause polarization in social media has become a profitable source of revenue in some countries, like Macedonia [25].

Catuscia feels strongly about the divisive impact of social networks in our society. We wish to honor her with a paper on a preliminary model of cognitive behavior for the analysis of group polarization in social networks.

One of our future directions, in collaboration with the INRIA team Comète and the PUJ team AVISPA, is to conduct interdisciplinary research in economy and social sciences for modeling human cognitive and social processes. Our long term goal is to develop a model for the dynamics of polarization allowing for the derivation of formal results that explain and predict how the phenomenon arises, sustains itself, or fades away.

Nevertheless, formalizing all elements necessary for such a goal is not a trivial task. Although there is an extensive body of work on how to represent individuals' beliefs and the way they interact to update these beliefs, most of that work assumes individuals are rational and optimally use all information at their disposal. However, a growing body of work by psychologists, sociologists, and economists has challenged these assumptions, and several common, relevant cognitive biases have been cataloged [8,21,34]. Experimental work shows the effect of such biases, and indicate they should be considered when modeling polarization.

A major shortcoming in current models is that they either investigate the effect of each bias individually, and independently from how people interact in society, or assume the same type of interaction occurs among every pair of agents. The particular way different individuals interact is also relevant, and to the best of our knowledge no study has been done to analyze the combination of biases and interactions in the rising of polarization. Furthermore the models do not formally define these biases and interactions in a way that allows for the precise statement, and proofs, of theorems about polarization.

Contributions. In this work we take a first step towards a formal model for the dynamics of polarization. We have singled out the fundamental issues discussed above in the context of social networks and made them parameters of our basic model. This has allowed us to observe some interesting and even counter-intuitive behavior. More precisely, the contributions of this paper are the following:

- We develop a simple model for agents' beliefs that allows for two key components: belief updates that can be biased (confirmation bias, backfire effect), and interaction models that can allow for several different ways in which society can be interconnected (cliques, disconnected groups, strong influencers).
- We show how our model, however simple, can capture interesting, realistic scenarios, through a few case studies.
- We show simulations exploring how the combination of interaction graphs and cognitive biases can lead to polarization. More specifically, we identify with our experiments counter-intuitive situations in which polarization arises or not in surprising ways. In particular we shall see that polarization is not necessarily monotonic, and that sometimes people with different opinions interacting more strongly may lead to more polarization.
- We organize the insights from our experiments in a series of precisely formulated conjectures about how the phenomenon of polarization comes about.

We believe that our model, experiments and observations contribute to the study of polarization in our society. It also paves the way for the design of more robust computational models for human cognitive and social processes. Perhaps one of our most interesting findings is that polarization is not necessarily monotonic: it can increase substantially, only to then decrease and fade away within a group. This insight alone should bring Catuscia some hope in our future.

Plan of the Paper. The remainder of this paper is organized as follows. In Sect. 2 we provide an overview of related work. In Sect. 3 we introduce our preliminary model for the dynamics of polarization that combines cognitive biases with the way agents interact. In Sect. 4 we present a series of case studies and simulations that help us gain insight on the dynamics of polarization, and we enumerate a list of conjectures inspired by these results. Finally, in Sect. 5 we discuss directions of future work.

2 Related Work

Social networks are fairly new and are becoming ubiquitous. Their wide-ranging effects on the world have motivated a great deal of research. Here we provide an overview of the range of relevant approaches to the problems we are considering, and put into perspective the novelty of our approach to the problem.

Polarization. Polarization was first rigorously defined by the economists Esteban and Ray [10]. Their measure of polarization, discussed in Sect. 3, is influential, and is the one we adopt in this paper. Li et al. [32] were the first to model consensus and polarization in social networks. Like most other work, they did not quantify polarization, but focused on when and under what conditions a population reaches consensus. Proskurnikov et al. [37] investigated the formation of consensus or polarization in social networks , but considered polarization as lack of consensus, rather than a phenomenon in its own right.

Formal Models. Sîrbu et al. [40] use a model somewhat similar to ours, which updates probabilistically. They investigate the effects of algorithmic bias on polarization by counting the number of opinion clusters, interpreting a single opinion cluster as consensus, rather than directly measuring polarization itself. Leskovec et al. [13] develop simulated social networks and observe group formation over time. Their work is not concerned with a formal measure of polarization.

Data-Based Approaches. There has been work on analyzing data from social networks and other systems with human users to understand polarization and related phenomena. Crandall et al. [7] were among the first to investigate changes in user agreement over time in such systems, based on data from Wikipedia and LiveJournal. Although they do not explicitly mention polarization, their concerns are similar to ours, as they focus on users becoming more similar over time and the formation of distinct communities, referred to as "Balkanization." Leskovec et al. [31] delve into measuring and representing users' status and positive or negative opinions of one another in social networks, using data from Epinions, Slashdot, and Wikipedia. Status and mutual opinions are not the same as influence but they share some similarities. Flaxman et al. [11] use data on readers' visits to news websites to measure their political segregation: the average distance between the political beliefs of two randomly selected users. This measure is related to polarization and it will be interesting to compare the two measures. Fletcher and Nielsen [12] analyze users' access to news outlets both online and offline in different countries. Their ideas on fragmentation are relevant to our work and helped us develop test scenarios concerning news outlets.

Logic-Based Approaches. Liu et al. [33] use ideas from doxastic logic and dynamic epistemic logic to qualitatively model influence and belief change in social networks. Christoff [6] develops several logics for social networks. This work is also non-quantitative, but it is concerned with problems related to polarization, such as information cascades. Seligman et al. [38] introduce a basic "Facebook logic." This logic is non-quantitative, but its interesting point is that an agent's possible networks are different social networks. This is a promising approach to formal modeling of epistemic issues in social networks. Hunter [22] introduces a logic of belief updates over social networks where closer agents in the social network are more trusted and thus more influential. While beliefs in this logic are non-quantitative, it includes a quantitative notion of influence between users.

In [15–19] logic-based semantic structures are used to reason about beliefs, lies, and group epistemic behaviour inspired by social networks. A quantitative extension of these structures seems to be a natural direction to deal with more complex epistemic phenomena such as group polarization.

Other Related Work. The seminal paper Huberman et al. [20] is concerned with determining which friends or followers in a user's network have the most influence on the user. Although this paper does not quantify influence between users, it still addresses an important question to our project.

3 A Preliminary Model for the Dynamics of Polarization

As we have discussed, our long-term goal is to develop a model for the dynamics of polarization allowing for the derivation of formal results that explain and predict how the phenomenon arises, sustains itself, or fades away. To ensure the tractability of our model, we must identify the minimal features allowing sufficient expressiveness for the results we envision. In this section we provide a preliminary model of polarization, including what we see as its most essential elements: a representation of agents' beliefs, of how agents interact with and influence each other, and of how they update their beliefs in face of new evidence.

A central concern of our model is that the majority of the (quite extensive) literature on the representation of agents' beliefs –as well as of the way these agents interact to update such beliefs– assumes that agents are rational and optimally use all information at their disposal. But a growing body of work by psychologists, sociologists, and economists has challenged these assumptions, and several common, relevant cognitive biases have been catalogued [8,21,34].

Our preliminary model incorporates such cognitive biases, as well as allows several different configurations for how agents in a social network may interact. The model is employed in Sect. 4 in a series of case studies and simulations. We divide the elements of our model into static and dynamic, described next.

3.1 Static Elements of the Model

Static elements of the model represent a snapshot of a social network at a given point in time. This includes the information needed to assess the level of polarization of agents' beliefs according to the desired measure. Our model is general enough to represent both opinions about subjective matters, and beliefs about factual topics, so we discuss both in the rest of the paper. Our model includes the following static elements:

- A (finite) set $\mathcal{A} = \{a_0, a_1, \ldots, a_{n-1}\}$ of $n \geq 1$ *agents*.
- A *proposition* p of interest, about which agents can hold beliefs.
- A *belief state* $\mathrm{Bel}_p : \mathcal{A} \to [0,1]$. Each value $\mathrm{Bel}_p(a)$ is the instantaneous confidence of agent $a \in \mathcal{A}$ in the veracity of proposition p. The extreme values 0 and 1 represent a firm belief in, respectively, the falsehood or truth of the proposition in question.
- A *polarization measure* $\mathrm{Pol} : [0,1]^{\mathcal{A}} \to \mathbb{R}$ mapping belief states to real numbers. The value $\mathrm{Pol}(\mathrm{Bel}_p)$ indicates how polarized the belief state Bel_p is.

There are several polarization measures described in the literature. In this work we focus on the influential measure proposed by Esteban and Ray [10].

Definition 1 (Esteban-Ray (ER) polarization measure). *Consider a (finite) discrete set $\mathcal{Y} = \{y_0, y_1, \ldots, y_{k-1}\}$ of size k, s.t. each $y_i \in \mathbb{R}$. Let $(\pi, y) = (\pi_0, \pi_1, \ldots, \pi_{k-1}, y_0, y_1, \ldots, y_{k-1})$ be a* distribution *on $\mathcal{Y}$ s.t. π_i is the frequency of value $y_i \in \mathcal{Y}$ in the distribution. W.l.o.g. we can assume the values* *of* π_i *are all non-zero and add up to 1 (as in a standard probability distribution). The* Esteban-Ray (ER) polarization measure *is defined as*

$$\text{Pol}_{\text{ER}}(\pi, y) = K \sum_{i=0}^{k-1} \sum_{j=0}^{k-1} \pi_i^{1+\alpha} \pi_j |y_i - y_j|,$$

where $K > 0$ *is a constant, and typically* $\alpha \approx 1.6$.

The Esteban-Ray measure captures the intuition that polarization is accentuated by both intra-group homogeneity and inter-group heterogeneity. Moreover, it assumes that the total polarization of a group of agents is the sum of the effects of individual agents on one another.

Note, however, that Pol_{ER} is defined on a discrete distribution, whereas in our model a general polarization metric is defined on a belief state $\text{Bel}_p : \mathcal{A} \to [0, 1]$. To properly apply Pol_{ER} to our setup we need to convert the belief state Bel_p into an appropriate distribution (π, y) as expected by the measure. We can do this as follows. Given a number $k > 0$ of *(discretization) bins*, define $\mathcal{Y} = \{y_1, y_2, \ldots, y_k\}$ as a discretization of the range $[0, 1]$ in such a way that each y_i represents the interval $y_i = [{}^i/_k, {}^{i+1}/_k]$, and define the weight π_i of value y_i, corresponding to the faction of agents having belief in the interval y_i, as $\pi_i = (\sum_{a \in \mathcal{A}} \mathbb{1}_{y_i}(\text{Bel}_p(a)))/|\mathcal{A}|$, where $\mathbb{1}_{\mathcal{X}}(x)$ is the *indicator function* returning 1 if $x \in \mathcal{X}$, and 0 otherwise.

3.2 Dynamic Elements of the Model

Dynamic elements of the model capture the information necessary to model the evolution of agents' beliefs in the social network. Although these aspects are not directly used to compute the instantaneous level of polarization, they determine how polarization evolves. The dynamic elements of our model are the following:

- An *influence graph* $\text{In} : \mathcal{A} \times \mathcal{A} \to [0, 1]$ s.t. $\text{In}(a_i, a_j)$ represents the influence of agent a_i on a_j, for $a_i, a_j \in \mathcal{A}$. A higher value means stronger influence.
- A *time frame* $\mathcal{T} = \{0, 1, 2, \ldots, t_{max}\}$ representing the discrete passage of time. Each $t \in \mathcal{T}$ is called a *time step*, and t_{max} is the duration of a simulation.
- A *family of belief states* $\{\text{Bel}_p^t\}_{t \in \mathcal{T}}$ parametrized by the time frame, s.t. each Bel_p^t is social network's state of beliefs w.r.t. proposition p at time step $t \in \mathcal{T}$.
- An *update function* $\text{Updt}_p : \text{Bel}_p \times \text{In} \to \text{Bel}_p$ mapping a belief state and influence graph to a new belief state. $\text{Bel}_p^{t+1} = \text{Updt}_p(\text{Bel}_p^t, \text{In})$ models the evolution of agents' beliefs from time step t to $t + 1$.

There are many ways of defining update functions; we discuss ours below.

Classic Update-Function. This update function incorporates the essential idea that the impact of agent a_j's belief on agent a_i's is proportional to the influence $\text{In}(a_j, a_i)$ of a_j on a_i, and to the difference $\text{Bel}_p^t(a_j) - \text{Bel}_p^t(a_i)$ in their beliefs. Formally, the *classic update* of agent a_i's belief given the interaction with agent a_j at time t is given by

$$\text{Bel}_p^{t+1}(a_i \mid a_j) = \text{Bel}_p^t(a_i) + \text{In}(a_j, a_i) \left(\text{Bel}_p^t(a_j) - \text{Bel}_p^t(a_i)\right) . \tag{1}$$

The *overall classic update* of agent a_i's belief given the interaction with all other agents in $\mathcal{A}$ at time t is defined as the expected belief of agent a_i w.r.t. all interactions in the group, as if they all happened in parallel:

$$\mathrm{Bel}_p^{t+1}(a_i) = \frac{1}{|\mathcal{A}|} \sum_{a_j \in \mathcal{A}} \mathrm{Bel}_p^{t+1}(a_i \mid a_j) . \tag{2}$$

Confirmation-Bias Update-Function. This function incorporates the insight that agents may be prone to *confirmation bias*, i.e., tend to give more weight to evidence supporting their current beliefs than to evidence that contradicts them. In our model we capture that as follows. When agent a_j presents agent a_i with evidence for or against proposition p, the update function will still move agent a_i's belief toward the belief of agent a_j, proportionally to the influence $\mathrm{In}(a_j, a_i)$ that a_j has over a_i, but with a caveat: the move is stronger when a_j's belief is similar to a_i's than when it is dissimilar. We formalize the *confirmation-bias update* of agent a_i's belief given the interaction with agent a_j at time t as:

$$\mathrm{Bel}_p^{t+1}(a_i \mid a_j) = \mathrm{Bel}_p^{t}(a_i) + f_{confbias} * \mathrm{In}(a_j, a_i) \left(\mathrm{Bel}_p^{t}(a_j) - \mathrm{Bel}_p^{t}(a_i)\right) ,$$

where $f_{confbias} \in [0,1]$ is a *confirmation-bias factor* proportional to the difference in the agents' beliefs, defined as $f_{confbias} = 1 - |\,\mathrm{Bel}_p^t(a_j) - \mathrm{Bel}_p^t(a_i)|$.

The *overall confirmation-bias update* of agent a_i's belief given the interaction with all other agents in $\mathcal{A}$ at time t is defined similarly to (2).

Backfire-Effect Update-Function. The *backfire effect* occurs when, in the face of contradictory evidence, agents' beliefs are not corrected, but rather remain unaltered or even get stronger. The effect has been demonstrated in experiments by psychologists and sociologists [8,21,34]. The backfire effect is considered an important component of polarization, as it indicates that trying to refute false information may not be as effective as one may expect [9,42].

The backfire effect may occur because an agent tend to scrutinize evidence more thoroughly and critically when that evidence (i) is contradictory to the agent's original belief; and (ii) comes from a source the agent respects at least to a minimum degree [42].

To formally capture these intuitions, we define the *belief gap* between agents a_i and a_j at time t as the distance between their beliefs

$$\mathit{bel\_gap}_p^t(a_i, a_j) = \left|\mathrm{Bel}_p^t(a_i) - \mathrm{Bel}_p^t(a_j)\right| ,$$

and consider that, when agents a_j and a_i interact, the effect is triggered if:

(i) the belief gap $\mathit{bel\_gap}_p^t(a_i, a_j)$ of the two agents is above the *backfire belief-gap threshold*, denoted $\mathit{bf\_bel\_th} \in [0,1]$, and
(ii) agent a_j's influence on agent a_i, $\mathrm{In}(a_j, a_i)$, is above the *backfire minimum-interaction threshold*, denoted $\mathit{bf\_int\_th} \in [0,1]$.

Outside of the conditions triggering the backfire effect, belief update will occur according to the classic update function (1). But when the effect is triggered, agent a_i's belief will get stronger in its original bias[1] exactly because a_j disagrees with a_i. More precisely, the mapping $\text{Bel}_p^t(a_i) \mapsto \text{Bel}_p^{t+1}(a_i)$ must satisfy:

$$\text{Bel}_p^{t+1}(a_i) \leq \text{Bel}_p^t(a_i), \qquad \text{if } \text{Bel}_p^t(a_i) < 0.5 \text{ , } and \tag{3}$$

$$\text{Bel}_p^{t+1}(a_i) \geq \text{Bel}_p^t(a_i), \qquad \text{if } \text{Bel}_p^t(a_i) \geq 0.5 \text{ .} \tag{4}$$

The expected behavior of an update function respecting (3) and (4) above is depicted in Fig. 1. Note that the backfire update follows a form of *step function*, which can be approximated by the *logistic function*

$$\text{Bel}_p^{t+1}(a_i) = \frac{1}{1 + e^{-2 \cdot bf\_factor \cdot z}} \text{ ,} \tag{5}$$

where $z = 2 \cdot \text{Bel}_p^t - 1$ is a transformation to ensure the step-curve lies within the interval $[0, 1]$, and the parameter $bf\_factor \geq 0$ controls how sharp the step is at 0.5. The greater bf_factor, the sharper the step, and the stronger the backfire effect. We use the smoothed approximation of the backfire effect because intuitively, belief update should be continuous with a continuous derivative, since the backfire effect does not begin at a discrete point, rather it becomes stronger as beliefs are more strongly opposed.

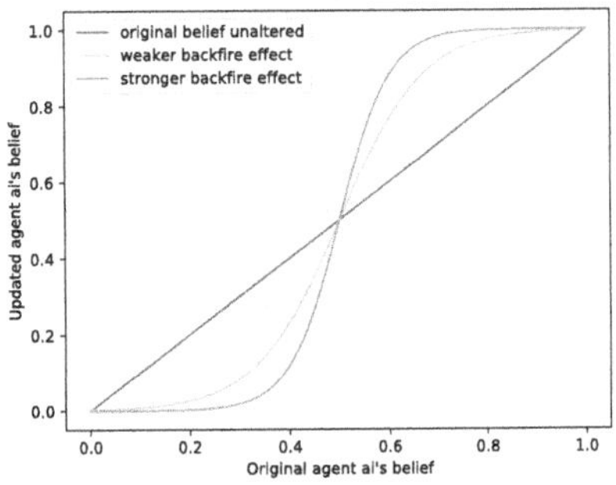

Fig. 1. Desired behavior of belief update with backfire effect, when presented with new information strongly contradicting the agent's current beliefs.

To tune the value of bf_factor, we specify two further conditions. First, the larger the belief gap $bel\_gap_p^t(i, j)$, the more pronounced is the strengthening of $\text{Bel}_p(a_i)$. Second, the lower the influence $\text{In}(a_j, a_i)$ of agent a_j on agent a_i, the more pronounced is the strengthening of $\text{Bel}_p(a_i)$. These can be formalized as

$$bf\_factor = \frac{bel\_gap_p^t(i, j)}{\text{In}(a_j, a_i)} \text{ .} \tag{6}$$

Finally, we put it all together to get our definition of backfire effect update function based on three cases: classic belief update if the belief gap threshold and strength of influence are below the thresholds, backfire effect in the negative

[1] We say agent a_i has *negative* or *positive bias* w.r.t. proposition p if $\text{Bel}_p(a_i)$ falls, respectively, in the intervals $[0, 0.5)$ or $[0.5, 1]$.

direction if the thresholds are met and the agent's initial belief in p is below 0.5 (the agent initially tended not to believe p, so this belief becomes stronger), and similarly, backfire effect in the positive direction if the agent's initial belief in p is above 0.5. Altogether, this gives us the *backfire-effect update* of agent a_i's belief given the interaction with agent a_j at time t, formalized as $\mathrm{Bel}_p^{t+1}(a_i \mid a_j)$ to be

- $\mathrm{Bel}_p^t(a_i) + \mathrm{In}(a_j, a_i)\left(\mathrm{Bel}_p^t(a_j) - \mathrm{Bel}_p^t(a_i)\right)$, if either $bel\_gap_p^t(i,j) < bf\_bel\_th$ or $\mathrm{In}(a_j, a_i) < bf\_int\_th$;
- $\min\left(\mathrm{Bel}_p^t(a_i), \frac{1}{1+e^{-2\cdot bf\_factor\cdot z}}\right)$, if the following holds $bel\_gap_p^t(i,j) \geq bf\_bel\_th$ and $\mathrm{In}(a_j, a_i) \geq bf\_int\_th$ and $\mathrm{Bel}_p^t(a_i) < 0.5$;
- $\max\left(\mathrm{Bel}_p^t(a_i), \frac{1}{1+e^{-2\cdot bf\_factor\cdot z}}\right)$, if the following holds $bel\_gap_p^t(i,j) \geq bf\_bel\_th$ and $\mathrm{In}(a_j, a_i) \geq bf\_int\_th$ and $\mathrm{Bel}_p^t(a_i) \geq 0.5$;

where $z = 2 \cdot \mathrm{Bel}_p^t(a_i) - 1$, and bf_factor is given by (6).

The *overall backfire-effect update* of agent a_i's belief given the interaction with all other agents in $\mathcal{A}$ at time t is defined similarly to the overall classic update of (2), by summing the effects of all the agents.

4 Case Studies and Simulation

In this section we present a series of simulations of the dynamics of polarization using the model introduced in Sect. 3. Our main goal is to gain insight into the interplay between the components of the model in bringing about polarization. We investigate the phenomenon under various configurations of (i) the initial belief state of agents' opinions in a social network; (ii) the influence graph modeling how these agents interact; and (iii) the belief update function dictating how they incorporate new information. Our experiments uncover a series of (sometimes counter-intuitive) facts about the dynamics of polarization, and provide valuable insight into the results we may prove about it, as well as ideas for refining our model.

4.1 Instantiation of Static and Dynamic Elements of the Model

In the following, let $i, j \in \{0, 1, \ldots, n-1\}$ range over the indexes of the set $\mathcal{A} = \{a_0, a_1, \ldots, a_{n-1}\}$ of n agents, and let p be the proposition of interest (about which agents may hold beliefs).

In our simulations we consider the following configurations for the initial belief state. Figure 2 presents instantiations of these belief states for $n = 10$ of agents, and the corresponding Esteban-Ray polarization value.

- A *uniform belief state* $\mathrm{Bel}_p^{uniform}(a_i)$ representing a set of agents whose beliefs are as varied as possible, all equally spaced in the interval $[0, 1]$: $\mathrm{Bel}_p^{uniform}(a_i) = i/(n-1)$.

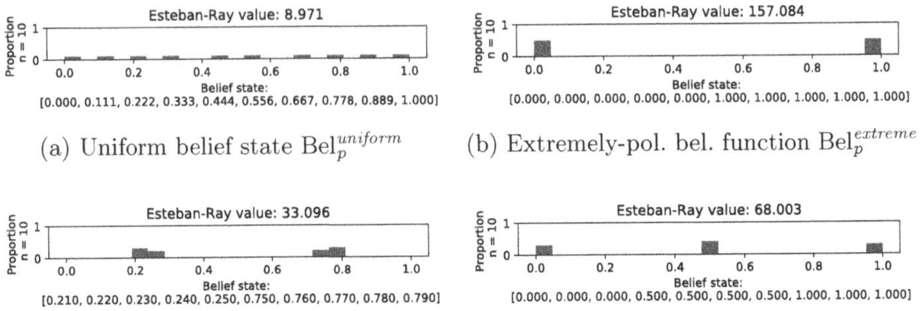

(a) Uniform belief state $\mathrm{Bel}_p^{uniform}$ (b) Extremely-pol. bel. function $\mathrm{Bel}_p^{extreme}$

(c) Mildly-polarized belief state Bel_p^{mild} (d) Tripartite belief state $\mathrm{Bel}_p^{tripartite}$

Fig. 2. Instantiations of belief states (for 10 agents) as histograms, and the corresponding Esteban-Ray polarization, with $K = 1000$, and 21 discretization bins.

- A *mildly polarized belief state* Bel_p^{mild} representing a set of agents clustered into two groups with rather different inter-group belief levels compared to their intra-group belief levels: $\mathrm{Bel}_p^{mild}(a_i) = \max(0.25-0.01 \cdot (\lceil n/2 \rceil - i - 1), 0)$, if $i \leq \lceil n/2 \rceil$; or $\mathrm{Bel}_p^{mild}(a_i) = \min(0.75-0.01 \cdot (\lceil n/2 \rceil - i), 1)$, otherwise.
- An *extremely polarized belief state* $\mathrm{Bel}_p^{extreme}$ representing the acute situation in which half of the agents are certain of the falsehood of the proposition, and the other half are certain of its truth: $\mathrm{Bel}_p^{uniform}(a_i) = 0$, if $i \leq \lceil n/2 \rceil$; or $\mathrm{Bel}_p^{uniform}(a_i) = 1$ otherwise.
- A *tripartite belief state* $\mathrm{Bel}_p^{tripartite}$ representing a set of agents divided into three clustered groups of similar size, where all agents in each group hold the same belief: $\mathrm{Bel}_p^{uniform}(a_i) = 0$, if $0 \leq i < \lfloor n/3 \rfloor$; $\mathrm{Bel}_p^{uniform}(a_i) = 0.5$, if $\lfloor n/3 \rfloor \leq i < \lceil 2n/3 \rceil$; or $\mathrm{Bel}_p^{uniform}(a_i) = 1$, if $\lceil 2n/3 \rceil \leq i < n$.

As for influence graphs, we consider the following configurations:

- A *clique* influence graph In^{clique} representing an idealized totally connected social network in which every agent exerts considerable influence on every other agent: $\mathrm{In}^{clique}(a_i, a_j) = 0.5$, for every i, j.
- A *disconnected* influence graph In^{disc} representing a social network sharply divided into two groups in such a way that agents within the same group can considerably influence each other, but agents from different groups do not influence each other at all: $\mathrm{In}^{disc}(a_i, a_j) = 0.5$, if i, j are both $\leq \lceil n/2 \rceil$ or both $> \lceil n/2 \rceil$, and $\mathrm{In}^{disc}(a_i, a_j) = 0$ otherwise.
- A *faintly-connected* influence graph In^{faint} representing a social network divided into two groups that evolve mostly separately, with only faint communication between them. More precisely, agents from within the same group influence each other much more strongly than agents from different groups: $\mathrm{In}^{disc}(a_i, a_j) = 0.5$, if i, j are both $\leq \lceil n/2 \rceil$ or both $> \lceil n/2 \rceil$, and $\mathrm{In}^{disc}(a_i, a_j) = 0.1$ otherwise. This represents a small, ordinary social net-

work, where some close groups of agents have strong influence on one another, and all agents communicate to some extent.

- An *unrelenting influencers* influence graph $\text{In}^{unr\text{-}inf}$ representing a social network in which two agents have strong influence on every other agent, but are not influenced at all by anyone else. We consider the influence of these two main agents (a_0 and a_{n-1}) to be the same: $\text{In}^{unr\text{-}inf}(a_i, a_j) = 0.6$ if $i = 0$ and $j \neq n{-}1$ or $i = n{-}1$ and $j \neq 0$; $\text{In}^{unr\text{-}inf}(a_i, a_j) = 0.0$ if $i \neq j$ and either $j = 0$ or $j = n{-}1$; and finally $\text{In}^{unr\text{-}inf}(a_i, a_j) = 0.1$ if $0 \neq i \neq n{-}1$ and $0 \neq j \neq n{-}1$. This could represent, for instance, a social network counting with the presence of two totalitarian TV networks dominating the news market, both with similarly high levels of influence on all agents. The two networks have clear agendas to push forward, and are not influenced in a meaningful way by other agents. Moreover, agents have a small influence on one another.
- A *malleable influencers* influence graph $\text{In}^{mall\text{-}inf}$ representing a social network in which two agents have strong influence on every other agent, but are barely influenced by anyone else. This scenario could represent a situation similar to the "unrelenting influencers" scenario above, with two differences. First, one TV network has a higher influence than the other. Second, the networks are slightly influenced by all the agents (e.g., by checking ratings and choosing what news to cover accordingly). More precisely, we consider the influence of one main agent (a_0) to be higher than the other's (a_{n-1}): $\text{In}^{mall\text{-}inf}(a_i, a_j) = 0.8$ if $i = 0$ and $j \neq n{-}1$; $\text{In}^{mall\text{-}inf}(a_i, a_j) = 0.5$ if $i = n - 1$ and $j \neq 0$; $\text{In}^{mall\text{-}inf}(a_i, a_j) = 0.1$ if $i \neq j$ and either $j = 0$ or $j = n{-}1$; and finally $\text{In}^{mall\text{-}inf}(a_i, a_j) = 0.2$ if $0 \neq i \neq n{-}1$ and $0 \neq j \neq n{-}1$.

We considered some other situations, such as a single major influencer, or two equally strong influencers who are also slightly influenced by all other agents. We chose the configurations with the most interesting results to present here.

4.2 Insights from Simulations

In this section we discuss the results of our simulations, and present the insights we obtain from them. In all simulations we consider a set of $n = 100$ agents, a time frame of 100 steps, and for the computation of the Esteban-Ray measure we use 201 discretization bins, setting parameters to $K = 1000$ and $\alpha = 1.6$.

We begin our discussion by considering the cases in which our simulations agree with the "expected" behavior of polarization. For this task, we focus on a scenario in which agents start off with varied opinions (initial belief state $\text{Bel}_p^{uniform}$), and all interact with each other (interaction graph In^{clique}), incorporating new information either in a classic way, or with confirmation bias. These "expected" results are shown in Fig. 3. In particular, Fig. 3a meets our expectation that social networks in which all agents can interact eventually converge to a consensus (i.e., polarization disappears), even if agents start off with very different opinions. Figure 3b shows that for the same update function and initial

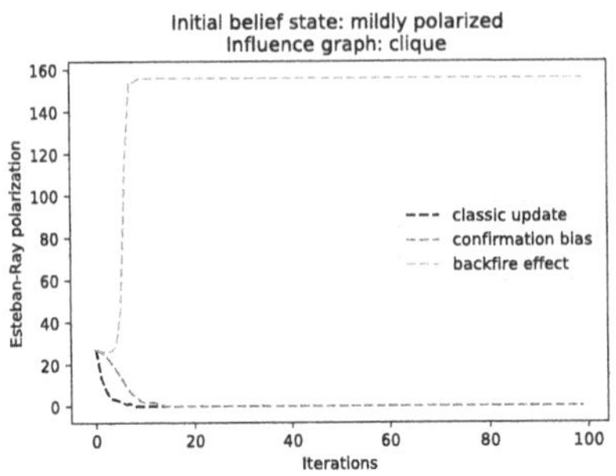

(a) Comparison of update functions

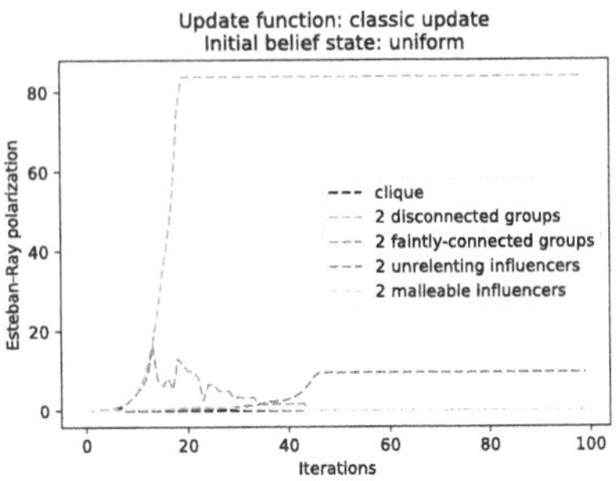

(b) Comparison of influence graphs

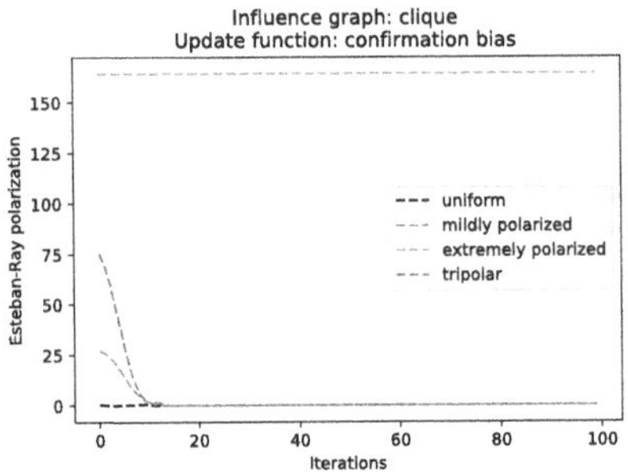

(c) Comparison of initial belief states

Fig. 3. Examples of "expected" behavior of the dynamics of polarization.

belief state, different interaction graphs may lead to very different evolutions of polarization: it grows to a maximum if agents are disconnected, achieves an intermediate value in the presence of 2 unrelenting influencers, and disappears in all other cases (i.e., when agents can influence each other). Finally, Fig. 3c shows that even when agents can all interact with one another, an extremely polarized belief state will remain polarized, since confirmation bias guarantees that agents will not properly update their beliefs.

We now turn our attention to other interesting cases, in which the simulations help shed light on perhaps counter-intuitive behavior of the dynamics of polarization. In the following we organize these insights into several topics.

Polarization is not Necessarily a Monotonic Function of Time. At first glance it may seem that for a fixed social network configuration, polarization either only increases or decreases with time. Perhaps surprisingly, this is not so.

Figure 4 shows examples of non-monotonic behavior of the evolution of polarization. In particular, Fig. 4a shows that for a uniform initial belief state and an interaction graph with two faintly connected groups, no matter the update function, polarization increases before decreasing again and stabilizing at a lower level. This is explained by the fact that in such a set-up agents within the same group will reach consensus relatively fast, which will lead society to two clear

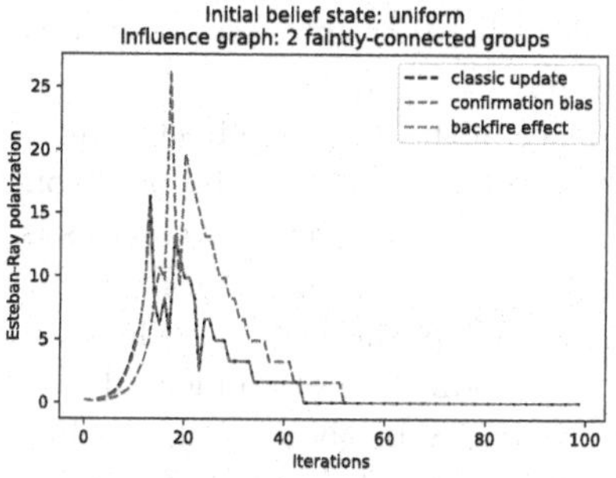

(a) Comparison of update functions

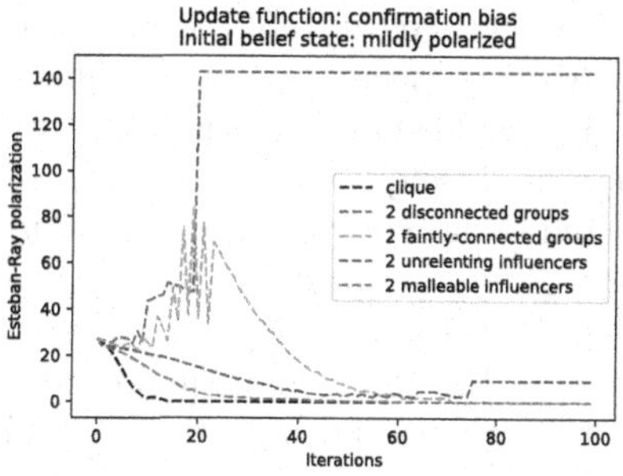

(b) Comparison of interaction graphs

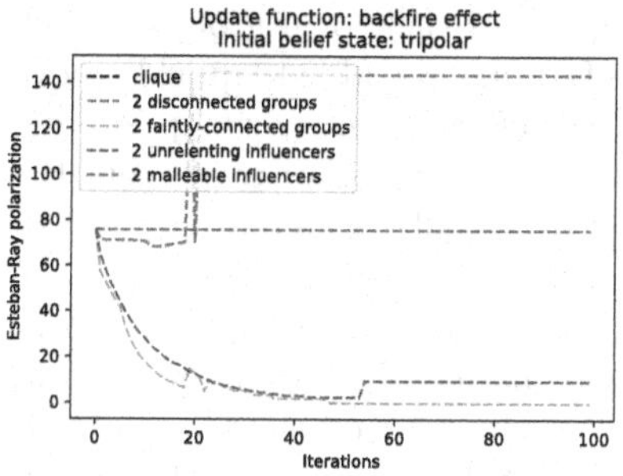

(c) Comparison of interaction graphs.

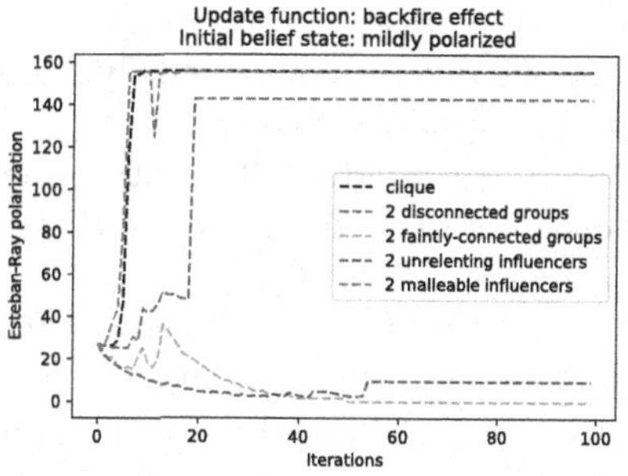

(d) Comparison of interaction graphs.

Fig. 4. Examples of cases in which the evolution of polarization is not monotonic.

poles and increase polarization at an initial stage. However, as the two groups continue to communicate over time, even if faintly, a general consensus is eventually achieved. Figure 4b shows that a mildly polarized social network in which agents have confirmation bias, a similar phenomenon occurs for all interaction graphs. The only exceptions are the cases of disconnected groups and unrelenting influencers, in which polarization stabilizes (as expected). In this case, this happens because the disconnected groups reach internal consensus but remain far from the other group, since they do not communicate. This represents a high level of polarization. The case of two unrelenting influencers retains a low level of polarization simply because the opinions of the unrelenting influencers never change. The other interesting case is the two faintly connected groups: here polarization first increases as the groups reach internal consensus, but then the two groups move toward one another and polarization decreases. The other two interaction graphs also stabilize to zero polarization.

Figure 4c shows that with the backfire effect taken into account, polarization increases under two malleable influencers. This occurs because the backfire effect actually pushes the malleable influencers apart, and then all the other agents also tend to one or the other major influencer, leading to a high level of polarization. Starting with two disconnected groups, polarization remains constant since the two groups do not communicate, and under all other interaction graphs, polarization decreases and stabilizes at or near zero.

Figure 4d shows another interesting consequence of the backfire effect: a fully-connected (clique) social network may polarize the most, followed by a disconnected social network, and only a faintly connected one may reach a consensus. This occurs because, as the social network's initial belief state is mildly polarized, if agents are exposed to every other agents' beliefs the backfire effect will be immediately triggered and ensure strong polarization. In a disconnected case, on the other hand, agents will achieve local consensus in their groups, which leads to a less extreme situation, even if still quite polarized. Most interestingly, however, a faintly connected network will be the only one to fully de-polarize: as agents interact more strongly with other agents in their own groups (which have different yet not extremely dissimilar beliefs), they are able to quickly move toward a local consensus in each group. When the effects of inter-group interaction are meaningful, all agents' beliefs are close enough together to prevent the backfire effect to be triggered, and consensus is eventually achieved.

The Effect of Different Update Functions. Figure 5 shows a comparison of different update functions in various interesting scenarios. Figure 5a shows that, as expected, polarization can permanently increase in a disconnected social network, with little difference between the behavior of different update functions. Figure 5b depicts the effects of the three different update functions beginning from a uniform belief state, with two unrelenting influencers as the influence function. In all three cases, all agents except the influencers eventually reach 0.5 belief (the middle of the belief spectrum, between the two extreme agents), representing an increased but still fairly low level of polarization. The classic belief update function achieves this equilibrium fastest, since in the other two cases agents are less influenced by others whose beliefs are far from their own, so their beliefs change more slowly. Figure 5c shows that without the backfire effect, even under two extreme unrelenting influencers, consensus is eventually nearly reached, since everyone except the influencers eventually reaches a belief state between the beliefs of the two influencers. However, with confirmation bias, polarization does not decrease, since agents give high credence to the influencer who agrees with their own beliefs and mostly discount all the agents who do not agree with their beliefs. Thus, in this extreme situation, no one's beliefs change.

The Effect of Different Interaction Graphs. Figure 6a shows a comparison of different interaction graphs in various scenarios. Interestingly, Fig. 6 shows that polarization can increase in a social network in which all agents are connected, as long as agents are prone to the backfire effect. Interestingly, in this case a social network with two disconnected groups will become less polarized than a clique social network, because it is exactly the interaction among agents who disagree that exacerbates polarization. Note also that Figs. 6a, b, and c show that a faintly connected graph leads to a temporary peak in polarization, which is reversed in all cases. As we discussed, this is explained by the fact that agents

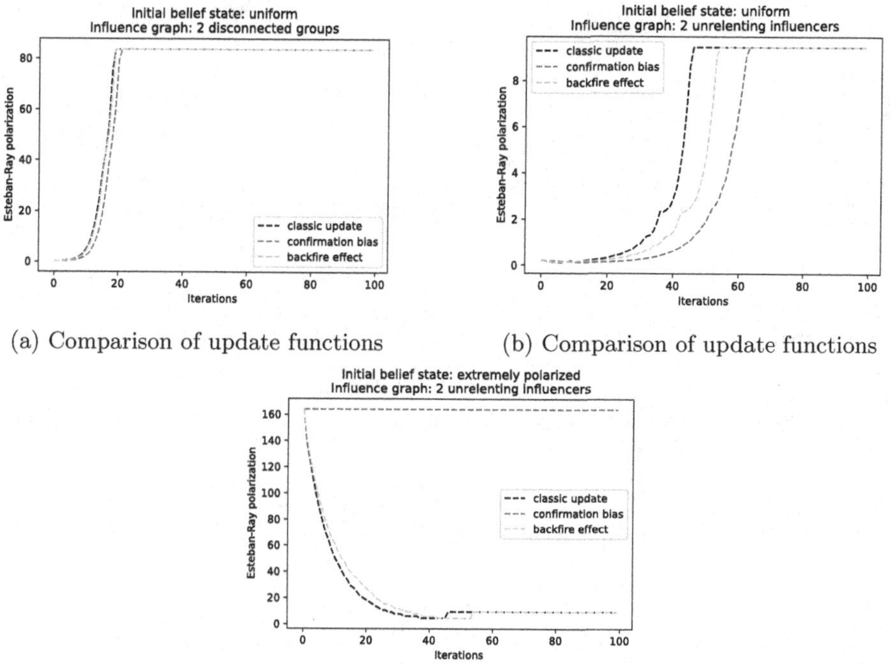

(a) Comparison of update functions

(b) Comparison of update functions

(c) Comparison of update functions

Fig. 5. Comparison of different update functions.

within the same group achieve consensus faster than agents in different groups, leading to a temporary increase in polarization.

The effect of Different Initial Belief States. Figure 7 compares different belief states in various scenarios. Figures 7a and c show that belief states that start off as less polarized can evolve to a more polarized state than situations that began as less polarized. For instance, Fig. 7a shows that a mildly polarized initial belief state will end up more polarized than a tripolar distribution, even though the tripolar distribution started out more polarized. Figure 7c shows that tripolar initial belief state starts out more polarized than uniform and mildly polarized situations, but eventually becomes the least polarized situation, since its polarization increases less than that of uniform and mildly polarized situations. This happens because in all cases, all the agents in the two disconnected groups move toward the center of the group they are in, and beginning with a tripolar distribution means the centers of the groups are moderately polarized.

4.3 Insights of Results to Prove

Altogether, these experiments show us that even in this simplified model, polarization is a complex and nuanced phenomenon which does not always behave

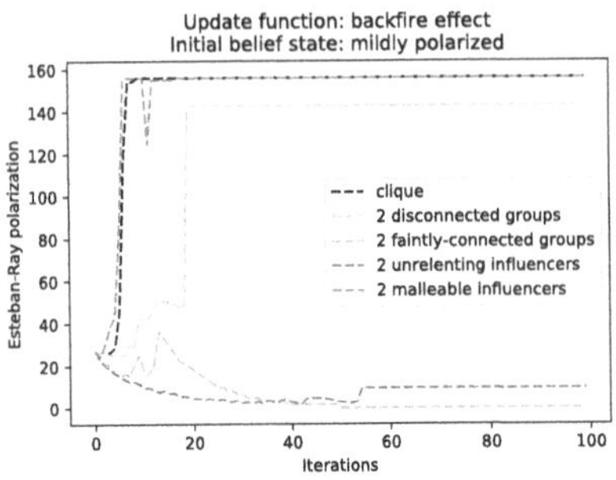

(a) Comparison of interaction graphs

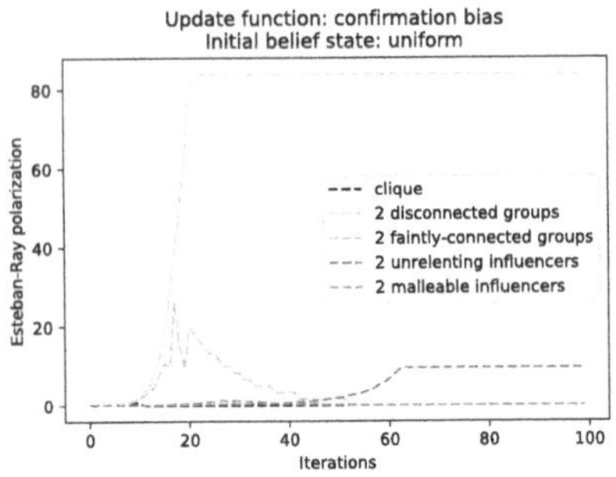

(b) Comparison of interaction graphs

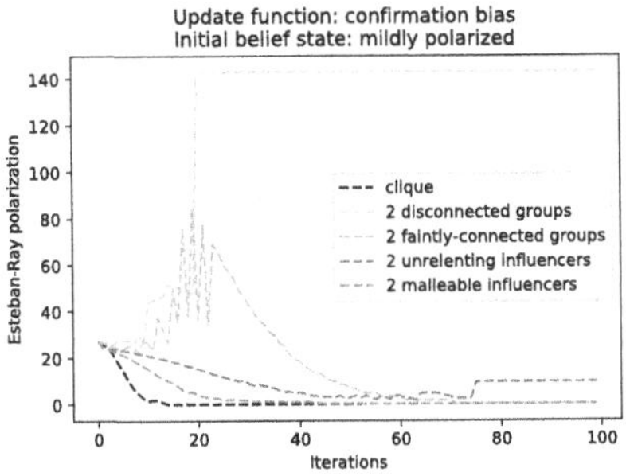

(c) Comparison of interaction graphs

Fig. 6. Comparison of different interaction graphs.

in the most obvious ways. However, we have some general conjectures about predictable behavior of polarization, which we plan to prove in the future. In all of the following, when we discuss convergence and limits we mean them in the standard mathematical sense, considering the updated beliefs or level of polarization as a sequence over $\mathbb{N}$, representing discrete time steps. We start with the conjecture that beliefs and polarization eventually reach a stable level.

Conjecture 1. *Under all conditions, each agent's belief eventually converges. (Note that different agents' beliefs may converge to different limits.)*

Conjecture 2. *Under all conditions, polarization eventually converges.*

For the remaining conjectures we first we need some definitions.

Definition 2. *We say an influence graph In is* strongly connected *if every agent has nonzero influence on every other agent: for all $a_i, a_j \in \mathcal{A}$, $In(a_i, a_j) > 0$.*

Definition 3. *For influence graphs In_1 and In_2 over a fixed set of agents $\mathcal{A}$, we say that $In_1 \sqsubseteq In_2$ iff for all $a_i, a_j \in \mathcal{A}$, $In_1(a_i, a_j) \leq In_2(a_i, a_j)$.*

Notice that in our experiments, the clique, faintly connected, and two malleable influencers interaction graphs are strongly connected.

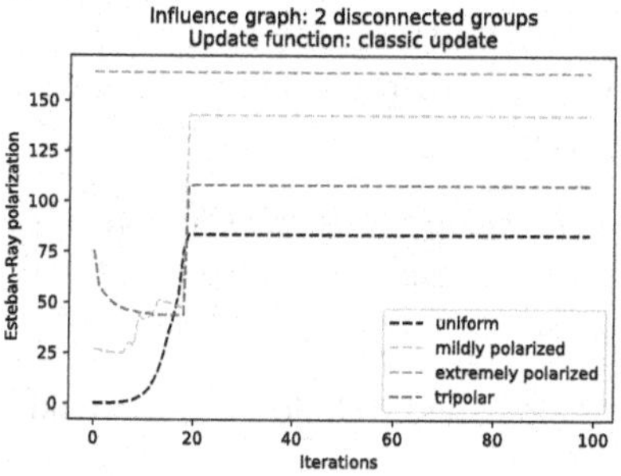

(a) Comparison of initial belief states

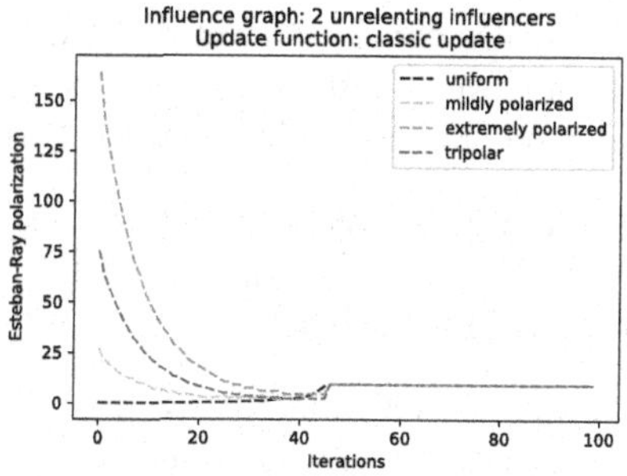

(b) Comparison of initial belief states

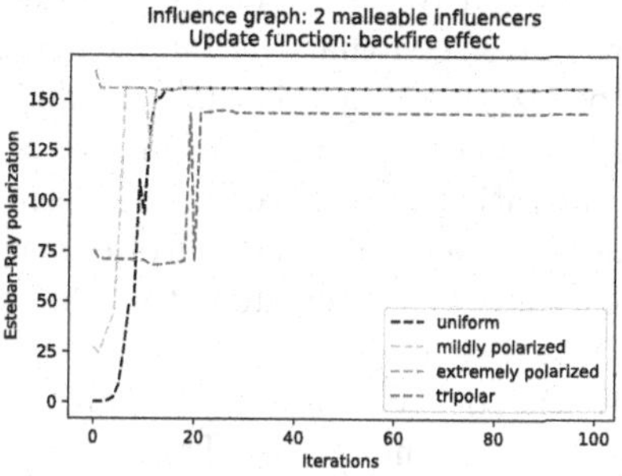

(c) Comparison of initial belief states

Fig. 7. Comparison of different initial belief states.

Conjecture 3. *With classic belief update and a strongly connected influence graph, increased influence does not increase polarization in the limit: if* In_1 *is strongly connected and* $In_1 \sqsubseteq In_2$*, then with any initial belief state, the polarization limit of* In_1 *is at least as high as the polarization limit of* In_2.

Conjecture 4. *With classic belief update and a strongly connected influence graph, all agents eventually converge to the same level of belief in the proposition of interest: for all* $a_i, a_j \in \mathcal{A}$, $\lim_{t\to\infty} Bel_p^t(a_i) = \lim_{t\to\infty} Bel_p^t(a_j)$.

Conjecture 4 implies the following result, which is stronger than Conjecture 3.

Conjecture 5. *With classic belief update and a strongly connected influence graph, with any initial belief state, polarization eventually converges to zero.*

Thus, we conjecture that under classic update, with connected influence graphs, polarization is bound to reach zero eventually, no matter what agents' starting beliefs are. However, we have seen in several of our experiments that this is not the case when update functions include confirmation bias or the backfire effect. This raises the question of whether there are methods for reducing polarization under non-classic belief update.

We also consider a notion of weakly connected influence graphs. The idea is that for every pair of agents there is a path from one to the other where every agent on the path influences the next agent on the path.

Definition 4. *We say an influence graph In over agents $\mathcal{A}$ is weakly connected if for all $a_i, a_j \in \mathcal{A}$, there exist $\{a_{k_1}, a_{k_2}, ..., a_{k_l}\} \subseteq \mathcal{A}$ such that $In(a_i, a_{k_1}) > 0$, $In(a_{k_l}, a_j) > 0$, and for $m = 1, ..., l-1$, $In(a_{k_m}, a_{k_{m+1}}) > 0$.*

We believe Conjectures 3, 4, and 5 can be strengthened to hold even in weakly (rather than strongly) connected graphs.

Finally, we consider the question of under what conditions it is best for agents who disagree with each other not to communicate in order to avoid polarization. In several simulations, we have seen that the faintly connected influence graph first leads to increased polarization, but eventually leads to low or zero polarization, even with the backfire effect or confirmation bias. This suggests that one strategy for reducing polarization is to encourage extreme agents to communicate first with those whose beliefs are not far from their own, moving toward a within-group consensus, and then later allow the different groups to come to a consensus. Note that this idea is different from creating "bubbles" in which agents communicate only with others who already agree with them, and it could play out differently depending on the setting. For example in a social network a user's feed could be tailored so that they see communications from other users with whom they share similar, yet sufficiently distinct beliefs. In real world situations this strategy could mean encouraging people to join carefully selected groups, for example. Such manipulation of users' beliefs would of course need to be carefully considered in terms of ethics, but perhaps it could also lead to positive effects in counter-balancing polarization in society.

5 Possible Directions of Future Work

In the current work, we have analyzed how group polarization can arise under certain configurations of agents' interactions. This analysis is a first step in a general model of polarization which takes more possibilities into account in order to model group polarization in the full complexity of the real world. Here we discuss how to formalize our model, how to enrich it, and how to apply it.

5.1 Possible Formalizations of the Model

Logic. One of our major goals is to develop a quantitative logic for reasoning about beliefs in social networks. Some interesting work has been done on logic in social networks, for example [6,22,38,39], but all current work on this topic does not consider quantitative level of beliefs for agents, nor a quantified influence of one agent on another. On the other hand, some work has been done on quantified or probabilistic epistemic logic outside of the setting of social networks, e.g. [29]. We plan to combine these two approaches to develop a logic for our models of social networks. Our logic would provide a formal way of making statements about agents' knowledge and beliefs and how these change as communication occurs. The major difficulty with creating such a logic is that in our model as currently described, agents do not have any information about one another; they only have beliefs or information about propositions. In order to develop an epistemic logic for these systems, we plan to introduce Kripke style possible worlds models so that agents have a view or information about one another. Then we will be able to develop a logic with doxastic or belief-based as well as epistemic or knowledge-based modalities. We also plan to include operators representing the effects of a message or announcement being sent, in the style of [35,41]. We further hope to incorporate ideas from *subjective logic* [24]. Subjective logic is a multi-agent logic which includes information both about agents' beliefs as well as their certainty about their beliefs. It is clear that in a logic of social networks, it would be beneficial to be able to discuss both how likely a fact is according to a user, and how certain they are about that likelihood.

Process Calculus. We plan to develop a process calculus which incorporates structures from social networks, such as communication, influence, posted messages, and individual opinions and beliefs. Since process calculi are an efficient, powerful, well studied formalism for reasoning about concurrent multi-agent systems, using them to represent social networks would simplify our models and give us access to already developed tools. A first step toward this goal was taken in [28] where we developed a process calculus with information about agents' beliefs and knowledge.

5.2 How to Refine the Elements of the Model

Dynamic Influence Revision. In the current work, we consider one agent's influence on another to be static. For modelling short term belief change, this is a reasonable decision, but in order to model long term belief change in social networks over time, we need to include dynamic influence. In particular, agent a's influence on agent b should become stronger over time if agent a sees b as reliable, which means that b mostly sends messages that are already close to the beliefs of agent a. Thus, we will enrich our model with *dynamic* influence between agents: agent a's influence on b becomes stronger if a communicates messages that b agrees with, and it becomes weaker if b mostly disagrees with the beliefs that are communicated by a. We expect that this change to the model will tend to increase group polarization, as agents who already tend to agree will have increasingly stronger influence on one another, and agents who disagree will influence one another less and less. This will be particularly useful for modelling *filter bubbles* [11], and we will consult empirical work on this phenomenon in order to correctly include influence update in our model.

Beliefs About Multiple Issues. The current paper considers the simple case where agents only have beliefs about one proposition. This simplification may be useful for modelling some realistic scenarios, such as beliefs along a liberal/conservative political spectrum, but we believe that it will not be sufficient for modelling all situations. To that end, we plan to develop models where agents have beliefs about multiple issues. This would allow us to represent, for example, Nolan's two-dimensional spectrum of political beliefs, where individuals can be economically liberal or conservative, and socially liberal or conservative [4], since more nuanced belief situations such as this one cannot be modelled in a one-dimensional space. Since Esteban-Ray polarization considers individuals along a one-dimensional spectrum, we will extend Esteban-Ray polarization to more dimensions, or develop a new, multi-dimensional definition of polarization.

Sequential and Asynchronous Communication. In our current model, all agents communicate synchronously at every time step and update their beliefs accordingly. This simplification is a reasonable approximation of belief update for an initial study of group polarization, but in real social networks, communication may be both asynchronous and sequential, in the sense that messages are sent one by one between a pair of agents, rather than every agent sending every connected agent a message at every time step as in the current model. The current model is deterministic, but introducing sequential messages will add nondeterminism to the model, bringing new complications. We plan to develop a logic in the style of dynamic epistemic logic [35] in order to reason about sequential messages. Besides being sequential, communication in real social networks is also asynchronous, making it particularly difficult to represent agents' knowledge and beliefs about one another, as discussed in [26,27]. Eventually, we plan to include asynchronous communication and its effects on beliefs in our model.

5.3 How to Apply This Model

Integrating Our Approach with Data. The current model is completely theoretical, and in some of the ways mentioned above it is an oversimplification. Eventually, when our model is more mature, we plan to use data gathered from real social networks both to verify our conclusions and to allow us to choose the correct parameters in our models (e.g. a realistic threshold for when the backfire effect occurs). Fortunately, there is an increasing amount of empirical work being done on issues such as polarization in social networks [2,5,30]. This will make it easier for us to compare our model to the real state of the world and improve it.

Avoiding Polarization. One of the long term goals of our work is learning how to avoid polarization. While the current work is only an analysis of when group polarization occurs, eventually we plan to develop a model which is sufficiently sophisticated to be able to prove facts about when polarization can and cannot occur. This would allow us to develop guidelines for what kind of communication and group formation should be avoided or prevented in social networks, in order to stop the incidence of group polarization, or at least know when it is likely to occur, without disregarding ethical issues. Similarly, we hope to develop guidelines to understand and prevent the occurrence of *radicalization*, a phenomenon which is related to but distinct from group polarization, in which an agent's beliefs become more extreme over time, and further from the beliefs of most other agents. We also plan to model the effects of filtering false information. When we are dealing with factual information, rather than opinions, we may incorporate a notion of outside truth, which is absent from the current model. We believe that by blocking even a small proportion of false messages (messages about p which are far from the true value of p) we may avoid polarization and reach consensus, depending on communication conditions. This would be an application of ideas presented in [14].

Applications to Issues Other than Polarization. Our model of changing beliefs in social networks certainly has other applications besides group polarization. One important issue is the development of false beliefs and the spread of misinformation. In order to explore this issue we again need the notion of outside truth. One of our goals is to learn the conditions under which agents' beliefs tend to move close to truth, and under what conditions false beliefs tend to develop and spread. In particular, we hope to understand whether there are certain classes of social networks which are resistant to the spread of false belief due to their topology and influence conditions. The spread of false beliefs is related to the issue of *information cascades* as discussed for example in [6]. In the future we also plan to model and analyze information cascades.

Acknowledgments. We would like to thank Alexis Elder, Camilo Rueda, and Valeria Rueda for thoughtful discussions and pointers to references.

References

1. Aronson, E., Wilson, T., Akert, R.: Social Psychology, 7th edn. Prentice Hall, Upper Saddle River (2010)
2. Bail, C.A., et al.: Exposure to opposing views on social media can increase political polarization. Proc. Natl. Acad. Sci. **115**, 9216–9221 (2018)
3. Bozdag, E.: Bias in algorithmic filtering and personalization. Ethics Inf. Technol. **15**, 209–227 (2013)
4. Bryson, M.C., McDill, W.R.: The political spectrum: a bi-dimensional approach. Rampart J. Individ. Thought **4**, 19–26 (1968)
5. Calais Guerra, P., Meira Jr, W., Cardie, C., Kleinberg, R.: A measure of polarization on social media networks based on community boundaries. In: Proceedings of the 7th International Conference on Weblogs and Social Media, ICWSM 2013 (2013)
6. Christoff, Z., et al.: Dynamic logics of networks: information flow and the spread of opinion. Ph.D. thesis, Institute for Logic, Language and Computation, University of Amsterdam (2016)
7. Crandall, D., Cosley, D., Huttenlocher, D., Kleinberg, J., Suri, S.: Feedback effects between similarity and social influence in online communities. In: Proceedings of the 14th ACM SIGKDD International Conference on Knowledge Discovery and Data Mining. ACM (2008)
8. Ditto, P.H., Lopez, D.F.: Motivated skepticism: use of differential decision criteria for preferred and nonpreferred conclusions. J. Pers. Soc. Psychol. **63**, 568 (1992)

9. Ditto, P.H., Scepansky, J.A., Munro, G.D., Apanovitch, A.M., Lockhart, L.K.: Motivated sensitivity to preference-inconsistent information. J. Pers. Soc. Psychol. **75**, 53 (1998)
10. Esteban, J.M., Ray, D.: On the measurement of polarization. Econometrica **62**, 819–851 (1994)
11. Flaxman, S., Goel, S., Rao, J.M.: Filter bubbles, echo chambers, and online news consumption. Public Opin. Q. **80**, 298–320 (2016)
12. Fletcher, R., Nielsen, R.K.: Are news audiences increasingly fragmented? A cross-national comparative analysis of cross-platform news audience fragmentation and duplication. J. Commun. **67**, 476–498 (2017)
13. Gargiulo, F., Gandica, Y.: The role of homophily in the emergence of opinion controversies. arXiv preprint arXiv:1612.05483 (2016)
14. Garrett, R.K.: The "echo chamber" distraction: disinformation campaigns are the problem, not audience fragmentation. J. Appl. Res. Mem. Cogn. **6**, 370–376 (2017)
15. Guzmán, M., Haar, S., Perchy, S., Rueda, C., Valencia, F.: Belief, knowledge, lies and other utterances in an algebra for space and extrusion. J. Log. Algebraic Methods Program. **86**, 107–133 (2016)
16. Guzmán, M., Knight, S., Quintero, S., Ramírez, S., Rueda, C., Valencia, F.: Reasoning about distributed knowledge of groups with infinitely many agents. In: CONCUR 2019–30th International Conference on Concurrency Theory. ACM SIGPLAN (2019)
17. Guzmán, M., Perchy, S., Rueda, C., Valencia, F.D.: Deriving inverse operators for modal logic. In: Sampaio, A., Wang, F. (eds.) ICTAC 2016. LNCS, vol. 9965, pp. 214–232. Springer, Cham (2016). https://doi.org/10.1007/978-3-319-46750-4_13
18. Guzmán, M., Perchy, S., Rueda, C., Valencia, F.: Characterizing right inverses for spatial constraint systems with applications to modal logic. Theor. Comput. Sci. **744**, 56–77 (2018)
19. Haar, S., Perchy, S., Rueda, C., Valencia, F.: An algebraic view of space/belief and extrusion/utterance for concurrency/epistemic logic. In: 17th International Symposium on Principles and Practice of Declarative Programming (PPDP 2015). ACM SIGPLAN (2015)
20. Huberman, B.A., Romero, D.M., Wu, F.: Social networks that matter: twitter under the microscope. arXiv preprint arXiv:0812.1045 (2008)
21. Hulsizer, M.R., Munro, G.D., Fagerlin, A., Taylor, S.P.: Molding the past: biased assimilation of historical information 1. J. Appl. Soc. Psychol. **34**, 1048–1074 (2004)
22. Hunter, A.: Reasoning about trust and belief change on a social network: a formal approach. In: Liu, J.K., Samarati, P. (eds.) ISPEC 2017. LNCS, vol. 10701, pp. 783–801. Springer, Cham (2017). https://doi.org/10.1007/978-3-319-72359-4_49
23. Jackson, J.: Eli pariser: activist whose filter bubble warnings presaged trump and brexit. The Guardian, January 2017. https://bit.ly/32sFT2O
24. Jøsang, A.: Subjective Logic - A Formalism for Reasoning Under Uncertainty. Artificial Intelligence: Foundations, Theory, and Algorithms. Springer, Cham (2016). https://doi.org/10.1007/978-3-319-42337-1
25. Kirby, E.: The city getting rich from fake news. BBC News, May 2017. https://www.bbc.com/news/magazine-38168281
26. Knight, S., Maubert, B., Schwarzentruber, F.: Asynchronous announcements in a public channel. In: Leucker, M., Rueda, C., Valencia, F.D. (eds.) ICTAC 2015. LNCS, vol. 9399, pp. 272–289. Springer, Cham (2015). https://doi.org/10.1007/978-3-319-25150-9_17

27. Knight, S., Maubert, B., Schwarzentruber, F.: Reasoning about knowledge and messages in asynchronous multi-agent systems. Math. Struct. Comput. Sci. **29**, 127–168 (2019)
28. Knight, S., Palamidessi, C., Panangaden, P., Valencia, F.D.: Spatial and epistemic modalities in constraint-based process calculi. In: Koutny, M., Ulidowski, I. (eds.) CONCUR 2012. LNCS, vol. 7454, pp. 317–332. Springer, Heidelberg (2012). https://doi.org/10.1007/978-3-642-32940-1_23
29. Kooi, B.P.: Probabilistic dynamic epistemic logic. J. Logic Lang. Inf. **12**, 381–408 (2003)
30. Lee, J.K., Choi, J., Kim, C., Kim, Y.: Social media, network heterogeneity, and opinion polarization. J. Commun. **64**, 702–722 (2014)
31. Leskovec, J., Huttenlocher, D., Kleinberg, J.: Signed networks in social media. In: Proceedings of the SIGCHI Conference on Human Factors in Computing Systems. ACM (2010)
32. Li, L., Scaglione, A., Swami, A., Zhao, Q.: Consensus, polarization and clustering of opinions in social networks. IEEE J. Sel. Areas Commun. **31**, 1072–1083 (2013)
33. Liu, F., Seligman, J., Girard, P.: Logical dynamics of belief change in the community. Synthese **191**, 2403–2431 (2014)
34. Nyhan, B., Reifler, J.: When corrections fail: the persistence of political misperceptions. Polit. Behav. **32**, 303–330 (2010)
35. Plaza, J.: Logics of public communications. Synthese **158**, 165–179 (2007)
36. Plous, S.: The Psychology of Judgment and Decision Making, 7th edn. McGraw-Hill, New York (2010)
37. Proskurnikov, A.V., Matveev, A.S., Cao, M.: Opinion dynamics in social networks with hostile camps: consensus vs. polarization. IEEE Trans. Autom. Control **61**, 1524–1536 (2016)
38. Seligman, J., Liu, F., Girard, P.: Logic in the community. In: Banerjee, M., Seth, A. (eds.) ICLA 2011. LNCS (LNAI), vol. 6521, pp. 178–188. Springer, Heidelberg (2011). https://doi.org/10.1007/978-3-642-18026-2_15
39. Seligman, J., Liu, F., Girard, P.: Facebook and the epistemic logic of friendship. arXiv preprint arXiv:1310.6440 abs/1310.6440 (2013)
40. Sîrbu, A., Pedreschi, D., Giannotti, F., Kertész, J.: Algorithmic bias amplifies opinion fragmentation and polarization: a bounded confidence model. Plos One **14**, e0213246 (2019)
41. Van Ditmarsch, H., van Der Hoek, W., Kooi, B.: Dynamic epistemic logic, vol. 337. Springer, Heidelberg (2007). https://doi.org/10.1007/978-1-4020-5839-4
42. Wood, T., Porter, E.: The elusive backfire effect: mass attitudes' steadfast factual adherence. Polit. Behav. **41**, 135–163 (2019)

Make Puzzles Great Again

Nicolás Aristizabal[1], Carlos Pinzón[1], Camilo Rueda[1], and Frank Valencia[1,2(✉)]

[1] Pontificia Universidad Javeriana Cali, Cali, Colombia
[2] CNRS & LIX, École Polytechnique de Paris, Palaiseau, France
frank.valencia@lix.polytechnique.fr

Abstract. We present original solutions to four challenging mathematical puzzles. The first two are concerned with random processes. The first, here called *The President's Welfare Plan*, can be reduced to computing, for arbitrary large values of n, the expected number of iterations of a program that increases a variable at random between 1 and n until exceeds n. The second one, called *The Dining Researchers*, can be reduced to determining the probability of reaching a given point after visiting all the others in a circular random walk. The other two problems, called *Students vs Professor* and *Students vs Professor II*, involve finding optimal winning group strategies in guessing games.

1 Introduction

Motivation. Puzzles have been a source of insights in many areas of mathematics. They have even led to the development of new fields as shown by the impact on graph theory of Euler's impossibility result to the *Seven Bridges of Königsberg Problem* [2].

Fig. 1. The Seven Bridges of Königsberg

The problem is to find a way to walk through each of the bridges in Fig. 1 exactly once. Also witnessed by this paradigmatic example is the fact that puzzle

This work have been partially supported by the ECOS-NORD project FACTS (C19M03) and the Colciencias project CLASSIC (125171250031).

M. S. Alvim et al. (Eds.): Palamidessi Festschrift, LNCS 11760, pp. 442–459, 2019.
https://doi.org/10.1007/978-3-030-31175-9_25

solution usually derives from its elegant and imaginative formalization. Although its solution may involve nontrivial mathematics, a defining feature of good puzzles is the simplicity of enunciation. Even non-experts are usually able to understand it and form some straightforward intuition about its answer (although finding a way to go about confirming that intuition is also usually recognized to be non-trivial). This simplicity greatly motivates undertaking the task of solving it. Nevertheless, it is also a feature of many good puzzles that their solutions end up contradicting that first common intuition. Simplicity is thus deceptive, which makes room for a variety of imaginative strategies to thrive. In fact, observing the way the puzzle solution defies common intuitions usually provides new insights into techniques for solving similar problems popping up in different realms.

One of most compelling examples of deceiving simplicity of enunciation is that of the *Collatz* problem [8]. The problem can be viewed as a puzzle asking us whether the simple function COLLATZ(n) defined below in Algorithm 1 terminates for every positive integer n. Paul Erdős, who had a passion for puzzles, said about the Collatz problem: "*Mathematics may not be ready for such problems*" [7]. In fact, the problem was introduced in 1937 and despite its simplicity of enunciation it is still open.

Algorithm 1. Collatz Function

```
function COLLATZ(n)                              ▷ n is a positive integer
    if n = 1 then
        return 1
    else if n MOD 2 = 0 then
        COLLATZ(n/2)
    else
        COLLATZ(3n + 1)
```

Erdős himself came up with severals puzzles that have led to surprisingly deep mathematical discoveries. One of remarkable simplicity of enunciation is his *Discrepancy Problem* [13]: Is it possible to find an infinite sequence $a_1, a_2, \ldots$ over the alphabet $\{-1, 1\}$ and a constant C, such that $\mid \sum_{i=1}^{n} a_{id} \mid \leq C$ for each n and d ? The problem was introduced in 1932 and despite many attempts to solve it, it was open until 2015 when Terence Tao, who as a child met Erdős, gave a negative answer to it [13].

The *secretary problem* [5] is another puzzle that has been studied extensively in the fields of applied probability, statistics, and decision theory. Suppose that a hiring committee wishes to select the best candidate out of n rankable applicants for a position. The applicants are interviewed one by one in random order. During the interview, the committee gains enough information to rank the applicant among all applicants interviewed so far, but does not know about the quality of the applicants that have not been interviewed. Immediately after each interview, the interviewed applicant is either the one selected for the job or rejected, and the decision is irrevocable. The question is about the optimal strategy to maximize the probability of selecting the best applicant. The problem was introduced in 1950 and solved in 1964 [4]. The optimal strategy is always rejecting the first n/e applicants that are interviewed (i.e., about 37% of all the applicants), where e is Euler's number, and then selecting the first applicant who is better than every applicant interviewed so far (or continuing to the last applicant if this never occurs). Researchers in social psychology and economics have studied the decision behavior of actual people in secretary problem situations (e.g., when buying online tickets). They have found that people tend to stop searching too soon.

Nevertheless, not all nice puzzles take several years to be solved. Some of them can be regarded as *brainteasers* that may take just a few minutes thought. E.g., suppose that a hiring committee wishes to select one person for a job from a group of three candidates. The only fact the committee knows about this group is that there is one that always tells the truth (A), there is another that never tells the truth (B) and there is a third one that sometimes tells the truth and sometimes does not (C). They would rather select (A), or even (B) because with that person at least they know where they stand. To make their choice the committee can ask one and only one of the candidates a single question whose answer should be yes or no. What question can they ask to avoid choosing (C)?

This Work. As part of the social activities of our research groups, we share puzzles of different complexity but significantly more difficult than brainteasers. The subsequent exchange of creative and elegant mathematical solutions to these problems after persistently trying to solve them gives our PhD students a glimpse into actual research problems. However, we mainly engage in solving puzzles because, as mentioned above, the appeal that lies in the simplicity of their enunciation, the fact they seem to contradict common intuition and the intellectual challenge they represent.

Catuscia is also well-known for having a great passion for puzzles. In fact she is the main contributor of puzzles in her team Comète. The challenging puzzles she takes up with great resolve and the sheer beauty of her solutions are a faithful reflection of the brilliance, tenacity and mathematical elegance that she displays in her own research.

In this paper we wish to honor Catuscia with a collection of some of her favorite and more challenging puzzles. Our solutions to these problems are original and based on basic principles and standard techniques from computer science.

The paper is structured as follows: There is a section per puzzle in addition to a section with concluding remarks. The *President's Welfare Plan* is a problem about a populist financial aid program by a fictional president. The *Dining Researchers* is a problem about a fair strategy to choose at random a researcher that should pay for dinner. The *Students vs Professor* problem is about a guessing challenge given by a professor to some students. Finally the *Students vs Professor: II* problem is about another guessing challenge, allegedly more difficult, by a professor to a selected group of students.

2 The President's Welfare Plan

Computer scientists often discuss about politics. The following problem is a parody on populism's political promises. Let us consider the following economic plan by a fictional populist president.

Description 1 (President's Welfare Plan). *A populist president of a populous country tweeted a popular plan: "I have a tremendous welfare plan to make our country great again: I will allocate a large amount of money n that would make anyone* very rich. *Every day, each person will receive a random amount in cents, uniformly chosen, between one cent and n until the day the person has accumulated an amount that exceeds n".*[1]

Given the president's welfare plan, some people may wonder about its implementability though not as many as one would like. In particular, it is natural to ask about the expected number of days after most people will accumulate more than the large amount n. More precisely:

Question 1. What is the expected value (or mean) of the number of days for a person to become *very rich* under the president's welfare plan?

For any given person and a large value of n, the president's plan can be simulated by the function $f(n)$ in Algorithm 2.

Algorithm 2. President's Plan.

```
function f(n)                                  ▷ n is a positive integer
    days, money ← 0
    while money ≤ n do
        money ← money + RANDOM(1, n)
        days ← days + 1
    return days
```

The algorithmic function $f(n)$ repeatedly increases *money* with a value between 1 and n chosen uniformly at random. It returns the number of iterations stored in *days* when *money* $> n$. Clearly, every outcome of $f(n)$ is between 2 and n.

To solve the question above it suffices to find the expected (or weighted average) output of $f(n)$ for an arbitrary large value of n. Such an expected output can be determined by calculating the limit when n approaches infinity of the expected value $E(F(n))$ where $F(n)$ is the random variable whose samples are drawn by calling $f(n)$. From the definition of expected value for a random

[1] It would seem that in real life this kind of plan is often implemented except that the random amounts of money are not uniformly chosen and often include 0.

variable [9] and the above observation about the outcomes of $f(n)$ we ought to calculate the following limit.

$$\lim_{n\to\infty} E\left(F(n)\right) = \lim_{n\to\infty} \sum_{i=2}^{n} ip_i \tag{1}$$

where p_i is the probability of $f(n) = i$.

At first glance it is not clear what the exact expected output of $f(n)$ would be. We have observed that many people tend to think that the expected value should be 2. However, after running simulations using $f(n)$ for large values of n one observes that the average output is in fact near 2.7. E.g., we implemented the algorithm in Python and run it 10^8 times for $n = 10^8$. The weighted average output was 2.71828205. Similar simulations with different values of n suggest that as n grows, the weighted average value approaches a ubiquitous constant in mathematics; *Euler's irrational number e*. In fact, we shall prove the following theorem.

Theorem 2. $\lim_{n\to\infty} E\left(F(n)\right) = e$

Notice that the above claim answers Question 1; e is the average number of days (i.e., about 2 days, 17 h and 14 min) for a person to become very rich under the president's plan.

Recall that the constant e equals the following limit and infinite series.

$$e = \lim_{n\to\infty} \left(1 + \frac{1}{n}\right)^n = \sum_{k=0}^{\infty} \frac{1}{k!} \tag{2}$$

In what follows we present two rather different solutions proving the claim, one using the infinite series representation with a counting argument and the other using the limit representation with a recursive version of the function $f(n)$.

2.1 Counting Solutions

Consider a constraint satisfaction problem with (random) variables $x_1, x_2, \ldots, x_i$ taking on values from $\{1, 2, \ldots, n\}$. Suppose that each x_i takes on the random value generated at the i-th iteration of $f(n)$. It is easy to see that $f(n)$ returns i (notice that always $i \geq 2$) iff the constraints C_1 and C_2 below are satisfied:

$$C_1 \equiv \sum_{j=1}^{i-1} x_j \leq n \text{ and } C_2 \equiv \sum_{j=1}^{i} x_j > n \tag{3}$$

Let $n \geq 1$ be fixed. We shall use $S(i)$ to denote the set of all *solutions* to the constraints C_1 and C_2. E.g., if $n = 10^8$ then the assignment $x_1 = 1$, $x_2 = 5$ and $x_3 = n$ is a solution in $S(3)$ but $x_1 = 1$, $x_2 = n$ and $x_3 = 5$ is not. Notice that we can express the probability p_i that $f(n) = i$ (see Eq. 1) as the ratio $\frac{|S(i)|}{n^i}$ since n^i is the number of all possible assignments for $x_1, x_2, \ldots, x_i$ and the random values are uniformly chosen.

Proposition 1. *We have* $p_i = \frac{|S(i)|}{n^i}$ *where* p_i *is the probability that* $f(n) = i$ *in Eq. 1.*

We shall now determine the value $|S(i)|$ for any $i \geq 2$. A nice characterization of $|S(i)|$, for reasons of space left as an exercise, is given by the following nested summation:

$$|S(i+1)| = \sum_{j_i=i}^{n} \sum_{j_{i-1}=j_i}^{n} \cdots \sum_{j_1=j_2}^{n} j_1 \tag{4}$$

The reader can verify that the above summation gives us $|S(2)| = n(n+1)/2$ and $|S(3)| = (n-1)n(n+1)/3$. However, obtaining the explicit value of $|S(i)|$ from the nested summation for arbitrary values of i seems a little involved.

Nevertheless $|S(i)|$ can also be determined as the total number of solutions to the following family of constraints. For every $k = i-1, \ldots, n$, define the constraint $C(k)$ as follows:

$$C(k) \equiv \sum_{j=1}^{i-1} x_j = k \wedge \sum_{j=1}^{i} x_j > n \tag{5}$$

Intuitively we are replacing "$\leq n$" in C_1 with "$= k$" for $k = n, ..., i-1$. Let $S_k(i)$ be the set of all solutions for $C(k)$. Notice that $S_k(i) \cap S_{k'}(i) = \emptyset$ whenever $k \neq k'$. Furthermore, $S(i) = \bigcup_{k=i-1}^{n} S_k(i)$. It follows that $|S(i)| = \sum_{k=i-1}^{n} |S_k(i)|$.

From combinatorics we know that for any pair of positive integers s and t, the number of t-tuples of positive integers whose sum is s is given by the binomial coefficient $\binom{s-1}{t-1}$ [3]. We can use this fact to show that the number of solutions for $\sum_{j=1}^{i-1} x_j = k$ is given by $\binom{k-1}{i-2}$. Furthermore, each solution for $\sum_{j=1}^{i-1} x_j = k$ can only be extended to k solutions for $\sum_{j=1}^{i} x_j > n$. Therefore $|S_k(i)| = k\binom{k-1}{i-2}$. This and some basic algebraic manipulations lead us to the following result.

Lemma 1. *Let* $i \geq 2$. *We have*

$$|S(i)| = \sum_{k=i-1}^{n} k\binom{k-1}{i-2} = \frac{(n+1)(n-i+2)\binom{n}{i-2}}{i} \tag{6}$$

By expanding $\binom{n}{i-2}$ in the above lemma and distributing some products to express $|S(i)|$ as a polynomial of degree i, we obtain the following corollary.

Corollary 1. *Let* $i \geq 2$. *Then* $|S(i)| = \frac{n^i}{(i-2)!i} + r(n)$ *where* $r(n)$ *is a polynomial of degree* $m < i$.

We now have all the elements to prove Theorem 2. Using distributive properties of limits at infinity, Proposition 1, Eq. 2 and the above corollary we obtain e as wanted.

$$\lim_{n\to\infty} E(F(n)) = \lim_{n\to\infty} \sum_{i=2}^{n} ip_i = \lim_{n\to\infty} \sum_{i=2}^{n} i\frac{\frac{n^i}{(i-2)!i} + r(n)}{n^i} = \sum_{i=2}^{\infty} \frac{1}{(i-2)!} = e. \tag{7}$$

2.2 Recursive Solution

For our second solution we first consider a recursive algorithm for computing $f(n)$, as shown in Algorithm 3, by introducing a parameter $m \leq n$ that represents the difference $m := n - money$ where $money$ is the program variable in Algorithm 2. Notice that $A(n, n)$ is equivalent to $f(n)$ for all $n \geq 1$.

Algorithm 3. Recursive President's Plan

```
function A(n, m)
    if m < 0 then
        return 0
    else
        return 1 + A(n, m − RANDOM(1, n))
```

Let $n \geq 1$ be fixed, and let $h(m)$ with $-\infty < m \leq n$ denote the expected output of the algorithm $A(n, m)$. Clearly, $h(m) = 0$ whenever $m < 0$. Now, for all $0 \leq m \leq n$, the function $A(n, m)$ yields uniformly at random one of $\{1 + A(n, m-1), 1 + A(n, m-2), ..., 1 + A(n, m-n)\}$, thus we have

$$\begin{aligned} h(m) &= \tfrac{1}{n} \textstyle\sum_{k=1}^{n} (1 + h(m-k)) \\ &= 1 + \tfrac{1}{n} \textstyle\sum_{k=1}^{n} h(m-k) \\ &= 1 + \tfrac{1}{n} \textstyle\sum_{k=m+1}^{n} h(m-k) + \tfrac{1}{n} \textstyle\sum_{k=1}^{m} h(m-k) \\ &= 1 + 0 + \tfrac{1}{n} \textstyle\sum_{k=0}^{m-1} h(k) \\ &= 1 + \tfrac{1}{n} \textstyle\sum_{k=0}^{m-1} h(k) \end{aligned}$$

We conclude that $h(0) = 1$, and derive the following formula whenever $1 \leq m \leq n$.

$$\begin{aligned} h(m) &= 1 + \tfrac{1}{n} \textstyle\sum_{k=0}^{m-1} h(k) \\ &= \left(1 + \tfrac{1}{n} \textstyle\sum_{k=0}^{m-2} h(k)\right) + \tfrac{1}{n} h(m-1) \\ &= h(m-1) + \tfrac{1}{n} h(m-1) \\ &= h(m-1)\left(1 + \tfrac{1}{n}\right) \\ &= h(m-2)\left(1 + \tfrac{1}{n}\right)^2 && \text{(recursively)} \\ &\vdots && \text{(recursively)} \\ &= h(0)\left(1 + \tfrac{1}{n}\right)^m \end{aligned}$$

This implies that the expected output of $f(n) = \mathrm{A}(n, n)$ is given by $h(n) = \left(1 + \frac{1}{n}\right)^n$, and since $e = \lim_{n\to\infty} \left(1 + \frac{1}{n}\right)^n$ (Eq. 2), it follows that, indeed,

$$\lim_{n\to\infty} E\left(F(n)\right) = e.$$

3 Dining Researchers

Computer scientists often dine with their colleagues. Despite their mathematical skills, however, they often find it difficult to split the bill. In fact, they would rather select one of them at random to pay for the entire bill. The following problem is concerned with a fair strategy for such selection assuming they are sitting at a round table.

Description 3 (Dining Researchers). *There are $n \geq 2$ players (i.e., the researchers) sat at a round table with only one of them having a (fair) coin. The coin will be flipped and passed around the table using the following rules for each player:*

- *If they have or receive the coin, they should flip it randomly.*
- *If they obtain heads, they should pass the coin to the player on their left.*
- *If they obtain tails, they should pass the coin to the player on their right.*

Once the coin has reached everyone, the last player to receive the coin loses the game.

Question 2. Is there a place that a player, other than the owner of the coin, can choose at the table so that they are less likely to lose the game?

Many people confronted with this problem consider a good strategy to sit closest to owner of the coin, the idea being to have at least the opportunity to receive the coin after the first flipping. To get a better understanding of the game, let us consider it for the case of 4 players, that we identify in clock notation as players 12 (or 0 if you prefer), 3, 6 and 9. Let p_{12}, p_3, p_6, p_9 be their respective probabilities of *losing* the game. Then,

- $p_{12} = 0$, since player 12 has the coin at the beginning.
- Starting from 12 and considering the first two moves, $p_3 = \overbrace{\frac{1}{4}}^{12\to 9\to 6} + \overbrace{\frac{1}{4}p_3}^{12\to 9\to 12}$, where each probability value is decorated with the corresponding path of the coin. Thus, $p_3 = \frac{1}{3}$.
- By symmetry, $p_9 = p_3 = \frac{1}{3}$.
- Finally, since someone must lose, $p_6 = 1 - p_{12} - p_3 - p_9 = \frac{1}{3}$.

So, it seems the game is fair for all players regardless their place in the table. We go on to prove this for the general case thus answering negatively Question 2.

First, notice that p_3 can be seen as composed of the probabilities of the coin reaching each of the other players first without having reached player 3. In the general case, each of these could be denoted by a function $f(l, r)$ giving the probability of the coin reaching some player r for the first time before it reaches some other player l.

The proof follows three steps:

1. Finding a recursive formula for function $f(l, r)$.

2. Proving that $f(l,r) = \frac{l}{l+r}$.
3. Using f to prove that the probability of losing is $\frac{1}{n-1}$ for all players except the initial one.

Step 1:

Let n be the number of people in the table, and let $0, 1, ..., n-1$ be their positions, with 0 being the starting position.

Let $f(l,r)$, with $0 \leq l$, $0 \leq r$, and $1 \leq l + r < n$, denote the probability of the coin reaching for the first time the position r without having reached the position $-l \mod n$. Starting from 0, r denotes *right* (clockwise) positions whereas l denotes *left* (counterclockwise) positions up to position $r+1$. For example, in the left circle of Fig. 2 below a table for 12 players is represented:

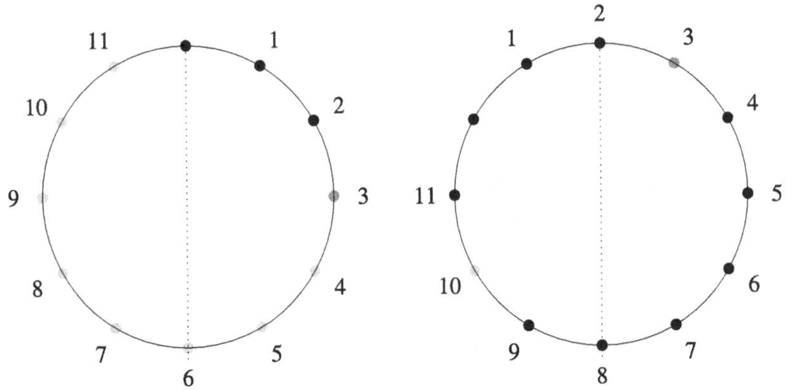

Fig. 2. (left) A state in a 12 players table (right) Rotating for a new first player (Color figure online)

For $r = 3$, $f(l,r)$ can refer to the probabilities of player 3 (in blue) receiving the coin before any of the players in the left half circle, and also players 5, 4 (in red).

Consider first the cases where r or l are equal to zero:

- when $r = 0$ then $l > 0$ (since $1 \leq l + r$) and, for all $1 \leq l < n$, we have that $f(l,r) = 1$, since the coin starts at 0.
- If $l = 0$, then $r > 0$ and, for all $1 \leq r < n$, we have that $f(l,r) = 0$, since the coin starts at 0.

Let us consider now the case $(l, r > 0)$:

In order for the coin to reach r without reaching $-l \mod n$, it must first reach $r-1$ (without reaching $-l \mod n$). After the coin reaches position $r-1$, what is left is that it move one additional position to the right, either directly in one step, or after many steps with the restriction that these steps do not reach $-l \mod n$.

Notice that once $r-1$ is reached, we can consider position $r-1$ to be the initial position, from which $f(l+r-1,1)$ would represent the probability of reaching position r, that is 1 position to the right of $r-1$, without reaching position $-l \mod n$, which is $l+r-1$ to the left of $r-1$. For instance, in a game with 12 players, if $r=3$ and $l=2$, $f(l,r)$ is the probability of reaching 3 without reaching player 10 (i.e., $-2 \mod 12$), which is located $l+r-1=4$ positions to the left of player $r-1=2$, as shown in the right circle of Fig. 2 above.

Therefore, the following holds:

$$f(l,r) = f(l,r-1)\, f(l+r-1,1)$$

Moreover, taking into account that the converse of $f(l,r)$ is the probability of the coin reaching for the first time position $-l \pmod m$ without having reached position r, we also have that

$$f(l,r) = 1 - f(r,l)$$

Combining these two, we finally have that,

$$f(l,r) = f(l,r-1)\ (1 - f(1,l+r-1)))$$

Step 2:

We now prove that

$$f(l,r) = \frac{l}{l+r}$$

For $l=0$ or $r=0$, clearly the equation holds. For $l=1$, consider two cases,

1. If $r=1$, then $f(1,1) = f(1,0)\ (1-f(1,1)) = 1 - f(1,1) = \frac{1}{2}$.
2. If $r>1$, assume $f(1,r-1) = \frac{1}{r}$. Then, $f(1,r) = f(1,r-1)\ (1-f(1,r))$ and, by the assumption, $f(1,r) = \left(1 + (f(1,r-1))^{-1}\right)^{-1} = (1+r)^{-1}$.

Finally, let $l>1$. We then have,

$$\begin{aligned}
f(l,r) &= f(l,r-1)\ (1 - f(1,l+r-1))) \\
&= f(l,r-1)\left(1 - \tfrac{1}{l+r}\right) \\
&= \tfrac{l+r-1}{l+r}\, f(l,r-1) \\
&= \tfrac{l+r-1}{l+r}\,\tfrac{l+r-2}{l+r-1}\, f(l,r-2) && \text{(recursively)} \\
&\ \vdots && \text{(recursively)} \\
&= \tfrac{l+r-1}{l+r}\,\tfrac{l+r-2}{l+r-1}\cdots\tfrac{l+1}{l+2}\,\tfrac{l}{l+1}\, f(l,0) \\
&= \tfrac{\cancel{l+r-1}}{l+r}\,\tfrac{\cancel{l+r-2}}{\cancel{l+r-1}}\cdots\tfrac{\cancel{l+1}}{\cancel{l+2}}\,\tfrac{l}{\cancel{l+1}}\, f(l,0) \\
&= \tfrac{l}{l+r}
\end{aligned}$$

Step 3:

Going back to the Dining Researchers Problem, let p_i denote the probability of player i **losing** the game, where $0 \leq i < n$.

For the initial player, by definition $p_0 = 0$. We now show that $p_i = \frac{1}{n-1}$ for all $i \neq 0$. Let $i > 0$. Notice that there are only two ways in which player i can **win**:

1. The coin reaches position i for the first time without having reached position $i+1$. This is the same position as $-(n-i-1) \mod n$. The probability of reaching i before reaching that position is given by $f(n-i-1, i)$.
2. The coin reaches position i for the first time without having reached position $i-1$. Equivalently, it does not occur that the coin reaches for the first time the position $i-1$ without having reached $-(n-i) \mod n$, or simply $1 - f(n-i, i-1)$.

Since the two scenarios are non-overlapping, we can conclude that

$$\begin{aligned} p_i &= 1 - (f(n-1-i, i) + 1 - f(n-i, i-1)) \\ &= 1 - \left(\tfrac{n-1-i}{n-1} + 1 - \tfrac{n-i}{n-1}\right) \\ &= \tfrac{1}{n-1} \end{aligned}$$

4 Students vs Professor

Computer science professors often challenge and reward well the creativity of their students. The following problem asks some students to devise a group strategy to win in the following guessing game.

Description 4 (Guessing Challenge). *A computer science professor is giving their n students an unusual exam. Each student undergoes the following. An arbitrary number from $\{0, \ldots, n-1\}$ is drawn on the student's forehead. A close up photo of the student is then taken. The student does not know their own number. (2 or more students may have the same number.) The student is put in a room and given the photos of all the other students. The students must guess their own number.*

- *If at least one student gets their number right, everybody gets full marks. Otherwise, everyone fails.*
- *The students may not cheat by somehow communicating or seeing the number on their own forehead during the exam, but they may all confer before the exam and work out a strategy.*

Question 3. Assume $n \geq 2$. Is there a strategy for the students to ensure getting full marks ?

A strategy amounts to a function g that takes two pieces of information, a student number s and a tuple $(x_1, x_2, ..., x_n)$, where $1 \leq x_i \leq n$ represents the number on the forehead of student i, and outputs a number $1 \leq y \leq n$ whose computation does not involve x_s (i.e. no cheating). For the strategy to be successful, there must be some $1 \leq s \leq n$ s.t. $g(\boldsymbol{x}, s) = x_s$

For $n = 2$ (say, players 0 and 1), a winning strategy is simple: one students writes the opposite number to what she sees and the other student writes the number he sees, i.e., $g((x_0, x_1), 0) = 1 - x_1$ and $g((x_0, x_1), 1) = x_0$. Since it cannot happen that both $1 - x_1 \neq x_0$ and $x_0 \neq x_1$ hold, the strategy works.

For the general case, let n be the number of students and $\mathbb{Z}_n$ be the set $\{0, 1, ..., n-1\}$. Enumerate the students from 0 to $n - 1$, and consider the following algorithm for each student k:

- **Input:** vector $\boldsymbol{x}$ with the numbers drawn on the students foreheads.
- **Output:**

$$g(\boldsymbol{x}, k) := \left(k - \sum_{i \neq k} x_i\right) \mod n$$

Notice that the computation of the guess $g(\boldsymbol{x}, k)$ does not involve the value x_k, thus the algorithm complies with the restriction that no student can see their number.

The group strategy for an input $\boldsymbol{x}$ will be that each student k will guess $g(\boldsymbol{x}, k)$. In order to prove that this strategy solves the given problem, we must prove that for all $\boldsymbol{x} \in \mathbb{Z}_n^n$, there exists $k \in \mathbb{Z}_n$ such that $g(\boldsymbol{x}, k) = x_k$.

To prove this, let $\boldsymbol{x} \in \mathbb{Z}_n^n$ be an arbitrary vector. To simplify the notation, assume that all additions and subtractions are done modulo n, and consider the following definitions:

- $\text{set}(\boldsymbol{x})$ represents the set of elements in a vector, i.e. $\text{set}(\boldsymbol{x}) := \{x_i : i \in \mathbb{Z}_n\}$.
- $\boldsymbol{1}$ represents the $n \times 1$ *unit* vector $[1, 1, ..., 1]$, so that $\text{set}(\boldsymbol{1}) = \{1\}$.
- $\boldsymbol{r}$ represents the $n \times 1$ *range* vector $[0, 1, ..., n-1]$, so that $\text{set}(\boldsymbol{r}) = \mathbb{Z}_n$, and more generally, $\text{set}(\boldsymbol{r} + m\boldsymbol{1}) = \mathbb{Z}_n$ for all $m \in \mathbb{Z}$ because additions are modulo n.
- $\sigma := \sum_{i=0}^{n-1} x_i$, so that $g(\boldsymbol{x}, k) = k - \sigma + x_k$.
- $\boldsymbol{y} := \boldsymbol{r} - \sigma\boldsymbol{1} + \boldsymbol{x}$, so that $g(\boldsymbol{x}, k) = y_k$.

Given that $g(\boldsymbol{x}, k) = y_k$ for all $k \in \mathbb{Z}_n$, the proof reduces to showing that there exists $i \in \mathbb{Z}_n$ with $y_i = x_i$ or, equivalently, that $0 \in \text{set}(\boldsymbol{y} - \boldsymbol{x})$. This is true since

$$\begin{aligned} \text{set}(\boldsymbol{y} - \boldsymbol{x}) &= \text{set}(\boldsymbol{r} - \sigma\boldsymbol{1} + \boldsymbol{x} - \boldsymbol{x}) \\ &= \text{set}(\boldsymbol{r} - \sigma\boldsymbol{1}) \\ &= \mathbb{Z}_n. \end{aligned}$$

5 Students vs Professor II

Since the students won the challenge in the previous section, the professor will now given them another, allegedly, more difficult guessing challenge.

Description 5 (Guessing Challenge II). *The professor gives their n students a new unusual test: A random number from $\{0, 1\}$ is drawn on the student's forehead. As before, a close up photo of the student is then taken. The student does not know their own number. The student is put in a room and given the photos of all the other students. The students* can *(but do not have to; they may abstain/pass) guess their own number.*

- *If somebody guesses correctly and nobody guesses incorrectly, everybody gets full marks. Otherwise, everyone fails.*
- *The students may not cheat by somehow communicating or seeing the number on their own forehead during the exam, but they may all confer before the exam and work out a strategy.*

Question 4. We consider the following questions:

1. How to win with probability strictly greater than $1/2$ for any $n \geq 3$?
2. Prove that probability $15/16$ can be achieved for $n = 15$.
3. Prove that, for any number n the form $2^k - 1$, there is an optimal strategy with probability $(2^k - 1)/2^k$

For $n = 3$, all triple combinations but two have exactly two equal numbers. A good strategy would then be for each student seeing two equal numbers to guess the opposite number and to abstain otherwise. This gives probability of success equal to $3/4$. For any $n > 3$ by choosing three students to play the same strategy and having the rest to abstain we obtain the same probability of success $3/4$. This answers positively Question 4(1). Below we prove Question 4(3) which also implies the special case in Question 4(2).

Solution for the General Case
Let $k > 0$ and let $n - 1$ be the number of students, where $n = 2^k$, and let $\boldsymbol{x}$ be a vector. We use the following notation:

- $\mathbb{B}$ is the boolean set $\{0, 1\}$.
- $e_i \in \mathbb{B}^{n-1}$ is the vector whose single non-zero entry is the i-th.
- $\boldsymbol{x} \uparrow_i$ denotes the entry-wise *or* between $\boldsymbol{x}$ and e_i (set bit i to 1).
- $\boldsymbol{x} \updownarrow_i$ denotes the entry-wise *xor* between $\boldsymbol{x}$ and e_i (invert bit i).
- $\boldsymbol{x} \downarrow_i$ denotes $\boldsymbol{x} \uparrow_i \updownarrow_i$(set bit i to 0).

We also denote the Manhattan-distance between vectors $\boldsymbol{x}, \boldsymbol{y}$ as $d(\boldsymbol{x}, \boldsymbol{y})$. Since vectors are binary, this distance also corresponds to the Hamming-distance (i.e. the number of positions at which the vectors differ).

As shown in Lemma 2, there is a set $H \subseteq \mathbb{B}^{n-1}$ of size 2^{n-1-k} such that $d(\boldsymbol{x}, \boldsymbol{y}) \geq 3$ for any two different binary vectors $\boldsymbol{x}, \boldsymbol{y} \in H$.

The students should precompute the set H (or an algorithm for checking $\boldsymbol{x} \in H$) and enumerate themselves from 0 to $n - 2$. For a given input $\boldsymbol{x}$, the strategy is:

- **Input:** vector $\boldsymbol{x}$ with the numbers drawn on the students foreheads.
- **Action:** (for each student, being i their number)
  - Guess $g(\boldsymbol{x}, i) = 1$ if $\boldsymbol{x} \downarrow_i \in H$.
  - Guess $g(\boldsymbol{x}, i) = 0$ if $\boldsymbol{x} \uparrow_i \in H$.
  - Otherwise, do not guess.

Notice that the computation of the guess $g(\boldsymbol{x}, i)$ does not involve the value x_i, thus the strategy complies with the restriction that no student can see their own number.

As shown in Lemmas 3 and 4, the following propositions are true:

1. "If $\boldsymbol{x} \notin H$ and a student makes a guess, he/she will be correct" (Lemma 3).
2. "If $\boldsymbol{x} \notin H$ at least one student will guess" (Lemma 4).
3. "If $\boldsymbol{x} \notin H$ the group of students will win" (consequence of the two previous points).

Moreover, since $|H| = 2^{n-1-k}$, the students will win in at least $2^{n-1} - 2^{n-1-k}$ cases out of 2^{n-1} possibilities, which yields a probability P of winning of

$$P \geq \frac{2^{n-1} - 2^{n-1-k}}{2^{n-1}} = 1 - \frac{1}{2^k}.$$

However, it turns out that the students fail when $\boldsymbol{x} \in H$, thus $P = 1 - \frac{1}{2^k}$ as wanted in Question 4(3).

Lemma 2. There is a set $H \subseteq \mathbb{B}^{n-1}$ of size 2^{n-1-k} *such that* $d(\boldsymbol{x}, \boldsymbol{y}) \geq 3$ *for any two different binary vectors* $\boldsymbol{x}, \boldsymbol{y} \in H$, *where* d *represents the Hamming distance.*

What is required here is basically what is known in the literature as a *hamming code*[12]. A hamming code is an encoding that adds redundancy to the message to enable so called "Forward Error Correction", that is, if an error occurs due to an external source of noise, the receiver might fix the error without asking the sender to resend the data. Specifically, a hamming code with distance 3 enables the receiver to fix one-bit errors, i.e. if just one bit of the encoded message was inverted because of noise in the transmission channel, the receiver can detect it, invert it, and decode the original message. We proceed to show how to construct such a set H while proving that the requirements in the statement hold.

Let $h := n - 1 - k$. We will define a boolean matrix M of size $k \times h$ and, based on it, we will let

$$H = \left\{ \mathrm{concat}(\boldsymbol{x}, M\boldsymbol{x}) \mid \boldsymbol{x} \in \mathbb{B}^h \right\},$$

where all the additions involved in the matrix product are considered modulo 2.

Notice that $|H| = 2^h$ as desired, since $|\mathbb{B}^h| = 2^h$ and the mapping $\boldsymbol{x} \mapsto \mathrm{concat}(\boldsymbol{x}, M\boldsymbol{x})$ is injective no matter how M is chosen. Since the elements of H result of a concatenation, we have that

$$d(\mathrm{concat}(\boldsymbol{x}, M\boldsymbol{x}), \mathrm{concat}(\boldsymbol{y}, M\boldsymbol{y})) = d(\boldsymbol{x}, \boldsymbol{y}) + d(M\boldsymbol{x}, M\boldsymbol{y})$$

and, therefore, the restriction $d(\boldsymbol{x}, \boldsymbol{y}) \geq 3$, *for any two different binary vectors* $\boldsymbol{x}, \boldsymbol{y} \in H$, can be reduced to the requirement that, for all $\boldsymbol{x}, \boldsymbol{y} \in \mathbb{B}^h$:

1. If $d(\boldsymbol{x}, \boldsymbol{y}) = 1$, then $d(M\boldsymbol{x}, M\boldsymbol{y}) \geq 2$.
2. If $d(\boldsymbol{x}, \boldsymbol{y}) = 2$, then $d(M\boldsymbol{x}, M\boldsymbol{y}) > 0$ or, what amounts to the same, $M\boldsymbol{x} \neq M\boldsymbol{y}$.

Or, equivalently, for all $\boldsymbol{x} \in \mathbb{B}^h$, and all $i, j \in \mathbb{Z}_h$ with $i \neq j$,

1. $d(M\boldsymbol{x}, M(\boldsymbol{x} \updownarrow_i)) \geq 2$, and
2. $M\boldsymbol{x} \neq M(\boldsymbol{x} \updownarrow_i \updownarrow_j)$.

These two conditions can be further simplified to the following::

1. Every column in M has at least two ones.
2. No two columns in M are equal.

The reason for this is that,
1. on the one hand, if every column in M has at least two ones, then for all $i \in \mathbb{Z}_h$, there are at least two different $r, s \in \mathbb{Z}_k$ such that $M_{r,i} = M_{s,i} = 1$; and this implies $d(M\boldsymbol{x}, M(\boldsymbol{x} \updownarrow_i)) \geq 2$ because $(M\boldsymbol{x})_r \neq (M(\boldsymbol{x} \updownarrow_i))_r$ and $(M\boldsymbol{x})_s \neq (M(\boldsymbol{x} \updownarrow_i))_s$ and,
2. on the other, if no two columns are equal, then for all $i, j \in \mathbb{Z}_h$ with $i \neq j$, there is $r \in \mathbb{Z}_k$ with $M_{r,i} \neq M_{r,j}$; therefore, $M(e_i + e_j) \neq \boldsymbol{0}$ (addition is assumed modulo 2), and

$$\begin{aligned} M\boldsymbol{x} + M(\boldsymbol{x} \updownarrow_i \updownarrow_j) &= M\boldsymbol{x} + M(\boldsymbol{x} + e_i + e_j) \\ &= M\boldsymbol{x} + M\boldsymbol{x} + M(e_i + e_j) \\ &= M(e_i + e_j) \\ &\neq \boldsymbol{0}. \end{aligned}$$

Thus $M\boldsymbol{x} + M(\boldsymbol{x} \updownarrow_i \updownarrow_j) \neq 0$ or, what is the same, $M\boldsymbol{x} \neq M(\boldsymbol{x} \updownarrow_i \updownarrow_j)$.

As a conclusion, the matrix M is any matrix whose h columns are vectors of length k, such that all of them have al least two ones. We will proceed to show how to compute such a matrix and how many of these are there.

Let $C \subseteq \mathbb{Z}_{2^k}$ be the set of numbers whose binary representation contains at least two ones. Since C is just $\mathbb{Z}_{2^k}$ minus the set of numbers whose binary representation has exactly one or zero ones, then

$$C = \mathbb{Z}_{2^k} \setminus \left\{0, 2^0, 2^1, 2^2, ..., 2^{k-1}\right\}.$$

Consider each number in C as the vector of its binary representation. From this perspective, the act of choosing the columns of M reduces to drawing h different numbers from C. Moreover, since $|C| = 2^k - 1 - k = h$, then there is only one way of choosing the columns of M (up to the order).

In summary, the algorithm for computing H would be:

1. List the numbers in $\mathbb{Z}_{2^k}$ skipping 0 and those of the form 2^i (there will be $2^k - 1 - k = h$ such numbers). Call C this list.

2. Let M be a $k \times h$ boolean matrix whose columns are the binary representation of the numbers in C. The order of the columns does not matter, nor does the order of the binary representation, which can be LSB or MSB.
3. Compute $H = \{\text{concat}(\boldsymbol{x}, M\boldsymbol{x}) \mid \boldsymbol{x} \in \mathbb{B}^h\}$.

Lemma 3. If $\boldsymbol{x} \notin H$ and a student makes a guess, the guess will be correct.

Suppose that $\boldsymbol{x} \notin H$ and that student i makes a guess. Then,

- if $x_i = 0$ then $\boldsymbol{x} \downarrow_i = \boldsymbol{x} \notin H$. Thus, since the student guessed, it must be that $\boldsymbol{x} \uparrow_i \in H$ and, therefore, the guess was $g(\boldsymbol{x}, i) = 0 = x_i$.
- if $x_i = 1$ then $\boldsymbol{x} \uparrow_i = \boldsymbol{x} \notin H$. Thus, as before, since the student guessed, it must be that $\boldsymbol{x} \downarrow_i \in H$, and the guess was $g(\boldsymbol{x}, i) = 1 = x_i$.

In both cases, the guess is correct.

Lemma 4. If $\boldsymbol{x} \notin H$ at least one student will guess.

For each $i \in \mathbb{Z}_{n-1}$, let S_i correspond to the set of input vectors for which student i makes a guess, i.e. $S_i := \{\boldsymbol{x} \mid \boldsymbol{x} \downarrow_i \in H \vee \boldsymbol{x} \uparrow_i \in H\}$.

Since $\boldsymbol{x} \downarrow_i \neq \boldsymbol{x} \uparrow_i$ necessarily holds for any vector $\boldsymbol{x}$, then $|S_i| = 2|H|$. Additionally, notice that $H \subseteq S_i$, since all $\boldsymbol{x}$ must be either $\boldsymbol{x} \downarrow_i$ or $\boldsymbol{x} \uparrow_i$.

Let $i, j \in \mathbb{Z}_{n-1}$ with $i \neq j$. We will show that $S_i \cap S_j = H$.

Towards a contradiction, suppose that $\boldsymbol{x} \in S_i \cap S_j$ and $\boldsymbol{x} \notin H$. Then, by definition of S_i and S_j, it must hold that $\boldsymbol{x} \updownarrow_i \in H$ and $\boldsymbol{x} \updownarrow_j \in H$. But this contradicts that $d(\boldsymbol{x}, \boldsymbol{y}) \geq 3$ for any two different binary vectors $\boldsymbol{x}, \boldsymbol{y} \in H$ because $d(\boldsymbol{x} \updownarrow_i, \boldsymbol{x} \updownarrow_j) = d(\boldsymbol{0} \updownarrow_i, \boldsymbol{0} \updownarrow_j) = 2$.

Thus, given that for different i, j, $S_i \cap S_j = H$, it must hold that

$$\begin{aligned}
\left|\bigcup_{i \in \mathbb{Z}_{n-1}} S_i\right| &= H \cup \left|\bigcup_{i \in \mathbb{Z}_{n-1}} S_i \setminus H\right| \\
&= |H| + \sum_{i \in \mathbb{Z}_{n-1}} |S_i \setminus H| \\
&= |H| + \sum_{i \in \mathbb{Z}_{n-1}} |H| \\
&= n|H| \\
&= 2^k 2^{n-1-k} = 2^{n-1}.
\end{aligned}$$

Therefore

$$\bigcup_{i \in \mathbb{Z}_{n-1}} S_i = \mathbb{B}^{n-1},$$

and as a consequence, for all $\boldsymbol{x} \in \mathbb{B}^{n-1}$, at least one student will make a guess. This is true in particular if $\boldsymbol{x} \notin H$.

6 Concluding Remarks

We have presented original solutions to four challenging problems. The first problem was given to us by Catuscia Palamidessi. There are solutions to an equivalent problem that involve stochastic techniques using integration and probability distribution functions (e.g., [10]). Instead, our first solution uses a constraint satisfaction problem and a simple argument from combinatorics. Our second solution to this problem is even simpler and it uses a program transformation that allows us to derive e in a few steps. The second problem was also given to us by Catuscia Palamidessi. The problem can also be solved using the general theory of random walks on symmetric structures [11]. Our simple solution uses the recursive function $f(l, r)$ giving the probability of the coin reaching some player r for the first time before it reaches some other player l.

The third problem is defined in [6] as a hat problem but no solution is given. The fourth problem arose as a hat problem in the context of computational complexity [1]. It is another compelling example of a problem that arose a puzzle and it has found growing interest in Mathematics, Statistics and Computer Science. Our solution to this problem constructively builds the set H while proving its main properties. Previous authors that tackled this problem, such as Guo et al. [6], focused on a more general version of the problem in which more than two numbers (hats in their case) are allowed in the students foreheads. Their solution for the general case, which is based on the same principle as ours, is not constructive and much less detailed.

Acknowledgements. We would like to thank Thomas Given-Wilson and Bartek Klin for bringing the third and fourth puzzles to our attention.

References

1. Ebert, T.: Applications of recursive operators to randomness and complexity. Ph.D. thesis, University of California, Santa Barbara (1998)
2. Euler, L.: Solutio problematis ad geometriam situs pertinentis. Commentarii Academiae Scientiarum Imperialis Petropolitanae **8**, 128–140 (1736)
3. Feller, W.: An Introduction to Probability Theory and Its Applications, 2nd edn. Wiley, Hoboken (1971)
4. Ferguson, T.S.: Who solved the secretary problem? Stat. Sci. **4**, 282–289 (1989)
5. Gardner, M.: Martin Gardner's New Mathematical Diversions from Scientific American. Simon and Schuster, New York (1966)
6. Guo, W., Kasala, S., Rao, M., Tucker, B.: The Hat Problem and Some Variations, pp. 459–479. Birkhäuser, Boston (2006)
7. Guy, R.K.: Unsolved Problems in Number Theory, 3rd edn. Springer, Berlin (2004). https://doi.org/10.1007/978-0-387-26677-0
8. Lagarias, J.: The Ultimate Challenge: The 3x+1 Problem. American Mathematical Society, Providence (2010)
9. Ross, S.M. (ed.): Introduction to Probability Models. Academic Press, Cambridge (2007)
10. Russell, K.G.: Estimating the value of e by simulation. Am. Stat. **45**, 66–68 (1991)

11. Sbihi, A.M.: Covering times for random walks on graphs. Ph.D. thesis, McGill University (1990)
12. Solov'eva, F.I.: Perfect binary codes: bounds and properties. Discrete Math. **213**(1–3), 283–290 (2000)
13. Tao, T.: The Erdos discrepancy problem. arXiv e-prints arXiv:1509.05363 (2015)

Author Index

Printed in the United States
By Bookmasters